CONTROL SYSTEM TRENDS IN TURBO-MACHINERY

CONTROL SYSTEM TRENDS IN TURBO-MACHINERY

Part I – Control System Basics
Part II – Measurement and Regulation
Part III – Turbo-Machinery Operation and Control

A.S. Rangwala
International Turbo-Machinery Consultant

PUBLISHING FOR ONE WORLD

NEW AGE INTERNATIONAL (P) LIMITED, PUBLISHERS

New Delhi • Bangalore • Chennai • Cochin • Guwahati • Hyderabad
Jalandhar • Kolkata • Lucknow • Mumbai • Ranchi
Visit us at **www.newagepublishers.com**

Published by New Age International (P) Ltd., Publishers
First Edition: 2009

Branches:

- 36, Malikarjuna Temple Street, Opp. ICWA, Basavanagudi, **Bangalore**. ✆ (080) 26677815
- 26, Damodaran Street, T. Nagar, **Chennai**. ✆ (044) 24353401
- Hemsen Complex, Mohd. Shah Road, Paltan Bazar, Near Starline Hotel, **Guwahati**. ✆ (0361) 2543669
- No. 105, 1st Floor, Madhiray Kaveri Tower, 3-2-19, Azam Jahi Road, Nimboliadda, **Hyderabad**. ✆ (040) 24652456
- RDB Chambers (Formerly Lotus Cinema) 106A, Ist Floor, S.N. Banerjee Road, **Kolkata**. ✆ (033) 22275247
- 18, Madan Mohan Malviya Marg, **Lucknow**. ✆ (0522) 2209578
- 142C, Victor House, Ground Floor, N.M. Joshi Marg, Lower Parel, **Mumbai**. ✆ (022) 24927869
- 22, Golden House, Daryaganj, **New Delhi**. ✆ (011) 23262368, 23262370

ISBN : 978-81-224-2666-3

Rs. 795.00

C-09-04-3567

Printed in India at Glorious Printers, Delhi
Typeset at Bharti Composers, Delhi

PUBLISHING FOR ONE WORLD

NEW AGE INTERNATIONAL (P) LIMITED, PUBLISHERS
4835/24, Ansari Road, Daryaganj, New Delhi - 110002
Visit us at **www.newagepublishers.com**

Preface

Reliable and continuous operation of turbo-machines at high speeds calls for means to control the unit's performance characteristics in the least amount of time. An airliner's pilot or the operator in charge of an electric power generating system may be alert, but the speed at which the machine operates is so high and the time available to react to a possible emergency situation so small that the human interface is mostly inadequate. A high level of automation and sophistication is required to detect the source and cause of a fault, and to take appropriate corrective action within the system to forestall a deteriorating condition. The automatic control system built inside modern turbo-machines is capable of sensing and measuring the value of operating parameters while the engine is running. If one or more parameters are beyond set limits because of a malfunctioning component, the control system has the capability to detect the fault, establish a course of corrective action, issue instructions, and execute the commands that will rectify the situation. The entire sequence of detecting, evaluating, and taking remedial action occurs over a period that is timed in milliseconds.

Modern turbo-machines rely on the marvelous devices, instruments, and fly-by-wire control systems to ensure smooth and trouble-free operation at all times. Instead of manipulating an array of switches, valves, and mechanisms of the various electrical, hydraulic, pneumatic, and mechanical actuators, the pilot or the operator in charge of the engine instructs a computer of his intentions. The central processing unit of the computer activates the necessary protocol and issues command instructions to achieve the desired objective.

Automatic control theory and its methods can be put to use to improve efficiency by reducing fuel consumption, and in the process also reduce exhaust gas emissions. As an example, during a take-off climb an airplane's engines go through rapid acceleration from ground idle to maximum take-off speed before the aircraft settles into a steady mode of cruise condition at a selected altitude. The microprocessor-based controller determines the ratio of air and fuel to be burned in the combustor of the turbine at each stage, and constantly adjusts the ratio to obtain optimum performance as parameters such as rotor speed, power output, compressor discharge pressure, and exhaust gas temperature vary with the load demand. Thus, the pilot can pay his full attention to navigating and controlling the overall airplane, and does not have to tinker with the air-fuel ratio. Oxygen sensors measuring the content of oxygen in the exhaust path also play a deciding role in establishing the amount of fuel to be consumed for a given set of conditions in the turbine.

This book is written to meet the needs of students in engineering colleges and practicing engineers in a large number of industries where turbo-machines are used. All of the material has been specifically tailored for college undergraduate and graduate level design engineering of rotating machine courses. Electronic spreadsheet type of calculation is used in example problems to calculate control system natural frequencies, dynamic response, stability and other related design parameters. It is expected that the reader is familiar with basic to medium level calculus offered at the college undergraduate level.

The book is split into three parts. The first part focuses on the fundamentals of automatic control systems. The four chapters in the section look into the theory of control methods, and provide the basics of electricity and electronics, and of fluid power. The second part investigates the measurement and regulation aspects of sensors, manipulators, actuators, and programmable logic controllers. The third part discusses operation and control techniques of various types of turbo-machines, and deals with topics such as rotor vibrations, clearances between rotor and stator, combustion and emissions, and engine diagnostics.

The first chapter of part I provides the basics of automatic control system theory as applied to turbo-machinery. The second chapter delves into different control methods, the third chapter into the basics of electrical and electronic forms of control, and the foruth chapter into the fundamentals of fluidic control. In thc sccond part of thc book, chapter five looks at various types of sensors that are capable of measuring parameters such as pressure and temperature, chapter six considers the role of valves, chapter seven at manipulation and actuation methods, chapter eight at programmable logic controllers, and chapter nine at the design and analysis of controllers. Part III starts with a description of various types of turbo-machines in the tenth chapter, and the eleventh and twelfth chapters delve into the control techniques for aircraft power plants and marine gas turbines. Chapter thirteen considers vibrations arising from the rotating system of the engine, chapter fourteen goes into control of clearance between the rotating and stationery parts of the engine, and chapter fifteen looks at control of combustion of fuel and production of emissions. Chapter sixteen considers ways of enhancing the performance of turbo-machinery, chapter seventeen looks into diagnostics and trouble shooting in turbines, chapter eighteen into control of manufacturing processes and chapter nineteen into economic aspects of operating turbo-machinery.

A list of symbols is provided mostly to facilitate identification with commonly used parameters in the equations and the associated text. However, because of the considerable number of topics the corresponding variables are adequately defined within each section. Oftentimes it is found necessary within the sections to redefine many of the symbols for convenience and better understanding of the subject matter. Thus, the list of symbols may be used only as a general guideline.

I gratefully remember and appreciate past students of the course on this topic who have sent in comments and reported errors, and express my hope that those who work with this treatise will do likewise. I am indebted to Mr. Mark Belloni and Dr. Fred Ehrich of General Electric Company for performing a vast amount of computational work in the finite element analysis and for valuable advise on the text and layout of the book. I greatly appreciate comments provided by Dr. Ahmad Kamel, Mr George Robinson and Dr. Raj Subbiah of Siemens-Westinghouse Power Corporation for checking the problems and reading the proof.

Orlando, FL **A. S. Rangwala**

Contents

Part I

Control System Basics

1

Automatic Control Theory in Turbo-Machines

1.1 INTRODUCTION

Turbo-machines perform a plethora of services, come in every conceivable size and shape, and are found in the air, on the ground, and on the high seas. Aircraft turbine engines have transformed the manner in which people and goods are transported at speeds that hitherto were considered unimaginable just a few decades earlier. Industrial power generation steam and gas turbines provide cheap and reliable electricity to illuminate the remotest corners of the world. Marine turbine engines are built taking advantage of the lighter weight characteristics of their aviation counterpart to satisfy huge power requirements of offshore oil platforms and onboard ships. Compressors provide clean and moisture-free compressed air for underwater breathing and for food processing. Turbo-chargers improve the performance of reciprocating compression ignition engines using diesel or other easily available hydrocarbon fuels to provide large amounts of mechanical power for railroad locomotives, construction machinery, and factories, all the while operating at exceptionally high levels of efficiency.

Reliable and continuous operation of the turbo-machines at high speeds calls for means to control the unit's characteristics in the least amount of time. In spite of high levels of diligence on the part of the airliner's pilot or the operator in charge of the system, the speed at which the processes occur within the unit is so high and the time available to react to a possible emergency situation so small that the human interface is mostly inadequate. A high level of automation and sophistication is required to detect the source and cause of a fault and to take appropriate corrective action within the system to forestall a deteriorating condition. The automatic control system is capable of measuring operating parameters within the engine's system, establishing the appropriate course of action and executing command instructions, all occurring over a period that is timed in milliseconds. An increase in the temperature of lubricant oil departing from a bearing sump, for instance, could be due to a leak in the oil circuit, defect within the bearing, or from imposition of a higher than normal load on the rotating system. If corrective action is not promptly taken, the elevated temperature can cause the components to thermally expand and seize the bearing, forcing a shutdown of the engine.

Modern turbo-machines rely on fly-by-wire control systems. Instead of manipulating an array of switches, valves, and mechanisms to the various electrical, hydraulic, pneumatic, and mechanical actuators, the pilot or the operator in charge of the engine instructs a computer of his intentions. The central processing unit of the computer activates the necessary protocol and issues command instructions to achieve the desired objective. Maximum power and performance rely on operation of the engine close to aerodynamically unstable limits that, if exceeded, can potentially cause the compressor to stall and set up excessive vibrations, so control by conventional methods is virtually impossible. Computerized control is essential.

Automatic control theory and methods can be put to use to improve efficiency by reducing fuel consumption, and in the process also reduce exhaust gas emissions. As an example, during a take-off climb an airplane's engines go through rapid acceleration from ground idle to maximum take-off speed before the aircraft settles into a steady mode of cruise condition at a select altitude. The microprocessor-based controller determines the ratio of air and fuel to be burned in the combustor of the turbine, and constantly adjusts the ratio to obtain optimum performance as parameters such as rotor speed, power output, compressor discharge pressure, and exhaust gas temperature vary with load demand. Thus, the pilot can pay his full attention to navigating and controlling the overall airplane, and does not have to tinker with the air-fuel ratio. Oxygen sensors measuring the content of oxygen in the exhaust path also play a deciding role in establishing the amount of fuel to be consumed for a given set of conditions in the turbine.

Rapid advances in electronics and microelectronics, coupled with materials technology, have made miniaturization a major factor in advancing the state of the art in automatic control of equipment and processes. Sensors to measure operating parameters, actuators, and linear motors that develop thrust through a combination of magnetic and electro-dynamic forces are indispensable in modern turbo-machines. The concept has been extended to meet the demands of a production-line machine that can carry out assembly of miniature components in an electrical or electronic subassembly; or for machining of components to demanding specifications; or has the capability and intelligence of a very precise position fixing for a quality inspection process. The application of the theoretical concepts extends to position fixing and measuring devices, and has spread into a number of areas as their cost versus performance ratio is improved by reducing the number of configuring elements.

1.2 OBJECTIVES AND BENEFITS OF CONTROL SYSTEM

A control system essentially maintains a controlled variable to a set value, irrespective of all demand parameters. The system ideally must respond to a change before an error is detected, but the inherent operating principle of the feedback system precludes it. A load change must initially trigger a variation in the controlled variable to produce an error. The controller then acts on the error information to generate a change in the manipulated variable. The sequence of events is completed when the shift in the manipulated variable drives the controlled variable towards the preset limit.

Achieving a nearly perfect operation where the objective is defined in terms of response to the change thus has to suffice. The objective may be based on minimizing the maximum value of the error signal, which could eventually be reduced to zero in a system with an integral mode. But a residual error that returns to a steady and non-changing value may persist in other systems. The setting time required to accomplish the minimization process provides another objective. Still another objective arises from the residual error after the system settles out. Simultaneous minimizing of the maximum value of the error, settling time and residual error may thus be considered the primary focus of any control system (figure 1.1).

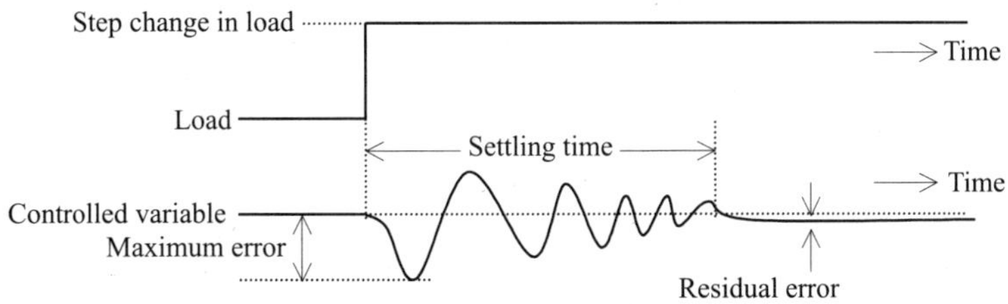

Fig. 1.1 Objectives of Control System

But the objectives as defined are inherently incompatible with each other. A reduction in residual error can be obtained if the gain of the controller is increased, so that a smaller residual error will produce the required corrective action from the controller. However, the increase in gain will also tend to increase the settling time. A trade-off is consequently necessary between the rate at which the error is eliminated, or sensitivity of the control mechanism, and the settling time, or stability of the system that ensures the controlled variable does not stray too far in both positive and negative directions as it attempts to zero in on the target value.

A case in point is the speed of the rotor of a power generation steam turbine. For 60 Hz electric power the shaft must spin at 3,600 rpm. The rotor will tend to accelerate in the event of a load shed. The speed control mechanism will respond by decreasing the flow of steam through the inlet valve, which will lower the speed. The reduced flow of steam continues till the rotor's speed reaches and even falls below the target speed. The speed control mechanism will then be forced to respond by increasing the steam flow to bring it back up in speed. The turbine's rotor will experience 'hunting' as the speed keeps oscillating a few rpm below and above the necessary 3,600 rpm, as long as variations in the steam flow ordered by the speed controlling mechanism persist.

Evaluation of a control system requires setting up a set of criteria and specifying a test to establish its acceptability. A step change, such as may occur in the example of the power generating steam turbine when a customer's large electric furnace suddenly trips, is a common test. Some other criteria call for determination of decay rate for a quarter of the amplitude, integral of the absolute error, and critical damping.

1.3 EVOLUTION OF CONTROL MECHANISMS

Control system technology has constantly evolved and improved from imposition of newer and stricter process or machine needs and from advancement in the design of components. Process control traces its origins from sensing elements connected to a controller, which signaled a control valve. The loop was placed in close proximity of the process and produced reasonably good control of the variables. However, the method suffered from lack of simultaneous monitoring of different phases of the process spread out over a large system, so measured data of other critical variables could not interact to produce a more intelligent command instruction.

When pneumatic transmitters, controllers and valve actuators made their appearance, use of control loops for the full process collected in a single control area became a reality. Monitoring of a large number of variables in the process was now possible. But the system required a large number of pneumatic lines between the various control loops, each of which required a signal from the measuring transmitter to the controller and another signal from the controller to the activating valve. Also, the signals experienced transmission delay since they travel at the relatively low speed of about 1,000 ft/second.

The arrival of electronic analog transmitters, controllers and electro-pneumatic converters replaced the pneumatic signals with electric signals. The switch considerably reduced the transmission delay problem, since the electric signals travel nearly at the speed of light. But the electric lines in a closed loop form an antenna, and another problem crept up in the form of noise signals picked up by the closed loops from the surroundings. The noise interferes with the transmitted signal to reduce its accuracy, and creates problems in the controller's algorithm.

The digital computer turned out to be a landmark development when it was first used in controlling the process of a polymerization unit of a Texaco plant in 1959. The digital computer permitted implementation of all process variables into the memory and making it available for use in the control algorithm. A single digital computer effectively replaced the analog controllers. Lack of reliability, however, still persisted when the computer crashed, bringing the entire control process to a standstill. A backup system relieved the problem, where a standby computer takes over in the event of failure in the primary system. Analog controllers for loop control are also used in tandem with the digital computer playing a supervisory role.

Control of operation of a turbo-machine initially started with hard-wired panels of relay logic, motor starters, fluid actuators, and solenoid valves. Vacuum tubes were employed in servo controls, and variable speed drives required a dc generator driving a dc motor. Numerical control of machines using digital technology appeared in the 1950's, with programming carried out off-line using punched paper tape for storage of command instructions. Programmable controllers with re-programmable computer logic came on the scene around 1970. This was followed by the microprocessor in the 1980's, which took advantage of semiconductor switching elements and memory systems.

The microprocessor is chiefly responsible for integrating the various processes occurring in the different sections of the machine to provide overall control of the turbine. The microprocessor enables communication of the distributed data gathering system to provide advantages of localized control and of the fully informed centralized control system. Local control modules have access to all inputs from throughout the system, permitting freedom to apply and satisfy ever more demanding control needs. The shorter control loops are relieved of extraneous signals. Smarter measuring transmitters may contain their own microprocessor to convert an analog signal to a digital signal, linearize, eliminate noise, convert the signal to engineering units, and also store data for trend analysis.

1.4 BLOCK DIAGRAM AND TRANSFER FUNCTION

A block diagram provides a schematic of how the components are connected in a control system. Any control system can be represented by a set of mathematical equations that define the characteristics of the component. A wide range of control problems – coolant flow and temperature variation of a surface over which it flows, engine throttle lever position changing power output – can be analysed by the same mathematical methods, as shown in figure 1.2.

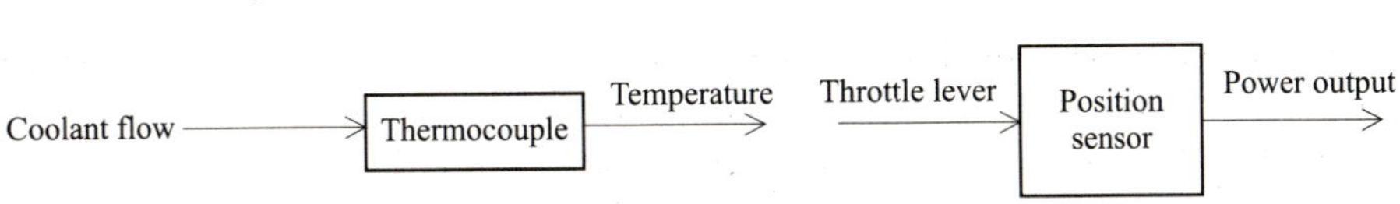

Fig. 1.2 Examples of Block Diagrams for Control Components

The block diagram retains the significant feature of the effect each component has on the system. Signal lines indicate the input and output signals of the component. The signals can take the form of electric current or voltage, air pressure, flow rate, temperature, pressure, speed, acceleration, position, direction, or some other parameter. The path traversed by the signal can be through electric wires, pneumatic or hydraulic lines, mechanical linkages, or related mechanisms that are capable of transferring signals.

The relationship between the input and output signals is of considerable interest and significance. The ratio of the output and input signals expresses the 'transfer function' of the component. Referring to the first example of figure 1.2, if input Q is the coolant flow and T the output temperature, then the transfer function C is given by $C = T/Q$, and $T = CQ$. Besides the magnitude of the input and output signals, the transfer function also incorporates the timing relation between the two signals.

1.5 GAIN, BANDWIDTH, FREQUENCY RESPONSE

If a linear component is subjected to an input signal of sinusoidal waveform, the output signal will also be sinusoidal, as illustrated in figure 1.3. The ratio of the magnitude of the output and input signals provides the 'gain' of the component. Thus, if the amplitude of the output signal is B and that of the input signal A, then gain $G = B/A$. The input signal becomes zero when $\omega t + \alpha = 0$, or $t = -\alpha/\omega s$. Similarly, the output signal approaches zero value when $\omega t + \beta = 0$, or $t = -\beta/\omega s$. Hence, the time difference between the input and output signals is $t = (-\alpha - [-\beta])/\omega = (\beta - \alpha)/\omega s$. The phase difference of the component is defined as the difference in the phase of the output and input signals, or $\alpha - \beta$.

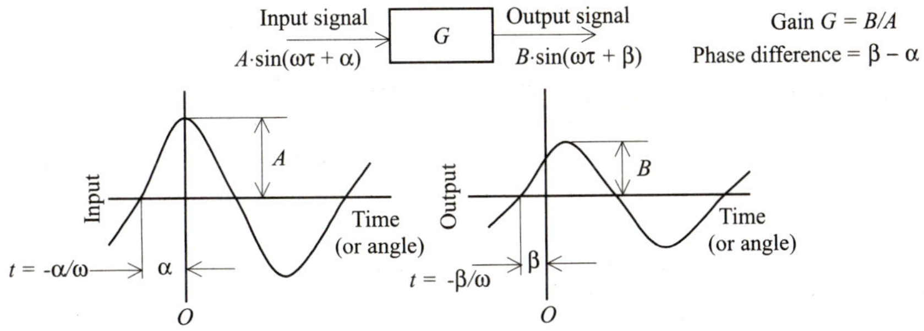

Fig. 1.3 Gain and Phase Difference of Linear Component Subjected to Sinusoidal Signal

Complex numbers in the polar form are commonly used in control theory calculations for convenience. In complex notation the input signal is $A\angle\alpha$ and the output signal is $B\angle\beta$. The transfer function is obtained by dividing the output signal by the input signal, or $G = [B\angle \beta]/[A\angle \alpha] = (B/A)\angle(\beta - \alpha)$. Thus, the transfer function of the component is provided by the gain B/A and by the phase angle difference of the output and input signals.

The gain and phase difference jointly provide the 'frequency response' of the component at the given frequency. As an illustration, the gain of a certain control system may be 0.875 and the phase difference is 37° at a frequency of 20 Hz.

1.6 OPEN-LOOP CONTROL

An open-loop control system relies on a calibrated setting previously obtained from a known procedure or from calculations. Comparison of the actual output with the desired value is thus absent.

The characteristics of a needle valve used in an open-loop control system are shown in figure 1.4. The calibration curve is obtained from tests that measure the flow rate through the valve for various valve openings. Multiple calibration curves are obtained for different values of pressure drop across the valve.

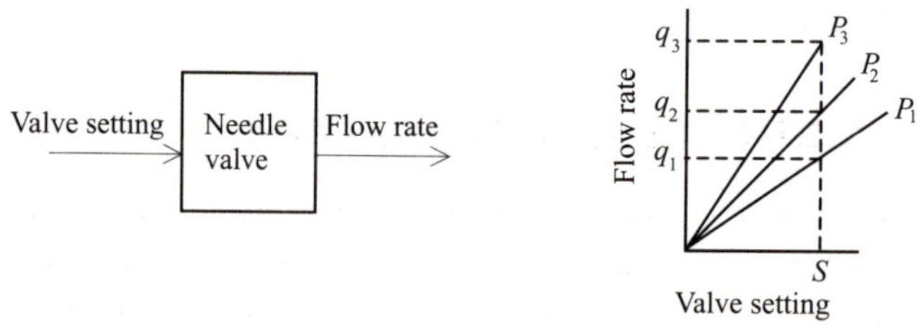

Fig. 1.4 Open-Loop Control System – Needle Valve

The open-loop controller offers the advantage of simplicity of hardware and installation, since it eliminates the need for measuring the output parameter value and for subsequent corrective action based on the error. But if unexpected disturbances in the system cause the output signal to go out of bounds, manual adjustment may be necessary to limit the error.

1.7 FEEDBACK AND CLOSED-LOOP CONTROL

In the example of the coolant flow over a hot surface, the temperature of the object may be used to increase or decrease the rate at which the coolant is made to bring down its temperature to a specified level (figure 1.5). This is the basis of 'feedback', and control systems using feedback are referred to as 'closed-loop control systems'. The corrective action is not present in the absence of feedback, and the block diagram loop is then open.

The principle of feedback relies on evaluating the difference between the actual and desired values of the output signal. The output signal traverses back and modifies the input signal before entering the controller block. In addition to measuring the controlled variable, the feedback control system computes the error (desired value minus measured value) and uses the error to determine the control action. The control action manipulates some process variable to reduce the error. In figure 1.5, reference S is the input signal to the mechanism, and C is the output variable. The transfer function C/S is determined as follows.

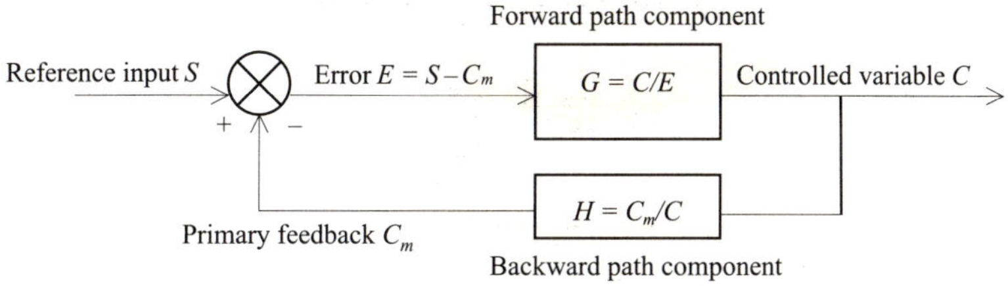

Fig. 1.5 Closed-Loop System with Feedback

$$\text{Error } E = (\text{reference input}) - (\text{primary feedback}) = S - C_m$$

$$\text{Controlled variable } C = (\text{error}) \times (\text{system transfer function}) = EG$$

$$\text{Measured control variable } C_m = (\text{controlled variable}) \times (\text{feedback transfer function}) = CH$$

Now: $$C = (S - C_m) \times G = SG - CHG, \quad \text{or:} \quad C(1 + GH) = SG$$

Hence transfer function $$C/S = G/(1 + GH)$$

Sometimes the primary feedback signal is inverted, which calls for adding it to the reference signal to determine the error signal. The transfer function in this case would be $C/S = G/(1 - GH)$. Also, the forward path may contain a control mode, manipulating element, and process element with transfer functions G_C, G_m, and G_p, respectively. The overall transfer function then becomes $G = G_C G_m G_p$.

1.8 ELEMENTS OF MECHANICAL VIBRATING SYSTEM

A mechanical vibrating system consists of a spring for storing potential energy, mass and inertia for kinetic energy, and damper for dissipating mechanical energy. The vibration process alternatively converts energy between its potential and kinetic forms. In its general sense, the vibration is a periodic motion that repeats itself in all its details after a certain interval of time called the period of vibration. Some energy must be replaced in each cycle of vibration from an external source to maintain the vibration.

An electric motor mounted on a flexible but mass-less beam supported at its ends is an example of a vibrating system. The single degree of freedom vibration system may be schematically represented by figure 1.6. The electric motor is of a certain mass m. A mass is a rigid body; so the force acting on the mass according to Newton's second law of motion is proportional to its acceleration, or md^2x/dt^2. The elastic beam has its own stiffness k, the constant of proportionality that relates the force in the spring with its deflection, or $f = kx$. Mass m is suspended by the spring from a fixed location. The ideal spring has no mass, and so the forces acting at its opposite ends are equal and opposite. Damping or energy-dissipating forces arise from friction at the end supports and from the molecules rubbing against each other within the beam. In a viscous damper, the applied force is proportional to the relative velocity of its connecting points. The damping coefficient, c, provides the proportionality constant, and is the characteristic parameter of the damper. The ideal damper also is without mass, and so the force at one end is equal and opposite of the force at the other end.

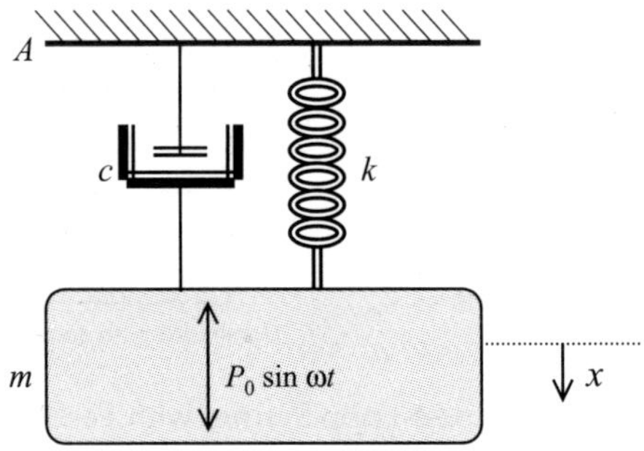

Fig. 1.6 Single Degree of Freedom Vibrating System

The damper is also placed between the mass and the fixed wall. The damper does not transmit force to the mass when at rest, but as soon as it moves, it exerts a damping force of cdx/dt.

Free vibrations are initiated when the mass is displaced from its position at rest and then released. In the absence of external excitation the forces acting on the system may be expressed as:

$$md^2x/dt^2 + cdx/dt + kx = 0 \quad \text{...(1.1)}$$

When external forces are not acting on the system, the free motion is suppressed by the energy dissipating damping mechanism. For sustained motion an external alternating force of the type $P_0 \sin(\omega t)$ is required, where P_0 is the amplitude and ω is the frequency. The governing differential equation of motion then takes the form:

$$md^2x/dt^2 + cdx/dt + kx = P_0 \sin \omega t \quad \text{...(1.2)}$$

1.9 RESONANT FREQUENCY

The role of 'resonant frequency' cannot be overemphasized in automatic control systems theory. A resonant frequency is one where the mechanical vibrating system acquires a portion of its operating energy through a more or less self-sustaining feedback loop or circuit. The single degree of freedom vibration model provides a good example. In equation (1.1), if the damping coefficient is increased, the maximum displacement x will be reduced, while if the damping is reduced (meaning low damping), the amplitude of motion will increase. In the limiting case of no damping, the equation of motion reduces to:

$$md^2x/dt^2 + kx = 0 \quad \text{...(1.3)}$$

The most general solution of this equation is:

$$x = C_1 \sin \sqrt{\frac{k}{m}} t + C_2 \cos \sqrt{\frac{k}{m}} t \quad \text{... (1.4)}$$

where C_1 and C_2 are arbitrary constants. Physically, the result as it stands is indefinite, since the constants can take any value. This is because the problem itself was not fully defined. If it is specified that the mass is pulled out of its equilibrium position to $x = x_0$, then released without initial velocity, two conditions are specified:

At $t = 0 : x = x_0$ and $dx/dt = 0$.

The first condition gives $C_2 = x_o$. For the second condition, (1.4) must be differentiated, then get $C_1 = 0$. Substitution into (1.4) leads to the specific solution:

$$x = x_0 \cos \sqrt{\frac{k}{m}}\, t \qquad ...(1.5)$$

This represents an undamped vibration. Denoting the period of vibration by T, then:

$$T = 2\pi \sqrt{\frac{m}{k}} \qquad ...(1.6)$$

Customarily, $(k/m)^{1/2}$ is denoted by ω_n, called the natural circular frequency. The natural frequency f_n is given by:

$$f_n = \frac{1}{T} = \frac{1}{2\pi}\sqrt{\frac{k}{m}} = \frac{\omega_n}{2\pi} \qquad ...(1.7)$$

measured in cycles per second. If the negligibly damped system is subjected to the external exciting force of the type $P_0.\sin \omega t$, and if the exciting force frequency ω coincides with the natural frequency ω_n, then a condition of resonance prevails. The amplitude of motion then theoretically reaches infinite without damping even if P_0 is small in magnitude, and the natural frequency of vibration then also becomes the resonant frequency.

1.10 DAMPING AND STABILITY

Free vibrations without damping persist indefinitely, which cannot happen; all free vibrations eliminate with time. The term viscous damping is associated with the expression cdx/dt since it represents the conditions of damping due to the viscosity of oil in a dashpot. Other types of damping exist, but their solution cannot be determined with as much ease. Consider the function $x = e^{st}$, where t is the time and s an unknown constant. Upon differentiation, the same function multiplied by a constant results. This function when substituted in the equation:

$$md^2x/dt^2 + cdx/dt + kx = 0 \qquad ...(1.8)$$

permits division by e^{st} and leads to an algebraic equation instead of a differential equation. Thus, assuming e^{st} as the solution, then the above equation becomes:

$$(ms^2 + cs + k).e^{st} = 0 \qquad ...(1.9)$$

This is a quadratic in s, with two values that will satisfy the equation:

$$s_{1,2} = -\frac{c}{2m} \pm \sqrt{\left(\frac{c}{2m}\right)^2 - \frac{k}{m}} \qquad ...(1.10)$$

The most general solution of the problem is:

$$x = C_1 e^{s_1 t} + C_2 e^{s_2 t} \qquad ...(1.11)$$

where C_1 and C_2 are arbitrary constants. The physical significance of this equation will depend on whether the expressions for s are real or complex. When $(c/2m)^2 > k/m$, both values of s will be real. Also, they are both negative since the square root is smaller than the first term $c/2m$. Thus, (1.11) describes a solution consisting of the sum of two decreasing exponentials, as shown in figure 1.7. The case of $C_1 = 1$ and $C_2 = -2$ is shown as a dashed line.

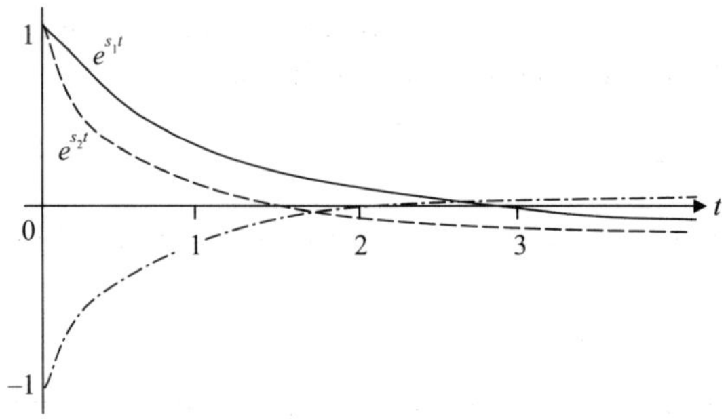

Fig. 1.7 Motion of Single Degree of Freedom System with $c/c_c > 1$

This figure shows that the motion is not oscillatory, but rather a creeping back to the original position. This is due to the fact that when $(c/2m)^2 > k/m$ the damping c is large. For smaller values of c pertaining to more practical cases, (1.10) gives complex values for s. The damping c at which this transition occurs is called the critical damping c_c:

$$c_c = 2m\sqrt{\frac{k}{m}} = 2\sqrt{mk} = 2m\omega_n \qquad \text{...(1.12)}$$

If the damping is less than critical damping, the solution consists of two factors, the first a decreasing exponential and the second a sine wave. The combined result is an exponentially decreasing sine wave, lying in the space between the exponential curve on both sides of the phase angle axis (figure 1.8). The smaller the damping constant c, the flatter will be the exponential curve and the more cycles it will take for the vibrations to be eliminated.

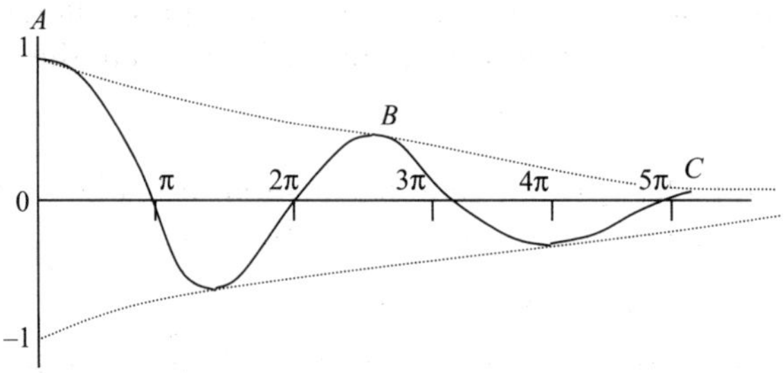

Fig. 1.8 Free Vibration of a System with $c/c_c < 1$

The rate of this dying down of the vibrations is of interest, and can be calculated by considering two consecutive maxima of the curve, *A–B, B–C,* and so on. During the time interval between any two such maxima, the amplitude of vibration diminishes from:

$$e^{-\frac{c}{2m}t} \text{ to } e^{-\frac{c}{2m}\left\{t+\frac{2\pi}{q}\right\}} \text{ where } q = \sqrt{\frac{k}{m} - \frac{c^2}{4m^2}} \qquad \text{...(1.13)}$$

The latter of these two expressions is equal to the first one multiplied by the constant factor:

$$e^{-\frac{\pi . c}{mq}} \qquad \text{...(1.14)}$$

which factor is smaller than unity. This factor is the same for any two consecutive maxima, and is independent of the amplitude or of time. Thus, the amplitudes decrease in a geometric series.

The frequency of vibration diminishes with increasing damping. If written in a dimensionless form with the help of (1.13), it becomes:

$$\frac{q}{\omega_n} = \sqrt{1 - \left\{\frac{c}{c_c}\right\}^2} \qquad \text{...(1.15)}$$

This equation is plotted in figure 1.9, where the ordinate q/ω_n is the ratio of the damped to the undamped natural frequency, and the abscissa is the ratio of the actual to the critical damping. The figure is a circle; for critical damping the natural frequency is zero. Due to the horizontal tangent of the circle at $c = 0$, the natural frequency is nearly constant and equal to $\sqrt{(k/m)}$ for all typically encountered values of damping ($c/c_c < 0.2$).

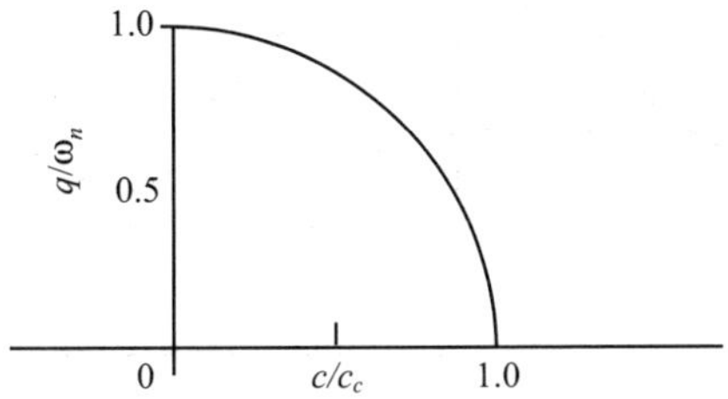

Fig. 1.9 Natural Frequency as a Function of Damping

1.11 TRANSIENT RESPONSE

The complete equation of motion is:

$$md^2x/dt^2 + cdx/dt + kx = P_o \sin(\omega t) \qquad \text{...(1.16)}$$

The full solution consists of the complete solution of the equation with the right hand side set to zero and a particular solution of the whole equation (1.16), or:

$$x = e^{-\frac{c}{2m}t}\{C_1 \sin(qt) + C_2 \cos(qt)\} + \text{particular solution} \qquad \text{...(1.17)}$$

Thus, only the particular solution needs to be determined. Assuming a solution of the type $x = x_0 \sin(\omega t - \varphi)$, the particular solution can be determined by finding x_0 and φ. Newton's law requires that the sum of all the forces be zero at all times, implying that the sum of the vertical and horizontal components must also be zero. From these conditions the unknowns x_0 and φ are solved:

$$x_0 = \frac{P_0/k}{\sqrt{\left\{1 - \frac{\omega^2}{\omega_n^2}\right\}^2 + \left\{2\frac{c}{c_c}\frac{\omega}{\omega_n}\right\}^2}} \qquad \text{...(1.18)}$$

$$\tan(\varphi) = \frac{2\dfrac{c}{c_c}\dfrac{\omega}{\omega_n}}{1-(\omega^2/\omega_n^2)} \qquad \text{...(1.19)}$$

The expressions for the amplitude and the phase angles are in terms of ratios of frequency and damping, where c_c is the critical damping. P_0/k may be interpreted as the static deflection of the spring due to P_0. These relations are plotted in figure 1.10. The amplitude diagram contains a family of curves, one for each value of damping. All curves lie below the one for zero damping, since

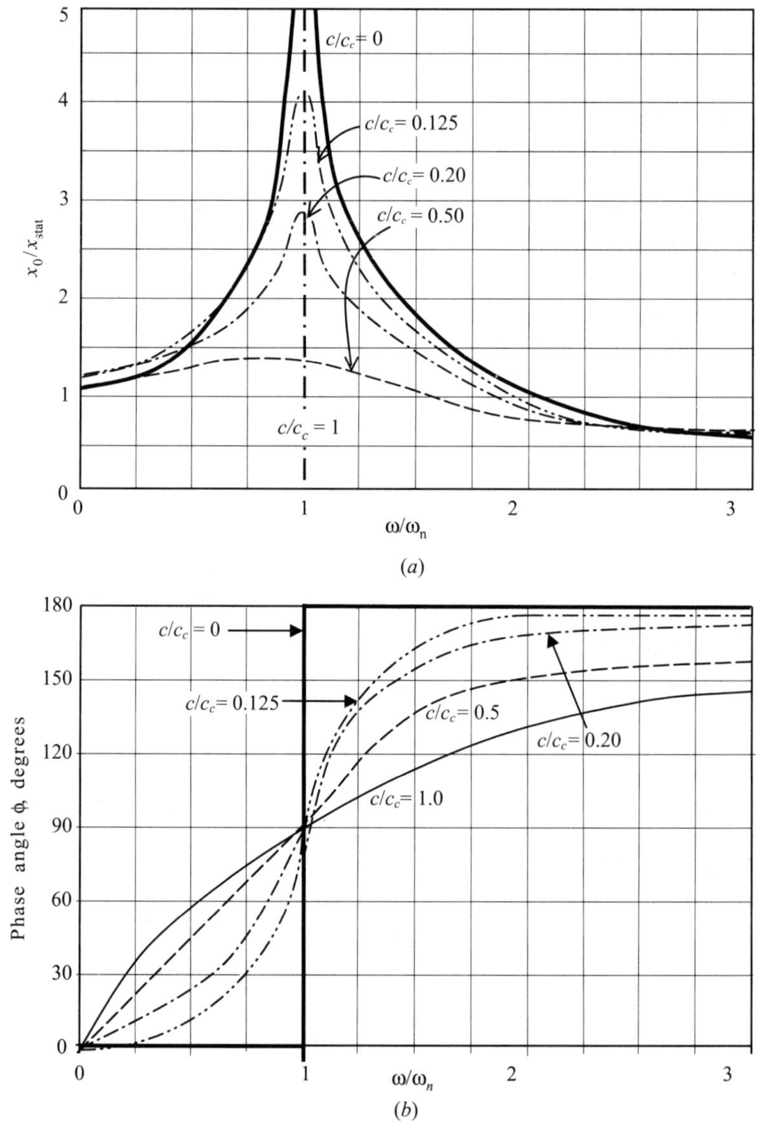

Fig. 1.10 (*a*) Amplitudes of Forced Vibrations, (*b*) Phase Angle Between Force and Displacement as a Function of Frequency for Various Damping Values

damping reduces the amplitude of forced vibration. The maxima of the different curves do not occur any more at $\omega/\omega_n = 1$ but at a smaller frequency. In fact, three different frequencies need to be distinguished, all of which coincide for $c = 0$:

- $\omega_n = \sqrt{(k/m)}$ = undamped natural frequency
- $q = \sqrt{\frac{k}{m} - \left\{\frac{c}{2m}\right\}^2}$ damped natural frequency
- The frequency of maximum forced amplitude, or the resonant frequency.

The phase angle diagram is also of interest. For no damping, it was seen that the force and displacement are in phase below resonance and 180° out of phase above resonance. So the phase angle shows a discontinuous jump at the resonance point. For damping values different from zero the other curves in the phase angle diagram are plotted. In general, damping tends to smooth out the sharpness of the undamped diagrams for both amplitude and phase.

Returning to the assertion pertaining to the particular solution, the general solution consists of the damped free vibration superposed on the forced vibration. After a short time the damped free vibration disappears and the forced vibration alone persists. For this reason the forced vibration may also be referred to as sustained vibration and the free vibration becomes transient vibration. The values of the constants C_1 and C_2 depend on the conditions at the start and can be calculated from these conditions using the analytical process described earlier. It is possible to describe the whole process by physical reasoning. For example, consider a mass suspended by a spring and acted upon by an external harmonic force of frequency eight times as slow as the natural frequency of the system. The mass is clamped while the external force is acting. Suddenly the clamp is withdrawn. Assume the damping in the system is such that the free vibration decreases by 10 per cent for each cycle. Assume also that the clamp is released when the forced vibration has its maximum amplitude. From the initial conditions it follows that at the instant of release the mass has no deflection and no velocity. The prescribed forced vibration starts with $x = x_0$ and $dx/dt = 0$. These two conditions can only be satisfied by starting a free vibration with $x = -x_0$ and $dx/dt = 0$. Then the combined motion will start at zero with zero velocity. Figure 1.11*a* shows the free vibration, 1.11*b* shows the forced vibration and 1.11*c* the combined motion.

Note that the transient vibration disappears quickly and that the maximum amplitude at the start is nearly twice as great as the sustained final amplitude. If the difference between free and forced frequencies is small and damping is also small, the free and the forced vibration vectors would retain nearly the same relative motion during one revolution, the included angle between them changing only slightly. The vectors can then be added geometrically, and during one revolution of the two vectors the motion will practically be a sine wave of frequency $\omega_1 \approx \omega_2$ and amplitude the arithmetic sum of the two amplitudes. After a large number of cycles, the relative position of the vectors varies, and the magnitude of the sum vectors also changes. The resulting motion can then be described as an approximate sine wave with frequency ω_1 and an amplitude varying slowly between the sum and the difference of the amplitudes of the free and forced vibrations. This phenomenon is known as beats. The free vibrations will be eliminated after a while due to energy dissipation from the damping. In order to have sustained beats it is necessary to have two sustained, or forced, sources of vibration.

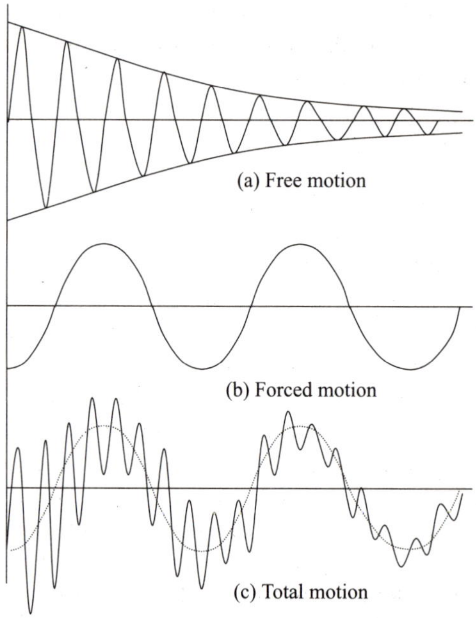

Fig. 1.11 Starting Transient Motion of a Forced Vibration

1.12 BLOCK DIAGRAM SIMPLIFICATION

Although block diagram schematics are more useful in explaining the mechanics of a process and flow of information within an automatic control system, a mathematically rigorous approach exists to simplify the system by modifying and moving the blocks. Each block representing a component has its own transfer function, which must work with other transfer functions in a systematic manner to generate the overall desired output.

Figure 1.12 shows a rather simplified version of a gas turbine system. Input signals from the operator in charge of the unit is required to set the variable inlet guide vanes and the fuel throttle lever to control the overall functioning of the machine. Each component has its own unique transfer function. For instance, the fuel throttle lever has the transfer function described by fuel flow through the flow valve divided by the lever position setting.

An arbitrary swapping of the components is not permitted. If the position of the combustor block were to be exchanged with the compressor stages 1 – 6 block, as an example, the system would be rendered inoperable. Looking at it mathematically, the transfer function of the first six stages of the compressor is air pressure at sixth stage divided by air pressure at the intake. On the other hand, the transfer function of the combustor is obtained by dividing temperature of gases exiting the combustor with the temperature of air entering it. The two transfer functions are radically different from each other, and swapping the two component blocks would lead to a breakdown in compatibility between the output and input information. Combining or swapping of the component blocks for the sake of simplification is permissible only if the output and input results consistently remain unaffected.

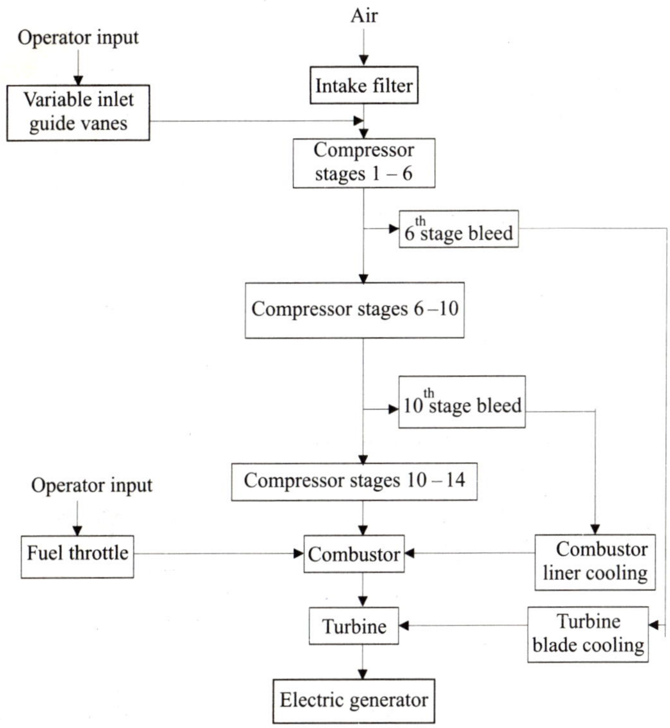

Fig. 1.12 Block Diagram of Simplified Representation of Gas Turbine

Some basic rules to be followed in simplifying or modifying the block diagram are provided below.

- Replacing two cascading blocks with a single block.

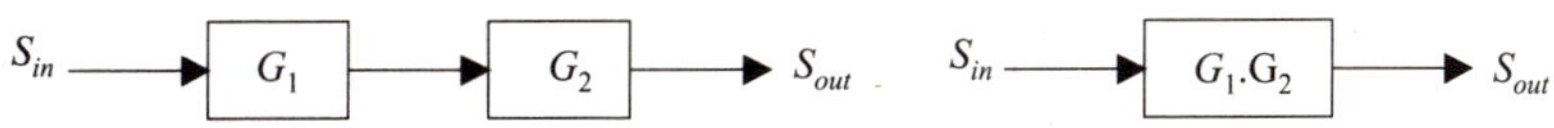

- Eliminating forward loop.

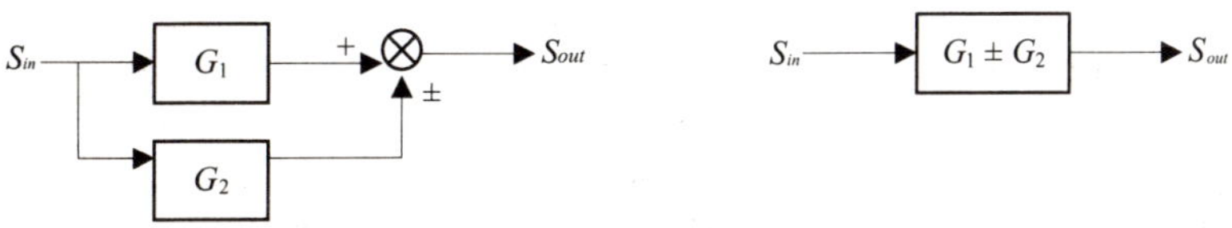

- Moving take-off point from input to output side.

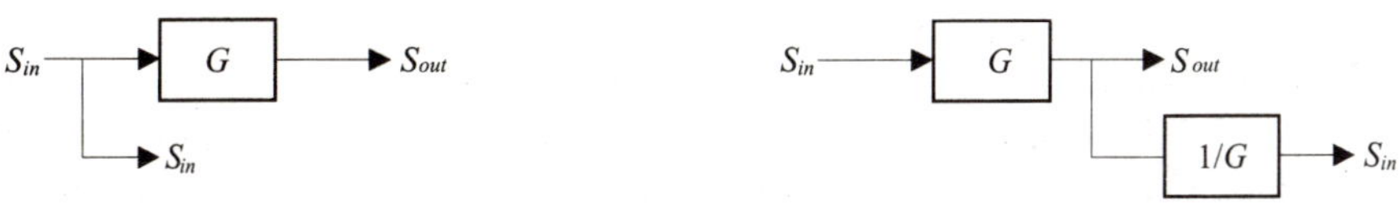

- Moving take-off point to a block ahead.

- Moving summation point after a block.

- Moving summation point from output to input block.

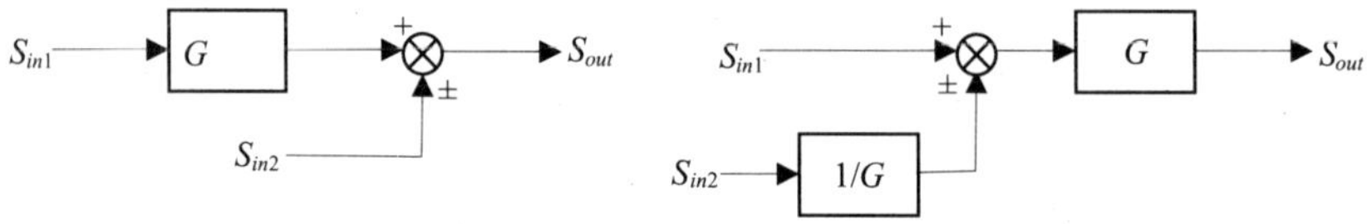

- Eliminating a feedback loop.

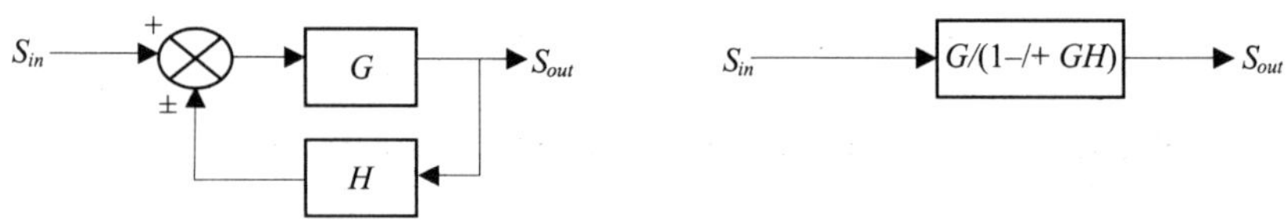

- Switching blocks.

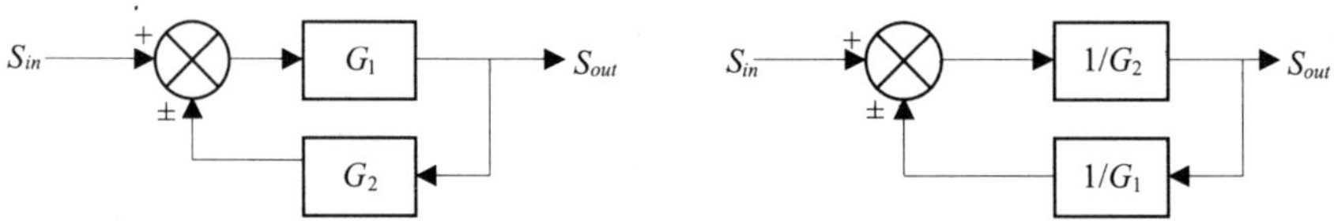

1.13 LAPLACE TRANSFORMS

Laplace transforms find extensive usage in control systems theory for description, evaluation and design. The mathematical equations for describing the time related connection between the input and the output parameters require derivative and integral functions. The Laplace transform theory offers the distinct advantage of converting a differential/integral equation into an algebraic function of frequency, which inherently is an easier process than operational calculus. The process effectively takes the problem from the time domain to the frequency domain.

The frequency-domain equations also establish the relation between the output (or the frequency response) and input parameters, the ratio of the two also defining the transfer function. After solving the algebraic equation, an inverse Laplace transformation yields the output as a function of time. Most control system evaluation and design problems avoid the inverse Laplace transformation by working directly with the algebraic equations. But the inverse transformation aids in understanding the situation and for answering stability related issues.

Note that the Laplace transformation is only one of a number of useful transforms. For instance, the logarithmic transformation converts numbers into their logarithms, thus replacing multiplication and division by the easier addition and subtraction. Other related methods such as the Fourier transform also find widespread usage.

A useful aspect of a component's characteristics is its response to sinusoidal inputs of more than one frequency. Knowledge of gain and phase difference of the component due to different frequencies is oftentimes of special interest. The analysis can be accomplished by letting the transfer function of the component to be a function of the frequency parameter s, where s has the unit of $(\text{second})^{-1}$. If the input is sinusoidal of frequency ω radians/second, and j is the imaginary operator $(\sqrt{-1})$, then the frequency parameter in the transfer function $s = j\omega$, yielding a complex number with magnitude M and angle θ, or $M\angle\theta$. Magnitude M provides the gain of the component and angle θ is the phase difference of the component.

Split into two categories, the functional Laplace transform relates to particular functions such as $\sin(\omega t)$ or $e^{-\alpha t}$, while operational transforms are concerned with the result of an operation involving one or more functions. The notation $\mathcal{L}\{f(t)\}$ and $F(s)$ are commonly used to indicate the Laplace transform of $f(t)$, which is defined by the expression:

$$\mathcal{L}\{f(t)\} = F(s) = \int f(t)e^{-st}dt \qquad ...(1.20)$$

where the upper and lower limits of integration are 0 and ∞. As an illustration, the Laplace transform of a constant K is:

$$F(s) = \int K.e^{-st}dt = -(K/s)(e^{-s\infty} - e^{-0}) = -(K/s)(0-1) = K/s \qquad ...(1.21)$$

When the Laplace transform of two functions acting independently are known, then operational transforms provide a means to obtain the Laplace transform of the two functions added together, or the derivative or integral of a single function. Table 1.1 provides Laplace transform pairs of commonly encountered functions in control systems theory. The tabulation aids in finding the Laplace transform of a time-domain function or the inverse Laplace transform of a frequency-domain function.

The inverse Laplace transform converts a frequency-domain function of s into a time-domain function of t. If the frequency domain function is available in the tabulation of Laplace transform pairs, the inverse transform is the associated time-domain function in the table. But the frequency-domain expressions in control system analyses mostly occur as a ratio of polynomials made up of integer powers of s, so the task is more complex than a simple conversion. For instance, consider a function with both polynomials factored in the form:

$$X(s) = 5[(s+3)(s+7)]/[s(s+2)(s+4)] \qquad ...(1.22)$$

If the ratio of polynomials is split into a sum of terms where each term is in the Laplace transform table, the task of obtaining the inverse transform is facilitated, and indeed, the 'method of partial fractions' does precisely the same. The above polynomial may be expanded using partial fractions to assume the form:

Table 1.1 Laplace Transform Pairs

Time Domain, $f(t)$, $t > 0$	Frequency Domain, $F(s)$
1. K	K/s
2. Kt	K/s^2
3. Ke^{-at}	$K/(s+a)$
4. Kte^{-at}	$K/(s+a)^2$
5. $K.\sin(\omega t)$	$K\omega/(s^2+\omega^2)$
6. $K.\cos(\omega t)$	$Ks/(s^2+\omega^2)$
7. $Ke^{-at}.\sin(\omega t)$	$K\omega/[(s+a)^2+\omega^2]$
8. $Ke^{-at}.\cos(\omega t)$	$K(s+a)/[(s+a)^2+\omega^2]$
9. $K(1-e^{-t/\tau})$	$K/[s(\tau s+1)]$
10. $2Ke^{-at}.\cos(bt+\theta)$	$\{K\angle\theta/[s+a-jb]\}+\{K\angle-\theta/[s+a+jb]\}$
11. $K.f(t)$	$K.F(s)$
12. $f_1(t)+f_2(t)-f_3(t)$	$F_1(s)+F_2(s)-F_3(s)$
13. $df(t)/dt$	$sF(s)-f(0)$
14. $d^2f(t)/dt^2$	$s^2F(s)-sf(0)-df(0)/dt$
15. $d^3f(t)/dt^3$	$s^3F(s)-s^2f(0)-sdf(0)/dt-d^2f(0)/dt^2$
16. $\int f(t)dt$	$F(s)/s$
17. $f(t-a)$	$e^{-as}.F(s)$

Note: K represents a numerical constant.

$$X(s) = 5[(s+3)(s+7)]/[s(s+2)(s+4)] = K_1/s + K_2/(s+2) + K_3/(s+4) \quad ...(1.23)$$

where K_1, K_2 and K_3 are constants to be determined. In general, a term is present on the right side for each root of the polynomial in the denominator of the left side. Multiple roots for factors such as $(s+2)^n$ will have a term for each power of the factor from 1 to n. Thus:

$$Y(s) = 8(s+1)/[(s+2)^2] = K_1/(s+2) + K_2/[(s+2)^2] \quad ...(1.24)$$

Complex roots are commonly encountered, and will occur in conjugate pairs. The corresponding constants in the numerator will also be complex conjugates. If $j = \sqrt{-1}$, K is a complex constant and K^* is its complex conjugate, then:

$$Z(s) = 7.5/(s^2+2s+5) = K/(s+1-j2) + K^*/(s+1+j2) \quad ...(1.25)$$

A two-step approach is useful in determining each distinct non-multiple real or complex root. To determine the value of the constants, for instance K_2 in equation (1.23), multiply both sides of the equation by the denominator of K_2, $(s+2)$. Next, replace s on both sides by the root of the factor that was multiplied (or, replace s by -2). The remaining terms on the right side will be zero, leaving only the constant term K_2. The left side reduces to a numerical value, that of the constant K_2. The process is repeated for each term on the right side of the equation. In the case of complex roots, the reduction of the left side will be a complex term, as may be expected. Additional steps are needed in the event of multiple roots. The inverse transformation completes the execution. The procedure is illustrated in example problems # 1.8 to 1.10.

To ensure that the analytically predicted solution to a problem is correct, a quick check of the initial and final values is helpful. The initial and final value theorems are available for answering the question, and may even eliminate the need to find the inverse transformation of the frequency-domain solution. The initial value theorem states that multiplying $X(s)$ by s and taking the limit as $s \to \infty$ will yield the initial value of $x(t)$, or $x(0)$. The final value theorem states that multiplying $X(s)$ by s and taking the limit as the parameter $s \to 0$ will yield the final value of $x(t)$, or $x(\infty)$.

1.14 REVIEW PROBLEMS

Example problem # 1.1: The input signal to a vibration measurement instrumentation is 0.75 Hz sinusoidal signal with a peak amplitude of 13×10^{-3} V and a phase angle of 45°. The output has a peak amplitude of 10×10^{-3} ins. and a phase angle of 27°. Calculate the gain, phase angle and transfer function of the instrument.

Solution:

$$\begin{aligned} \text{Gain} &= \text{output/input} = 10 \times 10^{-3}\ \text{in.}/13 \times 10^{-3}\ \text{V} \\ &= 0.769\ \text{in./V} \\ \text{Phase difference} &= (\text{output angle}) - (\text{input angle}) = (27 - 45)^\circ \\ &= -18^\circ \\ \text{Transfer function} &= 0.769\angle{-18^\circ}\ \text{in./V} \end{aligned}$$

Example problem # 1.2: A pressure probe transmitter has an input range of 2.5 to 150 lbs/in² and output range of 5 to 25 mA. If the transmitter is assumed to be linear, determine (*i*) the transmitter gain $H = C_m/C$, (*ii*) expression for output C_m in terms of input C, (*iii*) output of transmitter when input is 70 lbs/in², and (*iv*) input required when output is 15.5 mA.

Solution:

(*i*) Gain H = (output range)/(input range) = [(25 – 5) mA]/[(150 – 2.5) lbs/in²]
= 0.136 mA/lbs/in²

(*ii*) Expression for slope/intercept equation of straight line is $C_m = mC + b$

Slope is given by the gain, so $m = H = 0.136$

To determine intercept b, C_m = 25 mA when C = 150 lbs/in².

Hence: $25 = 0.136 \times 150 + b$, so $b = 4.6$

The required expression is: $C_m = 0.136C + 4.6$

(*iii*) If C = 70 lbs/in², then $C_m = 0.136 \times 70 + 4.6 = 14.12$ mA

(*iv*) If C_m = 15.5 mA, $C = (15.5 - 4.6)/0.136 = 80.15$ lbs/in².

Example problem # 1.3: A spring of stiffness 75 lbs/in has one end fixed and the other end attached to a mass weight of 35 lbs. Assuming no damping or external forces acting on it, write the expression describing the free motion of the mass.

Solution: The term $\sqrt{(k/m)} = \sqrt{[(75\ \text{lbs/in})/(35\ \text{lbs}/386\ \text{in/sec}^2)]}$
= 28.76 radian/sec.

Using equation 1.4, the expression for the free motion of the mass is:

$$x = C_1 \sin 28.76t + C_2 \cos 28.76t$$

Example problem # 1.4: In example problem # 1.3 assume that the mass is pulled out of its stationary position at time $t = 0$ with $x_0 = 0$ and released with an initial velocity $dx_0/dt = 1.25$ ins/second. Write the equation of motion.

Solution: At $t = 0: x = 0$ and $dx_0/dt = 1.25$

Substituting in the general expression: $0 = 0 + C_2$, or, $C_2 = 0$

Differentiate the general expression: $dx/dt = 28.76 \times C_1 \cos 28.76t$

At $t = 0$: $1.25 = 28.76 \times C_1$

Or: $C_1 = 0.0435$

The equation of motion is: $x = 0.0435 \times \sin 28.76t$

Example problem # 1.5: In example problem # 1.3 determine the period of vibration and the natural circular frequency of vibration.

Solution: The period of vibration T is given in equation 1.6:

$$T = 2\pi \times \sqrt{(m/k)} = 2\pi \times \sqrt{[(35 \text{ lbs}/386 \text{ in/sec}^2)/(75 \text{ lbs/in})]}$$

$$= 0.218 \text{ sec.}$$

The natural circular frequency $f_n = 1/T = 1/0.218 = 4.58$ cycles/sec.

Example problem # 1.6: A vibration system of the type shown in figure 1.5 has a mass $m = 5.5$ lbs, spring stiffness $k = 11.75$ lbs/in. and damping $c = 0.375$ in./sec. The system is subjected to a sinusoidal force of amplitude 4.25 lbs and of frequency 30 radians/sec. Determine (*i*) critical damping, (*ii*) damped natural frequency, (*iii*) amplitude of motion and corresponding phase angle.

Solution:

(*i*) From equation 1.12 critical damping:

$$c_c = 2\sqrt{(mk)} = 2 \times \sqrt{[(5.5 \text{ lbs}/386 \text{ in/sec}^2) \times (11.75 \text{ lbs/in})]} = 0.818 \text{ in./sec.}$$

(*ii*) From equation 1.7 natural frequency:

$$\omega_n = \sqrt{(k/m)} = \sqrt{[(11.75 \text{ lbs/in})/(5.5 \text{ lbs}/386 \text{ in/sec}^2)]} = 28.716 \text{ radian/sec.}$$

From equation 1.15 damped natural frequency:

$$q = \omega_n \times \sqrt{[1 - (c/c_c)^2]} = 28.716 \times \sqrt{[1 - (0.375/0.818)^2]} = 25.524 \text{ radian/sec.}$$

(*iii*) From equation 1.18 amplitude of motion:

$$x_0 = (P_0/k)/\{[1 - (\omega/\omega_n)^2]^2 + [2(c/c_c \times \omega/\omega_n)^2]\}^{1/2}$$

$$= (4.25/11.75)/\{[1 - (30/28.716)^2]^2 + [2 \times (0.375 \times 30/0.818 \times 28.716)^2]\}^{1/2}$$

$$= 0.529 \text{ in.}$$

From equation 1.19 corresponding phase angle:

$$\tan(\varphi) = 2(c/c_c \times \omega/\omega_n)/[1 - (\omega/\omega_n)^2] = -10.477, \quad \text{or} \quad \varphi = -84.5°$$

Example problem # 1.7: Using the block simplification technique, reduce the control system shown below into an equivalent single block.

Solution: The variables may be expressed as: $E = S - T$

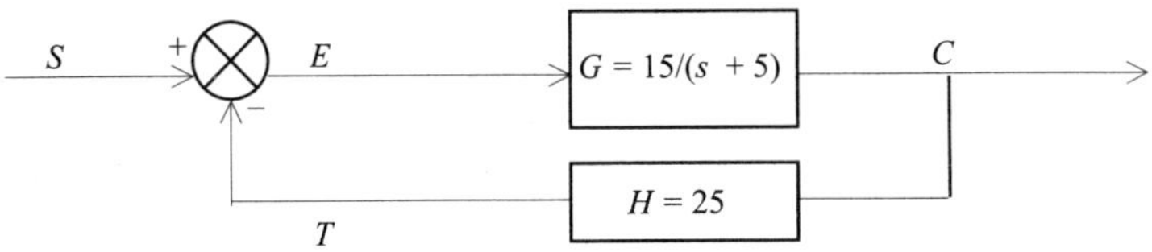

$$C = GE$$

$$T = CH$$

Hence: $C = G(S - T)$

And: $C = G(S - CH)$

Simplifying: $C = GS - GCH$

Or: $C/S = G/(1 + GH)$

Substituting values of G and H: $C/S = [15/(s + 5)]/[1 + \{15 \times 25\}/(s + 5)]$

The single block representation of the control system is given below.

$$\frac{15}{(s+5)\times[1+375/(s+5)]}$$

Exmple problem # 1.8: Apply the partial fraction expansion and inverse transformation method to the following equation:

$$X(s) = 5[(s+ 3)(s + 7)]/[s(s + 2)(s + 4)] = K_1/s + K_2/(s + 2) + K_3/(s + 4)$$

Solution:

(*i*) Find K_1: Let $s = 0$. Then: $K_1 = [5(0 + 3)(0 + 7)]/[(0 + 2)(0 + 4)] = 13.125$

(*ii*) Find K_2: Let $s = -2$. Then: $K_2 = [5(-2 + 3)(-2 + 7)]/[-2(-2 + 4)] = -6.25$

(*ii*) Find K_3: Let $s = -4$. Then: $K_3 = [5(-4 + 3)(-4 + 7)]/[-4(-4 + 2)] = -1.875$

Then: $X(s) = 13.125/s - 6.25/(s + 2) - 1.875/(s + 4)$

The inverse Laplace transformation from table 1.1 is:

$$x(t) = 13.125 - 6.25e^{-2t} - 1.875e^{-4t}$$

Example problem # 1.9: Provide the partial fraction expansion and inverse transformation of the equation:

$$Y(s) = 5(s + 1)/[(s + 2)^2] = K_1/(s + 2) + K_2/[(s + 2)^2]$$

Solution: Multiply both sides of the equation by $(s + 2)^2$. Then:

$$5(s + 1) = K_1(s + 2) + K_2$$

Replace s by -2 and solve for K_2:

$$5(-2 + 1) = K_1(-2 + 2) + K_2$$

or:

$$K_2 = -5$$

Next, substitute value of K_2 into the equation $\{5(s + 1) = K_1(s + 2) + K_2\}$ and determine the value of K_1.

$$5(s + 1) = K_1(s + 2) - 5$$

$$K_1 = [5(s + 1) + 5]/(s + 2)$$

Letting $s = 0$, then $K_1 = 5$.

The partial expansion of the original equation is:

$$Y(s) = 5/(s + 2) - 5/[(s + 2)^2]$$

The inverse Laplace transform is obtained from table 1.1:

$$y(t) = 5e^{-2t} - 5te^{-2t}$$

Example problem # 1.10: Provide the partial fraction expansion and inverse transformation of the equation:

$$Z(s) = 7.5/(s^2 + 2s + 5) = K/(s + 1 - j2) + K^*/(s + 1 + j2)$$

where K and K^* are complex conjugates, and $(s^2 + 2s + 5) = (s + 1 - j2)(s + 1 + j2)$.

Solution: Multiply left side by: $[7.5/(s^2 + 2s + 5)] \times (s + 1 - j2) = 7.5/(s + 1 + j2)$

Then at $s = -1 + 2j$, $K = 7.5/(s + 1 + j2) = 7.5/(-1 + 2j + 1 + 2j)$

$$= 7.5/4j$$

$$= -1.875j = 1.875\angle -90°$$

and:

$$K^* = 1.875\angle 90°$$

The partial fraction expansion is:

$$Z(s) = 1.875\angle -90°/(s + 1 - j2) + 1.875\angle 90°/(s + 1 + j2)$$

The inverse Laplace transform is:

$$z(t) = 2 \times 1.875 \times e^{-t} \times \cos(2t - 90°)$$

$$= 3.75 \times e^{-t} \times \cos(2t - 90°)$$

Example problem # 1.11: A control valve has the following step response:

$$X(s) = 7.5/s(s^2 + 2s + 5)$$

Determine the initial $x(0)$ and final $x(\infty)$ positions of the valve.

Solution: Initial position: $x(0) = \text{limit } [sX(s)]$ as $s \to \infty$, $[7.5s/\{s(s^2 + 2s + 5)\}] = 0$

Final position: $x(\infty) = \text{limit } [sX(s)]$ as $s \to 0$, $[7.5s/\{s(s^2 + 2s + 5)\}] = 1.5$ mm.

1.15 EXERCISES

1.1 Name the input and output parameters of the type shown in figure 1.2 for a vibration measuring accelerometer.

1.2 An integrator of a signal processor can convert an electronic signal from an accelerometer. What will be the engineering parameters and units if the signal is integrated the first time, then reintegrated a second time?

1.3 A temperature measurement probe's transmitter has a linear input/output characteristic of the type shown in figure 1.4. If the temperature range is 100 to 1500°F and the output range is 5 to 25 mA, find (*i*) the transmitter gain $H = C_m/C$, (*ii*) the equation for output C_m in terms of input C, (*iii*) output when input is 750°F, and (*iv*) input temperature when out signal is 12.75 mA.

1.4 If the spring stiffness k is 7.5 lbs/in of an undamped single degree of freedom vibrator and has period of oscillation of 0.135 minutes, find the magnitude of the attached mass.

1.5 When the stiffness of the spring in a mass-spring-damper vibration model is decreased, what is its impact on (*i*) the period of oscillation, (*ii*) c/c_c ratio, (*iii*) undamped circular frequency ω_n, and (*iv*) damped natural frequency q?

1.6 If the damping ratio $c/c_c < 1$ and the frequency ratio $\omega/\omega_n > 1$ in a mass-spring-damper vibration model subjected to a sinusoidal excitation on the mass, what is the effect of increasing the mass on (*i*) the amplitude of motion x_0, (*ii*) phase angle φ?

1.7 A mass m of 45 lbs is attached to a spring of stiffness k = 135 lbs/in. System damping is estimated to be 5.5 in./sec. The system is subjected to a sinusoidal force of amplitude 57.5 lbs and frequency 27.5 cycles/sec. If the vibration response is to be halved, calculate the necessary equivalent spring stiffness.

1.8 A mass m of 2.5 lbs is attached to a spring of stiffness k = 4.5 lbs/in. Determine the damping coefficient if the system when subjected to a sinusoidal force of amplitude 5.25 lbs and frequency 33 cycles/sec has a response of 0.175 in.

1.9 Determine the transfer function of the following control system.

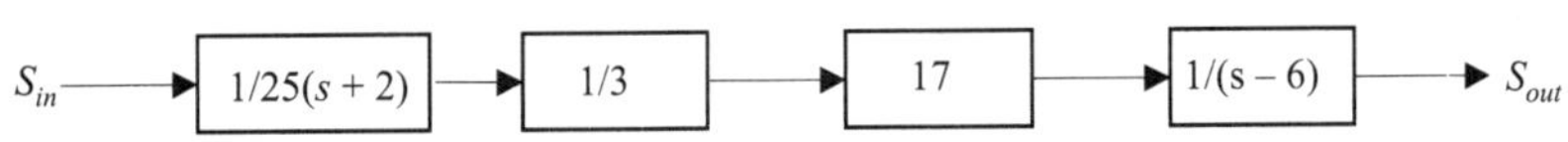

1.10 Reduce the system shown below to an equivalent single block system.

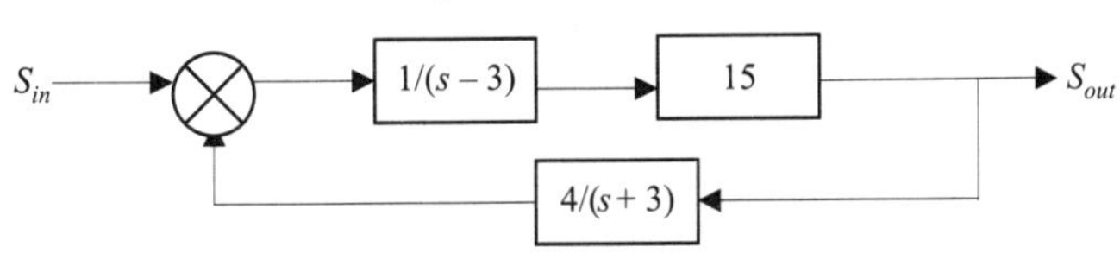

1.11 Determine the transfer function of the following control system.

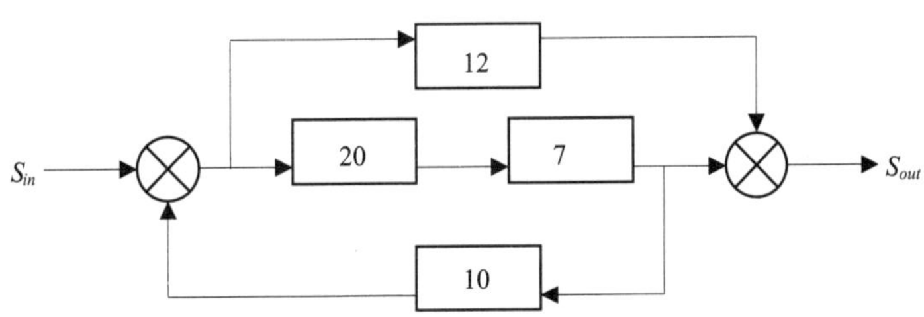

1.12 Reduce the control system shown in the diagram to a single block.

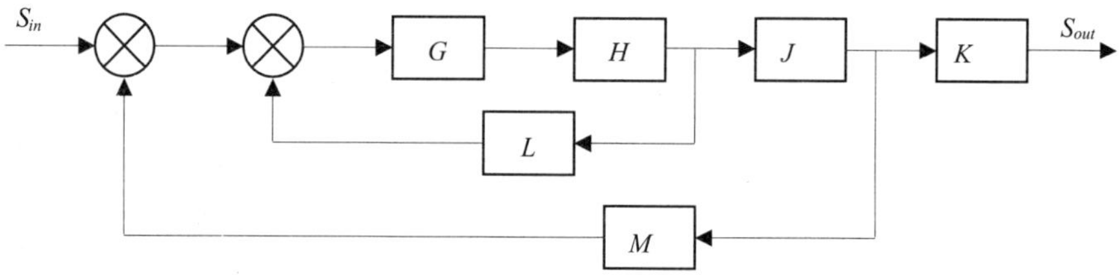

1.13 Find the Laplace transform of the following functions:

(*i*) $f(t) = 7.5$ (*ii*) $f(t) = 45 \cos(37.5t)$

(*iii*) $f(t) = 55t$ (*iv*) $f(t) = 2.5e^{-2t}$

(*v*) $f(t) = 35\, e^{-2t} \cos(37.5t)$

1.14 Apply the partial fraction expansion and inverse transformation method to the following equation assuming all initial conditions are zero:

$$f(t) = 3.25\, dx(t)/dt + 7.25x(t) + 3.5\int x(t)dt$$

1.15 Determine the initial $x(0)$ and final $x(\infty)$ positions of the valve with the following step response:

$$Y(s) = 5(s + 1)/[(s + 2)^2]$$

References & Bibliography

Bateson, R.N., "Introduction to Control System Technology", Prentice-Hall, Clifton, NJ, 1999.
Boyce, M., "Handbook for Cogeneration and Combined Cycle Power Plants", ASME Press, New York, 2002.
Carstens, J.R., "Automatic Control Systems and Components", Prentice-Hall, Clifton, NJ, 1990.
DeRoy, B.E., "Automatic Control Theory", John Wiley & Sons, New York, 1996.
Funakubo, H., "Actuators for Control", Gordon and Breach Science Publishers, New York, 1991.
Kiameh, P., "Power Generation Handbook", McGraw-Hill, New York, 2002.
Miller, R.W., "Servo-mechanism Devices and Fundamentals", Reston Publishing Co., Reston, VA. 1977.
Patrick, D.R. and Fardo, S.W., "Industrial Process Control System", Prentice-Hall, Clifton, NJ, 1979.
Rangwala A.S., "Turbo-machinery Dynamics: Design and Operation", McGraw-Hill, New York, 2005.
Sante, D.P., "Automatic Control System Technology", Prentice-Hall, Clifton, NJ, 1980.

2

Control Methods

2.1 INTRODUCTION

Automatic control systems may be divided into a number of groups. The split may be based on usage of feedback, the nature of the signals received and issued, values required for the set point, and the environment in which the application occurs.

The open or closed configuration of the loop is determined by the need for feedback. If the feedback mechanism is inserted, the loop will be closed. The analog or digital method of control will be governed by the continuous or discrete characteristics of the signals. Continuous signals will call for analog method of control, while discrete signals need the digital approach. If the set-point changes are seldom, the regulator system will suffice. However, frequent changes in the set point will require a follow-up method of control.

Control of a process in a processing environment may take the form of continuous or batch systems. Manufacturing of discrete parts and assembling of components requires machine control, either through a numerical or a robotic system. Depending on the flow of the product in the manufacturing circuit and on the assembly line, the process may be classified as continuous or intermittent and periodic.

The controllers may be located in one central area or may be distributed in close proximity of the sensors and actuators. Some other classifications may be based on the use of servo-mechanisms, control of process related to sequence either of event or of time, and on the type of programmable controllers.

The controllers rely on mechanical, electrical, electronic, pneumatic, hydraulic or thermal elements for operation. A combination of two or more element types is also extensively used.

2.2 ANALOG AND DIGITAL CONTROL

Figure 2.1 shows an example of variation of compressor discharge pressure during a slow acceleration, and is depicted in two different forms. The two forms are the 'analog' and the 'digital' signals. The continuously varying analog signal of the pressure parameter is capable of assuming any value between set limits. The digital version of the signal, on the other hand, takes discrete values between the limits.

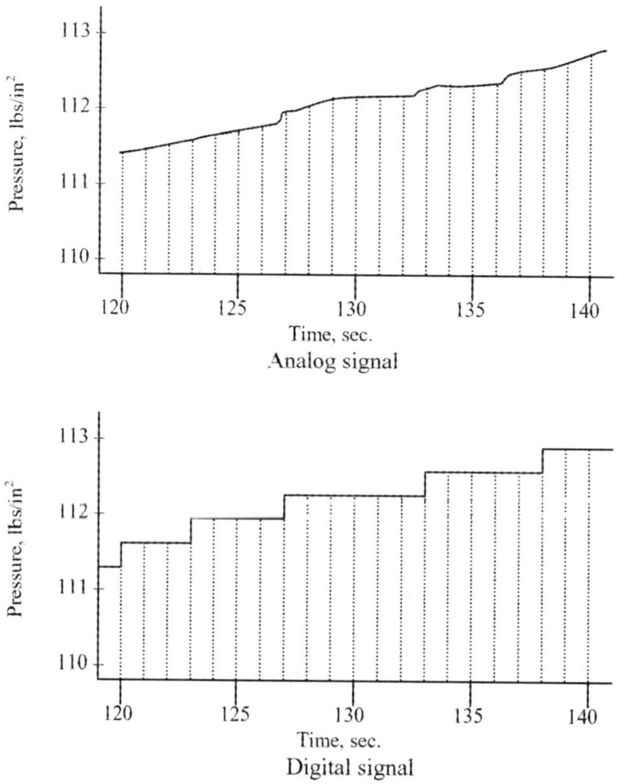

Fig. 2.1 Different Forms of Signal Indicating Compressor Discharge Pressure

A graphical representation of the analog signal with time is displayed as a continuous line. The discrete values of the digital signal causes its graph with time to appear disjointed, with straight-line segments connected by time steps. The signal's magnitude does not change during the interval, but may jump to a different value for the succeeding interval.

2.3 REGULATOR AND FOLLOW-UP METHODS

A regulator in a feedback control system is used to maintain the controlled variable at a constant set point, even in the presence of unwarranted changes in the load acting on the system. A pressure or a temperature regulator in a process control to maintain the pressure or temperature at a fixed level for a processing condition is an example. The speed controller for a power generation steam turbine needs the regulator method to maintain the speed of the rotor at 3,600 rpm for 60 Hz electric power, or 3,000 rpm for grids operating at 50 Hz.

If the reference set point of the controlled variable in a feedback control system needs to be modified frequently, the follow-up method is essential. The follow-up system keeps the controlled variable in close correspondence with fluctuations in the set point. Some examples are exhaust gas temperature, bearing vibration level, and variable stator vane angle setting of an aircraft engine during take-off.

2.4 PROCESS CONTROL

The principle objective of a process control system is to regulate the variables associated with it through systematic exploitation of the combination of available equipment and materials. The desired end results are brought about through changes in energy, physical properties, or chemical properties.

To understand the concept, consider the closed-loop block diagram shown in figure 2.2 used in the design of a typical system for control of a process. The set point (SP) acts as the input and controlled variable C is the output. A measuring transmitter of transfer function H is inserted in the feedback path, while the forward path has a control mode, manipulator, and the process element, their respective transfer functions being G_c, G_m and G_p. Thus, the combined forward transfer function is $G = G_c G_m G_p$. The blocks represent all of the necessary equipment, and procedures are carried out through which the output function is controlled. The measuring sensor's transmitter indicates the value of the controlled variable and converts it into a useable signal. The controller is composed of the error detector to compute the difference between the measured and required (or set-point) values of the controlled variable and a unit to implement the control modes. The control modes convert the error signal into command instructions for reducing the error. Some commonly used control modes are the proportional mode P, the integral mode I, and the derivative mode D. The control modes are discussed at length in section # 2.10. The manipulating element executes the command instructions issued by the controller's output to regulate the output variable. The manipulating system comprises of an 'actuator' to translate the requirements of the controlled variable into an actionable instruction and a 'final controlling element' that carries out the change in the output signal. The manipulating elements may be in the form of valves, dampers, fans, pumps, and other associated components.

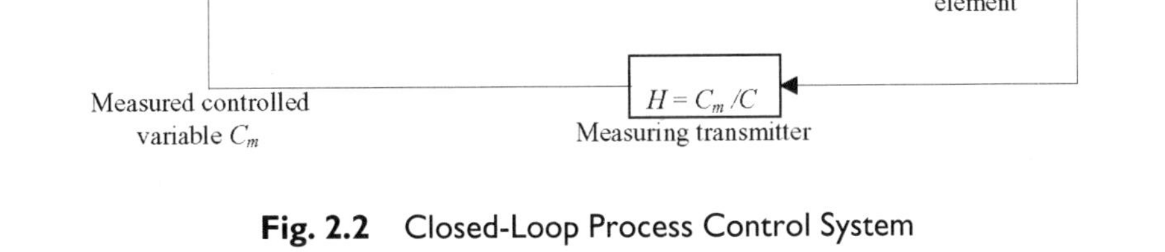

Fig. 2.2 Closed-Loop Process Control System

Since it is available, the measured variable C_m may be used instead of C to compare with the input signal SP, and thus establish the performance of the control system. The transfer function C_m/SP of the overall process control system may be obtained in the following manner:

$$E = SP - C_m$$

$$C = EG$$

$$C_m = CH$$

$$C_m = EGH$$
$$C_m = (SP - C_m)GH$$
$$C_m + C_mGH = SPGH$$

Hence: $$C_m / SP = GH/(1 + GH) \quad ...(2.1)$$

As a further illustration of the functioning of a process control system, figure 2.3 shows the instrumentation diagram of a compensated flow control loop used in a gas turbine engine using natural gas as fuel. The rate of flow of natural gas is required to be in proportion by weight to the demand signal. The flow meter produces a signal that is proportional to the square of the volume flow rate Q of the gas. If the output of the flow meter is D and meter proportionality constant is k_f,

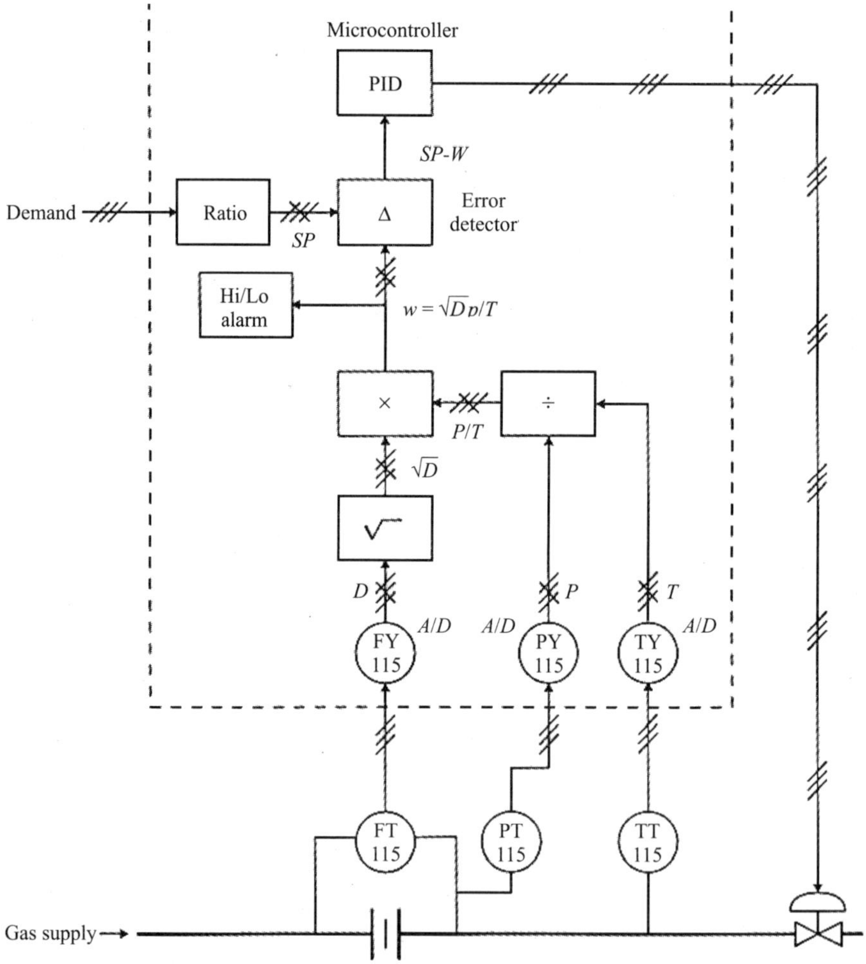

Fig. 2.3 Instrumentation Schematic of Compensated Mass Flow Control Loop (*Courtesy*: Foxboro Company)

then $Q = k_f \times \sqrt{D}$. Conversion of volume V to mass m depends on absolute pressure p, absolute temperature T, universal gas constant R = 8.314 J/K·mol and gram molecular weight M of the gas, and is in accordance with the following equation:

$$m = (MVp)/(10^3 RT) \quad ...(2.2)$$

Dividing both sides by time t, and noting flow rate $Q = V/t$, then:

$$w = m/t = (MVp/t)/(10^3 RT) = (MQp)/(10^3 RT) \quad ...(2.3)$$

Further substitution of Q in terms of flow meter parameter leads to:

$$w = k \times \sqrt{D} \times (p/T) \quad ....(2.4)$$

where $k = Mk_f/(10^3 R)$ = mass flow proportionality constant, D = flow meter differential pressure, pascal, p = absolute pressure of the gas, pascal, and T = absolute temperature of the gas, kelvin.

The output signal from the flow meter, D, is converted into mass flow rate signal, w, in accordance with equation 2.4. The mass flow rate acts as an input signal to the PID controller, and its output is applied to the control valve to regulate the mass flow rate of natural gas to the fuel injection assembly in the gas turbine combustor.

2.5 SERVO-MECHANISM

Servo-mechanism controls are predominantly used to obtain a specific physical motion or position of a controlled parameter in a feedback system. Acting as a part of another system, the follow-up servo-mechanism is capable of maintaining an output position as it tracks the input reference signal with the aid of electro-mechanical devices. The stylus of a vibration recording on a strip chart, for example, requires a servo-mechanism to establish its position as the chart travels at a regulated pace.

Although the development of closed-loop process control and servo-mechanism followed their own path, the mathematical tools required in evaluating the two are identical. Differences arise in the design techniques. Also, smaller time steps, oftentimes less than 1 second, are the norm for servo-mechanisms since the requisite procedures are inherently faster. Time periods may be measured in minutes or even hours for the slower moving process control methods. Another distinguishing feature of servo-mechanism components is better definition of the analytical techniques. As a consequence, a more precise design of the controller is possible with little or no adjustments necessary in the field. Because of a general lack of mathematical accuracy, process controls necessitate more elaborate fine-tuning to obtain optimum results.

Many criteria for evaluating operating machinery vibrations are based on velocity, and it is often preferable to express it in terms of displacement. It may also be desirable to determine the total lateral motion of a rotating shaft by adding displacement of the bearing housing, measured with an accelerometer, to the relative displacement of the shaft in its bearing, measured with proximity probes. Acceleration signals can be electronically integrated to yield velocity and displacement. Assuming a sinusoidal representation, a typical displacement signal may be expressed by the equation:

$$x(t) = X_0.\sin(\omega t) \quad ...(2.5)$$

where X_0 is the amplitude of motion and ω is the speed of the rotating shaft. Differentiating once against time t yields velocity:

$$v(t) = dx(t)/dt = X_0.\omega.\cos(\omega t) \quad ...(2.6)$$

Differentiating a second time provides the acceleration term:

$$a(t) = d^2x(t)/dt^2 = -X_0.\omega^2.\sin(\omega t) \quad ...(2.7)$$

A phase angle of 90° separates the displacement signal from the velocity signal, and a further 90° separates the velocity from the acceleration. Also, the amplitude of velocity equals the product of the amplitude of displacement and angular speed, and the amplitude of acceleration is ω^2 times the amplitude of displacement.

Detectors are used to extract parameters that characterize a signal, typical ones being arithmetic average, mean square and root mean square values. A full wave rectifier obtains the signal magnitude, and with a smoothing circuit, produces the arithmetic average. Mean square and root mean square values are independent of phase relationships, and are preferred for describing stationary signals. Mean square values are directly additive when two signals of different frequency bands are added together.

Vibration meters receive signal from a transducer and process it to display relevant vibration parameters. For measurement on rotating machines, a frequency range of 10 Hz to 10 kHz is desirable. The lower frequency limit includes any sub-harmonic components due to factors such as oil whip in plain bearings. The upper limit includes higher order frequencies arising from blade passing, tooth meshing frequencies and their harmonics in gears and frequencies associated with faults in rolling element bearings.

Evaluation of vibration signatures can be done at a later date if the signal is stored on instrumentation tape recorders using direct, frequency modulation, or digital techniques. Ability to record low frequency signals is the prime difference between the methods. Direct recordings cannot record down to a direct current signal, 25 Hz being the lower limit. In contrast, the frequency modulation method can go down to a direct current signal, which represents a constant deviation of the carrier frequency.

Figure 2.4 provides a schematic diagram of a typical vibration measuring system consisting of preamplifier, signal conditioner, and an error detector. The set point input signal is obtained from an accelerometer mounted on a bearing housing or on an outer turbine casing flange. Many accelerometers operate on the piezo-electric effects of certain crystals. When the crystal is physically shaken, or oscillated, it generates small electric charges whose frequency and amplitude correspond to the physical motion. The charge is collected from the surface of the crystal and amplified for processing. The raw signal after charge amplification is not of much value in evaluating the machine's vibration characteristics. The signal requires processing before it can be put to use, and must be passed through a number of aliasing, band-pass, and tracking filters, and further amplifiers before it can be used to evaluate the vibration characteristics of the machine component on which the accelerometer is mounted. Vibration meters often combine all of these elements in a single unit.

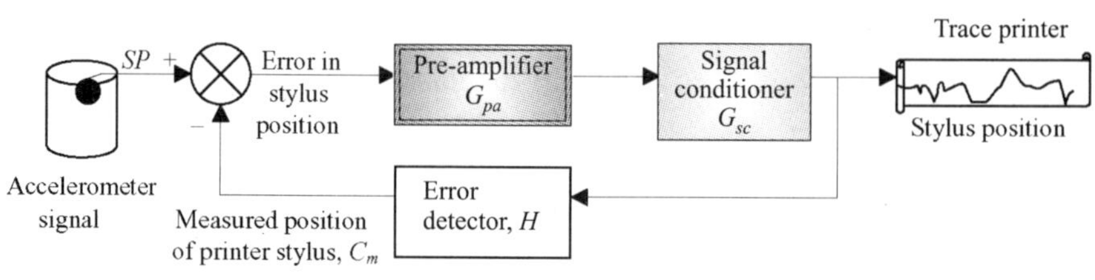

Fig. 2.4 Vibration Measurement System

2.6 SEQUENTIAL CONTROL

A 'sequential process' calls for execution of a sequence of one or several operational steps in a prescribed order. The orderly execution of the steps ensures production of a set quantity of the finished material. The final product may be a turbine blade or some fluid used during the manufacturing cycle or a turbine blade.

Lubrication oil returning from a bearing sump of an aircraft engine needs to be processed before it can be returned to carry on the task of lubricating the rolling element bearings. The oil goes through a metal chip detector, a filter with a coarse mesh, another filter with a fine mesh, oil cooler, excess oil reservoir, and finally pumped back to the oil sprayer in the sump. Loose metal chips hanging around the edges of a component left over during manufacture often get disengaged. Rubbing between mating rotating and stationery components in the seals is another source of loose particles. The suspended particles are washed away by the flow of lubricant, and must be removed before getting caught in the tight clearance areas between mating components and scoring the surfaces. Contamination of the oil from moisture condensate, gas fumes, and other hydraulic system fluids cannot be readily taken care of since it is not expedient in aviation applications. Most engine manufacturers require the lubricant oil to be replaced after the engine accumulates a certain number of operating hours or cycles.

The demands placed on turbine blades operating at high temperatures require the application of directional solidification, or columnar grain, form of alloys. This technology is preferred to the conventional casting method since grain boundaries can be oriented normally to the stress axis. This considerably improves ductility, and also deletes the grain boundaries where failures may be expected to initiate. Single crystal alloys go a step further by totally eliminating all grain boundaries. Substantial improvements in the resistance to thermal fatigue is also available as a consequence of the solidification process in a specific direction, a feature of major significance for turbine blades operating at elevated temperatures.

Since the array of grain boundaries run parallel to the principal axis of stress in the direction of solidification, the stresses at the weaker grain boundaries are minimized (*ref*: Ver Snyder et al., 1970). The initiation of failure is then prolonged, and creep rupture life is improved. Orientation of the primary and secondary axes in specific directions is possible by using seed crystals. Columnar grains may also be obtained in the solid state from the growth of secondary grains (*ref*: Hughes et al., 1978), but is easier to achieve with transformation in the liquid phase. Essentially a similar casting method is required for obtaining the columnar grain and the single crystal structures.

The process calls for the transition from the liquid to the solid phase according to a controlled thermal gradient to produce elongation of the grains in the direction parallel to the temperature variation. Thus, to obtain elongated grains along the principal stress axis, the thermal gradients are aligned parallel to it. In figure 2.5 the molten alloy is poured into a preheated ceramic shell in a vacuum, with the temperature maintained above that of the super alloy in the liquid form. A starter block of columnar grains is placed directly between the mold and cooled copper chilling plates for producing single crystal castings. The mold is open at the bottom to cause the material in the vicinity of the chilling plate to solidify upon contact and form a layer of equi-axed grains. The grains then proceed to form, aligning themselves with the direction parallel to the thermal gradient to create a columnar array. Induction coils are placed inside the casing of the furnace to permit exchange of heat between the walls and the mold. The temperatures along the height are closely controlled. Once the work-piece begins solidifying, the ceramic mold assembly is gradually retracted from the heated furnace base. Extraction of heat then proceeds by radiation between the

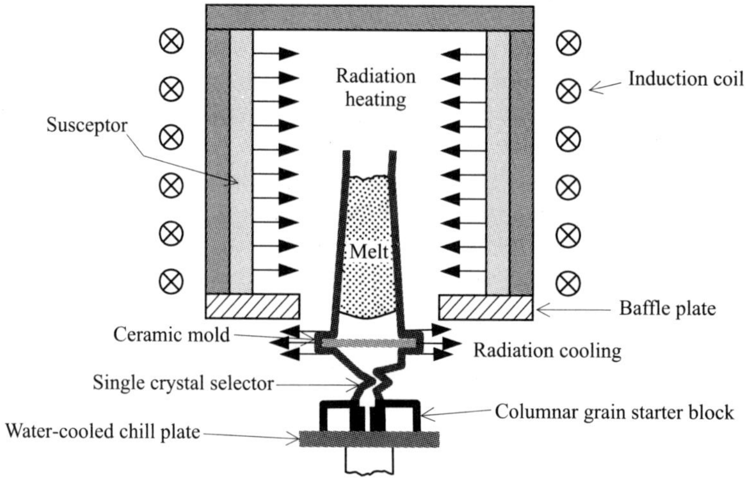

Fig. 2.5 Directional Solidification Process Schematic

mold and the cooler vacuum chamber walls. Thermal gradient is controlled with a baffle plate at the base of the furnace. The grain structure continues to grow from the base to fill the cavity in the mold to produce a complete airfoil casting with the desired grain structure.

The single crystal casting requires a helical constriction between the mold and the starter block, allowing only a single grain to pass through it. Solidification takes place in the form of dendrites growing in all three directions, so the continuously changing direction of the helix then restricts the growth of the dendrite to a single grain releasing at the top of the helix. The selected grain then expands to fill the cavity in a manner similar to the growth in columnar grain castings. Depending on the design of the shell, the process can produce hollow single crystal and columnar grain airfoils with cooling passages. For single crystal castings, the seed crystal is assumed to be of the same alloy with the same or higher melting temperature, and must be oriented to allow its duplication in the alloy material in the mold. The seed crystal does not need to melt completely, and hence it is placed on the chilled plate.

Casting defects may appear in the form of cracks at the grain boundaries in columnar grain components. A ceramic core is needed for hollow blades that will be subsequently leached out, and the difference in the coefficient of thermal expansion between the ceramic and the alloy builds up hoop stresses during cooling as the alloy shrinks around the core. This results in cracks at the grain boundaries. The problem is avoided by adding a small amount of hafnium to the alloy.

2.7 NUMERICAL CONTROL

A set of defined instructions is used in the 'numerical control' method for carrying out an operation. Processes such as boring, drilling of holes, milling of flat surfaces, grinding for smooth finish, and turning of circular features are performed on numerically controlled machining centers. Numerical values stored in a computer's database specify many different parameters. In the machining of the casing of a turbine, parameters such as location of a salient feature, machining speed, depth of tool cut and supply of coolant flow at the machining point are contained within the program of the numerical control machine to produce the part to desired tolerances.

A considerable amount of automation can be achieved in a modern computerized numerically controlled (CNC) machining center, and is essential in the case of a large component such as the casing for a turbine engine where the necessary dimensions are obtained from a computer-aided design (CAD). The mathematically defined features of the component provide total definition of the part, which is then used in the machining process.

Starting with the engineering drawings, the process uses the specifications and the sequence of operations as defined by the manufacturing engineer in the form of a computer program. At each step the programmer specifies data such as the *x*, *y* and *z* coordinates of the starting point of the machining operation on the part, machine running speed, and feed rate using a special programming language. Figure 2.6 illustrates a numerically controlled machining center, where the *x*-controller moves the work-piece along the axis of the table. The machining head travels along the *y*-axis, while the cutting tool with its attachment holder is carried along the vertically oriented *z*-direction. Position sensors measure and send the information to the central machine control unit. The controller compares the measured and required motion and sends electronic pulse signals. Individual machine actuators rotate lead screws to advance or retract the position after receiving appropriate command instructions from the control unit.

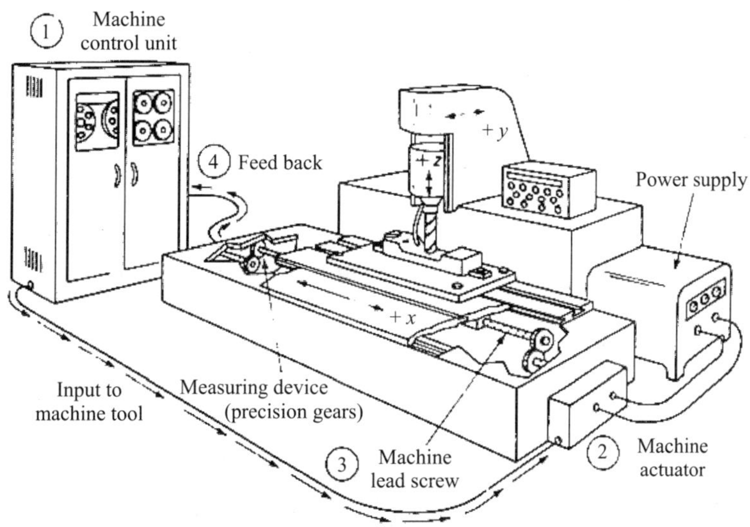

Fig. 2.6 Computerized Numerical Controlled Machining Center (*ref:* Olesten, 1970)

The automation can be extended even to relatively mundane tasks such as changing of cutting tools. Since a single unit can perform a number of tasks, the necessary cutting tools must be changed at the end of one process and the beginning of the following step. This calls for removal of the cutting tool used in the process that has just ended after the machining center is commanded to come to a pause, carrying of that tool to a storage rack, retrieving the proper tool from the storage rack for the next manufacturing process, transporting it to the tool holding mechanism in the machining center, mounting of the cutting tool, and restarting the machine to proceed with the next step.

2.8 DISCRETE PROCESS

A discrete process consists of a series of distinct operations executed sequentially, with each step bearing the hallmark of a defining condition for its initiation, a controlled form, and final conclusion. Discrete processes vary in complexity from a single output (such as on and off, or high and low) to complicated processes made of a number of operations, with each operation combining a number of steps. The steps built into each process are required to be performed in a definite order. In a manufacturing cycle of an item, for example, the completion of the requisite steps is essential before the finished product is ready. The sequence of manufacturing operations and the built-in steps are then repeated for production of additional items.

The event-driven sequential process will be used to explain the electro-discharge machining method, abbreviated EDM. The technique is used extensively for drilling of cooling holes in turbine blades. The procedure is essentially a controlled burning of the metal substrate on the surface of the blade through a shaped electrode. The process offers the advantage of avoiding potentially damaging forces and consequent hairline cracks around the hole that are encountered during a conventional metal chipping form of drilling procedure. The work-piece may be immersed in an oil bath to negate dielectric effects. The procedure can be accelerated by using a precisely sized shaped electrode, usually made of copper, which can machine a number of holes in a single pass. At the end of the pass the electrode is spent, and a new electrode is attached for drilling of more holes.

Figure 2.7 explains the machine setup. A hydraulic cylinder moves the platform and discharge head in the vertical direction between the 'Reset' position (up) and 'Hole Machined' position (down). In the 'Reset' position the limit switch LS1 is actuated, and in the 'Hole Machined' position the limit switch LS2 is actuated. The oil bath and the work-piece are mounted on the stationary table. A single-acting pneumatic cylinder with a return spring clamps the turbine blade against a bracket bolted on the support table. Limit switch LS3 is actuated when the clamping cylinder is in the 'Unclamped' position, while limit switch LS4 is actuated when the clamping cylinder is in the 'Clamped' mode. A double-acting pneumatic cylinder ejects the blade after all the holes are machined. Limit switches LS5 and LS6 are actuated when the cylinder is in the 'Ejector Retracted' and in the 'Ejector' modes.

Four solenoid valves are required to control the two cylinders. A three-position, four-way valve controls the hydraulic cylinder. Solenoid *A* is energized when the pump is running, which causes the cylinder to move the platform down, while solenoid *B* makes the cylinder move the platform up. A two-position, three-way valve controls the pneumatic cylinder. The cylinder is vented and the spring returns it to the unclamped position when solenoid *C* is de-energized. Energizing solenoid *C* supplies air to the cylinder and moves it into the clamped mode. The ejector is pushed into the retracted position when solenoid *D* is de-energized. Energizing solenoid *D* ejects the turbine blade as the cylinder moves to the ejected mode.

The sequential function flow chart of figure 2.8 describes the operational steps required for this operation. The progression through the ordered numbers of the steps is from top to bottom. The condition for advancing to the next step is shown next to the cross transfer path between intermediate steps.

2.9 CONTINUOUS PROCESS

An uninterrupted flow of inputs and outputs to regulate a desired variable parameter within certain bounds is the distinguishing feature of a continuous process. The output of continuous control

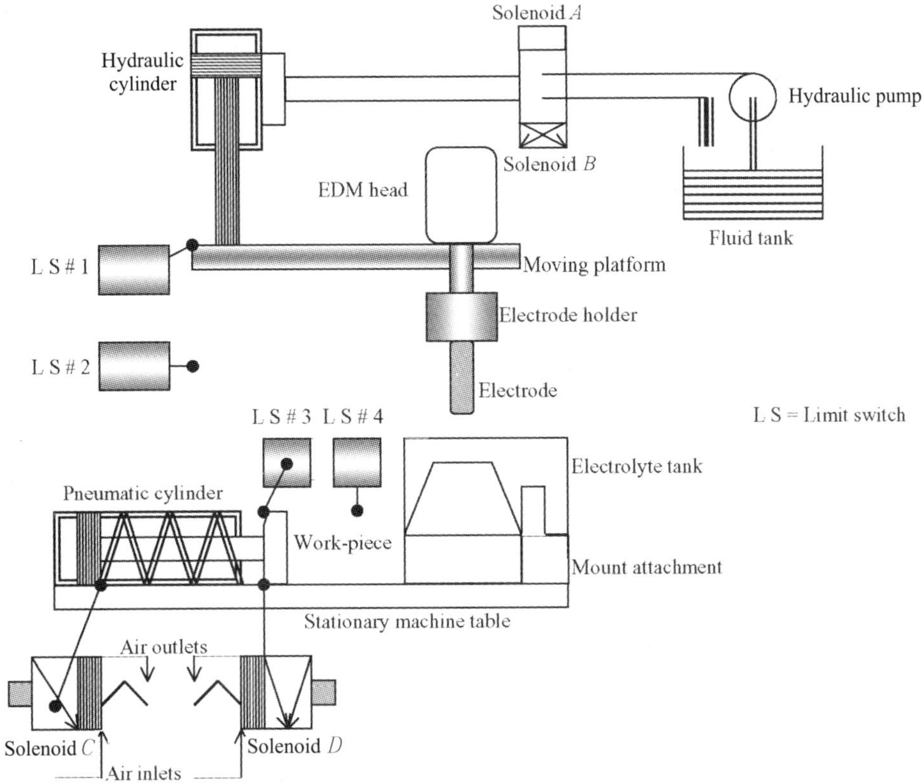

Fig. 2.7 Automated Electro-Discharge Machining of Cooling Holes in Turbine Blades

processors with either a constant or varying set point is determined by one or more modes of control. Commonly used control modes are the two-position, floating, proportional, integral, and derivative modes. The control function is achieved through measurement of the controlled variable, computation of the error between the desired and measured values, and manipulation of some variable in the process system to minimize the error.

Pneumatic and electronic digital and analog circuits are widely employed for controlling continuous processes. Pneumatic controllers require an operational equivalent of an amplifier that depends on compressed air. Electronic analog control is accomplished with a resistive circuit to establish error and an amplifier to develop controlling action. Digital controllers rely on a microprocessor and relevant algorithm. Advanced controlling methods go beyond a single loop form of feedback control that adjusts just one variable. Controllers arranged in cascade require two loops in which output of the primary controller acts as input signal to the secondary controller. The feed-forward method produces modifications in the output signal in response to substantial changes in an important load variable without waiting for the error to occur. Adaptive control of parameters may be achieved to 'adapt' to variations in the process. Artificial intelligence uses lessons learnt from expert systems for superior control of an adaptive system to manipulate multiple variables.

The concept of first-order lag process may be understood in a continuous system where the process curve is developed. The compressed air storage tank in figure 2.9 (*upper*) is filled with air

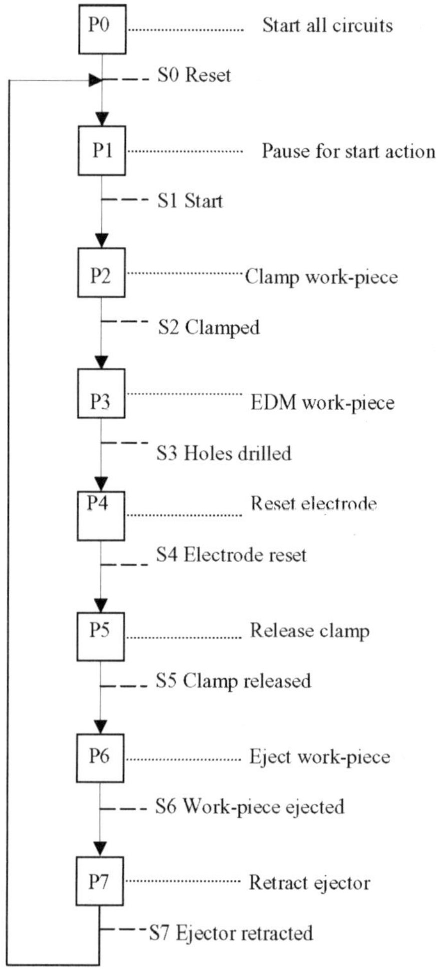

Fig. 2.8 Sequential Operation Chart for Cooling Holes in Turbine Blades

maintained at a constant temperature and at varying pressure p. The density ρ and pressure of air in the tank depend only on the mass of air inside the tank. The inflow valve has resistance R_m and the outflow valve has resistance R_L. The input valve acts as the control element, pressure p is the controlled variable, inflow rate m and the outflow rate q are the manipulated variables, and outflow resistance R_L is the disturbance variable. A state of equilibrium is achieved by adjusting the m until it balances q. If the flow is assumed to be laminar, then $q = (\rho/R_L)p$. If R_L is held at a constant value and inflow valve resistance R_m is modified, then q is proportional to p.

For a given pressure level p, there is a specific inflow rate m and outflow rate q for the necessary balanced condition. A specific flow rate m depends on the position of the control element R_m, which provides the basis for the developed load curves shown in figure 2.9 (*lower*). As inflow valve resistance R_m increases, the net flow rate decreases, while a reduced resistance R_m increases the flow rate into the tank.

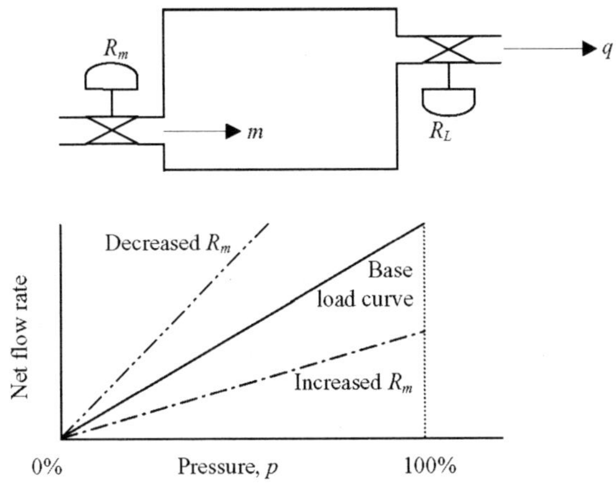

Fig. 2.9 Pressure-Mass Load Curve for Air Storage Tank

2.10 MODES OF CONTROL

The simplest mode of control is the 'two-position' type with two (on or off) possible outcomes, and depends on the error signal. A neutral region for a limited range around zero value helps to eliminate chatter, as indicated in figure 2.10. The two-position controller is suitable for processes possessing adequate capacitance to counteract the combined effects of dead-time lag and load variation capability. The mode provides energy pulses for cycling the controlled parameter, and is the preferred method if the cycling is reduced to acceptable levels.

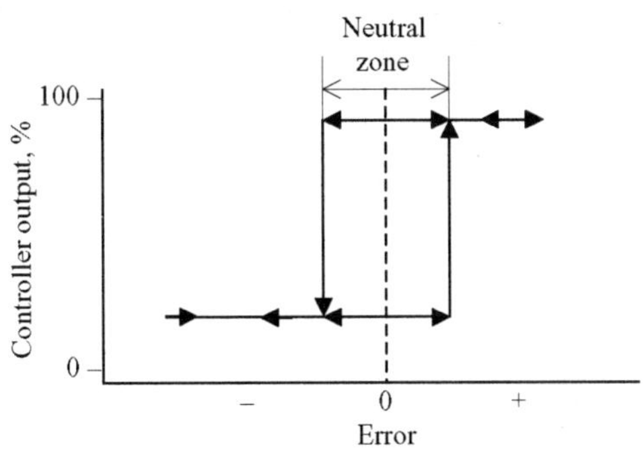

Fig. 2.10 Two-Position Controller Output

The temperature control of a domestic baking oven can be accomplished with a two-position controller. The air inside the oven has a large capacitance, and the reaction time lag is relatively small. Energy from the heating elements is mostly adequate to raise the temperature.

If the two-position mode is applied to achieve stationary control, then it is referred to as a 'floating' control mode. If the error exceeds the bounds of the neutral zone, the control element proceeds at a constant rate until it reaches an extreme position, as illustrated in figure 2.11.

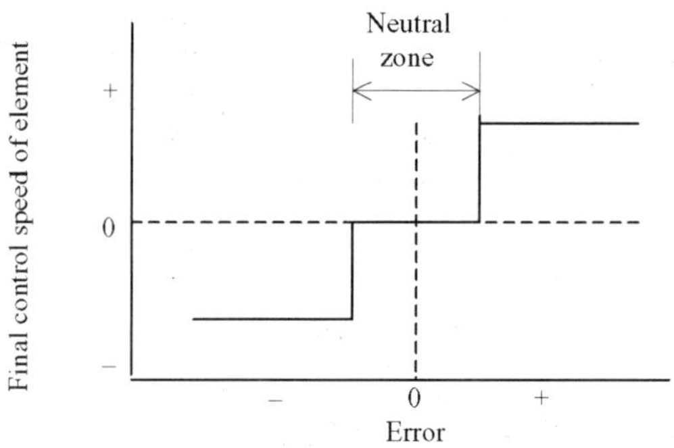

Fig. 2.11 Operation of Floating Control Mode

The tendency to cause cycling of the controlled variable increases in the floating control mode, with the amplitude of cycling dependent on dead-time lag and capacitance of the process, and the speed at which the controller changes the final control element.

The floating control mode offers the benefit of handling load changes through gradual adjustments of the final control element, and finds applications where substantial load variations are anticipated. Electric motor and hydraulic cylinder form of actuators operated by on/off relays find widespread use of the floating control mode.

The 'proportional' control mode may be used in applications requiring output from the controller to be linearly proportion to the error. Consider, for example, the level of a tank holding a fluid. The tank is provided with a float to measure the level, a valve to manipulate the level, and a horizontal lever rotating about a fixed fulcrum point to control the operation. One end of the lever is attached to the valve, and the float is attached at its other end. When the error is nil, the level stands at height h_0, and the corresponding position of the valve is v_0. The position of the valve v at any level h is given by: $(v - v_0)/a = (h_0 - h)/b$, error signal $e = h_0 - h$, and valve position $v = (a/b)e + v_0$. Here a is the distance of the valve from the fulcrum on the beam, and b is the distance from the fulcrum to the float mount point (figure 2.12).

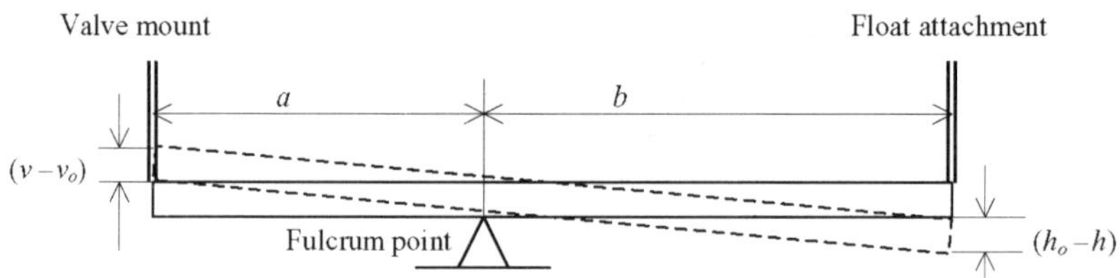

Fig. 2.12 Proportional Control Mode Using Lever for Liquid Level

The gain P of the proportional controller is obtained from the ratio of shift in the position of the valve and the associated change in the level. Thus, if a shift in the valve of 0.25 ins. leads to a change in the fluid level in the holding tank of 18 ins., then gain is determined to be

$$P = [(v - v_0)/0.25]/[(h_0 - h)/18.0] = 72 \times (a/b).$$

Moving the location of the fulcrum point on the horizontal beam will modify the ratio of proportion for the control mode. The action effectively changes the values of beam lengths a and b, and hence, the proportional ratio.

The output from the control mechanism may be changed in proportion to the integral of the error signal in an 'integral control mode'. The output will keep varying at a rate of change that is proportional to the magnitude of the error with this control mode, as shown in figure 2.13. The integral control mode is generally used in conjunction with the proportional control mode.

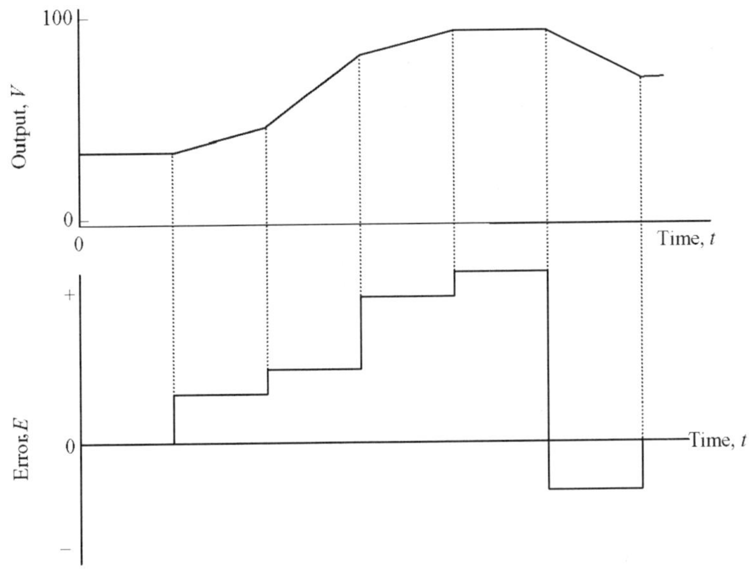

Fig. 2.13 Integral Control Mode Response to Error Signal

The integral control mode is defined mathematically by the following expressions:

- $v = I\int edt + v_0$ where e and v are the time domain error and output, I is the integral action rate, and t is the time variable
- $V = (I/s)E$ where E and V are the frequency domain error and output, I is the integral action rate, and s is the complex frequency parameter
- or $(V/E) = (I/s)$ in the form of a transfer function.

In the associated Bode diagram of the integral control mode (figure 2.14), the gain decreases at a rate of 16 dB per decade increase in frequency, and passes through 0 dB at a radian frequency equal to I, the integral action rate. The phase angle remains fixed at –90° for all frequencies.

An automatic reset action that cancels the proportional effect is available if the integral mode works in tandem with the proportional mode of control. Referred to as the 'proportional plus integral (or the PI) control mode', the controller's output is continuously changed until the error signal is eliminated by the rest action (figure 2.15). The proportional mode alters the output in proportion to the error signal, while the integral mode imposes a further alteration that is in proportion to the integral of the error signal.

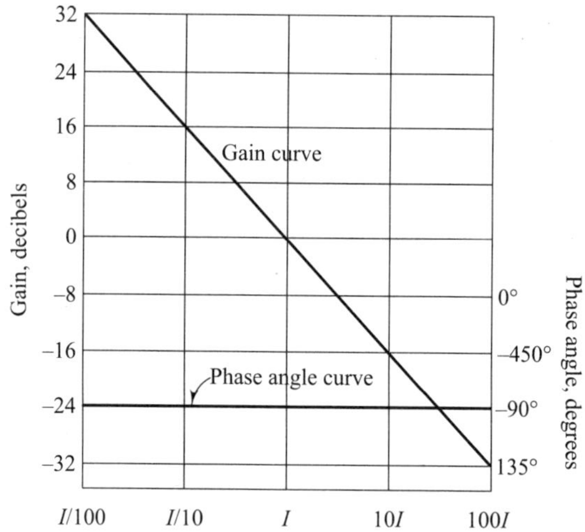

Fig. 2.14 Integral Control Mode Bode Diagram

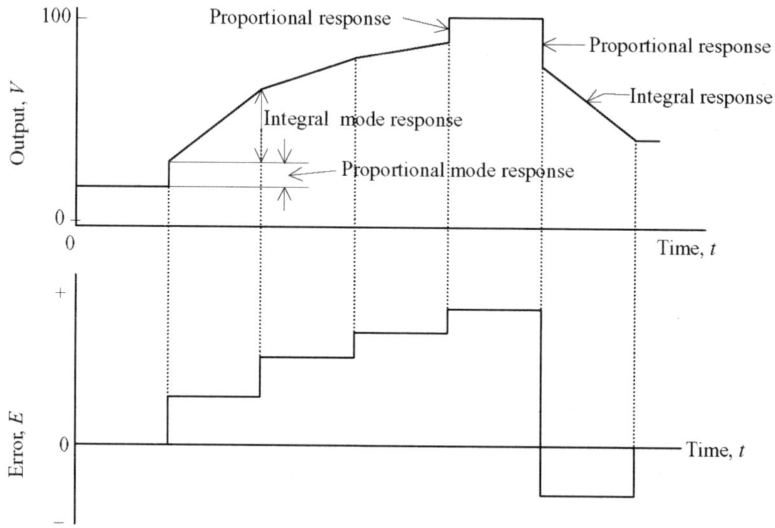

Fig. 2.15 Proportional Plus (PI) Integral Control Mode Response to Error Signal

Referring to the example of pressure control in the air storage tank of figure 2.9, the proportional mode establishes the slope of the load curve, while the integral mode of control raises or lowers its position without affecting the slope.

The integral mode plays the role of adjusting the placement of the curve to accommodate load variations. But the integral mode suffers from the drawback of increased tendency for oscillations in the controlled variable, so the gain of the proportional mode needs to be decreased when functioning in tandem with the integral mode. The ability of the controller to respond to quick shifts in the load is then compromised. Presence of a large dead-time lag in the process will not affect the error immediately, but the delay can cause the integral mode to provide excessive correction, even after the error signal is eliminated.

The proportional plus integral control mode may be defined mathematically by the following expressions:

- $v = Pe + PI\int edt + v_0$ where e and v are the time domain error and output, P is the proportion gain, I is the integral action rate, and t is the time variable
- $V = PE + (PI/s)E$ where E and V are the frequency domain error and output, I is the integral action rate, and s is the complex frequency parameter
- or $(V/E) = P[(I + s)/s)]$ in the form of a transfer function.

The associated Bode diagram of the proportional plus integral control mode (figure 2.16) is split into two halves due to the integral action breakpoint frequency, which is equal to integral action rate, or $\omega_i = I$. On the left side ($\omega < \omega_i$), the gain decreases at a rate of 16 dB per decade increase in frequency and the phase angle is –90°. On the right side of the diagram ($\omega > \omega_i$), the proportional action dominates with phase angle 0° and magnitude established by the proportional gain P. A transition occurs between $\omega_i/10$ and 10 ω_i. If $P = 1$, the magnitude on the Bode diagram is 0 dB. For other values of P, the gain curve is shifted without affecting the phase angle curve. A gain value of 10, for instance, will shift the gain curve upwards by 20 dB.

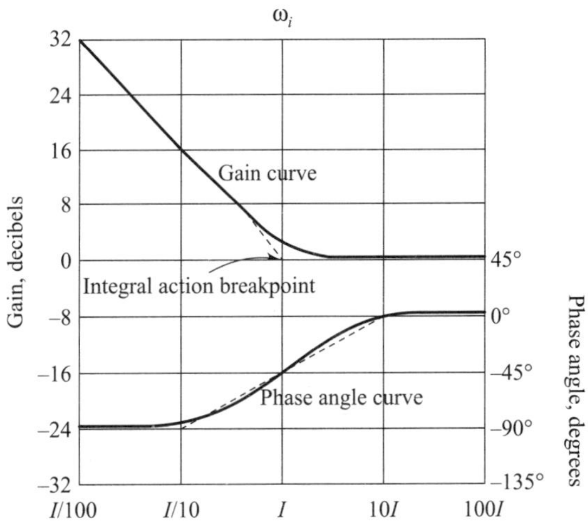

Fig. 2.16 Proportional Plus Integral Control Mode Bode Diagram

The 'derivative' control mode modifies the output variable in proportion to the rate of change of the error, the change occurring as a consequence of variations in the measured variable or the set point or a combination of the two. Used in combination with either the proportional or the proportional plus integral modes of control, the derivative mode tries to predict the error through observation of speed with which the error signal is taking place. Extrapolation of this data then permits developing a control action to reduce the error.

Since the output of the derivative control mode is proportional to the rate of change of the error signal, response to a step-up function will be $+\infty$ and $-\infty$ to a step-down function, which explains its lack of acceptability in practical applications. Response to the step and ramp type of functions is shown in figure 2.17.

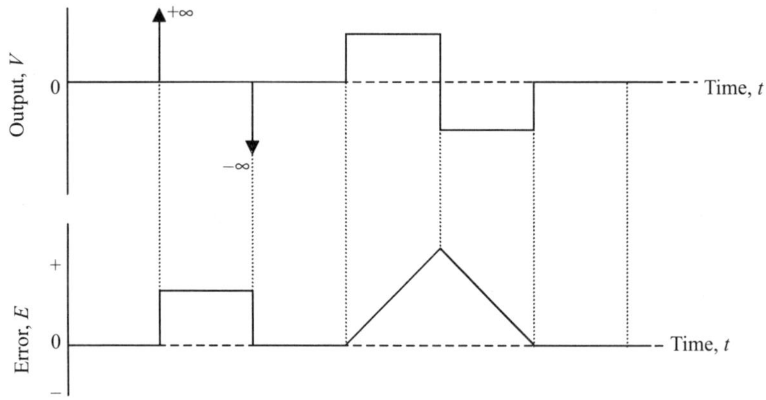

Fig. 2.17 Derivative Control Mode Response to Step and Ramp Functions

The Bode diagram of the derivative control mode is the opposite of the integral mode shown in figure 2.14. The gain increases at a rate of 16 dB per decade increase in frequency, and passes through 0 dB at a radian frequency equal to $1/D$. The phase angle remains constant at +90° for all frequencies.

The ideal derivative control mode may be defined mathematically by the following expressions:

- $v = D(de/dt)$ where e is the time domain error, D is the derivative action time constant, and t is the time variable
- $V = DsE$ where E and V are the frequency domain error and output, and s is the complex frequency parameter, or $(V/E) = Ds$ in the form of a transfer function.

The derivative control mode may be implemented together with a proportional mode to reduce tendencies towards oscillation and to permit higher gain setting. Termed as 'proportional and derivative (PD)' control mode, the combination controller provides the benefit of an output signal that is proportional to the error and an additional output that is proportional to the rate of change of the error signal. The anticipatory nature of the derivative mode of control renders it useful in handling sudden load shifts in the process and its attendant potential to generate excessively high errors. In such situations the proportional mode by itself lacks the capability to maintain the error within acceptable bounds.

The proportional plus derivative control mode may be defined mathematically by the following expressions:

- $v = Pe + PD(de/dt) - \alpha D(dv/dt) + v_0$ where e and v are the time domain error and output, P is the proportion gain, D is the derivative action time constant, α is the derivative limiter coefficient ($0 < \alpha < 1$), and t is the time variable
- $V = PE + PDsE - \alpha Dsv$ where E and V are the frequency domain error and output, I is the integral action rate, and s is the complex frequency parameter
- or $(V/E) = P[(1 + Ds)/(1 + \alpha Ds)]$ in the form of a transfer function.

The Bode diagram of a combination PD control mode is shown in figure 2.18. On the left side of the diagram ($\omega < \omega_d = 1/D$), the proportional mode is in control, with the proportional gain shifting the curve up or down. The derivative mode causes the gain curve to slope at 16 dB per decade at the

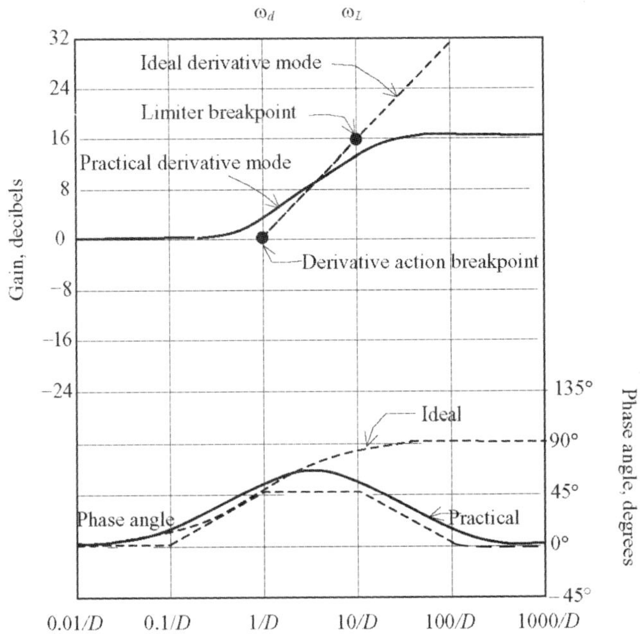

Fig. 2.18 Proportional Plus Derivative Control Mode Bode Diagram

derivative action breakpoint. The gain returns to a horizontal profile due to the action of the derivative limiter at the limiter breakpoint. High-frequency signals are amplified by the derivative mode, but the limiter limits the amplification at the same time. Also, the limiter causes the phase angle to 0° at higher frequencies. The action effectively causes the PD control mode to provide a phase lead over a band of frequencies.

The proportional, integral and control modes may also be combined to produce a triple 'proportional plus integral plus control (PID)' mode of control. The integral mode neutralizes the proportional offset arising from large load changes, while the derivative mode diminishes the oscillatory tendencies to produce a controlling action that predicts changes in the error signal.

Sudden imposition of a large load cannot be accommodated in one and two-control modes due to unacceptable errors. The maximum error is kept within bounds by the anticipatory nature of the derivative mode in the event of a sudden load shift. The integral mode plays the role of resetting the action by eliminating the proportional effect.

The proportional plus integral plus derivative control mode may be defined mathematically by the following expressions:

- $v = Pe + PI\int edt + PD(de/dt) - \alpha D(dv/dt) + v_0$ where e and v are the time domain error and output, P is the proportion gain, D is the derivative action time constant, I is the integral action rate, α is the derivative limiter coefficient ($0 < \alpha < 1$), and t is the time variable
- $V = PE + PIE/s + PDsE - \alpha Dsv$ where E and V are the frequency domain error and output, I is the integral action rate, and s is the complex frequency parameter
- or $(V/E) = P[(I + s + Ds^2)/(1 + \alpha Ds^2)]$ in the form of a transfer function.

2.11 REVIEW PROBLEMS

Example problem # 2.1: In the control system of figure 2.19, determine the transfer function.

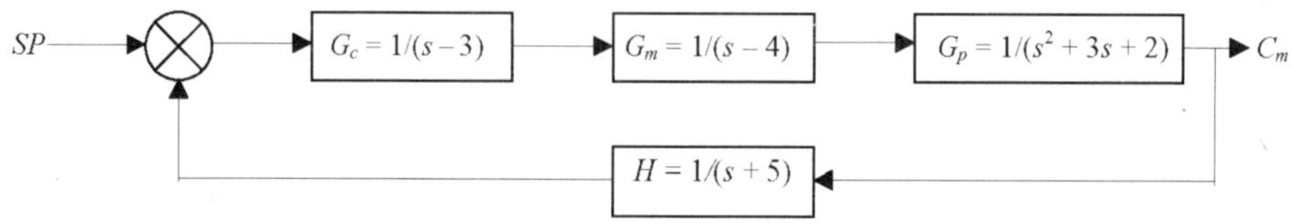

Fig. 2.19

Solution: The closed-loop control system has the forward transfer function:

$$G = G_c G_m G_p = 1/[(s-3)(s-4)(s^2 + 3s + 2)]$$

From equation 2.1, the system transfer function is:

$$C_m/SP = GH/(1 + GH)$$

$$= \{1/[(s-3)(s-4)(s^2 + 3s + 2)(s + 5)]\}/\{1 + 1/[(s-3)(s-4)(s^2 + 3s + 2)(s + 5)]\}$$

Example problem # 2.2: The flow measurement device of a gas metering system produces a signal proportional to the square of the volume flow rate Q of the gas. If the output of the flow meter is 750 m^3/second and meter proportionality constant is 0.725, determine the flow rate Q.

Solution: If the output of the flow meter is D and meter proportionality constant is k_f, then the flow rate:

$$Q = k_f \times \sqrt{D}$$

$$= 0.725 \times \sqrt{750}$$

$$= 19.85 \text{ m}^3\text{/second.}$$

Example problem # 2.3: In example problem # 2.2 assume the gram molecular weight M of the gas is 22, gas pressure p is 350000 pascal, and absolute temperature T is 500 K. Find the flow rate by weight.

Solution: Conversion of volume V to mass m depends on absolute pressure p, absolute temperature T, universal gas constant R = 8.314 J/K·mol and gram molecular weight M of the gas, and is in accordance with the equation 2.4:

$$w = k \times \sqrt{D} \times (p/T),$$

where $k = Mk_f/(10^3 R) = 22 \times 0.725/(1000 \times 8.314) = 0.001918$

Then: $w = 0.001918 \times \sqrt{750} \times (350000/500) = 36.78$ kilogram/second.

Example problem # 2.4: The displacement signal from a vibration transducer mounted on a turbine bearing running at 5400 rpm has double-amplitude reading of 1.85×10^{-3} in. Express the vibration reading as velocity and acceleration.

Solution: Operating speed = $2\pi N/60 = 2\pi \times 5400/60 = 565.49$ radian/second. Single-amplitude vibration reading is: $1.85 \times 10^{-3}/2 = 0.925 \times 10^{-3}$ in. From equation 2.5, the displacement reading is:

$$x(t) = X_0.\sin(\omega t) = 0.925 \times 10^{-3} \times \sin(565.49t) \text{ ins.}$$

From equation 2.6 the velocity reading is:

$$\nu(t) = X_0.\omega.\cos(\omega t) = 0.925 \times 10^{-3} \times 565.49 \times \cos(565.49t)$$
$$= 0.523 \times \cos(565.49t) \text{ ins/second}$$

From equation 2.7 the acceleration reading is:

$$a(t) = -X_0.\omega^2.\sin(\omega t) = -0.925 \times 10^{-3} \times (565.49)^2 \times \sin(565.49t)$$
$$= -295.80 \times \sin(565.49t) \text{ ins/second}^2$$

Example problem # 2.5: Express the acceleration amplitude of example problem # 2.4 in terms of g's (or the gravitational constant).

Solution: Gravitational constant g = 386 ins/second^2. Then:

$$a(t) = 295.80/386 = 0.766\, g.$$

Example problem # 2.6: The level of a lubricating oil reservoir is controlled with a float and lever mechanism of the type shown in figure 2.12. If the gain is 3.25 and the error is 4%, calculate the valve displacement $(v - v_0)$.

Solution: Gain P is defined by the ratio of difference in valve displacement from the position where the error signal is zero, or:

$$P = (v - v_0)/e.$$

Hence: $(v - v_0) = Pe = 3.25 \times 4 = 13\%$

Example problem # 2.7: In example problem # 2.6, what is the fluid level height h_0 if h = 15 cm and beam segment b = 25.5 cm long?

Solution: When the error is zero, the oil level stands at height h_0 and the corresponding position of the valve is v_0. The position of the valve v at any level h is given by:

$$(v - v_0)/a = (h_0 - h)/b$$
$$(v - v_0)/a = 13\% = (h_0 - h)/b$$

or: $(h_0 - 15)/25.5 = 13/100 = 0.13$

Hence: $h_0 = 0.13 \times 25.5 + 15$
$= 18.315$ cm

Example problem # 2.8: In example problem # 2.7, what is the zero error valve position v_0 if the beam segment a = 4.5 cm and b = 25.5 cm long and v = 0.15 cm?

Solution: If the error is 4%, the position of the valve v at any level h is given by:

Valve position: $v = (a/b)e + v_0$

or: $0.15 = (4.5/25.5) \times 0.04 + v_0$

Hence: $v_0 = 0.15 - (4.5/25.5) \times 0.04$
$= 0.1429$ cm

Example problem # 2.9: The integral control mode characteristic shown in figure 2.13 has the following values: $I = 1.5\ s^{-1}$, $v = 1.25$, $t = 0$. Error $e = 1.75$ when $0 \le t \le 1$ s. Express the straight-line by an equation.

Solution: The integral control mode is defined mathematically by the expressions:

$$v(t) = I \times \int (e)dt + v(t)$$

$$= 1.5 \times \int (1.75)dt + 1.25$$

$$v(t) = 2.625t + 1.25, \text{ when } 0 \le t \le 1 \text{ s.}$$

Example problem # 2.10: In example problem # 2.9, determine the straight-line equation assuming $I = 2.5\ s^{-1}$, $v(1) = 2.25$, error $e = 1.25$ when $1 \le t \le 2$ s.

Solution: The straight-line equation is:

$$v(t) = I \times \int (e)dt + v(t)$$

$$= 2.5 \times \int (1.25)dt + 2.25, \text{ when } 1 \le t \le 2 \text{ s}$$

or,

$$v(t) = 3.125(t - 1) + 2.25, \text{ when } 1 \le t \le 2 \text{ s}$$

Example problem # 2.11: The PI control mode characterized by the error and output curves in figure 2.20 has a gain $P = 2.5$ and integral action rate $I = 0.0325\ s^{-1}$. At time $t = 0$ the value $v_0 = 27\%$. Calculate the value of the output at time $t = 0$ and at $t = 50$s.

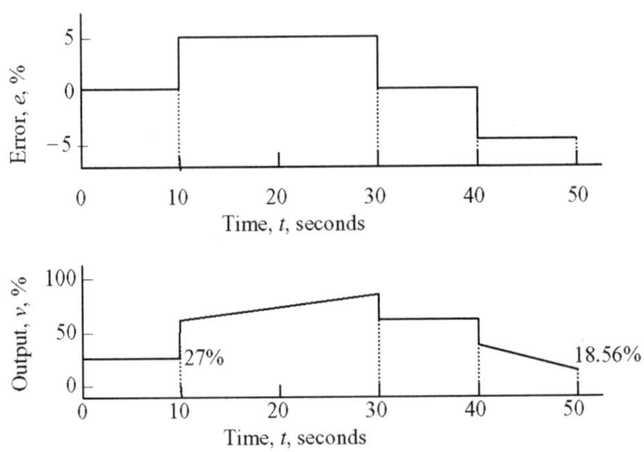

Fig. 2.20

Solution: The output of a PI controller is:

$$v = Pe + PI\int edt + v_0$$

$$= 2.5e + 2.5 \times 0.0325 \times \int edt + 27$$

$$= 2.5e + 0.08125 \times \int edt + 27,$$

where the integral equals the net area under the error curve between 0 and ts.

At $t = 0$ s, $e = 0$, and net area $= 0$. Hence:

$$v = 2.5 \times 0 + 27 = 27\%$$

At $t = 50$ s, $e = -5\%$, and net area $= 5 \times (30 - 10) - 5 \times (50 - 40) = 50$

$$v = 2.5 \times (-5) + (2.5 \times 0.0325 \times 50) + 27 = 18.56\%.$$

The output curve is shown in the lower half of figure 2.20.

Example problem # 2.12: A PID controller mode with gain $P = 1.25$ has $I = (1/8)\,s^{-1}$, a derivative action time constant $D = 1.15$ s, and an initial output $v_0 = 12\%$. Assume negligible delimiter effects on the system. The error curve is shown in figure 2.21. Calculate the output at time $t = 5$, 10, 20 and 25 s.

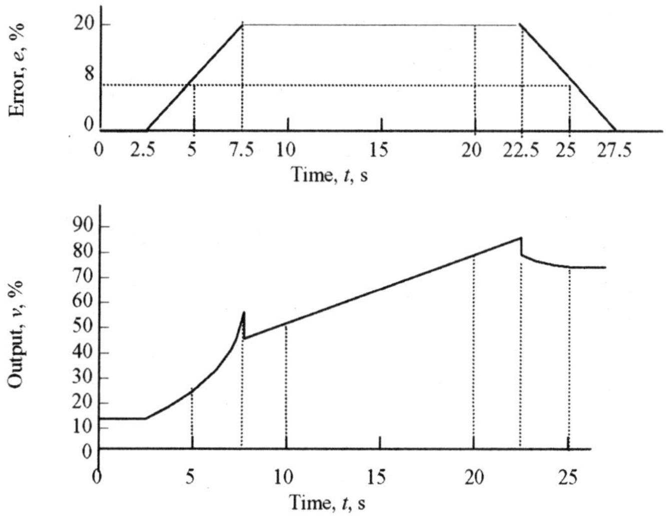

Fig. 2.21

Solution: The output from a PID controller is:

$$v = Pe + PI\int edt + PD(de/dt) + v_0 \text{ when } \alpha = 0.$$

$$v(t) = 1.25e + [1.25 \times (1/8) \times \int edt] + [1.25 \times 1.15 \times (de/dt)] + 12,$$

where the integral equals the net area under the error curve between 0 and t s and (de/dt) is equal to the slope of the curve at time t.

At $t = 5$ s, $e = 8$, $(de/dt) = 20/(7.5 - 2.5) = 4$, and net area $= (8 \times 2.5/2) = 10$. Hence:

$$v(t = 5) = 1.25 \times 8 + [1.25 \times (1/8) \times 10] + [1.25 \times 1.15 \times 4] + 12 = 29.31\%.$$

At $t = 10$ s, $e = 20\%$, $(de/dt) = 0$, and net area $= (20 \times 5)/2 + 20 \times 2.5 = 100$. Hence:

$$v(t = 10) = 1.25 \times 20 + [1.25 \times (1/8) \times 100] + [1.25 \times 1.15 \times 0] + 12 = 52.63\%.$$

At $t = 20$ s, $e = 20\%$, $(de/dt) = 0$, and net area $= (20 \times 5)/2 + 20 \times (20 - 7.5) = 300$. So:

$$v(t = 20) = 1.25 \times 20 + [1.25 \times (1/8) \times 300] + [1.25 \times 1.15 \times 0] + 12 = 83.88\%.$$

At $t = 25$ s, $e = 8\%$, $(de/dt) = -20/(27.5 - 22.5) = -4$, and net area

$= (20 \times 5)/2 + 15 \times 20 + [(20 + 8)/2] \times 2.5 = 385$. Hence:

$v(t = 25) = 1.25 \times 8 + [1.25 \times (1/8) \times 385] + [1.25 \times 1.15 \times (-4)] + 12 = 76.41\%$.

2.12 EXERCISES

2.1 The velocity signal from a vibration transducer mounted on a turbine bearing running at 3600 rpm has double-amplitude reading of 0.775 in/s. Express the vibration reading as displacement and acceleration.

2.2 The acceleration signal from an accelerometer mounted on a turbine bearing running at 5400 rpm has double-amplitude reading of 0.625 *g*'s. Express the vibration reading as displacement and velocity.

2.3 The flow measurement device of a gas metering system produces a signal proportional to the square of the volume flow rate Q of the gas. If the output of the flow meter is 125 m^3/second and meter proportionality constant is 0.575, determine the flow rate in terms of mass.

2.4 In example # 2.3 assume the gram molecular weight M of the gas is 24, gas pressure p is 75000 pascal, and absolute temperature T is 550 K. Find the flow rate by volume.

2.5 Replace the two-position on/off controller in figure 2.9 by a valve that moves between two partially open positions. The intake flow m rate varies between 0.05 and 0.35 m^3/sec when the valve is in the minimum and maximum flow positions. Determine the amplitude and period of oscillation of the level h.

2.6 Sketch the proportional controller's input and output graphs for the following conditions:

(*i*) $P = 5, v_o = 33\%$ (*ii*) $P = 0.75, v_o = 13\%$ (*iii*) $P = 3.25, v_o = 23\%$

2.7 Calculate the necessary proportional controller's offset for obtaining $(v - v_o)$ at 8% for the following gain values:

(*i*) $P = 0.35$ (*ii*) $P = 0.75$ (*iii*) $P = 0.15$

(*iv*) $P = 1.25$

2.8 A PI controller has gain $P = 1.25$, integral action rate $I = 1/7\ s^{-1}$, and output at zero error $v_o = 20\%$. The error signal is as shown in figure 2.20. Calculate the output of the controller at $t = 10$ and 40 s.

2.9 A PD controller has gain $P = 1.5$, $D = 7$s, $\alpha = 0$, and output at zero error $v_o = 27\%$. The error signal is as that shown in figure 2.20. Calculate the output of the controller at $t = 10$ and 30 s.

2.10 A PID controller has gain $P = 2.5$, $I = 0.125\ s^{-1}$, $D = 5$ s, $\alpha = 0$, and output at zero error $v_o = 22\%$. The error signal is as that shown in figure 2.21. Calculate the output of the controller at $t = 5$, 15 and 25 s.

2.11 A PI controller is defined by the transfer function: $(V/E) = P[(I + s)/bI + s)]$, where b in the denominator aids in limiting the gain to a maximum value of $1/b$. Plot the Bode diagram when:

(*i*) $P = 2.5,\ I = 0.125\ s^{-1},\ b = 1$ (*ii*) $P = 2.5, I = 0.125\ s^{-1}, b = 2$.

References & Bibliography

Bateson, R.N., "Introduction to Control System Technology", Prentice-Hall, Clifton, NJ, 1999.

Carstens, J.R., "Automatic Control Systems and Components", Prentice-Hall, Clifton, NJ, 1990.

DeRoy, B.E., "Automatic Control Theory", John Wiley & Sons, New York, 1996.

Foxboro Company, "Compensated Mass Flow Control Loop", Bulletin # C–40A, Foxboro, MA, November, 1985.

Miller, R.W., "Servo-mechanism Devices and Fundamentals" Reston Publishing Co., Reston, VA. 1977.

Olesten, N.O., "Numerical Control", John Wiley & Sons, New York, 1970.

Patrick, D.R. and Fardo, S.W., "Industrial Process Control System", Prentice-Hall, Clifton, NJ, 1979.

Rangwala, A.S., "Turbo-Machinery Dynamics: Design and Operation", McGraw-Hill, New York, 2005.

Sante, D.P., "Automatic Control System Technology", Prentice-Hall, Clifton, NJ, 1980.

3

Basics of Electricity and Electronics

3.1 INTRODUCTION

Electrically and electronically operated control system components are extensively used either by themselves or in conjunction with mechanical, liquid flow, gas flow and thermal elements. Transducers are used to sense a number of physical parameters, with each transducer specifically designed to change a particular quantity into a small current or voltage signal that is proportional in magnitude to the parameter being sensed. The information is then transmitted in the form of electronically equivalent signals to a control circuitry for producing the desired response.

A pipe carrying flowing water has been used to explain the flow of electricity through a conductor, but the comparison is not adequate in a number of respects since the characteristics of the electrons are not addressed. If, however, the water is replaced by tightly packed small marbles traveling through the pipe, the many characteristics of the flow of electrons may be easier to grasp. A marble pushed at one end of the pipe will emerge from the other end of the pipe, although it will not be the same marble placed at the first end. The flow of electricity through a conductor may be described similarly. The electrons bump, or displace, neighbouring electrons along a continuous path inside the conductor. The cause of the bumping is of special interest. An electric charge exists on the electron, which causes attractive and repulsive forces between adjacent electrons and protons. Just as a pressure differential is required to force water through a conduit, an electrical potential difference is needed to force the electrons to flow through a conductor. The electrical form of pressure may be derived through an electrochemical process, as in batteries, or can be created electronically from a power supply. In both cases, a reservoir of electrons is produced opposite a similar reservoir that has a deficiency of electrons. The unstable situation will lead to the over-abundant number of electrons striving to find a path to the deficient electron area.

3.2 ELECTRICAL CONCEPTS

The electrical charge carried by a fixed number of electrons past a point in an electrical conductor in one second defines the flow of current, or $I = Q/t$, where I is the current in amperes (or amps), Q is the charge in coulombs, and t is the time elapsed in seconds. The electrical pressure, or potential difference required to force the electrical flow represents the expended energy or work. If the unit of expended energy, W, is joule, and the unit of voltage, V, or potential difference, is volt, then $V = W/Q$.

An important aspect of DC circuits is the direction of flow of electrons (or current) resulting from the pressure difference. Since the electrons are negatively charged, it is assumed the electrons flow from the negative to the positive side of the voltage source. Conventionally, however, the flow of electrical current occurs from the positive to the negative terminals. The polarity of the voltage drop across a device can then be established as shown in figure 3.1, where the circuit devices are resistors.

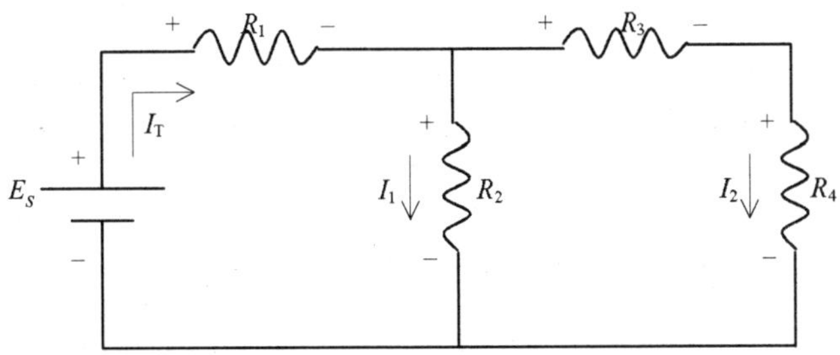

Fig. 3.1 Current Flow Through Circuit with Resistors and Resulting Voltage Polarities

The relation between the flow of current charge and the voltage that forces it down the conductor depends on the extent of impediment that prevails in the conductor's material. The internal resistance is a form of friction, and dissipates the energy required to overcome it in the form of heat. The relationship is referred to as 'Ohm's law', and is expressed as: $I = E/R$, where I is the current in amps, E is the voltage in volts, and R is the resistance to the flow in ohms.

The electrical resistance of a conducting material can be expressed by four parameters: the material's characteristics (or chemical makeup), its temperature, length and cross-sectional area. The factors may be combined into the equation: $R = kl/(A_{CM})$, where R is the resistance (ohms), k = temperature dependent resistance proportionality constant (units: ohms-circular mils/ft) and A_{CM} is the cross-sectional area (circular mils). Resistors for control system applications are specifically designed for insertion into a circuit to impede the flow of current through the conductor.

Two forms of resistive circuits, the 'series circuit' and the 'parallel circuit', are in common use. When the resistors are connected in series with each other (figure 3.2), the total current flows through all of the resistors R_1, R_2, R_3, R_4 present in the circuit. The flow of current I_T originates in the voltage source E_s and flows back into its negative terminal. A voltage drop occurs across each resistor as the current I_T flows through it. The voltage drop in each resistor (V_1, V_2, V_3, V_4) varies with its resistance value. The sum of the voltage drops must equal E_s. Hence, $E_s = (V_1 + V_2 + V_3 + V_4)$. Also, the total equivalent resistance in the circuit $R_T = (R_1 + R_2 + R_3 + R_4)$.

In a parallel resistance circuit (figure 3.3), the resistors are connected across a voltage source such that the positive and negative terminals are common to all the devices. The total current flow I_T is divided at point a as it flows from the voltage source through all the resistor branches, combines at point b and goes back to the negative terminal. However, the potential difference E_s from the voltage source is identical across each of the four resistors. The total equivalent resistance, R_T, of the resistors in parallel connection with each other is given by the expression:

$$1/R_T = 1/R_1 + 1/R_2 + 1/R_3 + 1/R_4$$

where R_1, R_2, R_3, R_4 are individual resistors, (ohms).

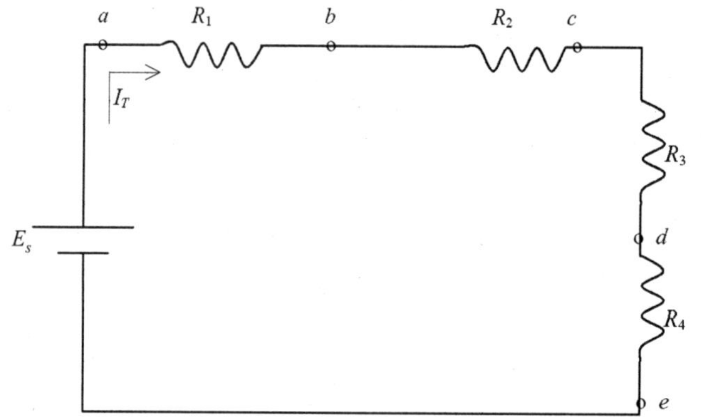

Fig. 3.2 Series Resistive DC Circuit

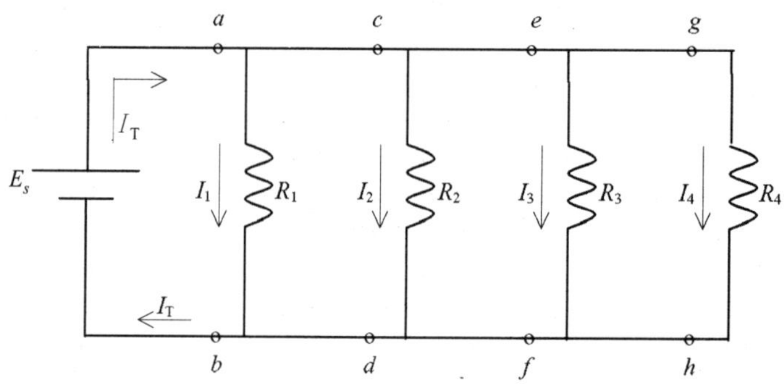

Fig. 3.3 Parallel Resistance Circuit

A commonly encountered combination of series and parallel connection of resistors is shown in figure 3.4. To determine the combined equivalent resistance of all the resistors, note that R_1 and R_2 are in parallel, hence $R_{1-2} = R_1 \cdot R_2/(R_1 + R_2)$. Resistors R_6 and R_7 are in series, hence $R_{6-7} = (R_6 + R_7)$, R_3 and R_{4-5} are in series, hence $R_{3-4-5} = [R_3 + \{R_4 \cdot R_5/(R_4 + R_5)\}]$. The other connections can be similarly combined. Example problems # 3.2 to 3.6 illustrate the sequence of calculations.

3.3 KIRCHHOFF'S LAWS

Kirchhoff's voltage and current laws regulate the behaviour of circuit voltages and currents. The voltage law states that the algebraic sum of all voltage rises and drops around a closed-loop in a circuit must be zero.

To illustrate, resistors R_1, R_2 and R_3 in figure 3.5 are connected in series with the voltage source E_s in a closed loop. Proceeding in an anticlockwise direction, the voltage drop and polarities are based on the direction and magnitude of the total current I_T. Starting at the voltage source, then:

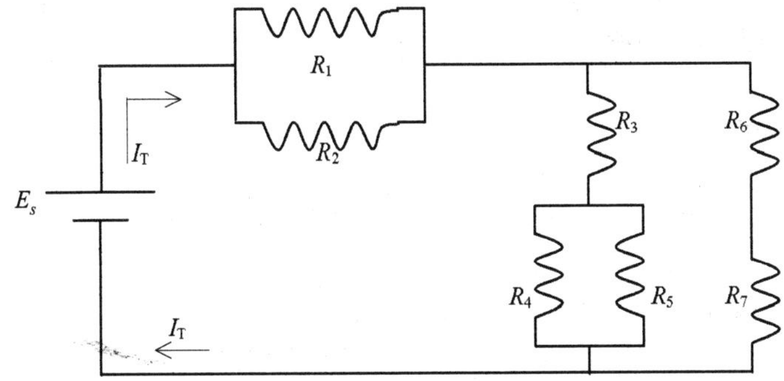

Fig. 3.4 Combination of Series and Parallel Resistors in Circuit

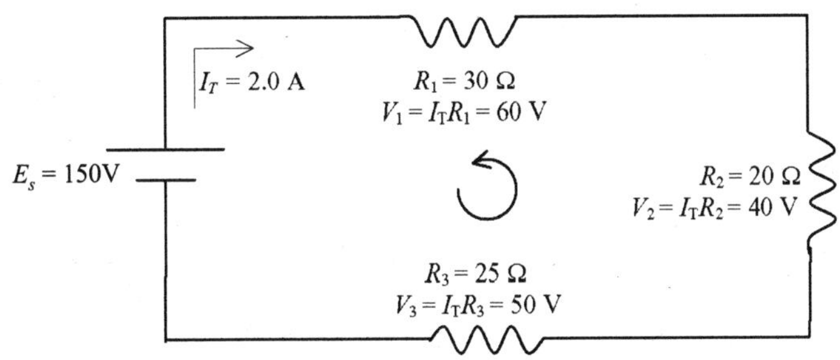

Fig. 3.5 Application of Kirchhoff's Voltage Law

$$-150\text{V (fall)} + 50\text{V (rise)} + 40\text{V (rise)} + 60\text{V (drop)} = 0$$

A comparable law pertains to the current flow. It states: all currents flowing into and out of a junction must equal zero. Four conductors meet at junction point *a* in figure 3.6.

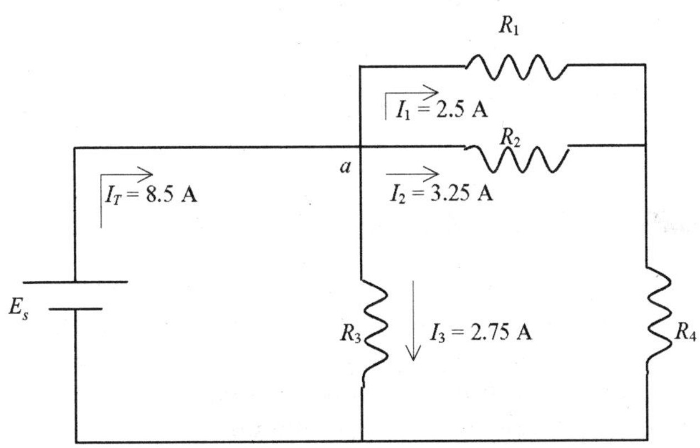

Fig. 3.6 Application of Kirchhoff's Current Law

The current flows into and out of the junction have been determined. Using Kirchhoff's law and the convention that currents coming into the joint are positive and leaving the joint are negative, then:

$$+ 8.5 \text{ A (entering)} - 2.5 \text{ A (leaving)} - 3.25 \text{ A (leaving)} - 2.75 \text{ A (leaving)} = 0$$

A convenient method to calculate voltage drops in resistive network circuits is through the use of the 'voltage divide rule'. For instance, in figure 3.7 the voltage rule states:

$$V_{out} = V_{in} (R_{out}/R_{in})$$

where V_{out} = divider's output voltage, V_{in} = divider's input voltage, R_{out} = total resistance across V_{out}, and R_{in} = total resistance across V_{in}. In the example shown, R_{out} comprises of R_5 in parallel connection with $(R_3 + R_4)$, R_{in} comprises of R_5 in parallel connection with $(R_3 + R_4) + R_1 + R_2$.

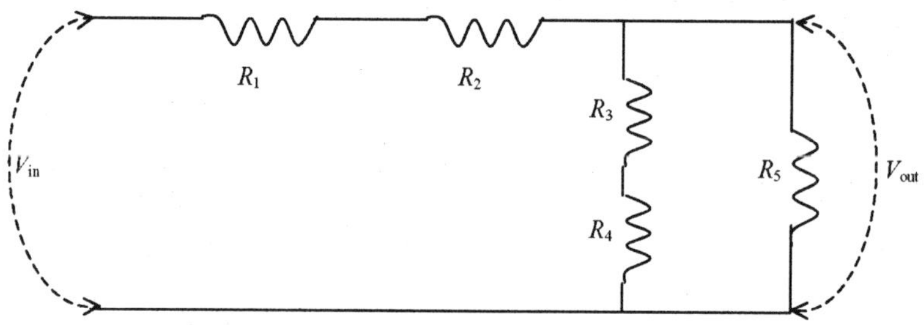

Fig. 3.7 Applying Voltage Divider Rule

Electrical power such as batteries have an internal resistance that is inherent and unavoidable because of their construction features. A voltage drop occurs across the internal resistance just as if it were a resistance external to the power source, leaving a lesser available voltage at its terminals. Thus, if a current of 8 amps is demanded from a 14 V battery with an internal resistance of 0.5 Ω, the internal voltage drop is 8 × 0.5 = 4 V and the available voltage is reduced to 14 – 4 = 10 V. Thus, the battery's voltage regulation is not good because of the internal resistance. The 'per cent of voltage regulation' is obtained with the aid of the equation:

$$\% \, VR = 100 \times (V_{NL} - V_{FL}/ V_{FL})$$

where VR = voltage regulation, V_{NL} = no load voltage at terminals, and V_{FL} = full load voltage at terminals.

3.4 ELECTRICAL ELEMENTS OF CONTROL SYSTEM

When the electrical resistance discussed in section 3.2 is non-linear with changes in the flow of current, it may be necessary to split it into two components. The 'static resistance' is the slope of a straight line from the origin to the operating point, or $R = \Delta E/\Delta I$, where ΔE is the change in voltage and ΔI is the change in current. Operation in a control system is usually confined to small variations around the operating point. 'Dynamic resistance' is obtained from the slope of the tangent line at the operating point. Dynamic resistance may be expressed mathematically as $R = dE/dI$.

'Electrical capacitance' is the quantity of electrical charge (Q coulombs) required to create a unit increase in the electrical potential (E volts). The unit of capacitance is 'farad' (F). Capacitance is defined by the expression: $C = \Delta Q/\Delta E$. But since $I = \Delta Q/\Delta t$, then current $I = C.\Delta E/\Delta t$, where Δt = time in seconds needed to obtain ΔQ change in charge and I is the average current during that time period. If Δt approaches zero, then current $I = dQ/dt = C.(dE/dt)$. Thus, if $E = A.\sin(\omega t)$, then $dE/dt = A\omega.\cos(\omega t)$, and current $I = CA\omega.\cos(\omega t)$. Thus, the current I leads voltage E by 90° in a capacitor. Capacitors installed in parallel with each other may be combined to provide an equivalent capacitor by the expression: $C_T = C_A + C_B + C_C + ... + C_X$, where C_A, C_B, C_C, are individual capacitors. Capacitors installed in series with each other may be combined to provide an equivalent capacitor by the expression: $1/C_T = 1/C_A + 1/C_B + 1/C_C + \cdots + 1/C_X$.

'Electrical inductance' is the voltage required to produce a unit increase in the electric current per second. The unit of electrical inductance is 'henry' (H). The voltage-current relation for an inductor is: $E = L.(dI/dt)$. Hence, if $I = A.\sin(\omega t)$, then $dI/dt = A\omega.\cos(\omega t)$, and voltage $E = LA\omega.\cos(\omega t)$. Thus, the voltage E leads the current I by 90° in an inductor. Inductors installed in series with each other may be combined to provide an equivalent inductor by the expression: $L_T = L_A + L_B + L_C + \cdots + L_X$, where L_A, L_B, L_C, are individual capacitors. Inductors installed in parallel with each other may be combined to provide an equivalent capacitor by the expression: $1/L_T = 1/L_A + 1/L_B + 1/L_C + \cdots + 1/L_X$.

'Electrical dead-time delay' is the time period required for a signal to travel from the source to the destination. In spite of its high speed, the travel time of an electric signal and the consequent time delay has an impact in many systems. The delay in the transmission line may also be advantageously used in a digital computer circuit to accomplish the desired logic function. But in control systems the effect of the electrical dead time is negligible because the delay is small when compared to that in other system components. The dead time delay of the line is obtained by dividing the distance traveled (D) by the velocity of propagation, v_p, which varies from 2×10^8 and 3×10^8 m/s, or: $t_d = D/v_p$ s. Mathematical expressions for electrical elements of control systems are shown in table 3.1.

Table 3.1 Equations for Electrical Elements of Control Systems

Resistance	$R = E/I = \Delta E/\Delta I = dE/dI$
Capacitance	$C = \Delta Q/\Delta E$, $I = C.dE/dt$
Inductance	$E = L.\Delta I/\Delta t = L.dI/dt$
Dead-time delay	$t_d = D/v_p$

3.5 AC CIRCUIT

An alternating current or voltage is generated by rotation of an electric coil in a magnetic field or by using an electronic oscillator. Figure 3.8 explains the principle on which an electric generator, or alternator, works. The direction of rotation (F) of the wire loop and magnetic field strength (B) determines the amount of current (I) generated. The right hand rule governs the direction of F, I and B. The speed at which the loop turns in the magnetic field establishes the frequency of current reversals in the circuit attached to the generator rotor. Hence, if the loop rotates at 3,600 rpm, the number of current reversals is 3,600 cycles per minute, or 60 cycles per second (or hertz).

The electronic oscillator provides the other method of producing an AC signal. Many different oscillator designs are in use. The frequency of oscillation of the electric signal is determined by the value of a resistor, capacitor, and inductor, or by an oscillating piezo-electric crystal.

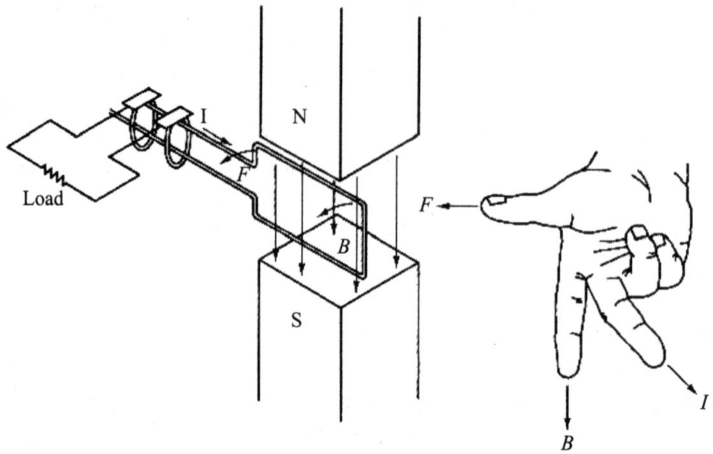

Fig. 3.8 Induction of Electric Current in Loop Rotating in Magnetic Field

The sinusoidal waveform is commonly used in the discussion of alternating electric currents, but any periodic or repeatable trace or path made by a measurable quantity such as voltage or current that is plotted as a function of time or angular position is also valid (figure 3.9). Since the role of the electric generator is widely recognized, it is common practice to plot current or voltage against an angular function such as degrees or radians. The instantaneous magnitude of the parameter measured at any instant in time provides a snapshot of the waveform's height of an otherwise fleeting value. Amplitude (E_{max}) represents the peak value of the waveform, while period (T) is the elapsed time between identical adjacent points of the repeating waveform. The portion of the waveform contained in one period is referred to as the cycle. Frequency refers to the number of cycles that occur in one second, or $f = 1/T$, and has the units of Hz, kHz or MHz. Angular velocity $\omega = 2\pi/T = 2\pi f$ radians/second.

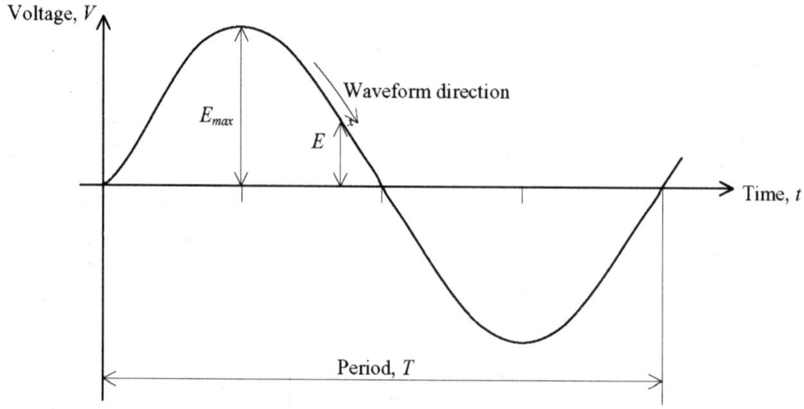

Fig. 3.9 Cyclic Waveform

When an electric signal is shifted to the left or right of another signal, the two signals are referred to as being out of phase with each other. A sinusoidal waveform shifted by angle $+\theta$ to another waveform is conventionally referred to as leading, and is represented by the expression: $y = A_{max}.\sin(\omega t + \theta)$, and a lagging waveform by the expression: $y = A_{max}.\sin(\omega t - \theta)$, where θ is the phase angle shift.

3.6 TRANSFORMER AND ITS CHARACTERISTICS

A transformer is a device for changing the voltage and current of AC in a circuit.

Comprising of two inductors placed close to each other, the magnetic flux surrounding the one encompasses the windings of the second inductor. Thus, the voltage that has developed across the 'primary' winding of the transformer inductively couples with the 'secondary' winding (figure 3.10). However, an alternating current is needed for the flux to develop and the inductive coupling to take place.

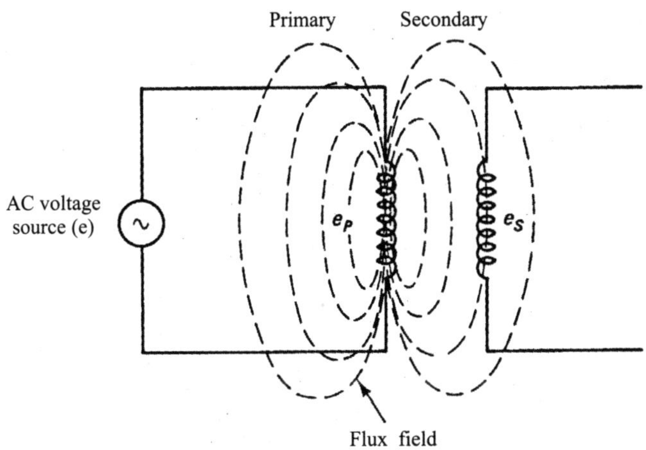

Fig. 3.10 Magnetic Coupling of Transformer Coils

Based on their style of construction, transformers may be split into three groups. Transformers with an iron core are used mostly for transforming large amounts of current and voltage at audio, or line, frequencies. Transformers with an air core are used for communication applications, while variable core transformers may be used for both power and communications applications.

The voltage induced in the secondary coil of the transformer is dependent on the number of coils in the primary and secondary windings, and is governed by the equation:

$$(E_p/E_s) = (N_p/N_s)$$

where E_p and E_s are the primary and secondary winding voltage and N_p and N_s are the number of turns on the primary and secondary coils. An alternate form relies on the current flow:

$$(I_p/I_s) = (N_s/N_p)$$

where I_p and I_s are the primary and secondary winding current (amps) and N_p and N_s are the number of turns on the primary and secondary coils.

A significant point to be noted is that the power rating of the transformer remains unaffected by the step-up or step-down of the voltage or the current flow. Often rated as VA rating, or volt-amp rating, it is the product of the voltage and current values in the primary or the secondary windings.

3.7 TRANSIENT BEHAVIOUR

Because of their ability to store signals, capacitors and inductors exhibit charging (or filling) and discharging (or emptying) characteristics for a short duration of time. The 'transient behaviour' may be explained as follows.

In the 'Resistor-Capacitor' (*R-C*) circuit of figure 3.11, the intent is to charge the capacitor through a resistor and slow down the charging rate. In the absence of the resistor, the capacitor would be charged instantly. With the resistor, the charge current I flows through R to reach C after the switch is turned on (position A). The greater the resistive value of R, the slower is the charging rate, or the voltage buildup v_C of C. It takes in theory at least six time constants to fully charge the capacitor, but in practice it is commonly agreed that five time constants is a more accurate representation for most applications.

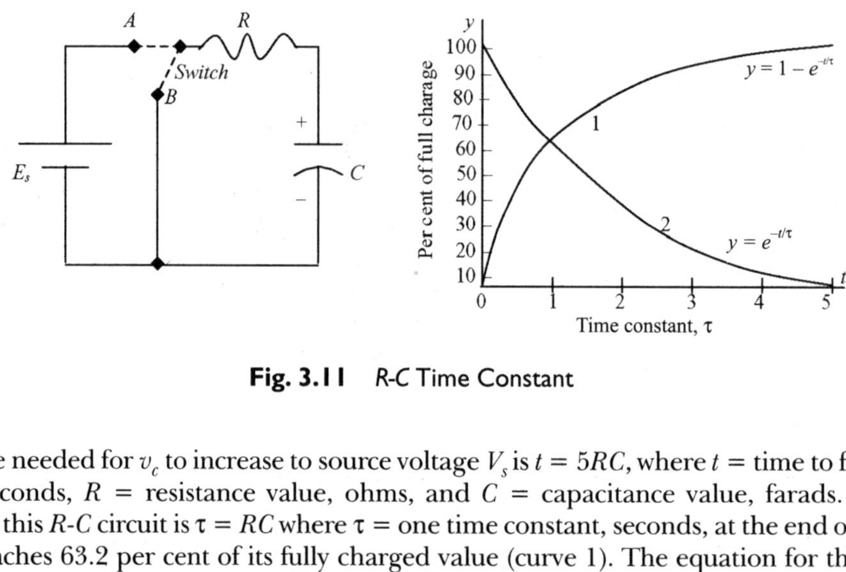

Fig. 3.11 *R-C* Time Constant

The time needed for v_c to increase to source voltage V_s is $t = 5RC$, where t = time to fully charge capacitor, seconds, R = resistance value, ohms, and C = capacitance value, farads. The 'time constant' for this *R-C* circuit is $\tau = RC$ where τ = one time constant, seconds, at the end of which the capacitor reaches 63.2 per cent of its fully charged value (curve 1). The equation for the charging curve shown in figure 3.11 is $y = 1 - e^{-t/\tau}$.

If the switch were to be placed in the B position, discharge curve 2 is mathematically expressed as: $y = e^{-t/\tau}$. After one time constant the capacitor will lose 63.2 per cent of its full voltage.

A combination of 'Resistance-Inductance' (*R-L*) circuits follows a similar trend as *R-C* circuits. Substituting an inductor in place of the capacitor in figure 3.11 will give voltage v_L across the inductor. Placing the switch in position A causes voltage v_L to equal supply voltage E_s. The time period for the flux field surrounding the inductor when the switch is closed to reduce to zero is $t = 5L/R$, where t = time to fully eliminate flux field, or for V_L to reduce from E_s to zero, seconds, R = resistance value, ohms, and L = inductance value, henry. The 'time constant' for the *R-L* circuit is $\tau = L/R$ where τ = one time constant, seconds.

Similar to the R-L circuit, at the end of the time constant the inductor's voltage reduces to 63.2 per cent of its initial value, E_s. The curve representing this deterioration of voltage is given by curve 2 in figure 3.11. Note that according to Kirchhoff's voltage law, while the V_L approaches zero, the voltage across the resistor R is building up to E_S. Thus, the inductor performs as a short.

When the switch is moved to position B, $V_R = E_S$ for a short duration, since the inductor's current does not change instantly. The voltage across the inductor changes in polarity. The flux field around the inductor begins to collapse, the characteristic of this reduction being similar to curve 2. Thus, curve 2 may be flipped over so that it shows reduction from a negative value up to zero as time elapses.

3.8 REACTANCE AND IMPEDANCE

The frequency of an AC electrical circuit plays a role in controlling the effect of a component's capacitive or inductive nature, the reaction being in the form of a blockage or a pseudo-resistance. The presence of 'reactance and impedance' in a DC electric circuit in comparison may be considered to be relatively simple.

An ideal resistor reacts to an AC signal by posing the same resistance to it as a DC signal. When an alternating current waveform passes through a resistor, it develops an alternating voltage waveform (figure 3.12 *upper*). The current and voltage waveforms will appear in phase with each other when viewed together on an oscilloscope. If a capacitor is substituted for the resistor, the current and voltage waveforms will appear out of phase with each other, with the voltage lagging the current waveform by 90° (figure 3.12 *middle*). When the current flow into the capacitor is at a peak, the voltage developed across it is at a minimum. Only when the current is at a minimum, or zero, does the capacitor develop the full voltage. Thus, when the capacitor is fully charged, a form of resistance is created that causes the current to stop flowing. The action is referred to as capacitive reactance. The size of capacitance determines the time needed to charge the capacitor, and the AC waveform's frequency controls the rate of filling to develop the resistive reactance. Expressed mathematically, the relation is: $X_C = -j/(2\pi.fC)$, where X_C = capacitive reactance, ohms, f = frequency, Hz, C = capacitance, farads, $-j = -\sqrt{(-1)}$, which causes the 90° phase shift between current and voltage. A phase angle shift invariably occurs when a reactance is present in the circuit, and is –90° for a capacitor. Since $\omega = 2\pi f$, then $X_C = (1/\omega C)\angle{-90°}$.

If an inductor is substituted for the resistor in figure 3.12 (*lower*), the current and voltage waveforms will appear out of phase with each other, but with the voltage leading the current waveform by 90°. When the flux field is expanding around the inductor, the lines of flux cut across its windings to develop an induced voltage of opposite polarity. The ensuing backpressure tends to diminish any further current or voltage buildup. When the flux field is at its peak, the current begins to flow again since the backpressure is relieved. The voltage across the inductor becomes zero, the flux field collapses and once again cuts across the turns of the inductor. This generates a forward pressure to push the current through the inductor, while also reversing the polarity across it.

Expressed mathematically, the inductive reactance is given by: $X_L = +j2\pi fL$, where X_L = inductive reactance, ohms, f = frequency, Hz, L = inductance, henry, $+j = \sqrt{(-1)}$, which causes the current to lag the voltage by 90°.

Combining the inductive and capacitance reactance with resistance provides the 'impedance' in a circuit. The central problem is to determine the current i and keeping track of the associated phase

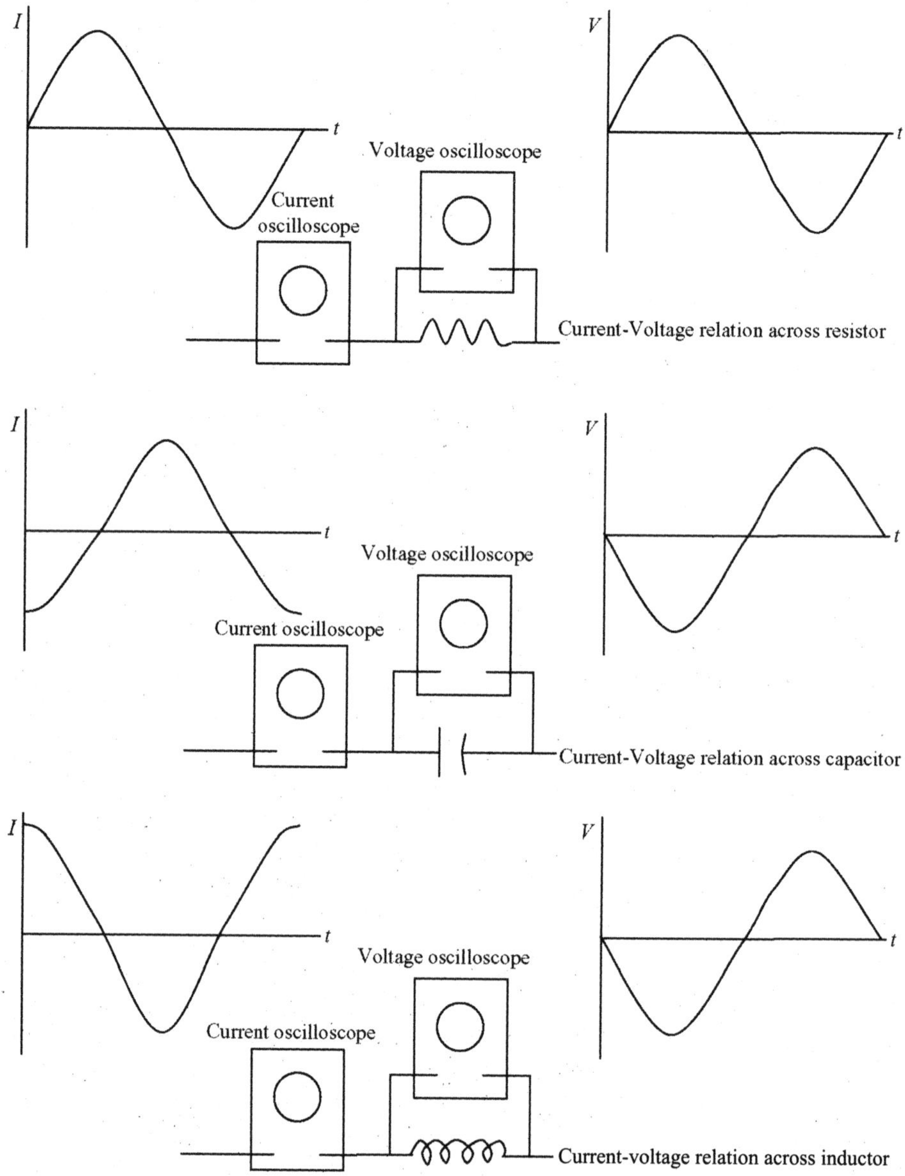

Fig. 3.12 Current-Voltage Relation for Resistor (*upper*), Capacitor (*middle*), Inductor (*lower*)

shifts between the output voltage and current of a circuit containing resistors and either capacitors or inductors or both. The solution is to devise a way to combine all three components into an equivalent impedance with a phase angle, and which permits using Ohm's law to solve for the current i. The impedance will have units of ohms just as do reactance and impedance. Ohm's law may then be formulated as: $I = E/Z\angle\theta$, where I = current, amps, E = voltage, volts, and $Z\angle\theta$ = impedance of circuit in polar form.

Reactance and impedance values are commonly considered to be vectors in electronics with both magnitude and direction. In the complex plane, the reactance and impedance may be represented by a real number (along the horizontal axis) and by its imaginary counterpart (along the vertical axis). The phase angle for any complex number representation is: $\theta = \text{Arctan}\ (jB/A)$, where A and jB are the components of the complex number. The magnitude of the complex number is $Z = \sqrt{(A^2 + B^2)}$. Note that the A term is the resistive value, and is plotted along the real axis. The B term is the reactive or impedance value is plotted along the imaginary axis.

The instantaneous values of any current (i) or voltage (v) waveform is given by the expression: $y = A_m \cdot \sin(\omega t \pm \theta)$, or $y = A_m \cdot \cos(\omega t \pm \theta)$. Another designation calls for root-mean-square (or, RMS) values. For a sinusoidal waveform the RMS value equals 0.707 times the maximum value of the waveform.

The apparent power in an AC circuit is expressed as $P_a = VI$, where V and I are the RMS values of voltage and current. Taking into account the phase angle between the voltage and current waveforms, the effective or real power $P = P_a \cos\theta$, where θ is the phase angle. The term $\cos\theta$ is called the power factor. Electrical components are generally rated in volt-amperes (VA) instead of watts, since the electrical load or impedance of the attached equipment is not known.

3.9 RESONANT FREQUENCY

Resonant frequencies play a significant role in electronics and automatic control theory. A 'resonant frequency' is one developed in an electrical circuit that acquires a part of its driving energy through a more or less self-sustaining feedback loop or circuit. Consider, for example, the simplified regenerative receiver (or, Armstrong oscillator) for a radio circuit shown in figure 3.13. Feeding in some of the amplified radio signal back to the antenna provides further amplification of the weak signals detected and heard.

But if excessive re-amplification were to occur, then the entire circuit oscillates, or resonates, at a particular frequency established by the radio's tuning circuits. The undesirable consequences from the standpoint of reception from the device become apparent.

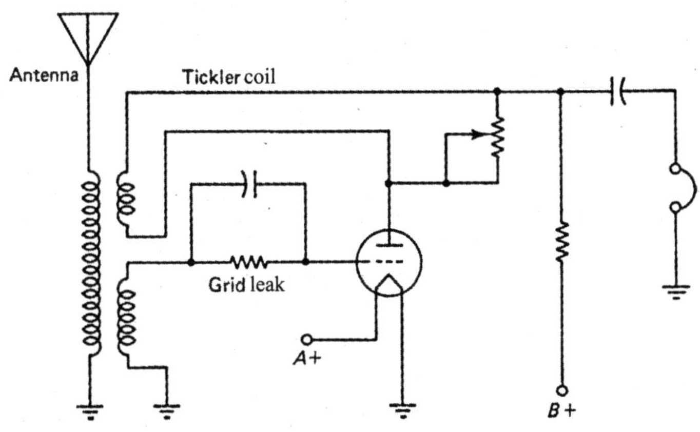

Fig. 3.13 Regenerator Receiver Circuit

Any electrical circuit containing an inductance and capacitance is susceptible to resonance at a particular frequency. Since the capacitor is permitted to discharge into an inductor, the inductor's field builds up and then collapses to reintroduce a flow of charge back into the capacitor. The event is followed by the capacitor charge building up and discharging back into the inductor, and repeating the process. The specific frequency of this back and forth oscillation is established by the amount of capacitance and inductance in the circuit according to the equation: $f_r = 1/[2\pi\sqrt{(LC)}]$, where f_r = resonant frequency, Hz, L = inductance, henry, and C = capacitance, farad. Internal resistance within the circuit eventually dampens out the electrical oscillations.

3.10 OSCILLATOR

In spite of internal resistance, the electrical oscillations as a consequence of resonant conditions mentioned in the previous section may persist if a fresh charge of electrons from an external source is injected at the right instant. This forms the principle behind the functioning of an oscillator. The oscillator has useful applications in the area of timing, becoming an electronic clock to synchronize or coordinate functions inside a computer and control circuits. Their performance is further enhanced through the application of precisely machined piezo-electric crystals made from certain materials that vibrate at a resonant frequency.

The time constant of an *R-C* or an *R-L* combination may also be put to use in an oscillator. Varying the resistance value in the combination over a wide range of values simulates a resonant condition. This acts as the basis for producing the necessary timing frequency for control circuitry applications.

3.11 SOLID-STATE DEVICES

Electronic solid-state devices find widespread applications in automatic control circuits. Figure 3.14 identifies many of the components and the symbols used in designs.

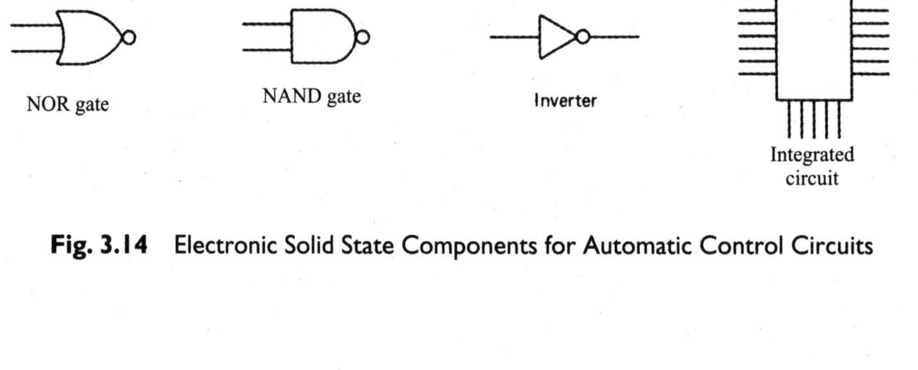

Fig. 3.14 Electronic Solid State Components for Automatic Control Circuits

A 'diode' is used to let the current flow only in one direction. The symbol's pointer indicates the flow direction. Similar to it is the 'Zener diode', which can maintain or regulate the level of voltage in a loop. The 'NPN and PNP transistor' is used to amplify current signals. The NPN or PNP designation identifies the direction of flow of current in the emitter circuit, as shown by the indicator. The 'FET transistor' amplifies voltage signals, and has characteristics that are similar to those of a vacuum tube. Variations are available in FET transistors, the one shown in figure 3.14 being a dual gate MOSFET (MOS standing for metal oxide semiconductor). Frequently used to mix two signals, the MOSFET is marked by its small physical size. 'NOR, VAND and INVERTER gates' are logic devices for logical decision making from a number of choices. 'Integrated circuits (IC)' are spread over a wide array of devices that are too numerous for mentioning. IC's are electronic circuits with various combinations of components packaged into a small container that may be soldered or attached to a larger circuit. Made by a number of manufacturers, the behaviour of the components can be obtained from their reference manuals.

3.12 ELECTRO-MECHANICAL COMPONENTS

Electro-mechanical devices used for control systems are based on the concept of development of electro-dynamic force. If an element is moving at a speed v within a magnetic flux of density $\boldsymbol{B}$, a magnetic field $\boldsymbol{E}$ is generated that may be mathematically expressed as: $\boldsymbol{E} = v\boldsymbol{B}$. If the moving element is carrying an electric charge e, the force $\boldsymbol{F}$ is: $\boldsymbol{F} = e\boldsymbol{E}$, so that: $\boldsymbol{F} = ev\boldsymbol{B}$, where $ev = \boldsymbol{i}$, and hence: $\boldsymbol{iB} = \boldsymbol{F}$. The expression is referred to as Fleming's left hand rule, and is applied in electric motors.

If the magnet placed near the body were to be moved, the magnetic flux associated with the body changes and a voltage is induced. Application of electro-magnetic induction and Fleming's right hand rule yields the relation: $\boldsymbol{F} = \boldsymbol{iB}$.

A linear motor is a device inside which the rotor of a turning motor, or the slider in a translating motor, and the stator are arranged such that the electro-magnetic energy is transformed into mechanical energy to generate the thrust. Miniature motors may be broadly classified into linear DC motors, stepping motors, induction motors, oscillatory motors and synchronized motors. The motors have a simplified structure, and in spite of their low power factor and efficiency, are rapidly permeating many different industrial arenas. The linear DC motors and induction motors may be classified as electro-dynamic force type of motors, and the stepping motors and oscillatory motors are magnetic force type of motors.

3.13 ANALOG AND DIGITAL ELECTRONIC MODES OF CONTROL

An electronic analog controller consists of an error detector and a control mode unit. The error detector compares the output of a measuring transmitter with a set-point signal. The two voltage signals are connected in opposition, so the voltage between two points equals the set-point signal, sp, minus the measured value, c_m, signal or $e = sp - c_m$.

The electronic analog controller uses a single operational amplifier to act as a function generator, and some resistors and capacitors to implement the transfer function of the desired control mode. Thus, the analog proportional controller has three resistors to form an inverting amplifier, the proportional plus integral controller uses two resistors and a capacitor to provide the

PI transfer function, and the proportional plus derivative controller uses four resistors and a capacitor to provide the PD mode.

An electronic PID controller may be obtained either by forming the derivative action on the input side and the integral action on the output side, or by doing the opposite and switching to form the integral action on the input side and the derivative action on the output side. The transfer function of the analog PID controller is a modified version of the expression obtained in section 2.10, where an interaction term *PIDe* is inserted in the time domain:

$$v = Pe + PIDe + PI\int edt + PD(de/dt) - \alpha D(dv/dt) + v_o$$

The Laplace transformation of the equation with $v_o = 0$ yields the frequency-domain equation:

$$V = PE + PIDE + PIE/s + PdsE - \alpha DsV$$

The transfer function is given by the expression:

$$V/E = P[(I + s)/s]\,[(1 + Ds)/(1 + \alpha Ds)].$$

Microprocessor based digital controllers have found wide acceptance in industrial control systems because of their advanced features such as self-tuning, multi-variable control, and expert systems. Digital controllers may be adapted for closed-loop control to provide the PI, PD or PID modes of control. Controlled variables are measured after specific time intervals called the sampling time, Δt. Measured samples of the control variables are converted into a binary form for input to a digital computer, which subtracts the sample from the set-point to calculate a set or error samples. The digital PID controller follows an algorithm to determine the output of the controller based on the error samples. Two versions of the algorithm are widely used. The 'positional' version determines the position of the actuation valve based on error signals, while the 'incremental' version determines the change in valve position based on error samples. The incremental algorithm is better suited for incremental output devices such as stepper motors, while the positional algorithm has the advantage that the controller acts from the previously set position of the valve. Computational problems that produce unsatisfactory results can develop when the controller algorithm is used in the integral (I) and derivative (D) modes of control, especially if the value of the term $PI\Delta t$ is small. The computer may then ignore the relatively large errors because of inadequate resolution. One solution to this integral offset problem is to increase the precision by increasing the expression length in the computer. Alternatively, the unused portion of the sum of the error samples may be added to the current error sample.

3.14 REVIEW PROBLEMS

Example problem # 3.1: A 14.5 foot long copper bar has a width of 3.25 inches and height of 2.5 inches. Find its resistance if copper has resistivity constant of 10.375 ohms-circular mils per foot at room temperature.

Solution: The cross-sectional area of bar in mils is:

$$\text{Area} = 3.25 \times 10^3 \times 2.5 \times 10^3 = 8.125 \times 10^6 \text{ mils}^2$$

Equivalent mils^2 area in circular mils:

$$\text{Area}_{CM} = 8.125 \times 10^6 \times (4/\pi) = 10.345 \times 10^6 \text{ CM}$$

Then resistance $R = [(10.375 \text{ ohms-CM per ft}) \times 14.5 \text{ ft}]/(10.345 \times 10^6 \text{ CM}) = 14.542 \times 10^{-6}\ \Omega$

Example problem # 3.2: Determine the total resistance current flow and voltage drop across each element of the circuit shown in figure 3.15.

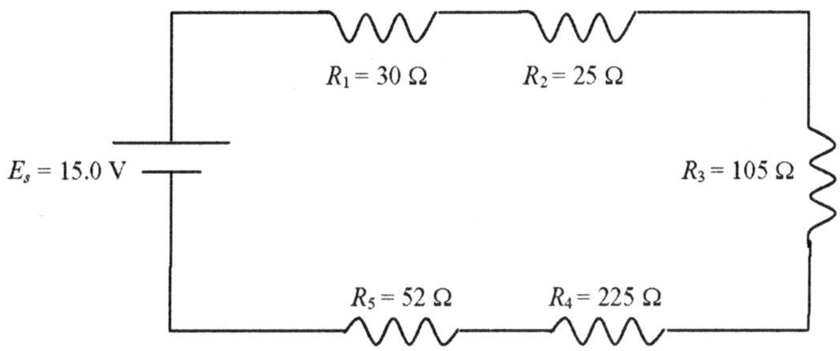

Fig. 3.15

Solution: Since the total current flows through all the resistors back to the voltage source, the resistances are connected in series. Hence, the total equivalent resistance in the circuit $R_T = (R_1 + R_2 + R_3 + R_4 + R_5) = 30 + 25 + 105 + 225 + 52 = 437\ \Omega$.

Current $I_T = E/R_T = 15.0/437 = 0.0343$ amps $= 34.3$ milliamps.

Voltage drop across R_1: $I_T R_1 = (0.0343 \text{ amps}) \times 30\ \Omega = 1.0297$ V.

Voltage drop across R_2: $I_T R_2 = (0.0343 \text{ amps}) \times 25\ \Omega = 0.858$ V.

Voltage drop across R_3: $I_T R_3 = (0.0343 \text{ amps}) \times 105\ \Omega = 3.604$ V.

Voltage drop across R_4: $I_T R_4 = (0.0343 \text{ amps}) \times 225\ \Omega = 7.723$ V.

Voltage drop across R_5: $I_T R_5 = (0.0343 \text{ amps}) \times 52\ \Omega = 1.785$ V.

According to Kirchhoff's voltage law, all five voltage drops must equal supply source voltage, or: $1.0297 + 0.858 + 3.604 + 7.723 + 1.785 = 15.0$ V, hence the solution is acceptable.

Example problem # 3.3: In figure 3.3, $R_1 = 30\ \Omega$, $R_2 = 45\ \Omega$ and $R_3 = 60\ \Omega$. If $E_s = 75.0$ V, calculate the total resistance, current flow and voltage drop across each element of the circuit.

Solution: The total equivalent resistance, R_T, of the resistors in parallel connection with each other is: $1/R_T = 1/R_1 + 1/R_2 + 1/R_3 = 1/30 + 1/45 + 1/60 = 0.0722$.

Hence: $R_T = 13.846\ \Omega$.

Total current $I_T = E_S/R_T = 75.0/13.846 = 5.417$ amps.

For resistors connected in parallel, the voltage drop in each resistor is the same as source voltage. Hence: $I_1 = E_s/R_1 = 75.0/30.0 = 2.50$ amps.

$I_2 = E_s/R_2 = 75.0/45.0 = 1.667$ amps.

$I_3 = E_s/R_3 = 75.0/60.0 = 1.25$ amps.

According to Kirchhoff's current law, the flow of current into a grid point must equal the flow of current out of it, or: $2.50 + 1.667 + 1.25 = 5.417$ amps, hence the solution is acceptable.

Example problem # 3.4: Using the following values for the resistors shown in figure 3.4, determine the combined equivalent resistance for the circuit.

$R_1 = 8\,\Omega,\ R_2 = 7\,\Omega,\ R_3 = 6\,\Omega,\ R_4 = 10\,\Omega,\ R_5 = 12\,\Omega,\ R_6 = 9\,\Omega,\ R_7 = 11\,\Omega.$

Solution: Parallel: $R_{1-2} = R_1.R_2/(R_1 + R_2) = 8 \times 7/(8 + 7) = 3.73\ \Omega$

Parallel: $R_{4-5} = R_4.R_5/(R_4 + R_5) = 10 \times 12/(10 + 12) = 5.45\ \Omega$

Series: $R_{3-4-5} = (R_3 + R_{4-5}) = 6 + 5.45 = 11.45\ \Omega$

Series: $R_{6-7} = (R_6 + R_7) = 9 + 11 = 20\ \Omega$

Parallel: $R_{3-4-5-6-7} = R_{3-4-5}.R_{6-7}/(R_{3-4-5} + R_{6-7}) = 11.45 \times 20/(11.45 + 20) = 7.28\ \Omega$

Series: $R_{1-2-3-4-5-6-7} = (R_{1-2} + R_{3-4-5-6-7}) = 3.73 + 7.28 = 11.01\ \Omega.$

Example problem # 3.5: Assume the supply voltage E_s = 22.5 Volts in example problem # 3.4. Calculate the magnitude of current I_T.

Solution: A single 11.01 Ω resistor can replace the group of resistors in figure 3.4. Total current flow then may be calculated to be:

$$I_T = E_s/R_T = 22.5/11.01\ \Omega = 2.044 \text{ amps.}$$

Example problem # 3.6: Find the output voltage in figure 3.16.

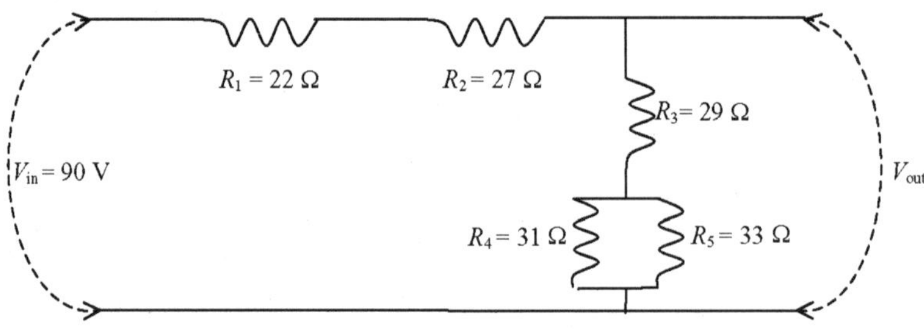

Fig. 3.16

Solution: Resistors R_4 = 31 Ω and R_5 = 33 Ω in parallel are equivalent to:

$$R_{4-5} = 31 \times 33/(31 + 33) = 15.98\ \Omega.$$

Then R_1, R_2, R_3, R_{4-5}, are connected in series, so total equivalent resistance:

$$R_E = 22 + 27 + 29 + 15.98 = 93.98\ \Omega.$$

Now, R_3, R_{4-5}, are connected in series, so total equivalent resistance:

$$R_{3-4-5} = 29 + 15.98 = 44.98\ \Omega.$$

According to voltage divider rule:

$$V_{out} = V_{in} \times (R_{3-4-5}/R_E) = 90 \times (44.98/93.98)$$

$$= 43.077 \text{ volts.}$$

Example problem # 3.7: A sine curve starting at origin ($t = 0$) has maximum amplitude of 3.75 amps and frequency of 50 Hz. If the curve is shifted by 45°, or $\pi/4$ radians, write the expression for the new curve. Also, rewrite the equation of shifted curve as a cosine curve.

Solution: The sine curve is: $y = A_0 \times \sin(\omega t + \alpha)$, where maximum amplitude $A_0 = 3.75$ amps, frequency $\omega = 2\pi f = 2\pi \times 50 = 314.16$ radians/second, and phase angle $\alpha = 0$. After shifting, $\alpha = \pi/4$ radians.

The expression for the new curve is: $y = 3.75 \times \sin(314.16t + \pi/4)$.

A cosine waveform is 90°, or $\pi/2$ radians, out of phase with a sine curve. Rewriting as a cosine curve implies subtracting $\pi/2$ radians, or:

$$y = 3.75 \times \cos(314.16t - \pi/4).$$

Example problem # 3.8: A capacitor made of two parallel, flat aluminium plates each of 75 cm^2 are separated by a glass sheet of 0.20 cm thickness. Find the capacitance if the dielectric constant of glass is $e_r = 7.5$. If three such plates are connected in parallel with each other, what is their equivalent capacitance?

Solution: The capacitance value, C, is related to plate area, distance between plates, and dielectric constant through the expression:

$$C = (8.85 \times 10^{-12} A e_r)/d. \text{ Area } A = 0.0075 \text{ m}^2 \text{ and } d = 0.0020 \text{ m}.$$

Hence:

$$C = (8.85 \times 10^{-12} \times 0.0075 \times 7.5)/0.0020$$
$$= 24.89 \times 10^{-11} \text{ farads.}$$

When three such plates are connected in parallel with each other, then:

$$C_T = 3 \times 24.89 \times 10^{-11} \text{ farads.}$$
$$= 74.67 \times 10^{-11} \text{ farads.}$$

Example problem # 3.9: A coil of wire with 350 turns is wrapped on a core of length 0.85 m and diameter 0.008 m. Assume permeability of air is 12.60×10^{-7} Wb/At-m. Calculate the inductance of the coil. If three such coils are connected in series with each other, what is their equivalent inductance?

Solution: The inductance value, L, is related to core area A, length of coil l, number of turns N, and air permeability constant, μ, through the expression:

$$L = (N^2 \mu A)/l.$$

$$N = 350 \text{ turns, } l = 0.85\ \mu, \text{ Area } A = (\pi/4) \times (0.008)^2 = 5.027 \times 10^{-5} \text{ m}^2$$

Then:

$$L = (350^2 \times 12.60 \times 10^{-7} \times 5.027 \times 10^{-5})/0.85$$
$$= 912.76 \times 10^{-7} \text{ henry}$$
$$= 91.28\ \mu\text{H}.$$

If three such coils are connected in series with each other, then:

Equivalent inductance:

$$L_T = 3 \times 91.28\ \mu\text{H}$$
$$= 273.84\ \mu\text{H}.$$

Example problem # 3.10: A transformer has 30 turns on its primary windings, and is supplied with 125 VAC. Determine the number of turns required on the secondary winding to obtain 25 VAC. Also, if the primary current is 10 amps, what is the secondary current?

Solution: $N_p = 30$ turns, $E_p = 125$ V, $E_s = 25$ V.

Now: $E_p/E_s = N_p/N_s$.

Hence: $N_s = N_p \times (E_s/E_p) = 30 \times (25/125)$

$= 6$ turns.

The relation between a transformer's turns and current is: $I_p/I_s = N_s/N_p$.

If $I_p = 10$ amps, then $I_s = 10 \times (30/6) = 50$ amps.

Example problem # 3.11: What is the reactance of a 0.1 μf capacitor through which passes a current of 750 mA at a frequency of 300 Hz?

Solution: Reactance $X_C = -j/(2\pi fC) = -j/[2\pi \times 300 \times 0.1 \times 10^{-6}]$

$= -5305.17j$ ohms

or: $X_C = 5305.17\angle{-90°}$.

Example problem # 3.12: Solve for the current, I, in the circuit of figure 3.17.

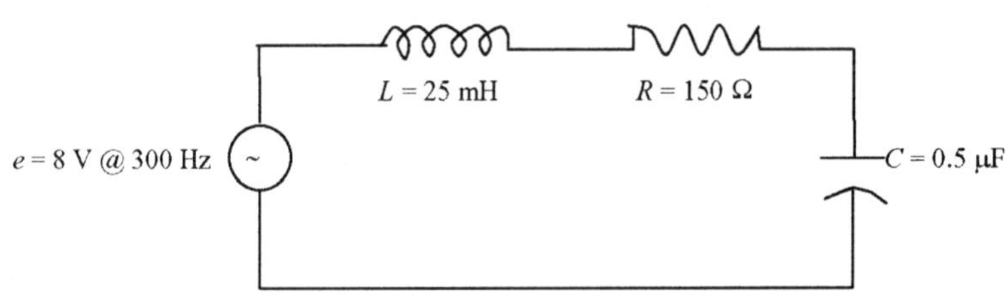

Fig. 3.17

Solution: Capacitance $X_C = -j/(2\pi fC) = -j/[2\pi \times 300 \times 0.5 \times 10^{-6}]$

$= -1061.03j\ \Omega$

or: $X_C = 1061.03\angle{-90°}$.

Inductance $X_L = j(2\pi fC) = j[2\pi \times 300 \times 25 \times 10^{-3}]$

$= +\ 47.12j\ \Omega$

or: $X_C = 47.12\angle 90°$.

Total reactance $Z = -1061.03j + 47.12j + 150$

$Z = 150 - 1013.91j\ \Omega$

$= 1024.95\angle{-81.6°}$

Then current $I = V/Z = 8/(1024.95\angle{-81.6°})$

$= 0.0078\angle 81.6°$ amps.

3.15 EXERCISES

3.1 A copper bar has a width of 2.25 inches and height of 2.75 inches. Find the length of the bar if its resistance is $5.5 \times 10^{-6}\ \Omega$. Copper has resistivity constant of 10.50 ohms-circular mils per foot.

3.2 Determine the total resistance, current flow and voltage drop across each element of the circuit shown in figure 3.15, but with the resistors connected in parallel with each other.

3.3 Using the following values for the resistors shown in figure 3.4 and assuming the supply voltage E_s = 12.5 volts, calculate the magnitude of current I_T.

$$R_1 = 8\ \Omega,\ R_2 = 7\ \Omega,\ R_3 = 6\ \Omega,\ R_4 = 10\ \Omega,\ R_5 = 12\ \Omega,\ R_6 = 9\ \Omega,\ R_7 = 11\ \Omega.$$

3.4 Find the input voltage in figure 3.16 if R_1 = 18 Ω, R_2 = 20 Ω, R_3 = 24 Ω, R_4 = 27 Ω, R_5 = 9 Ω, R_6 = 30 Ω, R_7 = 32 Ω and V_{in} = 25 volts.

3.5 A capacitor made of two parallel, flat aluminium plates each of 90 cm^2 are separated by a glass sheet. Find the thickness of the glass sheet if the dielectric constant of glass is e_r = 7.5 and the reactance of the capacitor is 20×10^{-11} farads. If three such plates are connected in series with each other, what is their equivalent capacitance?

3.6 A coil of wire is wrapped on a core of length 0.25 m and diameter 0.012 m. Assume permeability of air is 12.65×10^{-7} Wb/At-m. If the inductance of the coil is 150 μH, calculate the number of turns in the wire coil. If three such coils are connected in parallel with each other, what is their equivalent inductance?

3.7 A transformer has 240 turns on its primary windings, and a rating of 4,400 volts. Find the number of turns required on the secondary winding to obtain 1,100 volts. Also, if the secondary current is 250 amps, what is the primary current?

3.8 Calculate the reactance of a 0.125 μf capacitor through which passes a current of 7.5 amps at a frequency of 300 MHz?

3.9 Determine the capacitance if a capacitor has a reactance of 11,500 Ω from a current of 2.5 A at a frequency of 250 Hz?

3.10 An inductor has a reactance of 150 Ω when a current of 12.5 A at a frequency of 390 Hz flows through it. Find its inductance.

3.11 In figure 3.17 assume the generator voltage e = 18.75 V at 3500 Hz. Find the current in the circuit.

3.12 In figure 3.17 assume current is 0.85 amps, and frequency is 350 Hz. Find the supply source voltage.

References & Bibliography

Bateson, R.N., "Introduction to Control System Technology", Prentice-Hall, Clifton, NJ, 1999.

Carstens, J.R., "Automatic Control Systems and Components", Prentice-Hall, Clifton, NJ, 1990.

DeRoy, B.E., "Automatic Control Theory", John Wiley & Sons, New York, 1996.

Miller, R.W., "Servo-mechanism Devices and Fundamentals", Reston Publishing Co., Reston, VA, 1977.

Patrick, D.R. and Fardo, S. W., "Industrial Process Control System", Prentice-Hall, Clifton, NJ, 1979.

Sante, D.P. "Automatic Control System Technology", Prentice-Hall, Clifton, NJ, 1980.

4

Fluid Power Review

4.1 INTRODUCTION

Fluid power is required to energize equipment and devices in industries ranging from manufacturing to transportation and construction. Heavier applications call for hydraulically activated systems, while pneumatic systems are ideal for lighter loads, shorter strokes, and higher speed applications. Manual controls suffice in a number of situations, but fluid power in combination with electrical and electronic command, control, and sensing finds far more extensive usage. Electronically activated digital devices have the considerable advantage of providing nearly instantaneous reaction to limiting situations. Components such as a turbine flow meter, tachometer-generator, linear variable differential transformer, accelerometer, and piezo-electric pressure sensor offer continuous sensing with electrical output that permits precise control of linear and rotational position, speed, acceleration, and force. In conjunction, the fluid power system delivers versatility, accuracy, and reliability. As an example, manual manipulation of large valves located on or near a power generation steam turbine may be deleted and replaced by electro-hydraulically operated actuators from a remote station, thus reducing operational fatigue and enhancing safety.

Employing an efficient merger between two clear and well-defined disciplines, electro-hydraulic systems can cause a cylinder to stop at a precisely specified position, reciprocate automatically, or operate in a certain sequence. Solenoid valves activated by limit switches and other digital devices are capable of achieving the most rigorous set of operating events under demanding conditions.

Many different applications of electro-hydraulic control systems exist, including:

- Variable inlet and stator guide vanes for the compressor of an aircraft engine. As the speed of the rotor changes, the stator vanes must be opened or closed to maintain peak operating efficiency.
- Roll control system for airplanes during turning manoeuvres.
- Pitch and speed control of propellers in a turbo-prop engine.
- Control of steam chest valves for maintaining rotor speed in turbines for electric power generation.
- Vectoring of thrust from exhaust nozzle of fighter aircraft engines.

Although a relatively mature technology, fluid power in combination with electro-pneumatics and electro-hydraulics continues to see evolutionary advancements. Improvements in solenoid design provide higher forces and faster operation. Proportional solenoids have reached a level of control and frequency response that rival many servo-valves. Refinements in transducers, digital electronics, and fibre optic devices are also finding their way into the world of electro-hydraulics. Feedback elements and servo electronics combine with smart components such as servo-cylinder, servo-pump, and servo-motor in a single package. The only external electrical connections are a power lead and a single command line from the command computer.

4.2 CONCEPTS IN FLUID POWER

A fluid power system consists of a hydraulic pump to convert mechanical energy into fluid energy, conduits and associated control devices to transmit the pressurized fluid with minimum resistance, and a hydraulic cylinder or motor to re-convert into mechanical energy. An electric motor is needed to drive a positive-displacement hydraulic pump to pressurize the fluid.

The velocity of flow through a cylindrical device is determined from the expression: $v = Q/A$, where v = flow velocity, Q = flow rate, and A = effective area. The pressure resulting from application of force depends on the area over which the force is exerted, or: $p = F/A$, where p = pressure, F = force, and A = effective area.

The speed of a hydraulic motor may also be determined by the flow rate, being a function of volumetric displacement of the motor, or: $N = Q\eta_v/V_M$, where N = rotor speed, Q = flow rate through the motor, η_v = volumetric efficiency of motor, and V_M = volumetric displacement of the hydraulic motor.

A relationship between pressure developed and the required driving torque for hydraulic motors may also be developed. The expression is: $T = V_M\,\Delta p\eta_m/(2\pi)$, where T = torque, V_M = volumetric displacement of motor, Δp = pressure differential built up in the motor, and η_m = mechanical efficiency of the motor.

The power required to operate the hydraulic pump, motor, cylinder, and to overcome losses is of interest. Horsepower for a rotating machinery is denoted by the expression: HP = $(2\pi \times N \times T)/33000$, where N = turning speed, rpm, and T = torque, ft-lbs. In SI units, power is measured in kilowatts: $P_{in} = (N \times T)/9550$, where N = turning speed, rpm, and T = torque, N-m. Power requirement for a pump is termed hydraulic horsepower (HHP), given by the expression: HHP = $(p \times Q)/1714$, where p = system pressure, lbs/in^2, and Q = flow rate, gpm. In SI units, hydraulic power is measured in kilowatts: $P_{hyd} = (p \times Q)/60000$, where p = pressure, kPa, and Q = flow rate, lpm. Overall pump efficiency is governed by the ratio of the output power to the input power: η_o = HHP/IHP.

The output of a hydraulic motor is OHP = $(2\pi \times N \times T)/33000$, where N = turning speed, rpm, and T = torque, ft-lbs. In SI units, power is measured in kilowatts, or: $P_{out} = (N \times T)/9550$, where N = turning speed, rpm, and T = torque, N-m. Input horsepower requirement for a motor is termed hydraulic horsepower (HHP), given by the expression: HHP = $(\Delta p \times Q)/1714$, where p = system pressure, lbs/in^2, and Q = flow rate, gpm. In SI units, hydraulic power is measured in kilowatts: $P_{hyd} = (\Delta p \times Q)/60000$, where $\Delta p = p_{inlet} - p_{outlet}$, kPa, and Q = flow rate, lpm. Overall motor efficiency is governed by the ratio of the output power to the input power: η_o = OHP/HHP.

Cylinder horsepower can be computed from the same fundamental equations, or:

OHP = $(F \times d)/(550 \times t)$, where F = force, lbs, d = distance, feet, and t = time, seconds. Since velocity $d/t = v$ (ft/second), the alternate expression is: OHP = $(F \times v)/(550)$.

4.3 LIQUID AND GAS FLOW ELEMENTS IN CONTROL SYSTEM

Resistance to flow of a fluid through a pipe, fittings, valves and other restrictive devices is measured in terms of loss of pressure. The resistance is measured as a relation between pressure drop and flow rate. The SI unit for flow resistance is pascal-s/m^3, and in English units it is lbs/in^2-gpm.

The characteristics of flow depend on the density of the fluid, ρ, inside diameter of the pipe, d, the absolute viscosity of the fluid, μ, and the average velocity of flow, v, which equals $Q/(\pi d^2/4)$. The four parameters may be arranged in dimensionless grouping called the Reynold's number: Re = $(\rho v d)/\mu$. Laminar flow is said to take place if the Reynold's number is less than 2000, and the flow is turbulent if the Reynold's number exceeds 4000. A transition between the laminar and turbulent regime prevails when the number is between 2000 and 4000.

The flow equation for laminar conditions in a round pipe is given by Hagen-Poiseuille law: $p = R_L Q$, Pa, where p = pressure loss from friction, laminar flow resistance $R_L = 128\mu l/(\pi d^4)$, Pa-s/m^3, μ = absolute viscosity, d = diameter of pipe, and l = length of pipe.

The flow equation for turbulent conditions in a round pipe is given by Fanning's equation: $p = K_t Q^2$, Pa, where turbulent flow coefficient $R_t = 2K_t Q$, Pa-s/m^3, where:

$K_t = 8fl/(\pi^2 d^5 \rho)$, Pa-s^2/m^6, ρ = density of fluid, d = diameter of pipe, and l = length of pipe, and f = friction factor that varies with pipe size and condition. Typical values for f are shown in table 4.1.

'Liquid flow capacitance' is defined in terms of increase in volume of liquid in a container that is required to obtain a unit increase in pressure at the outlet of the container, or: $C_L = \Delta V/\Delta p$, where C_L = capacitance, m^3/Pa, ΔV = increase in volume, m^3, and Δp = corresponding increase in pressure, Pa. Increase in pressure is brought about through increase in level of fluid in the container (ΔH), acceleration due to gravity (g), and density of the liquid, ρ. The relation may be expressed as follows: $\Delta p = \rho g \Delta H$. The increase in level of the liquid depends on the change in volume divided by average surface area A of the liquid in the container, or: $\Delta H = \Delta V/A$. Hence, $\Delta p = \rho g \Delta V/A$, and capacitance $C_L = A/\rho g$ m^3/Pa.

'Liquid flow inertance' is defined in terms of pressure drop in a pipe required to increase flow rate by 1 unit/second, or: $I_L = p/(\Delta Q/\Delta t)$, where I_L = inertance, m^3/s^2, ΔQ = change in flow rate, m^3/s, p = pressure drop in pipe, Pa, and Δt = time interval, s. A more practical equation is obtained

Table 4.1 Friction Factor for Pipes

Type	Diameter, cm	Reynold's Number					
		4×10^3	10^4	10^5	10^6	10^7	10^8
Smooth tubing	1 – 2	0.039	0.030	0.018	0.014	0.012	0.012
	2 – 4	0.039	0.030	0.018	0.013	0.011	0.010
	4 – 8	0.039	0.030	0.018	0.012	0.010	0.009
	8 – 16	0.039	0.030	0.018	0.012	0.009	0.008
Commercial steel	1 – 2	0.041	0.035	0.028	0.026	0.026	0.026
	2 – 4	0.040	0.033	0.024	0.023	0.023	0.023
	4 – 8	0.039	0.030	0.022	0.020	0.019	0.019
	8 – 16	0.039	0.030	0.020	0.018	0.017	0.017

Reference: Moody, L.F., "Friction Factors for Pipe Flow".

by letting the pressure drop (p) acting on the cross-sectional area of the pipe (a) produce a force (F) equal the force due to the acceleration of the fluid in the pipe ($\Delta v/\Delta t$). The expression for inertance then takes the form: $I_L = \rho l/a$, Pa/m^3/s^2.

'Gas flow resistance' is that characteristic of pipes, fittings, valves and other restrictive devices that impede the flow of gas. The resistance is measured as a relation between pressure drop and flow rate of 1 kg/s. The SI unit for flow resistance is pascal-s/kg.

Flow of gas through a pipe may be either laminar or turbulent. In laminar flow the pressure drop varies directly with the flow velocity, while in turbulent flow the pressure drop varies with the square of the flow velocity. The Reynold's number for gas flow is: $\text{Re} = 4W/\pi\mu d$. In practice, the flow is mostly turbulent. If the pressure drop is under 10 per cent of the initial gas pressure, the equation for incompressible flow gives reasonably good accuracy. The expression is similar to that used for turbulent liquid flow, with volume flow rate (Q) replaced by mass flow rate (W): $p = p_1 - p_2 = K_g W^2$, Pa, where p_1 = pressure at inlet of pipe, Pa, p_2 = pressure at outlet of pipe, turbulent flow coefficient $R_g = 2\ K_g W$, Pa-s/kg, where: $K_g = 8fl/(\pi^2 d^5 \rho)$, Pa-s^2/kg^2, ρ = density of fluid, kg/m^3, d = inside diameter of pipe, m, and l = length of pipe, m, and f = friction factor. Note that gas properties such as density and viscosity vary with pressure and temperature.

'Gas flow capacitance' is defined in terms of increase in the mass of gas in a pressure container that is required to obtain a unit increase in pressure, with the temperature remaining constant. Then: $C_g = \Delta m/\Delta p = MV/(10^3 RT)$, where C_g = gas flow capacitance, kg/Pa, M = gram molecular weight of the gas, g/mol, V = volume of the gas, m^3, R = universal gas constant = 8.314 J/K-mol, T = absolute temperature, K, of the gas.

4.4 THERMAL ELEMENTS

'Thermal resistance' is a property of materials that impedes the flow of heat, and is measured as a difference in temperature needed to produce a heat flow rate of 1 W (joule per second). Heat generally flows through a wall separating two liquids or gases from the fluid at higher temperature to the fluid at lower temperature. Thus, if T_o is the higher temperature on one side and T_i is the lower temperature on other side of the wall, then heat flow rate, Q, from the hotter to the cooler fluid is $Q = (T_o - T_i)/R_i$, W, where R_i is the thermal resistance of the separating wall. Thermal resistance is often expressed as resistance of 1 m^2 of the wall, or unit resistance, R_u. If the area of the exposed wall between the two fluids is A, then: $Q = (T_o - T_i)A/R_i$. If the separating wall is a composite wall made of several layers (figure 4.1), each material contributes to the total resistance. The two outer layers are thin films of stagnant fluid that act as insulating blanket, while the two inner layers are of different materials. The thickness of the films depends on the motion of the fluid. For instance, a strong circulating forced draft reduces the thickness of the film. The resistance of the film varies with the thickness and thermal resistance of the fluid. Empirical formulas for calculating the conductance, h, (reciprocal of resistance) of the fluid film under different conditions are shown in table 4.2. The unit resistance of the film is equal to the reciprocal of the film coefficient, or: R_u (film) = $1/h$, K-m^2/W. The unit resistance of a typical layer of thickness x, thermal conductivity k of material of the composite wall is: R_u (composite layer) = x/k, K-m^2/W.

'Thermal capacitance' is defined as increase in the amount of heat required to raise temperature by one unit. The SI unit of thermal capacitance is joule/kelvin (J/K).

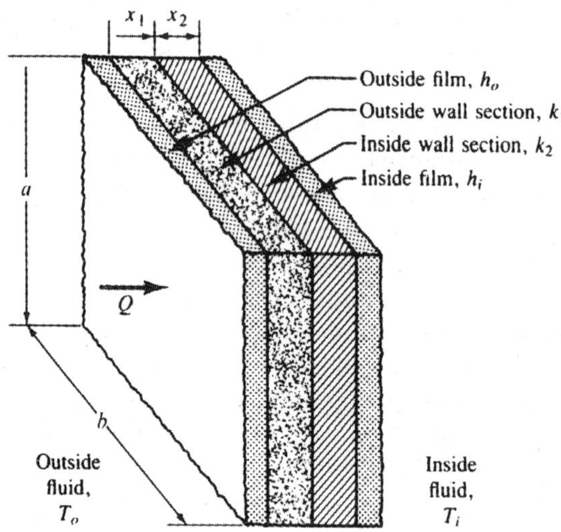

Wall surface area = $A = a \times b$, m_2
Total thermal resistance = $R_T = [(1/A) \times (1/ho + x_1/k_1 + x_2/k_2 + 1/h_i)]$ kelvin/watt
Outside fluid temperature = T_o, °C
Inside fluid temperature = T_i, °C
Heat flow rate = $Q = (T_o - T_i)/R_T$, watts

Fig. 4.1 Thermal Resistance of Composite Wall

The heat capacity, or the specific heat (S_h), of a substance is the amount of heat required to raise the temperature of 1 kg of the substance by 1 K. Hence, thermal capacitance (C_T) of an item is the product of mass and the specific heat of the substance, or: $C_T = mS_h$.

4.5 CIRCUIT DIAGRAMS

A fluid power circuit diagram may be represented by ISO symbols, which when put together convey important information about the system. The basic elements of a hydraulic pump symbol are a circle and a filled triangle (figure 4.2). Combining the two gives a fixed displacement, single-acting pump. Adding a diagonal arrow represents a variable displacement, single-acting pump. In the open-loop, open-center circuit all fluid returns at the low pressure reservoir, where the pump picks it up for re-circulation. A tandem center directional control valve is also included. When the piston in the cylinder is not acting, the valve is centered and so pump output flow passes through the valve at low pressure and returns to the reservoir. In the open-loop, closed-center mode, the circuit uses a closed-center directional control valve. When the valve is centered, the cylinder is stopped and held in place with the tandem center valve. This causes the fixed-displacement pump to operate under its maximum load condition, and dumps its full output flow across the relief valve at a high pressure. Energy wastage and heat generation is the consequence, but it provides the ability to maintain full system pressure.

Table 4.2 Thermal Resistancquations

Resistance	$R_u = 1/h_o + x_1/k_1 + x_2/k_2 + \cdots x_n/k_n + 1/h_i$
	$R_T = R_u/A$
	$Q = (T_o - T_i)/R_T$
	$Q = (T_o - T_i) \times A/R_u$
Film Coefficients – Natural convection in still air.	
a. Horizontal surfaces facing up:	$h = 2.50 \times T_d^{0.25}$
b. Horizontal surfaces facing down:	$h = 1.32 \times T_d^{0.25}$
c. Vertical surfaces:	$h = 1.78 \times T_d^{0.25}$
Natural convection in still water.	$h = 2.26 \times (T_w + 34.3) \times T_d^{0.5}$
Natural convection in oil.	$h = 7.0T_d^{0.5}/\mu^{0.4}$
Forced convection of air against smooth surfaces and inside pipes.	
a. Air velocity $v_a < 4.6$ m/s:	$h = 4.54 + 4.1 \times v_a$
b. Air velocity $v_a > 4.6$ m/s:	$h = 7.75 \times v_a^{0.75}$
Forced convection for turbulent water flow in straight pipes.	$h = 20.93(T_w + 68.3) \times v_w^{0.8}/d^{0.2}$
Capacitance	$C_T = mS_h$
where: A = area of wall surface, m^2 C_T = thermal capacitance, J/K d = pipe inside diameter, m h = film coefficient, W-m^2/K k = thermal conductivity, W-m/K m = mass, kg Q = heat flow rate, W R_u = unit thermal resistance, K-m^2/W R_T = total thermal resistance, K/W S_h = heat capacity, J/kg-K T_d = temperature difference between fluid and wall surface, K or °C T_i = inside fluid temperature, °C T_o = outside fluid temperature, °C T_w = water temperature, °C v_a = velocity of air, m/s v_w = velocity of water, m/s x = thickness of layer, m μ = absolute viscosity, Pa-s	

Reference: Salisbury, J.K., "Kent's Mechanical Engineer's Handbook: Power".

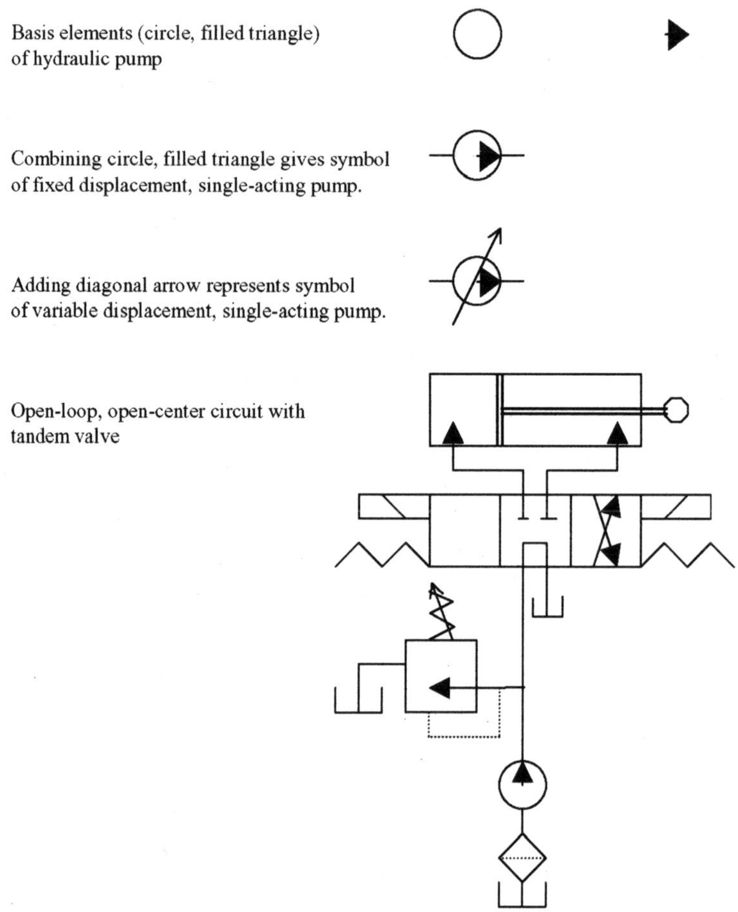

Fig. 4.2 Fluid Power Circuit Graphic Symbols

4.6 ELECTRO-HYDRAULIC CONTROL SYSTEM

Many applications require rapid cycling. For example, fatigue testing of turbine super alloys requires a test machine with a hydraulic cylinder to produce cyclic stress on the specimen. The test specimen is cyclically loaded to a predetermined state of stress and relaxed to zero load at the highest possible frequency. The on-off operation can be carried out with a solenoid valve and a suitable switching circuitry, but regulating the rate at which the loading is increased and decreased at a higher cycling frequency can be better achieved with either a proportional or a servo-valve.

The primary concern for this application is to ensure that the frequency response of the valve is appropriate. Examination of the valve manufacturer's data sheet and frequency response curves, usually in the form of a Bode plot, provides valuable information. A typical Bode plot indicating the components amplitude response and phase lag for a range of input signal frequencies is shown in figure 4.3.

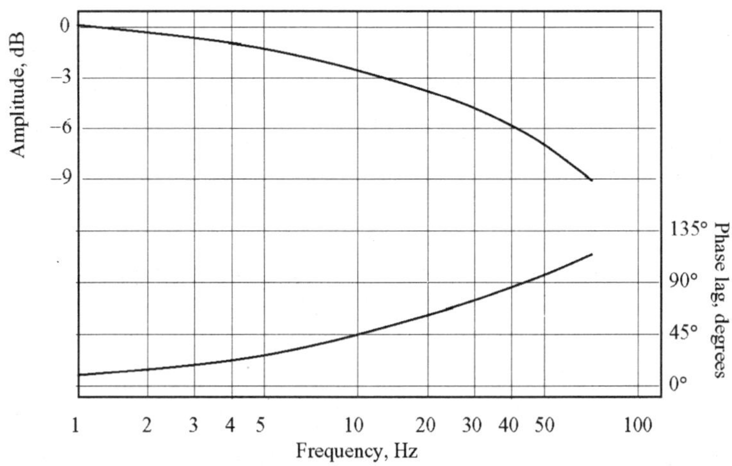

Fig. 4.3 Typical Bode Plot for Proportional Valve

Electrical responses are far faster than mechanical responses because motion in mechanical devices experience greater drag. Thus, a phase lag occurs between the electrical signal command and the response of the mechanical device. The phase lag tends to increase with the frequency of the cycling rate. Increasing the command frequency in figure 4.3 to 44 Hz causes the valve's spool movement to lag by 90°. When the phase lag exceeds 180°, the valve becomes unstable since the command signal is in a direction opposite that of the valve travel.

By convention, proportional valves and servo-valves are rated at their 90° phase lag frequency. Thus, the valve illustrated in figure 4.3 is rated as a 90° phase lag at 44 Hz, and is sometimes expressed as $f_{90} = 33$.

Motion of the valve spool in response to the input command is defined by the amplitude response curve in figure 4.3. The amplitudes may also be defined in terms of spool position. If X_a is the actual peak position and X_c is the commanded peak position, the amplitude response = 20 × log (X_a/X_c). Valves are conventionally rated at their −3 dB amplitude response. In figure 4.3, the valve has −3 dB amplitude response at 12 Hz, which may be expressed as $f_{-3} = 12$.

If the intent in the fatigue testing of the super alloy specimen is to completely relax, but not reverse, the loading, valve performance about the zero position must be known. The manufacturer's data sheet provides information of the type shown in table 4.3 concerning step response of the valve. The spool reaches 90 per cent of its full flow in 34 ms in response to a step input command for 100 per cent flow. The valve closes in 31 ms after the command signal is removed. Thus, for one full cycle (zero to maximum and back to zero) will require 65 ms, which equates to around 15 Hz. This cycle time may be reduced if the valve is not required to be fully open in each cycle. A practical circuit for the fatigue test equipment using a proportional valve is shown in figure 4.4.

Just as valves are not capable of responding to the maximum frequency of the electronic control circuitry, the overall system is incapable of responding to the valve's maximum frequency capability. This is due to the natural (or resonant) frequency of the hydro-mechanical system, which is usually lower than the maximum operating frequency of the valve. Attempts to operate the system at a frequency higher than the natural frequency will result in erratic and unstable behaviour. The equation for natural frequency of a hydraulic cylinder is: $\omega_o = \{(E/M) \times [(A_p^2/V_p) + (A_N^2/V_N)]^{1/2}\}$,

Table 4.3 Response to Step Input (at Δp = 72 lbs/in^2 per metering path)

Required Flow Step	*Time to Reach 90% of Required Step*
0 to 100%	34 ms
100 to 0%	31 ms
+90 to −90%	52 ms

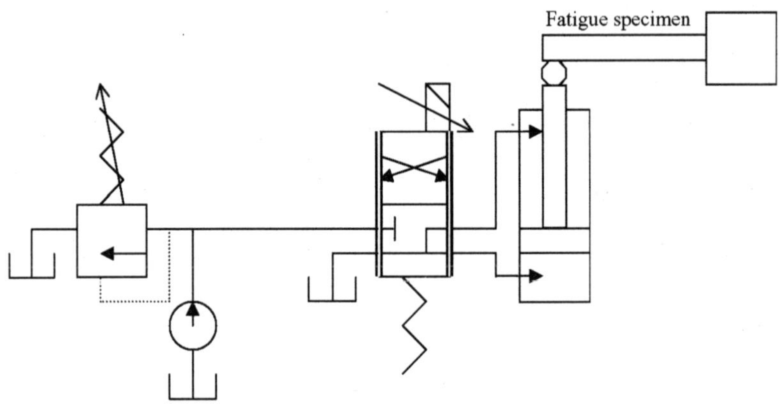

Fig. 4.4 Cyclic Fatigue Test Circuit Diagram Using Proportional Valve

where ω_o = natural frequency, radian/s, E = fluid bulk modulus, M = mass of load, (weight/g), A_p = piston area, V_p = total volume of oil between valve and piston face, A_N = annular piston rod area, and V_N = total volume of oil between annular face of piston face and valve. Fluid bulk modulus E = $[-\Delta p/(\Delta V/V)]$, where Δp = pressure change, ΔV = change in volume due to applied pressure, and V = original volume. Values of V_p and V_N depend on the position of the piston as it moves through the stroke. Reducing the volume of oil between the valve and cylinder can increase the frequency of the system. One possibility is to attach valves directly on the cylinders and motors and eliminate connecting lines.

The natural frequency of a hydraulic motor circuit is given by the expression:

$\omega_o = \{[2E \times [(V_M/\pi)^2]/(V_T/I_T)\}^{1/2}$, where V_M = hydraulic motor displacement, V_T = effective trapped volume of fluid = $V_M + V_L$, I_T = total reflected inertia = Mr^2R^2, V_L = volume of fluid in pressure and return lines between the valve and motor, r = radius of gyration, R = gear ratio of input and output gear teeth, if used, and E = fluid bulk modulus.

An important aspect of a servo system is its ability to continuously monitor its output and make corrections to ensure output remains in the command range. This is achieved through feedback from a transducer that monitors the output parameter. The forward loop gain, G, and the backward loop gain, H, determine the system gain. Increasing the amplifier gain in an electro-hydraulic control system can saturate and push the valve wide open until the cylinder is very close to its stopping point. The time required for moving from one position to another is decreased, and the system stiffness would improve. This results in a stiff and fast acting system. But extremely high pressures can result from a rapidly closing valve, and lead to plumbing components to rupture. Still another consequence can be 'water hammer' due to the sudden stopping of a flowing fluid by rapid closure of a valve. System stability can also be adversely affected due to high amplifier gain. Sensitivity of the system to disturbances is greatly enhanced as the gain is increased, resulting in a tendency to overcorrect, then abruptly reverse and overcorrect in the opposite direction. An unstable system will oscillate wildly, which can hold potential to cause serious damage to itself and the machine on which it is installed. The maximum allowable gain of a system can be estimated mathematically if data about the components of the system is available.

On the other hand, reducing the amplifier gain would increase the stopping distance and result in a sluggish and spongy system that will tend to 'hunt' for the commanded position.

'Repeatable error' defines the ability of the system to return to the same position every time a specific input command is provided. The repeatable error is a function of the dead-band of the valve and the electronic gain of the system, or: R_e = (dead-band, mA)/[(G_A, mA/V) × (H, V/in.)], where G_A = amplifier gain and H = feedback gain.

When the cylinder is actuated, its position lags the error signal command by a small distance, and is referred to as the 'tracking error'. The error is at its peak when the valve is fully open. The error diminishes as the valve begins closing, and is reduced to zero when the cylinder reaches the commanded position.

4.7 ANALOGY BETWEEN COMMON ELEMENTS OF SYSTEM COMPONENTS

The governing differential equation for mechanical, electrical, liquid flow, gas flow and thermal elements takes the form: $a.d^2x/dt^2 + b.dx/dt + c.x = f(x)$. As an illustration, in a mechanical mass-spring-damper system, coefficient a = value of the oscillating mass, coefficient b = damper constant, and coefficient c = spring constant.

The analogy may be extended towards similar physical identification of the parameters for electrical, liquid flow, gas flow, and thermal elements.

In a further extension of the analogy, the mathematical analysis for all control elements is going to be identical, but the mathematical parameters for the input and calculated items assume their own distinctive physical characteristics. Table 4.4 provides a summary of equations and a description of their physical characteristics.

4.8 REVIEW PROBLEMS

Example problem # 4.1: A double-acting single ended hydraulic cylinder has a 2.5 in. bore and 1.25 in. diameter rod. The system flow rate is 3 gallons per minute. Calculate the extension and retraction velocities of the cylinder.

Solution:

Piston diameter $A_p = \pi D_p^2/4 = \pi \times (2.5)^2/4 = 4.91$ in^2.

Extension velocity $v_e = Q/A_p$ = (3 gal. per min.) × (231 in^3/gal)/(4.91 in^2)

= 141.18 in./minute

Rod area $A_r = \pi D_r^2/4 = \pi \times (1.25)^2/4 = 1.23$ in^2.

Retraction velocity $v_r = Q/(A_p - A_r)$

= (3 gal. per min.) × (231 in^3/gal)/(4.91 – 1.23 in^2)

= 188.24 in./minute

Table 4.4 Equations for Common Elements of System Components

Element and Type	*Description*	*Equation*	*Units*
1. *Resistance*:			
Mechanical	Force needed to increase velocity by 1m/s.	$R_m = F/v$	N-s/m
Electrical	Voltage needed to increase current by 1 A.	$R_e = e/i$	ohm
Liquid flow	Pressure needed to increase flow by 1 m^3/s.	$R_e < 2000$, (laminar), $R_L = 128\mu/\pi d^4$	Pa-s/m^3
		$R_e > 2000$, (turbulent), $R_L = 2K_T Q$	Pa-s/m^3
		(where $K_T = 8\rho/\pi^2 d^5$)	Pa-s^2/m^6
Gas flow	Pressure needed to increase flow by 1 kg/s.	$R_g = 2K_g W$	Pa-s/kg
Thermal	Temperature needed to increase heat flow by 1 W.	$R_T = 1/A[1/h_o + x_1/k_1 + \dots x_n/k_n + 1/h_i]$	K/W
2. *Capacitance*:			
Mechanical	Spring displacement needed to increase force by 1N.	$C_m = 1/K = F/v$	m/N
Electrical	Electric charge needed to increase potential by 1 V.	$C = q/e$	Farad
Liquid flow	Volume needed to increase pressure by 1 Pa.	$C_L = A/\rho g$	m^3/Pa
Gas flow	Mass needed to increase pressure by 1 Pa (T constant).	$C_g = MV/RT$	kg/Pa
Thermal	Heat needed to increase temperature by 1 K.	$C_T = mS_h$	J/K
3. *Inertia/Inductance/Inertance*:			
Mechanical	Force needed to increase acceleration by 1m^2/s.	$m = F/(dv/dt)$	kg
Electrical	Voltage needed to increase current rate by 1A/s.	$L = e/(di/dt)$	Henry
Inertance	Pressure drop needed to increase liquid flow rate by 1 m^3/s/s.	$I_L = \rho l/a$	Pa-s^2/m^3

where:

a = area of flow through pipe, m^2.	A = tank cross section area, m^2.
C = electrical charge, farad.	C_g = gas flow capacitance, kg/Pa
C_L = liquid flow capacitance, m^3/Pa	C_m = mechanical capacitance, m/N
C_T = thermal capacitance, J/K	d = pipe diameter, m
e = voltage, volts	F = force, N
g = gravitational constant, m/s^2	i = current, Amps
I_L = liquid flow inertance, Pa-s^2/m^3	K_g = turbulent gas flow coefficient, Pa-s
K_T = turbulent liquid flow coefficient, Pa-s^2/m^6	l = length, m
L = electrical inductance, henry	m = mass, kg
M = molecular weight, g/mol	q = electrical charge, coulomb
Q = gas flow rate, m^3/s	R = gas constant = 8.314 J/K-mol.
R_e = electrical resistance, ohms	R_e = Reynold's number
R_g = gas flow resistance, Pa-s/kg	R_L = liquid flow resistance, Pa-s/m^3
R_m = mechanical resistance, N-s/m	S_h = heat flow rate, J/kg-K
R_T = thermal resistance, K/W	t = time, s
T = temperature, K	v = velocity, m/s
v = voltage, volt	V = volume, m^3
ρ = density, kg/m^3	

Example problem # 4.2: Determine the speed of a hydraulic motor with a displacement of 3.2 in^3/rev for a flow rate of 7 gallons/minute. Assume the volumetric efficiency of the motor is 93 per cent.

Solution: Speed $N = \eta_v Q/V_M = [(7 \text{ gal/min}) \times (231 \text{ in}^3\text{/gal})) \times 0.93]/3.2 \text{ in}^3\text{/rev}$

$= 470$ rpm.

Example problem # 4.3: A hydraulic motor with a displacement of 3.2 in^3/rev has a mechanical efficiency of 92 per cent. If the pressure at the pump inlet is 1200 lbs/in^2, find the torque produced by the motor.

Solution: Assume no pressure at the outlet port. Torque developed is:

$$\text{Torque} = [V_M \Delta p \eta_m]/(2\pi) = [(3.2 \text{ in}^3\text{/rev}) \times (1200 \text{ lbs/in}^2) \times 0.92]/(2\pi)$$

$$= 562.26 \text{ in-lb.}$$

Example problem # 4.4: A hydraulic pump operating at 1250 lbs/in^2 and 11.5 gallons/minute is driven by an electric motor at 1760 rpm. If the motor develops 325 in-lbs of torque, find the overall efficiency of the pump.

Solution: The horsepower supplied into the pump is:

$$\text{IHP} = (2\pi NT)/33000$$

$$= (2\pi \times 1760 \text{ rpm} \times 325 \text{ in-lbs})/(33000 \times 12) = 9.08 \text{ HP}$$

The horsepower supplied into the pump is:

$$\text{HHP} = (p \times Q)/1714 = (1250 \text{ lbs/in}^2 \times 11.5 \text{ gallons/minute})/1714$$

$$= 8.39 \text{ HP.}$$

$$\text{Overall efficiency } \eta_o = \text{HHP/IHP} = 8.39/9.08$$

$$= 0.924$$

Example problem # 4.5: A hydraulic cylinder can push a 3750 lb load at 2.5 ft/s. Find its output horsepower.

Solution: Output horsepower $= (F \times v)/550 = (3750 \text{ lbs} \times 2.5 \text{ ft/s})/550 = 17.05$ HP

Example problem # 4.6: The resistance of an electric light bulb with a non-linear volt-ampere graph can be approximated from a small voltage increment resulting in small current increment. If 5.95 V results in 0.450 A and 6.05 V results in 0.453 A, find its resistance at 6 V.

Solution: Voltage increment $\Delta e = 6.05 - 5.95 = 0.10$ V

Current increment $\Delta i = 0.453 - 0.450 = 0.003$ A

Then resistance $R = \Delta e/\Delta i = 0.10 \text{ V}/0.003 \text{ A} = 33.3\ \Omega$

Example problem # 4.7: A current pulse of amplitude 0.25 mA and duration of 0.12 s is applied to an electrical capacitor. The action increases the voltage in the capacitor from 0 to 24 V. Determine the capacitance of the capacitor.

Solution: Capacitance $C = I \times (\Delta t/\Delta e) = (0.25 \times 10^{-3} \text{ A}) \times [0.12 \text{ s}/(24 \text{ V} - 0 \text{ V})] = 1.25 \times 10^{-6}\ \mu\text{F}$

Example problem # 4.8: Calculate the dead-time delay of a signal sent from a space vehicle located 3500 km from a receiving station if the signal travels at 3×10^8 m/s.

Solution: Time delay t_d = (distance, d)/(speed of propagation)

$$= (3500 \times 10^3 \text{ m})/(3 \times 10^8 \text{ m/s})$$

$$= 11.67 \times 10^{-3} \text{ s} = 11.67 \text{ ms}.$$

Example problem # 4.9: A lubricating oil at 17°C temperature flows through a 1 cm diameter and 12 m long horizontal tube at a flow rate of 8.5 litres/minute. Calculate the Reynold's number, resistance and pressure drop in the line if density ρ of oil is 875 kg/m³ and viscosity μ = 0.155 Pa-s.

Solution: Diameter of tube d = 1 cm = 0.01 m

Flow rate, $Q = 8.5 \text{ litres/min} = 8.5/(60 \times 10^3) \text{ m}^3/\text{s} = 1.417 \times 10^{-4} \text{ m}^3/\text{s}$

Flow velocity $v = (4Q)/(\pi d^2) = (4 \times 1.417 \times 10^{-4} \text{ m}^3/\text{s})/(\pi \times 0.01^2)$

$= 1.80$ m/s

Reynold's number $= \rho v d/\mu = [(875 \text{ kg/m}^3) \times (1.80 \text{ m/s}) \times (0.01 \text{ m})]/(0.155 \text{ Pa-s})$

= 101.85, hence flow is laminar.

Resistance $R_L = (128\mu l)/(\pi d^4) = (128 \times 0.155 \text{ Pa-s} \times 12 \text{ m})/(\pi \times 0.01^4)$

$= 7.578 \times 10^9 \text{ Pa-s/m}^3$.

Pressure drop $p = R_L Q = (7.578 \times 10^9 \text{ Pa-s/m}^3) \times (1.417 \times 10^{-4} \text{ m}^3/\text{s})$

$= 1.074 \times 10^6$ Pa.

Example problem # 4.10: A holding tank is of diameter 2.25 m and height 8 m. Determine the capacitance of the tank if it contains a liquid of density ρ = 885 kg/m³.

Solution: Cross section area of tank $A = (\pi d^2)/4 = (\pi \times 2.25^2)/4 = 3.98 \text{ m}^2$.

Capacitance $C_L = A/\rho g = (3.98 \text{ m}^2)/(885 \text{ kg/m}^3 \times 9.81 \text{ m/s}^2)$

$= 4.58 \times 10^{-4} \text{ m}^3/\text{Pa}$.

Example problem # 4.11: Calculate the liquid flow inertance of water inside a pipe of diameter 2.25 cm and length 75 m.

Solution: Cross sectional area of pipe $a = (\pi d^2)/4 = (\pi \times 0.0225^2)/4 = 3.98 \times 10^{-4} \text{ m}^2$.

Density of water $\rho = 1000 \text{ kg/m}^3$.

Inertance $I_L = \rho l/a = (1000 \text{ kg/m}^3 \times 75 \text{ m})/(3.98 \times 10^{-4} \text{ m}^2)$

$= 1.89 \times 10^8 \text{ Pa/m}^3/\text{s}^2$.

Example problem # 4.12: If the flow rate in example problem # 4.11 is 0.001275 m³/s, calculate the dead-time delay.

Solution: Velocity of flow $v = 4Q/(\pi d^2) = (4 \times 0.001275 \text{ m}^3/\text{s})/(\pi \times 0.0225^2)$

$= 3.21$ m/s

Time delay $t_d = D/v = (75 \text{ m})/(3.21 \text{ m/s}) = 23.39 \text{ s}$.

Example problem # 4.13: A section of a wall similar to figure 4.1 has two inner layers, a 1.25 cm thick steel plate and a 2.5 cm thick insulation layer. On one side there is water, and on the other side there is air. Surface temperature of wall on the water side is 55°C, water temperature is 45°C, and temperature difference T_d = 10°C. Air on the other side has a velocity of 5 m/s and temperature of 95°C. Wall dimensions are a = 2.5 m, b = 3.5 m. Calculate the unit resistance, total resistance, total heat flow and direction of heat flow. Assume k_1 = 50 W/m-K for steel and k_2 = 0.035 W/m-K for insulation.

Solution: Area of wall $A = 2.5 \times 3.5 = 8.75 \text{ m}^2$.

$$\text{Inside film coefficient} = h_i = 2.26(T_w + 34.3)T_d^{0.5} = 2.26 \times (45 + 34.3) \times 10^{0.5}$$

$$= 566.74 \text{ W/m}^2\text{-K.}$$

$$\text{Outside film coefficient} = h_o = 7.75 v_a^{0.75} = 7.75 \times 5^{0.75}$$

$$= 25.91 \text{ W/m}^2\text{-K.}$$

$$\text{Unit thermal resistance } R_u = 1/h_o + x_1/k_1 + x_2/k_2 + 1/h_i$$

Hence,

$$R_u = 1/(25.91 \text{ W/m}^2\text{-K}) + 0.0125 \text{ m}/(50 \text{ W/m-K}) + 0.025 \text{ m}/(0.035 \text{ W/m-K}) + 1/(566.74 \text{ W/m-K})$$

$$R_u = 0.755 \text{ K-m}^2\text{/W}$$

$$\text{Total thermal resistance } R_T = R_u/A = (0.755 \text{ K-m}^2\text{/W})/(8.75 \text{ m}^2)$$

$$= 0.0863 \text{ K/W}$$

$$\text{Total heat flow } Q = (T_o - T_i)/R_T = (95 - 45)\text{K}/(0.0863 \text{ K/W})$$

$$= 579.55 \text{ W.}$$

4.9 EXERCISES

4.1 A cylinder has a 3.0 in. bore and a 1.25 in. rod. If the oil pressure is 2750 lbs/in^2, what is the (*i*) maximum retraction force, (*ii*) minimum retraction force the component can produce?

4.2 A 3.0 in. cylinder is required to extend at 4 in./s. What is the flow rate required in gallons/minute?

4.3 A 2.5 cm cylinder can move a 5000 N resistance. Calculate the pressure needed.

4.4 If a 4.5 cm cylinder has a flow rate of 22 litres per minute, what is the extension velocity?

4.5 A hydraulic motor with displacement of 3.4 in^3 operates at a pressure of 800 lbs/in^2, and has mechanical efficiency of 0.925. Find the available torque from the motor.

4.6 A pump delivers 22 gallons/minute at 2500 lbs/in^2. Find its hydraulic horsepower.

4.7 A hydraulic motor produces 130 ft-lbs torque at 725 rpm. Find its horsepower output.

4.8 Calculate the thermal resistance of the following boundary layer film conditions:

(*i*) Natural convection in still air on a vertical surface, if T_d = 13°C.

(*ii*) Natural convection in still water, if T_d = 17°C and T_w = 25°C.

(*iii*) Natural convection in oil, if T_d = 14°C.

(*iv*) Forced convection in air, if air velocity is 2.5 m/s.

(*v*) Forced convection in water, if T_w = 35°C, pipe diameter is 4 cm and velocity is 3 m/s.

4.9 An automobile weighing 2500 kg accelerates from standstill to 75 km per hour in 6 s. Find the average force required.

4.10 A 25 m long belt conveyor has a speed of 2.25 m/s. Calculate the delay time between the loading and delivery ends of the belt.

4.11 A composite wall section has an area of 4.5 m^2, and has an inner aluminium layer 0.70 cm thick. The inside fluid is oil and outside fluid is water, both of which are still. Temperature difference T_d = 2°C for the water film and 18°C for the oil film. Temperature of oil is 92°C, and of water is 67°C. Calculate the unit resistance, total resistance, total heat flow and direction of heat flow.

4.12 Ethyl alcohol at 15°C has density of 800 kg/m^3 and absolute viscosity of 0.0013 Pa-s. Calculate the Reynold's number, resistance, and pressure drop if the alcohol flows through a 15 m long commercial pipe with diameter 2.20 cm. Flow rate is 20 litres per minute.

4.13 An electric current of 0.45 A is applied to a capacitor for 0.22 s. If the voltage across the capacitor increased from 0 to 11.5 V, calculate its capacitance.

4.14 A holding tank is 1.75 m in diameter and 2.5 m tall. Determine the capacitance of the tank for the following fluids:

(*i*) Water – density: 1000 kg/m^3	absolute viscosity: 0.001 Pa-s
(*ii*) Oil – density: 880 kg/m^3	absolute viscosity: 0.160 Pa-s
(*iii*) Kerosene – density: 800 kg/m^3	absolute viscosity: 0.0024 Pa-s
(*iv*) Gasoline – density: 740 kg/m^3	absolute viscosity: 0.0005 Pa-s
(*v*) Turpentine – density: 870 kg/m^3	absolute viscosity: 0.0015 Pa-s

References & Bibliography

"Industrial Hydraulics Technology", Parker-Hannifin Corp., Cleveland, OH, 1997.

Moody, L.F., "Friction Factors for Pipe Flow", ASME Transactions, Volume 16, No. 8, page 671, New York, NY. 1944.

Norvelle, F.D., "Electro-hydraulic Control Systems", Prentice-Hall, Upper Saddle River, NJ, 2000.

Norvelle, F.D., "Fluid Power Technology", West, St. Paul, Minn., 1995.

Salisbury, J.K. "Kent's Mechanical Engineer's Handbook: Power", Page 3-17 to 3-20, John Wiley & Sons, New York, 1950.

"Vickers Industrial Hydraulics Manual", Vickers, Inc., Rochester Hills, Mich., 1993.

Part II

Measurement and Regulation

5

Sensors for Measurement

5.1 INTRODUCTION

Among the many desirable characteristics of measurement sensors, accuracy and speed of measuring a rapidly changing signal play the leading role in controlling the output variable in an automatic control system. Control of a varying parameter first requires measurement of its value before the controller and the manipulating element execute their task. Feedback information serves as the basis of a control system. In the absence of measurement, the controller cannot decide on the necessary course of action for achieving the stated objective, and the manipulating device cannot act without a decision from the controller.

Response from the primary element (or sensor) relies on the properties of its material and construction details. The measured value is converted into an electrical, mechanical, or fluidic signal. The output signal may then take the form of an electrical voltage, force, displacement, pressure, or some other physical parameter. Regardless of the form, the output of the sensor depends on the value of the measured parameter.

Although rarely achieved, the measuring instrument must obtain the true value of the variable. A certain level of uncertainty is present in the measurement, but that may also prevail in the standards of measurement used for calibrating the measurement devices. Statistical evaluation is essential in this context. Uncertainties occur when repeated measurements under identical conditions give different results. Statistics cannot determine the true value of the variable, but it helps in the understanding and confronting of the uncertainties. Individual measurements of a variable must be obtained to form a collection sample. The simplest statistical measure of the sample is the arithmetic average, or the mean. The sample mean is an estimate of the expected value of the next observation of the variable. Insight into the dispersion, or variability, of the observations can be obtained from deviation between each observation and the mean. If the errors in measurement are truly random, the standard deviation may be computed to determine the variability of the observations in the sample.

The selection of the measurement instrument for a given application can be a complex task. As an illustration, in selecting a temperature transmitter consideration must be given to a range of primary elements: thermocouple, RTD, thermistor, IC sensor, liquid-filled element, gas-filled element, vapour-filled element, or radiation pyrometer. A set of criteria must be established prior to choosing a measurement instrument. Operating, static, and dynamic characteristics of the measuring instruments must be weighed for in determining its applicability.

5.2 SIGNAL CONDITIONING

A signal conditioner is generally required to convert the sensor's output signal into an electrical or pneumatic signal suitable for use by a controller or a display device. Conditioning of the raw signal obtained from a sensor is essential prior to its use by another component. Typical operations performed by a signal conditioner include isolation and impedance conversion, noise reduction through filtering, amplification, analog-to-digital and digital-to-analog conversion, data sampling, and linearization.

The operational amplifier, a high-gain dc amplifier, plays the central role in most modern signal conditioners. The gain ranges from 10^4 to 10^7, with 10^5 being a typical value. The operational amplifier requires a positive and a negative voltage supply and two more inputs to offset null adjustments. Operational amplifiers come in a variety of IC packages, with their operating characteristics almost entirely dependent on components external to the amplifier. Since the gain, input and output impedance, and frequency response depend on external resistors and capacitors in the circuit and not on the operational amplifier, the behaviour of the assembled circuitry can be made to fit a particular application by proper selection and placement of the resistors and capacitors.

The two inputs are the 'inverting input' (v_1) and the 'non-inverting input' (v_2) voltages. Output terminal voltage is v_{out}. Corresponding currents are: i_1, i_2, and i_{out}. The gain, A, of the operational amplifier determines the output voltage according to the expression: $v_{out} = A(v_2 - v_1)$. Figure 5.1 shows a plot of the open-loop response of an operational amplifier. The sloping straight-line segment is the linear operational region, and the horizontal segments at both ends make up the saturation region.

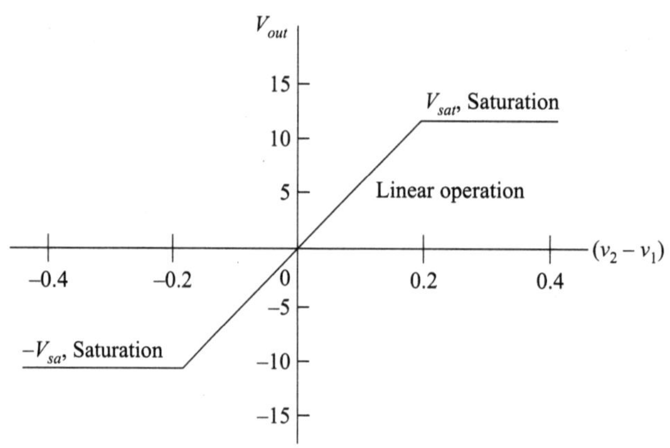

Fig. 5.1 Open-Loop Response of Operational Amplifier

Operational amplifier circuits may be configured as 'comparator' to compare two voltages and indicate the larger voltage. (The amplifier is then in the open-loop mode without any feedback or input resistors.) Another application is the 'voltage follower' to check on the impedance in a circuit containing the primary sensing element and the signal conditioning amplifier. The 'inverting amplifier' changes the sign and level of the input signal. Based on the value of two resistors, one between the input signal and the inverting input and the other between the inverting input and the output terminal, the inverting amplifier can increase, decrease, or maintain the size of the signal. The 'non-inverting amplifier' circuit can increase the size of the signal, but cannot decrease it. The position of the input voltage and ground connection are reversed from their position in the inverting amplifier circuit. A 'summing amplifier' adds two or more input signals forming an output that is the inverse of the sum. Each input may also be multiplied by a weighting factor that is formed by the ratio of two resistors. An 'integrator' circuit produces an output that is proportional to the integrator of the input voltage. An inverting amplifier can be converted to an integrator by replacing the resistor between the inverting input and the output terminal by a capacitor. A 'differentiator' circuit produces an output that is proportional to the rate of change, or derivative of the input voltage. An integrator can be converted to a differentiator by replacing the resistor between the input signal and the inverting input by a capacitor. A 'differential amplifier' circuit amplifies the difference between two input voltages, neither of which is zero. An 'instrumentation amplifier' (IA) circuit is essentially a differential amplifier with high input impedance, high common mode rejection, balanced differential inputs, and gain determined by a user selected external resistor. A 'function generator' circuit provides an output that is a non-linear function of its input. The inverting amplifier may be converted into a function generator by replacing one of its two resistors with a component that has a non-linear volt-amp relationship.

Conditioning of an analog signal requires changing its level, which may be accomplished with an inverting amplifier or a non-inverting amplifier or a differential amplifier. Sometimes it may be necessary to convert the signal from a voltage to a current, or from a current to a voltage.

Many sensors convert changes in the measured variable into small changes in the resistance of the element. Strain gage force and pressure transducers are prime examples. A bridge circuit is the traditional method of measuring small changes in resistance using the balanced or unbalanced forms of operation. The 'current balance bridge' is useful for balancing to provide a useful signal for control systems. The 'unbalanced Wheatstone bridge' is another method for obtaining a signal suitable for control systems.

Electric signal lines are prone to operate in an environment that is filled with high levels of electro-magnetic energy. The effect is to produce noise in the electric signal lines. Noise preventative measures such as careful grounding of lines and shielding of cables are successful to a point, and some noise is bound to creep in. Special circuits, referred to as filters, help reduce the level of noise in the signals. Filters may be used to decrease unwanted signals in a specific band of frequencies, giving them names such as 'low-pass filter', 'band-pass filter', and 'notch-pass filter'.

Some primary elements, such as a platinum resistance detector, produce a linear signal calibration curve so the output goes from 0 to 100 per cent as the input goes from 0 to 100 per cent. But other primary elements produce a non-linear signal calibration curve, and the signal must be linearized to produce a nearly ideal calibration curve. The preferred method of handling the signal is to condition it by linearizing small increments in the input variable. The linearizer mathematically is defined by the inverse of the function that defines the sensor element.

Most physical parameters in a process are analog signals, which must be converted into digital form for processing by computers. This is a function of the 'data acquisition system'. After processing in the computer, the digital code is converted back into analog voltage levels for input to the analog actuators, and is the job of the data distribution system. The acquisition and distribution tasks are closely associated with sampling and conversion of the data. Data sampling is a process in which a switch connects momentarily to an analog signal in a sequence of pulses separated by evenly spaced increments of time called the 'sampling interval'. Values of the sampled signal give an approximation of the original signal. Data converters are the interface between the analog and digital domains. The physical parameter measured by the primary sensing element is converted to produce a digital code with the aid of an analog-to-digital (*A/D*) converter to produce a code that accurately represents the level of the analog input voltage. The digital-to-analog (*D/A*) converter converts digital codes into analog voltage levels for input to the analog actuators, and is the job of the 'data distribution system'. Figure 5.2 illustrates the various roles played by the components of data acquiring and distributing systems.

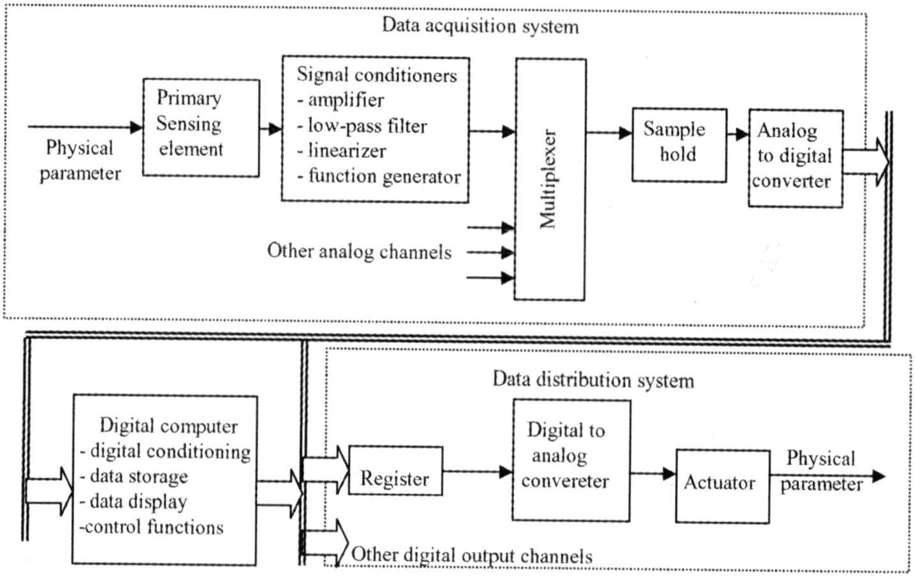

Fig. 5.2 Components of Data Acquiring and Distributing Systems

5.3 SENSOR CHARACTERISTICS

The impact of measurement procedures and environmental effects on the operational characteristics of measuring instruments must be understood.

When the measured variable is continuously changed in its upper and lower range limits, the measuring device sometimes changes the signal in discrete steps. The 'threshold' of the measuring instrument defines the smallest increment in the measured variable that will result in a shift in the indicator output, while its 'sensitivity' is the ratio of change in input and the corresponding shift in output. Environmental effects of a measurement instrument include ambient temperature and pressure, fluid temperature and pressure, mounting method and location, and electro-magnetic effects. Operating conditions dictate the environment to which a measurement instrument is subjected.

Static characteristics of a measurement instrument at room conditions describe its precision when the measured value is constant or changing slowly. The instrument output's 'accuracy' provides the degree of conformity to the ideal value of the measured variable as determined by an established standard. 'Error' is defined by the difference in the measured and the ideal values. 'Repeatability' establishes the dispersion, or standard deviation, of a measurement instrument. A repeatable but inaccurate instrument will produce a tight pattern that is not centered at the ideal value, the offset being attributable to 'bias' or 'systemic error'. Thus, an automatic controller that is repeatable but inaccurate may still be useful if adjustment to compensate for bias is made.

Calibration data may be presented in a graphical form. Figure 5.3 shows two forms of calibration curves. Both curves form a loop as the upscale and downscale curves follow different path. The loop is formed from a combination of threshold and hysteresis effects. 'Hysteresis' defines the dependence of the value of the variable on the history of past variables and current direction of traverse.

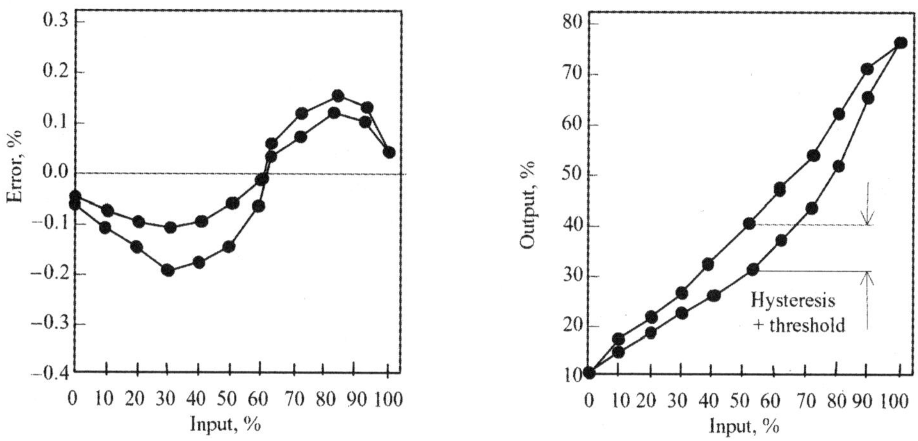

Fig. 5.3 Examples of Calibration Curves for Measuring Instrument

The 'dynamic characteristics' of a measuring instrument are of interest under rapidly changing conditions. Generally, sensors do not provide immediate and complete response to a sudden change in the measured variable. Depending on the electronic characteristics of the instrument, a certain period of time elapses before the total response is received. Step, ramp and frequency responses of the measuring instrument are needed to determine its dynamic characteristics. A step change in the measured variable, such as moving a temperature probe from ice water to boiling water, produces a worst-case response, and is often used to define its dynamic characteristics. The response may be classified as over-damped, critically damped or under-damped. Response and rise time are good indicators for step response over-damped and critically damped systems. Step response of under-damped systems is stated in terms of rise time, peak percentage of overshoot and settling time (figure 5.4).

Although not widely used, response to a ramp function (when the measured variable is gradually increased) provides additional insight into the dynamic characteristics of the measuring instrument. 'Dynamic lag' indicates the time period required by the indicated value to catch up behind the input variable, and 'dynamic error' is the difference in the variable's ideal and measured values at any time.

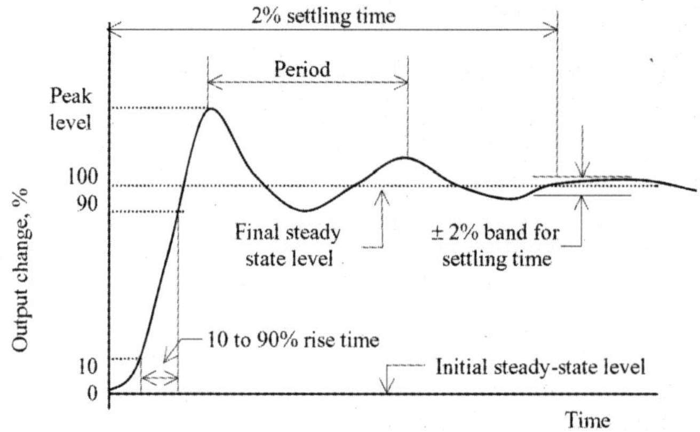

Fig. 5.4 Step Response Curve for Under-Damped Measuring Instrument

Frequency response and transfer function are most useful in defining the dynamic characteristics of the measuring instrument. Chapters 1 and 4 discuss the methods of determining the two items. Some measuring instruments have a more complex transfer function than the first order lag. If there are two layers of protective sheath, for instance, the two resistances and two capacitances form a two-capacity interacting system, the differential equation has a first-order and a second-order derivative term. Hence, the system is called a 'second-order lag'. The frequency response of a second-order system is also more complex than that of a first-order, especially if the damping ratio is less than 1. Figure 5.5 shows the Bode diagram of a second-order lag for values of damping ratio between 0.05 and 20.

In selecting the measuring instrument for an application, certain questions must be addressed. The questions pertain to: (*i*) identification of measured variable, (*ii*) accuracy required at various levels of input, (*iii*) nature of item being measured, (*iv*) upper and lower range limits, (*v*) over-range, and (*vi*) occurrence of dynamic changes (rate of change and frequency) in measured variable. Questions concerning the sensor must also be answered: (*i*) primary sensor identification, (*ii*) effect of primary sensor on measured variable, (*iii*) intent of output, (*iv*) analog or digital signal, (*v*) output signal type, (*vi*) provision for signal conditioning, (*vii*) transmitter operation, (*viii*) environmental conditions, (*ix*) thermal sensitivity, and (*x*) economic considerations.

5.4 DISCRETE SENSORS AND TRANSDUCERS

Most common sensors and transducers for motion and sequence control are on-off switches. Actuated by a number of system parameters, the switches respond by completing or breaking the control circuit, or provide an input to the programmable logic controller.

Electro-mechanical limit switches are the most common type, but inductive, capacitive, light sensitive, and optical types are also used. A mechanical actuated lever or pushbutton triggers the electro-mechanical switch. Depending on the function required to be performed, the switch may be normally in the open or closed position. Switching characteristics of a particular switch design must take into consideration actuating force, pre-travel, acceptable over-travel, movement distance, actuating point, releasing position and repeatability (*ref*: Balluff, 1987).

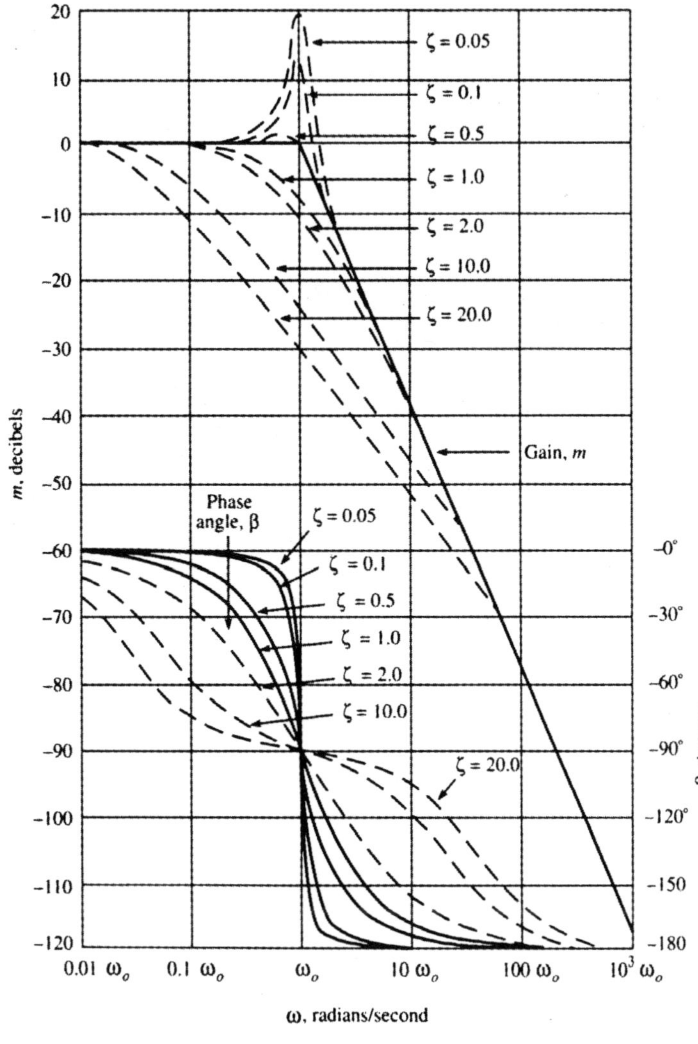

Fig. 5.5 Bode Diagram of Second-Order Components

More sophisticated than the mechanical type, inductive and capacitive proximity switches and photoelectric sensors accomplish non-contact discrete position sensing. Inductive probes (figure 5.6) are activated by the presence of a metal in the sensing zone, and have four functional sections – sensing head, oscillator, detector, and output amplifier (*ref*: Balluff, 1988). An inductive alternating radio frequency field is developed near the sensing surface by the resonant circuit coil of the oscillator. Energy is absorbed when a metal piece traverses the radio frequency field, which causes the oscillation to stop.

If a metal is not available for a target, capacitive proximity switches may be used. The device has the same four functional units of the inductive sensor, but the oscillator does not operate continuously. When an object approaches the active surface of the sensor, capacitance in the region increases. When a preset value is exceeded, the oscillator starts to produce a signal that activates a trigger. The control circuit then uses the output signal. Figure 5.7 shows some typical proximity switches.

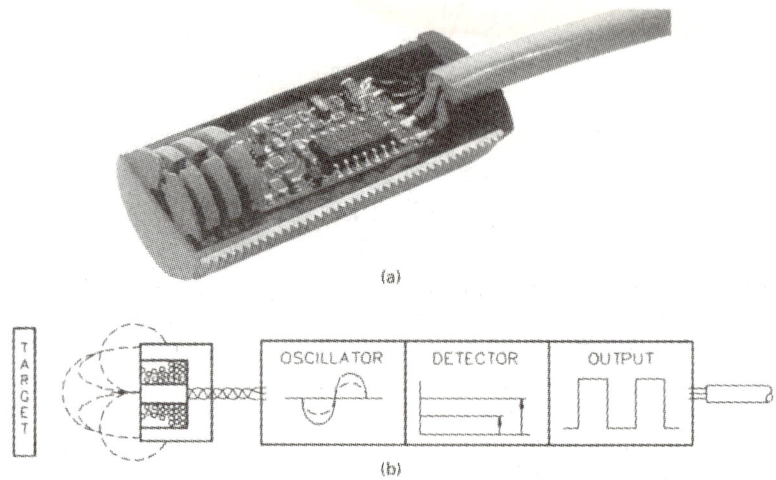

Fig. 5.6 Inductive Proximity Switch - (*a*) Cutaway, (*b*) Circuit Block Diagram (*Courtesy*: Turck, Inc.)

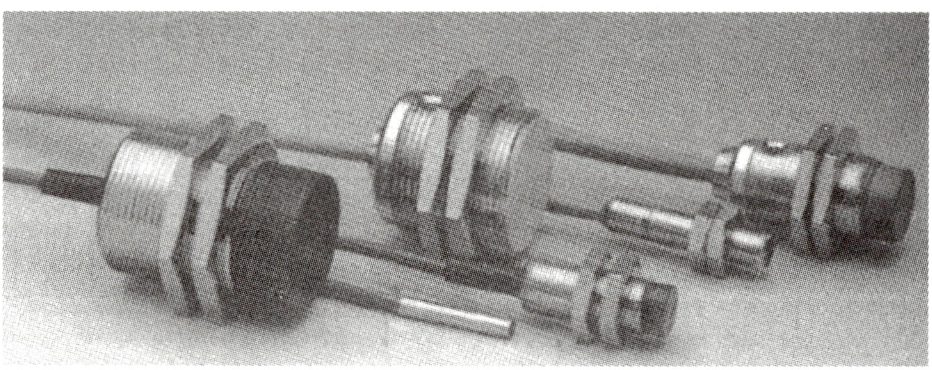

Fig. 5.7 Typical Proximity Switches (*Courtesy*: Balluff, Inc.)

Magnetic reed switches are limit switches that are usually mounted on cylinders. A magnetic ring is needed on the piston, and the cylinder barrel must be made of a magnetic permeable material. The switch armature is attracted when the piston's magnet moves close to the reed switch, causing the switch to close and complete the circuit. As the piston later moves away, the switch opens and breaks the circuit. However, normal hysteresis will cause a repeatable error at both opening and closing positions of the switch.

Hall effect switches are functionally similar to magnetic reed switches but do not rely on a moving mechanism. A solid-state element is used instead that is activated by the magnetic field. If the magnet is directly aligned with the sensor, the current distribution is distorted. Known as the

'Hall effect', the distortion causes a small potential difference across the output and used as a control signal. The device is well suited for interfacing with programmable controllers but cannot be used to directly switch a circuit.

Photoelectric sensors respond to a change in the intensity of light falling on them. Using a 'light emitting diode' (LED), the property allows the sensors to detect an object. Applications for photosensors include counting, sizing, timing, and level measuring. They are also used as limit switches in fluid power circuits when a specific position is attained. LED's are solid-state semiconductors that are similar to electrical diodes, except that they emit a small amount of light when electric current flows through them in a specific direction. Besides colour shades in the visible spectrum, infrared LED's are also used. Being solid-state devices, LED's can last the life of the sensor. To overcome the disadvantage of low emission, LED's can be made to switch on and off at a high frequency to provide a pulsed or modulated LED. The receiver, usually a photo-transistor, is tuned to respond only to light received at the modulated frequency, and ignore other ambient lights except direct sunlight and very bright lights (figure 5.8). In the 'retro-reflective' mode of operation, the target reflects the light beam back to a single unit containing both the emitter and the receiver, as illustrated in figure 5.9. The retro targets are made of tiny corer cube prisms that reflect the light from the perpendicular

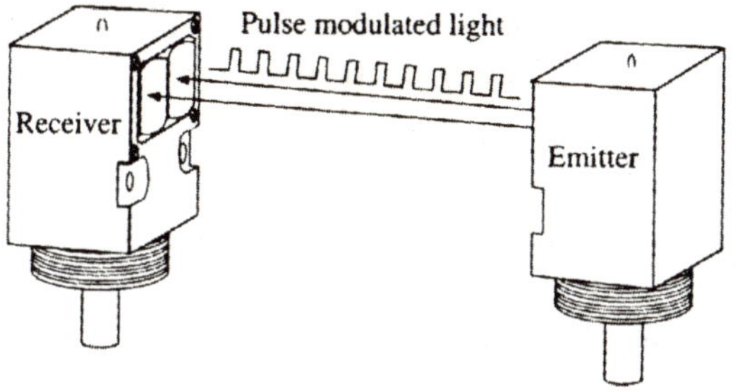

Fig. 5.8 Modulated Light Emitting Diode (*Courtesy*: Banner Engineering Corp.)

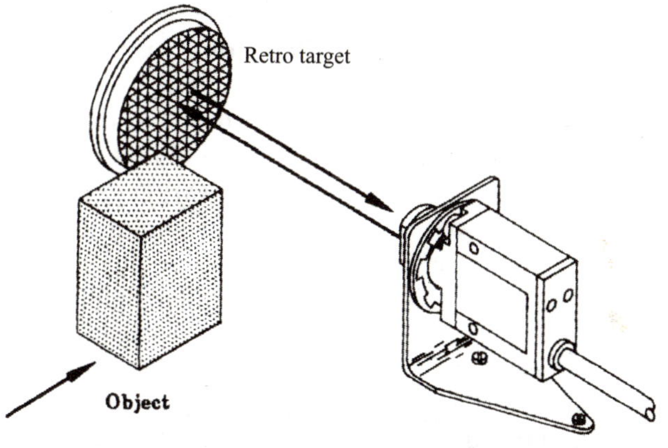

Fig. 5.9 Retro-reflective Sensing Mode of Operation (*Courtesy*: Banner Engineering Corp.)

surfaces and exit the prism parallel to the path of incident. The 'proximity-sensing mode' uses reflected light from the object itself to trigger the output. Diversity of object sizes and surface reflectivity will require special arrangement of the proximity sensors. 'Ultrasonic' sensors are similar to photoelectric types except that they emit and detect sound waves instead of light. Sound frequencies are above 20 kHz, the upper range of human hearing. The receiver is tuned to filter out sound at frequencies other than that of the transmitter.

Pressure switches permit detection of prescribed pressure situations and switch the control circuit in response. Response may be in the form of changing a contact or initiating an output signal to activate a control device. Pressure switches come in the form of bellows, pistons, tubes, and diaphragm mechanisms. The switches are labeled by terms such as 'make on rise' and 'break on rise' to describe the operation of the contact in response to variations in the pressure level. Pressure switches are electrically similar to other contact type of switches.

5.5 CONTINUOUS SENSORS

Unlike discrete sensors to detect individual events, continuous sensors are active throughout the operation for monitoring a parameter and producing an output to represent the parameter.

Devices for continuous sensing include flow parameter sensors and vibratory characteristic transducers. The output from the measuring instruments may be displayed to indicate the operating condition of a machine or a process plant, or may be used as feedback and control signals.

5.6 POSITION AND DISPLACEMENT

Linear positioning and feedback plays a major role in manufacturing and quality inspection checks of turbo-machinery components. The 'linear variable differential transformer' (LVDT) is noted for sensing position accurately, and works with a fluid power cylinder. The 'linear potentiometer', or 'pot', has a tightly packed coil of wire wound around a supporting core. The terminals of the coil are connected to a known voltage potential. Another lead is connected to a brush that moves with the cylinder rod along the coils. The brush taps a voltage that is proportional to its distance along the coil. Location of the movable contacting brush on the core determines the magnitude and phase of the signal in the secondary windings of the LVDT. Depending on the core's relative position, the secondary signal is in phase for one direction of travel and 180° out of phase in the opposite direction. The voltage output at any point is: $V = ED/L$, where V is the voltage output, E is source voltage, D is distance from the zero end, and L is total length of potentiometer, ins.

A 'magneto-strictive linear displacement transducer' (MLDT) consists of a non-magnetic stainless steel probe welded to an enclosed base. The probe contains a stretched magneto-strictive wire, and the base contains the control electronics. A permanent magnet ring slides over the outside of the probe and serves as a movable position marker, as illustrated in figure 5.10. The control circuit periodically sends a current pulse of short duration in the wire. When pulse's magnetic field intersects that of the magnetic ring, magneto-striction twists the wire. The torsional strain is detected by the electronics. The time elapsed between the emission of the pulse and receipt of the strain signal provides a precise indication of the position of the magnetic ring, and hence of the cylinder rod.

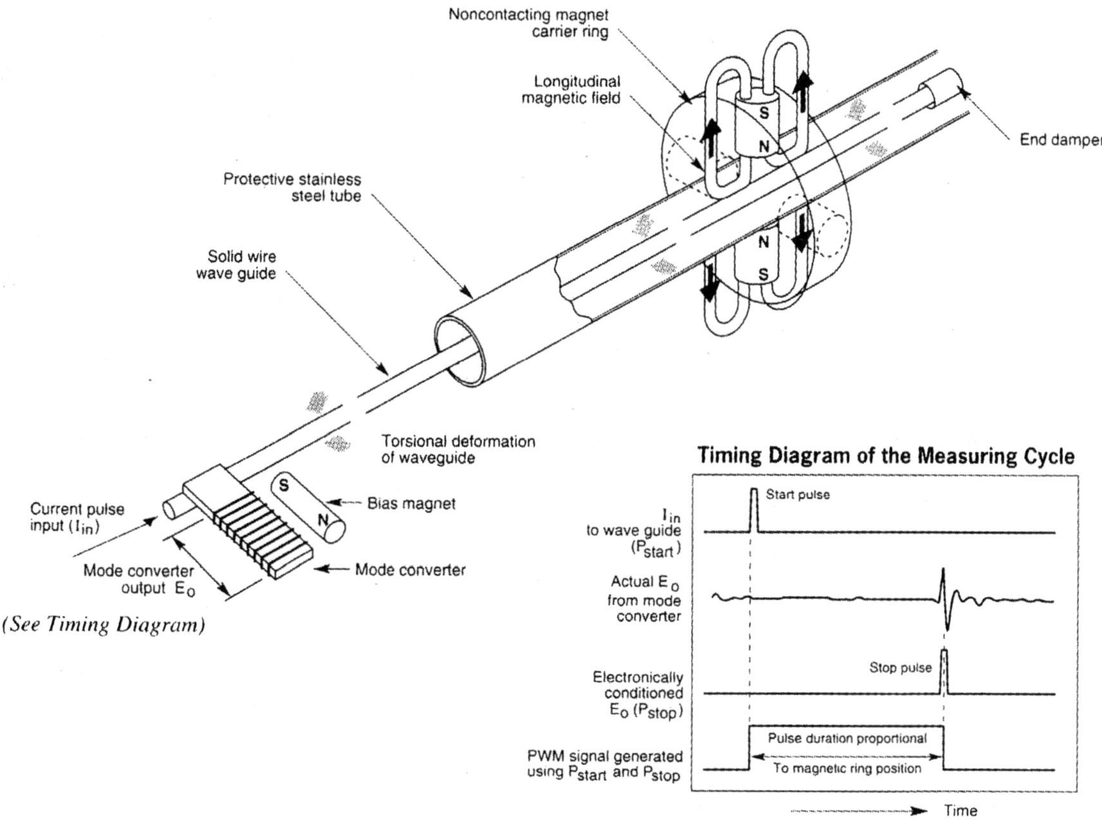

Fig. 5.10 MLDT for Sensing Position and Velocity
(*Courtesy*: Lucas Control System, Schaevitz Sensors)

Tracking the position of rotating elements is required in many machines. Rotary potentiometers, resolvers, encoders, synchros and 'rotary variable differential transformer' (RVDT) are commonly used for the purpose. As in a linear pot, a rotary potentiometer consists of a tightly packed coil with a brush. The output voltage, proportional to the angle of rotation, may be mathematically expressed: $V = E\theta/\theta_T$, where V is the voltage output, E is source voltage, θ is angle of rotation from the zero end and θ_T is total rotation of the potentiometer. The encoder establishes the rotary position with the aid of a pair of photo-sensors and a disk with rows of slits. As the disk turns, the light beam is interrupted and detected by the photo-detectors. Each interruption causes a pulse to be generated and sent to the computer. A 'resolver', essentially a rotary transformer, uses the output signals resolved by the computer to indicate the precise angular position of the shaft.

5.7 VELOCITY AND ACCELERATION

Velocity and acceleration can be computed from the output of linear and rotary position transducers by including the time rate of change of displacement, or time derivatives, calculations in the computer software.

Angular velocity may also be computed with digital devices producing a pulsed output. Typical of these are magnetic pickups that sense the passing of indicators on a rotating disk, and electro-optical units that detect pulses of light through slits in a rotating disk. Photo-reflective devices are also used to detect passing of a strip of reflective tape on a rotating shaft.

Analog measurement of shaft speed may be obtained with a tachometer-generator. The output from the generator is mostly linearly proportional to the rotor speed up to its maximum, or saturated, speed.

5.8 FORCE

Newton's second law of motion does not imply that there are no forces acting on a body if it is not accelerating, only that there is no net unbalanced force. Two methods for force measurement rely on producing a measurable balancing force, the null balance and the displacement method. A beam balance is an example of the null balance method, where the unknown mass is placed in one pan and a calibrated mass in the other pan until the beam is balanced. A spring balance is an example of the displacement method. The unknown mass is placed in a pan supported from a calibrated spring. The stretch in the spring balances the force of gravity on the unknown mass. The spring stretch provides a measure of the mass.

The 'strain gage load cell' is a displacement type of force sensor. If an unknown force acts on an elastic member of length L, the bar is stretched to length $L + \Delta L$. The strain ε is $\Delta L/L$. A strain gage converts the elastic strain into an electrical resistance, based on the premise that the resistance of a wire changes when it is stretched, or strained. Bonded strain gages are cemented directly on an elastic member at the point where the strain is to be measured (figure 5.11). Unbonded strain gages measure the total displacement of an elastic member, with a mechanical linkage stretching a strain wire. Change in resistance of the strain wire gives a measure of the displacement. The gage factor, G, of a strain gage is given by: $G = (\Delta R/R)/(\Delta L/L)$, where ΔR = change in resistance, Ω, R = resistance, Ω, ΔL = change in length, and L = length of strain gage. Gage factor is usually between 2 and 4,

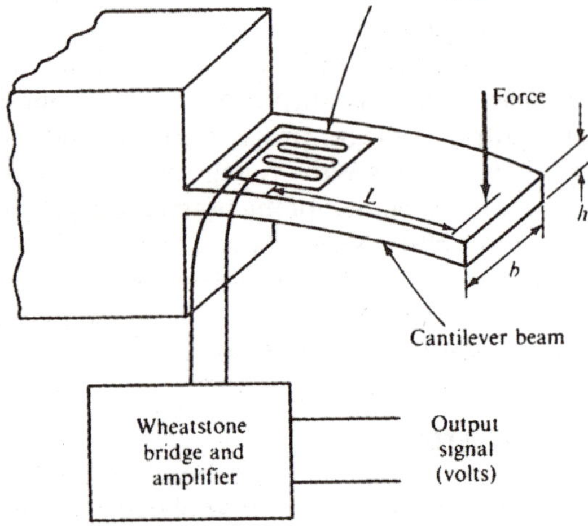

Fig. 5.11 Strain Gage Load Cell for Measuring Stress

length L ranges from 0.5 cm to 4 cm, and resistance R ranges from 50 to 5000 Ω. Stress is defined from Hooke's law for elastic materials: $E = \sigma/\varepsilon$, where E = modulus of elasticity, N/m^2, σ = stress, N/m^2, and ε = strain, m/m.

Figure 5.12 illustrates a pneumatic force transmitter as an example of a null balance force sensor. The unbalanced force f is balanced by air pressure acting on the diaphragm. The ball and nozzle are arranged to balance the two forces. Thus, as the force f increases, the force rod moves upward to reduce the opening between the ball and nozzle, pressure in the diaphragm chamber increases, and the balance is restored. Pressure of air, p, in the chamber is determined from: $f = (p - 3)A$, where A is effective area of diaphragm. Signal range is between 3 and 15 lbs/in^2.

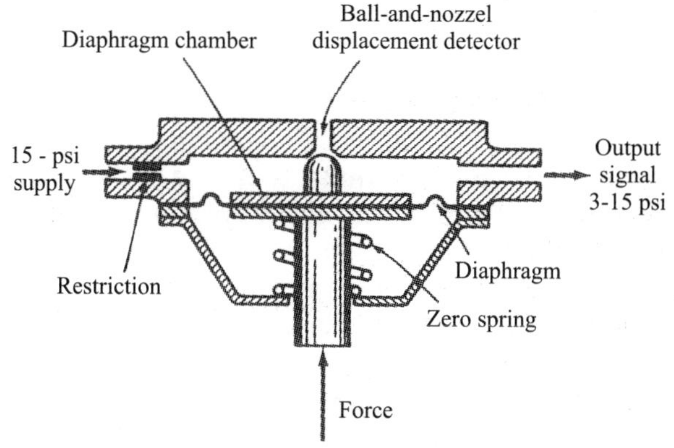

Fig. 5.12 Pneumatic Force Transmitter

5.9 FLOW RATE

Three types of sensing methods are used for measuring the flow rate: differential pressure, vortex shedding and turbine flow. 'Differential pressure flow meters' operate on the principle that a restriction placed in a flow line produces a pressure drop, h, that is proportional to the square of the flow rate. The method requires a differential pressure transmitter to measure the pressure drop due to the restriction. Flow rate, q, is given by: $q = K\sqrt{h}$. The restriction may be in the form of an orifice plate with a small hole, and the line is set up such that all of the flow must pass through the orifice plate. A 'vortex shedding flow meter' uses an un-streamlined obstruction in the flow stream to cause pulsations in the flow. The pulsations result from vortices, or eddies, that are alternately formed and shed on one side and then on the other side of the obstruction. The resulting pulsations are measured using a piezo-electric crystal. The frequency of the pulses is directly proportional to the fluid velocity, thus forming the basis for a volumetric flow meter.

The most widely used flow rate measurement device is probably the 'turbine flow meter' (figure 5.13). It consists of a turbine (made of magnetic grade of stainless steel) and a magnetic sensor. The flowing fluid turns the turbine. As the blades of the turbine enter the magnetic field, the losses in the field path decrease and magnetic field strength increases.

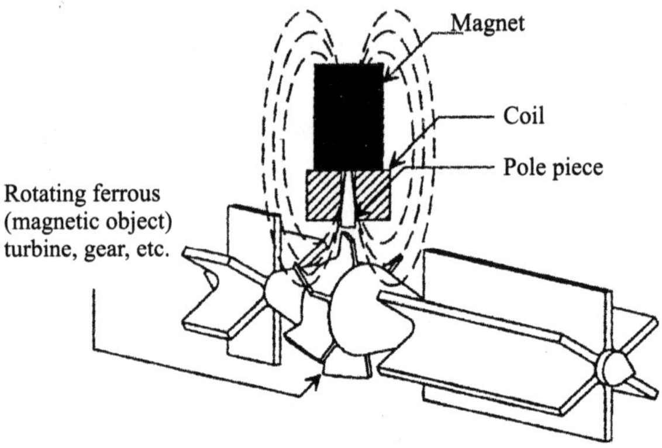

Fig. 5.13 Turbine Flow Meter Operation (*Courtesy*: Racine Federated, Hedland Division)

The operation of the turbine flow meter is similar to that of an AC generator, with the spinning ferro-magnetic turbine interrupting the magnetic field to produce an AC voltage. The assembly is generally calibrated for a specific fluid type, so application for another fluid can lead to errors. Non-linearity problems become pronounced at low flow rates when the meter is out of the linear range. A flow rate transducer for liquid flow measurements in a pipe is in figure 5.14.

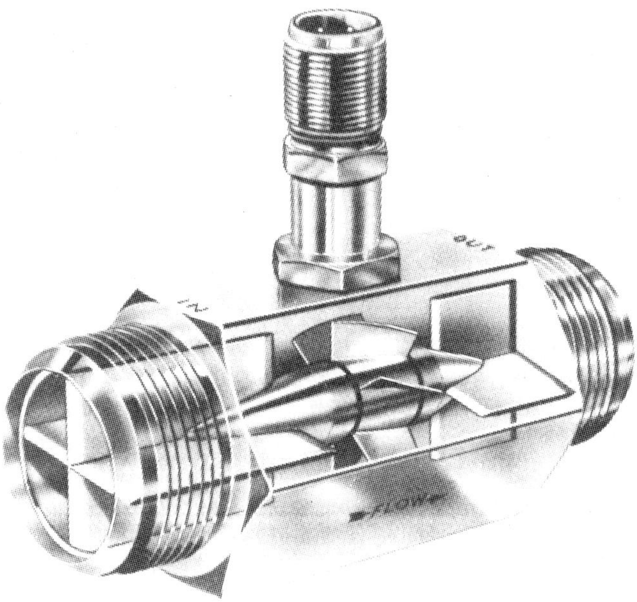

Fig. 5.14 Flow Rate Transducer for Liquid Flow in Pipes (*Courtesy*: Flow Technology, Inc.)

5.10 LIQUID LEVEL

Measurement of level of a liquid may be done directly by tracking the surface or indirectly by measuring some variable related to the level. Direct methods include sight glasses and floats with external indicators. Indirect methods rely on measurement of static pressure at some point in the liquid, and the fact that static pressure is proportional to the product of density of the liquid and height of the liquid column above the point of measurement, or: $p = \rho gh$, where p is static pressure, Pa, ρ is density, kg/cm^3, h is height, m, and g is gravitational constant, 9.81 m/s^2. If the top of the tank is open to atmospheric pressure, an ordinary pressure gauge may be used to measure at some point in the tank. If the top of the tank is not vented, the pressure at the liquid surface increases the static pressure. A differential pressure measurement is then needed between the pressure in the dome of the tank and at some point in the liquid to measure the level.

A displacement float level sensor relies on the buoyancy force on a stationary float being in proportion to the level of liquid around the float. Net force is: $f = Mg - \rho gAh$, where f = net force, N, M = mass of float, kg, g is gravitational constant, 9.81 m/s^2, ρ = liquid density, kg/cm^3, A = horizontal cross section area of float, m^2, and h = length of float below the surface of liquid, m. A bonded strain gage load cell applies a balancing force on the force beam connecting the float with the load cell and pivoted in the middle. The load cell is a strain gage force transducer that changes its resistance in proportion to the applied force. A signal condition in the form of Wheatstone bridge and amplifier converts the resistance into a usable electrical signal.

The capacitance probe method is based on the fact that the capacitance between the stationary probe and the vessel's wall depends on the level of liquid around the probe. Another method, the gamma-ray system, relies on the premise that the number of gamma rays that penetrate a layer of liquid depends on the thickness of the layer.

5.11 PRESSURE

A typical pressure measurement system calls for a pressure transducer, signal conditioning amplifier and one or more readout units. A frequently used pressure transducer depends on a strain gage, where a force or a displacement causes a change in the length or position of the sensing element. The gage's sensing element may be of the unbonded, bonded, thin film, or diffused semiconductor type. The bonded strain gage is applied to a diaphragm (figure 5.15). Pressure acting on the diaphragm causes it to flex, resulting in a strain in the gage element.

'Piezo-electric pressure transducers' operate on the concept of strain applied on an asymmetric crystalline material such as quartz or tourmaline generates an electrical charge. The small charge needs an amplifier installed close to the crystal, with a low-noise cable connecting the two.

The 'semiconductor gage' requires a silicon chip as a substrate into which an impurity, usually boron, is diffused to produce a piezo-resistive element. The device has a high gage factor for sensitive pressure change measurements, meaning that it produces a large signal for a relatively small strain. The signal experiences little distortion from electrical noise, and has better overall performance, stability, linearity, and hysteresis than other strain gages.

The linear variable differential transformer (LVDT), shown in figure 5.16, may be used for pressure measurements. System pressure deflects the diaphragm (or a bellows element), which moves the core of the LVDT. This results in a signal that is proportional to the pressure.

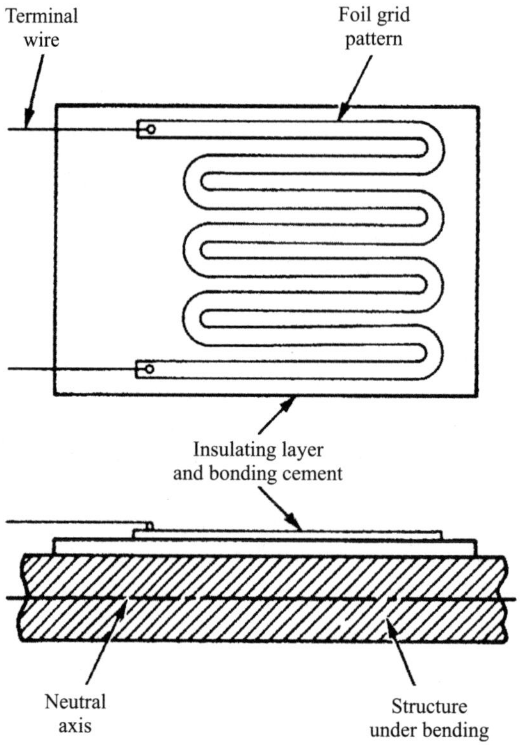

Fig. 5.15 Bonded Foil Strain Gage for Pressure Diaphragm (*Courtesy*: Omegadyne, Inc.)

5.12 TEMPERATURE

Generally not a major operating parameter, temperature nevertheless plays an important role in controlling a process. The operation of the temperature sensors may be transparent to the operator, so the operation of the machine changes without the knowledge of the operator. But in many situations the actual fluid temperature is of interest, so such sensors are required to provide a continuous readout. For instance, if the temperature of oil returning to the oil tank from the sump of a turbine bearing exceeds a certain level, the oil loses its viscosity and hence its capacity to provide hydrodynamic action. The system may then trigger a valve to divert the flow of oil through a cooler. On the other hand, if the oil is too cold, the sensor may turn on a fluid heater. Either response may turn on an indicator light, or annunciator, on the control panel to inform the operator about the action taken.

Thermocouples, thermistors and resistance temperature detectors (RTD) are devices commonly used for measuring temperature. A thermocouple consists of two wires made from dissimilar metals joined at the ends. A difference in temperature between the two junctions creates an electric potential due to the 'thermoelectric' or the 'Seebeck' effect. Measuring the electrical potential between the junctions is interpreted as a temperature differential, and if the temperature at one junction is known, the temperature at the other junction can be established. The response to the

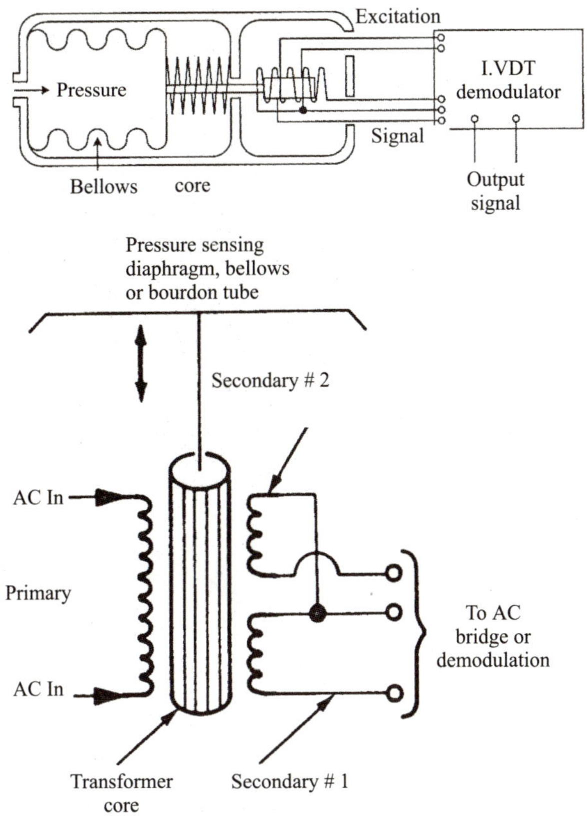

Fig. 5.16 Bellows-Inductance Pressure Sensor with LVDT (*Courtesy*: Omegadyne, Inc.)

temperature differential is non-linear, requiring a compensation of the temperature to be included as a feature in the signal conditioning process. The lead wires from the thermocouple need to be adequately supported and protected. Commercially available thermocouples have lead wires encased in a protective sheath, as shown in figure 5.17.

'Thermistors' have a large change in electrical resistance in response to a change in temperature, although the variation is considerably non-linear. The associated software must include a lookup table in the memory to convert the observed resistance to the right temperature level. The devices are mostly used to measure temperatures over a limited range, and are restricted to temperature alarm and switching circuit applications.

A 'resistance temperature detector' (RTD) works on the principle that electrical resistance of a metal wire changes with temperature. Virtually any metal shows this property, but nickel, tungsten, copper and platinum are more commonly used. Although not as sensitive as thermistors, the change in resistance due to temperature is more linear. The feature makes the RTD more convenient from the perspective of signal conditioning.

Platinum RTDs are more widely used because of their linear characteristics over a wider temperature range. The metal wire may be shaped as a flat strip, disk, cylinder, or other configuration to accommodate installation and application needs. The element must be protected from stresses, physical damage and hostile environment, since that would change its resistance. Many types of electronic circuits accept the RTD signal. A common one uses the RTD as a leg of the Wheatstone bridge.

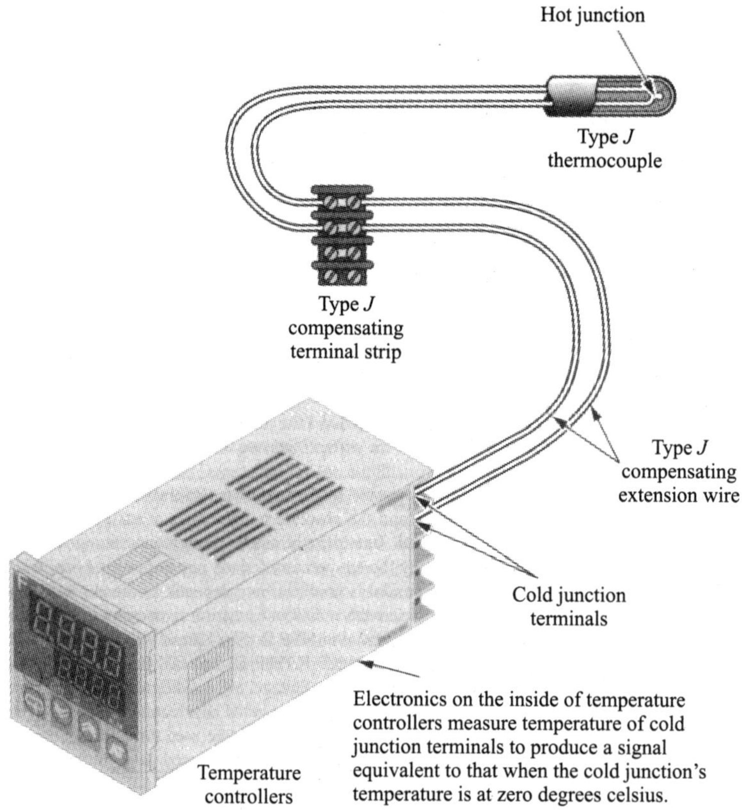

Fig. 5.17 Thermocouple with Digital Readout (*Courtesy*: Womack Machine Supply)

5.13 SOUND

Noise is not only an annoying irritant. Medical science has proved that prolonged exposure to high levels of noise results in hearing impairment, sometimes even in a total loss of hearing. High levels of noise are developed in the engines of an airplane while taking off. Power generation turbines are required to satisfy near and far field noise requirements. The need for accurate measurement of noise is thus readily discernible.

Sound waves travel as pressure fronts emitted from a source. The pitch of the sound is determined by the frequency of the sound waves, and its intensity is measured in units of the decibel.

Sound is converted into electrical signal in the sensors. The earliest microphones used a diaphragm fastened rigidly to a container holding dried carbon granules (figure 5.18). The diaphragm was free to vibrate when struck by sound waves, and the granules would be alternately compressed and decompressed with each impinging sound pressure wave front. A small electric signal passed through the granules experienced proportional increase and decrease in electrical resistance as the granules experienced compression and decompression. The varying voltage and current provided a fair representation of the frequency and intensity of the sound picked by the microphone.

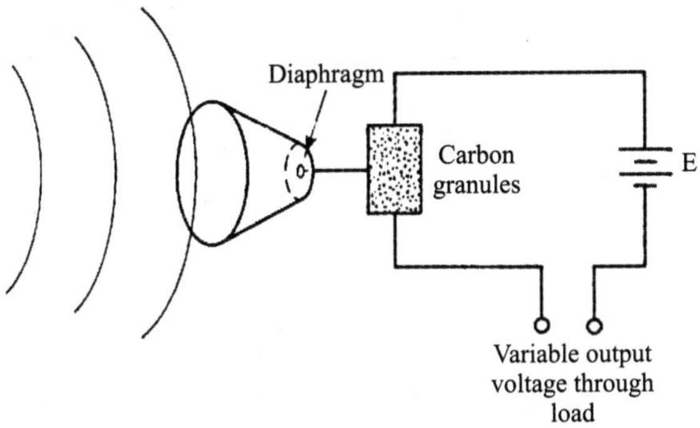

Fig. 5.18 Carbon Button Microphone

Piezo-electric crystals may be used to convert sound waves into a varying output voltage. Sound waves strike the diaphragm, which activates the attached piezo-electric crystal. The ability of the crystal material to generate voltage for a given input of force is often stated in coulombs per newton.

Other microphones using different processes function in a similar way. A capacitor microphone has one plate of the capacitor attached to the vibrating diaphragm, creating a varying capacitance between the two plates. A voltage supplied to the capacitor varies the charge. A proportional voltage may also be obtained with a combination of a magnet and a coil. The magnet is attached to the diaphragm, and is permitted to move inside the coil. The lines of magnetic flux sweeping past the coiled wire then generate a voltage that is proportional to the sound waves reaching the microphone (figure 5.19).

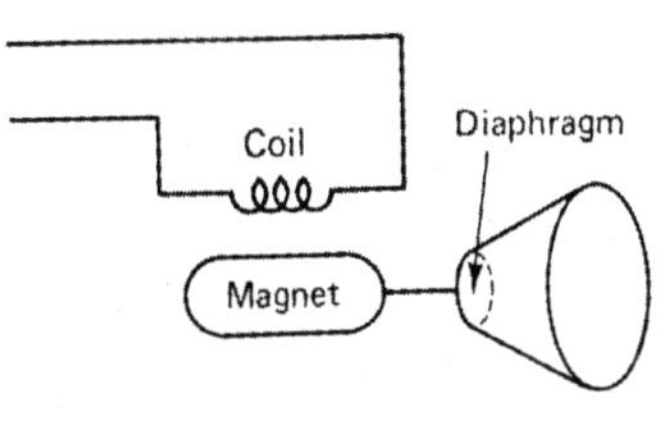

Fig. 5.19 Electro-Magnetic Microphone

5.14 LIGHT

The unit of illumination defined as the flow rate of light is based on the concept of light being comprised of lines of flux, just as in a magnetic field. The unit in metric system for flow rate of light is the 'lumen'.

Depending on the type of sensor, sensing of a light beam may be split into three categories: photoresistive sensor, photoemissive sensor, and photovoltaic sensor. The photoresistive sensor (figure 5.20) varies its internal DC resistance according to the light intensity striking the device. The sensor's chemistry bears some resemblance with that of the thermistor, except for the wavelength of the electromagnetic energy response being shorter. The variation in resistance response of the sensing transducer is converted to change in voltage that is correlated to light intensity, using the following transfer function: $E_{out}/Q = K_{photo}/(1 + s\tau)$, where E_{out} = voltage output, Q = input flow rate of light, lumens, K_{photo} = transfer function constant for the transducer, measured experimentally,

s = Laplace transform operator, and τ = time constant, s, also measured experimentally for the light sensitive material. Lead selenide sensors have a time constant of around 10 μs, while lead sulfide cells are in between 10 and 100 μs.

Fig. 5.20 Photo-Resistive Sensor

Photoemissive sensors have the characteristic of being extremely sensitive to weak levels of light. Its transfer function is the same as that of photoresistive sensors. Photovoltaic sensors offer the advantage of electric potential being linear with the amount of light flow. Typical time constant is 20 μs, and its transfer function is also the same as that of photoresistive sensors.

5.15 REVIEW PROBLEMS

Example problem # 5.1: Determine the value $(v_2 - v_1)$ that will saturate an operational amplifier of gain $A = 10^6$ and supply voltages are –25 V and +25 V.

Solution: The amplifier will saturate at 80 per cent of supply voltages, or about –20 V and +20 V. Then at +20 V, $(v_2 - v_1) = 20/10^6$

$$= 0.02 \text{ mV}$$

$$\text{At } -20 \text{ V}, (v_2 - v_1) = -20/10^6$$

$$= -0.02 \text{ mV}$$

Example problem # 5.2: The strain gage force sensor of the type shown for the cantilever beam in figure 5.10 has the following data:

Beam: material = steel, $E = 2 \times 10^{11}$ N/m^2

Maximum allowable stress = 3.5×10^8 N/m^2

Dimensions: b = 1.5 cm, h = 0.25 cm, L = 8 cm

Strain gage: Gage factor = 2, nominal resistance = 150 Ω

Determine the maximum force that can be measured and the variation in the resistance produced by that force.

Solution: The maximum force that can be accepted by the cantilever beam may be determined from the maximum stress developed:

$$f_{max} = (S_{max} \times bh^2)/6L$$
$$= (3.5 \times 10^8 \times 0.015 \times 0.0025^2)/(6 \times 0.08)$$
$$= 68.36 \text{ N}$$

The change in resistance corresponding to this stress level is obtained from the definition for the gage factor: $G = (\Delta R/R)/(\Delta L/L)$. Hence:

$$\Delta R = GRL/(\Delta L)$$
$$= 6GRL \times f/(bh^2E)$$
$$= (6 \times 2 \times 150 \times 0.08 \times 68.36)/(0.015 \times 0.0025^2 \times 2 \times 10^{11})$$
$$= 0.525\ \Omega.$$

Example problem # 5.3: The pneumatic force transmitter in figure 5.11 has an effective area of 2.5 in^2. If the air supply is at 15 lbs/in^2, determine the range of force of the transmitter.

Solution: Expression for force is: $f = (p - 3)A$

$$= (15 - 3) \times 2.5$$
$$= 30.0 \text{ lbs.}$$

Example problem # 5.4: The resistance of a resistance temperature detector may be expressed as: $R = R_o(1 + a_1T + a_2T^2)$, where R = resistance at T°C, R_o = resistance at 0°C, and T = temperature, °C, and a_1, a_2 are constants. If the resistance is 101.1 Ω at 0°C, 120.2 Ω at 50°C, and 141.8 Ω at 100°C, find the values of R_o, a_1 and a_2. What is the resistance at 75°C?

Solution: Three equations are needed for the three unknown values. Substitute each pair of values in the controlling equation.

$$101.1 = R_o(1 + 0a_1 + 0a_2)$$
$$120.2 = R_o(1 + 50a_1 + 2500a_2)$$
$$141.8 = R_o(1 + 100a_1 + 10000a_2)$$

Then $R_o = 101.1\Omega$. Substituting this in the other two equations:

$$120.2/101.1 = (1 + 50a_1 + 2500a_2) = 1.1889$$
$$141.8/101.1 = (1 + 100a_1 + 10000a_2) = 1.4026$$

Then $a_1 = 0.004026$ and $a_2 = -4.96 \times 10^{-6}$.

At 75°C the resistance is: $R = 101.1(1 + 75 \times 0.004026 - 4.96 \times 10^{-6} \times 75^2)$

$$= 128.8\ \Omega.$$

Example problem # 5.5: A turbine flow meter has a value of $K = 17.4$ cm^3 per pulse. Determine the volume of liquid transferred for a pulse count of (*i*) 225 and (*ii*) 535. Also calculate the flow rate if each pulse count occurs in a time period of 125 s.

Solution: For a turbine flow meter, $V = KN$, where N = pulse count.

(*i*) Then: $V = 12.2 \times 225$

$= 2745 \text{ cm}^3$

and: $Q = V/\Delta t = 2745/125$

$= 21.96 \text{ cm}^3/\text{s}$

(*ii*) When: $V = 12.2 \times 535$

$= 6527 \text{ cm}^3$

and: $Q = V/\Delta t = 6527/125$

$= 52.22 \text{ cm}^3/\text{s}$

Example problem # 5.6: A bellows pressure element of the type shown in figure 5.15 has an effective area = 13.4 cm^2, and a spring rate = 15 N/cm. The spring has spring rate = 125 N/cm. Determine the pressure range of the sensor if the motion in the bellows is restricted to 1.75 cm.

Solution: The combined spring rate is 15 + 125 = 140 N/cm.

Force required to deflect the springs by 1.75 cm = 1.75 × 140 = 245 N.

Pressure corresponding to this force = 245/13.4 = 18.28 N/cm^2.

Range is between 0 and 18.28 N/cm^2, or 0 and 183 kN/m^2.

Example problem # 5.7: A linear pot is required to have an output of 14.5 volts for a positioning circuit when its output shaft moves through 2.75 ins. for a positioning circuit. Find the transfer function value of the pot

Solution: The pot's transfer function = (14.5 volts)/(2.75 ins.)

= 5.27 volts/in.

Example problem # 5.8: A fluid of density ρ = 940 kg/m^3 is to stored in an open tank. A static pressure measurement probe is placed 1.5 m below the top of the tank. Determine the range of a pressure transducer for 100 per cent output when the tank is full.

Solution: Expression for static pressure $p = \rho g h$

Maximum pressure $p_{max} = 940 \times 9.81 \times 1.5$

$= 13832.10$ Pa, or 13.83 kPa.

Example problem # 5.9: A displacement float level transducer has a float mass of 5.25 kg, area of float is 60 cm^2, length of float is 2.5 m, and liquid in the tank is water. If the equation for force on the float is given by $f = Mg - \rho g A h$, find the minimum and maximum values of force on the float.

Solution: Force on the float: $f = Mg - \rho g A h$

$= 5.25 \times 9.81 - 1000 \times 9.81 \times (60/1000)h$

Maximum force occurs when $h = 0$.

So $f_{max} = 5.25 \times 9.81$

$= 51.50$ N

Minimum force occurs when $h = L$

$$\text{So } f_{min} = 5.25 \times 9.81 - 1000 \times 9.81 \times (60/1000) \times 2.5$$
$$= -1420 \text{ N}$$

Example problem # 5.10: A differential pressure flow meter is used as a sensor for a liquid flow control system. In calibration tests, the following data was obtained while measuring the time required to fill a 0.5 gallon container: Controller setting – 25%, time – 6.63 minutes, setting – 50%, time – 4.78 minutes, setting – 75%, time – 3.98 minutes, setting – 100%, time – 3.42 minutes. Calculate the average flow rate in gallons/min and construct a calibration curve.

Solution: Flow rate for 25% controller setting = 0.5/6.63 = 0.0754 gal/min.

Flow rate for 50% controller setting = 0.5/4.78 = 0.1046 gal/min.

Flow rate for 75% controller setting = 0.5/3.98 = 0.1256 gal/min.

Flow rate for 100% controller setting = 0.5/3.42 = 0.1462 gal/min.

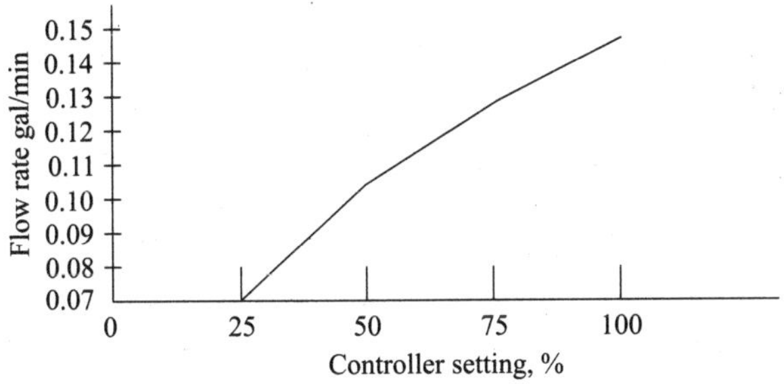

Example problem # 5.11: The flow control system in example problem # 5.10 may be described by the equation $q = K\sqrt{(sp)}$, where q = flow rate, gal/min., K = constant, sp = controller setting, %. Calculate K for each observation, find the average and use it to describe the flow condition.

Solution: At 25% controller setting: $0.0754 = K\sqrt{(25)} = 5K$, or $K = 5/0.0754 = 66.3$

At 50% controller setting: $0.1046 = K\sqrt{(50)} = 7.071K$, or $K = 7.071/0.1046 = 67.6$

At 75% controller setting: $0.1256 = K\sqrt{(75)} = 8.660K$, or $K = 8.66/0.1256 = 68.94$

At 100% controller setting: $0.1462 = K\sqrt{(100)} = 10K$, or $K = 10/0.1462 = 68.4$

Average $K = (66.3 + 67.6 + 68.94 + 68.4)/4 = 67.81$.

Hence: Flow rate $q = 67.81 \times \sqrt{(sp)}$

5.16 EXERCISES

5.1 A bellow pressure element has effective area of 5.25 cm^2, and spring rate of 33 N/cm. The spring has spring rate of 450 N/cm. If the bellows is to have a range of 0 to 80 N/cm^2, find the stroke required.

5.2 Determine the gain of an operational amplifier that will saturate at 75 per cent of supply voltages if supply voltages are –25 V and +25 V, and the maximum and minimum values of $(v_2 - v_1)$ are +0.03 mV and –0.03 mV.

5.3 An operational amplifier has gain $A = 5.25 \times 10^6$ and supply voltages of –12.55 V and +12.5 V. Assuming saturation at 75%, determine the value of $(v_2 - v_1)$.

5.4 The maximum force measured by a cantilever type strain gage is 75 N. If the beam material has $E = 2 \times 10^{11}$ N/m^2, maximum allowable stress = 3.5×10^8 N/m^2, width b = 1.5 cm, height h = 0.25 cm, gage factor = 2, nominal resistance = 150 Ω, calculate the variation in the resistance produced by the force.

5.5 A pneumatic force transmitter is required to develop a force of 72 lbs. What is the required air supply pressure if the effective piston area is 4.75 in^2?

5.6 The resistance of a resistance temperature detector may be expressed as: $R = R_o(1 + a_1T + a_2T^2 + a_3T^3)$, where R = resistance at T°C and T = temperature, °C, and a_1, a_2, a_3 are constants. If the resistance is R_o = 59.8 Ω at 0°C, 72.8 Ω at 20°C, 90.2 Ω at 70°C, and 122.4 Ω at 125°C, find the values of R_o, a_1, a_2, and a_3. What is the resistance at 100°C?

5.7 Determine the value K per pulse of a turbine flow meter that transfers a volume of liquid per pulse count of 365 cm^3/s if each pulse count occurs in 55 s.

5.8 The output shaft of a linear pot for a positioning circuit travels 2.29 cm. Find the voltage output if the transfer function value is 9.38 volts/cm.

5.9 Calculate the location of a static pressure measurement probe below the top of an open tank if the density of the fluid is 63.5 lbs/ft^3 and the range of the pressure transducer is 80 per cent output when the tank is full.

5.10 A displacement float level transducer has float area of 8.75 in^2, is 43 in long, and liquid in the tank is kerosene of density 57.5 lbs/ft^3. The equation for force on the float is $f = Mg - \rho gAh$. Find weight of the float mass if the maximum force on the float is zero and the minimum force is –3.85 lbs.

5.11 A sensor for a liquid flow control system is calibrated using the following data: Controller setting – 22.5%, time – 3.79 minutes, setting – 33%, time – 3.59 minutes, setting – 65%, time – 5.57 minutes, setting – 100%, time – 4.28 minutes. If the container has 0.70 gallon capacity, calculate the average flow rate in gallons/min.

5.12 The flow control system in the above problem may be described by the equation $q = K\sqrt{(sp)}$, where q = flow rate, gal/min., K = constant, sp = controller setting, %. Find the average value of the constant K.

References & Bibliography

Balluff, Inc., "Electro-mechanical Single and Double Limit Switches with Accessories", Catalogue # 103 E, Issue # 105, Florence, KY, 1987.

Balluff, Inc., "Non-Contact Switching", Catalogue # 321E, Edition 8806, Florence, KY, 1988.

Banner Engineering Corporation, "Handbook of Photo-Electric Sensing", 2nd edition, Minneapolis, MN, 1993.

Bateson, R.N., "Introduction to Control System Technology", Prentice-Hall, Clifton, NJ, 1999.

Carstens, J.R., "Automatic Control Systems and Components", Prentice-Hall, Clifton, NJ, 1990.

CEC Instruments, "Pressure Transducer Handbook", Pasadena, CA, 1989.

Funakubo, H., "Actuators for Control", Gordon and Breach Science Publishers, New York, 1991.

Norvelle, F.B., "Electro-Hydraulic Control Systems", Prentice Hall, Upper Saddle River, NJ, 2000.

Parr, E.A., "Industrial Control Handbook", Volume I, Industrial Press, Inc., New York, 1987.

Ryoji, O., "Intelligent Sensor Technology", John Wiley, New York, 1992.

6

Role of Valves

6.1 INTRODUCTION

The flow rate of fluid is the manipulated variable in many process control systems. A pneumatic control valve is commonly used as the final control element. An air pressure signal in the 3 to 15 lbs/in^2 is applied on the diaphragm of the valve, which transforms the pressure into motion of the valve stem. The valve body in the valve-seating region is shaped to vary the flow area as governed by the displacement of the stem. The pressurized air signal may originate from a pneumatic controller, a valve positioner or an electro-pneumatic transducer. The transducer converts the current signal from an electronic controller into a pressure signal.

The solenoid valve is the simplest type of electrically operated valve, and has been used over a large span of time with relatively few changes to the original concept. Electro-hydraulic proportional control valves (EHPV) represent a step up in the technology, moving up from the relatively less demanding electrical switching to the more sophisticated operation by electronics. Control of the valve through electronic circuitry offers the advantage of flexibility in system design and operation, as well as reduction in complexity of the system for processes requiring multiple speed and force outputs. Functions such as flow control and directional control may then be combined in a single valve. Modern EHPV also permit rapid cycling that is encountered in manufacturing and material handling systems.

Servo system components offer the ultimate control capability to continuously monitor an output parameter and make instantaneous corrections to the parameter whenever it deviates from a predetermined set point. Servo systems require continuous monitoring, feedback provision, and corrective action to improve efficiency, accuracy, and repeatability that are far beyond the capacity of the human operator. Applications for servo valves for turbo-machines include operation of variable stator vanes of an axial flow compressor and control of rotor speed for power generation turbines. Controlling the rate of flow of fuel to the injectors in the combustion system to ensure development of required power setting and managing the speed of the rotor to maintain air pressure at the compressor's discharge are some other examples where servo system components play a primary role.

6.2 SOLENOID VALVE CONSTRUCTION

A direct-acting solenoid actuated valve, or a 'solenoid valve', comprises of a sliding spool type valve section and a solenoid. The hydraulic section is similar to that of a manually operated valve, except that the spool does not extend through the end of the valve body for connection to a lever or plunger. Figure 6.1 shows a cutaway view of a direct-acting solenoid-operated control valve, with the symbol indicating a four-way, three-position valve with an open center spool and two solenoids. The assembly consists of a valve body, spool, fluid ports and a return spring at each end.

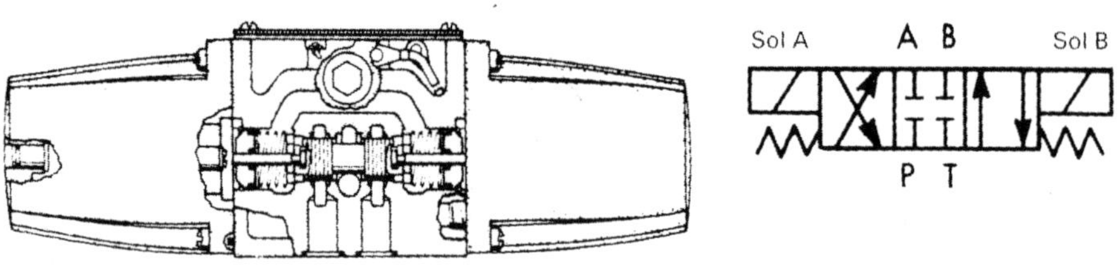

Fig. 6.1 Cutaway View of Direct-Acting Solenoid-Operated Control Value (*Courtesy:* Vickers, Inc.)

When the solenoids are not energized, the springs push the spool into the center position. All four ports are blocked, and no flow passes through the valve. Energizing solenoid A pushes the spool to the left and exposes two flow paths. One path allows flow from the pressure port P to the A port, and the other path allows flow from the B port to the tank port T. Energizing solenoid B pushes the spool to the right, again exposing two flow paths. This time the flow is from P to B and from A to T. The two positions permit extension and retraction of a cylinder, or running a hydraulic motor in the forward and reverse directions. The closed center position lets the cylinder stop anywhere in the stroke, or to stop the hydraulic motor.

The solenoid is mostly an electromagnet adapted for a specific application. When an electric current passes through its coil wrapped around a core made of a ferro-magnetic material, the coil and the core are magnetized. Increasing the number of turns strengthens the magnetic field. Providing the core with a proper shape allows the magnetic field to concentrate in a specific region, with the flux lines flowing through the center of the coil. Figure 6.2 shows a coil wound around a plastic spool called the bobbin placed inside a C-framed stack of steel laminations. Another piece of iron, termed the plunger, is placed in the hollow center of the coil of wire. When the coil is energized, the concentrated magnetic field draws the plunger into the C-frame until physical contact occurs between the plunger and the frame. When the solenoid is inactive and the plunger is resting partially out of the frame, a space is formed between the plunger and the base, which is called the 'air gap'. Although some magnetic flux lines flow through the gap, the air represents a large resistance that weakens the field. The magnetic field and the consequent force are at their peak strength when the plunger is in contact with the frame. The force is minimum when the plunger is out. But the maximum force requirement is usually at the start of the stroke and little force at the end of the stroke. This reversal of required and available force at the start of the stroke represents a drawback.

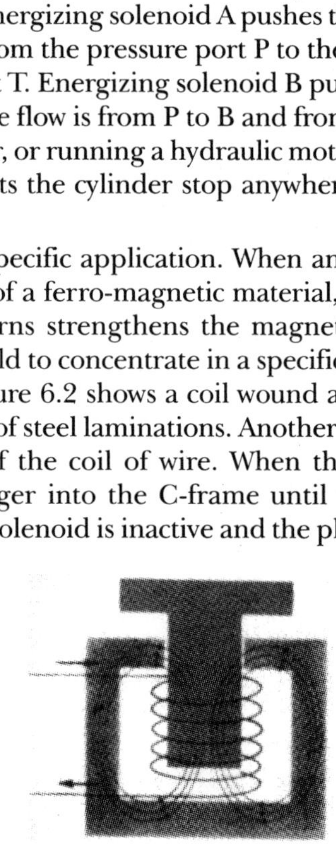

Fig. 6.2 C-Framed Solenoid with Plunger (*Courtesy:* Detroit Coil Co.)

The force available from a solenoid may be calculated from the following equation: $F = (NI_0)^2 \times \mu_0 A/(2X^2)$, where F = initial solenoid force, N, N = number of turns in coil wire, I_0 = initial current, A, μ_0 = permeability ($4\pi \times 10^{-7}$ H/m in air), A = plunger cross sectional area, cm^2, and X = air gap, cm.

The initial force may be increased to some extent by increasing the number of turns in the coil wire and the cross sectional area of the plunger. Another possibility is to increase the initial current, since the force varies with the square of the current. But heat generation in the solenoid also changes with the square of the current. This represents a problem, particularly in AC circuits, where the initial current is 4 to 10 times the holding current when the solenoid is closed. The problem is not as severe in DC solenoids, but heat production must be taken into consideration.

Rapid cycling of an AC solenoid, even with a 25 per cent duty cycle, can cause the solenoid to overheat (figure 6.3). Peak current is 3.25 A at the start of the cycle. As the plunger moves in, the current reduces until it reaches the holding value of 0.75 A. The area under each energizing pulse represents the heat generated. If this heat buildup from the continuous cycling cannot be dissipated, the solenoid will overheat and fail. Heat contributions from the operating fluid and the ambient must also be taken into consideration in calculating the total operating temperature and establishing insulation requirements. Usually an epoxy, the insulation serves the multiple purposes of electrical insulation, heat dissipation, and sealing out moisture, dirt, oil, and corrosive vapors.

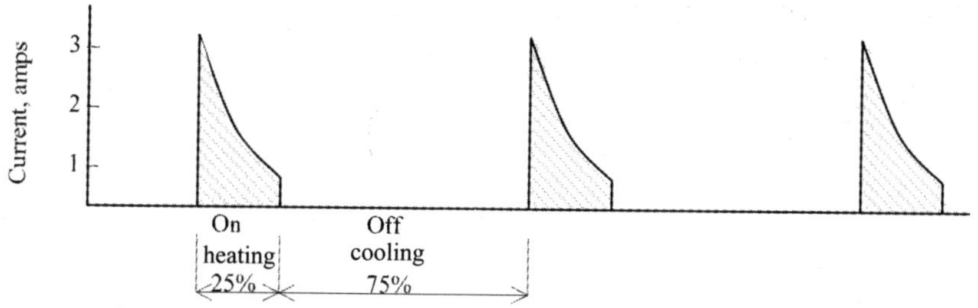

Fig. 6.3 Rapid Cycling of AC Solenoid with 25% Duty Cycle

In addition to the amperage rating from overheating considerations, the voltage determines the power of a solenoid, or: $P = VI$. Ideally, a solenoid must be designed for a specific power supply. A DC solenoid may be used for AC power only for on-off operation, when the unit is not held energized. Eddy currents do not pose a problem in DC circuits, so DC solenoids are made with solid iron parts. When operated on AC power, high losses will result in excess heat buildup. AC power in a DC solenoid may also cause solenoid buzz or chatter as the magnetic field increases and decreases sinusoidally at the frequency of AC power.

Since a solenoid is an inductive load, it acts as an open circuit when first energized and delays the increase in the current before it reaches steady state operation. When the inductive coil is disconnected, a high induced voltage spike can occur. The disconnect spike can be observed in the form of arcing in the contacts of the relay controlling a solenoid. Transient suppression is needed to protect sensitive components from the spikes, which may be achieved with a conductive device connected in parallel with the solenoid coil.

The operation of an air-gap solenoid can be understood from figure 6.4. The upper figure shows the solenoid energized and the plunger beginning to move. In the lower figure the plunger is pulled in.

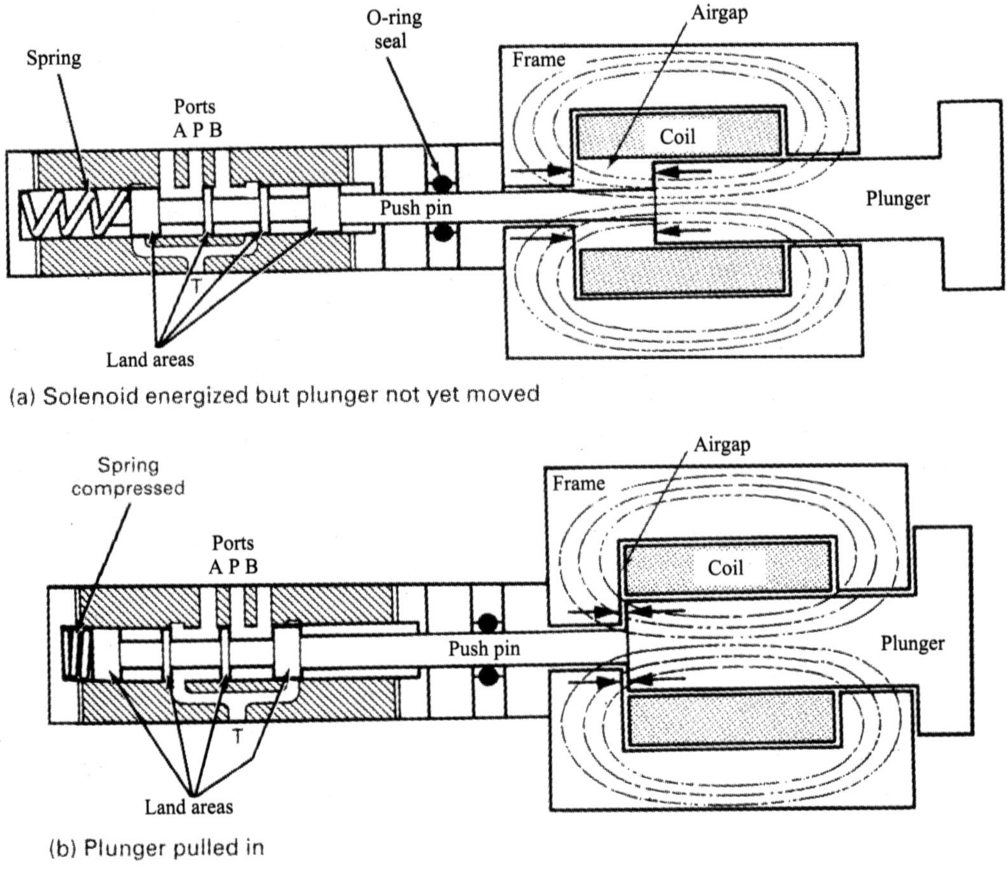

Fig. 6.4 Operation of Air-Gap Solenoid (*Courtesy*: Vickers, Inc.)

6.3 PILOT OPERATED SOLENOID VALVE

Standard solenoid designs are practically limited in terms of the amount of force developed to around 15 lbs, of which 8 to 10 lbs is expended to overcome spring resistance, friction and other related forces when moving the spool. Thus, there is little force left to overcome the effects of contamination, sludge formation and other problems.

Removing the cylinder and installing in its place a large valve resolve the problem. The lines from the solenoid valve are connected to the ends of the spool cavity just as in the cylinder. Operating the solenoid valve as before causes the spool of the larger valve to move in the same manner as the piston moved in the cylinder. Sliding the valve spool opens and closes the flow paths through the larger valve. Thus, the hydraulic pressure of the fluid is accomplishing what the solenoid valve alone could not achieve. In practical applications the solenoid valve is stacked on top of a larger main valve to make a single unit. The configuration results in a 'pilot-operated' directional control valve, and is illustrated in figure 6.5. The smaller direct-acting solenoid valve is called the 'pilot valve', and the larger valve is referred to as the 'main spool'.

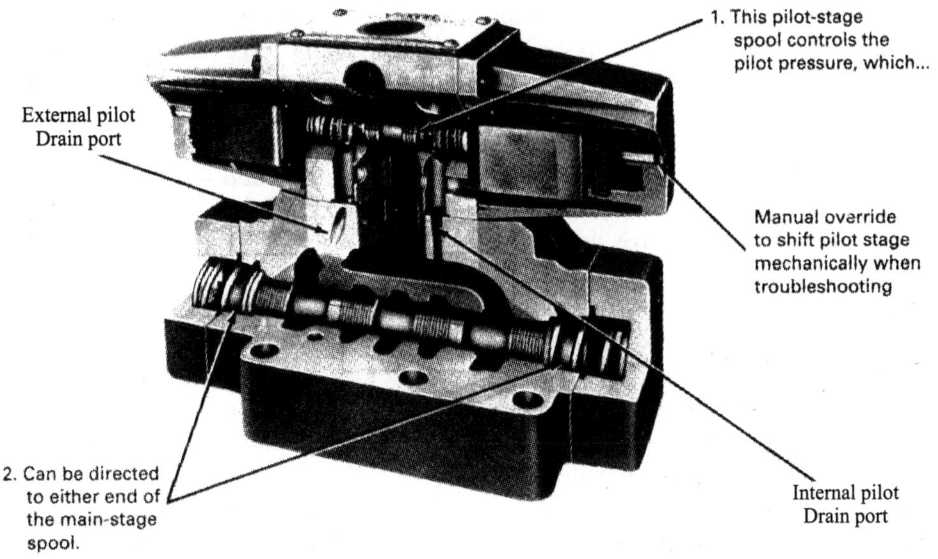

Fig. 6.5 Cutaway View of Pilot-Operated Directional Control Valve (*Courtesy*: Vickers, Inc.)

Pilot valves may be configured as four-way, two or three position types, with single or double solenoids. The main spool can be of any configuration. The pilot pressure, which is the same as the pressure of the fluid the pilot valve, directs to operate the main spool, which may be the same as the system operating pressure. In some cases the system pressure fluctuates widely, making the pilot pressure unpredictable and unreliable. Excessive pressure can also lead to rapid acceleration of the spool and cause a water hammer type of effect in the system. It may then be desirable to provide the pilot pressure from a controlled source that is isolated from the system pressure, which must be ported into the pilot valve through an external port. The same options must also be available for the drain, or tank, lines.

The speed of the pilot valve and of the main spool must be considered in determining the shifting time of a pilot-operated valve. The two speeds must be combined to establish total operating time.

6.4 SHIFTING TIME

Data about the solenoid and the pilot spool is required for determining the shifting time of the pilot spool. Dividing the initial and final solenoid forces by the mass yields acceleration. If the force/air-gap curve of the solenoid is approximately linear, the spool shifting time can be obtained when viscous damping forces are negligible. The assumption is mostly valid because of the short stroke and the relatively loose fit of the spool in the valve body.

The approximate solenoid force at any location of the plunger is given by the equation: $F_x = F_i + Bx$, where F_x = force in solenoid at any point x, N, F_i = initial force at maximum displacement, N, B = force gradient, N/cm, and x = displacement from position when solenoid is not actuated, cm. The approximate time period to reach a specific displacement x is given by: $t = \{(m/b) \times \ln[x + r + (2rx + x^2)^{0.5}\}^{0.5}$, where t = shifting time, s, m = spool mass, N-s^2/cm, $r = F_i/B$, cm, x = displacement from initial position, cm, and B = force gradient, N/cm.

The shifting time for a pilot-operated valve is significantly longer because of the additional dynamic system, larger mass of the main spool, and hydrodynamic forces resisting the motion of the spool. The hydrodynamic forces result from backpressure in the return line to the pilot section, viscous damping in the main valve section, and force from the incoming fluid stream. The speed at which the cylinder moves is determined by the incoming flow rate, or: $v = Q/A$, where v = velocity, Q = flow rate, and A = effective area. The flow rate can be estimated from a form of the orifice equation: $Q = CA \times [2\Delta p/\rho]$, where C = discharge coefficient (0.5 to 0.7), A = effective orifice area, Δp = pressure drop across pilot valve, and ρ = fluid density. Shifting time t is then given by: $t = S/v$, where S = main spool stroke, and v = velocity.

6.5 TRANSDUCERS FOR SOLENOID SYSTEM

Solenoid valves are mainly simple on-off type of devices, and are used to change linear or rotary position, turning speed, liquid level, pressure and similar other parameters into an electrical signal. The signal then controls the solenoid valve.

The transducers that are used as control elements are themselves switches that respond to a system parameter to either complete or break the control circuit. A majority of transducers sense a discrete position, and act as limit switches. Switches that sense high or low pressure or vacuum limits are called pressure switches. In hydraulic motors, centrifugal switches may be used to detect over-speed or under-speed conditions. Level switches may be used to monitor the level in reservoirs and tanks, and to open or close valves in response to the sensed level.

6.6 PROPORTIONAL CONTROL VALVE

A proportional control valve, also referred to as electro-hydraulic proportional control valve (EHPV), produces an output that is proportional to an electronic control input. The valve stands comfortably between the solenoid valve and the servo valve in performance, cost and complexity. Maximum frequency response is around 10 Hz, which is faster compared with the frequency capability of the solenoid valve, but considerably slower compared with that of the servo valve. The design of the actuating device, the proportional solenoid, allows the spool to be stopped at intermediate positions at the ends of the solenoid stroke. The associated electronics controlling the spool position offers a high degree of flexibility compared with the solenoid valve.

The performance can be improved and applicability extended by including a spool position feedback loop in the circuit. External feedback systems, higher frequency response, better performance in hysteresis, deadband, threshold and other features can also be included in proportional control valves. The enhanced performance and control features blur the distinction between servo valves and proportional control valves, but they increase the cost.

Aside from performance and control characteristics, physical features such as manufacturing process and design define the proportional control valve. For instance, proportional control valves require a proportional solenoid, whereas servo valves are operated by torque motor. The spool of a proportional control valve is mostly machine produced, while servo valve spools need a considerable amount of manual lapping and finishing due to the tighter clearances and tolerances.

6.7 DIRECTIONAL CONTROL

The directional control valve is the most commonly used type of proportional control valves. The directional EHPV bears some resemblance with a solenoid valve, but there are substantial differences in the solenoid and the spool.

The proportional solenoid differs from a standard solenoid in the design of the region near the end of the plunger stroke. The proportional solenoid eliminates the effect of the diminishing air-gap dimension by using a magnetically impervious material that to the solenoid appears to be a constant air gap. For a given current, this results in modifying the force-current curve as shown in figure 6.6. The flat portion of the curve occurs in the effectively constant air-gap portion of the stroke.

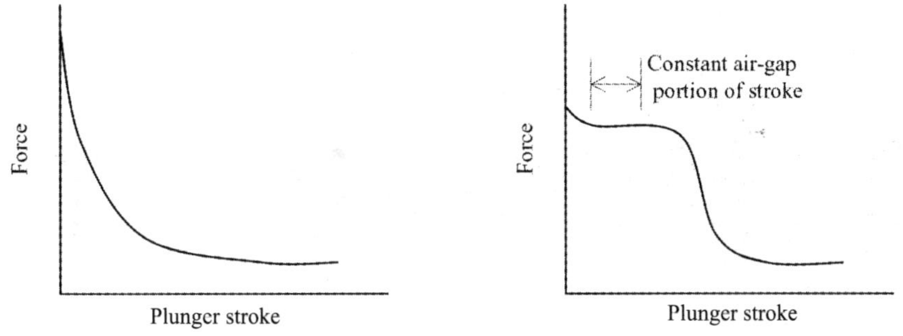

Fig. 6.6 Force-Current Curve Comparison Between Standard (*left*) and Proportional (*right*) Solenoids

The centering spring opposing the solenoid force needs careful calibration. Figure 6.7 shows a family of solenoid force versus stroke curve for various current loads, with the spring force curve also included. The spring force versus extension curve must intersect all the solenoid force lines in the flat region of the solenoid force lines. The physical consequence of this arrangement is that the position of the solenoid plunger, and subsequently of the spool, can be controlled by the current flow to the solenoid. A higher current produces a higher solenoid force. The force compresses the spring until the spring force balances the solenoid force, and the plunger then stops. This position determines the size of the flow path through the valve, which along with the pressure differential across the valve determines the fluid flow rate. Thus, both the direction of flow and the flow rate can be controlled with a single valve.

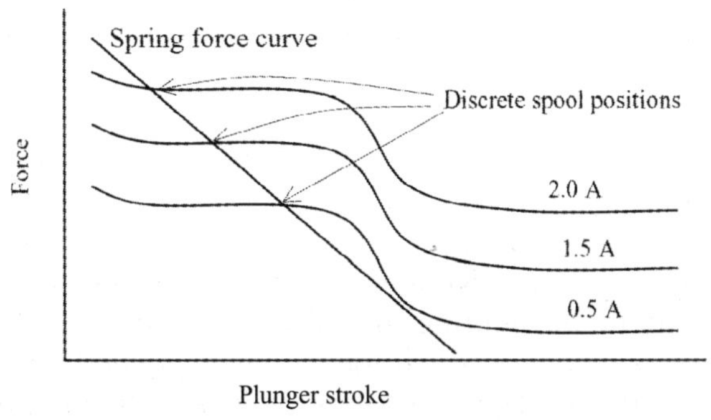

Fig. 6.7 Proportional Solenoid Force Curves with Overlaid Spring Force Curve

The directional control proportional valve has a valve body, spool and other parts as illustrated in figure 6.8. The primary difference between the hardware of the proportional and conventional valves is in the design of the spool. The surface finish, straightness and roundness are closely controlled. Metering notches on the spool land provide a greater degree of control throughout the travel of the spool than is possible with conventional square-edged spool lands. Identical control grooves on the land permit the flow to be metered evenly. Omitting some of the grooves on one of the lands permits differential flow rates in two directions. The concept is used to cause a single ended hydraulic cylinder to retract and extend at the same speed by reducing the input flow to the rod end on retraction.

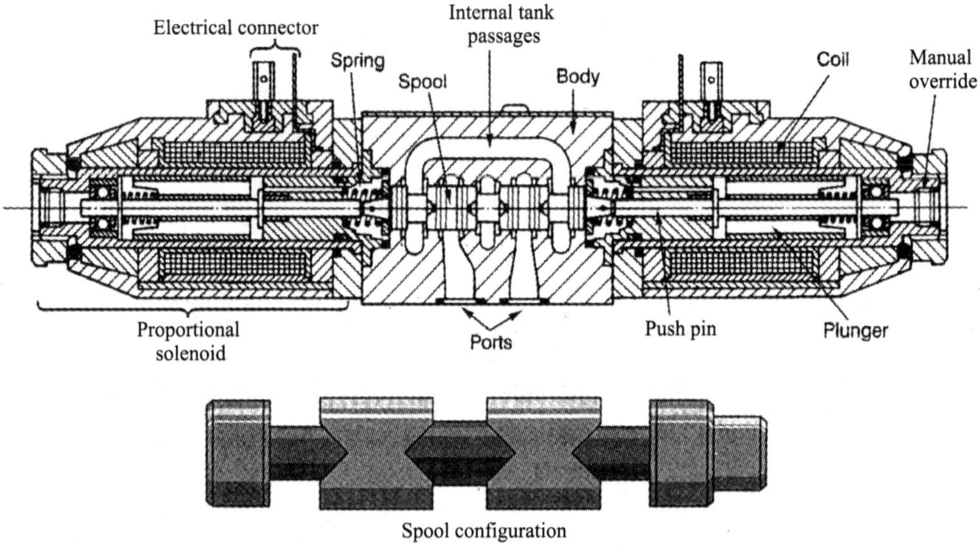

Fig. 6.8 Cutaway View of Directional Control EHPV (*Courtesy*: Vickers, Inc.)

6.8 FLOW AND PRESSURE CONTROL

EHPV's may be designed to provide only flow control, directional control being deleted. The valve then basically works in the same manner as a manually controlled flow valve.

The current input to the solenoid dictates the position of the spool, so the valve throttles the flow in proportion to the command signal. The pressure drop across the valve also determines the flow.

Proportional pressure control valves provide the same functions as adjustable spring type of pressure relief, pressure reducing and counterbalancing valves. Use of the ramping function ensures shock free yet rapid pressure changes. The adjustable solenoid permits remote control without the additional fluid piping for standard valves. The current going to the proportional solenoid determines the force exerted on the valve mechanism to hold the conical poppet on its seat to prevent flow. This may be done by omitting the valve spring and coupling the solenoid plunger directly with the poppet through a pushpin. If the spring is retained, the solenoid compresses the spring to provide the requisite force. Figure 6.9 shows a direct-acting poppet type of pressure relief valve using a spring. A linear variable differential transformer (LVDT) is used to provide feedback about the position of the valve.

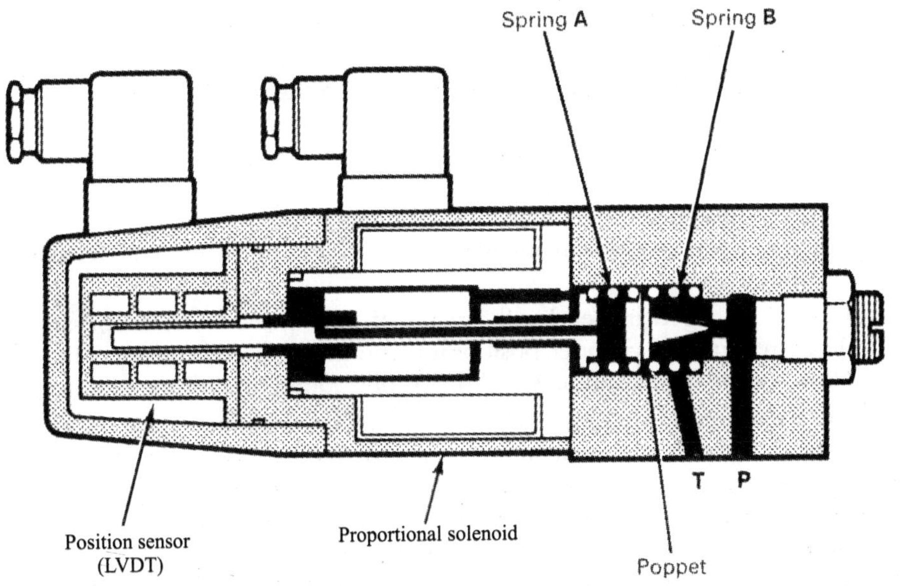

Fig. 6.9 Poppet Type Pressure Relief Valve with LVDT Feedback (*Courtesy*: Vickers, Inc.)

Since proportional solenoids typically produce a maximum force of around 15 lbs, the flow handling capability of a direct acting proportional relief valve is limited to mostly about half a gallon per minute. Only small poppet areas may be used for the high operating pressures needed for modern fluid power systems. Pilot-operated valves must be used for greater flow capability.

6.9 COMBINING SPEED AND DIRECTIONAL CONTROL

An important feature of the EHPV is its ability to provide flow, and hence speed, control, directional control and controlled acceleration and deceleration. A single EHPV can replace a series of solenoid and flow control valves to combine the varied control functions into a single unit.

Manufacturing and process operations relying on different machine speeds at various points can take advantage of the capabilities of the EHPV. A timed switching device, such as a programmable controller, may be used to sequence the switches and increment the drive motor speeds. Additional circuitry may be used by the switching control to activate flow of other process liquids for each stage of the operation. A ramping function helps to minimize shocks and flow surges in the system, and to flush out the liquids from prior processes before the next level of operation.

6.10 SERVO VALVES

Defined as 'an automatic device for controlling large amounts of power through very small amounts of power and automatically (and continuously) correcting the performance of the machine', the principle of servo-mechanism implies a return flow of information, or feedback, from the machinery. Although an EHPV system provided with feedback from the main system may be occasionally found, a servo valve operating without feedback is difficult to justify since it essentially becomes an expensive EHPV.

Numerous servo-mechanisms have been invented since the mid-1700's. Many were associated with raining steam in a boiler for controlling water level, water and steam flow, steam pressure and speed of machinery. Technological advances of the twentieth century in electrical power generation units, aircraft engines and automobile turbo-chargers spurred greater advances in servo technology. Significant contributions have been made in advancing the practical servo-mechanisms for controlling fluid power circuits prior to and after World War II by Siemens of Germany and Tiebel and Foxboro of United States.

Servo valves are engaged in nearly all aspects of fluid power system operations, the most common application being for cylinder positioning and motor speed control. The valves provide direction and flow control with a sliding spool that is positioned by a torque motor.

The torque motor is a low-power electro-magnetic component that is assembled from permanent magnets, two pole pieces, an armature and two coils, and is operated by low-voltage DC power. The coils may be connected in series, parallel, or in the more commonly used push-pull arrangement. Increasing the voltage input to one coil and reducing the voltage to the other coil by the same amount causes the armature to rotate, with its position determined by differential torque. The current input is expressed as a difference between the two coils, ΔI, and the control power P is expressed by the equation: $P = (\Delta I)^2 R$, where ΔI = differential current and R = resistance of one coil. If the coils are connected in parallel, the control power is: $P = (I_p)^2 R/2$, where I_p = total current through the circuit. For series connection, the control power is: $P = (I_s)^2(2R)$, where I_s is current in the series circuit.

The spool and the bore in which it travels are precision machined. Circumferential grooves are machined into the land areas to reduce transverse loading. The term 'lap' defines the width of lands relative to the width of the ports in the valve bore. The most commonly used is the zero overlap spool, where the land width matches the port width. When the spool is centered, there is no flow through the valve. The valve is useful for closed-loop position, speed and force control applications because of its precise metering characteristics about the neutral position. Overlapped spools have the advantage of providing lower leakage flow in the neutral position, but the dead band compromises the precision about the neutral position. Under-lapped spools provide rapid response to commands about the neutral position, but have non-linear flow characteristics around the region.

6.11 CONFIGURATION, CHARACTERISTICS AND PERFORMANCE OF SERVO VALVE

Single-stage servo valves are used where the flow requirement is low, under 5 gallons per minute. The valve has a sliding spool that directly connects to the torque motor armature. The force of the torque motor and the stroke of the spool dictate its flow capacity. Figure 6.10 shows a single stage servo valve. A stiff wire provides mechanical connection between the torque motor's armature and the spool. A clockwise turning of the armature pushes the spool to the left, opening the flow paths from P to B and A to T. A counterclockwise turn opens flow paths P to A and B to T.

Two and three stages are used to obtain higher flow rates. The second and third stages have a sliding spool that is pilot-operated from the preceding stage. Figure 6.11 shows a valve with a spool-type pilot stage for controlling the direction and speed of a hydraulic motor. A sleeve is used in the pilot stage, which is associated with the internal feedback mechanism of the valve. Two pressure sources are provided, p_c for piloting the main spool, and p_s for the system. The larger end of the spool has twice the area of the smaller end to obtain static balance when the spool is in the neutral position.

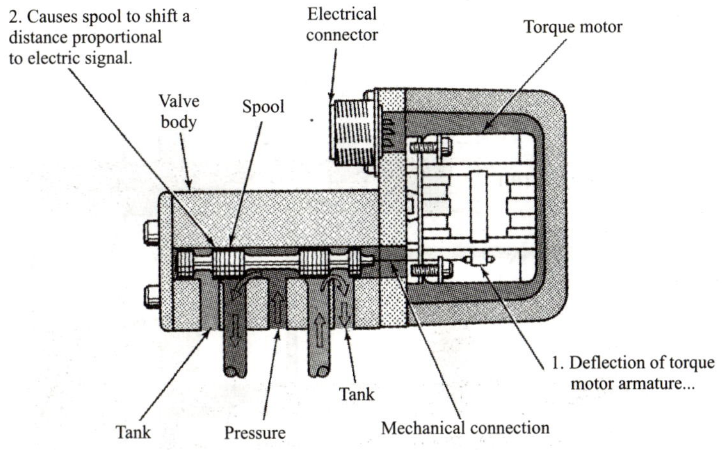

Fig. 6.10 Single-Stage Spool Type of Servo Valve (*Courtesy*: Vickers, Inc.)

A counterclockwise movement of the torque motor armature pushes the pilot spool to the left, which initiates an internal feedback mechanism. Flow to the large end of the main spool is terminated, with the stopping point determined by the deflection of the torque motor. De-energizing the torque motor returns the pilot spool to its neutral position, and the main spool is pushed back to the right. A similar but opposite sequence takes place when the torque motor armature is deflected clockwise.

The jet pipe is another form of pilot stage, where the nozzle of the jet is attached to, and moves with, the torque motor armature (figure 6.12). The pilot fluid is directed by the nozzle into two receiver ports that connect to the pilot chambers at the ends of the main spool. Deflection of the torque motor directs the jet into the receiver ports to unbalance the forces, which moves the main spool.

Servo valves are usually saturated at 3000 lbs/in^2, but most of them are capable of operating at 5000 lbs/in^2. The stated flow rate occurs at 1000 lbs/in^2 pressure drop across a fully open valve (figure 6.13). If the pressure drop across the valve differs from the optimum value, the flow rate changes proportionally. The non-optimum flow rate may be calculated from the equation: $Q = Q_R \times [\Delta p/\Delta p_R]^{0.5}$, where Q = adjusted flow rate, Q_R = rated flow, Δp = actual pressure, and Δp_R = rated pressure drop.

The substantial pressure drop in a servo valve will lead to a large amount of heat, which is determined from: HGR = HP × 42.4 Btu/min/HP, where HGR = heat generation rate, and HP = $(\Delta p \times Q)/1714$ = horsepower loss across the valve. Fluid temperature rise across the valve is: $\Delta T = \text{HGR}/(C_P W)$, where ΔT = temperature rise, C_P = specific heat, and W = weight of the fluid.

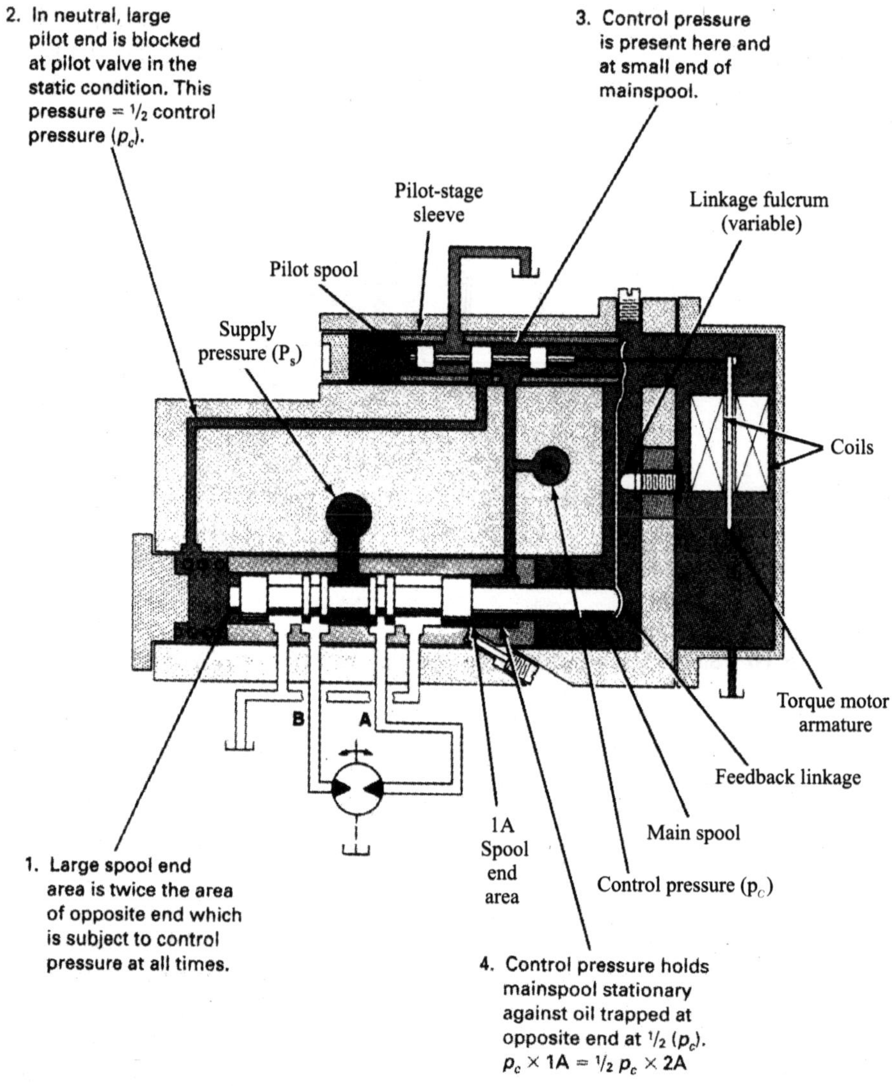

Fig. 6.11 Two-Stage Pilot Operated Servo Valve (*Courtesy*: Vickers, Inc.)

6.12 SERVO PUMP AND SERVOMOTOR

Servo controllers are built inside servo pumps and servomotors to provide variable displacement. The servo assembly includes a servo valve, a hydraulic cylinder and an internal feedback unit (figure 6.14).

Depending on the size and application, servo pumps can handle maximum flow rate of 500 gallons per minute and pressure of 5000 lbs/in^2. The piston pumps may be single or double-acting, and may be arranged as either inline or radial units. Servo pumps are mostly used to control the speed of large hydraulic motors to adjust pump displacement in response to feedback signals from the motor.

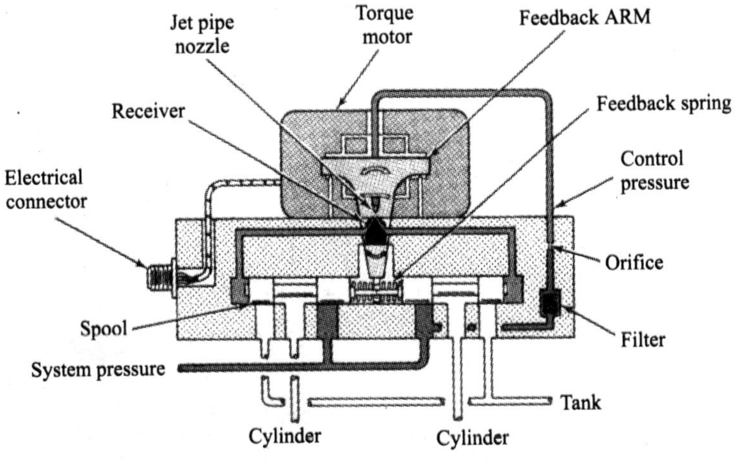

Fig. 6.12 Jet Pipe-Type Servo Valve (*Courtesy*: Vickers, Inc.)

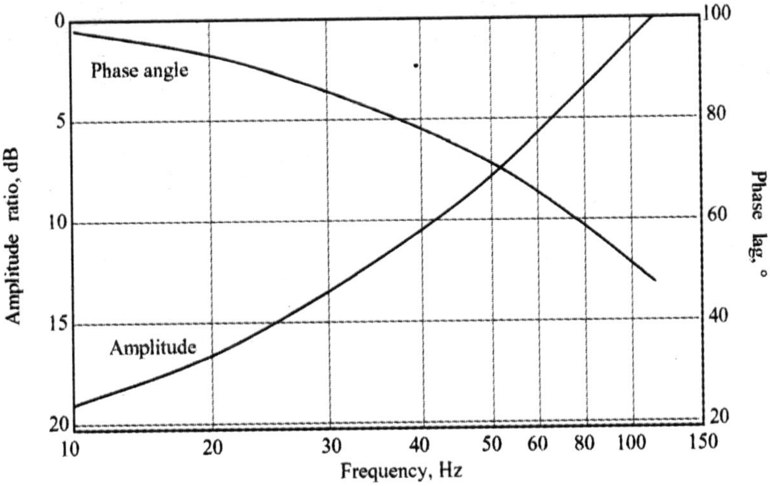

Fig. 6.13 Bode Diagram for Typical Servo Valve

Similar to servo pumps in many ways, servomotors can alter the displacement in response to command and feedback signals. Servomotors are used to control motor speed and output torque.

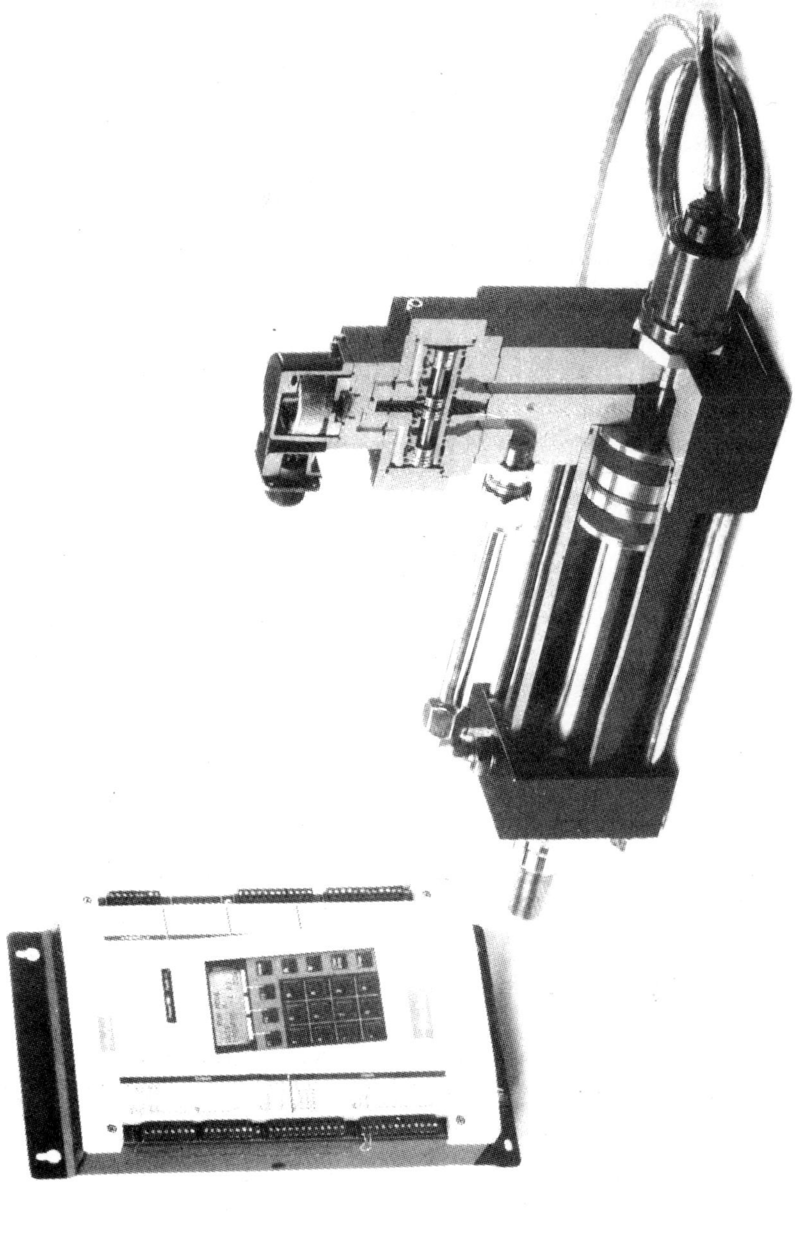

Fig. 6.14 Servo Valve with Internal Feedback Loop Mounted Directly on Cylinder (*Courtesy*: Parker Hannifin Corp.)

6.13 REVIEW PROBLEMS

Example problem # 6.1: The resistance of the coils connected in a push-pull circuit of a torque motor is 25 Ω, and rated at 225 mA. Find (*i*) the voltage of each coil when the armature of the torque motor is centered, (*ii*) the maximum value of ΔI, and (*iii*) the maximum control power of the motor.

Solution: (*i*) The maximum voltage in the coils is:

$$E_{max} = I \times R = 225 \text{ mA} \times 25 \ \Omega$$
$$= 5.625 \text{ V}$$

The armature is centered when: $E = E_{max}/2 = 2.81$ V

(*ii*) Differential current ΔI is maximum when maximum voltage is applied on one coil, and the other coil has no voltage.

Then: $$\Delta I = E_{max}/R - 0 = (5.625 \text{ V})/(25 \ \Omega) = 225 \text{ mA}$$

(*ii*) Maximum control power is:

$$P = (\Delta I)^2 \times R = (225 \text{ mA})^2 \times (25 \ \Omega)$$
$$= 1.27 \ \Omega$$

Example problem # 6.2: Assuming the coils in example problem # 6.1 are connected in parallel, repeat the calculations.

Solution: The voltage to each coil remains at 5.625 V, but the current to each coil will increase since the resistance has decreased. For a parallel circuit with two equal resistors, the equivalent resistance is $R/2$, or 12.5 Ω.

$$\text{Total current } I_P = E/R = (5.625 \text{ V})/(12.5 \ \Omega)$$
$$= 0.45 \text{ A} = 450 \text{ mA}$$

Maximum control power is:

$$P = (I_P)^2 \times R/2 = (0.45 \text{ A})^2 \times (12.5 \ \Omega)$$
$$= 2.53 \text{ W}$$

Example problem # 6.3: Assuming the coils in example problem # 6.1 are connected in series, repeat the calculations.

Solution: In a series circuit the currents are assisting the armature rotation. Total resistance is $2R = 50 \ \Omega$, and maximum current is 225 mA.

$$\text{Then maximum voltage} = E = IR = (225 \text{ mA}) \times (50 \ \Omega)$$
$$= 11.25 \text{ V}$$

Control power is: $$P = (I_S)^2 \times 2R = (0.225 \text{ A})^2 \times (50 \ \Omega)$$
$$= 2.53 \text{ W.}$$

Example problem # 6.4: A servo valve is rated at 17 gallons per minute and 1500 lbs/in^2. If it is to be operated in a 2500 lbs/in^2 system, what is the adjusted flow rate for optimum power transfer?

Solution: Optimum power transfer occurs when the pressure drop Δp across the valve is 2500/3 = 833 lbs/in^2. Then flow rate $Q = Q_R \times [\Delta p/\Delta p_R]^{0.5}$.

$= 17 \times [833/1500]^{0.5}$.

$= 12.67$ gallons per minute.

Example problem # 6.5: Calculate the horsepower loss, heat buildup rate and temperature rise across the valve in example problem # 6.4. Assume specific heat is 0.47 Btu/lb-°F and specific weight is 54.8 lbs/ft^3.

Solution: Horsepower loss across the valve:

$$\text{HP} = (\Delta p \times Q)/1714$$
$$= (833 \times 12.67)/1714$$
$$= 6.16 \text{ HP}$$

Heat generation rate: $\text{HGR} = \text{HP} \times 42.4$ Btu/min/HP

$$= 6.16 \times 42.4$$
$$= 261.08 \text{ Btu/min}$$

Temperature rise across valve:

$$\Delta T = \text{HGR}/(C_P W)$$

Weight flow rate: $W = \gamma Q = (54.8 \text{ lbs/ft}^3 \times 12.67 \text{ gal/min})/(7.49 \text{ gal/ft}^3)$

$$= 92.69 \text{ lbs/min}$$

Then temperature rise: $\Delta T = (261.08)/[(0.47) \times (92.69)]$

$$= 5.99°\text{F}$$

Example problem # 6.6: Find the amplitude response in decibels if a valve spool moves sinusoidally 0.225 in. in response to a 0.32 in. sinusoidal input.

Solution: Amplitude response $= 20 \times \log (\text{output/input})$

$$= 20 \times \log (0.225/0.32)$$
$$= -3.06 \text{ dB}$$

Example problem # 6.7: The forward loop in figure 6.15 contains a hydraulic motor with gain of 335 rpm/V. The feedback loop has a tach-generator with gain of 0.235 V/rpm. Calculate the speed of the motor for a 4.25 V input.

Solution: Relating the variables of the diagram to the problem data,

$$R = 4.25 \text{ V}, G = 335 \text{ rpm/V}, H = 0.235 \text{ V/rpm}$$

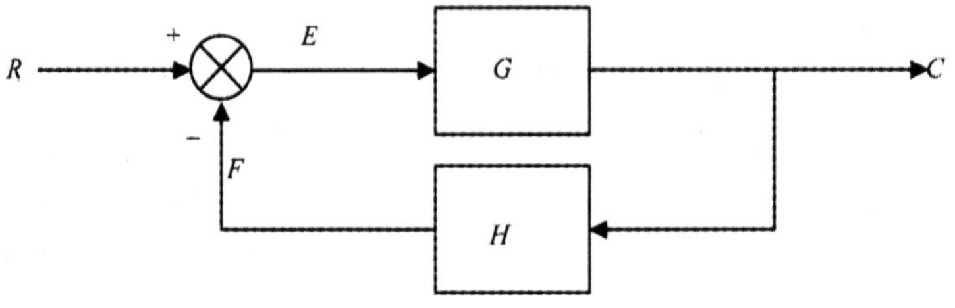

Fig. 6.15

Output: $C = GR/(1 + GH)$

$= (335 \text{ rpm/V} \times 4.25 \text{ V})/[1 + (335 \text{ rpm/V} \times 0.235 \text{ V/rpm})] = 17.86 \text{ rpm}.$

Example problem # 6.8: Calculate the speed of the hydraulic motor in the circuit shown in figure 6.16 if $R = 7$ V, $G_{amp} = 75$ mA/V, $G_{SV} = 0.15$ gpm/mA, $G_{HM} = 35$ rpm/gpm and $H_T = 0.25$ V/rpm.

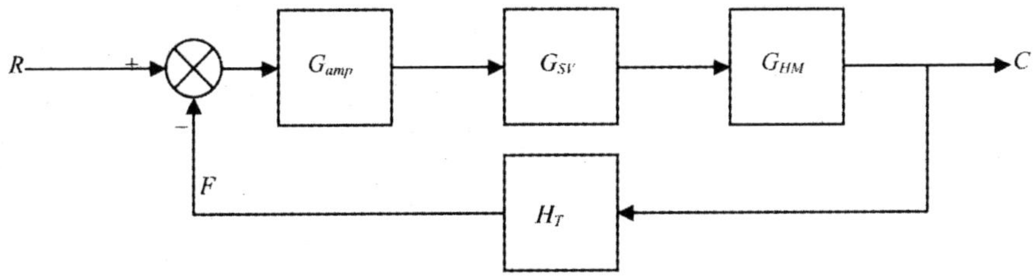

Fig. 6.16

Solution: The forward loop gain:

$$G = G_{amp} \times G_{SV} \times G_{HM}$$
$$= 75 \text{ mA/V} \times 0.15 \text{ gpm/mA} \times 35 \text{ rpm/gpm}$$
$$= 393.75 \text{ rpm/V}$$

Output: $C = GR/(1 + GH)$

$$= (393.75 \text{ rpm/V} \times 7.0 \text{ V})/[1 + (393.75 \text{ rpm/V} \times 0.25 \text{ V/rpm})]$$
$$= 27.72 \text{ rpm}$$

Example problem # 6.9: The input command from a joystick controller for a servo-controlled positioning system is 36 V, and is linearly distributed over a 90° throw (figure 6.17). The cylinder has a 4.75 in^2 piston, a 2.5 in^2 rod, and a stroke of 6.5 in. The position transducer is a linear resistive

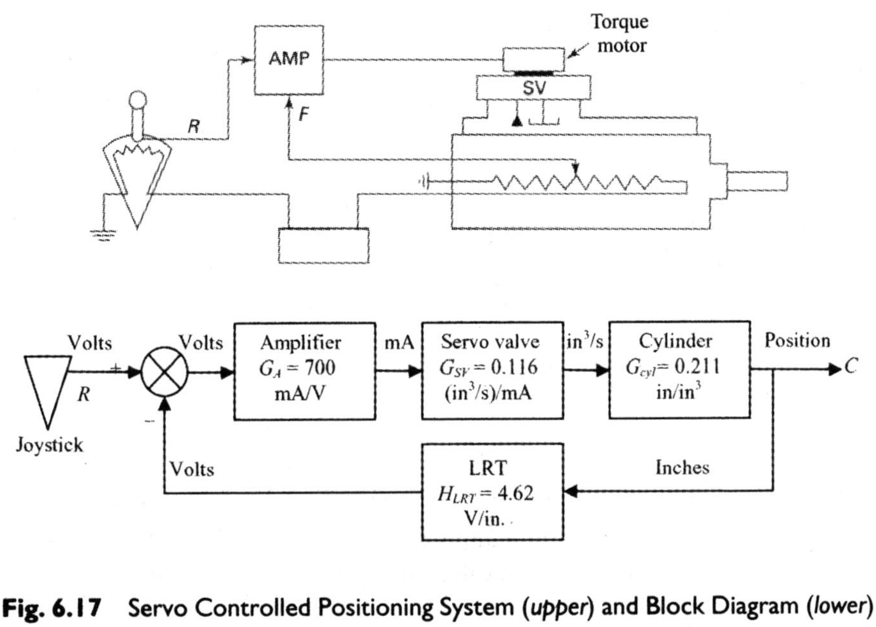

Fig. 6.17 Servo Controlled Positioning System (*upper*) and Block Diagram (*lower*)

transducer (LRT) that uses a 30 V DC power supply. The operational amplifier has an output of 700 mA for each input volt, but the torque motor on the servo valve is saturated at 250 mA, at which point the valve is fully open. Thus, the current limiter of the operational amplifier is set to a maximum of 250 mA, regardless of the input signal to protect the torque motor. The hydraulic system working pressure is 3200 lbs/in^2 with a flow rate of 7.5 gallons per minute. When the torque motor input is 12 mA, the pressure differential across the valve is 1250 lbs/in^2.

Calculate: (*i*) the gain of the joystick controller, (*ii*) LRT gain, (*iii*) operational amplifier gain, (*iv*) servo valve pressure gain, (*v*) valve pressure gain, (*vi*) cylinder position gain, (*vii*) control ratio of the system, (*viii*) the new position if joystick is moved through 2°.

Solution:

(*i*) Joystick controller: G_{js} = 36 V/90° = 0.4 V/degree.

(*ii*) LRT, or feedback gain: H_{LRT} = 30 V/6.5 in. = 4.62 V/in.

(*iii*) Operational amplifier gain: G_A = 700 mA/V

(*iv*) Servo valve gain: G_{SV} = (7.5 gpm/250 mA) × [(3.85 in^3/s)/gpm]

= 0.116 [(in^3/s)/mA]

(*v*) Servo valve pressure gain: G_p = 1250 lbs/in^2/12 mA = 104.17 lbs/in^2/mA

(*vi*) Cylinder position gain: $G_{cyl} = v/Q = 1/A$ = 1/(4.75 in^2) = 0.211 in/in^3

(*vii*) Control ratio of the system: $C/R = G/(1 + GH_{LRT})$,

where: $G = G_A \times G_{SV} \times G_{cyl}$

= (700 mA/V) × [0.116 (in^3/s)/mA] × (0.211 in/in^3)

= 17.02 in./V

Then: C/R = (17.02 in./V)/[1 + (17.02 in./V) × (4.62 V/in.)]

= 0.214 in./V

(*viii*) Cylinder position due to joystick motion:

$R = G_{js} \times 2.0°$

= 0.4 V/° × 2.0°

= 0.8 V

New position: $C = (C/R) \times R$ = (0.214 in./V) × (0.8V) = 0.171 in.

Example problem # 6.10: Calculate the open loop gain (also called velocity constant) in example problem # 6.9.

Solution: Open-loop gain = $G_A \times G_{SV} \times G_{cyl} \times H_{LRT}$

= (17.02 in./V) × (4.62 V/in.) = 78.63 s^{-1}

Open-loop gain = Velocity constant = K_V = 78.63 s^{-1}

Note: velocity constant is reciprocal of time constant, or: $K_V = 1/\tau$

Example problem # 6.11: Assume the piston of a hydraulic cylinder is in critical position, at which point V_p = total volume of oil between valve and piston face = 25 in^3 and V_N = total volume of oil between annular face of piston face and valve = 18 in^3. Take bulk modulus to be E = 200,000 lbs/in^3, and mass M = 10 lb-s^2/in. Calculate the natural frequency of the hydraulic cylinder.

Solution: The natural frequency of the hydraulic cylinder is discussed in section 4.6. Natural frequency $\omega_o = \{(E/M) \times [(A_p^2/V_p) + (A_N^2/V_N)\}^{1/2}$, where ω_o = natural frequency, radian/s, E = fluid bulk modulus, M = mass of load, (weight/g), A_p = piston area, A_N = (annular piston) – (rod area). Fluid bulk modulus $E = [-\Delta p/(\Delta V/V)]$, where Δp = pressure change, ΔV = change in volume due to applied pressure, V = original volume. Values of V_p and V_N depend on the position of the piston as it moves through the stroke.

$$\text{Then} \quad \omega_o = \{(200000/10) \times (4.75^2/25 + 2.5^2/18)\}^{0.5}$$
$$= 134.35 \text{ radian/s.}$$

6.14 EXERCISES

6.1 A torque motor's coils have a resistance of 17.5 Ω and are rated at 240 mA. Find the voltage of each coil when the armature of the torque motor is centered when the coils are connected (*i*) in push-pull arrangement, (*ii*) in series with each other, and (*iii*) in parallel with each other.

6.2 In the above problem find the maximum value of ΔI and the maximum control power of the motor.

6.3 A servo valve has an adjusted flow rate for optimum power transfer of 16.5 gallons per minute and 2250 lbs/in^2. If it is to be operated in a 1250 lbs/in^2 system, what is the rated flow?

6.4 In the above problem calculate the horsepower loss, heat buildup rate and temperature rise across the valve assuming specific heat is 0.475 Btu/lb-°F and specific weight is 55.5 lbs/ft^3.

6.5 Find the sinusoidal valve spool movement in response to a 0.275 in. sinusoidal input if the amplitude response is –3.75 dB.

6.6 Find the forward loop gain in figure 6.15 containing a hydraulic motor if the feedback loop has a tach-generator with gain of 0.325 V/rpm and the speed of the motor is 15.75 rpm for a 3.85 V input.

6.7 Determine the input signal voltage if the speed of the hydraulic motor in the circuit shown in figure 6.16 is 32.25 rpm, G_{amp} = 80 mA/V, G_{SV} = 0.175 gpm/mA, G_{HM} = 33 rpm/gpm and H_T = 0.275 V/rpm.

6.8 Determine the feedback H_T if the speed of the hydraulic motor in the circuit shown in figure 6.16 is 29.75 rpm, R = 5 V, G_{amp} = 65 mA/V, G_{SV} = 0.135 gpm/mA, and G_{HM} = 32.25 rpm/gpm.

6.9 Determine the servo valve gain G_{SV} if the speed of the hydraulic motor in the circuit shown in figure 6.16 is 28.5 rpm, R = 6.25 V, G_{amp} = 50 mA/V, G_{HM} = 29.75 rpm/gpm and H_T = 0.245 V/rpm.

6.10 Consider the piston of a hydraulic cylinder to be in critical position, at which point V_p = total volume of oil between valve and piston face = 19.75 in^3 and V_N = total volume of oil between annular face of piston face and valve = 12.875 in^3. Assume bulk modulus to be E = 215,000 lbs/in^3, and mass M = 9.35 lb-s^2/in. Calculate the natural frequency of the hydraulic cylinder.

6.11 The natural frequency of a hydraulic cylinder is 165.25 radians/s. Assume the piston of a hydraulic cylinder is in critical position, at which point V_p = total volume of oil between valve and piston face = 25 in^3 and V_N = total volume of oil between annular face of piston face and valve = 18 in^3. Take bulk modulus to be E = 209,000 lbs/in^3. Find the mass M.

6.12 In the above problem assume the bulk modulus E = 189,000 lbs/in^3 and the natural frequency of the hydraulic cylinder is 165.25 radians/s. Find the mass M.

References & Bibliography

Bateson, R.N., "Introduction to Control System Technology", Prentice-Hall, Clifton, NJ, 1999.

Carstens, J.R., "Automatic Control Systems and Components", Prentice-Hall, Clifton, NJ, 1990.

CEC Instruments, "Pressure Transducer Handbook", Pasadena, CA, 1989.

Funakubo, H., "Actuators for Control", Gordon and Breach Science Publishers, New York, 1991.

Norvelle, F.B., "Electro-Hydraulic Control Systems", Prentice-Hall, Upper Saddle River, NJ, 2000.

International Organization for Standardization, ISO 4572, Geneva, 1972.

International Organization for Standardization, ISO 4406, Geneva, 1992.

Johnson, J.E., "Electro-Hydraulic Servo Systems", Penton, IPC, Cleveland, OH, 1977.

Merritt, H.E., "Hydraulic Control Systems", John Wiley & Sons, New York, NY, 1967.

Mannesmann Rexroth, GmbH, "The Hydraulic Trainer: Proportional and Servo Valve Technology", Lohr am Main, Germany, 1986.

Maskrey, R.H. and Thayer, W.J., "A Brief History of Electro-Hydraulic Mechanisms", ASME Journal of Dynamic Systems Measurement and Control, New York, NY, 1978.

Tonyan, M.J., "Electronically Controlled Proportional Valves", Marcel Dekker, New York, NY, 1985.

Vickers Inc., "Vickers Industrial Hydraulics Manual", Rochester Hills, MI, 1992.

7

Manipulation and Actuation

7.1 INTRODUCTION

The controller interacts with the process through the input signals it receives at selected stages of the process, and through the output commands controls the process. Sensors and process data signals come form the input side. Switches, actuators, control valves and electric motors execute the command signals to provide the required output.

Switches make or break the connection in a circuit, either mechanically with an actuator or electro-mechanically with a solenoid. Control systems require manipulation of process tasks by switching on an electric motor or a heating element, regulate speed of the motor, actuate solenoid valves in pneumatic and hydraulic cylinders, and initiate sequential action in control systems. A set of switches may be assembled in a control relay to control multiple functions. Switching action in a control system may be delayed with the use of time-delay relays. Special relays with heavy-duty contacts are used where large amounts of electric power are required. For instance, the relay for switching a three-phase electric motor is called a motor starter.

Solid-state components are used extensively in control systems. Among other applications, they are used as electronic switches for control motors and solenoid valves, and to convert DC voltage to controlled AC voltage in full-wave bridge rectifiers and in servomotors.

Actuators provide the power for executing the final control element. Hydraulic and pneumatic cylinders provide linear motion, and hydraulic and pneumatic motors develop large torque to produce rotational motion. Electric motors are often used as actuators in a servo control system, for instance, in numerically controlled machining centers to position the work piece and to locate the cutting tool. Variable speed electric motors are also used to drive pumps for controlling the flow rate and pressure of the fluid.

Hydraulic and pneumatic cylinders are the preferred choice for producing linear motion from economic and maintenance considerations. Hydraulic actuators are better suited for slower, more precise and heavier load applications. Pneumatic actuators operate at lower pressure than the hydraulic ones, and are useful for lighter loads and where faster response is necessary. The speed of the hydraulic cylinders is limited by the maximum flow rate received from the pump.

7.2 DC MOTOR

DC motors are versatile drives that provide accurate control of speed over a wide range, and are capable of operating in the reverse direction. The high torque to inertia ratio gives them quick response to control signals, and deliver over three times their rated torque for short periods. The motors range from fractional to several hundred horsepower ratings. DC motors also have the unique capacity for dynamic and regenerative braking. The motor momentarily turns into a DC generator driven by the inertia of the load during braking.

The lighter weight and inertia and lower inductance of the armature permits DC servomotors to respond quickly to command signals for a shift in position or speed. Typically the servomotors have electrical time constants ranging from 0.1 to 6 ms and mechanical time constants ranging from 2.3 ms to around 40 ms.

The armature and field windings may be connected in series, shunt, compound, and separately excited configurations. Series connection provides the highest starting torque and the largest no-load speed of the four connections. The shunt motor has a comparatively lower starting torque and no-load speed. The compound motor expectedly has speed and torque characteristics that lie between those of the series and the shunt connections. The separately excited version is a special case of the shunt motor, allowing independent control of the armature and field voltages.

Permanent magnets provide the magnetic field instead of the field winding in a permanent magnet motor of low power rating. The feature offers advantages of simpler construction due to the elimination of field power supply, higher efficiency, and a more linear torque-speed curve.

The torque to inertia ratio may be increased in a DC servomotor with the moving coil combined with a stationary core. The armature winding spins in the annular space between the field poles on the outside and the iron core on the inside. The arrangement yields a torque to inertia ratio that is several times greater than the conventional iron core armature motor. A disk, or pancake, armature motor is another approach for obtaining high torque to inertia ratio. The short length and large diameter armature of non-magnetic material operates between the permanent magnetic poles on either side of the armature. The magnetic field passes from the magnetic pole on one side, through the armature, and to the pole on the other side.

The need to make electrical contact between the armature winding and the brush and commutator assembly of the DC servomotor is a considerable drawback. Limitations are posed by the need for replacing worn brushes, arcing caused by commutation, and the long path for heat dissipation from the armature. Brushless motors, illustrated in figure 7.1 overcome these problems. The rotor of the brushless motor is a permanent magnet, the winding is in the stator, and a solid-state circuit replaces the brush and commutator assembly. Heat from electrical resistance is confined to the field windings, resulting in quicker dissipation. Also, the lighter weight of the rotor results in better torque-inertia characteristics. A current amplifier modulated by pulse width provides smoother operation at any speed or torque.

The steady state characteristic of an armature controlled DC motor has a linear relation between the armature current (i_a) and the motor torque (T). The slope of the curve provides the torque constant (K_T). The relation between the parameters is given by: $T = K_T \times i_a - T_f$, N-m, where $K_T = \Delta T/\Delta i_a$, N-m/A, is the intercept on the current axis, and i_f, is the current needed to overcome friction torque, T_f. The back emf (e_b) has a linear relation with armature speed (ω_m). Hence, $e_b = K_E \times \omega_m$, V, where the slope of this curve is called the voltage constant, K_E, and is given by $K_E = \Delta e_b /\times \omega_m$,

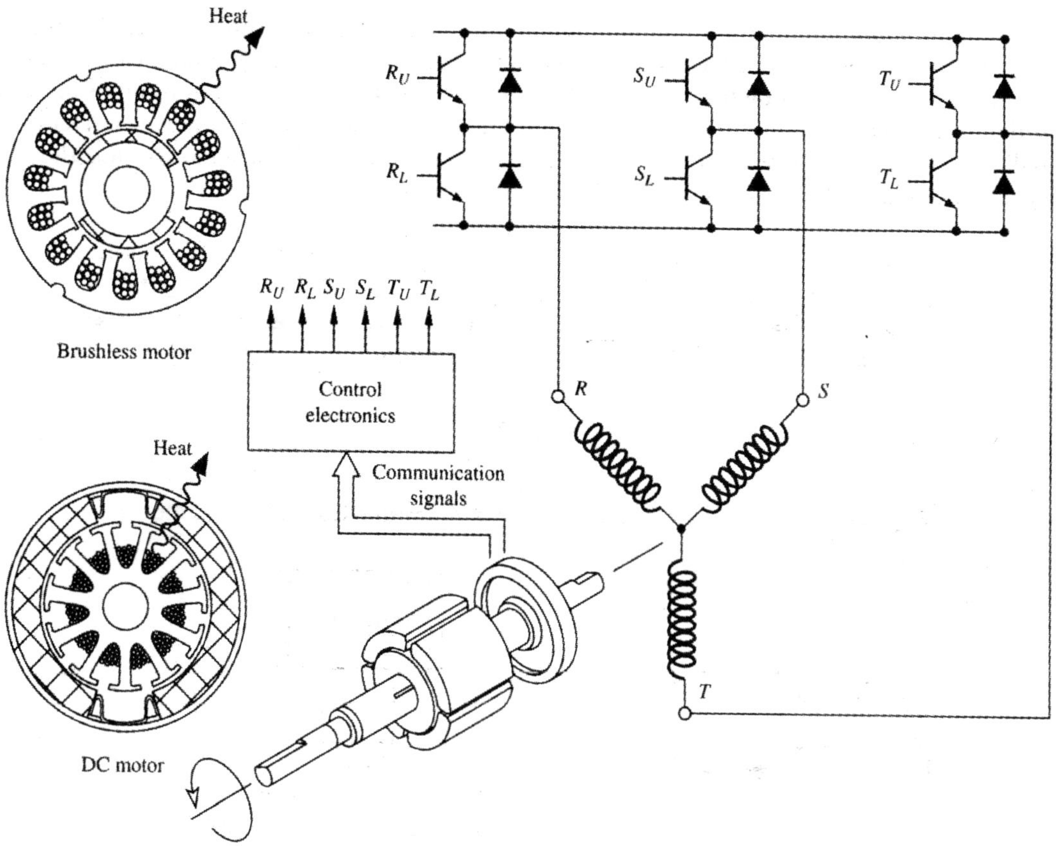

Fig. 7.1 Brushless Servomotor Construction (*Courtesy*: Reliance Motor Control)

V-s/radian. The armature voltage versus motor speed with constant torque curve has a constant armature current (i_a), and the armature voltage (e_a) is made up of the back emf (e_b) and voltage drop across the armature (i_aR), or: $e_a = e_b + i_aR$, V. The equation for torque versus motor speed with constant armature voltage curve may be expressed by: $i_a = (T + T_f)/K_T$ and by: $e_a = e_b + (T + T_f)R/K_T$. Two constants are defined: the speed regulation constant is given as $K_R = R/K_EK_T$, radians/N-m-s, and the no-load speed is given by: $\omega_{NL} = (e_a K_T - RT_f)/K_E K_T$, radians/s. Motor speed $\omega_m = -K_RT + \omega_{NL}$, radian/s, and power produced in the motor is determined from: $P = \omega_mT$, watts.

The block diagram of an armature controlled DC motor is given in figure 7.2. The electric circuit consists of resistance R and inductance L of the armature winding, plus a source for the back emf e_b induced in the armature winding. A variable voltage e_a is applied to the armature winding. The armature current i_a produces the torque T. Equations for the transfer functions for the DC motor are shown in table 7.1.

Brushless DC motors used for converting digital pulses into fixed incremental motion are called 'stepping motors'. They play a valuable role in machine tools, coordinate measuring machines and scientific instruments (figure 7.3). Pulse inputs may be provided from a computer or a programmable controller. The control unit converts the pulses into step motion of the motor shaft through power transistors. The lead screw converts the rotary motion of the motor shaft into linear motion of the work piece attached to the table. The motor's stator has an even number of equally spaced poles, or teeth, each with an electrical coil. The rotor has equally spaced external teeth with a small gap between the rotor and stator teeth. The number of teeth on the rotor and the stator determine the size of the step, also called the stepping angle.

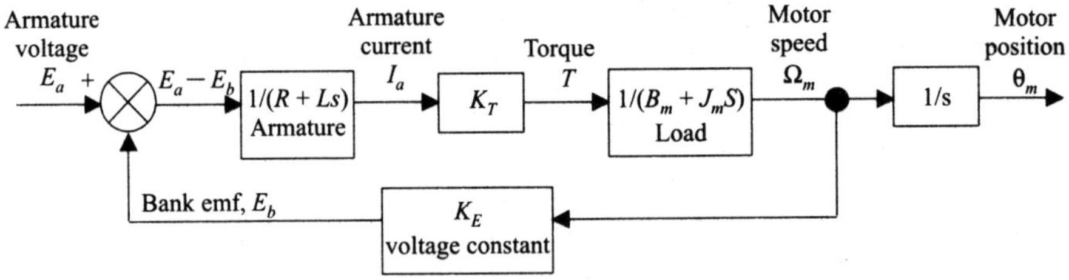

Fig. 7.2 Block Diagram Representation of Armature-Controlled DC Motor

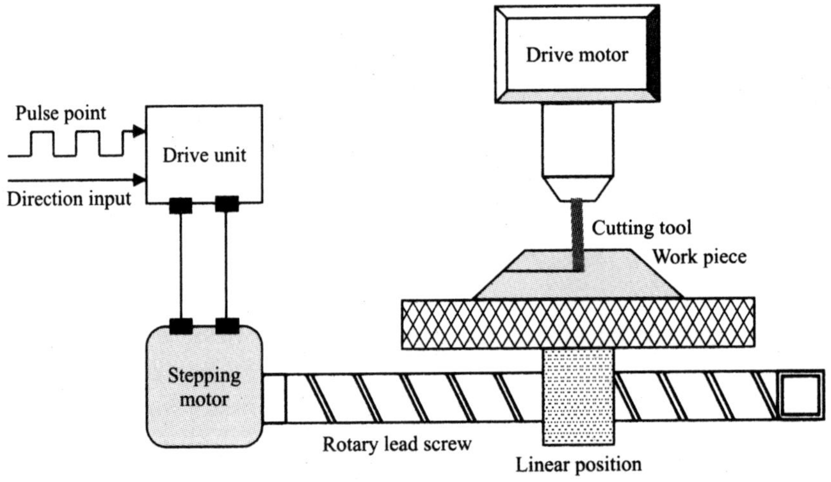

Fig. 7.3 Stepping Motor Operation in Numerical Controlled Machine

Depending on the construction features of the rotor, stepping motors may be classified as variable reluctance, permanent magnet, or hybrid types. Variable resistance stepping motors are used for larger step angles, permanent magnet motors have improved design and performance, and hybrid stepping motors combine the variable reluctance and permanent magnet features.

Table 7.1 Armature-Controlled DC Motor Transfer Functions

Note: Velocity transfer functions are provided below. Corresponding position transfer function may be obtained from the expression: $\theta_m(s)/E_a(s) = (1/s) \times \Omega_m(s)/E_a(s)$.

Voltage-driven motor

Velocity transfer function:

$$\Omega_m(s)/E_a(s) = K_T/[RB_m + K_E \times K_T + RB_m \times (\tau_m + \tau_e)s + RB_m\tau_m\tau_e s^2]$$

Simplified velocity transfer function

$$\Omega_m(s)/E_a(s) = K_s/(1 + \tau_s s)$$

where: $K_s = K_T/(RB_m + K_E \times K_T)$

$\tau_s = RJ_m/(RB_m + K_E \times K_T)$

Current-driven motor

Velocity transfer function:

$$\Omega_m(s)/I_a(s) = K_c/(1 + \tau_m s)$$

where: $K_c = K_T/B_m$

$\tau_m = J_m/B_m$

Velocity transfer function of motor, speed reducer and load:

$$\Omega_L(s)/E_a(s) = K_T \times (N_1/N_2)/[RB_T + K_E \times K_T + RB_T \times (\tau_T + \tau_e)s + RB_T\tau_T\tau_e s^2]$$

where: $J_T = J_m + (N_1/N_2)^2 \times J_L$

$B_T = B_m + (N_1/N_2)^2 \times B_L$

$\tau_T = J_T/B_T$

B_L = viscous friction from load, N-m-s/radian

B_m = viscous friction of motor, N-m-s/radian

B_T = friction from motor and load, N-m-s/radian

E_a = armature voltage, V

J_L = moment of inertia of load, N-m-s^2/radian

J_m = moment of inertia of motor, N-m-s^2/radian

J_T = inertia of motor and load, N-m-s^2/radian

K_c = current driven gain, radian/A-s

K_E = voltage constant, V-s/radian

K_s = simplified gain, radian/V-s

K_T = torque constant, N-m/A

L = armature inductance, H

N_1 = number of teeth in input gear

N_2 = number of teeth in output gear

R = armature resistance, Ω

s = frequency parameter, 1/s

τ_e = electrical time constant = L/R, s

τ_m = mechanical time constant = J_m/B_m, s

τ_s = simplified time constant, s

θ_L = position of load, radian

θ_m = position of motor, radian

Ω_L = speed of load, radian/s

Ω_m = speed of motor, radian/s

Torque $T = J(d\omega/dt)$

7.3 INDUCTION, SYNCHRONOUS AND SERVOMOTORS

AC motors dominate control system applications that require a single operating speed. Single-phase motors are used where power requirement is limited to 20 hp. Poly-phase motors extend the range to substantially higher horsepower ratings. Although DC motors are widely used in variable speed applications, improvements in solid state switching components and integrated circuits have made variable speed AC drives competitive.

AC motors may be split into single-phase and poly-phase groups, and further sub-divided into synchronous and induction categories, bringing to a total of four major groups.

The single-phase squirrel cage induction motor, the most widely used type, does not have brushes to wear out or to generate sparks. The simple construction of the motor has low initial and maintenance costs. Resembling a squirrel cage, the rotor winding is in the form of copper or aluminium conductors fitted in slots in the rotor and connected at the ends by continuous rings. The stator has one or two windings that are externally excited. The rotor winding is shorted, and obtains energy through electro-magnetic induction. The induction motor functions as a rotating transformer in which the stator winding is the primary and the rotor winding is the secondary.

Unlike a DC motor, the stator winding's magnetic field is not stationary. The poles of the field alternate between two positions, in effect rotating about the rotor's axis. The speed at which the field of the induction motor rotates is called 'synchronous speed'. A motor with a two-pole stator has the synchronous speed equal to the line frequency of 60 Hz, or 3600 rpm. For a four-pole motor the synchronous speed is one-half the line frequency, or 1800 rpm. The rotor, however, runs at less than this speed, and the difference between the synchronous and actual speeds is called 'slip'. The torque developed by the motor is proportional to the slip, and even at no-load some slip occurs to overcome wind resistance and friction losses.

A single stator winding induction motor is not self-starting, since the torque produced in both directions of rotation is equal. The rotor needs to be pushed in one direction by a starting device, after which the rotor gains speed. A second stator winding rotated 90° form the main winding and called the start winding provides the starting impetus, and is the preferred method for starting a single-phase induction motor (figure 7.4). A speed sensitive switch disconnects the start circuit when the rotor reaches a preset speed.

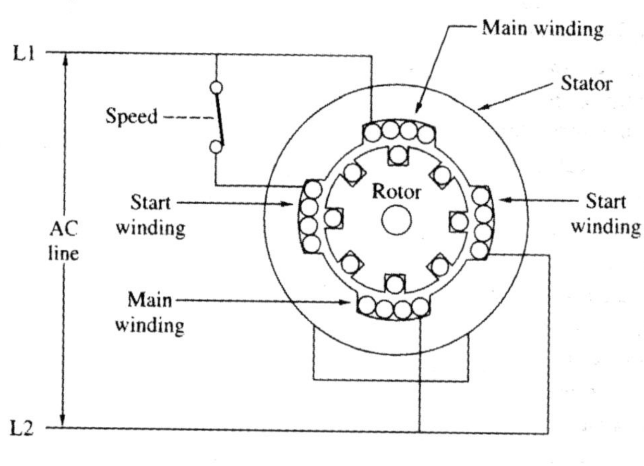

Fig. 7.4 Single-Phase Induction Motor with Start Winding

Poly-phase induction motors use three phases, and are either of the squirrel cage or of the wound rotor type. In a wound rotor the conductors are connected in series to form three windings, with the windings joined at one end in a Y connection. The other of the windings is connected to a slip ring, with one slip ring for each winding. Three brushes are in contact with the three slip rings to permit external resistances to be connected in series with the windings. The other end of the resistors is connected together in a Y connection. The resistors limit the rotor current during startup when the inertia load is large. In the absence of external resistors the wound motor is the same as a squirrel cage motor.

A two-pole three-phase induction motor's stator has three windings that are located 120° apart. The windings may be connected in either a delta or a Y connection. During each sixth of a cycle, the poles move to a new location that is rotated through 60°. The rotation of the field poles in six steps is smoother than the two-step rotation of the single-phase motor. An important advantage of the three-phase induction motor is that it is self-starting, eliminating the need for a starting device. A four-pole three-phase motor has six windings, located at 60° intervals. Each coil uses stator slots spaced 90° apart.

Synchronous motors are distinguished by their precise constant speed of operation. Ranging in size from fractional to several thousand horsepower, the motors are designed for single-phase and poly-phase operation. The stator winding of synchronous motors is mostly identical to that of the corresponding single or poly-phase induction motors. The rotor has fixed magnetic poles that lock into step with the rotating poles in the stator. The rotor may not be excited or may be excited by direct current.

Synchronous motors are not self-starting, so two methods are used for starting. A separate prime mover starts the motor and brings the motor to nearly synchronous speed, and when the rotor locks to the proper speed, the prime mover is disconnected. The other method calls for a squirrel cage winding to be added to the rotor. The motor starts as an induction motor and accelerates to near speed. When the motor reaches synchronous speed, the current in the squirrel cage conductor reduces to zero.

Often used in control systems that call for low power and variable speed, AC servomotors have an advantage over DC motors because of their ability to use the AC output of synchros, LVDT's, and other measuring units without demodulating the error signal. The AC servomotor is a modified two-phase reversible induction motor for servo operations. Amplified AC error signal, which is either in or out of phase, is applied to the control field. A constant AC voltage is applied to the reference field through a 90° phase shifting network. The two voltages are: $e_c = V_c \times \cos \omega t$ and $e_R = A \times \sin \omega t$, where e_c = control field voltage, e_R = reference field voltage, ω = operating frequency, V_c = variable amplitude of control voltage, and A = constant amplitude of reference voltage. A negative error signal yields a negative value of V_c, which is interpreted as phase shift of 180° in e_c. Linearized operating characteristics of an AC servomotor are shown in figure 7.5.

The position transfer function for an AC servomotor is given by the expression: $(\theta/V_c) = [K/s(1 + \tau s)]$ radians/V, and the velocity transfer function is given by the expression: $(\Omega/V_c) = [K/(1 + \tau s)]$ radians/s-V, where V_c = control voltage amplitude, Ω = motor speed, radians/s, θ = motor position, radians, $K = K_1/(b + K_2)$, radians/s-V, $\tau = J/(b + K_2)$, s, K_1 = stall torque/rated voltage, N-m/V, K_2 = stall torque/no-load speed at rated voltage, N-m-s/radian, J = load moment of inertia, N-m-s^2/radian, and b = load damping resistance, N-m-s/radian.

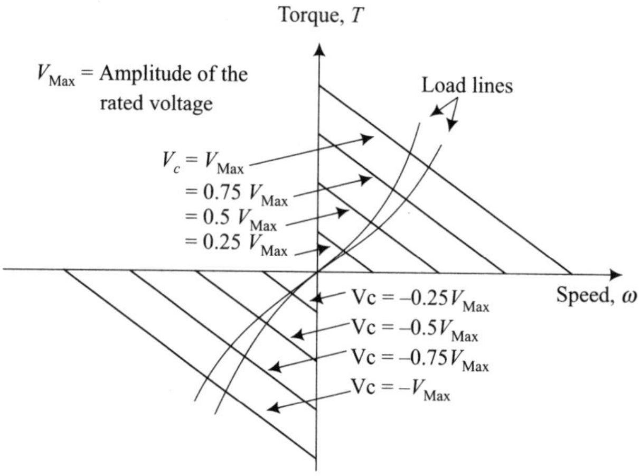

Fig. 7.5 Linearized Operating Characteristics of AC Servomotor

7.4 HYDRAULIC CYLINDER

The pressure energy of the hydraulic fluid acting on the piston is converted to mechanical energy in hydraulic cylinders. The force generated is derived either externally through the rod, or by keeping the rod fixed, through the cylinder. Single-acting cylinders are used for loads in a single direction, with a retracting spring returning the piston back to the start position at the end of the cycle. Double-acting cylinders are for two directional loads (figure 7.6). Telescopic cylinders are used for larger strokes in proportion to the axial length.

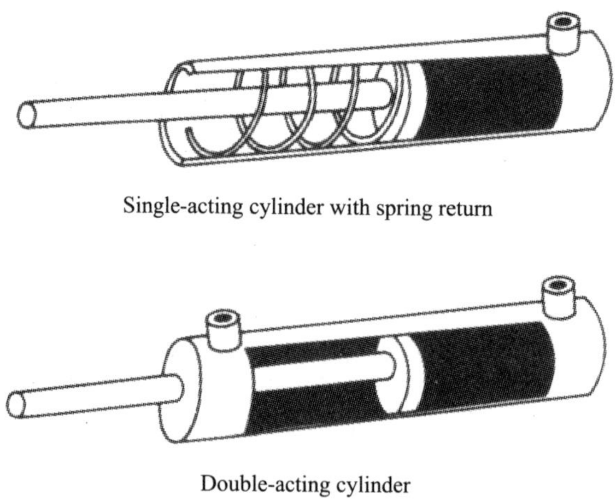

Fig. 7.6 Single and Double-Acting Hydraulic Cylinders

In the design of hydraulic cylinders, sealing and lubrication of the sliding parts, strength of the components in relation to the internal pressure, capacity to absorb shock and rust prevention must be considered. Non-contacting seals or hydrostatic bearings may be needed to reduce coulomb friction in units where high-pressure characteristics are called for.

Thrust of the hydraulic cylinder, $F_c = \Delta p A_c \lambda$, where Δp = pressure differential, A_c = internal cylinder area, and λ = thrust efficiency of hydraulic cylinder, which varies with construction features, friction characteristics of packing seals, size and operating pressure.

Seamless steel tubes are frequently used to meet pressure requirements of hydraulic cylinders. Internal diameter of the cylinders is stipulated in ISO 3320. Wall thickness, t, is calculated from the equation: $t = pD/(2S)$, where p = nominal pressure, MPa, D = inner diameter, $S = \sigma/5$, N/mm^2, and σ = minimum material tensile strength, N/mm^2. To prevent wear on the cylinder's inner surface, the piston is made from cast iron or copper alloys. Packing is mostly used to obtain piston sealing. Besides tensile and compression loads, buckling capacity of the piston rod must be determined. The rod must maintain a good rubbing surface.

The consequences of extremely high working pressures in hydraulic cylinders are wear and scuffing in the piston or the rod. Imposition of bending loads in the rod is chiefly responsible for a number of failures, hence aligning the direction of the external reacting force with the axis of the cylinder is of significance. Furthermore, reacting loads on the mount bolts must avoid shear. Leakage of oil may be prevented by ensuring absence of scratch marks on the sliding surfaces and maintaining purity of the oil. Non-metallic seals must be periodically checked and replaced.

It may not be possible to avoid mixing of air in hydraulic circuits, due to which air can also get into the cylinder. Accumulation of air leads to jerking motion, sticking, slipping and vibrations. Air must be continuously purged during operation to prevent the undesirable effects. Since air bubbles move upwards due to buoyancy, provision of a port where they collect helps in purging the air.

7.5 HYDRAULIC MOTOR

Hydraulic motors are required to produce large torque in a control element or load. The motors may be divided into four common types: piston motor, gear motor, vane motor and turbine motor. Gear and vane motors are more suitable for small to medium displacement, while piston motors are for larger displacement capacity.

External gear motors have a set of gears housed inside a two-piece casing. Fluid entering the space between the meshing gears causes the two gears to rotate in opposite directions. The gears transport the oil from the inlet to the outlet port. One of the gears delivers the power output. Internal gear hydraulic motors have an internal gear with circular teeth meshing with an external gear having one more tooth (figure 7.7). Keeping the internal gear stationary permits the external gear to walk around the internal gear. The design provides combined features of a hydraulic motor and a speed reducer.

In a vane motor, torque is produced from the action of pressurized oil on a number of radially sliding vanes fitted in slots on the rotor. The vanes are forced out towards the inner surface of the casing by spring action and by centrifugal force as the rotor spins at high speed. Friction and wear related problems occur mostly at vane tips. Instability may be encountered at low speeds, and the motors also suffer from low starting torque.

Radial piston motors have the pistons arranged around a drive shaft (figure 7.8). Fluid entering the cylinder forces the piston against an eccentric thrust ring attached to the shaft, which causes the rotor to spin. In axial piston motors, the axes of the cylinders run parallel with the cylinder axis (figure 7.9). A wobble plate attached to the rotor is made to rotate by the normal component of the reaction thrust that is produced from the motion of the piston.

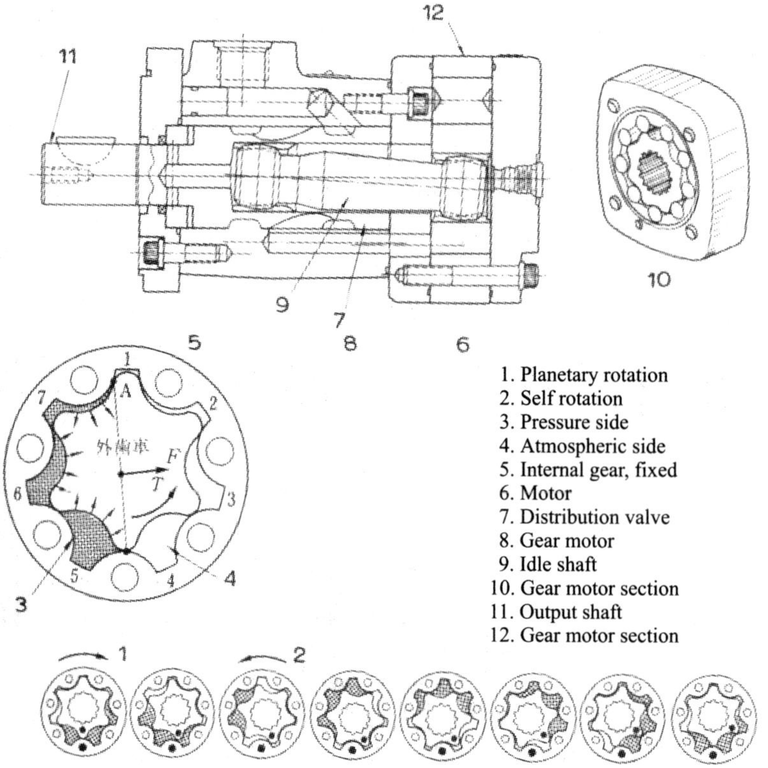

Fig. 7.7 Hydraulic Motor with Internal Gears

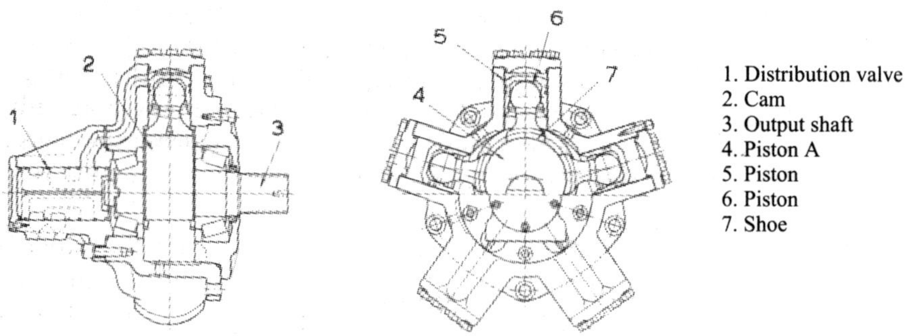

Fig. 7.8 Radial Piston Hydraulic Motor (*Courtesy*: Kawasaki Heavy Industries)

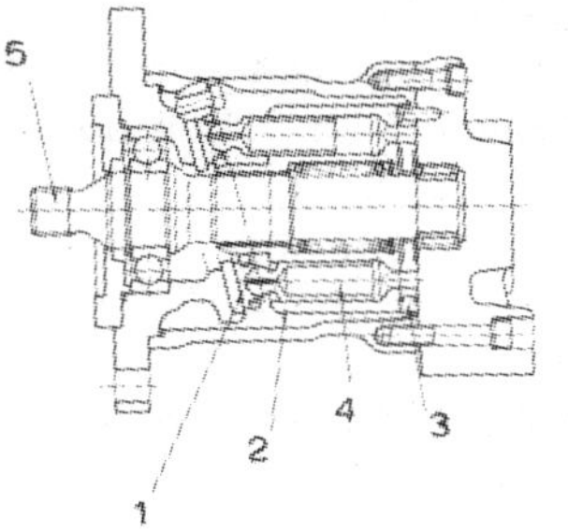

1. Piston shoe
2. Cylinder block
3. Valve plate
4. Piston
5. Output shaft

Fig. 7.9 Axial Piston Hydraulic Motor (*Courtesy*: Kawasaki Heavy Industries)

Turbine motors have a low starting torque, but operate at high speeds. The motor has a number of rotating blade sets attached to the rotor, separated by rows of stationary blades. Fluid entering the motor flows through the canted turbine blades and is deflected, which imparts a tangential force that spins the rotor. The stationary blades redirect the flow of the fluid towards the next row of rotating blades.

7.6 PNEUMATIC CYLINDER

Pneumatic systems are more suitable for light and medium loads than hydraulic systems. Two-stage reciprocating compressors are capable of delivering air above 1000 kPa, and the air may be readily stored in pressure tanks for regular production operation or for emergency operations during power failures.

Pneumatic actuators lend themselves to easier control of power and operation with the aid of pressure and flow control valves and directional valves in response to load demand. Compressibility characteristics of air make it difficult to obtain good accuracy in positioning and controlling speed, and operating efficiency is also relatively low. Their performance, however, may be improved by adding proportional valves in the system.

Pneumatic cylinders have a considerably simple construction, and are widely used for converting the energy of the compressed air into mechanical reciprocating form of motion. Cylinder designs vary with operational mode (single acting, double acting, or differential), form of piston (piston or ram), and method of positioning (multi-position, brake-attached, positioner). A double-acting pneumatic cylinder assembly is shown in figure 7.10.

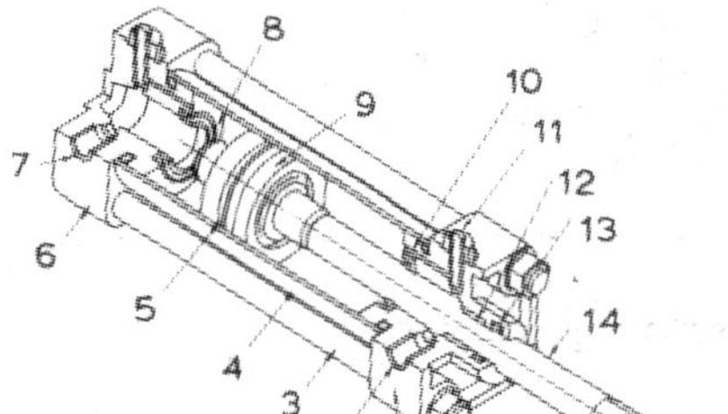

1. Rod cover
2. Port
3. Tie rod
4. Cylinder
5. Piston packing
6. Head cover
7. Port
8. Cushion packing
9. Piston
10. Cushion throttle
11. Cushion valve
12. Rod bush
13. Rod packing
14. Piston rod

Fig. 7.10 Double-Acting Pneumatic Cylinder

When the pneumatic cylinder is operating at no load, the volume of air flowing in and out of the cylinder is the same. At low speed the inertia force from the moving parts is low, and the force acting on both sides of the piston will equal the applied load. When the operating speed is high, the pressure inside the cylinder increases to the supply line pressure within the starting time to overcome the applied load, while the discharge pressure is close to atmospheric level. The acceleration rate depends on the pressure differential, and is given by: $a = g/W[\mu P\pi D^2/4 - F_c]$, where a = acceleration, m/s^2, g = acceleration due to gravity, m/s^2, W = weight of load, N, μ = thrust coefficient of cylinder, P = air supply pressure, Pa, D = cylinder bore, m, and F_c = actual load on cylinder, N.

7.7 OSCILLATING ACTUATOR

Oscillating actuators are employed where the angle of rotation of the shaft is limited. As opposed to the pneumatic cylinder, the energy of the compressed air is converted into reciprocating motion in rotary actuators.

The conversion mechanisms may be broadly split into vane and piston types (figure 7.11). In the vane type, the energy of the compressed air acts on the vanes and directly transmits the power to the shaft to produce torque. The surface area of the vane, number of vanes and applied air pressure determine the output. Single, double and triple vane oscillators are commonly used. The angle of oscillations increases, but the torque developed increases with the number of vanes. Complete shielding of the airflow in vane oscillators is difficult to obtain, and some leakage is inevitable. But the units are compact and have mechanically reliable functions.

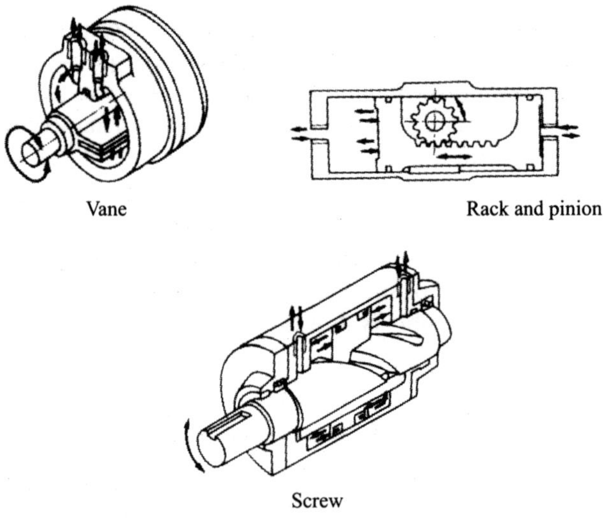

Fig. 7.11 Oscillating Actuators

Inside the piston of oscillating actuators, the linear motion of the pneumatic cylinder's piston is converted into rotary motion through screws or gears. With the rack and pinion arrangement, turning the pinion that is directly attached to the piston develops the required torque. Internal cushioning may be built in the cylinder. A small increase in friction is encountered due to the lateral thrust on the pinion and the piston, but the efficiency is still between 80 and 90 per cent. In screw type of oscillators the linear motion of the piston is converted into rotary motion through a screw mounted on the output shaft. The piston is provided with a device to prevent it from rotating. The rotary angle increases with the stroke, and may exceed 360°. Efficiency of the actuator is about 80 per cent due to the large friction losses in the anti-rotation device.

7.8 AIR MOTOR

The displacement type of air motor is designed for general industrial use, where only the pressure energy of the compressed air is utilized. Where high levels of rotation are the norm, velocity type of air motors are more suitable. Both the pressure energy and the velocity of the flowing air are put to use in velocity type of air motors.

One form of construction calls for vanes inserted in radial rotor slots, and the rotor is placed in an eccentric casing. The vanes slide inside the slots either by springs or by air pressure. The gear type of air motor employs double helical gears for expansion of the air to provide a low-speed, high-torque unit.

In piston type of motors, compressed air acts on the piston face to provide power through an inclined plate, cam and crankshaft. Figure 7.12 shows a radial piston air motor with between 3 and 6 cylinders arranged radially. Each piston has a connecting rod that links it to a crankshaft. The pistons operate through sequential supply and discharge of compressed air with the aid of a rotary distribution valve that is connected to the crankshaft. The reciprocating components are dynamically balanced and the outputs of the cylinders overlap, so vibrations are relatively low. Provision must be made for lubrication. Switching the connections to the supply and discharge ports and adjusting the distribution valve timing will reverse the rotation direction. Axial piston type of air motor generally has between 4 and 6 cylinders arranged concentrically around the output shaft. An inclined wobble plate is connected to the piston rod, which rotates the output shaft through a gear mechanism.

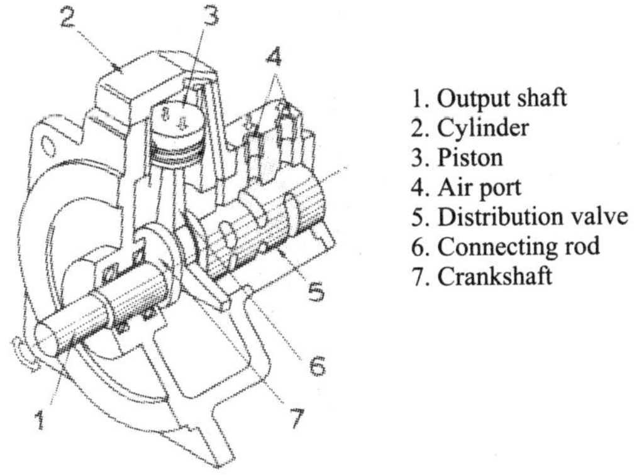

Fig. 7.12 Radial Piston Type of Air Motor

7.9 ACTUATORS EMPLOYING FUNCTIONAL MATERIALS

Mechanical conversion of energy through hydraulic, pneumatic and electrical media are not capable of providing precise micro-motions, and cannot be adapted to miniaturized sizes for use where space is restricted. Alternative methods must be considered to make up these shortcomings. Newer ideas look for the raw material of the actuator itself to posses the requisite energy conversion mechanism.

Certain dielectric substances such as solid-state crystals display the piezo-electric property of developing minute amounts of electricity when subjected to a mechanical force. Also, if an electrical field is added to a piezo-electric crystal, mechanical distortion is produced. This is the reversible effect of the piezo-electric phenomenon. Piezo-electric elements have long been in use as flexible elements in audio speakers and pick-ups. Advances in multi-layer construction of piezo-electric elements are recognized as actuators with micro-motion capabilities.

In a piezo-electric element actuator, varying the applied voltage controls mechanical displacement. Four forms of operational modalities occur. Longitudinal and lateral effects occur when the electric field is impressed in the polarized direction of the piezo-electric element, with the element stretching in the polarized direction and contracting in the perpendicular direction. The relation between voltage sensitivity, g, and electric load sensitivity, d, is expressed as $g = d/\varepsilon$, where ε is the dielectric efficiency. The longitudinal effect is twice as sensitive as the lateral effect.

For actuators made from built-up layers, materials possessing a high level of piezo-electric effect are used. Such crystals are made from barium-titanium oxide ($BaTiO_3$), lead-titanium-zircon oxide [$Pb(Zr,Ti)O_3$], and lead-titanium oxide ($PbTiO_3$). The longitudinal effects in lead-zircon-titanium (PZT) oxide ceramics receive a special mention.

When these materials are used as an ultra-sonic vibrator, a high power output can be obtained. The technique is used for ultra-sonic welding and cleaning. The electro-mechanical conversion efficiency in a laminated nickel plate oscillator, for instance, reaches over 90 per cent. Referred to as ultra-sonic motors, the actuators convert the oscillatory motion to rotary or linear motion.

An elastic body suspended in an ultra-sonic region will experience elastic waves propagating on its surface in the longitudinal and lateral directions. A point on the surface will travel on the surface along an elliptical track at a velocity that is proportional to the frequency.

Alloys exhibiting shape-memory effects are made from transformed thermo-elastic martensite. For use as an actuator, the martensite transformation should be crystallographically reversible and slippage must be absent during the shape change process. To function as an actuator, the shape-memory effects may be during heating only, or for both heating and cooling. The shape and the electrical resistance change during heating and cooling. In feedback systems, a displacement transducer is required for positioning. Being small and simple to operate, shape memory alloy actuators hold the potential for performing all the functions of a conventional actuator.

7.10 REVIEW PROBLEMS

Example problem # 7.1: An AC servomotor has the following specifications:

Rated voltage = 110 volts

Load inertia = 7.35×10^{-6} N-m-s^2/radian

Load damping $= 2.4 \times 10^{-5}$ N-m-s/radian

Stall torque $= 0.035$ N-m at 110 volt

No-load speed $= 3900$ rpm at 110 volt

Determine the motor's velocity and position transfer function.

Solution: At rated voltage and stall torque:

Constant $K_1 = 0.035$ N-m/110 V $= 3.18 \times 10^{-4}$ N-m/volt

No-load speed $= 3900 \times 2\pi/60 = 408.41$ radian/s

Constant $K_2 = 0.035$ N-m/408.41 radian/s $= 8.57 \times 10^{-5}$ N-m-s/radian

Solving for K and τ: $b + K_2 = 2.4 \times 10^{-5} + 8.57 \times 10^{-5} = 10.97 \times 10^{-5}$

Hence: $K = K_1/(b + K_2) = 3.18 \times 10^{-4}/10.97 \times 10^{-5} = 2.90$ radian/s-volt

And: $\tau = J/(b + K_2) = 7.35 \times 10^{-6}/10.97 \times 10^{-5} = 0.067$·s

Velocity transfer function:

$$(\Omega/V_c) = [K/(1 + \tau s)]$$
$$= 2.90/(1 + 0.067s) \text{ radian/s-volt}$$

Position transfer function:

$$(\theta/V_c) = [K/s(1 + \tau s)]$$
$$= 2.90/[s(1 + 0.067s)] \text{ radian/volt}$$

Example problem # 7.2: An armature controlled DC motor has friction torque $T_f = 0.014$ N-m, torque constant $K_T = 0.075$ N-m/A, maximum current $I_{max} = 2.5$ A, voltage constant $K_E = 0.08$ volt-s/radian, maximum motor speed $\omega_{max} = 475$ radian/s, and armature resistance $R = 1.4\ \Omega$. Find: (*i*) maximum output torque T_{max}, (*ii*) maximum power output P_{max}, (*iii*) maximum armature voltage E_{max}, and (*iv*) no-load motor speed.

Solution:

(*i*) Maximum output torque occurs when current is maximum.

$$T_{max} = K_T \times I_{max} - T_f = 0.075 \times 2.5 - 0.014 = 0.1735 \text{ N-m}$$

(*ii*) Maximum power output occurs when speed and torque are maximum.

$$P_{max} = \omega_{max} \times T_{max} = 475 \times 0.1735 = 82.41 \text{ W}$$

(*iii*) Maximum armature voltage occurs when current and speed are maximum.

$$E_{max} = I_{max} \times R + K_E \times \omega_{max} = 2.5 \times 1.4 + 0.08 \times 475 = 41.5 \text{ volt}$$

(*iv*) No-load motor speed occurs when $e = E_{max}$.

$$\omega_{NL} = (E_{max} \times K_T - R \times T_f)/(K_E \times K_T)$$
$$= (41.5 \times 0.075 - 1.4 \times 0.014)/(0.08 \times 0.075) = 515.5 \text{ radian/s.}$$

Example problem # 7.3: In example problem # 7.2 assume the motor runs at 375 radian/s at a load torque of 0.065 N-m. Calculate: (*i*) armature voltage and (*ii*) armature speed if torque rises to 0.08 N-m at the same armature voltage.

Solution:

(*i*) Torque $T = K_T \times i - T_f$

Armature current $i_a = (T + T_f)/K_T = (0.065 + 0.014)/0.075 = 1.053$ A

Back emf $e_b = K_E \times \omega_m = 0.08 \times 375 = 30$ volt

Armature voltage $e_a = i_a \times R + e_b = 1.053 \times 1.4 + 30 = 31.48$ V

(*ii*) $K_R = R/(K_E \times K_T) = 1.4/(0.08 \times 0.075) = 233.3$

$\omega_{NL} = (\omega_1 + K_R \times T_1)$ and $\omega_2 = (-K_R \times T_2 + \omega_1 + K_R \times T_1)$

Hence:

$$\omega_2 = (\omega_1 - K_R \times T_2 + K_R \times T_1) = 375 - 233.3 \times (0.08 - 0.065)$$
$$= 371.5 \text{ radian/s.}$$

Example problem # 7.4: A permanent magnet DC motor has following specifications:

Maximum operating speed ω_{max}	575 radian/s
Maximum armature current I_{max}	2.75 A
Voltage constant K_E	0.074 volt-s/radian
Torque constant K_T	0.064 N-m/A
Friction torque T_f	0.14 N-m
Armature resistance R	1.45 Ω
Armature inductance L	0.015 H
Armature moment of inertia J_m	7.5×10^{-4} N-m-s^2/radian
Armature viscous friction B_m	1.25×10^{-4} N-m-s/radian

Calculate the velocity and position transfer function of the voltage driven motor.

Solution: Mechanical time constant $\tau_m = J_m/B_m = (7.5 \times 10^{-4})/(1.25 \times 10^{-4}) = 6.0$ s

Electrical time constant $\tau_e = L/R = 0.015/1.45 = 0.0103$ s

$$RB_m = 1.45 \times 1.25 \times 10^{-4} = 1.81 \times 10^{-4}$$

$$RB_m + K_E \times K_T = 1.81 \times 10^{-4} + 0.074 \times 0.064 = 0.0049$$

$$RB_m \times (\tau_m + \tau_e) = 1.81 \times 10^{-4} \times (6.0 + 0.0103) = 0.00109$$

$$RB_m\tau_m\tau_e = 1.81 \times 10^{-4} \times 6.0 \times 0.0103 = 1.12 \times 10^{-5}$$

Voltage driven velocity transfer function:

$$\Omega_m(s)/E_a(s) = K_T/[RB_m + K_E \times K_T + RB_m \times (\tau_m + \tau_e)s + RB_m\tau_m\tau_e s^2]$$
$$= 0.064/[0.0049 + 0.00109s + 1.12 \times 10^{-5}s^2] \text{ radian/volt-s}$$
$$= 13.06/[1 + 0.222s + 0.00229s^2] \text{ radian/volt-s}$$

Voltage driven position transfer function:

$$\theta_m(s)/E_a(s) = (1/s) \times [\Omega_m(s)/E_a(s)]$$
$$= 13.06/[s + 0.222s^2 + 0.00229s^3] \text{ radian/volt-s}$$

Example problem # 7.5: If τ_e is neglected in example problem # 7.4, find the simplified velocity and position transfer function of the voltage driven motor.

Solution: Simplified velocity transfer function of a voltage driven motor is:

$$\Omega_m(s)/E_a(s) = K_s/(1 + \tau_s s),$$

where

$$K_s = K_T/(RB_m + K_E \times K_T) = 0.064/0.0049 = 13.06 \text{ radian/volt-s}$$

$$\tau_s = RJ_m/(RB_m + K_E \times K_T) = 1.45 \times 7.5 \times 10^{-4}/0.0049 = 0.222 \text{ s}$$

Then:

$$\Omega_m(s)/E_a(s) = 13.06/(1 + 0.222s)$$

Simplified position transfer function is:

$$\theta_m(s)/E_a(s) = (1/s) \times [\Omega_m(s)/E_a(s)]$$

$$= 13.06/(s + 0.222s^2)$$

Example problem # 7.6: Find the velocity and position transfer functions if the motor in example problem # 7.4 is current driven.

Solution: Velocity transfer function of a current driven motor is:

$$\Omega_m(s)/I_a(s) = K_c/(1 + \tau_m s),$$

where

$$K_c = K_T/B_m = 0.064/(1.25 \times 10^{-4}) = 512.0$$

$$\tau_m = J_m/B_m = (7.5 \times 10^{-4})/(1.25 \times 10^{-4}) = 6.0 \text{ s}$$

Then:

$$\Omega_m(s)/I_a(s) = 512.0/(1 + 6.0s) \text{ radian/A-s}$$

Position transfer function of a current driven motor is:

$$\theta_m(s)/I_a(s) = (1/s) \times [\Omega_m(s)/I_a(s)]$$

$$= 512.0/(s + 6.0s^2)$$

Example problem # 7.7: Find the velocity and position transfer functions if the voltage driven motor in example problem # 7.4 has the following speed reducer and load specifications:

$N_1 = 18$ teeth, $N_2 = 36$ teeth,

$J_L = 7.75 \times 10^{-4}$ N-M-s^2/radian, $B_m = 0.25 \times 10^{-5}$ N-M-s/radian.

Solution:

$$K_T \times (N_1/N_2) = 0.064 \times (18/36) = 0.032$$

$$J_T = J_m + (N_1/N_2)^2 \times J_L = 7.5 \times 10^{-4} + (18/36)^2 \times 7.75 \times 10^{-4}$$

$$= 9.44 \times 10^{-4} \text{ N-m-s}^2\text{/radian}$$

$$B_T = B_m + (N_1/N_2)^2 \times B_L = 1.25 \times 10^{-4} + (18/36)^2 \times 0.25 \times 10^{-5}$$

$$= 1.26 \times 10^{-4} \text{ N-m-s}^2\text{/radian}$$

$$\tau_T = J_T/B_T = (9.44 \times 10^{-4})/(1.26 \times 10^{-4}) = 7.51 \text{ s}$$

$$RB_T = 1.45 \times 1.26 \times 10^{-4} = 1.827 \times 10^{-4}$$

$$RB_T + K_E \times K_T = 1.827 \times 10^{-4} + 0.074 \times 0.064 = 0.004919$$

$$RB_T \times (\tau_T + \tau_e) = 1.827 \times 10^{-4} \times (7.51 + 0.0103) = 0.001374$$

$$RB_T \tau_T \tau_e = 1.827 \times 10^{-4} \times 7.51 \times 0.0103 = 1.41 \times 10^{-5}$$

Velocity transfer function:

$$\Omega_L(s)/E_a(s) = K_T \times (N_1/N_2)/[RB_T + K_E \times K_T + RB_T \times (\tau_T + \tau_e)s + RB_T\tau_T\tau_e s^2]$$
$$= 0.032/[0.004919 + 0.001374s + 1.41 \times 10^{-5}s^2] \text{ radian/volt-s}$$
$$= 6.51/[1 + 0.279s + 0.00287s^2] \text{ radian/volt-s}$$

Position transfer function:

$$\theta_L(s)/E_a(s) = (1/s) \times [\Omega_L(s)/E_a(s)]$$
$$= 6.51/[s + 0.279s^2 + 0.00287s^3] \text{ radian/volt-s}$$

Example problem # 7.8: An industrial process system requires a workpiece to be moved through a distance of 8.0 in. in 13.5 s with a force of 8500 lbs. A hydraulic cylinder of 4.25 in. diameter capable of exerting a force of 38500 lbs at a working pressure of 3000 lbs/in^2 is under consideration for the task. Operating and related system conditions, however, limit the working pressure to 1200 lbs/in^2. The 4.25 in. cylinder has a speed of 16.8 in./min for a flow rate of 1 gallon/minute. If the available hydraulic oil supply is 2.2 gallon/minute, determine the suitability of the cylinder.

Solution: Operating speed for a distance of 8.0 in. in 13.5 s is 8.0 × 60/13.5 = 35.56 in. per minute. Using an overload factor of 25 per cent, the required force is 8500 × 1.25 = 10625 lbs.

At 3000 lbs/in^2 working pressure the force is 38500 lbs. Then at 1200 lbs/in^2 working pressure, the available force is (1200/3000) × 38500 = 15400 lbs.

The required oil flow rate for producing the speed of 35.56 in./minute is determined to be 1 × (35.56/16.8) = 2.12 gallon/minute.

Thus, both available oil flow rate and force are within the capability of the given hydraulic cylinder.

Example problem # 7.9: If in example problem # 7.8 a 5.0 in. diameter hydraulic cylinder with a force capability of 57500 lbs is used, what is the working pressure of the oil?

Solution: For a 5.0 in. diameter cylinder the working pressure is 3000 × (10625/57500) = 554 lbs/in^2.

Example problem # 7.10: If in example problem # 7.8 a 5.0 in. cylinder needs a flow rate of 12.25 gallon/minute for full capacity force, determine the required flow rate for this application.

Solution: For a 5.0 in. diameter cylinder the required flow rate is 1 × (35.56/12.25) = 2.90 gallon/minute.

7.11 EXERCISES

7.1 An AC servomotor has the following specifications:

Rated voltage = 220 volts
Load inertia = 7.05×10^{-6} N-m-s^2/radian
Load damping = 2.475×10^{-5} N-m-s/radian
Stall torque = 0.042 N-m at 220 volt
No-load speed = 4000 rpm at 220 volt

Determine the motor's velocity and position transfer function.

7.2 An armature controlled DC motor has friction torque T_f = 0.0175 N-m, torque constant K_T = 0.0625 N-m/A, maximum current I_{max} = 2.75 A, voltage constant K_E = 0.11 volt-s/radian, maximum motor speed ω_{max} = 525 radian/s, and armature resistance R = 1.25 Ω. Find the maximum output torque T_{max}.

7.3 In problem # 7.2 determine the maximum power output P_{max}.

7.4 In problem # 7.2 determine the maximum armature voltage E_{max} and the no-load motor speed.

7.5 In problem # 7.2 assume the motor runs at 475 radian/s at a load torque of 0.072 N-m. Calculate: (*i*) armature voltage and (*ii*) armature speed if torque rises to 0.11 N-m at the same armature voltage.

7.6 A permanent magnet DC motor has following specifications:

Maximum operating speed ω_{max}	4255 radian/s
Maximum armature current I_{max}	3.25 A
Voltage constant K_E	0.078 volt-s/radian
Torque constant K_T	0.084 N-m/A
Friction torque T_f	0.17 N-m
Armature resistance R	1.30 Ω
Armature inductance L	0.0175 H
Armature moment of inertia J_m	5.25×10^{-4} N-m-s²/radian
Armature viscous friction B_m	1.135×10^{-4} N-m-s/radian

Calculate the velocity and position transfer function of the voltage driven motor.

7.7 Neglect τ_e in problem # 7.6, and find the simplified velocity and position transfer function of the voltage driven motor.

7.8 Find the velocity and position transfer functions if the motor in problem # 7.6 is current driven.

7.9 Find the velocity and position transfer functions if the voltage driven motor in problem # 7.6 has the following speed reducer and load specifications:

N_1 = 24 teeth, N_2 = 36 teeth,

$J_L = 2.75 \times 10^{-4}$ N-m-s²/radian, $B_m = 0.75 \times 10^{-5}$ N-m-s/radian.

7.10 A manufacturing process requires a workpiece to be moved through a distance of 9.0 in. in 12.0 s with a force of 8750 lbs. A hydraulic cylinder of 4.0 in. diameter capable of exerting a force of 37000 lbs at a working pressure of 3000 lbs/in² is under consideration for the task. Operating and related system conditions, however, limit the working pressure to 900 lbs/in². The 4.0 in. cylinder has a speed of 18.0 in./min for a flow rate of 1 gallon/minute. Calculate the working pressure of the oil and the required flow rate of the oil.

References & Bibliography

Bateson, R.N., "Motomatic Servo Control Course Manual", Electro-Craft Corporation, Hopkins, MN, 1971.

Bower, J.L. and Schultheiss, P.M., "Introduction to Design of Servo-Mechanisms", John Wiley & Sons, New York, NY, 1964.

Herman, S.L. and Alerich, W.N., "Industrial Motor Control", Delmar Publishers, Inc. Albany, NY, 1985.

Ichikawa, T., "Hydraulic Engineering", Asakura Shoten, 1979.

Kozoh, O., "Development of Multi-Cam High Torque Hydraulic Motor", Hitachi-Hyoron, Volume 56, Number 11, 1974.

Morris, N.M., "Control Engineering", McGraw-Hill Book Co., (UK) Ltd., 1968.

Reliance Motor Control, "Electro-Craft Handbook – Brushless Servo Systems", Racine, WI, 1992.

Shikano, M., "Application Technology on Control Use of Actuators", Encyclopaedia of Factory Automation, Volume 6, 1983.

Shimazu Manufacturing Co., "Shimazu Gear Motor", Shimazu Catalogue.

Takenaka, T., "Hydraulic Engineering", Yokendo, 1970.

Tuji, S., "Pneumatic Engineering", Asakura Shoten, 1973.

Zeines, B., "Automatic Control Systems", Prentice-Hall, Inc., Englewood Cliffs, NJ, 1972.

8

Programmable Logic Controller

8.1 INTRODUCTION

A programmable logic controller (PLC) is a special-purpose computer dedicated to controlling the operation of a machine and processes. The digital electronic device provides the control logic for industrial and manufacturing machinery, equipment or for complete processes. An entire manufacturing plant may also be controlled through a single master PLC. Specialized programming skills are not essential for writing the programs that provide detailed instructions for every aspect during the entire sequence of operations. Shop personnel can write program instructions into the computer system.

The turbo-machinery industry has played a leading role in originating and refining the PLC. Design criteria for the first logic controllers were initially developed in the 1950's and 1960's. The objective was to eliminate replacing the inflexible and hard-wired relay panels with a computer that can survive in the industrial environment. Initially the units were little more than elaborate switches, but their capabilities and sophistication grew rapidly.

From its modest beginnings, the modern PLC has developed into far greater than a substitute for relay logic panels. Individualized modules that execute a host of functions such as logic, timing, counting, sequencing and proportional-integrating-derivative (PID) control are included in the PLC's. The computers have the added advantage of carrying out arithmetical calculations, analyse data, and exchange data with other programmable controllers and host computers.

PLC's have expanded to include hierarchical systems for integrated manufacturing. A supervisory computer communicates with individual PLC's to coordinate their activities. The intent is to reduce the complexity in the computer program that is required to control a centralized system with one large PLC. Splitting the process into smaller and more manageable portions simplifies the software, and also permits use of smaller PLCs.

8.2 PLC ARCHITECTURE

PLCs range in size from small units performing the job of eight relays to large units for accommodating thousands of inputs and outputs. The controllers are assembled from a number of essential and some special-purpose components. The necessary components are the power supply, processor module, input and output modules and programming unit. The PID controller and data communicator unit are found only in certain PLC designs. Figure 8.1 illustrates the basic components.

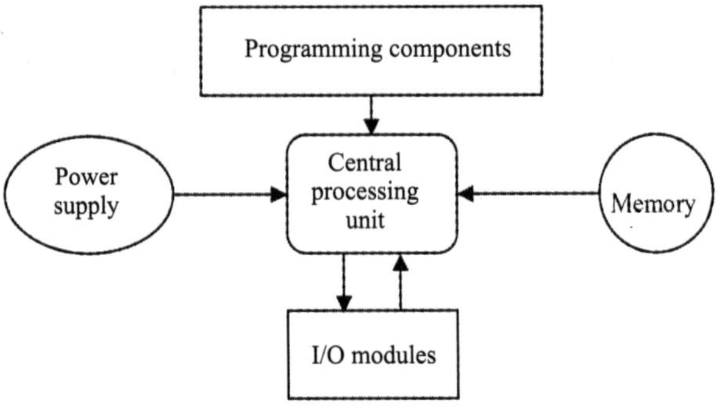

Fig. 8.1 Programmable Logic Controller Architecture

A PLC needs two power supplies. An external power supply is connected to a 120 V or a 240 V AC for the input devices and output loads. The other power source is an internal one for the central processor unit.

Most input devices are of the discrete type for bringing in data from pushbutton contacts and limit and pressure switches. The input and output modules may have up to 32 input terminals, plus a common and a ground terminal. Each input data is converted into a logic voltage. Input of analog signals is also used for control and data acquisition. The analog input module converts the data to a digital form with an analog-to-digital converter. Discrete output devices provide on or off signals for lamps, relays, solenoids and motor starters, among others. Analog outputs require 8 to 16 bits from the output section of the memory. A digital-to-analog converter converts the binary number into an analog voltage.

Programming of instructions may be done with a hand-held programmer, CRT terminal and personal computer. The programming unit allows the operator to examine, enter and edit programs residing in the memory. It also lets monitoring the status of input and output commands, display data and counter values.

8.3 CENTRAL PROCESSING UNIT

The central processing unit, or the CPU, forms the core portion of the PLC. The microprocessor-based unit contains the logic and control algorithm for the controller.

The CPU scans the total information package stored in the memory and the input and output devices continuously. During the scan the CPU also executes instructions based on input data, sends appropriate output responses to the output devices, updates data acquisition systems, and indicates condition changes.

Scan time for larger units depends on the size of the memory and on the configuration of the system. Smaller PLC's have longer scan times than larger units, and have less memory and reduced capacity for special instructions and peripheral equipment.

8.4 MEMORY

The memory in a PLC stores the digital control logic, the process program and the necessary instructions to operate the system. Stored in units called 'bytes', the memory may retain the information (called 'non-volatile memory') or may lose it when electrical power is turned off (called 'volatile memory'). User interface with the PLC mostly works through the volatile memory.

The digital memory of modern microprocessors is classified into four categories: random access memory (or RAM), read only memory (or ROM), programmable read only memory (or PROM), and erasable programmable read only memory (or EPROM). Only the RAM has a volatile memory, which is preferred since it is easy to program, provides large memory to store the programs and data, and is easy to access. ROM memory cells store information permanently, and is used to store the operating program of the computer to permit easy startup. They may also be used for permanent and unalterable process control. PROM's are manufactured with a full array of memory cells, each of which is electronically tied to a signal grid by a fusible link. Programming requires physically melting the fusible link to disable specific memory cells. EPROM's are more convenient to use than ROM or PROM because they can be erased and reprogrammed.

8.5 INPUT AND OUTPUT MODULES

Input and output modules provide interfacing between the CPU and the devices that provide and receive electrical signals. Digital input devices include toggle switches, pushbuttons, limit switches and pressure switches to provide discrete on-off signals. Touch screens that allow the operator to control functions are widely used for process control. Analog devices such as pressure sensors, thermocouples and accelerometers provide continuously changing signals that represent the state of the system.

Some level of signal conditioning may be assigned to the input module to ensure a usable signal is available for the CPU. Circuit protection capabilities will also be necessary in the module to protect the CPU circuitry from faults, surges or spikes in the incoming signal.

The output module reads digital signals transmitted from the CPU, and operates motor starters, solenoids and status monitors. The module essentially functions as a switching device, and provides on-off signals to the associated devices. If the output module is incapable of handling the power needed by the device, it sends a signal to an external relay, which connects with the actual power circuit.

Input and output devices are placed outside the PLC, with connecting wires attached to the terminal strips on the PLC. There are no connections between the input and the output. Each input and output terminal has a unique address. Where solid-state devices are employed as input or output signals for DC circuits, it is imperative that the correct device type be selected.

8.6 PLC PROGRAMMING AND OPERATION

The LD programming language used for a PLC has many similarities with the ladder diagram for relay logic. Contact and coil symbols are used to construct the diagrams. Contact and output symbols have a number written above or below it to identify the location in the RAM process memory where its image is stored.

A number of restrictions are placed on ladder diagram programs. Typical limitations include: (*i*) output must be at the right of the rung, (*ii*) power must flow from left to right, up or down, (*iii*) power must never flow from right to left, (*iv*) number of series contacts is limited to 11, (*v*) number of parallel contacts is limited to 11, (*vi*) a rung can have only one output, (*vii*) a single output must be on the top line in a rung. Sneak paths are not permitted in programming a PLC, which simplifies the programming. But adjustments are needed when converting a relay panel to a PLC program.

Figure 8.2 shows an example of the schematic and ladder diagrams of a relay logic circuit for controlling the action of a pneumatic cylinder. When the PLC is in the operating mode, the processor repeats the four-step cycle of: (*i*) scanning and storing the new input conditions, (*ii*) scan the program and derive new output conditions, (*iii*) transfer output information to output device, and (*iv*) communicate and carry out related routine tasks.

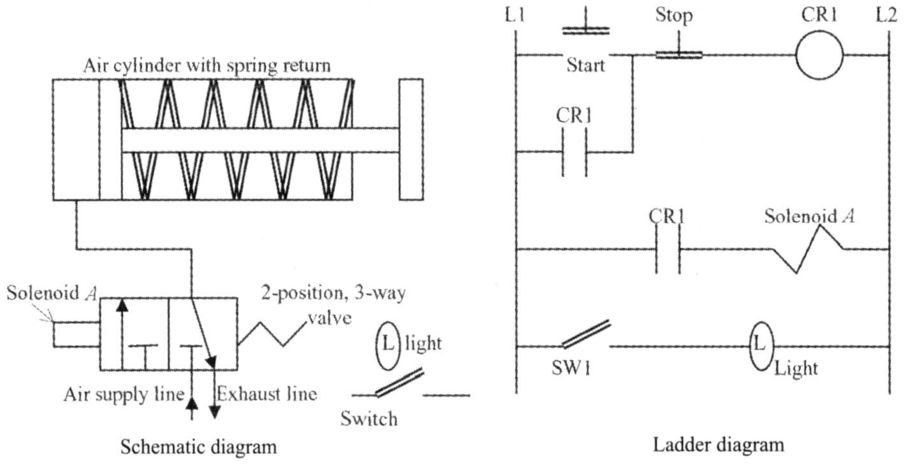

Fig. 8.2 Ladder Diagram Control Circuit of Pneumatic Cylinder for PLC Implementation

It may be observed that the order of operations is of significance, since the PLC evaluates the program one rung at a time. Also, a time delay occurs between the input action and the resulting change in output conditions, because the PLC takes time to execute one cycle.

8.7 PROGRAMMING FUNCTIONS

The open and closed contact symbols are instructions in the LD programming language. The instructions for event-driven processes with on and off inputs and outputs are the NO contact, the NC contact and the output instructions. Some control applications require the counting function, and others need the ability to compare and manipulate numbers with arithmetic functions. Control of a process variable requires the PID control function. All advanced functions are available in the PLC.

The 'timer' is frequently used in the PLC, and the 'time delay on' and 'time delay off' are the most common timing functions. Two types of timers are used, the one-input non-retentive timer and the two-input retentive timer.

The non-retentive timer has one output and a preset delay time. Thus, if the preset delay time is 12 s, the timer runs for 12 s following the turning on of the input before the output turns on. The output stays on until the input is turned off. The timer resets every time the input is turned on.

The retentive timer has two inputs (Enable/Reset and Run lines), one output (Done line), and two status bits (Enable and Run lines) that can be used as internal outputs. The timer runs only after turning on both the Enable/Reset and the Run lines, but stops when the delay time is satisfied. The timer is reset only when the Enable/Reset line is turned off.

The 'counter' of a PLC keeps track of the number of times a contact opens. The objective is to count the number of occurrences of an event in a control system. The counting sequence may run from zero to a preset value or may be from a preset value to zero. A memory register can store both the preset and the counted values. Multiple counters may be employed in a PLC if the incidence of more than one event is to be counted. For instance, in the filling of a storage tank with a fluid, one counter counts the number of gallons entering the tank at the inlet and another counter counts the number of gallons leaving the tank at the discharge. The two counters can use the same memory register for recording. For each gallon that enters or leaves the tank, the counter adds 1 to the count register for entering and subtracts 1 for leaving. Thus, at any given time the count register provides the number of gallons held in the tank.

The 'sequencer' function executes a series of steps to produce specific conditions in a number of on or off outputs at each step. The specified output pattern for each step is stored in the memory. The number of outputs that a sequencer can control depends on the word size of the PLC.

The generic sequencer is an event-driven function, where the event is the occurrence of an 'off' to an 'on' change on the step input line. Combined with a timer, it becomes a time-driven function. The generic sequencer shown in figure 8.3 controls eight outputs in module OM:1 in a five-step sequence. The desired output conditions are stored in memory cells RA:0151 through RA:0155, the contents of which are shown in the figure. Also shown are the lines connecting memory bits 0 through 7 to the eight terminals in the output module OM:1. When the step pointer selects terminal OM:1, the actual selection is memory cell RA:0150. Combined with a timer, as shown in figure 8.4, the sequencer becomes a time-driven function. When OM:1-1 is turned on, the timer output T-8 produces an off-on-off pulse every 30 seconds.

Most PLCs have the ability to compare two operands. The operands may be numbers stored in the memory, or one may be a constant and the other a memory register. A 'compare' function is programmed as a rung in the ladder diagram program with a contact that specifies the operands.

'Arithmetic' functions perform arithmetic calculations of one or two operands and store the result in a destination memory register. Addition, subtraction, multiplication, division and square root may be determined with the aid of the arithmetic function. This function is programmed as a rung in the ladder diagram program with a contact that specifies the operands.

A 'control' function changes the method of scanning the program. The function may be split into four separate operations. The 'skip' function permits the PLC to pass over a portion of the program as it is being scanned. The 'master control relay' (MCR) function is similar to the skip function in that a specified number of rungs immediately following the MCR rung are affected. The 'jump' function causes the PLC scan to go immediately to the next rung in the program. The 'jump to subroutine' (JSR) function causes the scan to go to another section of the program that ends with a rung that has a return (RET) label.

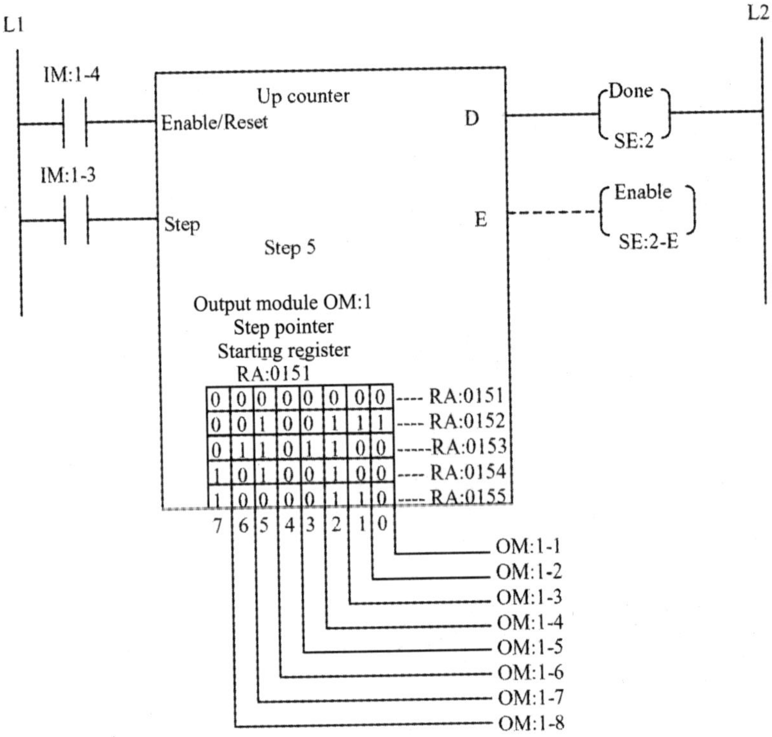

Fig. 8.3 Generic Sequencer

Medium and large PLCs have the ability to copy data from one place and place it in another area. A 'move' label copies data from the source register to the destination register. A 'block transfer' (BT) function copies data from a block of source registers into a block of destination registers.

'Bit manipulation' functions change the bits in a memory register in a specified manner. The functions permit changing the status of individual bits, moving and rotating the bits in a register to the left or to the right, and to perform logical operations on corresponding bits in two operands stored in the memory registers.

Many PLCs have the PID control function built into the input/output module. Other PLCs have the PID function programmed as a rung in the ladder diagram program. A generic PID controller function is shown in figure 8.5.

8.8 ANALOG CONTROL

Electronic 'analog control' calls for an error detector and a control mode unit. An analog controller requires a single operational amplifier as a function generator and some resistors and capacitors to implement the transfer function of the desired control mode or combination of modes.

The analog proportional controller uses three resistors to form an inverting amplifier. The circuit has two inputs, an error and an output offset. The proportional gain is obtained by dividing the feedback resistor by the error input resistor. The offset resistor must equal the feedback resistor

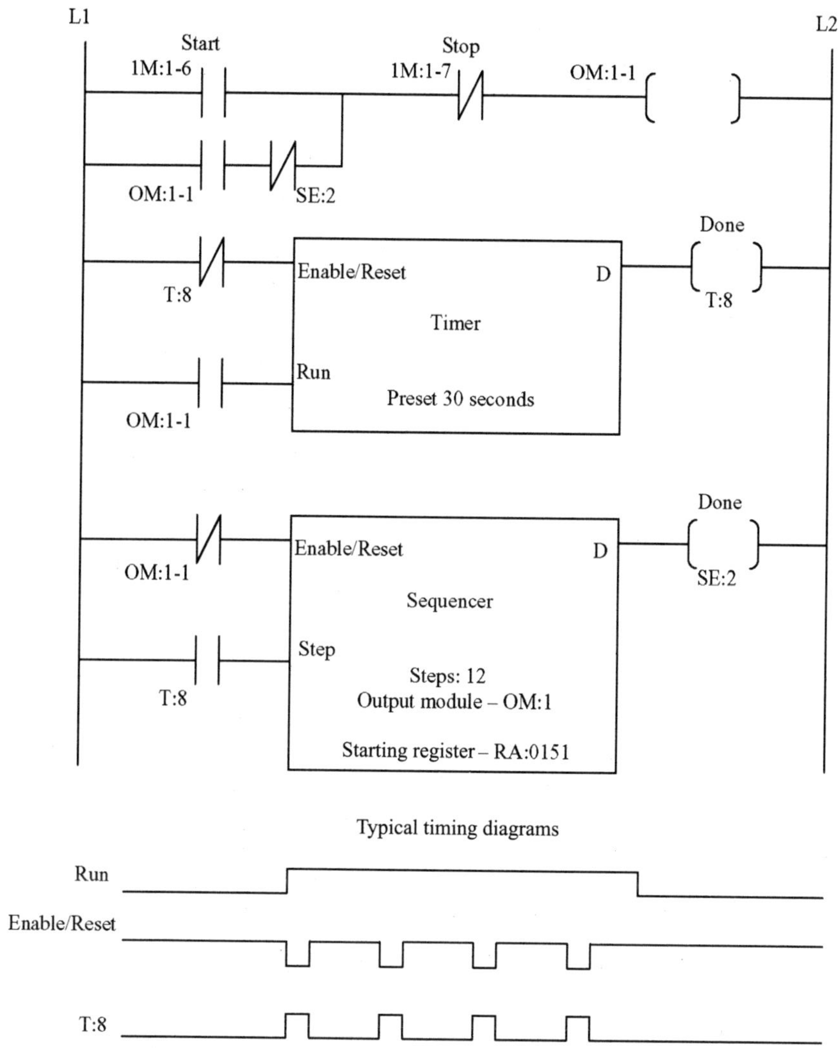

Fig. 8.4 Generic Sequencer with Timed Steps

to satisfy the time domain equation. The output lines may be reversed to make the output either positive or negative with respect to the error.

The proportional plus integral controller needs two resistors and a capacitor to implement the PI transfer function. The capacitor is connected in series with the feedback resistor. The gain is given by dividing the feedback resistor by the input resistor. The integral action rate is equal to the reciprocal of the product of the input resistor and the capacitor.

The proportional plus derivative controller uses four resistors and a capacitor to implement the PD transfer function. The circuit is a proportional controller with a parallel combination of a resistor and a capacitor placed in series with the input resistor.

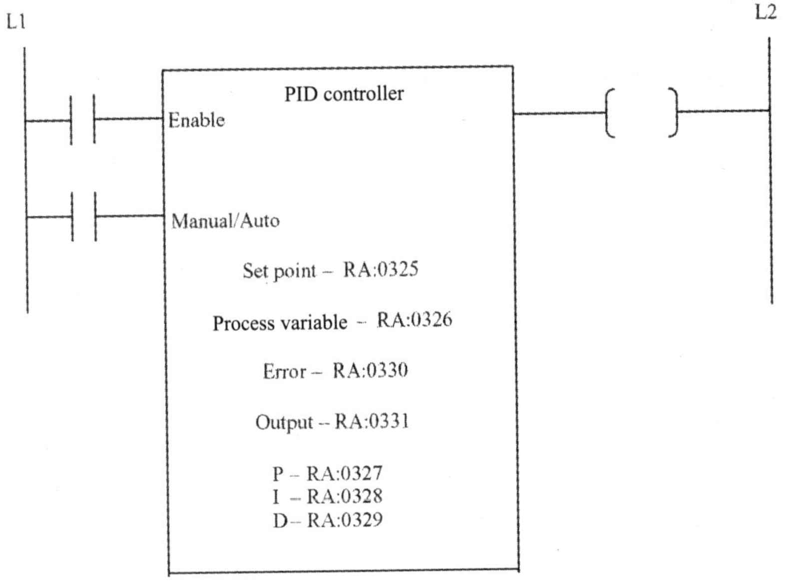

Fig. 8.5 Generic PID Controller Function

8.9 DIGITAL CONTROL

Microprocessor based digital controllers are widely used in industrial control systems since advanced tuning, multi-variable control and expert systems are available. Digital controllers in the closed-loop usually implement the PI, PD or PID modes.

A digital controller measures the controlled variable after specific intervals of time, called the sampling time, Δt. The measurement is converted to a binary number for input to a digital computer. The computer subtracts each reading of the sample from the set point value to obtain a set of error samples.

At the end of each error sample calculation, the digital PID controller follows a procedure called the PID algorithm to calculate the controller output based on the error samples. The positional version of the algorithm determines the actuator position, while the incremental version of the algorithm determines changes in the position of the actuator. The positional version of the algorithm has the advantage that the controller has the actuator position retained in its memory. The incremental version is suitable for output devices such as stepper motors.

The integral mode sometimes presents computational difficulties that can lead to unsatisfactory results. When the value of $PI\Delta t$ is less than 1, it is convenient to work with the reciprocal of $PI\Delta t$, which could be stored in the computer as an integer. If the value of $PI\Delta t$ is very small, the computer may ignore relatively large errors because of insufficient resolution. The change would then be lost unless a special provision is made to include the change in the calculations for the next sample. The end result is an error offset that the integral mode is unable to eliminate. One solution is to increase the precision by increasing the word length in the computer. A 16-bit word length, for example, has a precision of 1 part in 65,536. Another solution is to add the unused portion of the sum of the error samples to the current error sample before computing the integral mode change.

The derivative mode can also present computational difficulties that can lead to unsatisfactory results. A slowly changing signal, for instance, results in a fluctuating derivative action mode. The derivative mode estimates the rate of change of the error, *de/dt*. Since Δt is fixed by the sampling rate, attention is focused on the change in the error. To obtain a better estimate of Δe, estimators of previous samples may be used to smoothen the derivative term and thus improve the estimate.

8.10 REVIEW PROBLEMS

Example problem # 8.1: A batch blending of water and acid is to be controlled by a timer- and event-driven sequential control system, as shown in figure 8.6. Three levels are specified in the mixing tank. The process initiates with the drain valve open, water and acid valves closed, mixer motor is off, and tank is empty. The design criteria specifies the following steps:

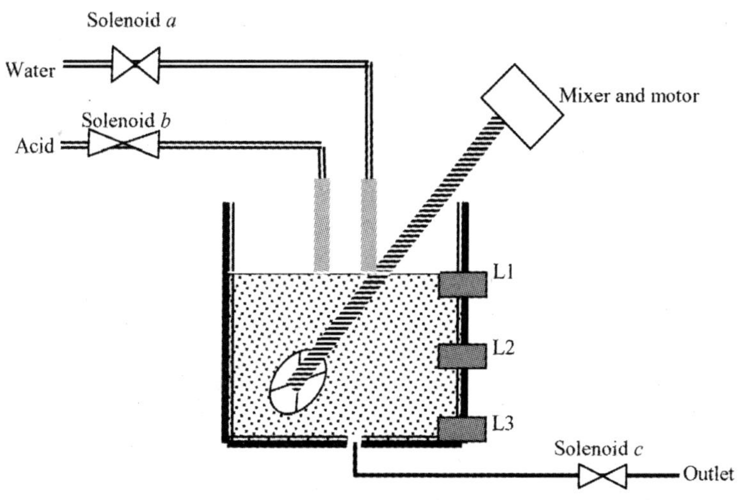

Fig. 8.6 Batch Mixer

Step 1 Begin when Start button is pressed, fill tank with water up to level 2, and ending with level switch L2 closed.

Step 2 Following step 1, fill tank with acid up to level 3, tank then fill with 60 per cent water and 40 per cent acid, close level switch 3. Turn on mixer motor.

Step 3 Mixer runs for 5 minutes, then stop mixer motor.

Step 4 Open switch L1 and drain tank.

Step 5 When tank is drained, close switch L1 and restart process for another batch.

Draw a PLC ladder diagram program for the process.

Solution: The PLC program has six rungs (figure 8.7). Rung 1 turns on output OM:1-4 when Start is pressed, until tank is empty at end of step 5. Rung 2 closes normally open drain valve, before timer T:4 activates. Rung 3 energizes solenoid *a*, L2 is closed and de-energizes it. Rung 4 energizes solenoid *b* when L2 closes, de-energizes it and closes L3. Rung 5 enables timer when cycle light is on, starts timer when L3 closes, turns on output T:4 when timer goes off. Rung 6 turns mixer on during steps 2 and 3.

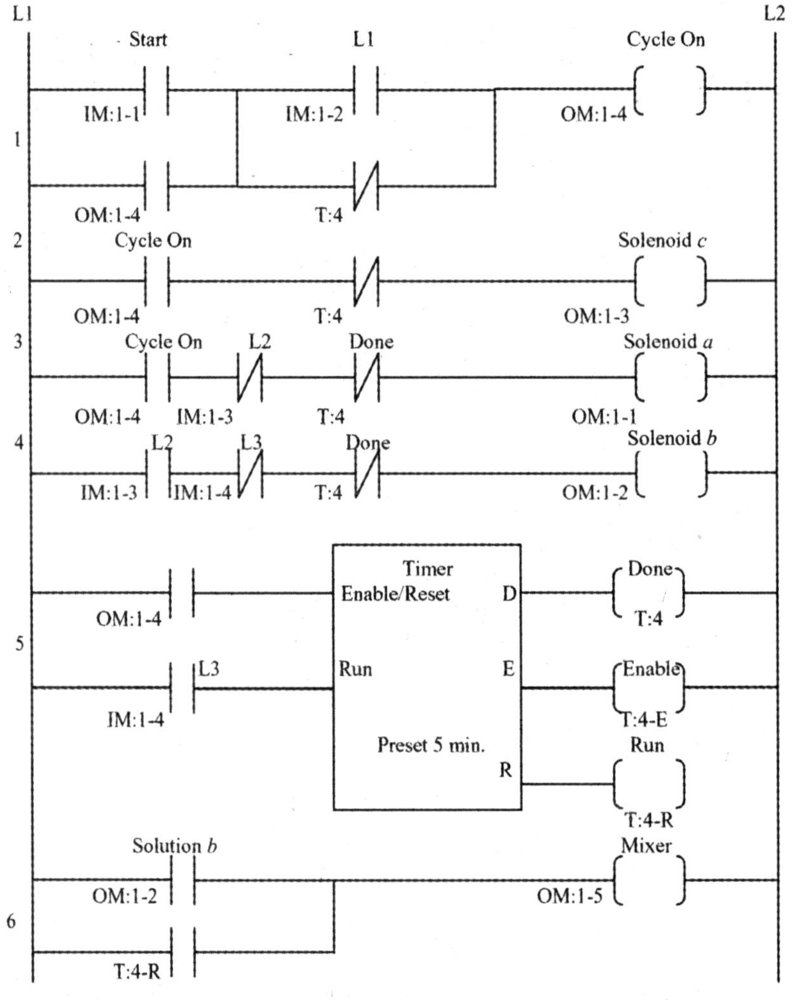

Fig. 8.7 PLC Ladder Diagram

Example problem # 8.2: Determine the values of R_1 and R_i for the electronic proportional plus integral controller shown in figure 8.8. Gain $P = 1.8$, and integral action rate $I = 0.03\ s^{-1}$. Use a 10 μF capacitor for C_i, and determine the transfer function.

Solution: Integral $I = 1/(R_i\ C_i)$

Hence $R_i = 1/(IC_i) = (1/0.03) \times (1/10^{-5}) = 3.33\ M\Omega$

Gain $P = R_i/R_1$.

Or $R_1 = R_i/P = (3.33 \times 10^6)/1.8 = 1.85\ M\Omega$

Transfer function $= V/E = 1.8 \times (1 + 33.33s)/(33.33s)$

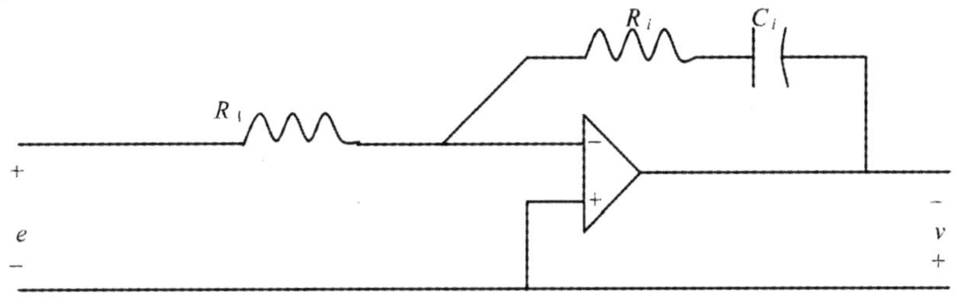

Transfer function = $V/E = P[I + S]/S$, $I = 1/R_iC_i$, gain $P = R_i/R_1$

Fig. 8.8 Analog Proportional Plus Integral Controller

Example problem # 8.3: Determine the value of R_1, R_i, R_d and C_d for the PID analog controller of figure 8.9. Gain P = 3.5, derivative action time constant D = 0.4 s, derivative limiter coefficient α = 0.1, and integral action rate $I = 1/6\ s^{-1}$. Use a 10 μF capacitor for C_i, and determine the transfer function.

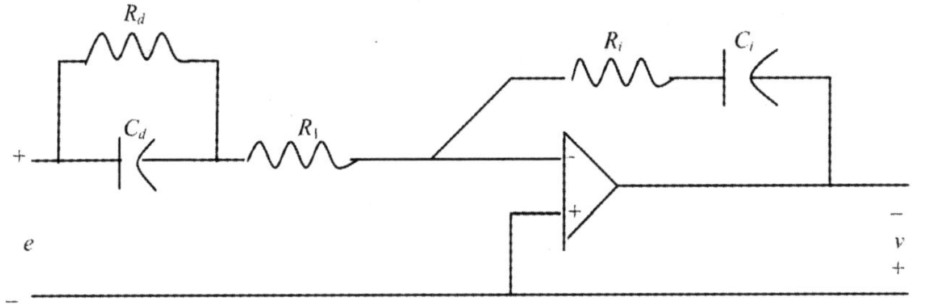

Transfer function = $V/E = P[(I + S)/S] \times [(1 + DS)/(1 + \alpha DS]$
$I = 1/R_iC_i$, gain $P = R_i/(R_1 + R_d)$, $D = R_dC_d$, $\alpha = R_1/(R_1 + R_d)$

Fig. 8.9 Analog Proportional Plus Integral Plus Derivative (PID) Controller

Solution: Integral $I = 1/(R_iC_i)$

Hence $R_i = 1/(IC_i) = 6 \times (1/10^{-5}) = 0.6\ M\Omega$

$\alpha = R_1/(R_1 + R_d)$

Hence $R_d = R_1(1 - \alpha)/\alpha = R_1(1 - 0.1)/0.1 = 9R_1$

$P = R_i/(R_1 + R_d) = R_i/(10R_1)$

Hence $R_1 = R_i/(10P) = 0.6\ M\Omega/(10 \times 3.5) = 17.1\ k\Omega$

And $R_d = 9R_1 = 9 \times 17.1 = 154.26\ k\Omega$

$D = R_dC_d$

Hence $C_d = D/R_d = 0.4/154.26 = 2593\ \mu F$

Transfer function $= V/E = 3.5 \times [1/6 + (1 + 0.4/6)s + 0.4s^2]/(s + 0.04s^2)$

$= (1 + 6.4s + 2.4s^2)/(1.714s + 0.0685s^2)$

8.11 EXERCISES

8.1 Draw a PLC ladder diagram for a system to turn on a light when the count reaches 25 and then switches off the light when the count reaches 45.

8.2 Draw a PLC ladder diagram for a system to turn on a solenoid actuated valve when count X reaches 20 or when count B reaches 30.

8.3 Draw a ladder diagram program for two comparators AB and BC to produce 5 comparisons as follows:

Rung 1 is TRUE when op1 equals op2 (AB).

Rung 2 is TRUE when op1 is not equal to op2 (NB).

Rung 3 is TRUE when op1 is greater than or equal to op2 (BC).

Rung 4 is TRUE when op1 is less than op2 (LT).

Rung 5 is TRUE when op1 is greater than op2 (CT).

8.4 An analog proportional plus integral mode controller of the type shown in figure 8.8 uses 50 kΩ resistor for R_1. Determine the value of R_i if the gain (*i*) $P = 0.35$, (*ii*) $P = 0.85$, (*iii*) $P = 2.5$, (*iv*) $P = 7.5$.

8.5 The analog PID controller of the type shown in figure 8.9 uses 80 kΩ resistor for R_i and $R_1 = 25$ kΩ. Determine the value of R_d if the gain (*i*) $P = 0.15$, (*ii*) $P = 0.75$, (*iii*) $P = 3.5$, (*iv*) $P = 8.5$.

References & Bibliography

Eckman, D.P., "Industrial Instrumentation", John Wiley & Sons, New York, 1980.

Floyd, T.L., "Digital Fundamentals", Macmillan Publishing Co., New York, 1990.

Herman, S.L. and Alerich, W.N., "Industrial Motor Control", Delmar Publishers, Albany, NY, 1985.

Hordeski, M.F., "Design of Microprocessor, Sensor and Control Systems", Reston Publishing Co., Englewood Cliffs, NJ, 1985.

Johnson, C.D., "Process Control Instrumentation Technology", Prentice Hall, Upper Saddle River, NJ, 1997.

Jones, C.T. and Bryan, L.A., "Programmable Controllers: Concept and Applications", International Programmable Controls, Inc., Atlanta, 1983.

Kilian, C.T., "Modern Control Technology Components and System", West Publishing Co., St. Paul, MN, 1996.

Stein, D.H., "Introduction to Digital Data Communications", Delmar Publishers, Albany, NY, 1985.

Webb, J.W. and Reis, R.A., "Programmable Logic Controllers", Prentice Hall, Upper Saddle River, NJ, 1995.

9

Controller Design and Analysis

9.1 INTRODUCTION

The design of a controller mechanism can only be selected after the control modes and their values are established. A process or a component is described by the relationship between the input and the output signals. It is this input and output relationship that determines the design requirements of the controller. If the input/output relationship is inadequately defined, the design must include a large adjustment of the controlled parameters so that optimum settings can be determined during start-up of the system.

The characteristics of a process are an expression of the relationship between the manipulated input variable and the controlled output variable. Process characteristics of a control system can be expressed in several ways, using the step response graph, the time-domain equation, the transfer function and Bode diagram. Characteristics of electrical, thermal, liquid flow, gas flow and mechanical processes may be determined through the integral, first-order lag, second-order lag, dead-time and first-order lag plus dead-time methods. Characteristic parameters include the steady state gain for all processes, time constant in first-order lag process, integral action time constant in integral process, dead-time lag in dead-time process, resonant frequency in second-order lag process and damping ratio in second-order lag process.

The proportional mode combined either with the integral or the derivative mode is usually suitable for continuous processes. The proportional gain P, the integral rate of action I and the derivative action time constant D characterize the three modes.

The analysis of a control system must focus on two aspects: stability and sensitivity. A stable system responds to any reasonable input with an error that diminishes with time. The error of a stable system may oscillate, and it may not be totally eliminated. The amplitude of the oscillating error will diminish exponentially with time. If the error is not suppressed, it eventually reaches a small and steady value. In contrast, an unstable system will respond to some reasonable inputs, but the error will usually oscillate and increase with time. For linear and continuous systems, an error oscillating at constant amplitude is considered to be unstable. Certain discontinuous systems are inherently oscillatory, but the error is considered stable as long as the amplitude does not increase with time.

The design of the control system must also answer the question concerning the rate at which the error diminishes. The system must be sensitive enough to quickly respond to an error signal.

However, a sensitive system is also likely to lead to excessive oscillations in the controlled variable, which defeats the objective of a stable system. At the same time, a considerably stable system will tend to be sluggish in its response. Thus, sensitivity and stability are conflicting aims. A combination of certain levels of the stability and sensitivity characteristics in the system would thus be desirable.

9.2 PROCESS CHARACTERISTICS

The relationship between the input and output signals may be defined with the aid of a step response graph, time domain equation, transfer function or the frequency response graph. Any or all of the methods generally serve the intent.

The step response versus time graph of an output signal follows a step change in the input signal. The time-domain equation, on the other hand, defines the magnitude versus time relation between the input and output signals. Expressed in terms of basic elements describing the system, the equation frequently contains derivative and integral terms. The transfer function is obtained after transforming the time-domain equation into the frequency domain, and then solving for the ratio of the output and the input. The transfer function provides gain and phase difference between the input and output signals as a function of frequency. The frequency response diagram, also referred to as Bode diagram, has twin plots of gain and phase difference as a function of frequency.

The characteristics of a process are independent of the system type, such as mechanical and electrical, but depend on the magnitude of the elements comprising the system (capacitance, stiffness, dead-time). Two different systems may have identical time-domain equation, transfer function, step response and Bode diagram. Hence, conclusions pertaining to the evaluation may be extended from one system to another.

9.3 RAMP PROCESS

The 'integral', or the 'ramp', process uses a single element, such as a capacitor, where the outflow of material or energy does not depend on the stored material or energy in the element. The stored quantity remains constant in the element if the inflow and outflow rates are equal.

Before a 'step change' of an integral process, the input and output flow rates are equal and the level remains fixed. After the step change in the input, the inflow rate exceeds the outflow rate, and the level increases at a constant rate. The sloping ramp-like response (figure 9.1) of an integral process is measured by the integral action time constant, which is the number of seconds needed for

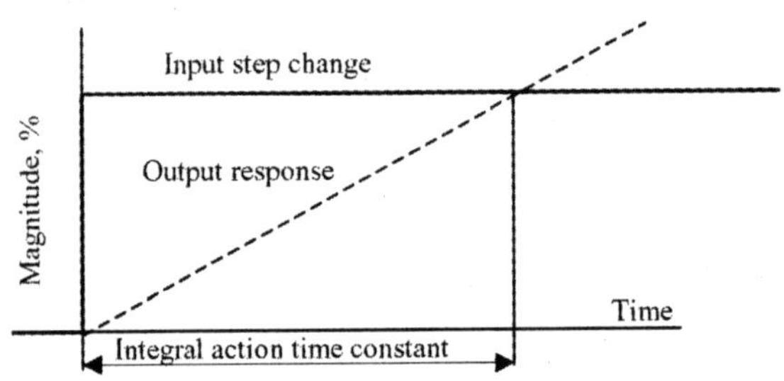

Fig. 9.1 Ramp Response of Integral Process to Step Change in Input Signal

the output to reach the same percentage change as the input. Equations defining the relationship between the input and output signals in the case of a tank being filled and discharged simultaneously with a liquid and the liquid level changing are shown in table 9.1.

Table 9.1 Integral Process

Time-domain equation: $h^*(t_1) - h^*(t_0) = 1/T_i \times [\int(q^*_{in} - q^*_{out})dt]$, $t_0 < t < t_1$

Transfer function: $H^*(s)/Q^*_{in}(s) = 1/T_i s$

Liquid flow integral process: $T_i = A[FS_{out}/FS_{in}]$

where: $h^*(t_0)$ = normalized output at time t_0, percentage of FS_{out}

$h^*(t_1)$ = normalized output at time t_1, percentage of FS_{out}

FS_{in} = full scale range of input, m^3/s

FS_{out} = full scale range of output, m

q^*_{in} = normalized input flow rate, percentage of FS_{in}

q^*_{out} = normalized output flow rate, percentage of FS_{in}

t = time, s

T_i = integral action time constant, s

A = cross-sectional area of tank at liquid's surface, m^2

Note: The equations refer to a situation where a tank is simultaneously being filled and discharged, with the liquid level changing.

9.4 FIRST-AND SECOND-ORDER LAG

The 'first-order lag' process may be illustrated in the form of a single capacitor configured to have the outflow of material or energy in proportion to the amount of material or energy stored in the capacitive element. The process regulates by itself, since the output flow rate is automatically adjusted to match the inflow rate. The integral process, on the other hand, is not a self-regulating mechanism.

The inflow and outflow rates are identical before a step change in the first-order lag process, with the level remaining constant. A step change causes the inflow rate to increase by a certain percentage, M. Outflow rate is proportional to the level, and does not change immediately. The higher inflow rate increases the liquid level at a rate in proportion to the difference. As the level increases, the difference between the inflow rate and the outflow rate decreases, which in turn reduces the rate at which the level increases (figure 9.2).

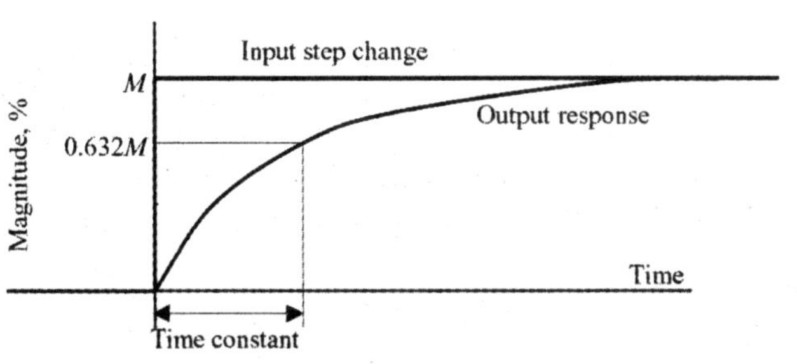

Fig. 9.2 Step Response of First-Order Lag

Response to a step change in a first-order lag is measured from the time constant that represents the time required for the output to reach 63.2 per cent of the total change. Each additional time constant interval will see the output reaching 63.2 per cent of the remaining change. In the first time interval the output changes by $0.632M$, and the remaining change is $M - 0.632M = 0.368M$. In the second time interval the output increases by $0.632 \times 0.368M = 0.232M$, and the output reaches $0.632 + 0.232M = 0.864M$. After five constants the output reaches $0.993M$.

Equations defining the relationship between the input and output signals for a first-order lag process for the liquid level of a tank-filling, current flow through an electrical circuit, heat flow through a thermal heat exchanger, gas pressure in a tank, and a blending mixer circuit are shown in table 9.2.

Table 9.2 First-Order Lag Process

Time-domain equation: $\tau \times dy/dt + y = Gx$

Transfer function: $Y/X = G/(1 + \tau s)$

where: G = steady state gain of process
t = time, s
x = input to process
y = output of process
τ = time constant, s

Liquid level first-order lag process: $\tau = R_L C_L$

$G = (R_L/\rho g)(FS_{in}/FS_{out})$

where: h^* = output level, percentage of FS_{out}, FS_{out} = output range, m
q^*_{in} = input flow rate, percentage of FS_{in}, FS_{in} = input range, m^3/s
R_L = liquid resistance, Pa-s/m^3
C_L = liquid capacitance, m^3/Pa
ρ = liquid density, kg/m^3
g = 9.81 m/s^2

Electrical first-order lag process: $\tau = RC$ (series and parallel R, C)

$\tau = L/R$ (series R, L)

where: e_{in} = input voltage
e_{out} = output voltage
i_{in} = input current
C = electrical capacitance, F
G = 1 (electrical series R, C, series R, L)
$G = R$ (electrical parallel R, C)
R = electrical resistance, Ω

Thermal first-order lag process: $\tau = R_T C_T$

θ_j = input jacket temperature, °C
θ_L = output liquid temperature, °C
G = 1
R_T = thermal resistance, K/W
C_T = thermal capacitance, J/K

Contd. Table 2

Gas pressure first-order lag process: $\tau = R_g C_g$

where: w^*_{in} = input gas flow rate, percentage of FS_{in}, FS_{in} = input range, kg/s

p^* = pressure in tank, percentage of FS_{out}, FS_{out} = output range, Pa

R_g = gas flow resistance, Pa-s/kg

C_g = gas flow capacitance, kg/Pa

$G = R \times (FS_{in}/FS_{out})$

Blending first-order lag process: $\tau = V/q$

where: c_i = blending input concentration, per cent

c_o = blending output concentration, per cent

q = blending liquid flow rate, m^3/s

$G = 1$

V = volume of blending tank, m^3

A 'second-order lag' process may comprise of two capacitance elements, one capacitance and one inertial element, or two inertial elements. Resonant frequency, ω_o, amount of damping present in the process expressed by damping coefficient, α, or damping ratio, ζ, and steady state gain, G, are the three parameters that characterize the response of a second-order system.

Depending on the value of the damping ratio, ζ, the step response of a second-order system may be split into three regions. If the damping ratio is greater than 1, the response is over damped, as shown in figure 1.6. If the damping ratio is less than 1, the response is under damped, as shown in figure 1.7. If the damping ratio is equal to 1, the response is critically damped. The under damped response has a rapid rise and oscillation, which contrasts with slow rise and lack of oscillation in the over damped system. The critically damped response has the fastest rise with no oscillation.

A mechanical mass, spring and damper system is shown in figure 9.3. The applied force, f, acts as the input signal, and the position of the moving mass, h, is the output signal. The spring exerts the force, $f_s = -h/C_m$, and the damping mechanism exerts force, $f_r = -R_m \times dh/dt$. The negative signs indicates that the direction of spring and damper forces is opposite to that of the motion, h. The force due to inertia of the mass is $M \times d^2h/dt^2$. The time-domain equation for the mass, spring and damper system then becomes:

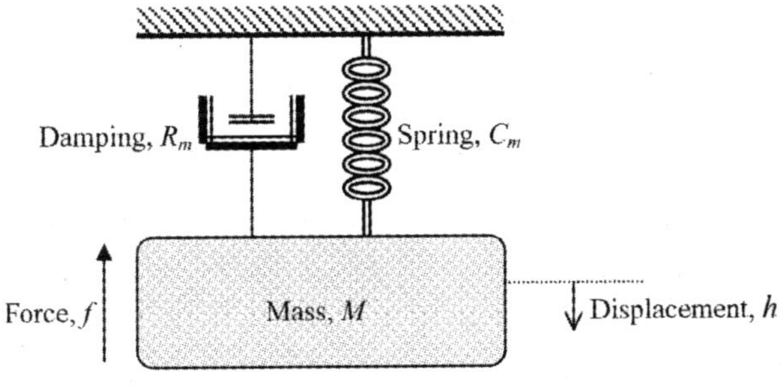

Fig. 9.3 Single Degree of Freedom Vibrating System

$$MC_m \times d^2h/dt^2 + R_mC_m \times dh/dt + h = C_m f$$

where: C_m = spring capacitance, m/N

f = externally applied force, N

h = position of mass, m

M = mass of moving body, kg

R_m = damper resistance, N-s/m.

The electrical analog of a mass spring, and damper is a series *RLC* circuit, which has the following expression:

$$LC \times d^2e_{out}/dt^2 + RC \times de_{out}/dt + e_{out} = e_{in}$$

A 'non-interacting second-order process' has two capacitive elements configured such that the elements have no effect on each other. The time-domain equation of a non-interacting process with input signal x and output signal y is:

$$\tau_1\tau_2 \times d^2y/dt^2 + (\tau_1 + \tau_2) \times dy/dt + y = Gx$$

where $\tau_1 = R_1C_1$ and $\tau_2 = R_2C_2$.

An 'interacting second-order process' has two capacitive elements configured such that the second element has an effect on the first one. The time-domain equation of an interacting process with input signal x and output signal y is:

$$A_2 \times d^2y/dt^2 + A_1 \times dy/dt + y = Gx$$

where $G = R_L/(R_1 + R_2 + R_L)$

$A_1 = G\{(\tau_1 + \tau_2) + \tau_1R_2/R_L + \tau_2R_1/R_2\}$

$A_2 = G\tau_1\tau_2$

The time-domain equation is obtained by substituting for i in the electrical equation:

$$A_2 \times d^2\omega/dt^2 + A_1 \times d\omega/dt + \omega = Ge$$

where $\tau_m = J/B$ = mechanical time constant

$\tau_e = L/R$ = electrical time constant

$A_2 = \tau_m\tau_eRB/(K_eK_t + RB)$

A DC motor-driven load serves as an example of a second-order process. The input signal is the voltage, e, applied to the armature, and the output signal is the speed of the load, ω. Three equations define the process. The electrical equation of the armature circuit is:

$$Ldi/dt + R_i + K_e\omega = e$$

The torque-current equation of the motor is: Torque = K_ti

The mechanical equation of the load is: $i = (J \times d\omega/dt + B\omega)/K_t$

Table 9.3 Second-Order Lag Process

Time-domain equation: $A_2 \times d^2y/dt^2 + A_1 \times dy/dt + y = Gx$

Transfer function: $Y(s)/X(s) = G/(1 + A_1s + A_2s^2)$

Parameters:

$$\omega_0 = \sqrt{(1/A_2)}$$
$$\alpha = A_1/2A_2$$
$$\zeta = \alpha/\omega_0 = A_1/2\sqrt{A_2}$$
$$A_2 = 1/\omega_0^2$$
$$A_1 = 2\zeta/\omega_0$$

where: G = steady state gain of process
t = time, s
x = input signal
y = output signal
α = damping coefficient, s^{-1}
ω_0 = resonant frequency, radian/s
ζ = damping ratio

Mechanical second-order process:

$$A_2 = MC_m$$
$$A_1 = RC_m$$
$$G = C_m$$

where: f = applied force, N
h = position of mass, m
M = mass of body, kg
R = damping resistance, N-s/m
C_m = spring capacitance, m/N

Electrical second-order lag process:

$$A_2 = LC$$
$$A_1 = RC$$
$$G = 1$$

where: e_{in} = input voltage, volt
e_{out} = output voltage, volt
L = electrical inductance, H
C = electrical capacitance, F
R = electrical resistance, Ω

Non-interacting second-order process:

$$\tau_1 = R_1C_1$$
$$\tau_2 = R_2C_2$$
$$A_2 = \tau_1\tau_2$$
$$A_1 = \tau_1 + \tau_2$$

Liquid level non-interacting: $G = (R_2/\rho g)(FS_{in}/FS_{out})$

where: q^*_{in} = input flow rate, percentage of FS_{in}, FS_{in} = input range, m^3/s
h^* = output level in tank 2, percentage of FS_{out}, FS_{out} = output range, m

Contd. Table 3

R_1, R_2 = liquid resistance, Pa-s/m^3
C_1, C_2 = liquid capacitance, m^3/Pa

Electrical circuit non-interacting: $G = 1$

where: e_{in} = input voltage, volt
e_{out} = output voltage across C_2, volt
R_1, R_2 = electrical resistance, Ω
C_1, C_2 = electrical capacitance, Ω

Interacting second-order process:

$$G = R_L/(R_1 + R_2 + R_L)$$
$$A_1 = G\{(\tau_1 + \tau_2) + \tau_1 R_2/R_L + \tau_2 R_1/R_2\}$$
$$A_2 = G\tau_1\tau_2$$

Liquid level interacting: Process input p_{in} = inlet pressure, Pa
Process output p_{out} = outlet pressure in tank 2, Pa

Electrical interacting: Process input e_{in} = input voltage, volt
Process output p_{out} = output voltage across R_L, volt

DC motor second-order process:

$$A_2 = \tau_m\tau_e RB/(K_eK_t + RB)$$
$$A_1 = (\tau_m + \tau_e)RB/(K_eK_t + RB)$$
$$G = K_t/(K_eK_t + RB)$$

where: e = armature voltage, volt
ω = output load speed, radian/s
$\tau_m = J/B$ = mechanical time constant
$\tau_e = L/R$ = electrical time constant
J = load moment of inertia, kg-m^2
B = damping resistance, N-m-s/radian
L = armature inductance, H
R = armature resistance, Ω
K_t = torque constant, N-m/A
K_e = EMF constant, volt-s/radian
$A_1 = (\tau_m + \tau_e)RB/(K_eK_t + RB)$
$G = K_t/(K_eK_t + RB)$
J = load moment of inertia, kg-m^2
B = damping resistance, N-m-s/radian
L = armature inductance, H
R = armature resistance, Ω
K_t = torque constant, N-m/A
K_e = EMF constant, volt-s/radian
e = armature voltage, volt and
ω = motor speed, radian/s

The Bode diagram of a second-order lag process, shown in figure 9.4, varies considerably for different values of the damping ratio. The frequency scale is normalized in terms of the resonant frequency, and process gain is 1. Damping ratio values range from 0.025 to 10.

Table 9.3 provides equations for the second-order lag process.

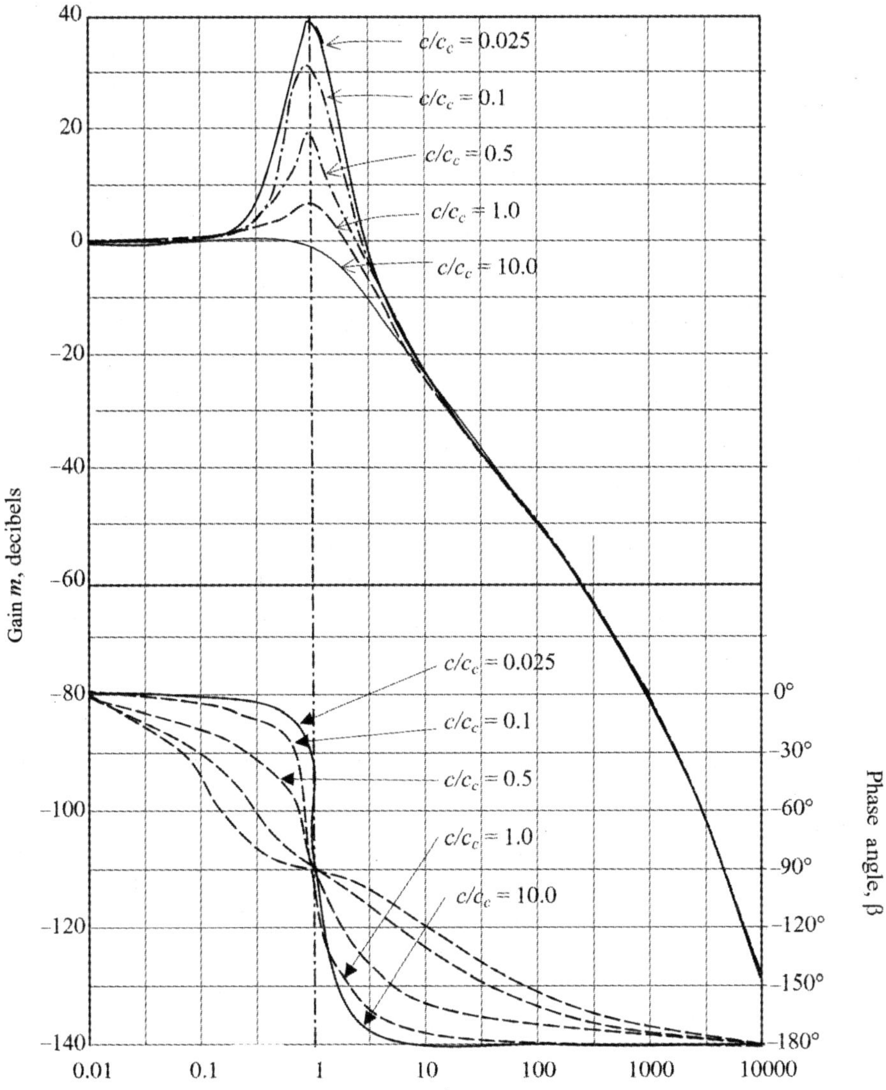

Fig. 9.4 Bode Diagram of Second-Order Lag Process

9.5 DEAD TIME

A 'dead-time' is associated with a process where mass or energy is transported from one point to another. The output and input signals are identical except for the time delay. The lag occurs due to the time required for the signal to travel from the input location to the output location, as shown in figure 9.5 for a step change. The time lapse that occurs between the input change and the corresponding output change is called the dead-time lag, and is denoted by t_d. Thus, if $f_i(t)$ and $f_o(t)$ represent the input and output signals, the equation representing the dead-time process is:

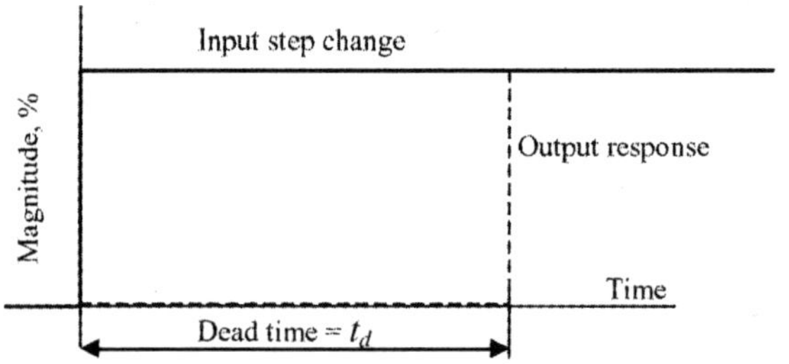

Fig. 9.5 Step Response of Dead-Time Process

$f_o(t) = f_i(t - t_d)$. The gain is 1 at all frequencies, which appears as a straight line on the 0 db line on the Bode diagram. Phase angle β is given by the equation: $\beta = -57.3\omega t_d$. The time-domain equation may also be expressed as $t_d = D/v$, and the transfer function is: $F_o(s)/F_i(s) = e^{-tds}$, where D = distance from input to output, m, $f_i(s)$ = input signal, $f_o(s)$ = output signal, t = time, s, and v = velocity of signal, m/s.

The first-order lag plus dead-time process is a combination of the first-order lag with a dead-time element (figure 9.6). The response is characterized by the first-order time constant (τ) and the dead-time lag (t_d), values of which may be determined from a step response test of the process. The process is frequently used as a first approximation of the model of more complex processes for design and analysis.

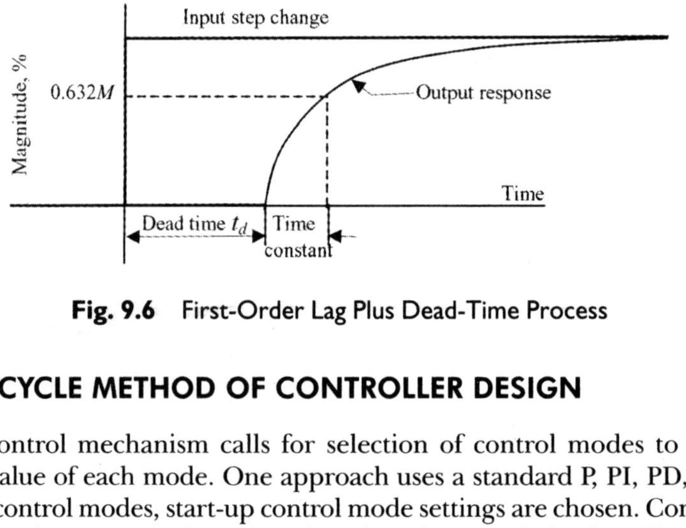

Fig. 9.6 First-Order Lag Plus Dead-Time Process

9.6 ULTIMATE CYCLE METHOD OF CONTROLLER DESIGN

The design of a control mechanism calls for selection of control modes to be employed and calculation of the value of each mode. One approach uses a standard P, PI, PD, or PID controller. After selecting the control modes, start-up control mode settings are chosen. Controller adjustment formulas provide optimum control mode values based on period of oscillation of the system. Another approach uses PID modes or compensating networks that are designed using I and D modes for a specific application. This method is frequently used if the transfer function of the process is known.

The 'ultimate cycle method' makes use of controller adjustment formulas to determine control settings. Two measurements from the process are needed for the formulas, the minimum controller gain that causes the control system to oscillate (ultimate gain, G_u) and the period of oscillation (ultimate period, P_u). The controller must be installed and the system ready to operate for using the ultimate cycle method.

The ultimate gain and ultimate period are determined first by setting the integral and derivative modes to the least effective setting. The process is then started with the controller gain at a low value. The gain setting is increased until the controlled variable begins to oscillate, which provides the ultimate gain G_u. The corresponding period of oscillation provides the ultimate period P_u.

System response may be optimized through modifications to the original ultimate cycle method. The gain setting is adjusted through a trial procedure until the desired response is obtained. This under-damped response is the quarter amplitude decay, where each succeeding positive peak value is one-fourth as large as the preceding positive peak value.

The measured P_u value is used to compute the integral and derivative mode settings. Next, the process is disturbed by a small change in the set point, the response of the controlled variable is observed, and the gain is adjusted. If the response is over-damped, the gain is increased; if the response is under-damped, the gain is decreased. The sequence of set point change, observation of response, and gain adjustment is repeated until one-quarter amplitude decay in the response is obtained. This method is deemed reliable for most fast processes where the waiting time is short, but is time consuming in slow processes.

Where a process can be approximated by a first-order lag plus dead-time model, the 'process reaction method' using controller adjustment formulas helps to determine the settings. Three parameters are required from an open-loop step response test of the method for the formulas. After operating the process on manual control (open-loop) until the measured variable stays constant, a step change is made in the manipulating element and the response in the controlled variable recorded in the form of a 'process reaction graph' (figure 9.7).

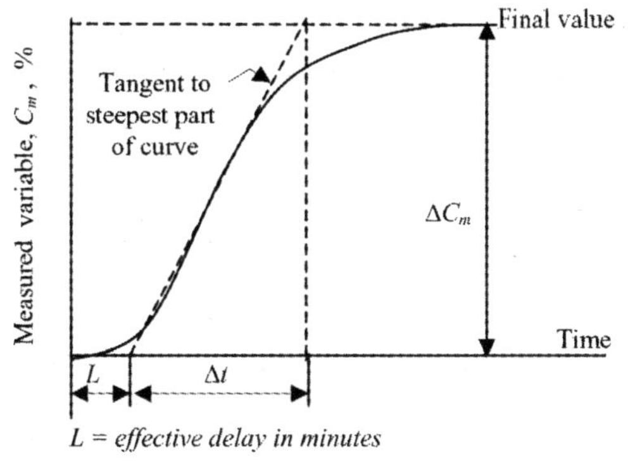

Fig. 9.7 Process Reaction Graph of Controlled Variable after Step Change in Manipulating Element

From the step response test, three variable are obtained for use in the reaction formulas. The process reaction graph provides the effective delay, L, minutes, and the change in value, ΔC_m, per cent. The third variable is ΔP, the per cent change in the manipulating element. Controller settings are determined from the slope of the tangent line $N = \Delta C_m/\Delta t$, %/minute. Proportional control $P = \Delta P/NL$. Proportional plus integral control (PI) $P = 0.9(\Delta P/NL)$, and $I = 0.3/L$ per minute. Proportional plus integral plus derivative control (PID) $P = 1.2(\Delta P/NL)$, $I = 0.5/L$ per minute, and $D = 0.5L$ minute.

An important advantage of the process reaction method over the ultimate cycle method is that once the model is obtained, the control modes can be determined and implemented immediately. This is of significance in slow processes where considerable time elapses in completing the ultimate cycle method.

9.7 SELF-TUNING ADAPTIVE CONTROLLERS

Difficulties arise in designing and tuning complex process models due to inaccuracies. Although elements such as resistance, inertia and dead-time are distributed in much the same way as a uniformly distributed load on a beam, the mathematical models assume the elements to be lumped at discrete locations. Sometimes the elements change in a process for reasons that vary from change in the production rate to lack of repeatability while tuning the parameters of a loop.

The intent in a 'self-tuning adaptive controller' is to maintain the desired control criteria when the load or the production rate changes in value. The adaptive controller has the capability to observe the response of the control system and automatically adjust the PID values in accordance. A conventional PID algorithm is provided in an outer loop within the adaptive controller to determine the values of P, I and D. The outer loop also serves as a feedback loop to observe the control system and adjust the values of the P, I and D elements for optimum performance.

A number of techniques are used to adapt the controller for changes in the process. A 'programmed adaptive controller' adjusts the proportional gain (P) as a function of process related variables such as set point, error signal, controller output or a remote input variable. The gain may also be adjusted in the adaptive algorithm using a combination of the variables.

A 'model based adaptive controller' uses an internal model of the process to determine optimum PID values. In this method, the controller introduces step changes above and below the set point and observes the reaction of the process to the changes. The set point changes and observations are repeated until adequate data points are obtained for establishing a process reaction curve similar to figure 9.7. Through this self-learned mode, the model is obtained from the process reaction curve, and the PID values calculated from the set of process reaction formulas.

A 'pattern recognition adaptive controller' uses a plot of error values that is similar to the quarter amplitude decay response type of curves to obtain optimum PID values. The adaptive algorithm examines the response of the control system continuously for disturbances caused by changes in the load or the set point. If the error magnitude exceeds a certain threshold, the adaptive algorithm searches for peaks in the magnitude of the error. When two or three peaks are detected, the algorithm uses the time period between the peaks to determine the decay ratio of the amplitude and the period of oscillation. Formulas similar to that used for the ultimate cycle method are then used to compute new optimum PID values.

9.8 COMPENSATION

In the design of a servo system to control position, velocity or mechanical loads, the component transfer function is well defined and dead-time delays are seldom encountered. This aspect differs from process control where transfer functions are not adequately known, and where dead-time delays are common.

Starting from a set of objectives (static accuracy, response time, overshoot, stability, among others), the servo system components (servo actuator, power supply, drive amplifier and feedback transducer) are selected. The components' transfer functions are then used to draw the open-loop Bode diagram or the root-locus plot.

If the simple loop-closure gain adjustment (proportional control mode) satisfies the defined objectives, adjusting the gain of one or more components or adding an amplifier provides the means to establish the required loop gain. If the performance objectives are not met, the open-loop frequency response of the system must be modified. This is done by changing or modifying components in the loop or adding components so that the open-loop frequency response can be altered to meet performance objectives through a simple gain adjustment. Addition of components is referred to as 'compensation'. The derivative and integral modes of a PID controller are examples of compensation.

Bode diagrams for some compensation networks may be constructed using the following equations:

(*i*) Integration-lead network compensation: $V_{out}/V_{in} = [(1 + T_i s)/T_i s]$

(*ii*) Lead-lag network compensation: $V_{out}/V_{in} = [(1 + T_{Lead} s)/T_{Lag} s]$, $T_{Lead} > T_{Lag}$

(*iii*) Lag network compensation: $V_{out}/V_{in} = [1/(1 + T_{Lag} s)]$

(*iv*) Lag-lead network compensation: $V_{out}/V_{in} = [(1 + T_{Lead} s)/(1 + T_{Lag} s)]$, $T_{Lead} < T_{Lag}$

9.9 OVERALL BODE DIAGRAMS

The 'frequency response' of a control system is usually displayed as a pair of curves, and is called Bode diagrams. One curve shows the overall gain (amplitude change from set point to output signal) versus frequency. The second graph indicates the overall phase angle change from the set point to the output signal. The same frequency scale is used for the two graphs, which are plotted on one diagram to facilitate reading the gain and phase angle at a specific frequency. As with transfer functions, two frequency responses are used for analysing a control system. The 'open-loop frequency response' is obtained with the feedback line disconnected at the error detector, and the 'closed-loop frequency response' is obtained with the feedback line connected.

Since a control system has a number of components connected in series to form a loop, the design and analysis of the system requires information about the overall transfer function and frequency response of the group considered as a unit. Consider, for instance, a group of three components with transfer functions TF_1, TF_2 and TF_3 connected in series. If signal S_1 is the input signal for component 1, the output signal of component 1 is the input signal S_2 of component 2. Similarly, the output signal of component 2 is the input signal S_3 of component 3. Signal S_4 is the output signal of component 3. Also, S_1 is the input signal and S_4 is the output signal of the assembled unit. Now, $TF_1 = S_2/S_1$, $TF_2 = S_3/S_2$, $TF_3 = S_4/S_3$, and the transfer function of the assembly is $TF = S_4/S_1$. Since $S_4/S_1 = (S_2/S_1)(S_3/S_2)(S_4/S_3)$, hence for the assembly the transfer function is $TF = (TF_1)(TF_2)(TF_3)$.

The frequency response is obtained by substituting $j\omega$ for s in the transfer function, which may then be transformed to a complex number in the polar form. The magnitude of the complex number is the gain, g, and the angle is the phase angle, β, of the frequency response. If the frequency response of the three components is represented as $g_1\angle\beta_1$, $g_2\angle\beta_2$ and $g_3\angle\beta_3$, and the overall frequency response of the assembly is $g\angle\beta$, then $g\angle\beta = (g_1\angle\beta_1)(g_2\angle\beta_2)(g_3\angle\beta_3) = g_1\, g_2\, g_3\angle(\beta_1 + \beta_2 + \beta_3)$. Thus, the overall gain of several components attached in series is equal to the product of the gains of the individual components, and the overall phase angle is the sum of the phase angles of the individual components.

The Bode diagram may be plotted using a logarithmic gain scale, so multiplication of the gain terms can be achieved by graphically adding the logarithms of individual gains terms. Since the decibel scale is a logarithmic scale, the addition of decibel values is equivalent to multiplying the associated gain values.

9.10 OPEN AND CLOSED-LOOP BODE DIAGRAMS

The 'open-loop Bode diagram' represents the frequency response of the control system when the measuring transmitter output is disconnected from the error detector (figure 9.8). The Bode diagram is obtained from the open-loop transfer function, given by the expression: $C_m/SP = [G(s)H(s)]$, where SP is the input signal, $G(s)$ is the transfer function for the forward path elements and $H(s)$ is the transfer function for the feedback path elements.

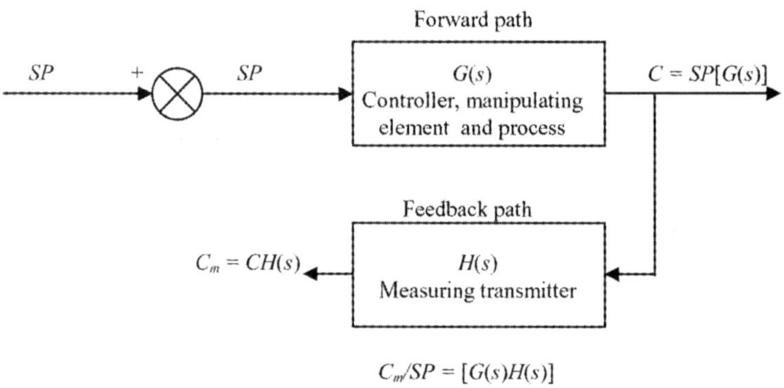

Fig. 9.8 Open-Loop Transfer Function

Multiplying the gains and adding the phase angles of the controller, manipulating element, process and measuring transmitter help in preparing the open-loop Bode diagram. Adding the decibel gains and phase angle graphs of the components may also draw the diagram.

The 'closed-loop Bode diagram' represents the frequency response of the control system when the measuring transmitter output is connected to the error detector, as shown in figure 9.9. The Bode diagram is obtained from the transfer function of the closed-loop control system.

Since the loop is closed, the output from the error detector diminishes the input to the forward path elements, or $E = (SP - C_m)$. The output of the forward path components is: $C = E[G(s)] = (SP - C_m)G(s)$. The output of the feedback path elements is: $C_m = CH(s) = (SP - C_m)G(s)H(s)$. The closed loop transfer function then takes the form: $C_m/SP = [G(s)H(s)]/[1 + G(s)H(s)]$.

The closed-loop values used for constructing the Bode diagram may be obtained by dividing the open-loop response by 1 plus the open-loop response. The open-loop gain and phase angle graphs may be graphically converted into closed-loop Bode diagram of the components with the aid of a Nichols chart.

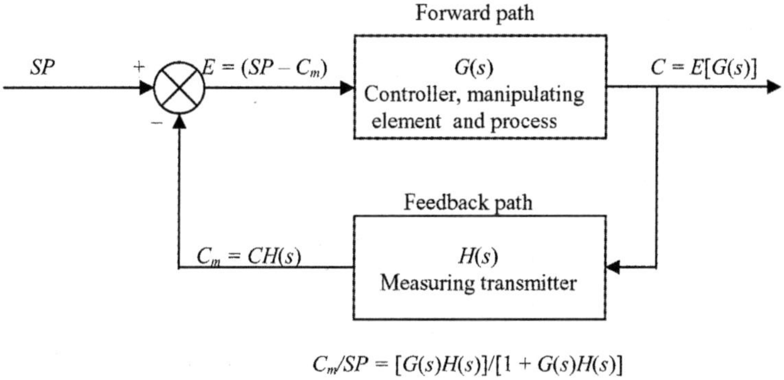

Fig. 9.9 Closed-Loop Transfer Function

9.11 ERROR AND DEVIATION RATIOS

Control systems are not capable of responding accurately to changes in the set point above a maximum frequency limit. The disturbances caused can prove to be annoying above this frequency limit, since the controller cannot reduce the error, and may even increase the error. The frequency limit can be established by comparing the errors with and without feedback. The 'error ratio' provides a useful method for comparison, and is given by the ratio of the closed-loop error magnitude over the open-loop error magnitude.

The error ratio graph in figure 9.10 is split into three frequency zones. In zone 1, the error ratio is less than 1, or 0 db, indicating that the closed-loop error is less than the open-loop error. The controller is capable of reducing the error in zone 1. In zone 2, the error ratio is greater than 1, the closed-loop error is greater than the open-loop error, and the controller is increasing the error. In zone 3, the error ratio is equal to 1, the closed-loop error is equal to the open-loop error, and the controller does not affect the error. The 'maximum frequency limit' is the frequency that separates zones 1 and 2.

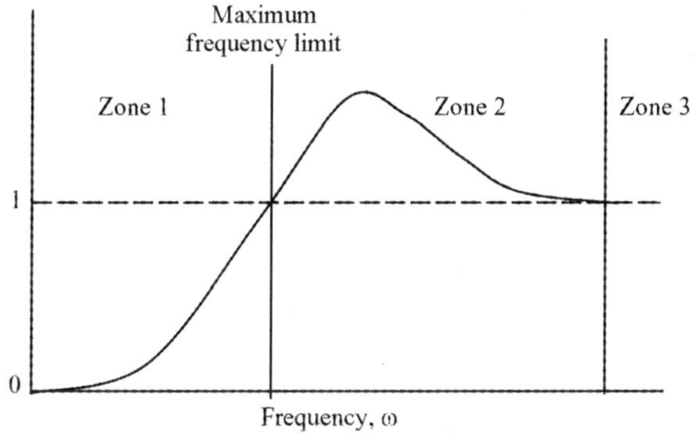

Fig. 9.10 Error Ratio Graph

To determine the mathematical expression for the error ratio, the closed-loop error is given by:

$$\begin{aligned}(SP - C_m) &= SP - SP(C_m/SP) \\ &= SP - SP[G(s)H(s)/\{(1 + G(s)H(s)\}] \\ &= SP[1/\{(1 + G(s)H(s)\}]\end{aligned}$$

and the open-loop error is given by the expression: $(SP - C_m) = SP[1 - G(s)H(s)]$. Hence: Error ratio $= |1/\{[(1 + G(s)H(s)][(1 - G(s)H(s)]\}|$. Since the expression is a complex number, the error ratio is its magnitude. In complex notation,

$$\text{Error ratio} = 1/[(1 + g_o\angle\beta_o)\,(1 - g_o\angle\beta_o)]$$

The 'deviation ratio' provides an indication of how accurately a control system follows a change in the set point, and is given by the ratio of the closed-loop error magnitude and the set point magnitude. The closed-loop error is given by the expression $SP[1/\{(1 + G(s)H(s)\}]$, and the deviation ratio is the magnitude of the ratio of closed loop error and the set point, or deviation ratio $= |1/[(1 + G(s)H(s)]|$. In complex notation, deviation ratio $= 1/(1 + g_o\angle\beta_o)$

9.12 GAIN AND PHASE MARGIN

Besides indicating the stability of a control system, the open-loop Bode diagram also reveals information about the extent to which the gain and phase angle can change before the stable system becomes unstable. The Bode diagram is also an indispensable design tool.

The stability of a closed-loop system may be defined by two conditions. The control system is stable only if the open-loop gain is less than 1, or 0 dB, at the frequency at which the phase angle is –180°, and the phase angle is greater than –180° (or phase lag is less than 180°) at the frequency for which the gain is 1, or 0 dB.

The conditions suggest two margins of safety for the stability of a closed-loop control system. The 'gain margin' at –180° frequency must be a safe amount less than 1, or 0 dB, and the 'phase margin' at 0 dB frequency must be a safe amount above –180°. Generally, a gain of less than 0.5, or –6 dB, at the 180° frequency and a phase greater than –140° at 0 dB is deemed acceptable. A control system is considered marginally stable if it satisfies the stability conditions but does not satisfy both the gain margin and phase margin criteria.

The gain and phase margins can be read directly from the open-loop Bode plot as illustrated in figure 9.11. The frequencies at which the open-loop gain is 0 dB and the phase angle is –180° must be located. The gain margin is obtained by reading the open-loop gain at –180° frequency, called $-m_{-180°}$. The phase margin is obtained by reading the open-loop phase angle at 0 dB to give β_0. The phase margin is $180 + \beta_0$.

9.13 NYQUIST STABILITY CRITERION

Where the function $[1 + G(s)H(s)]$ has positive roots, the 'Nyquist stability criterion' is a useful graphical approach to determine the control system's stability characteristics. The Nyquist diagram is a plot in the complex plane of the open-loop gain and phase angle as the frequency varies from 0 to infinity. The gain and phase angles are treated as magnitude and angle of a complex number in polar form. The decibel values must first be changed into gain values. The Nyquist diagram is a

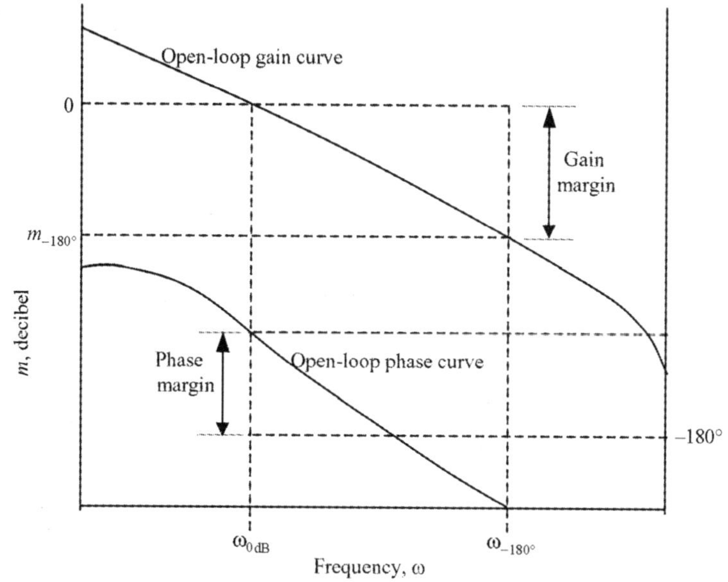

Fig. 9.11 Reading Gain and Phase Margins from Open-Loop Bode Plot

polar plot, but the two parameters may be converted to rectangular form and plotted on rectangular coordinates to yield the same result. Figure 9.12 shows the Nyquist plot of a stable closed-loop control system, with a close-up view showing the three points used for determining the stability of the system.

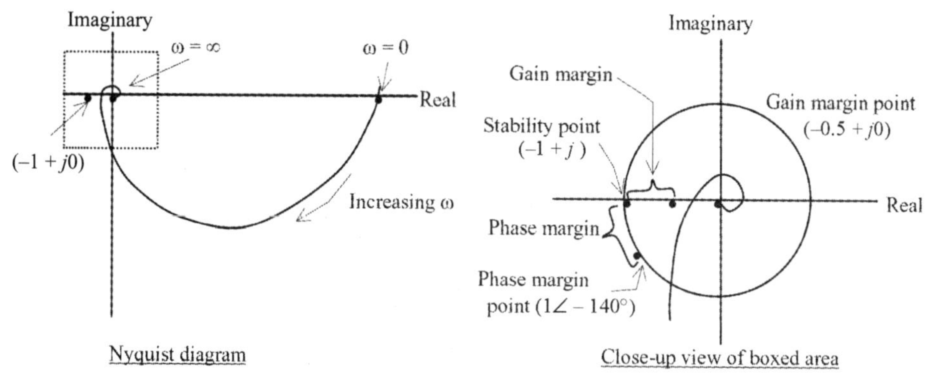

Fig. 9.12 Nyquist Diagram of Stable Closed-Loop Control System

When the control system's open-loop is stable, the Nyquist stability criterion reduces to observing whether or not the plot encloses the point $(-1 + j0)$. If the plot encloses the point, then $[1 + G(s)H(s)]$ has positive roots and the system is unstable. If on the other hand the plot does not enclose $(-1 + j0)$, then $[1 + G(s)H(s)]$ has only negative roots and the system is stable or marginally stable.

The Nyquist stability criterion may be stated as follows. If the open-loop is stable, the closed-loop control system is:

(i) Unstable when the Nyquist plot encloses the point $(-1 + j0) = (1\angle{-180°})$.

(ii) Marginally stable if the Nyquist plot does not encircle the stability point $(-1 + j0)$, but does encircle the gain margin point $(-0.5 + j0)$ or the phase margin point $(1\angle{-140°})$, or both the gain and phase margin points.

(iii) Stable if the Nyquist plot does not encircle the gain margin point $(-0.5 + j0)$ and also does not encircle the phase margin point $(1\angle{-140°})$.

Figure 9.13 shows examples of Nyquist stability plots of stable, marginally stable and unstable control systems. In the marginally stable systems, the gain or the phase or both margins are unacceptable.

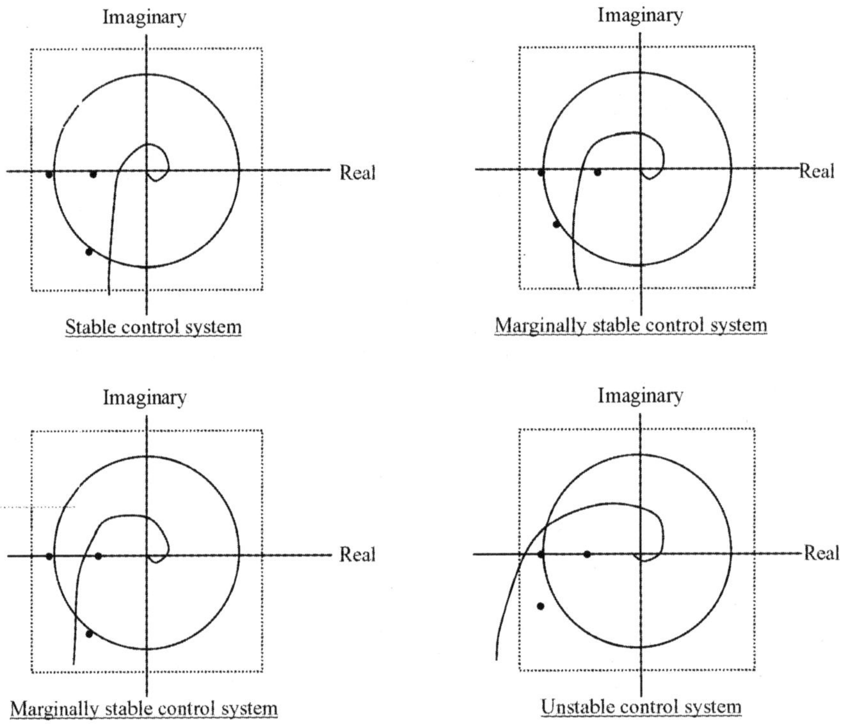

Fig. 9.13 Nyquist Diagrams of Stable, Marginally Stable and Unstable Systems

9.14 ROOT LOCUS

A condition for the stability of a closed-loop control system is that all roots of the 'characteristic equation' $1 + G(s)H(s) = 0$ must have all negative real parts. Note that $G(s)H(s)$ is the open-loop transfer function, while the term $1 + G(s)H(s)$ is the denominator of the closed-loop transfer function.

The 'root locus' is a graphical method for determining all possible roots of the closed-loop characteristic equation. The roots are plotted as the controller gain, K, varies from 0 to infinity. The graph provides an insight into the relation between gain, K, and the stability of the system. In addition, the method provides data about the damping constant. A graphing device called the spirule aids in the construction of the root locus plot.

As an illustration, consider the open-loop transfer function $G(s)H(s) = K/(s + 3)$. If s_0 represents the values of s that satisfies the characteristic equation, then:

$$1 + K/(s_0 + 3) = 0, \text{ or: } s_0 = -(K + 3)$$

The root locus will start at $s_0 = -3$ when $K = 0$ and moves to the left on the real axis as K increases, as shown in figure 9.14*a*.

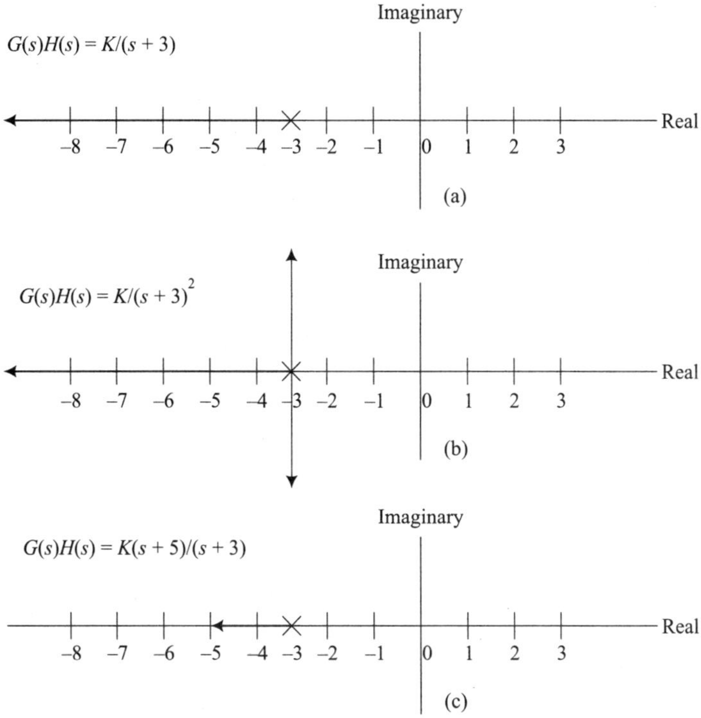

Fig. 9.14 Root Locus Examples

Consider another open-loop transfer function $G(s)H(s) = K/(s + 3)^2$, where the characteristic equation is: $1 + K/(s_0 + 3)^2 = 0$, or: $s_0 = -3 \pm j\sqrt{K}$. The root locus starts at $s_0 = -3$ when $K = 0$ and moves both up and down on a vertical line as K increases, as shown in figure 9.14*b*.

In another example, consider the open-loop transfer function:

$$G(s)H(s) = K(s + 5)/(s + 3).$$

The characteristic equation is: $1 + K(s_0 + 5)/(s_0 + 3) = 0$, or: $s_0(K + 1) + 5K + 3 = 0$. Hence: $s_0 = -(5K + 3)/(K + 1)$

To plot the root locus, note that s_0 has only negative values as K varies from 0 to infinity. When $K = 0$, $s_0 = -3$. When $K = \infty$, divide the numerator and denominator by K. Then: $s_0 = -(5 + 3/K)/(1 + 1/K)$, and $s_0 = -5$. Additional points are needed for this plot as K increases, as shown below.

K	½	1	4	40	400	4000	40000
s_0	–3.67	–4.0	–4.6	–4.90	–4.99	–4.999	–4.9999

9.15 REVIEW PROBLEMS

Example problem # 9.1: A liquid level integral process of the type shown in figure 9.15 has the following parameter values:

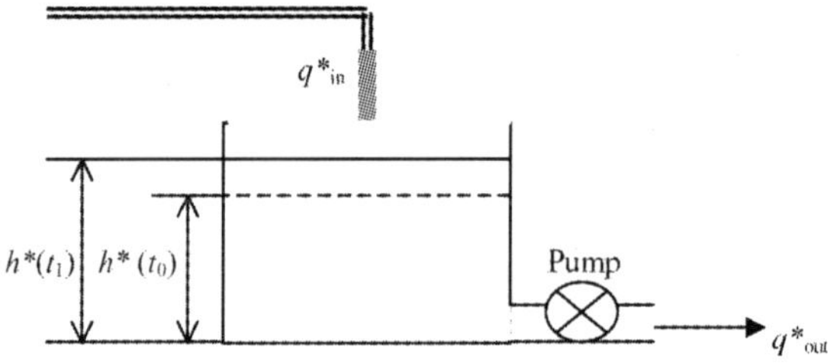

Fig. 9.15

Tank height = 4.5 m

Tank diameter = 1.75 m

$FS_{in} = 0.02\ m^3/s$

$FS_{out} = 5.0$ m

$h^*(t_0) = 24\ \%$ of FS_{out}, or 1.2 m

$q^*_{out} = 50\ \%$ of FS_{in}, or $0.01\ m^3/s$

$q^*_{in} = 70\ \%$ of FS_{in}, or $0.014\ m^3/s$

Determine the time domain equation, transfer function, integral action time constant and level at time $(t_0 + 80)$ s.

Solution: From table 9.1: Tank cross-sectional area $A = \pi d^2/4 = \pi(1.75)^2/4 = 2.405\ m^2$

$$T_i = A(FS_{out}/FS_{in}) = 2.405 \times (5.0/0.02) = 601 \text{ s}$$

Time-domain equation (t_0 to t_1): $h^*(t_1) = (1/601) \times \int[q^*_{in} - q^*_{out}]dt + h^*(t_0)$

Transfer function: $H^*(s)/Q^*_{in}(s) = 1/T_i s = 1/601\ s$

Change in level from t_0 to $t_0 + 80 = (70 - 50)(80)/601 = 2.66\ \%$ of FS

$$\begin{aligned} h^*(t_0 + 80) &= h^*(t_0) + \text{change in level} \\ &= 24 + 2.66 = 26.66\% \text{ of FS} \\ &= (26.66 \times 5.0)/100 \\ &= 1.333 \text{ m} \end{aligned}$$

Example problem # 9.2: A solution tank of the type shown in figure 9.15 has a diameter of 1.40 m and height of 3.2 m. The resistance at the outlet from the tank is 5.0×10^6 Pa-s/m^3. Temperature of the solution is 16°F, and density is 910 kg/m^3. Full-scale ranges are $FS_{in} = 0.00028$ m^3/s and FS_{out} = 3.0 m. Determine (*i*) the capacitance of the tank, (*ii*) the time constant of the process, (*iii*) the gain of the process, (*iv*) the time-domain equation, (*v*) the transfer function.

Solution: From table 9.2: Tank cross-sectional area: $A = \pi d^2/4 = \pi(1.4)^2/4 = 1.539$ m^2

Capacitance: $C_L = A/\rho = 1.539/910 = 16.92 \times 10^{-4}$ m^3/Pa

Time constant: $\tau = R_L C_L = (5.0 \times 10^6) \times (16.92 \times 10^{-4}) = 8458$ second.

Process gain: $G = (R_L/\rho) \times (FS_{in}/FS_{out})$

$= (5.0 \times 10^6/910) \times (0.00028/3.0) = 0.5128$

Time-domain equation: $\tau \times dh^*/dt + h^* = G \times q^*_{in}$

or: $8458 \times dh^*/dt + h^* = 0.5128 \times q^*_{in}$

Transfer function: $H^*(s)/Q^*_{in}(s) = G/(1 + \tau s) = 0.5128/(1 + 8458s)$

Example problem # 9.3: An electrical series *RC* circuit has a resistance of 7500 Ω and 50 μF capacitance. Find (*i*) the time constant and (*ii*) the transfer function.

Solution: From table 9.2: Time constant: $\tau = RC = (7500) \times (50 \times 10^{-6}) = 0.375$ s

Transfer function: $E_{out}(s)/E_{in}(s) = G/(1 + \tau s) = 1/(1 + 0.375s)$, where $G = 1$.

Example problem # 9.4: An oil bath of the type shown in figure 9.16 has a diameter of 1.25 m and height of 1.50 m. The inside film coefficient is 75 W/m^2, and outside film coefficient is 225 W/m^2. The steel (k = 45 W/m, heat capacity S_h = 2250 J/kg-K, density = 900 kg/m^3) wall of the tank is 1.0 cm thick. Determine (*i*) the thermal resistance, (*ii*) thermal capacitance, (*iii*) time constant, (*iv*) time-domain equation, (*v*) transfer function.

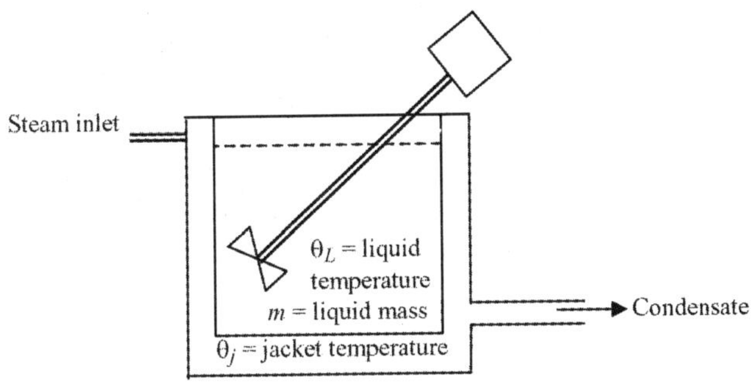

Fig. 9.16

Solution: Tank area: $A = \pi d^2/4 + \pi dh = \pi(1.25)^2/4 + \pi(1.25)(1.5) = 7.118$ m^2

Thermal resistance: $R_T = (1/h_i + x/k + 1/h_o)/A$, kelvin/watt

$= (1/75 + 0.01/45 + 1/225)/7.118 = 0.00253$ K/W

Thermal capacitance: $C_T = mS_h$

Mass of tank $m = \pi\rho hd^2/4 = \pi(900)(1.5)(1.25)^2/4 = 1656.7$ kg

Hence: $C_T = 1656.7 \times 2250 = 3.728 \times 10^6$ J/K

Time constant: $\tau = R_T C_T = (0.00253) \times (3.728 \times 10^6) = 9427$ s

Time-domain equation: $\tau \times d\theta_L/dt + \theta_L = \theta_j$, or: $9427 \times d\theta_L/dt + \theta_L = \theta_j$

Transfer function: $\theta_L(s)/\theta_j(s) = G/(1 + \tau s) = 1/(1 + 9427s)$, where gain $G = 1$.

Example problem # 9.5: A gas pressure process of the type shown in figure 9.17 has a volume of 1.5 m^3, temperature of 530 K, gas flow resistance of 2.25×10^5 Pa-s/kg, $FS_{in} = 1.9$ kg/s and FS_{out} = 225 kPa. Molecular weight of gas is 32. Calculate (*i*) the capacitance, (*ii*) time constant, (*iii*) process gain, (*iv*) time-domain equation, (*v*) transfer function for the pressure vessel.

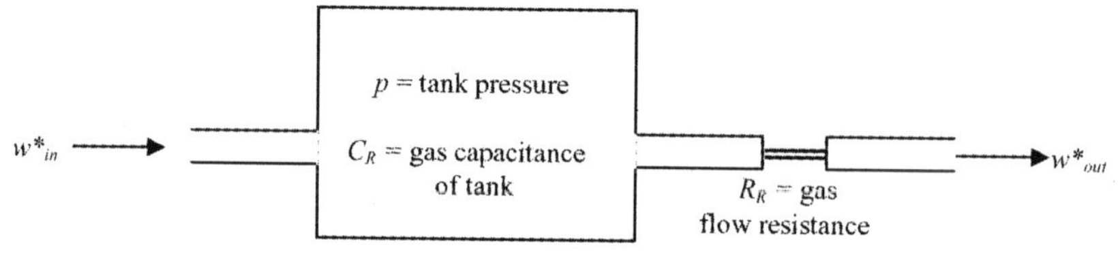

Fig. 9.17

Solution: From section 4.3, capacitance: $C_g = \Delta m/\Delta p = MV/(10^3 RT)$, where C_g = gas flow capacitance, kg/Pa, M = gram molecular weight of gas = 32 g/mol, V = volume of gas = 1.5 m^3, R = universal gas constant = 8.314 J/K-mol, T = absolute temperature = 530 K.

Hence: Capacitance: $C_g = [(32)(1.5)]/[(10^3)(8.314)(530)]$

$= 1.089 \times 10^{-5}$ kg/Pa

Time constant: $\tau = R_g C_g = (2.25 \times 10^5) \times (1.089 \times 10^{-5}) = 2.45$ s

Process gain: $G = R(FS_{in}/FS_{out}) = (2.25 \times 10^5) \times (1.9/225000) = 1.9$

Time-domain equation: $\tau dp^*/dt + p^* = Gw^*$, or: $2.45 \times dp^*/dt + p^* = 1.9w^*$

Transfer function: $P^*(s)/W^*(s) = G/(1 + \tau s) = 1.9/(1 + 2.45s)$

Example problem # 9.6: A blending tank of the type shown in figure 9.18 has a volume of 3.25 m^3, and flow rate of 0.00275 m^3/s. Determine (*i*) the time constant, (*ii*) time-domain equation, and (*iii*) transfer function.

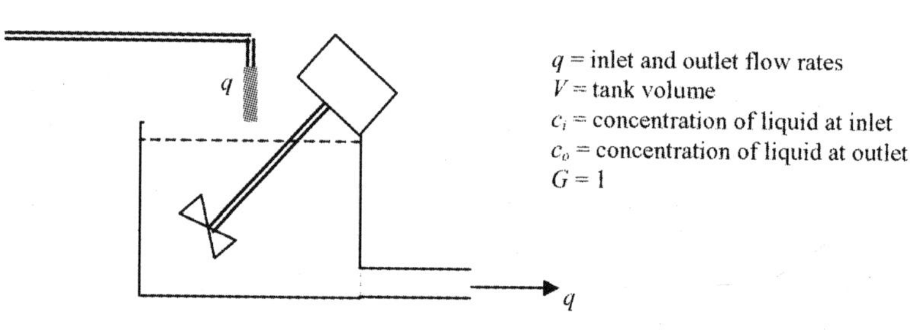

Fig. 9.18

Solution: From table 9.2: Time constant: $\tau = V/q = 3.25/0.00275 = 1182$ s

Time-domain equation: $1182 \times dc_o/dt + c_o = c_i$

Transfer function: $C_o(s)/C_i(s) = G/(1 + \tau s) = 1/(1 + 1182s)$, where $G = 1$.

Example problem # 9.7: A spring-mass-damper consists of mass $M = 12$ kg, spring capacitance $C_m = 0.0025$ m/N, and damping resistance $R = 25$ N-s/m. Calculate the (*i*) time-domain equation, (*ii*) transfer function, (*iii*) resonant frequency, and (*iv*) damping ratio. (*v*) Is the process under damped, over damped or critically damped?

Solution: From table 9.3: If h = output signal, f = input signal, time-domain equation for the mass, spring and damper system is:

$$MC_m \times d^2h/dt^2 + RC_m \times dh/dt + h = C_m f$$

where: $MC_m = 12 \times 0.0025 = 0.03$

$RC_m = 25 \times 0.0025 = 0.0625$

Time-domain equation: $0.03 \times d^2h/dt^2 + 0.0625 \times dh/dt + h = 0.0025f$

Transfer function: $H(s)/F(s) = 0.0025/(1 + 0.0625s + 0.03s^2)$

Resonant frequency: $\omega = \sqrt{(1/0.03)} = 5.774$ radian/s

Damping ratio: $\zeta = 0.0625/(2 \times \sqrt{0.03}) = 0.18$

Damping ratio is less than 1, so system is under damped.

Example problem # 9.8: A series electrical RLC circuit consists of inductor $L = 0.025$ H, capacitor $C = 12$ μF, and resistor $R = 250$ Ω. Calculate the (*i*) time-domain equation, (*ii*) transfer function, (*iii*) resonant frequency, (*iv*) damping ratio, and (*v*) damping type.

Solution: From table 9.3: Time-domain equation for the RLC circuit is:

$$LC \times d^2e_{out}/dt^2 + RC \times de_{out}/dt + e_{out} = e_{in}$$

where: $LC = 0.025 \times 12 \times 10^{-6} = 3.0 \times 10^{-7}$

$RC = 250 \times 12 \times 10^{-6} = 0.003$

Time-domain equation: $3.0 \times 10^{-7} \times d^2e_{out}/dt^2 + 0.003 \times de_{out}/dt + e_{out} = e_{in}$

Transfer function: $E_{out}(s)/E_{in}(s) = 1/(1 + 0.003s + 3.0 \times 10^{-7}s^2)$

Resonant frequency: $\omega = \sqrt{(1/3.0 \times 10^{-7})} = 1826$ radian/s

Damping ratio: $\zeta = (0.003 \times 1826)/2 = 2.74$

Damping ratio is greater than 1, so system is over damped.

Example problem # 9.9: A non-interacting two capacity liquid system has time constants τ_1 and τ_2 of 485 s and 890 s. The liquid of density $\rho = 1150$ kg/m^3 has $R_2 = 1.75 \times 10^6$ Pa-s/m^3. Full-scale ranges are $FS_{in} = 5.5 \times 10^{-3}$ and $FS_{out} = 1.75$ m. Calculate the (*i*) time-domain equation and (*ii*) transfer function.

Solution: From table 9.3: $\tau_1\tau_2 = 485 \times 890 = 431650\ s^2$

$\tau_1 + \tau_2 = 485 + 890 = 1375$

$G = (R_2/\rho g)(FS_{in}/FS_{out}) = [1.75 \times 10^6/(1150 \times 9.81)] \times (5.5 \times 10^{-3}/1.75)$

$= 0.488$

Time-domain equation: $431650d^2h^*/dt^2 + 1375dh^*/dt + h^* = 0.488q^*_{in}$

Transfer function: $H^*(s)/Q^*_{in}(s) = 0.488/(1 + 1375s + 431650s^2)$

Example problem # 9.10: A non-interacting two capacity electrical system has components $R_1 = 1200\,\Omega$, $R_2 = 1400\,\Omega$, $R_L = 330\Omega$, $C_1 = 14\ \mu F$, and $C_2 = 75\ \mu F$. Calculate the (*i*) time-domain equation and (*ii*) transfer function.

Solution: From table 9.3: $\tau_1 = R_1C_1 = 1200 \times 14 \times 10^{-6} = 0.0168$

$$\tau_2 = R_2C_2 = 1400 \times 75 \times 10^{-6} = 0.105$$

$$G = R_L/(R_1 + R_2 + R_L) = 330/(1200 + 1400 + 330) = 0.113$$

$$A_1 = G\{(\tau_1 + \tau_2) + \tau_1R_2/R_L + \tau_2R_1/R_2\}$$

$$= 0.113\{(0.0168 + 0.105) + 0.0168 \times 1400/330 + 0.105 \times 1200/1400\}$$

$$= 0.0319$$

$$A_2 = G\tau_1\tau_2 = 0.113 \times 0.0168 \times 0.105 = 0.000199$$

Time-domain equation: $0.000199 \times d^2e_{out}/dt^2 + 0.0319 \times de_{out}/dt + e_{out} = 0.113e_{in}$

Transfer function: $E_{out}(s)/E_{in}(s) = 0.113/(1 + 0.0319s + 0.000199s^2)$

Example problem # 9.11: An armature-controlled DC motor has $J = 7.35 \times 10^{-4}$ kg-m^2, $B = 1.40 \times 10^{-4}$ N-m-s/radian, $L = 0.025$ H, $R = 1.5\ \Omega$, $K_t = 0.065$ N-m/A, $K_e = 0.075$ volt-s/radian. Find the (*i*) mechanical time constant, (*ii*) electrical time constant, (*iii*) time-domain equation, and (*iv*) transfer function.

Solution: Mechanical time constant: $\tau_m = J/B = 7.35 \times 10^{-4}/1.40 \times 10^{-4} = 5.25$ s

Electrical time constant: $\tau_e = L/R = 0.025/1.5 = 0.0167$ s

$$A_2 = \tau_m\tau_eRB/(K_eK_t + RB)$$

$$= (5.25 \times 0.0167 \times 1.5 \times 1.40 \times 10^{-4})/(0.075 \times 0.065 + 1.5 \times 1.40 \times 10^{-4})$$

$$= 0.00361$$

$$A_1 = (\tau_m + \tau_e)RB/(K_eK_t + RB)$$

$$= (5.25 + 0.0167) \times 1.5 \times 1.40 \times 10^{-4}/(0.075 \times 0.065 + 1.5 \times 1.40 \times 10^{-4})$$

$$= 0.2175$$

$$G = K_t/(K_eK_t + RB) = 0.065/(0.075 \times 0.065 + 1.5 \times 1.40 \times 10^{-4})$$

$$= 12.78$$

Time-domain equation: $0.00361 \times d^2\omega/dt^2 + 0.2175 \times d\omega/dt + \omega = 12.78e$

Transfer function: $\Omega(s)/E(s) = 12.78/(1 + 0.2175s + 0.00361s^2)$

Example problem # 9.12: A dead-time process consists of a 15 m long conveyor belt traveling at 0.75 m/s. Determine the (*i*) dead-time lag, (*ii*) time-domain equation, (*iii*) transfer function, and (*iv*) the Bode diagram gain and phase line equations.

Solution: Dead-time lag: $t_d = D/v = 15.0/0.75 = 20.0$ s

Time-domain equation: $f_o(t) = f_i(t - 20)$

Transfer function: $F_o(s)/F_i(s) = e^{-20s}$

Bode diagram gain line is a constant 0 dB, and the phase line equation is:

$$\beta = -57.3\omega t_d = -57.3 \times 20\omega = -1146\omega$$

Example problem # 9.13: The derivative mode of a process control system tested at start-up is turned off, and the integral mode is at the minimum setting. Gain is gradually increased until the controlled variable begins oscillating. At gain = 2.5 the period of oscillation is 14 minutes. Calculated the PID controller settings using the ultimate cycle method.

Solution: $P = 0.5G_u$, or $I = 2.0/P = 2.0/14 = 0.143$ per minute

For a PID controller, $D = P/8 = 14/8 = 1.75$ minute

And $P = 0.6G_u = 0.6 \times 2.4 = 1.44$

Example problem # 9.14: The control valve of a process control system tested at start-up is maintained constant until the controlled variable reaches a constant value. The control valve position is then changed by 12 per cent, giving a response plot similar to figure 9.7 and the following values: $L = 7$ minutes, $\Delta t = 13$ minutes, initial value of $C_m = 43$ per cent, and final value of $C_m = 52$ per cent. Determine the settings using the process reaction method.

Solution: The process is approximated as a first-order lag plus dead-time model, hence the process reaction method is used to determine the settings. From section 9.6, the slope of the tangent line $N = \Delta C_m/\Delta t = (52 - 43)/(13 - 7) = 1.5\%$/minute.

Proportional control: $\Delta P = 12\%$, $L = 7$ minutes and $P = 1.2(\Delta P/NL)$. Hence:

$$P = 1.2 \times 12/(1.5 \times 7) = 1.37$$

Integral control: $I = 0.5/L = 0.5/7 = 0.071$ per minute.

$$D = 0.5L = 0.5 \times 7 = 3.5 \text{ minutes.}$$

Example problem # 9.15: A blending and heating process calls for three loops to control concentration, temperature and level in a tank, as shown in figure 9.19. Specifications are as follows:

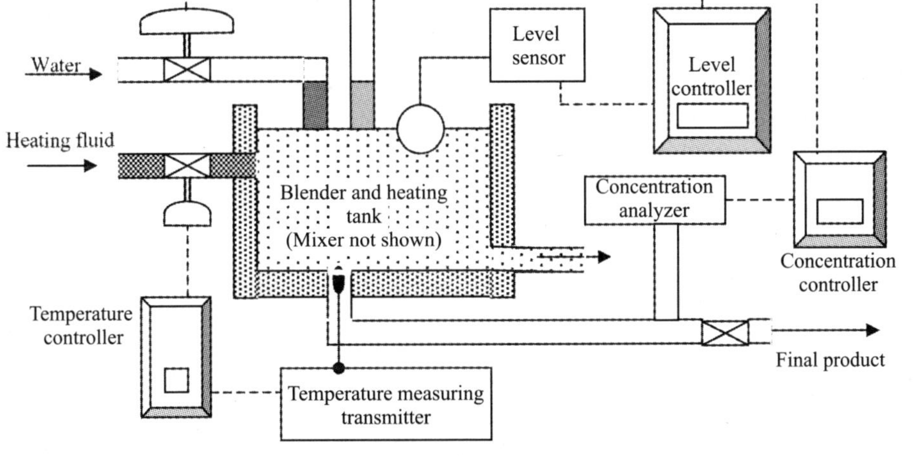

Fig. 9.19

Production rate: $Q = 2.25 \times 10^{-4}$ m^3/s, outlet flow velocity 0.32 m/s.

Blend properties: Composition = 92% water, 8% acid, density ρ = 1020 kg/m^3, viscosity μ = 0.1 Pa-s, specific heat H_h = 4240 J/kg-K, temperature = 65°C.

Tank: Diameter = 0.85 m, operating height = 1.2 m, wall thickness x = 0.015 m

Tank wall material: steel (conductivity K = 45 W/m-K, density ρ = 1020 kg/m^3, heat capacity S_h = 2250J/kg-K), operating level = 0.85 m.

Heating fluid temperature = 110°C, film coefficient $h_i = h_o$ = 1275 W/m^2-K.

Outlet length = 0.25 m.

Concentration measuring device: Model – first-order lag plus dead time.

Time constant ω_1 = 180 s, dead-time delay t_{d1} = 12 s, gain = G_C = 1.

Level measuring transmitter: Model – first-order lag plus dead time.

Time constant ω_2 = 3 s, dead-time delay t_{d2} = 0.8 s, gain = G_L = 1.

Temperature measuring transmitter: Model – over-damped second-order lag.

Time constant τ_3 = 48 s and τ_4 = 270 s, gain = G_T = 1.

Control valves: Model – under-damped second-order lag.

Water control valve: ω_{01} = 11.5 radian/s, ζ_1 = 0.70, gain $G(1)$ = 4.

Acid control valve: ω_{02} = 3.2 radian/s, ζ_2 = 0.85, gain $G(2)$ = 32.5.

Heating valve ω_{03} = 22.7 radian/s, ζ_3 = 0.95, gain $G(3)$ = 9.

Concentration process: Model (τ_5) – first-order lag, gain = G_P = 1.

Level process: Model (τ_5) – integral, FS$_{\text{in}}$ = 0.0025 m^3/s, FS$_{\text{out}}$ = 0.9 m.

Thermal process: Model (τ_6, τ_{d3}) – first-order lag plus dead time, gain =G_{Th} = 1

Design a PID controller for the concentration control loop in the blending and heating process.

Solution: Liquid volume in tank: $V = \pi d^2 h/4 = \pi(0.85)^2 \times (1.2)/4 = 0.6809$ m^3

Liquid flow rate $Q = 2.25 \times 10^{-4}$ m^3/s

Concentration process time constant $\tau_5 = V/Q = 0.6809/(2.25 \times 10^{-4}) = 3026$ s

Transfer coefficients: Acid valve $B(21) = 2\zeta_2/\omega_{02} = 2 \times 0.85/3.2 = 0.531$

$$B(22) = 1/(\omega_{02})^2 = 1/(3.2)^2 = 0.0977$$

The overall transfer function of concentration measuring device, concentration process and acid control valve:

$$\text{TF} = [G_C e^{-td1s}/(1 + \omega_1 s)][G_P/(1 + \tau_5 s)][G(2)/(1 + B(21)s + B(22)s^2)]$$
$$= [e^{-12s}/(1 + 180s)][1/(1 + 3026s)][32.5/(1 + 0.531s + 0.0977s^2)]$$

Example problem # 9.16: Design a PID controller for the temperature control loop in the blending and heating process in example problem # 9.15.

Solution: Thermal resistance of tank: $h_i = h_o$ = 1275 W/m^2-K

Tank area: $A = \pi d^2/4 + \pi dh = \pi(0.85)^2/4 + \pi(0.85)(1.2) = 3.772$ m^2

Tank material K = 45 W/m-K, wall thickness x = 0.015 m

Thermal resistance: $R_T = (1/h_i + x/K + 1/h_o)/A$

$= (1/1275 + 0.015/45 + 1/1275)/3.772 = 5.04 \times 10^{-4}$ K/W

Mass $m = \pi\rho d^2h/4 = \pi \times 1020 \times 0.85^2 \times 1.2/4 = 694.6$ kg

Thermal capacitance: $C_T = mS_h = 694.6 \times 4240 = 2.945 \times 10^6$ J/K

Time constant $\tau_6 = R_TC_T = 5.04 \times 10^{-4} \times 2.945 \times 10^6 = 1485$ s

Dead-time lag t_{d3} = length/velocity = 0.25/0.32 = 0.781 s

Temperature measuring transmitter transfer function:

$$B(1) = \tau_3 + \tau_4 = 48 + 270 = 318 \text{ s}, \; B(2) = \tau_3 \times \tau_4 = 48 \times 270 = 12960$$

Heating control valve transfer function:

$$B(31) = 2\zeta_3/\omega_{03} = 2 \times 0.95/22.7 = 0.0837, \; B(32) = 1/(\omega_{03})^2 = 1/22.7^2 = 0.001941$$

Overall transfer function of thermal process, acid control valve and heating control valve:

$$\text{TF} = [G_{Th}e^{-td3s}/(1+ \tau_6 s)][G(2)/\{1 + B(1)s + B(2)s^2\}][G(3)/\{1 + B(31)s + B(32)s^2\}]$$

$$\text{TF} = [e^{-0.781s}/(1 + 1485s)][32.5/(1 + 318s + 12960s^2)][9/(1 + 0.0837s + 0.001941s^2)]$$

Example problem # 9.17: Design a PID controller for the level control loop in example problem # 9.15.

Solution: Tank cross-section area: $A = \pi d^2/4 = \pi(0.85)^2/4 = 0.5675$ m^2

Integral action time constant for the level process:

$$T_i = A(\text{FS}_{out}/\text{FS}_{in}) = 0.5675 \times (0.9/0.0025) = 204.28 \text{ s}$$

Water control valve transfer function coefficients:

Water control valve $B(11) = 2\zeta_1/\omega_{01} = 2 \times 0.7/11.5 = 0.122$

$$B(21) = 1/(\omega_{01})^2 = 1/(11.5)^2 = 0.007561$$

Overall transfer function of level measuring transmitter, water control valve and integral process:

$$\text{TF} = [G_L e^{-td2s}/(1+ \omega_2 s)][G(1)/\{1 + B(11)s + B(21)s^2\}][1/(T_i s)]$$

$$\text{TF} = [e^{-0.8s}/(1 + 3s)][4/(1 + 0.122s + 0.007561s^2)][1/(204.28s)]$$

9.16 EXERCISES

9.1 A liquid level integral process of the type shown in figure 9.15 has the following parameter values:

Tank height = 11.25 ft

Tank diameter = 6.475 ft

FS_{in} = 0.155 ft^3/s

FS_{out} = 12.78 ft

$h^*(t_0)$ = 18% of FS_{out}

$q^*_{out} = 58\%$ of FS_{in}

$q^*_{in} = 82\%$ of FS_{in},

Determine the time domain equation, transfer function, integral action time constant and level at time $(t_0 + 65)$ s.

9.2 A solution tank has a diameter of 5.65 m and height of 11.35 m. The resistance at the outlet from the tank is 3.25×10^8 Pa-s/m^3. Temperature of the solution is 28°F, and density is 830 kg/m^3. Full-scale ranges are $FS_{in} = 0.00039$ m^3/s and $FS_{out} = 4.2$ m. Determine (*i*) the capacitance of the tank, (*ii*) the time constant of the process, (*iii*) the gain of the process, (*iv*) the time-domain equation, (*v*) the transfer function.

9.3 An electrical series *RC* circuit has a resistance of 3550 Ω and 80 μF capacitance. Find (*i*) the time constant and (*ii*) the transfer function.

9.4 An oil bath has a diameter of 4.30 ft and height of 5.25 ft. The inside film coefficient is 33 W/ft^2, and outside film coefficient is 85 W/ft^2. The steel (K = 16 W/ft, heat capacity S_h = 1070 J/kg-K, density = 0.25 lb/ft^3) wall of the tank is 0.6 in. thick. Determine (*i*) the thermal resistance, (*ii*) thermal capacitance, (*iii*) time constant, (*iv*) time-domain equation, (*v*) transfer function.

9.5 A gas pressure process has a volume of 1.85 m^3, temperature of 495 K, gas flow resistance of 3.47×10^5 Pa-s/kg, $FS_{in} = 2.23$ kg/s and $FS_{out} = 311$ kPa. Molecular weight of gas is 56. Calculate (*i*) the capacitance, (*ii*) time constant, (*iii*) process gain, (*iv*) time-domain equation, (*v*) transfer function for the pressure vessel.

9.6 A blending tank has a volume of 28.76 ft^3, and flow rate of 0.125 ft^3/s. Determine (*i*) the time constant, (*ii*) time-domain equation, and (*iii*) transfer function.

9.7 A spring-mass-damper consists of a 33.5 lbs mass, a spring capacitance of 0.125 in/lb, and a damping resistance of 28 lb-s/in. Calculate the (*i*) time-domain equation, (*ii*) transfer function, (*iii*) resonant frequency, (*iv*) damping ratio. (*v*) Is the process under damped, over damped or critically damped?

9.8 A series electrical RLC circuit consists of a 0.105 H inductor, a 9 μF capacitor and a 750 Ω resistor. Calculate the values of (*i*) time-domain equation, (*ii*) transfer function, (*iii*) resonant frequency, (*iv*) damping ratio, and (*v*) type of damping.

9.9 A non-interacting two capacity liquid system has time constants τ_1 and τ_2 of 320 s and 785 s. The liquid of density ρ = 49.5 lb/ft^3 has $R_2 = 21.5 \times 10^6$ Pa-s/ft^3. Full-scale ranges are $FS_{in} = 2.95 \times 10^{-3}$ and $FS_{out} = 1.125$ m. Calculate the (*i*) time-domain equation and (*ii*) transfer function.

9.10 A non-interacting two capacity electrical system has components R_1 = 473 Ω, R_2 = 539 Ω, R_L = 169 Ω, C_1 = 59 μF, and C_2 = 23 μF. Calculate the (*i*) time-domain equation and (*ii*) transfer function.

9.11 An armature-controlled DC motor has $J = 3.48 \times 10^{-5}$, lb-ft^2, $B = 2.13 \times 10^{-4}$, lb-ft-s/radian, L = 0.107 H, R = 2.1 Ω, K_t = 0.078 N-m/A, K_e = 0.081 volt-s/radian. Find the (*i*) mechanical time constant, (*ii*) electrical time constant, (*iii*) time-domain equation, and (*iv*) transfer function.

9.12 A dead-time process consists of a 45 ft long conveyor belt traveling at 2.47 ft/s. Determine the (*i*) dead-time lag, (*ii*) time-domain equation, (*iii*) transfer function, and (*iv*) the Bode diagram gain and phase line equations.

9.13 The derivative mode of a process control system tested at start-up is turned off, and the integral mode is at the minimum setting. Gain is gradually increased until the controlled variable begins oscillating. At gain = 4.85 the period of oscillation is 9 minutes. Calculated the PID controller settings using the ultimate cycle method.

9.14 The control valve of a process control system tested at start-up is maintained constant until the controlled variable reaches a constant value. The control valve position is then changed by 7 per cent, giving a response plot similar to figure 9.7 and the following values: $L = 5.2$ minutes, $\Delta t = 8.3$ minutes, initial value of $C_m = 59$ per cent, and final value of $C_m = 31$ per cent. Determine the settings using the process reaction method.

References & Bibliography

Floyd, T.L., "Digital Fundamentals", Macmillan Publishing Co., New York, 1990.

Herman, S.L. and Alerich, W.N., "Industrial Motor Control", Delmar Publishers, Inc., Albany, NY, 1985.

Johnson, C.D., "Process Control Instrumentation Technology", Prentice Hall, Upper Saddle River, NJ, 1997.

Jones, C.T. and Bryan, L.A., "Programmable Controllers: Concept and Applications", International Programmable Controls, Inc., Atlanta, 1983.

Kilian, C.T., "Modern Control Technology Components and System", West Publishing Co., St. Paul, MN, 1996.

Kozoh, O., "Development of Multi-Cam High Torque Hydraulic Motor", Hitachi-Hyoron, volume 56, Number 11, 1974.

Morris, N.M., "Control Engineering", McGraw-Hill Book Co., (UK) Ltd., 1968.

Reliance Motor Control, "Electro-Craft Handbook – Brushless Servo Systems", Racine, WI, 1992.

Shikano, M., "Application Technology on Control Use of Actuators", Encyclopaedia of Factory Automation, volume 6, 1983.

Shimazu Manufacturing Co., "Shimazu Gear Motor", Shimazu Catalogue.

Stein, D.H., "Introduction to Digital Data Communications", Delmar Publishers, Albany, NY, 1985.

Webb, J.W. and Reis, R.A., "Programmable Logic Controllers", Prentice Hall, Upper Saddle River, NJ, 1995.

Zeines, B., "Automatic Control Systems", Prentice-Hall, Inc, Englewood Cliffs, NJ, 1972.

Part III

Turbo-Machinery Operation and Control

10

Turbo-Machinery Applications

10.1 INTRODUCTION

Throughout the history of mechanical devices the power of rising hot air has been recognized in performing useful work, becoming apparent that efficiency is related to the use of high temperatures. This observation led to the development of the thermo-dynamic Brayton cycle with its basic physical tenet that higher operating temperatures (in conjunction with lower heat rejection temperatures) leads to enhanced efficiency. Applying the concept to rotating engines, improvements were introduced in steam turbines for power generation appearing in the 1800's and to gas turbines in the 1900's.

In the context of the evolution from piston to rotating engines, first exercised in steam turbines, the design target for the coming gas turbine becomes clearer: partial replacement of the Otto (or diesel) piston engine's linearly accelerating and decelerating pistons, connecting rods and cranks; and elimination of a steam plant's boiler, condenser with circulating water and condensate and feed water pumps. However, the first gas turbine patent of J. Barber in United Kingdom and the subsequent development work of F. Stolze in Berlin at the end of the nineteenth century indicated a decisive difficulty on this path: a gas turbine with a net useful power output is only possible if the total expansion work of the turbine exceeds the compression work.

The French R. Armengoud and C. Lemale in Paris built the first standalone power generating gas turbine in 1905–06. While Stolze searched for an axial design, the Frenchmen took advantage of a proven radial compressor design. The first experimental gas turbine consisted of 25 radial stages (system Rateau) with two intercoolers and a single stage turbine (also following Rateau's design). The unit achieved self-sustaining operation by injecting steam into the turbine, generated from cooling of the combustor. The efficiency of the unit should only be in the 2 to 3 per cent range, or 6 to 10 kW of equivalent power produced. The reason for the poor performance of this early design is due to the turbine inlet temperature of around 830 K at the nozzles. The radial compressor cannot furnish the large volume of compressed air required to reduce the combustion temperature of around 2250 K to the value admissible for the turbine blades. With water injected into the combustion chamber, the turbine was just able to supply merely the power for compressing the air.

First ideas for an advanced compressor design were formed through a series of tests on windmills of 8 to 10 m in diameter. C. Seippel at BBC carried out practical transfer of the single airfoil theory towards the design of a multistage axial compressor. The Gottingen airfoil no. 265 was selected as most appropriate based on lift/drag polar plots published by L. Prandtl, transformed by conformal mapping into a cascade flow and followed by experimental verification. The tests permitted reconstruction of the polar plots of the actual cascaded airfoil as a reliable design base. The world's first utility gas turbine went into operation at the Swiss town of Neuchatel, with a simple overall configuration of 23 stage axial compressor, single can combustor, 7 stage axial turbine and a synchronous generator mounted on the same shaft. This constant pressure design generated a maximum of 4 MW at a total thermal efficiency of 17 per cent, compressor pressure ratio of 4.4, adiabatic compressor efficiency of about 85 per cent, turbine inlet temperature of 820 K, turbine efficiency of 88 per cent and mass flow of 66.2 kg/s (*ref*: Eckardt et al., 2001).

Gas turbines, and more generally 'turbo-machinery', emerged in the wake of early electrification. In 1867 Werner von Siemens presented the first dynamo after the discovery of the principle of electrodynamics. In 1879 Thomas A. Edison invented the light bulb, which eventually led to the creation of the powerful General Electric Company in 1895 to manufacture power generation equipment. More than 100 years ago engineers at Switzerland's Brown-Boveri Company made significant contributions to the achievement of today's advanced gas turbine concept. And in 1891 Charles E.L. Brown succeeded in transmitting 220 kW of power from Lauffen/Neckar to Frankfurt/Main, a distance of 175 km. This offered the considerable prospects of not requiring power generation and consumption at the same site. An electrical cable could now link the source of energy with the place at which it was utilized; thus, the times of large mechanical power transmission trains were over. Interestingly, these pioneering companies are still today among the leading players in the industry.

Performance driven competition often characterizes the choice between steam and gas turbines. The simple cycle gas turbine in many cases may be the right selection when operation is restricted to meet peak loads in demand and where fuel price is competitive. Construction of the industrial gas turbine is simplified by using higher power density, a single shaft supported in two bearings and the generator located at the compressor end. The combustion turbines also offer the benefits of low initial installation cost and build time. Turbine inlet temperatures have gradually increased over the years by introducing thermal barrier coated super alloys with better corrosion resistance and by cooling of the airfoils through internal cooling ducts in the first turbine stages. Even with these increased temperatures, operation in the simple cycle mode cannot reach the efficiency of steam turbines.

Concurrent with the development for power generation, the application of gas turbine engines for propelling airplanes held even greater and exciting promise. In the time period between the two world wars piston engines were exclusively employed for propelling small aircraft; even today many light and medium-light aircraft use this engine. However, the requirement of low weight and the generation of inertia loads from the reciprocating parts caused excessive vibrations and consequent dynamic stresses in the engine and the airframe. The advanced axial compressor designs finally became the key to the successful realization of the aero gas turbine after 1940. Both England and Germany went through earlier jet engine configurations with radial compressor designs, although evidence indicates Germany made the transition to an all axial version faster. Flight Lieutenant Frank Whittle of Great Britain's RAF applied for a patent in 1930, where he describes a jet engine with a multistage axial compressor followed by a radial stage, annular combustor, single stage axial turbine and an exhaust nozzle.

Jet powered flight research was done in great secrecy due to the pressures and exigencies before and during WW II, reflecting the fact that fighters and large bombers played a pivotal role in the war machinery of both sides. On August 27, 1939, four days before the start of the Second World War, the first jet airplane took off in Germany to open a new era in aviation. Lasting a mere 6 minutes, the Heinkel He-178 aircraft reached a top speed of 400 mph, nearly 100 mph faster than conventional piston engine propelled fighters. The wailing HeS3B 1100 lb thrust turbojet engine was the brainchild of Hans von Ohain. Many difficulties remained in developing the military jets. Experience led to essential design advances such as replacing the standard tail wheel on piston engine aircraft with a nose wheel to stop the jet efflux from hitting the runway at take-off. But production of a sufficiently reliable jet engine proved time consuming, and remained essentially experimental. Later in the war, however, the jet engines were not without impact. In Germany the Messerscchmitt Me-262 twin-engined fighter powered by the Jumo 004A engine first flew in June 1942. By the spring of 1945 almost 6,000 Jumo 004A jet engines had been built.

A versatile energy converter, gas turbines produce thrust to propel most of the world's military and commercial aircraft. The largest commercial jet engines can produce as much as 120,000 lbs (or 534,000 newtons) of thrust. Engines for the multi-role Mach 1.5-plus Joint Strike Fighter will be made by Pratt and Whitney Company. The life of the US Air Force's fleet of giant four engine C-5 Galaxy transport aircraft is expected to double by replacing the turbofan engines.

Propulsive forces are generated by the reaction of exhaust gases from the nozzle in turbojet and turbofan engines. Turbofan engines bypass a certain portion of the airflow around the engine to generate thrust from the fan through the fan's exhaust duct. In turboprop engines the energy of the hot gas drives an additional turbine, and that in turn powers the propellers. A small percentage of the remaining hot gas energy also generates thrust.

Taking advantage of the lighter weight characteristics of aircraft engines, a new breed of derivative gas turbines has been developed expressly for ships, offshore oil platforms and land vehicles. By eliminating the fan, the power generated by the turbine drives the ships propellers, electrical generator or other equipment. Turbines rated at 1,100 kW are slated to power the US Army's M1A2 Abrams main battle tank and the next generation Crusader armoured vehicle. Rolls Royce manufactured recuperated and inter-cooled gas turbines propel destroyers for the Royal Navy. The gas turbine powered cruise ship *Millennium* went into service in 2001.

In desert, arctic and marine applications, gas turbine designs derived from aviation engine technology have demonstrated the flexibility to operate on a range of fuel and fuel/water combinations. The engines are also ideally suited for cogeneration and combined-cycle applications. Overall system efficiency can be greatly increased by using the exhaust gas heat to generate steam, which then may be used for other process requirements.

General Electric Company's LM6000 engine is developed from the highly successful CF6-80C2 commercial aircraft engine, generating 54,000 shaft horsepower (shp) from either end of the low-pressure rotor system at 3,600 rpm. This feature eliminates the need for a conventional power turbine. It provides the power and unprecedented efficiency needed by users at an installed cost that is competitive with any gas turbine. The LM6000 is suitable for a variety of marine applications, including fast ferry and high-speed cargo ship applications. GE's LM2500+ entered marine service in 2000 aboard the first cruise ship for Celebrity Cruises. Today, it is in service aboard Princess Cruises' Coral Princess. The Coral Princess's gas turbine is operated in a combined diesel and gas turbine configuration with four diesel engine-driven alternators. The LM2500+ has shaft horsepower of 33,600, specific fuel consumption of 0.373 lbs/shp-hr, 37% thermal efficiency and 6,860 BTU/shp-hr heat rate. All components incorporate corrosion-resistant materials and coatings to provide maximum parts life and reliability when operating in harsh environments. Its high efficiency and installation flexibility make it ideal for a wide variety of utility power generation and industrial applications, with high potential for marine applications.

Large vehicle and marine diesel engines are turbo-charged expressly with the intent of increasing the density of air or air-fuel charge prior to intake into the cylinders. Compared with a naturally aspirated engine, the increased amount of oxygen thus induced permits a greater quantity of fuel to be combusted following the compression stroke. As a consequence, the power delivered by the engine is increased. Turbo-charging is a special form of supercharging the air delivered to the cylinders in that the process uses a portion of the energy escaping into the atmosphere from the engine's exhaust to power a turbine wheel that is arranged in tandem with a centrifugal compressor stage on a single shaft. Extracting up to a third of the waste energy from the engine exhaust, the process results in improved volumetric efficiency by raising the incoming flow velocity and in higher cylinder pressure.

10.2 AXIAL AND RADIAL FLOW COMPRESSORS

Air flowing through the annulus of an axial compressor performs the relatively more difficult task of going against an adverse pressure gradient. As the compression ratio increases, the danger of stall troubles rises correspondingly, which is all too prevalent in axial compressors. In individual airfoils, stalling is encountered when the difference between the flow direction and the blade's angle of incidence becomes excessively large. The stability of the flow is jeopardized by the very fact that the movement is taking place in the same direction as that of the increase in pressure. Flow reversals occur under the right conditions of mass flow and rotor speed, and which vary from those for the blade's design.

Air is first imparted acceleration by the rotating blades and then decelerated in the stator passages, where the kinetic energy is converted to static pressure. A series of diffusions take place in both the rotor and stator passages. Absolute velocity of the working fluid is increased in the rotor, but the relative velocity will decrease. It is imperative for flow area to diminish moderately in a diffusing flow, hence a single axial compressor stage can only deliver a small increment in the pressure, considerably less then what a turbine can advantageously use with its efficient pressure gradient, converging blade passages and accelerating flow (figure 10.1). This explains why a few turbine stages can power many more stages of the compressor. As fluid density increases with progression of the flow, axial velocity does not change appreciably. During operation at lower speeds density in the latter stages deviates further from the design point, causing the flow velocity to reach a point where the blades may stall and the compressor to surge.

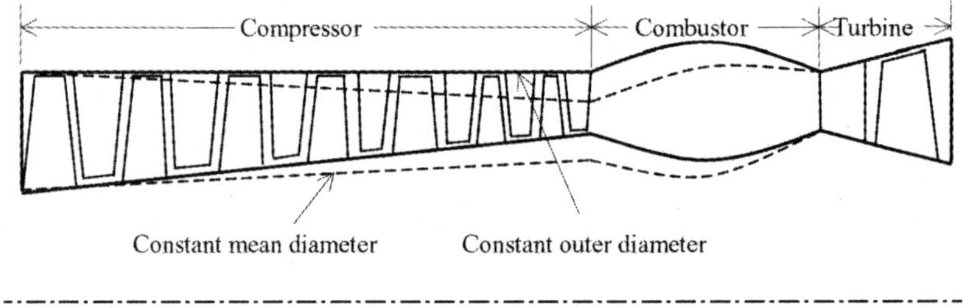

Fig. 10.1 Flow Path Configuration of Gas Turbine

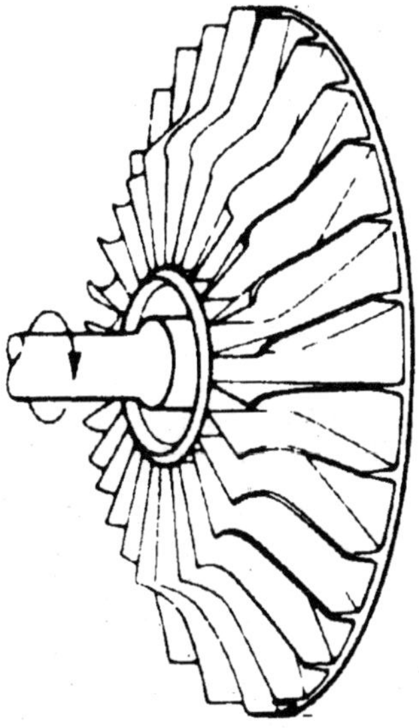

Fig. 10.2 Single Stage Centrifugal Compressor Rotor

Figure 10.2 shows the impeller of a single stage radial compressor. Air enters the eye of the impeller axially and then goes through a 90° rotation to exit at the periphery into the diffuser to go through another 90° turn before discharging into the manifold. Adequate precautions are needed to ensure flow separation does not occur in the flow passage and to avoid the associated losses. The diffuser transforms the high flow velocity into pressure head through the gradually increasing cross sectional area of the passages within.

Axial and radial centrifugal compressor types find wide acceptance in aircraft engines. Smaller engines can easily take advantage of the centrifugal impeller's requirement of reduced flow area at the inlet to obtain a pressure ratio as high as 5:1 in one stage, while operating at nearly 80 per cent efficiency. The rotating impeller's mechanical energy is used to create centrifugal forces in the air stream. Engines are also designed with a row of two to four axial stages followed by a centrifugal stage. Two impellers turning at a high speed, exceeding 30,000 rpm, may also be employed in an opposed arrangement to increase the flow capacity (*ref*: Hunecke, 1997).

Most engines, however, rely on the axial compressor for air pressurization to handle the large mass flow rate necessary for high levels of compression and thrust generation. Since flow direction is mostly axial, turning of flow direction is eliminated. Also, the consistent external dimensions in an axial flow machine help hold down aerodynamic drag. Axial compressors rely on the principle of lowering stream velocity to raise pressure, and hence are more susceptible to flow fluctuations.

A considerably larger number of parts go into the building of an axial compressor, thus increasing the complexity of design and manufacturing cost. Major components are support frames at the front and rear ends, an external casing with stator vanes attached to it and a bladed rotor. Figure 10.3 provides details of an axial compressor module.

The compressor case has a cylindrical geometry with an axial split to permit installation of the rotor. Made out of lighter weight titanium or a stainless steel, the material must be capable of maintaining radial clearance at blade tips due to thermal growth during operation. Fixed stage vanes are inserted as an assembly in T shaped grooves machined in the casing.

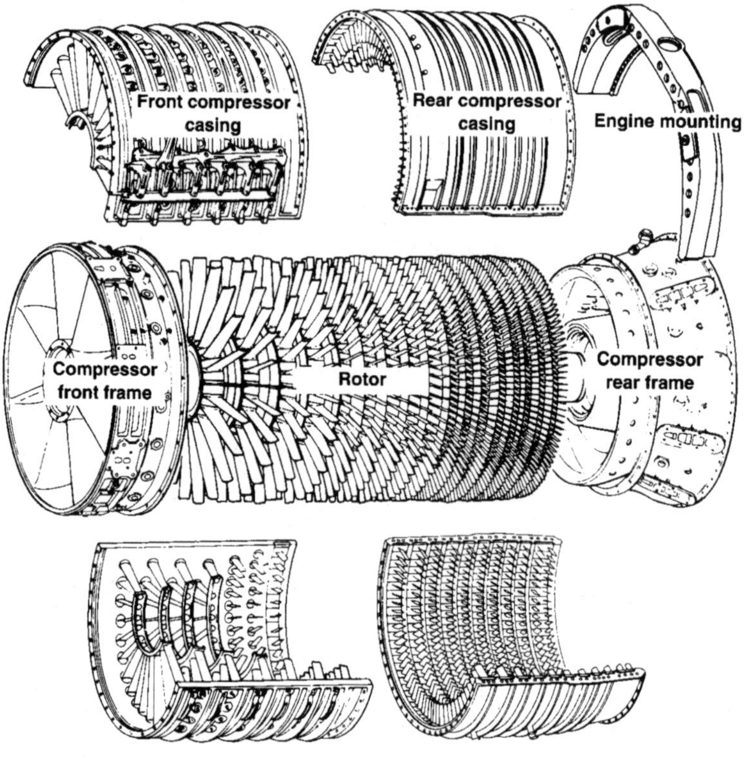

Fig. 10.3 Axial Compressor Module of General Electric J79 Engine (Hunecke, 1997)

A combination of a drum and discs is the common form of rotor construction, with a shaft segment at both ends to carry the bearings. Disks may be attached at either extremity of the drum by spacer rings. Rotor blades are twisted in order to get the right flow inlet angle along the full length. Variation in magnitude of the inlet angle arises because the blade has a much lower tangential velocity at the root than at the tip, while axial flow velocity must be maintained along the full face. Blade height reduces gradually as the airflow progresses through the stages, in nearly the same proportion as the pressure rises. Blade roots may be of the dovetail shape, or of the fir tree type if the loading is considerable. Attachment of the blade to the disk must be firm, but still permit room for growth during operation. Longer blades tend to move a little in the seat, but lock under the action of centrifugal forces during operation (*ref*: Rangwala, 2005).

Axial entry of the dovetail in the disk or drum grooves is common in forward stages, but circumferential insertion is favoured for the latter stage blades. Blade retainers may be in the form of a key or a fitted bolt.

10.3 AIRCRAFT POWER PLANT

Turbojet, turbofan, turboshaft and turboprop engines drive a plethora of aircrafts ranging from huge passenger and cargo planes to helicopters and cruise missiles, come in every conceivable size and shape and fly to the remotest corners of the world. Turbojet engines, with their relative simplicity of design and construction, are predecessors of the later breed of turbine propulsion systems for aircrafts. Basically comprising of a multi-stage compressor to pressurize air, a combustor to burn fuel and a multi-stage turbine, the jet engine function is facilitated by an intake to channel smooth flow at the front and by an exhaust system at the rear. Besides pressure the compression process also raises the density and temperature of the air as it enters the combustor, where burning of the fuel adds considerable thermal energy to create intensely hot and pressurized gases. The charged gases now have the right characteristics to convert this energy into a mechanical form to drive the turbine. The turbine and compressor modules are rigidly connected by a shaft, the former supplying the power to drive the latter and to a host of accessories required in the engine system. Energy level of the hot gases exiting from the turbine is not exhausted in turbojet engines; in fact, substantial pressure and thermal energy is still available to be expended in the exhaust nozzle for conversion into high velocity and consequent thrust generation.

Reciprocating engines for small aircraft have the distinct advantage of cheaper materials and initial cost, which is perhaps the overriding factor for the average private aircraft owner. The combination of a reciprocating engine with an exhaust gas driven turbine driving a supercharger is in wide use. On the other hand, the internal combustion turbine has relative mechanical simplicity, which makes it very attractive for aircraft propulsion. The absence of reciprocating parts gives freedom from vibration. It also has the significant advantage of lower weight to power ratio, making it the only power plant available for large military and commercial aircraft. Cooling the turbine blades helps to increase inlet gas temperature, which increases the efficiency and output from a given size unit. This requires sophisticated metals and manufacturing methods for the blades and elaborate pipes, sensors and valves to ensure adequate operational life of the components. Large engines are normally built for airplanes operated by the airlines or by the government; hence, the original cost of the engine is not as important to the buyer as the efficiency of its performance over a long period of time. The use of techniques such as direct fuel injection and supercharging on small engines has greatly improved their efficiency, weight to power ratio and specific fuel consumption.

A unique feature of all aviation related hardware, and especially the engines, is the weight requirement. In order for an aircraft to have the capacity to lift off from the ground and to stay aloft in the air, engine system weight must be as low as physically and operationally possible. This requirement poses interesting and considerable ramifications when compared with power generation and other land based turbines. All components, whether rotating or stationary, must be thin walled to maintain weight and mass inertia properties within set limits. From engineering, manufacturing and operation aspects this calls for unique solutions, even necessitating development of brand new technologies. Engineering methods such as finite element techniques accurately determine stresses and strength of the components. Because the parts have small wall thickness throughout, they lack adequate stiffness and have low natural frequencies of vibration, making them susceptible to metal fatigue. And since aircraft engines operate in a large speed range, synchronous vibrations must be determined and provided for to ensure structural integrity under many different load conditions. A number of super alloys possessing the right combination of strength and creep resistance are used extensively. Machining of thin-walled aircraft engine parts pose a special problem, requiring extra precautions to achieve the specified dimensions while still avoiding gouging due to an improper cut size, speed or feed rate. Special cutting tools to withstand the tougher machining traits of super alloys require frequent replacement due to excessive wear, even when liberal amounts of cooling fluid are dispensed. New machining and fabrication methods such as electro discharge machining, cold and hot rolling and inertia welding trace their origin in the production of aircraft engine components.

The philosophy behind the design and configuration of bearings used in aircraft engines very well illustrates the minimum weight concept. Hydrodynamic fluid film bearings are the norm in most other turbine designs. However, they tend to be axially long in order to provide the right length to diameter ratio. If they were to be used in aircraft engines, this requirement would translate into a longer and consequently heavier engine. Specially designed rolling element bearings are the rule for aircraft power plants. Ball bearings capable of withstanding axial thrust and radial loads are used near the fan. Roller bearings are used at other locations. Three bearings are commonly required for each spool of the engine; a two-spool system would then call for six main engine bearings. When compared with equivalent power generation turbines, the weight of an aircraft engine's rotor weight is a small fraction of its industrial counterpart. This factor certainly facilitates the design of rolling element bearings when static load carrying capacity is evaluated. Dynamic loads due to rotating unbalance in engines for larger applications can be considerable, mostly because of the presence of vibration modes within the operating speed regime and because they operate at high and varying speeds.

The turbine section converts more mechanical energy from the hot gases than is needed to drive the compressor, hence the excess energy powers the fan located upstream of the main compressor. A portion of the air entering the main intake splits from the main flow after passing through the fan to bypass the main engine, expanding in a separate diffuser to furnish its own thrust. The bypass ratio determines the split between airflow in the core engine and the fan's exhaust duct. Bypassing has the distinction of better fuel consumption characteristics over engines without this feature of comparable thrust. Lower values of bypass ratio, in the range of 0.2:1 to 1:1, are more convenient for supersonic combat airplanes. Higher values of 5:1 are more suitable for higher capacity wide body airplanes such as Boeing 747 and Airbus A300. The size of the C-5A Galaxy transport aircraft is so large that its stipulated propulsion characteristics of increased thrust at take-off and low fuel consumption at cruise can only be satisfied by the use of four high bypass thrust engines operating at high turbine inlet temperature. Higher thrust levels are achievable, particularly during take-off, as a consequence of acceleration of a larger mass of bypass air, since the flow through the remainder of the engine may contribute as little as 15 per cent of the combined total. Fuel is burned more efficiently by increasing the bypass ratio at the same flight speed, although the turboprop version is more economical to operate. Another benefit accrues from lower noise levels due to the low exit velocities of the propulsive jets. Figure 10.4 provides a layout of Pratt and Whitney's 2000 series of engine.

Propulsion power is obtained from the reactive forces of the exhaust gases in turbojet and turbofan engines. A large percentage of airflow in the turbofan principle does not pass through the main engine, and hence comes under the classification of bypass engines. When the energy of the escaping exhaust gases drives a separate turbine coupled to a propeller, the arrangement is called a turboprop engine. Kinetic energy yield from the escaping gases out of the nozzle is limited, since the turbine has already absorbed a considerable portion. Turboshaft engines, on the other hand, convert all of the thermal energy into mechanical power for a shaft, by means of a separate turbine, to drive, for example, a helicopter's propellers or a compressor or an electric power generator in an auxiliary power unit of an airplane.

The discussion on compressors relates to turbines in several ways, except for two special items: increased gas temperature at turbine inlet and falling pressures as flow progresses through a turbine, as opposed to rising pressures in a compressor. High gas temperature leads to lower tangential

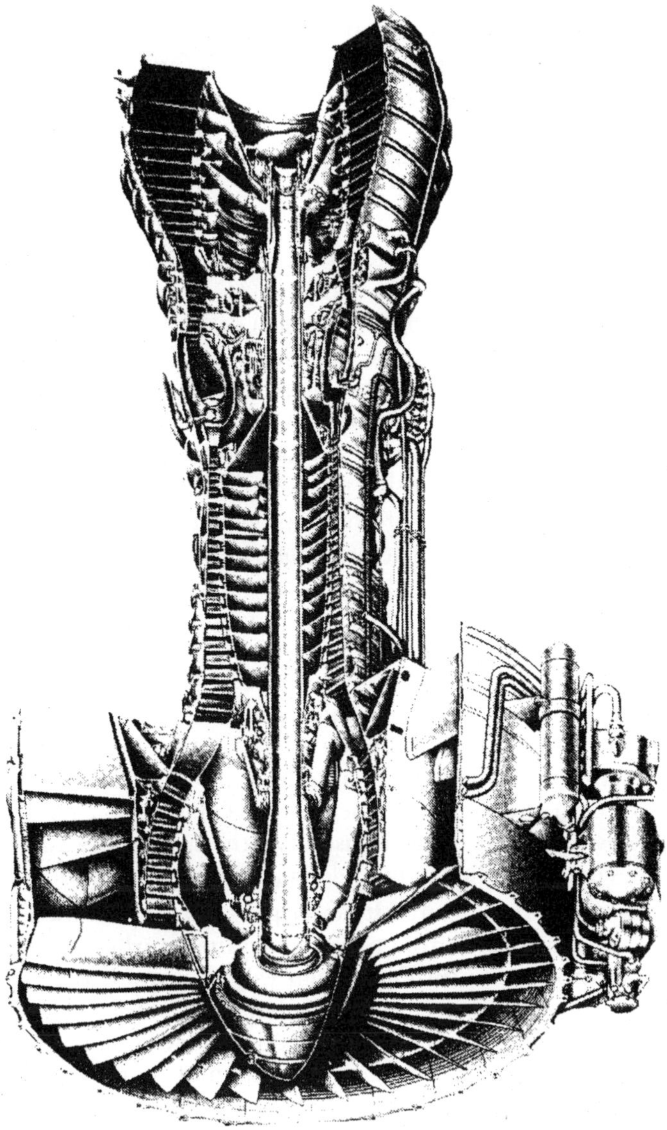

Fig. 10.4 Pratt & Whitney's Series 2000 High Bypass Turbofan Engine

Mach numbers for turbine blades than for compressor blades, relatively easing the aerodynamic problems. Also, the falling pressure in turbine thins the boundary layers to reduce separation concerns.

Compressor efficiency has a stronger impact on the overall engine system than turbine efficiency. Still, high bypass turbofan engines rely heavily on turbine efficiency. As noted before, increased turbine inlet temperature improves thermal efficiency, but generous provisions for cooling must be made. As a consequence, the previous definition of efficiency needs modification.

The ratio of actual turbine work per unit of total airflow, including cooling and ideal work that would be attained in expanding that flow through the defined pressure range, defines the turbine efficiency. Cooling flow may also be assumed to expand through the same pressure differential as the primary flow. Cooling flow impacts the turbine efficiency in multiple ways. Exiting cooling air from the blades causes a higher level of drag. It also suffers a pressure loss while traversing the cooling passage, so it has a lower stagnation pressure when mixed with the main downstream flow for a given τ_t. Note that entropy of the total flow increases due to the heat transfer from the hot primary flow to the cooling flow.

Of particular interest in high bypass turbofan engines are the fan blades. General Electric's CF6-6 engine fan uses 38 titanium blades to generate a considerable portion of the total thrust. Weight and stiffness (and hence natural frequency) control of the blades may be achieved by the simple expediency of drilling holes in the tip. Due to the large fan diameter the root of the blade may be operating in the subsonic regime, but tip peripheral velocity may exceed the speed of sound during maximum thrust conditions. A circular profile for the tip section of the airfoil aids in meeting the more rigorous demands of the supersonic flow, but conventional airfoil shapes are employed at the hub to serve the intended purpose.

A key consideration in the control of vibrations, dynamic stresses and operating life is the blade length. Mid-span shrouds promote structural integrity of the blades, and are also advantageous in the event of a bird strike. However, the shrouds come with a penalty in efficiency, since they act as an obstruction for the airflow. Erosion from rain, hailstone and ingestion of ice are other factors to be kept in mind in the design of fan blades. Close matching of dimensions, both in the blade and in the mating disk dovetail, is of special significance because of the high centrifugal loads. The problem becomes even more acute since the blades must permit easy removal in the field. Fan disks must be designed to withstand axial, radial and tangential direction loads arising from aerodynamic and centrifugal forces on the blades. Titanium has proved to be a reliable material for the disk because of its advantageous strength and weight characteristics. The fan stage may be followed by one to four low pressure compression stages, located just aft of the split in the air stream for bypass engines. Figure 10.5 provides details of the fan module in a high bypass turbofan engine.

Axial turbines are the norm in aircraft power plants because of the higher mass flow capability. Besides meeting mechanical loads from power generation, turbine components must operate in a high temperature environment. Turbine design will be controlled by the compressor's pressure ratio, quantity of energy to be extracted from the hot gases, rotor speed and maximum permissible diameter. Turbine stage designs tend to be based on a mixed combination of constant pressure and impulse principles. Selection of appropriate materials and cooling techniques are of fundamental importance in a successful turbine design (*ref*: Hunecke, 1997).

The rotor assembly consists of disks, blades and shaft segments at both ends. At the compressor end the shaft is of a conical design to permit attachment with the shaft driving the fan rotor (see figure 10.6). A splined joint with a lock nut is favoured for transmission of power from the turbine to the fan. The disks are bolted together, with spacers placed in between the disks. Varying thickness in the disks permits safe distribution of stresses caused by the blade centrifugal forces at the periphery, with the hub axial length being two to three times the length at the rim. Bore sizing is crucial in that hoop directional stresses peak in the region, and must also permit placement of a vent tube through the center hole. Blades are inserted in machined slots and held in place either by clamps or retaining plates, with passage for cooling air directed towards holes machined in the base of the blade's dovetail.

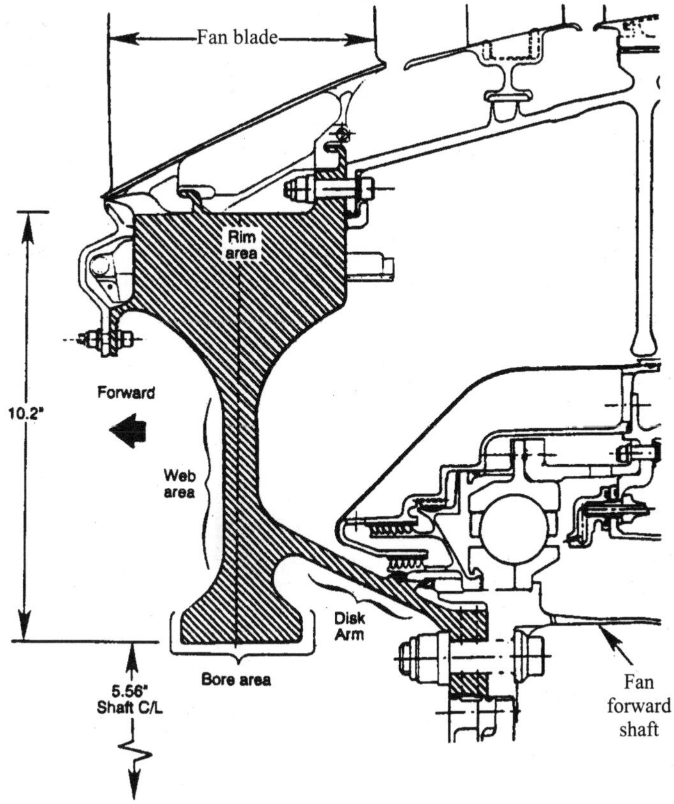

Fig. 10.5 CF6-6 Engine Fan Disk (NTSB/AAR-90/06 Report)

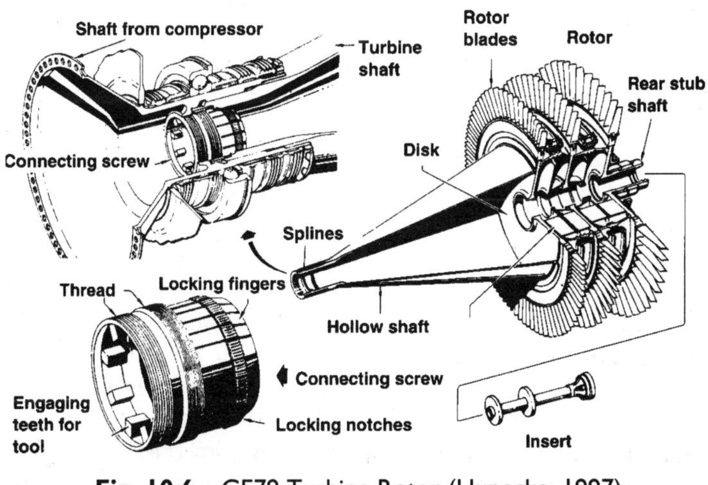

Fig. 10.6 GE79 Turbine Rotor (Hunecke, 1997)

A number of cooling methods are made use of to maintain metal temperatures well below those of the gas. Cooler air from the compressor flows through the inside of blades and vanes, exiting through holes drilled in the leading and trailing edges. Thermal barrier coating helps increase resistance of base metal to elevated gas temperature, while also providing protection from corrosion. Provision of cooling air to the rotating blades is at best a difficult proposition. Figure 10.7 illustrates the case for a gas generator turbine. Contamination of cooling air from dust particles poses a serious problem, in that the exit holes may get clogged, thus creating hot spots and premature metal fatigue. In this respect aircraft engines face a totally different situation when compared with industrial land based gas turbines provided with intake manifold filters to keep out the offending contaminants. One solution calls for cooling air to go through sharp turns before entering a swirl separator in which dust particles are whirled in the direction of shaft rotation, thus causing the particles to fly out by centrifugal action. Cooling air then proceeds through traps to separate the particles before entering the passages inside the blade.

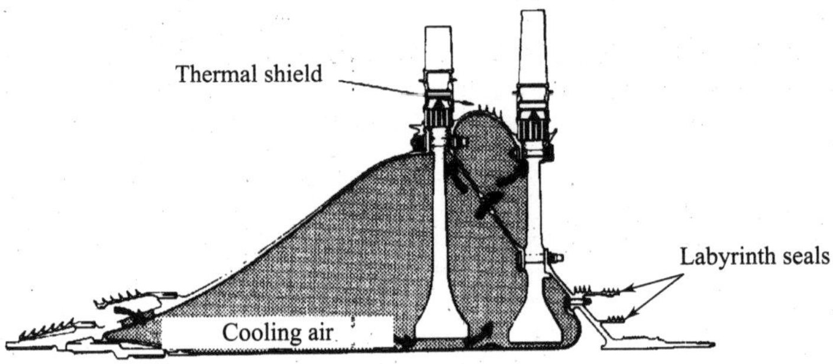

Fig. 10.7 Cooling Air Provision in Gas Generator Turbine (Hunecke, 1997)

10.4 INDUSTRIAL GAS AND STEAM TURBINES

Considering the dramatic increase in the cost of energy over the past few decades, acceptance of inefficient power generation is not an option. The implication of a turbine operating at 30 per cent thermal efficiency is that it wastes 70 per cent of the energy put into it. Energy losses are encountered as the fluid flows over nozzle vanes and moving blades, when fuel is not burnt completely in the combustion system and when frictional forces must be overcome between a rotating shaft and the bearings. Exhaust gases leaving the stack in a boiler take away considerable amounts of thermal energy into the atmosphere because of their elevated temperature.

Many parts of the world have abundant coal available, and consequently steam plants play a major role in power generation. By some estimates 42 to 45 per cent of power is derived from steam generating plants. Coal is a main prop of Kentucky's economy, trailing only Wyoming and West Virginia in production. The state generates more than 95 per cent of its electricity form coal burning. Electrostatic precipitation of suspended particles in effluent gases has made considerable strides in recent years to assure an ecologically friendly environment when compared with plants built in earlier years. East Kentucky's Power Cooperative employs circulating fluidized beds for clean burning of a variety of fuels, ranging from biomass to hard and soft coals, petroleum coke and coal washery wastes. The system includes a flash dryer absorber, and combines flue gas de-sulfurization to remove 98 percent of sulfur from the coal. Selective catalytic reduction in the pulverized coal boiler helps curb NO_x emission. Advanced coal technology has successfully demonstrated integrated gasification and combined cycle power generation.

Because of its clean burning characteristics and harmless residuals, natural gas is easily the preferred fuel. When the price is affordable natural gas as a fuel for power generation has the advantages of creating minimum pollution, while requiring minimal maintenance of machinery. Comparative prices of uranium for nuclear power operated and of coal for fossil fuel power plants have not changed substantially over the years, and are the lowest. Oil and natural gas generated electrical power is, on the other hand, costly in the simple cycle mode of operation. But when power generation in the combined cycle mode is used, where gas and steam turbines operate in conjunction, the cost proves to be quite effective. Natural gas fuels the turbine to produce electricity; turbine exhaust enters a heat recovery boiler for producing steam that runs another bank of turbines for a second kick from the first burn. Cogeneration of two or more forms of energy, such as steam and electric power, in a plant also finds wide application. Chemical and process plant facilities use exhaust gases directly from a gas turbine or other prime mover for pre-heating air in furnaces, absorption cooling systems and raising temperature of fluids during production. New and novel concepts for energy conversion in combined cycle plants are being developed for the modern day market place, resulting in constantly improving heat rates. Simple cycle gas turbines with firing temperature of 2400°F, recuperative gas turbines, combined cycle power plant, advanced combined cycle power plant and hybrid power plants represent some of the latest developments.

The gas turbine operates on the Brayton cycle. The Brayton principle consists of two isobaric and two isentropic processes, as shown in figure 10.8. The former take place in the gas turbine's combustor and in the steam generator's gas side, and the latter represents the compression of air and expansion of gases in the turbine. Gas turbine cycle efficiency depends on the compression ratio and turbine firing temperature, increasing with both parameters according to the relation shown in equation 3.1. It is assumed in the derivation that the pressure ratios in the compressor and the turbine remain the same, fuel flow rate is considerably smaller than that for air, specific heat of the gases stays constant and that all components operate without incurring any losses:

$$\eta = 1 - \left(\frac{1}{r^{\frac{\gamma - 1}{\gamma}}} \right) \qquad \text{...(10.1)}$$

where η is ideal cycle efficiency, r represents compression ratio and γ is ratio of specific heats at constant pressure and constant volume. Turbine and compressor efficiencies will have a moderating effect on the overall efficiency. Figures 10.8 and 10.9 demonstrate the impact of the two parameters. Work performed per pound of air occurs at a lower pressure ratio than the point of maximum efficiency for a given firing temperature. Overall cycle efficiency is improved with increased pressure ratio, cooler compressor inlet temperature and higher turbine inlet temperature (figure 10.10).

Evaporative coolers, direct water fogging and refrigerated cooling systems are some methods used to cool the air at the inlet. The impact of efficiencies in the compressor, combustor and turbine, as also system related pressure losses in a simple cycle turbine is shown in figure 10.11.

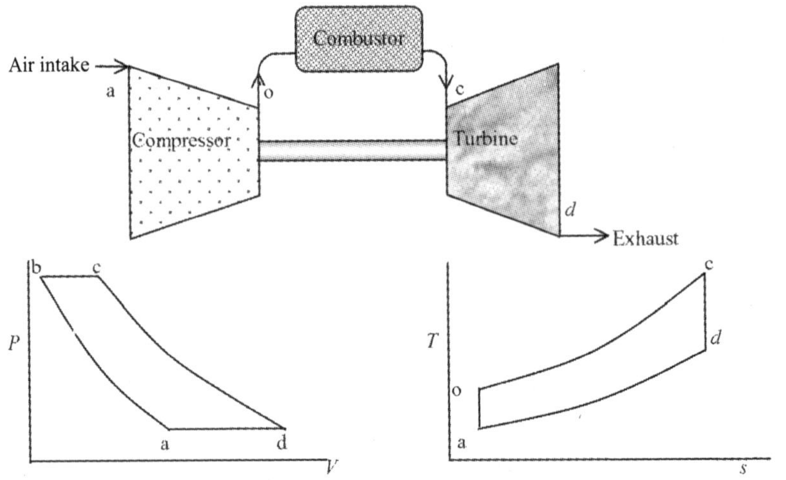

Fig. 10.8 Standard Brayton Cycle for Gas Turbine

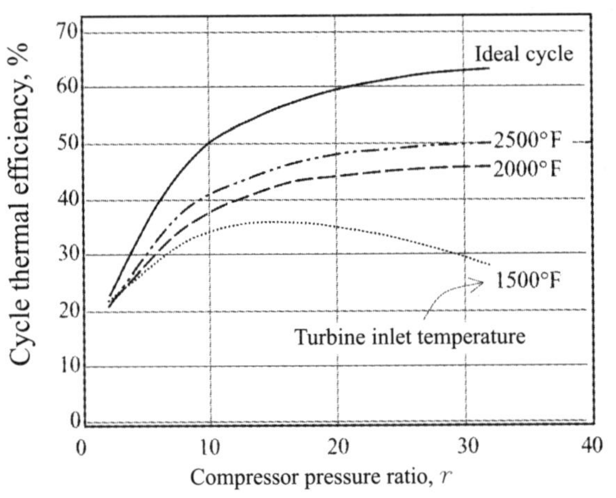

Fig. 10.9 Compression Ratio and Thermal Efficiency

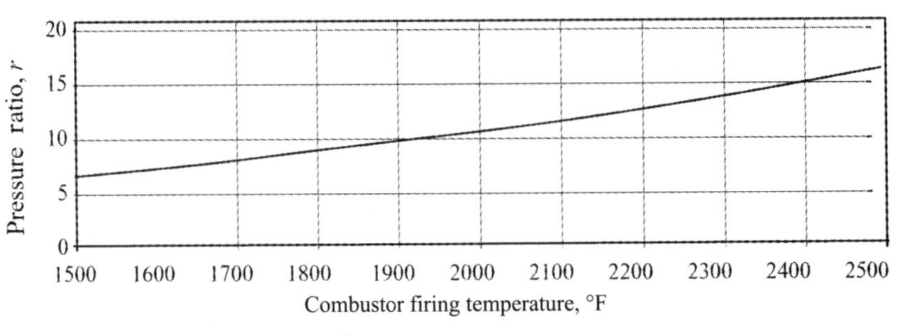

Fig. 10.10 Optimized Pressure and Temperature for Maximum Thermal Efficiency

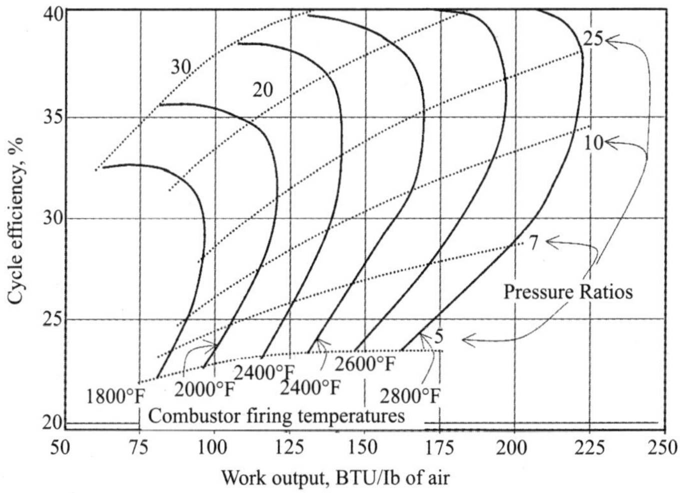

Fig. 10.11 Simple Cycle Gas Turbine Performance Map

Since gas temperature at the turbine exit is higher than that exiting from the compressor, sertion of a regenerator to preheat the air between the compressor and the combustor with the rbine's exhaust gases will reduce fuel requirement. If T_1, T_2, T_3 and T_4 represent inlet mperatures to compressor, regenerator, combustor and turbine and T_5 is the exit temperature om the turbine, ideal regenerator efficiency (assuming no pressure loss) may be expressed in the llowing form:

$$\eta_{Reg} = \frac{T_3 - T_2}{T_5 - T_2} \qquad ...(10.2)$$

Regenerator efficiency will depend on available surface area for heat transfer, but increased area ll call for higher cost, pressure loss and space. Heat exchangers may be of regenerative or cuperative type. Regenerative systems call for another medium to effect transfer of heat between rbine exhaust gases and compressed air.

Gas turbines are a practical and economic way for utility and industrial service. Advanced design its offer high firing temperatures, low NO_x and improved efficiency. For example, the 60 Hz 200 W class W501G engine has been jointly developed by Westinghouse Electric Corporation, itsubishi Heavy Industries and FiatAvio (*ref*: Southall et al., 1995). This machine continues a long e of large heavy-duty single shaft combustion turbines by combining the proven efficiency and iability concepts of the W501F with the latest advances in the aero technology. Designed for both nple and combined cycle applications, the turbine can operate on all conventional turbine fuels d on coal derived low BTU gas produced in an integrated gasification plant. Figure 10.12 ovides details of a single shaft gas turbine supported on two bearings.

The general configuration of the combustion turbine calls for a bolted construction rotor pported in two bearings. The 18 in. diameter bearings have compounded form, with tilting pads the lower half and a fixed arc element on the upper side. The compressor segment of the rotor is sembled from spigotted disks bolted together by 12 through bolts. Alignment and torque nsmission are assured by employing radial pins between the disks. The turbine rotor section is

Fig. 10.12 W501F Compressor and Turbine Components on a Single Shaft (*Courtesy*: Siemens Westinghouse)

Fig. 10.13 Turbine Rotor Disks with Curvic Coupling (*Courtesy*: Siemens-Westinghouse)

made up of disks provided with curvic clutches (see figure 10.13), and are bolted together by through bolts. The curvic clutch is machined in the form of uniformly spaced teeth protrudi axially from the flat face of the disk. The teeth engage and interlock with a similar pattern machin on the face of the adjacent disk, providing a slippage-free joint under the action of the clampi load from the through bolts.

The combustion system consists of 16 cans arranged in an annular pattern. Stability of the flame nd uniformity in the distribution of fuel flow between the combustors are monitored by ıermocouples located downstream of the last turbine stage. Malfunctions in the combustor when at ›ad and sensing of ignition during the start-up mode can also be detected with the aid of ultra-.olet sensors.

The casings are split horizontally to permit performance of maintenance with the rotor in place. 'he inlet and compressor casings are made of nodular cast iron and cast steel, while the combustor, ırbine and exhaust casings are of alloy steel. The inlet end bearing housing is supported from eight ıdial struts. At the exhaust end the bearing is supported by tangential struts that respond gradually uring transient conditions, and maintain alignment between the rotor and the bearing through ›tation of the housing about its axis to accommodate thermal growth. The fairings around the ruts are configured in the form of an airfoil to enhance aerodynamic performance. Individual lade rings are employed for each compressor and turbine stage to control leakage past the blade ps. The rings have a relatively high thermal response independent of the outer casing, and are rovided with features to obtain concentricity with the rotor to prevent rubbing from the blade tips, ninimize radial clearance and thus maximize performance.

Since the turbine operates at extremely high inlet temperatures, cooling air for the rotor is xtracted from the compressor discharge, and is externally cooled and filtered before returning to ıe torque tube casing to cool the disks and the first, second and third stage blades and vanes igure 10.14). Filtration is deemed essential for eliminating blockage by suspended particles of the ıtricate cooling passages inside the blades. Bleed air from the compressor is also used to cool the lade ring cavities and to cool and purge the inter stage disk cavities to prevent ingestion of hot lade path gases.

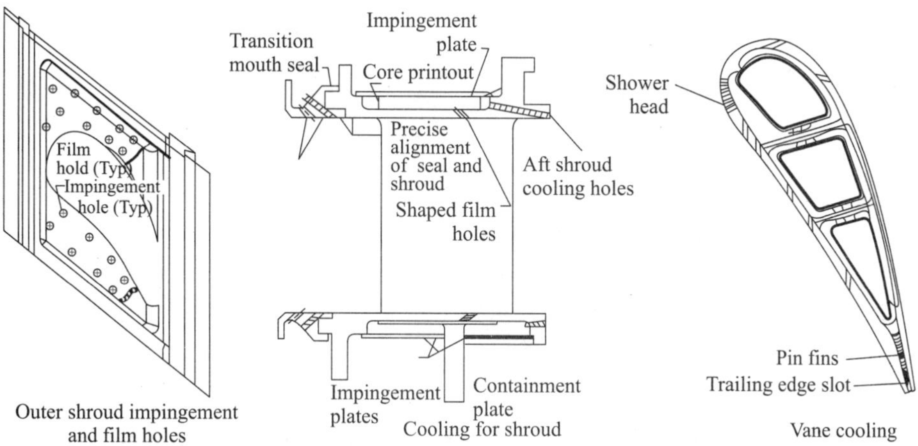

Fig. 10.14 Turbine Row I Vane Cooling Scheme (Southall, 1995)

Compressor diaphragms are coated to improve aerodynamic performance and to obtain rotection from corrosion. The stationary vanes and rotating blades for the first two stages of the turbine are provided with a thermal barrier coating. The operating firing temperature level 1425° C is selected to be commensurate with the capabilities of the super alloys for the componen in the hot path and with the cooling schemes.

The 17-stage axial flow compressor is patterned from the proven design for the W501F engin Flow and pressure coefficients are similar in the two designs, with mean diameter of the stag increased to accommodate the 25 per cent increase in flow. Bleeds for starting and cooling flows a located at the sixth and eleventh stages, and the fourteenth stage bleed is used for the hot pa components. The compressor is also equipped with variable inlet guide vanes (figure 10.15) f improving the low-speed surge characteristics and to enhance part-load performance in combine cycle applications. The rotating blades are controlled diffusion airfoils, made of multiple circular a forms. Stationary blades are fabricated in two 180° segments for easy removal, and provided wi sealing at the inner shroud.

Fig. 10.15 Compressor Inlet Variable Guide Vanes (*Courtesy*: Siemens Westinghouse)

Moderate aerodynamic loads are used for the four turbine stages operating at a high peripheral speed than the W501F engine. Aerodynamic airfoil shapes are obtained from a ful three-dimensional viscous analysis code. The third and fourth row blades are shrouded. The fir row stationary vanes are individual precision cast of IN939 alloy, and can be removed from acce man ways without lifting the cylinder cover. Inner shrouds are supported from the torque tul casing to limit flexural stresses and distortion. Vane segments for the other rows are supported in separate inner ring.

The cooling scheme maintains the NiCrMoV turbine disks below 400°C, within the creep range for long life. Row 1 vane cooling is by the methods of impingement, convection and film cooling, as shown in figure 10.14. Impingement inserts are used in combination with an array of film cooling holes and a pin fin at the trailing edge. Pin fins help to increase turbulence and the surface area. Film cooling is provided at the leading edge on the pressure and suction sides. This limits thermal gradients and external surface temperatures at the walls of the vane. Special attention is paid to the inner and outer shrouds because of the flat temperature profile from the dry low NO_x combustor. The shrouds are cooled by impingement plates, film cooling and by convection through drilled holes.

The cooling for the row 1 blades consists of serpentine passages with angled turbulators (figure 10.16). The film cooling uses fan shaped cooling holes, and is used extensively at the tip to reduce the metal temperature of the squealer tip. The airfoil is coated with a vapour deposited thermal barrier coating. For the row 3 blade the cooling is unique in that it positively cools the blade tip shroud (figure 10.17). Because of the flat profile from the combustor and because of leakage past the tips of row 1 and 2 blades, positive cooling for the tip shroud is deemed essential.

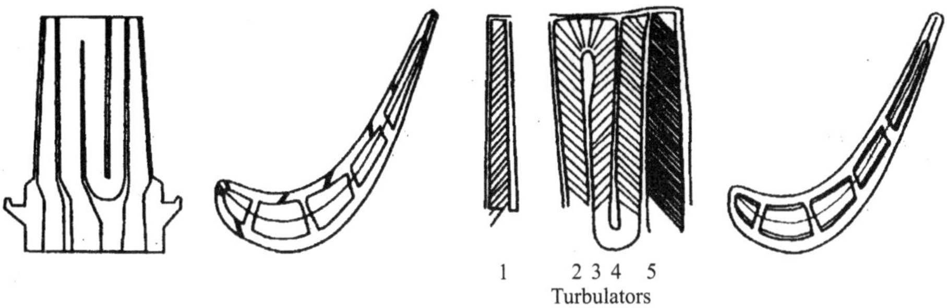

Fig. 10.16 Turbine Row 1 Blade Cooling Scheme (Southall, 1995)

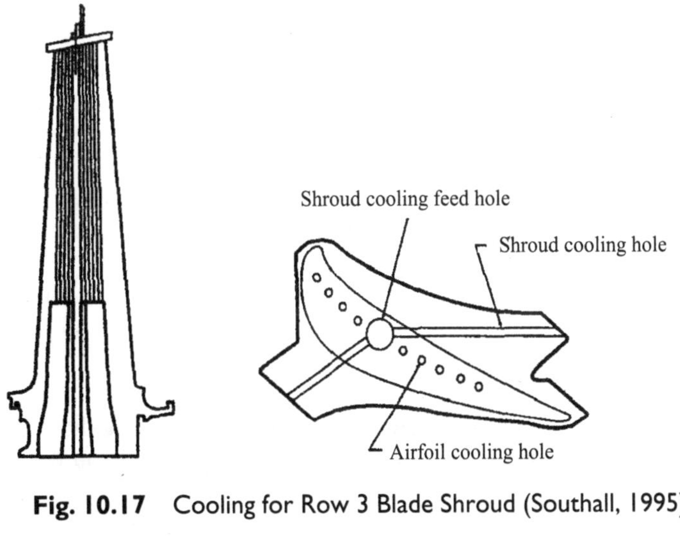

Fig. 10.17 Cooling for Row 3 Blade Shroud (Southall, 1995)

The rotating blades are made from CM247 for all four rows. The blades are provided with long root extensions, or transitions, in order to reduce the stress concentration as the load travels through the airfoil into the shank. The blade roots are of multiple serration type, with four serrations on the first three rows and five on the last stage blades.

The dry low NO_x combustor operates at 25 ppm NO_x level at 1260°C turbine inlet temperature. Steam cooling is used for reduced emissions at higher firing temperatures. By eliminating the transition cooling air virtually all the combustion air is introduced into the primary zone of the combustor to maintain the flame temperature at nearly the same level as in the W501F engine.

Steam turbines remain the workhorse of power generation worldwide. Hero of Alexandria is credited with developing the first steam turbine 2,000 years ago. Dr. de Laval demonstrated driving of a paddle attached to a shaft by expanding steam through a trumpet shaped steam jet in the latter half of nineteenth century. In 1894, Sir Charles Parsons invented the multi-staged steam turbine. Today, steam turbines are the favoured choice in the driving of electric generators, mostly due to extensive availability of coal.

A steam turbine converts thermal energy of the steam into kinetic energy by expansion in nozzles, the resulting jet then forcing rows of blades mounted on a rotor. Steam turbine power plants may be split into three groups: heat sources such as boilers or steam generators, feed water pump and heater; power generation components that include turbine and generator; and condensers and condensate pump. Rankine cycle is most commonly employed, with water-steam as the working medium. Water is pressurized isentropically by the feed-water pump before entering the boiler, where it evaporates to steam and eventually superheated. The cycle requires substantial amount of heat to raise the temperature of water at pump discharge to steam temperature at turbine inlet. Using high temperature exhaust gas for this purpose partially offsets the amount of external heat required by minimizing the temperature difference between the two points. This is the concept behind regenerative heating. A more common practice calls for intermediate pressure steam to perform the function of heating the feed water. Steam as it progresses through the stages of a turbine tends to increase in moisture content. Wet steam tends to affect the buckets, since the higher density of water impacting its leading edges causes erosion. A reheat of steam withdrawn after partial expansion is required to overcome the situation (*ref*: Boyce, 2002).

A regenerative and reheat steam arrangement is shown in figure 10.18 and the corresponding temperature-entropy diagram in figure 10.19. The compressed liquid at D is heated to saturation point (D1), evaporated to steam (D2) and finally superheated (D3). After isentropic expansion in the ideal engine to point E (point E′ after accounting for losses), some of the steam is extracted from an intermediate stage in the turbine to heat the pressurized condensate in the feed water heater. The rest of the extracted steam is reheated, and then returned to the turbine (point F). After further expansion to point G′ for the real case the steam passes to the condenser, converting it to saturated liquid at point A.

Heat rate, a modified reciprocal of thermal efficiency, defines the heat chargeable in BTU/kW-hour for a straight condensing or a non-condensing turbine, and may be expressed as:

$$\text{Heat Rate} = [(h_1 - h_{f2}) \times 3415]/W_{net} \text{ BTU/kW-hr} \quad ...(10.3)$$

where h_1 is enthalpy of throttle steam, h_{f2} is enthalpy of liquid water at exhaust pressure and W_{net} is available work output at the generator coupling, BTU/lb of steam at throttle.

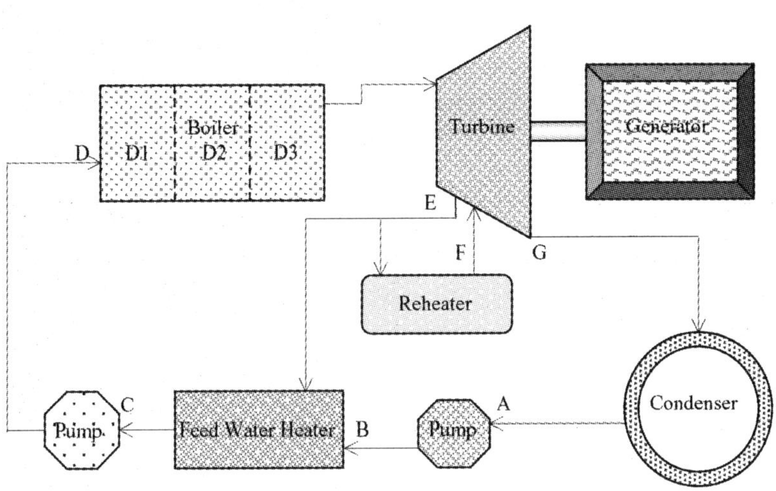

Fig. 10.18 Regenerative – Reheat Steam Turbine Plant Schematic

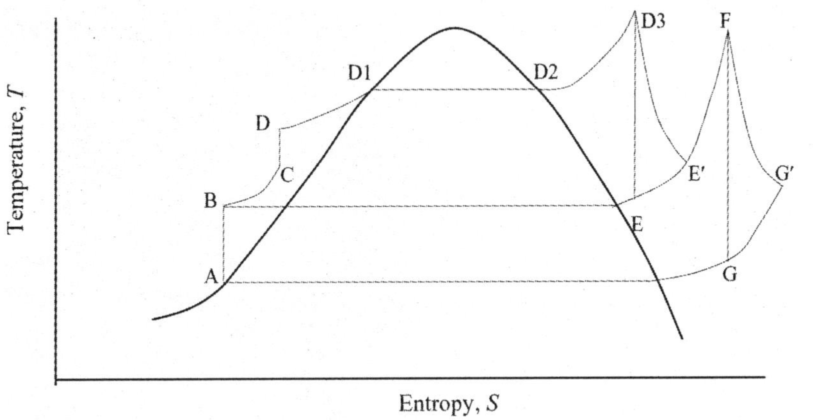

Fig. 10.19 Regenerative – Reheat Steam Turbine Temperature/Entropy Diagram

Among the many different ways of classifying steam turbines, fluid flow direction specifies the engine type as axial, radial or mixed flow. In axial flow machines, steam is flowing virtually parallel to machine axis; medium and large turbines are configured with this arrangement. Some small turbines have a radially outward flow among the blades. Axial flow machines may be split into two categories, impulse and reaction. The former uses a system in which all steam expansion occurs in fixed nozzles; pressure drop is absent in the passages between the moving blades. The change in enthalpy in the nozzles increases the steam's kinetic energy, which is then imparted to the blades. Reaction, or Parsons, turbine calls for expansion of steam in both the nozzle vanes and in the moving blades. Steam turbine manufacturers most commonly use a combination of the two methods. The initial stages are usually of the impulse type, while the subsequent stages are based on the reaction principle. The degree of reaction is defined as the ratio of change of enthalpy reduction in the blades to the total enthalpy change in the stage.

Two forms of losses play a major role in the design of steam turbines. For high flow Mach numbers the loss factor remains constant up to 1.0. To avoid choking of passages and severe shock, the Mach number is limited to 1.15. The other source of loss arises from the arrangement of nozzles along the periphery at the turbine's admission. At the high-pressure end the nozzles are placed in selected locations around the circumference. This form of partial arc admission facilitates operation of the turbine at part load, but it also causes loss in energy. As the flow progresses, the increased volume of steam necessitates larger circumferential arcs to be occupied by the nozzles. A balance between the flow Mach number and the partial arc length determines the design layout in steam turbines.

Most power plants employ turbine sections designated as high-pressure, intermediate-pressure and low-pressure. The number of stages in each section is based on pressure, temperature and amount of steam available from the boiler, as also the stage percentage reaction. Basic steam turbine configurations can be made in selected ways (see figure 10.20). Steam enters at one end, flows through the nozzles and blades parallel to shaft axis, and exits into the condenser at the other end in a single casing machine. Tandem compound turbines call for steam expansion in two or more separate units, with a single coupled shaft driving the generator. Exhaust from the high-pressure may be reheated before entering the intermediate section. Figure 10.21 provides a layout for a compound flow unit, with high and intermediate sections in one casing and low-pressure turbine in the second casing. Exhaust from the intermediate unit is carried to one or more low-pressure units by the crossover pipe. Path through the LP turbines is split into parallel flows because of limitations on the blade length and for the need to balance axial end thrust.

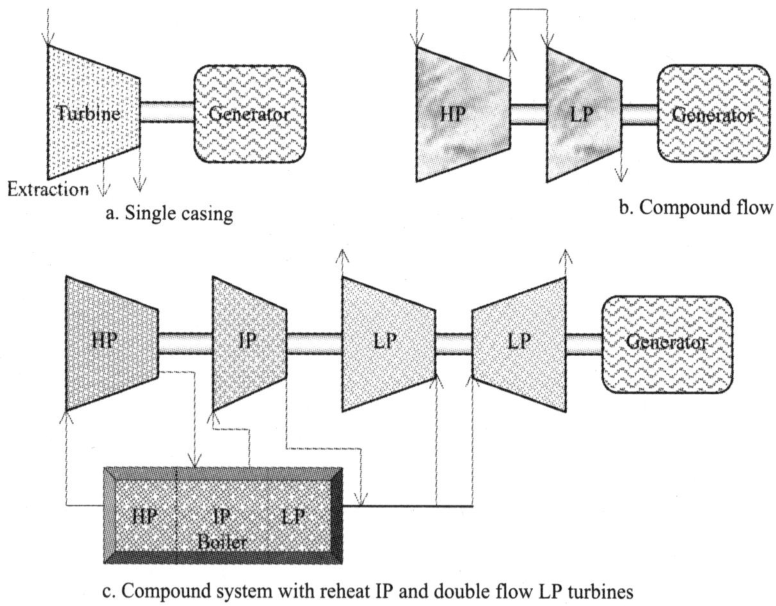

Fig. 10.20 Basic Steam Turbine Configurations

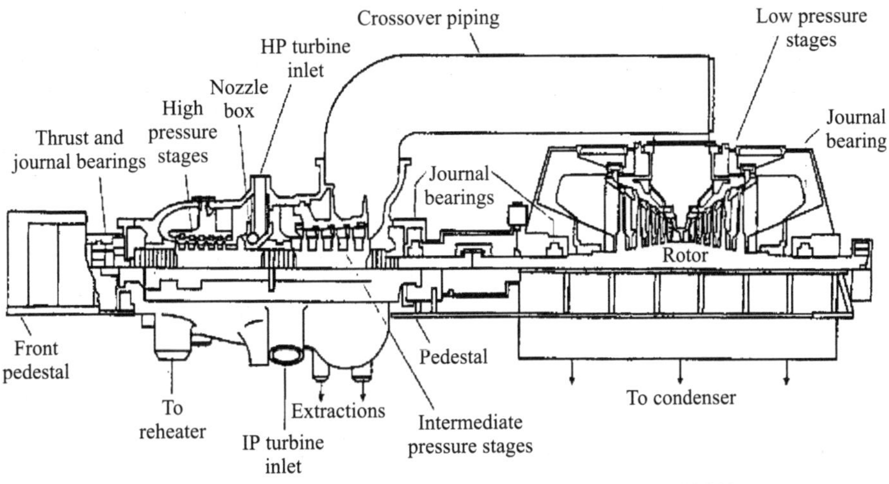

Fig. 10.21 Compound Turbine Layout (Boyce, 2002)

Low volumetric flow in the high and intermediate pressure turbines causes their blades and vanes to be shorter in length. Generally straight, HP and IP blades still tend to have some leaning and bowing, creating some three-dimensional characteristics. Shrouds at the outer tip help in sealing steam flow radially, thus contributing to improved efficiency. Structurally also the shrouds aid in constraining tip motion in axial and tangential directions; the feature reduces vibratory stresses and improves fatigue life by damping elastic motion of the blades. Low-pressure turbine blades are considerably longer and may require one or two mid-span interlocking constrain mechanisms (such as tie-wires) in addition to the outer shroud. LP blades may be connected in groups of between two to eight blades, or all blades in a row may be connected continuously. At the blade tip a protruding tenon aids in attaching the shroud in the form of a cover with a rivet. Tie wires may take the form of integral forged stubs welded or brazed together, or may be rods inserted through a hole with a boss in the blade.

Steam turbines for electric power generation offer some unique advantages. The work required to increase fluid pressure in the liquid phase in the feed water pump is considerably less than needed to compress gas. Steam boilers have the ability to burn a wide spectrum of fuels without the steam coming into direct contact with the products of combustion. Fuels range from natural gas to heavy residual fuel and solid fuel such as coal and refuse. Consequently, production costs compare favourably against that produced by gas turbines. However, a large amount of equipment is required, raising the initial capital investment and cost per installed kilowatt of power. Indirect heating in a furnace necessitates large heat exchangers to obtain pressurized and superheated steam. The boiler requires substantial land area for installation, involves many different auxiliary components and takes a long lead time for erection. Finally, it takes several hours for a boiler to raise steam and put the turbines on line; cool down sequence for the boiler also is considerably long. The system is designed to operate continuously for long periods of time rather than for meeting surges in demand of short duration.

The thermodynamic performance of a steam turbine is primarily determined by the design of the steam path components such as valves, inlet, nozzles, buckets, steam leakage control devices and exhaust. Because efficiency of the entire power plant cycle is highly dependent on the efficiency of the steam turbine, it is important to minimize aerodynamic and steam leakage losses in the path of both the rotating and stationary components. Nozzle and bucket aerodynamic profile losses, secondary flow losses and tip, root and seal leakage losses can be significant if the airfoil shapes are not optimized for the given set of operating conditions. Profile losses are driven by surface finish, total blade surface area, airfoil shape and velocity distributions. Attention to adequate matching between the nozzle vanes and buckets is also necessary to control incidence losses (*ref*: Cofer, 1995).

The last stage bucket is perhaps the most important contributor to the performance and reliability of the steam turbine, ranging in size from 20 in. for 3,000 rpm applications to 48 in. titanium for 3,600 applications. Some salient features of the blades are:

- Improved vane profile, including a convergent – divergent supersonic passage and high solidity
- Enhanced mass flow distribution
- Improved tip leakage control for moisture extraction
- Continuously coupled covers, with side or over and under entry
- Self-shielding erosion protection.

With the last stage buckets typically developing 10 per cent of the total unit output, and up to 15 per cent in combined cycle applications, improvements in their efficiency can considerably impact the total power generated by the unit. A bladed disk using 40 in. titanium last stage buckets for a 3,600 rpm turbine is shown in figure 10.22. Figure 10.23 shows bladed disk assemblies for a steam turbine rotor, and figure 10.24 provides examples of new steam turbine buckets.

Fig. 10.22 40 in. Titanium Last Stage Buckets (Cofer, 1995)

Fig. 10.23 Steam Turbine Rotor (*Courtesy*: General Electric/Toshiba)

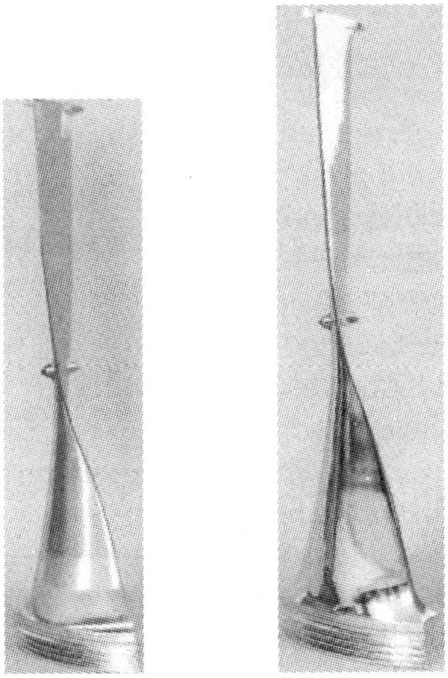

Fig. 10.24 Steam Turbine Buckets (*Courtesy*: General Electric/Toshiba)

Bucket tip leakage control mechanisms typically rely on a single spill strip mounted on the upstream side of the bucket, or two spill strips attached on either side of the bucket cover tenon. Leakage along the shaft is controlled by the packing, which contains multiple teeth located between the diaphragms and the shaft. Flow past the teeth is controlled due to the reduced radial clearance and by the torturous path between the rotating and stationary components. Last stage buckets may be provided with a radial spill band on the cover to maintain a tight clearance.

The Brayton-Rankine cycle combines the gas turbine with a steam turbine, and has many benefits for electric utilities and for process industries requiring steam. The hot gases from the exhaust of the gas turbine are employed in a fired boiler to generate superheated steam to drive a steam turbine (figure 10.25).

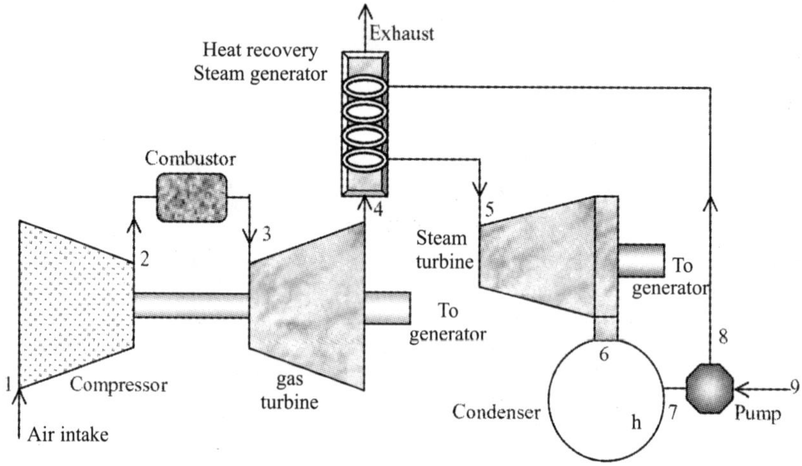

Fig. 10.25 Combined Gas and Steam Cycle Plant

The net work done is in the same range as the work done in a steam injection cycle but lesser fuel is required, which increases thermal efficiency. The cost of heat recovery steam generator (HRSG), steam turbine, condenser and pump tend to considerably increase the first cost associated with this cycle. NO_x content of the exhaust gases depends on the gas turbine. The main attraction of this cycle is the substantial advantages it offers in operating performance and fuel costs. Figure 10.26 provides performance curves for a typical combined cycle plant (*ref*: Boyce, 2002).

The power produced by a gas turbine can also be increased by injecting humidified and heated compressed air or steam into the compressor discharge or directly into the combustor, and by water injection in the mid compressor stages. Heating of air or water may be accomplished in a heat recovery steam generator, together with a separate compressor or pump. Flashing of water in the form of a spray mist in the middle rows of the compressor to cool the air aids in obtaining an isothermal compression process. Evaporation of the water through the consumption of latent heat lowers the temperature of the air stream as it enters the succeeding stage, and lowers the work done during compression. The technique has been successfully applied in many high-pressure gas turbines (for example, General Electric's LM6000 gas turbine engine), and may be combined with other methods described earlier. A more elaborate system is required to inject the external humidified and heated air at a point upstream of the combustor. A supplementary compressor for air pressurization, saturation column for humidification and preheating, water heater and balance of plant equipment in the form of pipes, valves and controls are required for the set up.

Water or steam may also be injected at the compressor discharge to augment generated power. For example, if injected steam is 12 per cent of the airflow, power increase will be in the neighbourhood of 25 per cent. If the steam can be generated from the gas turbine's exhaust, the procedure also yields higher efficiency.

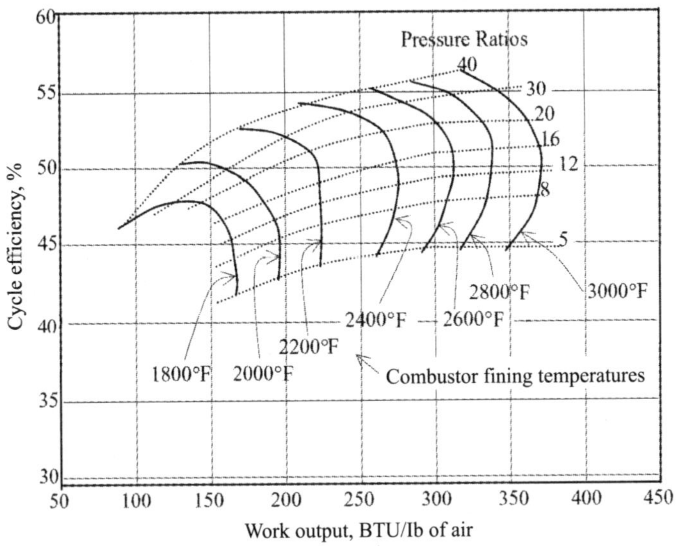

Fig. 10.26 Performance Curves for Combined Cycle Plant

When a gas turbine is equipped with nozzles to burn dual fuels, steam is injected into the combustor with the primary intent of controlling NO_x formation. Combustion concerns will limit the amount of steam injected, but 3 per cent of the airflow will provide a boost of 4 per cent in the power output. Steam would come from the HRSG.

A number of power plant operators are faced with inadequate power generation capacity during the summer, especially in the daytime when air conditioning and industrial requirements reach a peak. Peaking units designed to operate for only a few hours in the day and started and put on line quickly have to be installed expressly for the purpose. But the investment is not particularly attractive, since the return is not high. To alleviate the situation Alabama Electric Cooperative (*ref*: Brown, 2000) operates a plant based on a cycle involving storage of compressed air. During off-peak hours excess power drives a train of compressors, and stores compressed air into solution-mined underground caverns. The compressor train includes a combination motor and generator with clutch mechanisms. When compression is required the motor is engaged to pressurize and store air in the bunker, while also disengaging the expansion mechanism. To boost capacity the compressor may be split into two or more modules with inter-coolers in between them. Conversion from compressed air to electric power takes place in high and low pressure air expanders and generators. For this mode of operation the clutch disengages the compressor train. Air returning from the bunkers may be heated re-generatively in a recuperator using exhaust gas from the low-pressure expander and additionally burned in combustors before entering the high-pressure expander. From the high-pressure expander the air is further reheated in combustors before entering the low-pressure expander. Can combustors are used. With the aid of two combustors the high-pressure expander produces a quarter of the power, the remaining three quarters coming from the low-pressure expander with its eight combustors. The plant has dual fuel capacity, natural gas and No. 2 distillate fuel oil, and is capable of producing between 100 and 110 MW.

Many configurations for combined cycle plants are designed. In most applications the gas turbine cycle is the topping cycle and the steam cycle is the bottoming cycle. Thermal efficiencies reach close to 60 per cent in combined cycle, with the gas turbine producing 60 per cent of the power and the steam turbine contributing 40 per cent at design point. At off-design conditions the inlet guide vanes in the compressor control the amount of air entering in order to maintain temperature levels in the HRSG. Steam may be formed at one or at three different pressure levels. Steam generation at multiple pressure levels decreases losses at the exhaust stack.

Breakthroughs in energy efficiency continue to be made by the power generation industry. General Electric Power System's H turbine is designed to operate at 60 per cent thermal efficiency in a combined cycle mode, the equivalent of four-minute mile (*ref*: Valenti, 2002). The F series, for example, tops out at 57 to 58 per cent. Performance gains derive from advance materials and a new cooling system. This enables the H turbine to operate at 2,600°F firing temperature, nearly 200° F hotter than the F technology. Since fuel expense is easily the main component of cost of the electric power, the increase in efficiency has significant economic implications.

The H design calls for an 18-stage compressor with 23:1 pressure ratio and 1,500 lbs/second air flow. Design of the compressor uses many features of GE's CF6-80C2 aircraft engine compressor. A dry combustion system premixes air with fuel prior to ignition, creating under 9 ppm of NO_x. Steam is used to cool the first and second stage vanes and buckets (or blades). Steam cooling has the advantage over air-cooling of reducing the temperature gradient, and consequently thermal stress, in the cooled part. Closed loop steam used for this purpose is delivered to the stationary vanes by tubular seals, then recycled to the heat recovery steam generator. The third stage nozzles and blades are cooled by air, fourth stage not being cooled. Proprietary single-crystal super alloy with a dense, vertically cracked thermal barrier coating is used for first stage blades and vanes to increase heat resistance.

The gas turbines are teamed with a GE D-10 steam turbine. This three-admission reheat steam turbine has 33-inch last stage buckets. The gas turbine, steam turbine and electric generator are solidly coupled to generate 480 MW of power. The gas turbine is equipped to carry thrust created in both units. An auxiliary 2.3 MW diesel generator set is used to start the rotating train. Figure 10.27 shows a picture of the gas turbine rotor during assembly.

Fig. 10.27 GE Gas Turbine Rotor During Assembly (Valenti, 2002)

10.5 DERIVATIVE ENGINES FOR MARINE AND INDUSTRIAL USE

A number of industrial situations are present where the lightweight characteristics of an aircraft engine can be advantageously applied to generate mechanical power. The power, however, is mostly for driving another machinery instead of developing aerodynamic thrust. Hence, the fan is no longer needed, and the developed mechanical power drives an electric generator, compressor or a ship's propellers. Derivative engines based on aircraft engine technology have been developed expressly for such a situation for producing mechanical energy.

A good example is electrical power generation on an offshore oil platform. A tremendous amount of mechanical energy is required to operate drills, pumps, compressors and rigs on exploratory and production oil platforms. The geometry of a typical rig calls for a platform weighing several hundred thousand pounds located 50 to 100 feet above the surface of the water, with the super structure supported by four legs resting at the bottom of the sea. Strictly from stability considerations it is of vital significance to limit the weight of the overall structure above the water line, as also of the individual components located on the platform. The weight of a typical industrial gas turbine is prohibitively large, and hence a design derived from an aircraft engine technology would be ideal. Pipeline pumping and gas compression applications also rely on a large operating speed range, as opposed to a fixed speed requirement of a power generation gas turbine. Add to that the capability of a combustion system to burn liquid or gas fuels that are abundantly available on an oil platform, and the suitability of a derivative engine becomes readily apparent (figure 10.28).

Shipboard prime movers and many other marine applications have similar weight restrictions. Low speed reciprocating diesel engines traditionally have been employed for marine propulsion, but the engines tend to be physically large. For example, a supercharged and after cooled 12-cylinder direct drive diesel engine of 30 in. bore can produce 18,700 horsepower and weigh 74 lbs/brake horse power, for a total of 1,383,800 lbs. Physical dimensions of the engine are 25 ft high, 8 ft wide and 45 ft long, occupying a total of 9,000 ft^3. In comparison, an LM2500 derivative engine housed in its own 12 ft high, 11 ft wide and 32 ft long module can develop 19,500 horsepower and have a gross weight of 50,000 lbs, or 2.56 lbs/bhp.

By eliminating the fan of an aircraft engine a few stages may be added to the low-pressure compressor. The axial compressor is split into low and high-pressure modules, each powered by individual turbine sections mounted on the same shaft. The power turbine then is connected to the low-pressure compressor at the forward end and to a driven equipment at the other end. With concentric shafts the speed of both compressor sections offers more flexibility for optimizing. Three shaft derivative engines call for a separate power turbine, which is directly connected to the driven machinery. All three shafts operate in their designated speed range. Design innovations are incorporated to obtain the required long life characteristics of most industrial applications (figure 10.28).

Derivative engines offer a number of benefits. The size and weight of the complete engine lend them to assembly and packaging as a complete unit within the manufacturer's facility. A generator or a compressor may be included in the package, together with the accessories purchased by the customer. Installation may also proceed at the job site by factory personnel specially trained for debugging and performance matching. Because most customers strive to hold operational costs, the engines may be readily adapted for remote control and automation. Offshore and remote pipeline pumping stations are normally designed for unattended operation. When auxiliary systems are not

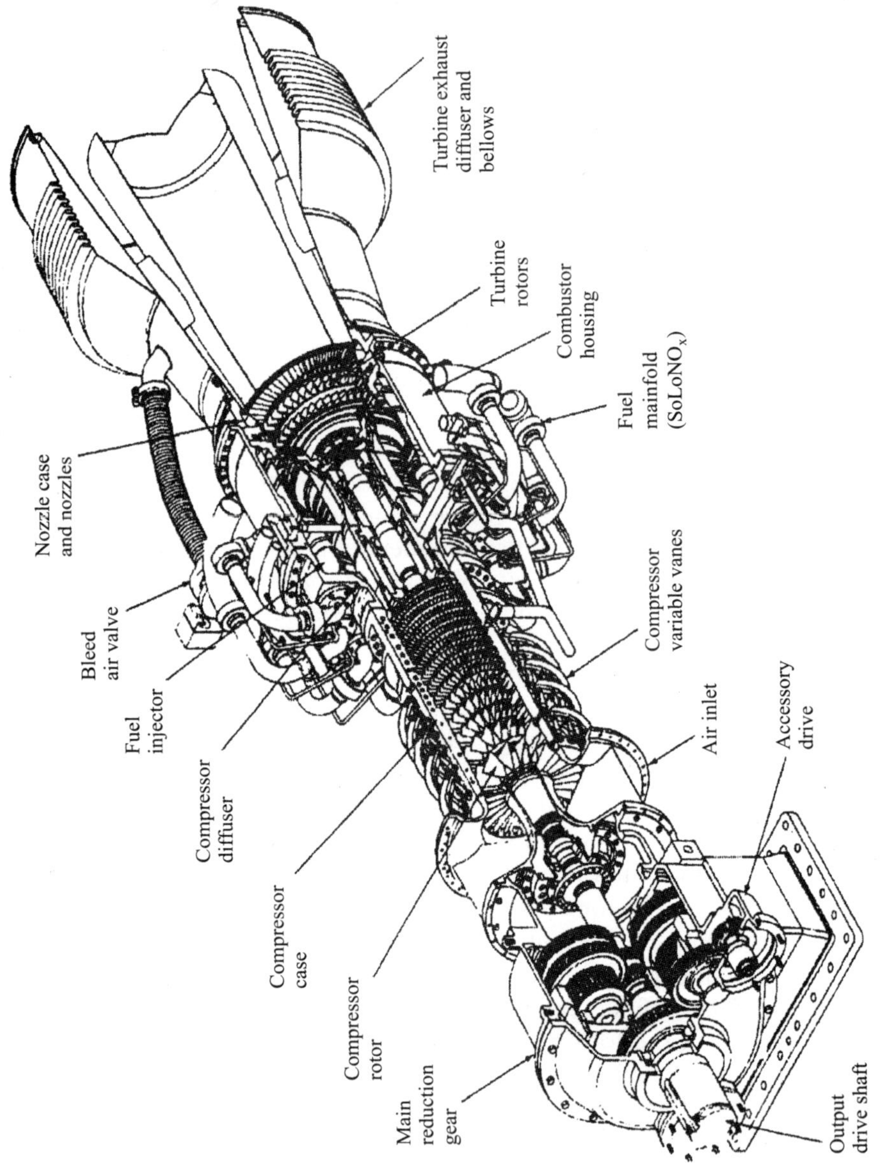

Fig. 10.28 Solar Turbines Taurus 70 Engine for Offshore Oil Production Platform

complicated, oil-to-air exchangers are used in place of water cooling. Starting devices requiring little energy are reliable. Hence, aviation technology derived engines lend themselves to automatic control from a distance. Aero-derivative engines can run continuously without inspection, until monitoring equipment indicates a fault or sudden performance variation. Such an incident can best be handled by removing and replacing the gas generator section with a spare, so the module can be inspected, evaluated and repaired more efficiently at the factory. Under these circumstances, offsite maintenance plans offered by a number of manufacturers and leading service organizations play a useful role. Technical servicing is then mostly restricted to conducting minor running adjustments and related routine tasks.

An integrated electric propulsion and power service system provides for greater flexibility, efficiency and survivability of naval ships. Examples of this concept include the type-45 destroyer program for the Royal Navy and DD (X) program for the US Navy (*ref*: Harvey, 2002).

The US Navy system comprises of a General Electric LM2500 gas turbine engine directly coupled to a Brush Electric Company synchronous generator. The unit is capable of producing 21.6 MW at 0.8 power factor. To determine system response during start-up a ramp load is conducted during system test to characterize power system interface, stability and control performance. Step loads are in the form of propulsion motor acceleration at an average of 4 per cent power per second up to steady state condition. System operating conditions are 25, 50, 75 and 100 per cent of shaft output power of the propulsion motor.

Step unloading of the power generating system represents a considerably more severe situation than a gradually unloaded ramp operation. The engine is driving at a constant mechanical power and the generator exciter is providing a constant excitation when the electrical load is suddenly lost. Frequency and voltage of the machine overshoot momentarily, hence mechanical power input to the generator must be reduced rapidly to prevent over speed and return the power turbine to normal speed. This is accomplished by adjusting the fuel-metering valve. In a similar manner the automatic voltage regulator must rapidly reduce excitation current to the generator to return the voltage to a normal level.

10.6 DIESEL ENGINE TURBO-CHARGER

An internal combustion engine cycle is a series of events that the engine goes through while it operates and delivers power. In a four-stroke, five-event cycle they may be described as intake, compression, ignition, combustion and exhaust events. Since the events occur in a certain sequence and at precise intervals of time, they are said to be timed. Most piston engines operate on the four-stroke, five-event cycle principle developed by August Otto, and is named after him as the Otto cycle. Other cycles for heat engines are the Carnot cycle, the Diesel cycle and the Brayton cycle; they differ in the particular engine theories developed by the scientists whose name is associated with the cycle.

The four strokes of a four-stroke cycle engine are the intake stroke, the compression stroke, the power stroke and the exhaust stroke. In a four-stroke-cycle engine the crankshaft makes two revolutions for each complete cycle, so the ignition of the fuel and air mixture takes place only once in two complete revolutions of the crankshaft. In a two-stroke engine, on the other hand, the complete cycle takes place during one revolution of the crankshaft. The sequence of events during the up and down strokes may be described as follows. During the intake stroke the piston starts at top dead center, the intake valve is open, the exhaust valve is closed, the piston moves downward, the fuel-air mixture is drawn into the cylinder, and at its end the intake valve closes. During the compression stroke both valves are closed, and the piston moves toward the top dead center while compressing the working fluid. Ignition takes place near the top of the stroke. In the power stroke both valves are closed, the ignited gases build up a lot of pressure and cause it to expand while forcing the piston toward the bottom dead center. The exhaust valve opens well before the bottom of the stroke. During the scavenge stroke the exhaust valve is open and the intake valve is closed, the piston moves toward top dead center, forcing the burned gases out through the open exhaust valve, and the intake valve opens near the top of the stroke.

Naturally aspirated reciprocating engines are designed to induce outside air, or a mixture of air and fuel, at nearly atmospheric pressure. The concept of supercharging, supplying pressurized air to an internal combustion engine, increases the mass flow rate of air induced into the cylinders. A greater amount of air crammed into the cylinders permits an increase in the fuel introduced into the system than would be possible if the air is aspirated into the cylinders without the benefit of pre-compression. This leads to an increase in the engine's power output as well as an improvement in thermal efficiency (*ref*: Heisler, 1995).

Turbo-charging is a particular form of supercharging in which a compressor is driven by an exhaust gas turbine to pressurize the incoming air. Turbo-chargers are extensively used on large compression ignition engines, and find increasingly greater applications on automotive engines as well. In diesel engines the process can reduce the specific fuel consumption from about 3 to 14 per cent in the engine's speed range. The reduction in fuel consumption becomes more marked as the engine's load is reduced, as can be seen in figure 10.29 from the family of constant load curves ranging from ¼, ½, ¾ and full engine load. However, at full load below 1,400 rpm and ¾ load below 1,000 rpm the specific fuel consumption is inferior to that of the naturally aspirated engine. Thus, the improvement in fuel consumed becomes more effective as the engine load is reduced. With the turbocharged engine, the level of exhaust smoke emission is considerably reduced with increasing engine speed, as excess air is supplied to the cylinders, which is in contrast to the naturally aspirated engine. In the upper speed range the naturally aspirated engine finds it difficult to clear and fill the cylinders with sufficient quantities of fresh air. Frictional losses rise rapidly with an increase in engine speed but do not rise in direct proportion to the engine load output.

Supercharging of spark ignition engines involves a compromise with efficiency, and can only be justified in a few cases, including:

- Aircraft engines: Supercharging is used here to provide both high specific power output for take-off and to compensate for lower air density at operating altitudes. All but small engines for light aircraft are supercharged. The problem of detonation is solved by use of high-octane fuels.

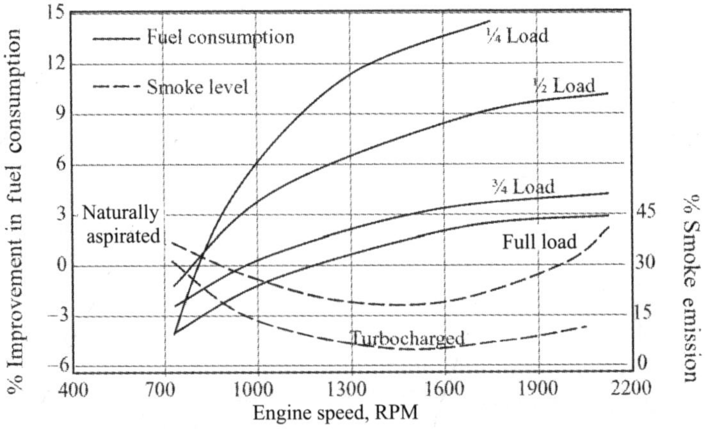

Fig. 10.29 Effect of Engine Speed on Exhaust Smoke and Fuel Consumption (Heisler, 1995)

- Automobile engines: With the advent of small, efficient turbo-superchargers and reliable electronic controls, many luxury and sports-type automobiles are supercharged. This usually lowers fuel economy in comparison with the same engine naturally aspirated. The decision to use supercharging is more one of marketing than one of utility.
- Racing engines: Here specific output has exaggerated its importance, and supercharging is used wherever it is allowed.
- Large natural gas engines: In this case the saving in weight and bulk by means of supercharging is great, and the fuel used has a high resistance to detonation. Nearly all natural gas engines above 500 horsepower are supercharged. No limit on supercharging in diesel engines is imposed by combustion.

The decision whether or not to use supercharging, and if so how much, depends on a balance between the relative simplicity of the non-supercharged engine together with its generally lower mechanical and thermal stresses and the smaller size of a supercharged engine with the same rating. The initial cost may be in favour of either type, depending on the cost of the supercharging system and the reduction in cost due to decreased engine size (*ref*: Taylor, 1985).

On the other hand, diesel engines use the compression ignition concept for trucks, buses and locomotives. Medium and large marine engines are almost always supercharged. The same is true for stationary applications, such as power generation, except in the smallest sizes. The allowable amount of supercharging in the diesel engines depends on the question of reliability and durability, plus economic factors such as the cost of the supercharging equipment as a function of its capacity and pressure ratio.

Compressors for pressurizing the intake air can be divided into two classes, positive displacement and dynamic, or non-positive, displacement types. Examples of positive displacement compressors include the Roots, sliding vane, screw, reciprocating piston and Wankel methods. These compressors can be more readily driven from the engine crankshaft, an arrangement often referred to as a 'supercharger'. Axial and radial flow compressors are of the dynamic type. Because of internal flow characteristics their rotational speed is of an order of magnitude higher than that of the internal combustion engine. Axial and radial compressors can be more adequately driven by a turbine to form a turbocharger. The turbine also can be of the axial or radial flow type. The advantage of turbochargers over superchargers stems from their use of recovered exhaust gas energy during the engine's blow-down stage.

The turbocharger relies solely on extracting up to a third of the wasted energy passing out from the engine's cylinders. However, this produces a penalty in the form of increased manifold backpressure, making it more difficult for each successive burnt charge to be expelled from the cylinders. The ideal available energy that may be used to drive the turbocharger comes from the blow-down energy transfer that occurs when the exhaust valve opens and the gas expands to atmospheric pressure (*ref*: Heisler, 1995).

Turbocharged engines produce higher cylindrical volumetric efficiencies compared with normally aspirated induction systems. Hence, there will be higher cylinder peak pressures. This increases mechanical loading of the components, and could cause detonation in gasoline engines. So it is customary to reduce engine's compression ratio from 10:1 for a naturally aspirated engine to 9:1 for a low boost pressure or 8:1 for a medium to high boost pressure turbocharged engine. In a direct injection diesel engine having a normal 16:1 compression ratio, the turbocharged engine compression ratio may be lowered to 15:1 or 14:1.

A correctly matched supercharger will raise the cylinder's brake mean effective pressure (b.m.e.p.) to well above that of a naturally aspirated engine without creating excessively high peak cylinder pressures; the actual increase in the brake mean effective pressure is basically determined by the level of boost pressure the supercharged system is designed to deliver. The advantage in raising the cylinder's b.m.e.p. is partly offset by having to transfer some of the engine's power to driving the supercharger (see figure 10.30). The power absorbed is dependent upon its displacement capacity per revolution and speed for a given boost pressure.

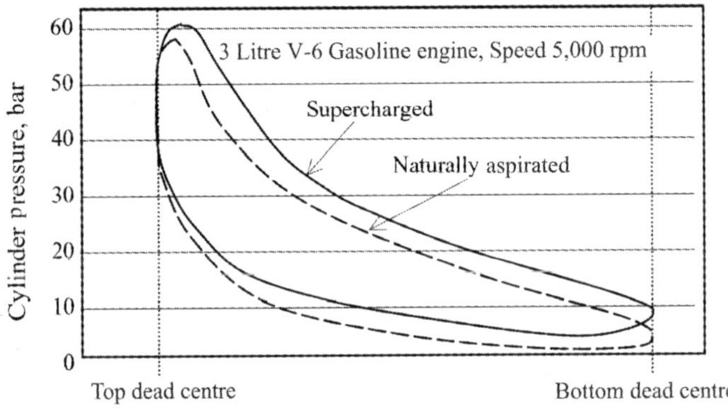

Fig. 10.30 Naturally Aspirated and Supercharged Engine *PV* Diagrams

A turbocharger comprises of an exhaust gas driven turbine and housing, a centrifugal compressor wheel with its housing and an interconnecting shaft mounted on a pair of fully floating plain bearings. The bearings are encased in a single bearing housing. Construction of a typical system is shown in figure 10.31.

The impeller, or compressor wheel, is an aluminium alloy casting in the form of a disk mounted at the hub, with curved radial blades (about 12) projecting from one side. This causes the air surrounding the impeller to be divided into an equal number of cells. The hub is contoured such that air enters the cells formed between adjacent pairs of blades axially. The enclosed air then travels along the flow path in the back wall of the impeller disk, causing the air to be expelled radially from the cells. Once air reaches the impeller periphery it passes to the parallel gap diffuser formed between the bearing housing and compressor wheel housing. From the diffuser gap the air flows into a circular volute shaped collector, which provides a constant expansion passage for the air (*ref*: Heisler, 1995).

Exhaust gas temperature at the inlet to the turbine wheel, under light-load to full-load high speed operating conditions, may range between 600°C and 900°C. Consequently, the turbine is usually made from a high-temperature heat resistant nickel-based alloy such as 'Inconel'. The turbine wheel (shown in figure 10.32) takes the form of a hub supporting a disk at one end and a number of radial blades projecting axially and radially from both the hub and the disc. The outer edges of the blades are curved backwards to trap the impinging exhaust gases. Exhaust gases from the manifold enter the spherical graphite cast-iron turbine housing flange entrance. The gases then flow around either a single or twin volute passageway surrounding the turbine wheel. The gases are then forced tangentially inwards from the throat of the turbine housing and also moves through a

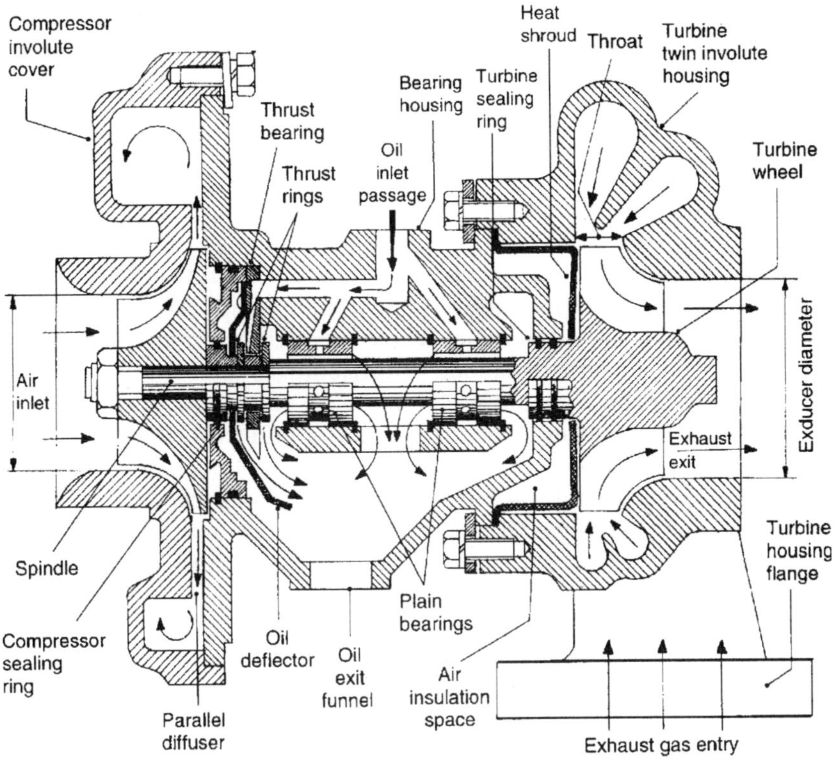

Fig. 10.31 Turbocharger (Heisler, 1995)

right angle, so that they come out axially from the center of the turbine hub before impinging on the blade faces. The flow path then directs the gases gradually through, and they are then expelled into the exhaust pipe system.

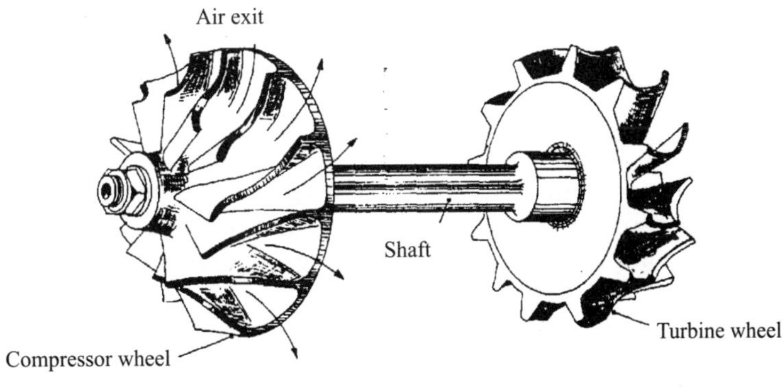

Fig. 10.32 Rotor Assembly (*ref*: Heisler, 1995)

In most turbocharger designs the turbine and spindle are joined together by a welding process (inert gas welding, resistance welding, electron beam welding). Inertia friction welding is also used, where the turbine and spindle are brought together under load, with one part revolving against the other so that frictional heat is generated at the interface. When the joint area is sufficiently plastic as a result of the increase in temperature the rotation is stopped and the end force increased to forge and consolidate the metallic bonds.

The turbine wheel and steel spindle can be welded together in a vacuum by an electron beam. The hollow space between the turbine wheel and the spindle prevents heat being transferred through the center of the spindle. Thus, heat is carried away along the outer section of the spindle, which can readily be cooled by the lubricating oil. The spindle is supported on a pair of free-floating phosphorus bronze plain bearings. Around the outside of each bearing shell are six radial holes, and there may be a circumferential groove machined to distribute the lubrication oil. The rotation of the spindle assembly, in conjunction with the speed and load conditions of the oil supply from the engine's lubrication system, produces a certain amount of end thrust as the spindle will want to move axially first in one direction and then the other. Both radial plain bearings and the axial end thrust bearing are supplied with ample oil from the engine's lubrication system by drillings made in the bearing housing and in the bearings themselves. The oil supply has two major functions, first, to lubricate the bearings so that a hydrodynamic oil film can be established, and second, to remove excess heat from the bearing assembly.

The impeller may be closed or shrouded; that is, the impeller is cast so that the cells or channels are completely enclosed. This construction eliminates direct leakage as the induced air is flung radially outwards in the cells. However, it is difficult to cast radial cells so that they curve backwards and also provide an axial angled inlet at the eye of the impeller. Other important disadvantages that must be considered are the mass of the shroud supported by the blades, such that at high rotational speeds the blades are subjected to severe centrifugal stresses. In addition, the shroud is away from the central hub, hence it raises the impeller wheel's moment of inertia, thus impeding its ability to accelerate or decelerate rapidly. This design has been virtually replaced by one without a shroud.

With an open impeller and scroll diffuser the impeller is cast with blades forming the walls of the cells. These blades can be shaped so as to provide the best inducement for the incoming air, and the radial flow path can be curved backwards to optimize the flow discharge at high rotational speeds. However, there is a clearance between the outer edges of the blades and the internal curved walls of the housing, which encloses the rotating cells. This gap will be responsible for leakage losses under high boost pressure operating conditions. With this arrangement, the kinetic energy of the air at the blade tips is converted into pressure energy by directly entering into the relatively large scroll volume. Thus, the air flung out at the rim of the impeller enters the scroll and into the circular passageway.

If a more positive method of converting the kinetic energy to pressure energy is required, a parallel annular space between the impeller and the volute or scroll will enlarge the circular passage from the entrance at the impeller rim to where it merges with the discharge volute. Thus, as the air moves outwards in a semi-spiral and radial direction the air will expand, thus causing its speed to reduce while its pressure rises. To reduce the maximum diameter of the volute or scroll housing, the circular volute passageway can be cast to one side of the parallel diffuser with a considerably reduced diameter. Looking at the sectional view of the compressor, the wall between the diffuser and volute resembles a tongue, and hence its name: parallel tongue diffuser. This produces a right-angled flow-path similar to the normal parallel wall diffuser design.

It is important for effective cylinder scavenging that pulsed exhaust gas energy is introduced to the turbine wheel, in contrast to a damped steady flow of exhaust gas. With a four-cylinder engine provided with a single-exhaust manifold, this is possible as there is an exhaust discharge every 180° so that there is very little exhaust gas interference between the cylinders. However, with more than four cylinders exhaust gas will discharge at shorter intervals than the 180°; that is, for 5, 6 and 8-cylinder engines the intervals between exhaust discharges will be 144°, 120° and 90° respectively. To overcome exhaust gas interference in the manifold, manifolds are sub-divided so that, in the case of an in-line six-cylinder engine, cylinders 1, 2 and 3 are grouped together and, similarly, cylinders 4, 5 and 6 are grouped together, and there is now an extensive exhaust interval between sub-divided manifolds of 240°. The exhaust discharge from each half-manifold is then fed to the turbine wheel through two separate passageways. If the exhaust gas in the branch pipes is permitted to discharge in the form of a pulse the initial blow-down from the open exhaust valve port will produce a rapid pressure rise until it peaks. The exhaust pressure then quickly decreases to a minimum value before the next cylinder, sharing the same common manifold gallery, discharges another batch of exhaust gas. This cycle of events will be continuously repeating. By reducing or even eliminating exhaust gas interference between the cylinders, by sub-dividing the manifold if need be, the exhaust pressure in the manifold will fall towards the end of the exhaust stroke to a value below the mean compressor pressure. Thus, during valve overlap near the end of the exhaust period and at the beginning of the inlet period, a positive pressure difference will exist between the cylinder intake and the cylinder exit, which will cause a blow-through of fresh charge from the intake manifold to the exhaust manifold. If there is sufficient pressure difference the fresh charge will rush into the cylinder and push out the residual exhaust gases still remaining in the non-swept combustion chamber space. The effectiveness of this scavenging action will also depend on the engine speed and the actual valve opening area during the time of valve overlap.

Transient response time is dependent upon the inertia of the rotating parts and the efficient projection of the exhaust gas onto the turbine blades. Immediately as the engine throttle is opened there will be an increased flow of mixture entering the cylinders with a corresponding exit of exhaust gas, which is directed onto the turbine blades, causing the wheel assembly to accelerate rapidly. The time taken for the turbine and compressor assembly to attain the maximum operating speed is dominated by the overall efficiency of the turbocharger and the polar moment of inertia of the rotating assembly.

Turbocharger lag will depend to some extent upon the excess torque available from the turbine wheel over that required to drive the compressor with the air flow and boost pressure existing at that instant. Therefore, a small turbine volute housing attached directly to a short and small diameter passageway manifold is desirable, as this will provide an undamped exhaust gas pulse directly and effectively to the turbine blades, thereby producing the least time lag.

The turbocharger has to be prevented from over-speeding and overheating since this can have two disastrous consequences: first, excessively high compressor and turbine-wheel rotational speeds, when subjected to high operating exhaust gas temperatures, can very quickly destroy the revolving components; and second, excessively high boost pressure will produce a correspondingly high cylinder pressure and temperature over a period of time, which can do considerable damage to the various supercharging components of the engine and in the case of a petrol engine, will certainly promote cylinder detonation during acceleration conditions.

To safeguard the turbocharger from over-speeding and overheating, a portion of the exhaust gas expelled from the cylinders under high engine load and/or speed conditions is deliberately made to bypass the turbine housing and instead flow directly to the exhaust pipe. Under extreme operating conditions, 30 to 40 per cent of the exhaust gas can be diverted away from the turbine throat with the effect that the turbine will not increase its speed and the output boost pressure will remain approximately constant with any further rise in engine speed. The exhaust gas bypass passage opening is controlled by a waste gate in the form of either a poppet or a swinging flap type of valve. Both types of waste gate valves are normally operated by a diaphragm actuator controlled by either the boost pressure from the volute impeller housing or by the exhaust manifold gas pressure. With the poppet valve waste gate the long stem of the valve is connected directly to the diaphragm actuator, with the stem usually enclosed in a finned housing to improve the heat dissipation from the valve and actuator assembly. Conversely, the swinging-flap type waste gate is operated by a short external lever, which is linked to the diaphragm actuator by a long push rod, so that the actuator is practically insulated from the exhaust gas heat. The waste gate and bypass passageways for small turbochargers can be integral with the turbine wheel housing or, for the larger turbochargers, the waste gate unit and the bypass passages can be mounted separately away from the turbine wheel housing.

Boost pressure can be controlled by blowing off either surplus exhaust gas from the turbine inlet through a waste gate valve or surplus air from the compressor delivery via a blow-off valve. Blowing off surplus air from the compressor discharge results in higher turbocharger speeds than the exhaust waste gate method. This is because the compressed air delivery load is reduced, but there will be very little change in the amount of gas energy passing through the turbine-wheel; consequently, the excess energy input to the turbine will raise the rotor assembly spin speed to a higher level. Since a portion of the compressed air delivery is discharged back into the atmosphere, and energy has been spent in driving the turbine wheel, it follows that there will be some reduction in the engine's thermal efficiency during this period when the compressed air is blowing off. Thus, because of the turbocharger's relatively low overall efficiency during the compressor discharge blow-off period (which may be prolonged under certain operating conditions) the waste gate method of diverting exhaust gas away from the turbine wheel has been universally adopted. However, the blow-off valve's simplicity has encouraged incorporation of this form of pressure-relief valve between the compressor and the inlet manifold as a secondary means of limiting boost pressure in the event of its build-up rate exceeding the waste gate's ability to divert sufficient gas energy from the turbine wheel.

10.7 CONTROL SYSTEM FOR AUTOMOTIVE TURBINE ENGINE

A gas turbine powering an automobile vehicle requires a number of components. For instance, the gas turbine engine developed by Chrysler Corporation drives a hybrid electric sports car (*ref*: Shortlidge, 1998). Each of the two spools consists of an integrated compressor, turbine and a three-phase induction motor/generator that produces 186 kW (250 hp) electric power output.

An inter-cooler is used between the stages of the compressor as shown in figures 10.33 and 10.34, and the combustor is placed between the high-speed and high-pressure compressor and the turbine. The turbine alternators supply power to a 373 kW (500 hp) flywheel and a 559 kW (750 hp) traction motor (figure 10.35). The 373 kW (500 hp) alternator flywheel, traction motor and a 9000 μF capacitor bank are electrically coupled through an 800 VDC power bus. Power flows electrically from one propulsion system component to the other. A microprocessor-based controller regulates the flow of current. The capacitor bank rejects the highest frequency bus voltage disturbances.

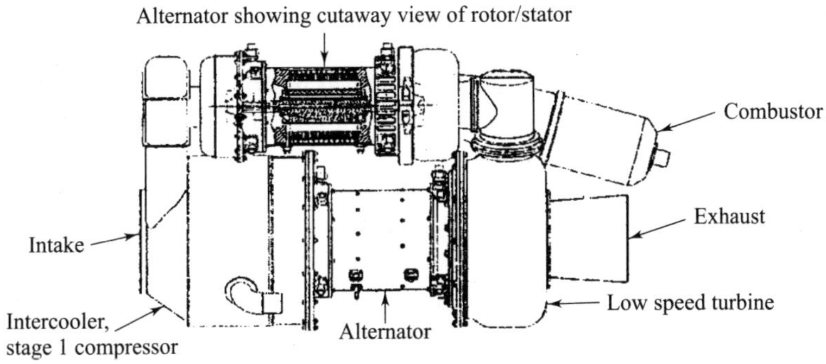

Fig. 10.33 Twin Spool Turbine Alternator

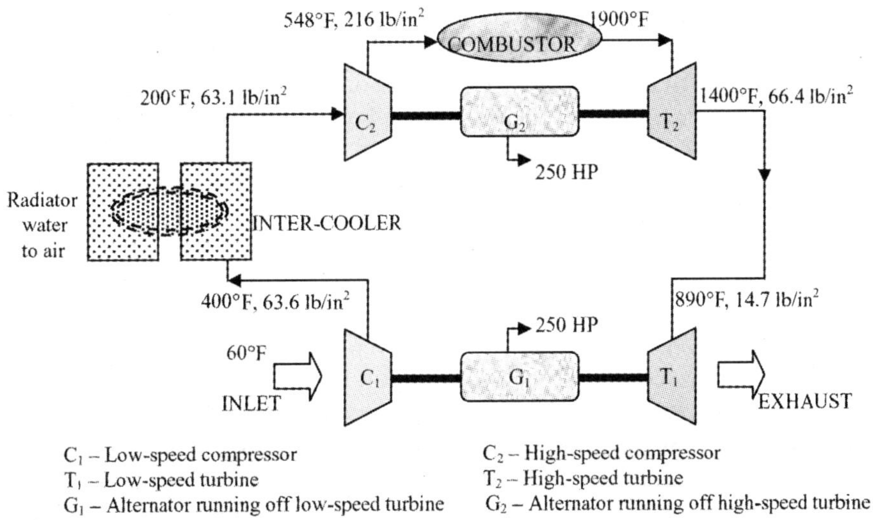

Fig. 10.34 Cycle Diagram for Turbine Alternator

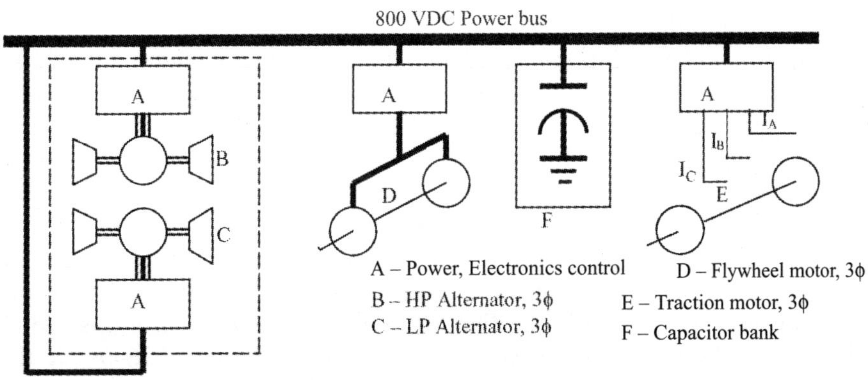

Fig. 10.35 Power Flow Architecture for Vehicle Propulsion System

The flywheel's permanent magnet motor controller is designed to yield the next highest bandwidth power capability in the system. This controller also acts as a bus voltage regulator in the event of disturbances introduced in the bus voltage by the traction motor during vehicle accelerations, and generates power during regenerative vehicle braking. The turbine alternator consumes power during starting and during operation at or near idle speed.

The flywheel provides power in the higher bandwidths and the turbine alternator provides high-efficiency power in the lower bandwidths. Proper power management requires maintaining the total kinetic energy in the system at or nearly constant level.

The two spools of the turbine alternator run at 100,000 rpm and at 60,000 rpm. Nominal pressure ratios for the low and high-speed compressors are 4.6 and 3.1, respectively. The intercooler cools 1.45 kg/s of air from 205°C to 94°C, while the cooling water flow rate is 2.27 kg/s and inlet and outlet temperatures of 68°C to 85°C respectively. The turbine alternator operates on an open Brayton cycle without a recuperator or a regenerator. The inlet temperature for the high-pressure turbine is 1038°C (1900°F), crossover point is 760°C, and low-pressure exhaust is at 477°C (890°F).

The alternator is centered between fluid film flexure pad bearings, with the compressor overhanging from one end of the shaft and the turbine on the other. Both rotating systems operate super-critically, with the low-pressure spool operating between the first and second critical speeds and the high-pressure spool operating between the third and fourth critical speeds. The thrust and radial bearings are located at the compressor end and are water lubricated. The water also serves as a coolant for the alternators. The stator of the alternator is cooled by an outer water jacket, while the rotor has an internal cooling system.

The engine burns natural gas, which is stored as a cryogenic liquid and vaporized in a separate unit prior to entering the combustor. For operation with a different fuel, only the combustor needs to be replaced. The vaporizer is essentially a helical coil passing natural gas through it, and warmer water flows outside in a cross counter-flow arrangement. Heat travels from the gas to the water in a convection-conduction-convection mode.

The power electronics for the induction type alternators consists of a three-phase inverter with a sine wave pulse modulated drive. The power electronics has its own microprocessor-based controller to obtain a closed-loop speed control for both turbine spools. The control loop effectively modulates alternator slip to achieve a target speed reference.

Power density is maximized and weight minimized through use of Inconel 100 for the turbines, titanium for the compressor, 303 Stainless steel for compressor shroud, and AerMet 100 for the rotor. The housing for the rotor is made from carbon composite.

The vehicle implements a distributed hierarchical control system (figure 10.36). The vehicle management computer (VMC), the power electronics controller (PC) and the turbine fuel system controller (TFC) are directly relevant. The VMC generates software commands based on driver control inputs such as throttle and brake positions. The PC receives commands from the vehicle's management computer through an optical serial data link. The power electronics and the turbine controllers are similarly linked.

Modulating fuel flow and torque in the low and high-pressure spools nominally controls the turbine's alternator. The primary control mode is implemented to achieve a requested engine power level by slewing the fuel flow servo loop. The alternator speed control loops are closed at a considerably higher bandwidth than the fuel flow servo loop. The speed loops are then able to

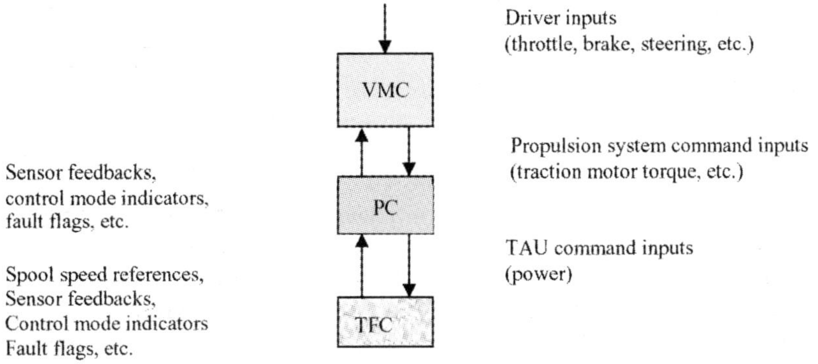

Fig. 10.36 Control System Architecture

follow the speed references accurately during relatively slower fuel flow transients. A consequence is that the turbine is generally operating on a predefined steady-state map of fuel flow and the speed of the two rotors. The map is specified in the turbine controller software in the form of corrected parameters.

10.8 REVIEW PROBLEMS

Example problem # 10.1: A variable length cantilever beam consists of a strip of spring steel 0.20 in. wide and 0.020 in. thick. It carries a weight of ¼ oz. at its free end. What should be the free length of the strip if it is required to have frequencies from 6 cycles/sec to 60 cycles/sec?

Solution: The spring constant of a cantilever beam is $3EI/l^3$. The moment of inertia of the cross section is:

$$I = (1/12)bh^3 = [0.2 \times (0.02)^3]/12 = 1.33 \times 10^{-7} \text{ in}^4.$$

$$\text{Bending constant } EI = 30 \times 10^6 \times 1.33 \times 10^{-7} = 4.0 \text{ lbs-in}^2,$$

$$\text{so spring constant } k = 12/l^3 \text{ lbs/in.}$$

$$\text{At the end, mass } m = 1/(4 \times 16 \times 386) = 4.05 \times 10^{-5} \text{ lbs-in}^{-1}\text{-sec}^2.$$

$$\text{Mass per inch length of strip } \mu = (0.2 \times 0.02 \times 0.28)/386 = 0.29 \times 10^{-5} \text{ lbs-in}^{-1}\text{-sec}^2.$$

$$\text{So total mass} = (4.05 + 0.29l) \times 10^{-5} \text{ lbs-in}^{-1}\text{-sec}^2.$$

Maximum length frequency is 6 cycles/sec, or, $\omega^2 = (2\pi \times 6)^2 = 1421$ (radians/sec)^2.

$$\text{Natural frequency: } \omega^2 = k/m, \text{ or } 1421 = 12 \times 10^5/\{(4.05 + 0.29l)l^3\}$$

By a method of trial and error, length of the beam is: $l = 5.33$ in.

Example problem # 10.2: A flywheel may be represented approximately by a solid steel disk of 4.5 in. diameter and 2 in. thickness. The shaft on which it is mounted has a torsional stiffness of 9.5 in-lbs/radian. Find the natural frequency of vibration of the system.

Solution: Weight of the flywheel is:

$$(\pi/4) \times (4.5)^2 \times 2.0 \times 0.28 = 8.9 \text{ lbs.}$$

and its moment of inertia is:

$$I = (mr^2)/2 = (8.9/386)(2.25^2)/2 = 0.0584 \text{ lbs-in-second}^2.$$

The natural frequency thus is:

$$\omega_n = \sqrt{(k/I)} = \sqrt{(9.5/0.059)} = 12.75 \text{ radians/second.}$$

or,
$$f_n = 12.75/2\pi = 2.03 \text{ cycles/second.}$$

Example problem # 10.3: A stroboscope is a device for producing intermittent flashes of light with which rapid vibratory motions can be made to appear to standstill, or to move very slowly. The flashes of light are of extremely short duration. When a vibrating object illuminated with this light is adjusted to the same frequency as the vibration, the object will be seen in a certain position; then it will be dark, and the object is invisible while traveling through its cycle. When it returns to the first position after one cycle another flash of light occurs. Thus the object appears to standstill. Now consider a point located 4 in. from the axis of a machine rotating at 10,000 rpm. If it is required to see the point with a blurring of less than 1/32 in., what should be the duration of the light flashes?

Solution: The point in question travels at:

$$10{,}000 \times 2\pi \times 4/60 = 4{,}189 \text{ in/second}$$

Flash time duration multiplied by blur velocity of the point will give blur size, which is

$$4189t = (1/32), \text{ so } t = 1/(4189 \times 32) = 1/134{,}000 \text{ second.}$$

Example problem # 10.4: Use Rayleigh's method for calculating the angular frequency of the fundamental mode of vibration of the beam with two masses, shown in figure 10.37. Assume $W_1 = W_2 = W$, flexural rigidity of the beam as EI and neglect the mass of the beam.

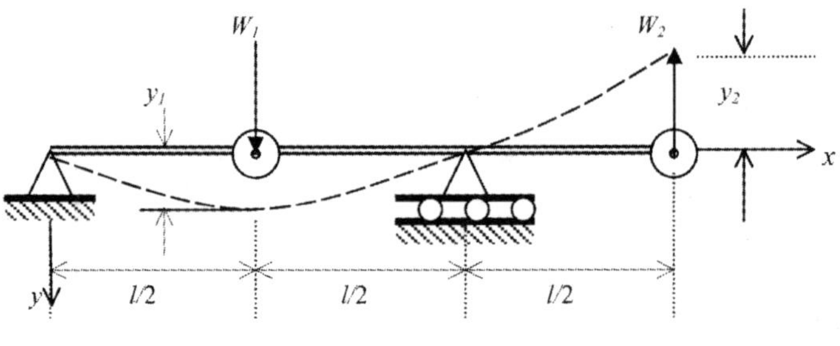

Fig. 10.37

Solution: From Rayleigh's method, for n masses on a beam the formula for angular frequency is:

$$p^2 = \frac{g\sum_{j=1}^{n} W_j y_j}{\sum_{j=1}^{n} W_j y_j^2}$$

It will be assumed that during dynamic motion the beam maintains a shape similar to the static deflection curve due to the two weights acting in opposite directions. The corresponding static deflections are:

$$y_1 = \frac{W_1 l^3}{48EI} + \frac{W_2 l^3}{32EI} = \frac{5Wl^3}{96EI}$$

$$y_2 = \frac{W_1 l^3}{32EI} + \frac{W_2 l^3}{8EI} = \frac{5Wl^3}{32EI}$$

Substituting these values into the expression for angular frequency gives:

$$p = \sqrt{192EIg/25Wl^3} = 2.77128 \times \sqrt{EIg/Wl^3}$$

Example problem # 10.5: If $W_1 = 3W_2$ in the above problem, what percentage change in the frequency is obtained?

Solution: Since $W_1 = 3W_2 = 3W$ and $W_2 = W$, static deflections are:

$$y_1 = (5/32) \times (Wl^3/EI) \text{ and } y_2 = (5/32) \times (Wl^3/EI)$$

Substituting in the expression for frequency:

$$p^2 = [W(5/32 + 5/32)]/[W\{(5/32)^2 + (5/32)^2\}] \times \sqrt{EIg/Wl^3}$$

$$p = 2.5298 \times \sqrt{EIg/Wl^3}$$

Hence the frequency is reduced by (2.7713 – 2.5298)/2.7713 = 8.71%.

Example problem # 10.6: A simply supported beam is impacted by a free falling sphere of weight W from height h, as shown in figure 10.38. Determine an expression for maximum momentary deflection produced in the beam.

Solution: When a load W falls a distance h on the beam, it causes a peak instantaneous deflection δ in the beam where the ball strikes. The total work done by the ball is $W(h + \delta)$. Assume a static equivalent load W_{eq} that produces an equal deflection δ at the same point on the beam, with the load applied gradually. The work done by this equivalent load is $(W_{eq}\delta)/2$. If the supports at the two ends of the beam are rigid, the work done by the falling and the stationary loads are the same, or:

$$W(h + \delta) = (W_{eq}\delta)/2$$

The static deflection δ may be expressed in terms of the equivalent static load W_{eq}. For example, if the load is applied centrally on the beam, $\delta = (W_{eq}l^3/48EI)$, where l is the length of the beam and EI represents the beam stiffness in flexure.

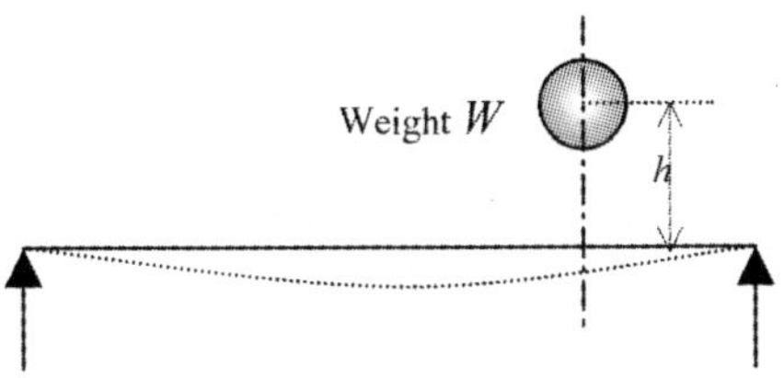

Fig. 10.38

Example problem # 10.7: If the ball weighs 8000 lbs, produces a static deflection of 0.75 in. and a maximum bending stress of 5500 lbs/in^2 in the above problem, from what height can a ball of 4000 lbs be dropped at the same point in the beam if the maximum permissible bending stress is 22000 lbs/in^2?

Solution: The required static load for a stress of 22000 lbs/in^2 is:

$$W_{eq} = (22000/5500) \times 8000 = 32000 \text{ lbs}$$

Since static stress and deflection are proportional to the load causing them, the corresponding deflection is: $\delta = (32000/8000) \times 0.75 = 3.0$ in. Substituting the values in the expression obtained from the previous example problem,

$$4000(h + 3.0) = (32000 \times 3.0)/2$$

Hence, $h + 3.0 = 12.0$ and the required height is 9.0 in.

Example problem # 10.8: Consider a prismatic bar rotating about an axis normal to the beam, resulting in a state of centrifugal stress in the bar (figure 10.39). If angular speed is 32,500 rpm, outer and inner radii are 5.5 in. and 2.5 in. and material density is 0.28 lbs/in^3, determine the maximum stress.

Solution: At any radius in the bar stress due to centrifugal force is:

$$\sigma = \int \rho\omega^2 r\,dr = (\rho\omega^2/2)(r_0^2 - r_1^2)$$

Stress at the inner radius near the axis of rotation is:

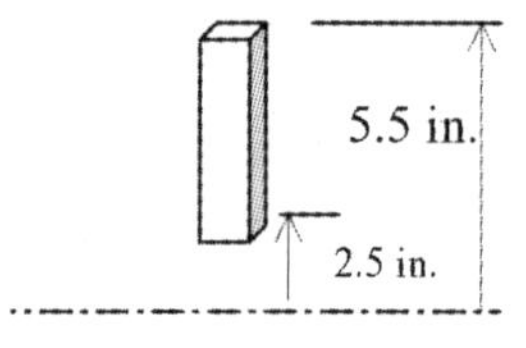

Fig. 10.39

$$\sigma = [(0.28/386) \times (2\pi \times 32500/60)^2](5.5^2 - 2.5^2)/2 = 100827 \text{ lbs/in}^2.$$

Example problem # 10.9: A freight ship is propelled by a 12-cylinder direct drive diesel engine of 30 in. bore giving 10,000 horsepower and weighing 120 lbs/bhp. The engine is 25 ft high, 8 ft wide and 45 ft long. It has been suggested that 8 cylinder engines of 12 in. bore be substituted, with gearing to the propeller shaft. The gearing is expected to weigh 20% as much as the engines and to add 15% to the floor space of the engine to which it is connected. Estimate the new dimensions and floor area required, assuming the smaller engines operate at the same stress level.

Solution: For the same output at the same piston speed and mean effective pressure, the number of 12 in. bore cylinders required will be $12 \times (30/12)^2 = 75$. Since there will be some losses in the gears, take 10 8-cylinder engines, or a total of 80 cylinders of 12 in. bore. Total engine weight is:

$$10{,}000 \times 120 \times (12/30) \times (80/75) = 512{,}000 \text{ lbs.}$$

This compares with 1,200,000 lbs for the single engine. To this must be added 20% for the gears, so the total weight of the engines and gears will be 512,000 × 1.20 = 614,400 lb. The height of the engines will be: 25 × (12/30) = 10 ft, which will allow two more useful decks over the engine room. The floor area covered by the engines plus gears will be: 15 × (80/75) = 16% greater than that of the single engine because floor area is proportional to the piston area with engines of similar design.

Example problem # 10.10: A diesel engine is rated at 2100 horsepower supercharged, with 12 cylinders of size 8.75 in. × 10.5 in. The engine weighs 39000 lbs. and is 5.33 ft wide and 16.1 ft long. Compute the weight and floor area required for the same total output from another engine type for marine duty rated at 235 horsepower each, and each 3.04 ft wide, 5.7 ft long and weighing 2740 lbs.

Solution: The number of smaller engines required will be 2100/235 = 9. Total weight will be $9 \times 2740 = 24660$ lb. Floor area = $9 \times 3.04 \times 5.7 = 156\ \text{ft}^2$, compared with $5.33 \times 16.1 = 85.8\ \text{ft}^2$ for the single engine. A single large engine of the V type occupies less floor area than the same power in 6 cylinder in-line engines. If the in-line engines were made in V type also, the floor area would be nearly the same as that occupied by the single large engine.

Example problem # 10.11: The large engine of example problem # 10.10 has 12 cylinders, 8.75 in. × 10.5 in., and is rated at 2100 hp at 850 rpm. The smaller engines have 6 cylinders, 4.25 in. × 5.0 in., and are rated at 235 hp at 1800 rpm for commercial marine purposes. Which engine has the higher stress due to inertia forces and which has higher stress due to gas loads?

Solution: If the engines are of similar design (which they usually are), inertia stresses will be approximately proportional to the square of piston speed.

For the larger engine piston speed:

$$V = \omega r = 2\pi \times 850 \times 10.5/(2 \times 12) = 2337 \text{ feet/minute}$$

And for the smaller engine:

$$V = \omega r = 2\pi \times 1800 \times 5.0/(2 \times 12) = 2356 \text{ feet/minute}$$

Inertia stresses, therefore, will not be substantially different. Gas loads will be nearly proportional to the indicated mean effective pressure (mep). If mechanical efficiency is assumed to be identical, then indicated mep will be equal to brake mep. For the larger engines:

$$\text{Piston area} = 12 \times (\pi/4) \times 8.75^2 = 720 \text{ in}^2. \text{ Therefore,}$$

$$\text{Bmep} = (2100 \times 33000 \times 2)/(720 \times 2337) = 82.4 \text{ lbs/in}^2$$

For the smaller engines:

$$\text{Piston area} = 6 \times (\pi/4) \times 4.25^2 = 85.17 \text{ in}^2$$

$$\text{Bmep} = (235 \times 33000 \times 2)/(85.17 \times 2356) = 77.3 \text{ lbs/in}^2$$

The larger engine will be carrying somewhat higher stress due to gas pressure. But the stress levels are quite close in the two engines.

10.9 EXERCISES

10.1 A variable length cantilever beam consists of a strip of spring steel 0.125 in. wide and 0.0275 in. thick, with a weight attached at its free end. If the free length of the strip is 4.75 in. and the spring is required to have frequencies ranging from 7 cycles/sec to 50 cycles/sec, calculate the weight at its free end.

10.2 A solid steel disk of 6.25 in. diameter and 1.875 in. thickness is attached to a shaft. If the natural frequency of vibration of the shaft and disk combination is 3.227 cycles/second, determine the torsional stiffness of the shaft.

10.3 A stroboscope is used on a rotating machine. If a point located 5.75 in. from the axis of the shaft is to be seen with a blurring of less than 1/36 in., and the duration of the light flashes is 1/136,000 second, calculate the speed of the rotating shaft.

10.4 Use Rayleigh's method for calculating the angular frequency of the fundamental mode of vibration of the beam with two masses, as shown in figure 10.40. Assume $W_1 = W_2 = W$, flexural rigidity of the beam as EI and neglect the mass of the beam.

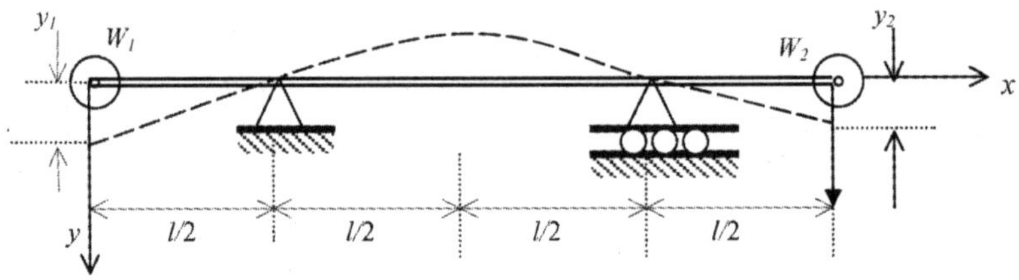

Fig. 10.40

10.5 If $W_1 = 2W_2$ in the above problem, what percentage change in the frequency is obtained?

10.6 Find the weight of the ball in figure 10.38 if it produces a static deflection of 0.625 in. and a maximum bending stress of 4750 lbs/in^2, when from a height of 8.25 in. a ball of 3750 lbs dropped at the same point in the beam produces a maximum bending stress of 29500 lbs/in^2?

10.7 A circular pipe rotating about an axis normal to the beam results in a state of centrifugal stress in the bar. Find the angular speed if the outer and inner radii are 4.75 in. and 2.255 in., material density is 0.28 lbs/in^3, Young's modulus is 27.5×10^6 lbs/in^2 and the maximum stress developed is 87500 lbs/in^2.

10.8 A freight ship is propelled by a 16-cylinder direct drive diesel engine of 27.5 in. bore giving 11,500 horsepower and weighing 112 lbs/bhp. The engine is 22.5 ft high, 9.5 ft wide and 42 ft long. It has been suggested that 6 cylinder engines of 14 in. bore be substituted, with gearing to the propeller shaft. The gearing is expected to weigh 24% as much as the engines and to add 18% to the floor space of the engine to which it is connected. Estimate the new dimensions and floor area required, assuming the smaller engines operate at the same stress level.

10.9 An 8-cylinder diesel engine rated at 1930 horsepower supercharged has 7.625 in. bore and 8.25 in. stroke. The engine weighs 32900 lbs. and is 4.84 ft wide and 14.9 ft long. Determine the weight and floor area required for the same total output from another engine type for marine duty rated at 262 horsepower each, and 3.18 ft wide, 5.27 ft long and weighing 3210 lbs.

10.10 The large engine of the previous exercise problem has 8 cylinders, 7.625 in. × 8.25 in., and is rated at 1930 hp at 680 rpm. The smaller engines have 6 cylinders, 5.75 in. × 4.85 in., and are rated at 265 hp at 700 rpm for commercial marine purposes. Which engine has the higher stress due to inertia forces and which has higher stress due to gas loads?

References & Bibliography

Blankinship, S., "Homeland Security: U.S. Brownfield", Power Engineering, page 28 – 32, June 2002.

Boeing Commercial Airplane Development, "High Speed Civil Transport Study", NASA Contractor Report # 4233, Seattle, WA, 1989.

Boyce, M.P., "Handbook for Cogeneration and Combined Cycle Power Plants", ASME Press, New York, 2002.

Chmielewski, R., Jacobucci, S., Harkins, W., Kuten, P., Wu, S., Berruti, A. and McArthur, J., "Unique Combined Cycle Design Caters to Plants with Cyclical Demand Profiles", Power Engineering, January, 2000.

Day, I.J., "Active Suppression of Rotating Surge and Stall in Axial Compressors", ASME paper # 91-GT-87, New York, 1991.

Douglas Aircraft Company, "Study of High Speed Civil Transport", NASA Contractor Report # 4235, Long Beach, CA, 1989.

Driscoll, M., McFetridge, E. and Arseneau, W., "Evaluation of At Sea Tested LM2500 Rainbow Rotor Blade Coatings", ASME paper # GT-2002-30263, New York, 2002.

Eckardt, D. and Rufli, P., "Advanced Gas Turbine Technology: ABB/BCC Historical Firsts", ASME paper # 01-GT-395, New York, 2001.

Fleeting, R. and Coats, R., "Blade Failures in the HP Turbine of RMS Queen Elizabeth II and Their Rectification", Transactions, Institute of Marine Engineers, volume 82, page 49, London, 1970.

Harvey, E., Kingsley, J. and Stauffer, M., "United States Navy Integrated Power System Gas Turbine Generator Set Test Experience", ASME paper # GT-2002-30260, New York, 2002.

Heisler, H., "Advanced Engine Technology", SAE International, Warrandale, PA, 1995.

Hoeft, R.F., "Heavy Duty Gas Turbine Operating and Maintenance Considerations", GER-3620B, General Electric I & PS, Schenectady, NY, 1993.

Hunecke, K., "Jet Engines – Fundamentals of Theory, Design and Operation", Motorbooks International Publishers, Osceola, WI, 1997.

International Civil Aviation Organization (ICAO), Annex 16, Volume II, 1981.

Kerrebrock, J.L., "Aircraft Engines and Gas Turbines", The MIT Press, Cambridge, MA, 2nd Edition, 1992.

Kurz, R. and Fozi, A. A., "Acceptance Criteria for Gas Compression Systems", ASME paper # GT-2002-30282, New York, 2002.

Langston, L., "Electrically Charged", Mechanical Engineering, page 50 – 52, June 2002.

Lefebvre, A.H., "Gas Turbine Combustion", Taylor and Francis, Philadelphia, 1999.

Nakazawa, N., Ogita, H., Takahashi, M., Yoshizawa, T. and Mori, Y., "Radial Turbine Development for the 100 kW Automotive Ceramic Gas Turbine", ASME paper # 96-GT-366, New York, 1996.

Pratt and Whitney Aircraft, "The Aircraft Gas Turbine Engine and its Operation (PWA Operating Instruction 200)", revised May, 1974.

Rangwala, A.S., "Structural Dynamics of Turbo-machines", New Age International, New Delhi, 2006.

Rangwala, A.S., "Reciprocating Machinery Dynamics – Design and Analysis", Marcel Dekker, Inc., New York, 2001.

Saravananamuttoo, H. I. H., Rogers, G. F. C. and Cohen, H., "Gas Turbine Theory", Prentice Hall, Harlow, England, 2001.

Shortlidge, C.C., "Control System for a 373 kW Intercooled Two-Spool Gas Turbine Engine Powering a Hybrid

Electric World Sports Car Class Vehicle", ASME paper # 96-TA-38, New York, 1996.

Smith, D.J., "Steam Turbines Remain the Workhorse for Power Generation", Power Engineering, page 44 – 46, August 2001.

Suo, M., "Turbine Cooling", The Aerodynamics of Aircraft Gas Turbine Engines, editor G. C. Oates, AFAPL TR-78-52, Wright Patterson Air Force Base, Ohio.

Thompson, B.D. and Wainscott, B., "Systematic Evaluation of US Navy LM2500 Gas Turbine Condition", ASME paper # 00-GT-667, New York, 2000.

Valenti, M., "Reaching for 60 Percent", Mechanical Engineering, page 35 – 39, April 2002.

Wadia, A.R., Wolf, D.P. and Haaser, F.G., "Aerodynamic Design and Testing of an Axial Flow Compressor with Pressure Ratio of 23.3:1 for the LM2500+ Gas Turbine", ASME paper # 1999-GT-210, New York, 1999.

11

Control Techniques for Aircraft Propulsion System

11.1 INTRODUCTION

Military and commercial engine controls have mostly been concerned with five or fewer engine-manipulated variables. Control of core engine functions such as fuel flow and stator vane settings has until recently relied on hydro-mechanical functions. In the early 1960's, engines like the J79 in the F4 Phantom used a single hydro-mechanical control for the fuel flow and for the variable stators. With the advent of transistorized circuitry, the 1960's and 1970's saw engines using analog electronic controls together with hydro-mechanical main fuel controls. The 1980's brought sufficient advances in data acquisition and microprocessor technologies to permit replacement of analog controls with a digital processor based unit on the GE F110 engine for the F16 Falcon. Commercial aircraft now fly with full authority digital engine controls, usually in a dual configuration to allow continued engine operation in the event of a single component failure. The Advanced Tactical Fighter and other advanced programs have identified performance advantages from further usage of electronics. Dual, Triplex and Quad control processing designs are all potential candidates that will let hydro-mechanical controls to be eliminated.

With the advent of vectored thrust, vertical lift and fly-by-wire aircraft, the complexity of aircraft gas turbine control systems has evolved to the point wherein they must approach or equal the reliability of current quad redundant flight control systems. Digital electronic controls with significantly expanded capabilities are required to handle multi-function control actuation systems, to integrate them with flight control systems, and to provide fail-operation capability. Areas that need careful scrutiny include digital electronic control capability and packaging, advanced fuel management system, high-pressure fuel hydraulic actuation systems for multi-function nozzles, integrated flight propulsion controls, and higher order language software development.

Preliminary design studies carried out in 1980's identified some challenging objectives for propulsion systems. The aim was to obtain increased thrust levels at reduced system weights, improved efficiency, and reduced specific fuel consumption. The ability to vector thrust and reverse thrust were projected to considerably improve overall aircraft manoeuvrability and capability (*ref*: Tilman, 1990). Besides improvements in the turbo-machine, capabilities of the control system to protect and operate the engine and the associated sub-systems would be essential. Maximum efficiency has to be attained at steady state operating conditions, while providing rapid and stall-free transient response to throttle or flight control inputs.

Thrust vectoring considerably enhances manoeuvrability of the aircraft, especially during take-off and landing. At the same time the addition of a thrust vector nozzle to the engine considerably increases control system requirements. The system must have convergent and divergent sections of the nozzle to maintain control of the engine pressure ratio during transient operation while simultaneously reacting to vectoring commands from the flight control. Besides placing significant demands on the control system computer, the addition of this capability requires a considerably larger overall horsepower that is available for actuation. To meet the control system requirements, engine control loops were doubled and electronic computing capability was tripled as compared to the current systems, as seen from the comparison between the F100-PW-220 engine in the F-15 aircraft and the Pratt and Whitney YF119 engine in the YF-22 flight prototype aircraft shown in figure 11.1. The F100 engine requires six modulated loops, while the YF119 requires fourteen. The capability to functionally integrate the engine control and the vectoring nozzle control with the flight control also introduces the requirement to communicate directly on the aircraft digital data bus, which also improves overall flight/propulsion capability.

11.2 MULTI-FUNCTION VECTORING NOZZLES

The ability to vector the thrust of an aircraft's gas turbine engine in a specific direction enhances the manoeuvrability of the aircraft. Thrust vectoring has been successfully demonstrated on the F-15,

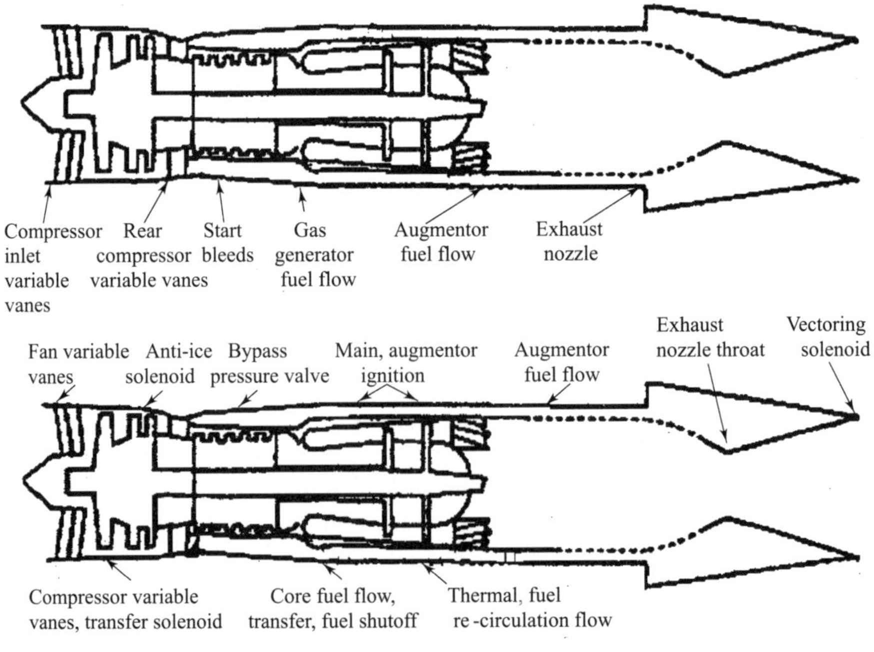

Fig. 11.1 Electronic Controls: F100-PW-220 Engine (*upper*), YF119 Engine (*lower*)

F-16, F-18, YF-22, YF-23 and other aircraft. Improvements have been demonstrated in overall aircraft performance, with significant benefits projected for advanced aircraft, both for conventional and for aircraft capable of taking off from short distance and vertical landing.

Mechanizing the multi-function nozzles poses a challenge to the hydraulic power and actuation systems. Present engine nozzle systems are either hydraulically or pneumatically powered, usually with self-contained systems that require 20 to 25 horsepower. Thrust vectored nozzles can double or even triple that requirement, depending on the functioning characteristics and the operational rate of the components of the system.

Component technology and material advances are essential in the pumps and the actuators to achieve the vectoring thrust nozzle design. The centrifugal pump requires features to permit operation of the pump at higher turndown ratios than historically demonstrated, while minimizing heat buildup and providing good stability at low flow conditions. Modified designs in gear pump provide the capability to operate over a large flow range at variable pressure, and to pump at higher pressure rise and flow range without imposing a harsh weight penalty. Use of titanium for actuator manifolds and barrels helps to minimize weight while operating in the high temperature environment at the aft end of the engine. Material coatings are employed to achieve durability and reliability of operation. Seals at the piston and rod ends have the capacity to tolerate high temperature and piston velocity.

11.3 ELECTRONIC CONTROL ACTUATION

The control of gas turbine engines with digital computers has been accompanied by marked improvements in overall engine performance, efficiency and operability. The opportunities for further benefits appear to be endless, although they result in increased complexity of the electronics and software. Figure 11.2 illustrates the growth in computational capability of commercial and military gas turbine controls. The primary difference between military and computer requirements lies in the additional logic needed to operate the augmentor, variable nozzle area and nozzle vectoring capability. The effect of growth in thrust and efficiency is best evidenced from the data for commercial engines.

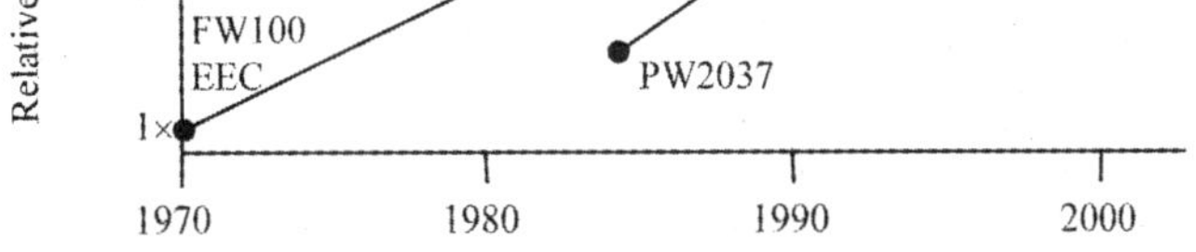

Fig. 11.2 Growth of Engine Control Software

Initial prototypes programs for vectoring and reversing nozzles utilized additional units for control of the dedicated nozzle functions, but this led to additional complexity of the data bus system and data delays in control loops. Integration of these functions into the engine's full authority digital electronic control (FADEC) eliminated the problems. Electronic components and circuit boards have been developed as highly integrated electronic packages to withstand elevated temperature and vibration environments. The integrated electronic packages accomplish integration of additional analog interfacing circuits into small board areas using custom monolithic microcircuits. Application specific integrated circuits have high board density through usage of surface mounted devices attached on both sides of the board. A copper invar laminate is used between the boards for thermal matching, and an efficient heat transfer path maintains acceptable component temperatures while on the engine. A comparison of the units for the YF119 FADEC and the F100 digital electronic engine control (DEEC) indicate the two are not far apart in size and weight, even though the former provides three times the latter's control function capability.

The digital electronic controls operating in the fleets are implemented using custom microprocessors and control logic programmed into these units by a process that includes defining the logic in flow diagrams. The diagrams are then programmed directly into assembly language code, and debugging the code through extensive code read.

11.4 FUEL MANAGEMENT

The control system primary effectors coordinated by the FADEC consist of fuel metering valves for the main combustor and augmentor. Hydraulic actuators are used to control variable geometry functions for the compression system, air bleed and cooling, and exhaust nozzle system.

The number of effectors has a considerable impact on the complicity of the FADEC and its software. Other substantial issues arise from the hydraulic system. The horsepower required to operate the vectoring nozzle constitutes a large amount. Also, the thermal impact on the heat load from the fuel combines with increased levels of heat from the aircraft system. Management of additional levels of heat requires architectural innovations. The imposed load, rate, temperature, weight limitation and fail-safe requirements of the nozzle actuator require new materials and design features for the actuator.

11.5 THERMAL MANAGEMENT

The usual approach to thermal management is isolation of the engine's fuel and oil systems from the aircraft's own system. Thus, fuel and oil specifications are defined at the delivery point to the engine, and the systems are managed independently. The practice is possible because the flow delivered to the engine is sufficient to absorb the heat generated from the inefficiencies of the engine oil and fuel pumping systems, and could be managed within the engine. The same is true on the aircraft side of the fuel system.

Advanced aircraft requirements for avionics and hydraulic power have changed substantially. At the same time, engine requirements for hydraulic power and oil system of advanced aircraft have increased. The capacity of the engine system to absorb and dissipate heat has decreased because engine cycle efficiencies have increased, thus decreasing burn flow at engine cruise points.

The integration is accomplished across the data bus between the aircraft's control systems and the engine's FADEC system. Either system can request changes that result in the best overall thermal efficiency through a set of integrated control laws. The engine burn flow temperature is essentially maintained at a maximum permissible level to transfer as much heat as possible into the combustor and to re-circulate excess levels through heat exchangers to cool the fuel before returning it to the aircraft system.

11.6 HYDRAULIC POWER FOR ADVANCED VECTORING NOZZLE

Demonstrator programs have shown that maximum vectoring rate of the vectoring surface can occur simultaneously with a rapid thrust command that requires rapid engine exhaust nozzle area changes to hold the desired pressure ratios in the engine. These maximum transient requirements, when combined with loads in the actuators, require considerably greater hydraulic power and flow capacity in the electro-hydraulic servo-valve system.

The additional hydraulic power is typically obtained from an additional closed-loop hydraulic system within the engine system that is dedicated to handle the nozzle, with the primary supply coming from a variable displacement piston type of pump. On most current engine configurations, the third pump, pulse damper, hydraulic reservoir and heat exchanger add to the weight. The arrangement also introduces a third hydraulic medium for maintenance of the engine.

Modifications in system architecture provide the additional hydraulic power without the addition of a third power system. The system is a re-configuration of current engine pumping technologies that uses a combination of gear and centrifugal pumps. The high flow centrifugal pump is used to boost the pressure for supplying fuel to the main combustor and the augmentor. This pump provides additional pressure to the gear pump that is used for actuation pressure to the main engine and nozzle actuators. A FADEC controlled pressurizing valve regulates the pressure rise required across the gear pump to minimize heat load by only providing high pressure during maximum nozzle demand periods. The open-loop hydraulic supply uses engine fuel for operation. This eliminates the need for a hydraulic reservoir and other support equipment or an air purging system after maintenance.

11.7 SUBSYSTEM INTEGRATION

Modern aircraft propulsion systems are complex and critical to mission performance. However, integrated design philosophies have not been fully exploited to optimize total system performance. Systems may be specified and designed independently, and assembled with little attention paid to interaction with other systems.

Integration during the design phase is a potential means of gaining overall system improvement. In an integrated design effort, performance tradeoffs among subsystems are made to optimize total system performance. Integration of the system necessitates common control to take advantage of sensitivities between subsystems. A single multivariable control structure may be applied to a network of associated subsystems. Robustness must be ensured since performance penalties may be needed to assure adequate safety margins.

Multivariable control methods do carry risks since they lack maturity for consideration in aircraft implementation. Experience and validation from existing systems may be insufficient predictors of in-service performance. Thus, the capability to evaluate potential benefits of integration during initial design is essential. Basic elements of this capability are methods for developing the mathematical equations, simulation software, control design tools and simulating presence of hardware within the loop.

As an illustration, consider a compressor, generator and a prime mover, with the devices designed and controlled independently (figure 11.3). Performance of the system at off design points will likely be away from peak values. A high-fidelity model of the system may be developed for evaluating the controls. Transfer functions or a series of coupled linear differential equations set up with the aid of state variables and system inputs and outputs may be used for simulating the components in the example. For a multivariable system the transfer function matrix is rearranged in block state-space form, from which a minimal realization is generated. The system structure is shown in figure 11.4. Control signal u_c is assumed constant over the interval between any two consecutive sampling intervals. The code calculates the resulting output for an arbitrary input set at each discrete time point to create a time history of the system states and outputs. Availability of all state variables at all times permits comparison of controls with partial or no state feedback.

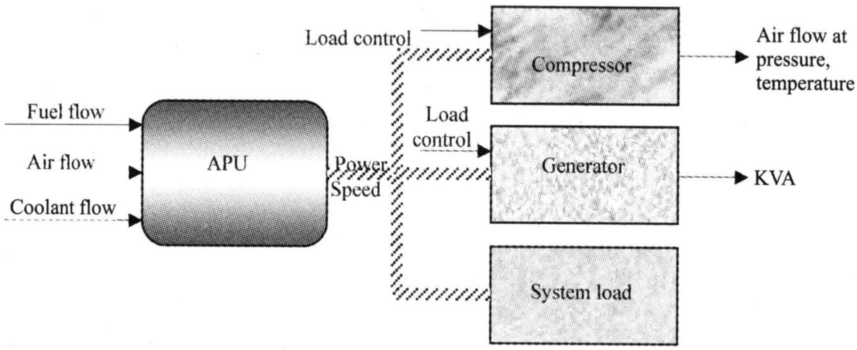

Fig. 11.3 Secondary Power System

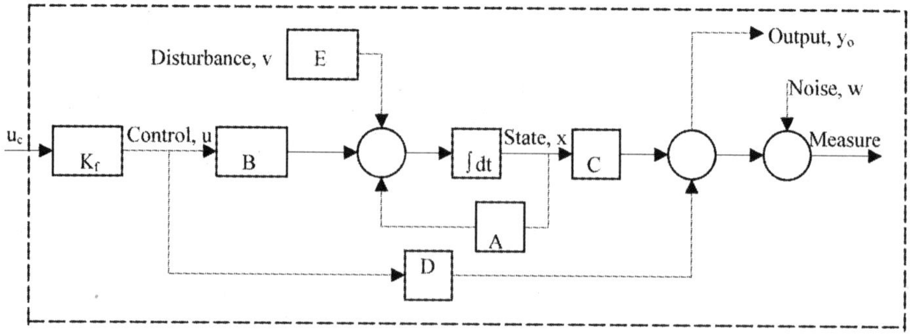

Fig. 11.4 Open-Loop Model Structure

11.8 RESILIENCE

The term resilience refers to the ability of a multi-channel control to accommodate an arbitrarily malicious fault. Thus, a resilient architecture containing the fault within a channel prohibits the fault from propagating into other control computations. Many fault tolerant control system concepts are subject to single point failures. As an example, commercial aircraft now in operation use two redundant computing channels wihin the engine control system. Such a system is sufficiently reliable, wherein use of multiple engines allows continued flight upon the loss of an entire engine control system. Over the water flights of aircraft with two engines are limited in duration primarily because reliability of the engine control system becomes a critical factor subsequent to loss of one of the engines. With only two operable channels, the health of an individual channel can only be assessed by itself. This severely limits the control's ability to provide coverage in the event of a total first failure.

Gas turbine control systems typically encompass five areas that require a systems design approach. The areas include sensors, actuators, control and computational hardware, and software and integration with the vehicle's management system. Each area requires a system design that stresses maintainability, reliability, performance and life cycle cost. The vehicle's management system incorporates that portion of the aircraft's system that is truly flight critical, and is useful in reducing component maintenance while improving the performance of the aircraft.

Integration of presently independent systems is essential in reducing pilot workload. An enabling technology for integration is a communications network with ability to perform high-speed data exchange. This ability will allow for a number of new functionality integrated modes, such as integrated flight and propulsion control and terrain and obstacle avoidance. Without a network for managing the vehicle, these modes could incur unacceptable data transfer delays arising from phase lag in the system. The delays would severely impact the system's transient response and performance.

11.9 RELIABILITY

Fault tolerance requirements for aircraft controls may generally be derived from the mean time between failures and repair and other associated mission needs of the engine.

The issue of fault tolerance in multi-channel computing systems has received considerable attention. The fault-tolerant core of one such computing system is a set of three tightly synchronized processors using hardware voting with fault containment and error detection mechanisms. Analytically derived synthesized values for various engine and effector readings are continuously updated based on engine operating parameters. In the event of loss of a sensor, comparison with synthesized values prepares for further information to the control. The availability of synthesized set of operating parameters allows full triplex control operation with only two sets of control input sensors. The synthesized set constitutes the third sensor set. The actuator set is equipped with triple coil drives. Each coil has sufficient drive capability to control each actuator independently.

11.10 ARCHITECTURE

Figure 11.5 illustrates the architecture used in the F100-PW-229 vectoring nozzle engine for a specific program in the modified F-15 aircraft. Addition of components or complexity in the data bus system may lead to modifications.

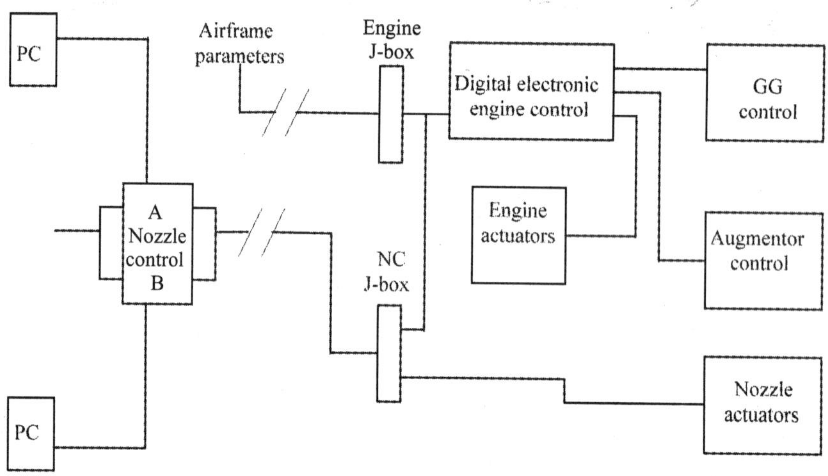

Fig. 11.5 Vectoring Nozzle Control Architecture

The triplex control design satisfies the requirement for 100 per cent first failure coverage. To assure operation during any single failure, a triplex set of all flight critical sensors and a duplex set of engine critical sensors is essential. In the event of mismatch between sensor values, the embedded engine model is used to arbitrate.

Advances in engine modeling techniques are leading to reduction in the number of minimal sensor sets. The reduction enables a set of dual sensors to be distributed across three channels, where each channel has sufficient sensors for normal engine operation. This method also provides for an increased capability for second failure operation. As modeling methods mature, a return to a single set of sensor may be feasible. This ability would reduce weight and cost while maintaining the capability for second failure operation.

An integrated architecture can provide the benefit of flexibility to operate the aircraft and engine systems at maximized thermal efficiency over most of the mission. Addition of a fuel return path to the aircraft tanker no longer restricts the heat sink available to the engine's burn flow, but allows the integrated system to adjust the flow of fuel for optimum thermal efficiency. The schematic diagram of the integrated approach is shown in figure 11.6.

11.11 INTEGRATING PROPULSION SYSTEM WITH AIRCRAFT FOR JET-BORNE FLIGHT

A conventional aircraft's pitch handling qualities are defined by its aerodynamics, pitching moment of inertia, location of centre of gravity, and flight control system, with the engine's characteristics not being a factor. The same is not true of a jet-borne flight that is capable of Vertical/Short Take-off and Landing (V/STOL). In a jet-borne flight, the aircraft is supported by propulsive lift rather than by aerodynamic lift. The characteristics of the engine do have an effect on the handling qualities (*ref*: Groll et al., 1989).

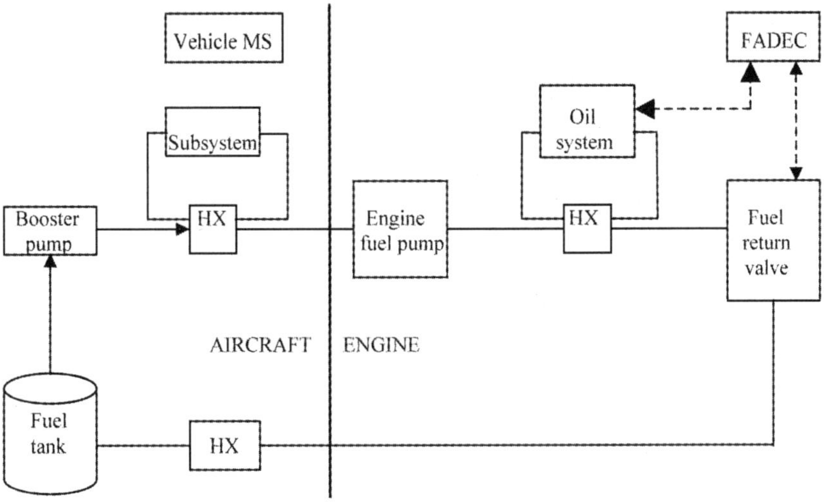

Fig. 11.6 Integrated Architecture Schematic

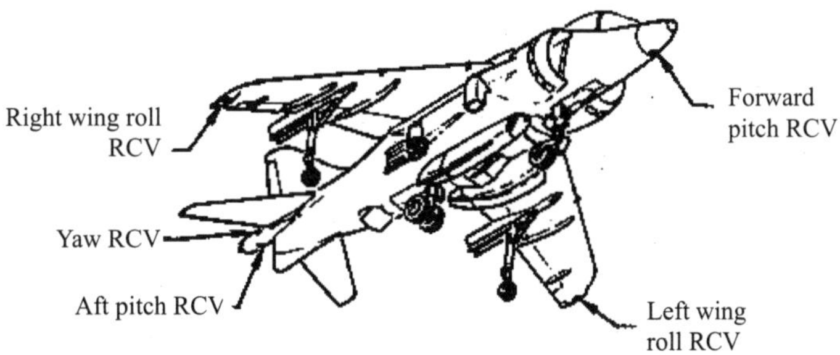

Fig. 11.7 AV-8B Harrier II Jet-Borne Aircraft for V/STOL (Groll, 1989)

The US Marine Corp's AV-8B Harrier II aircraft is an example of such flights (figure 11.7). The current production version is a single seat high wing aircraft powered by a single Rolls-Royce Pegasus F402-RR-406A engine. Propulsive thrust developed in the engine exits through four nozzles, two on each side of the fuselage (figure 11.8). Relatively cooler low-pressure fan bypass air is discharged through the pair of forward nozzles, while the hot turbine gases are exhausted through the aft nozzles. For jet-borne flights, all four nozzles rotate collectively downward.

Stabilization and manoeuvring during the jet-borne flight is achieved using a system of moment generating thrust valves, called 'reaction control system' (RCS). Two of the control valves provide pitch control that directs the thrust downward. One is located in the nose of the aircraft and the other is in the tail extension. The forward valve opens to develop a nose-up pitching moment, while the aft valve opens to provide nose-down pitching moment. To obtain roll control, a control valve is placed in each wing tip for generating both positive and negative rolling moments. Dual control valves are located in the tail for yaw control. The RCS is powered by bleed air from the engine's compressor.

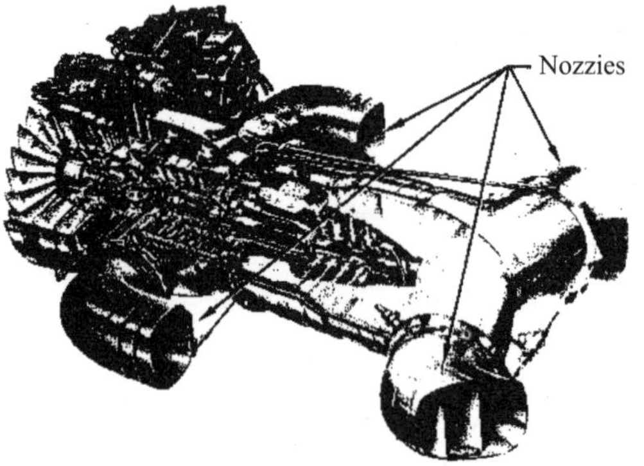

Fig. 11.8 Rolls-Royce Pegasus F402-RR-406A Engine (Groll, 1989)

A simulator type of evaluation is used to establish the forward centre of gravity location at which the pitch handling qualities are acceptable for variations in the pitching moment of inertia, weight and thrust centre. Testing is conducted on a manned flight simulator. Ground effects including suck-down and exhaust impingement on the tail are modeled, together with crosswinds and gusts. The aircraft has a conventional control stick and rudder pedals. Longitudinal stick movement controls both the stabilator and the pitch reaction control valves; lateral stick movement controls the ailerons and the roll reaction control valves; rudder pedal movement controls the rudder and the yaw reaction control valves.

Flight runs are conducted at sea level for a tropic day to cause the engine to run at higher temperatures than on a standard day. A consequence is that the maximum thrust available for vertical lift and for the RCS is reduced. Representative minimum, maximum and intermediate values of weight and moment of inertia and varying centres of gravity are used in different combinations for the flight tests. A forward centre of gravity is typical for this aircraft during a vertical landing because, as the fuel is expended, the centre of gravity moves forward from its full fuel aft location.

Simulation results are generalized in terms of the following pitch axis parameters: control power, pitch sensitivity, disturbance susceptibility and control margin. Of these, pitch sensitivity (pitch acceleration per unit linear stick deflection) dominated the pilot's judgment in determining a rating for a specific configuration. Pitch control power in terms of available pitch acceleration is an important factor affecting an aircraft's pitch handling qualities in hover. The exact value of pitching acceleration depends on the particular pilot task, magnitude of pitching moments to be trimmed in the event of changes in some state of the aircraft (such as thrust), and magnitude of pitching moments due to external disturbances.

For landing of the aircraft, the centre of gravity is forward of the thrust centre and a nose-down pitching moment is present, as shown in figure 11.9. To trim the aircraft in pitch, the pilot must command actuation of the forward reaction control valve to produce a nose-up moment. The amount of this thrust is limited and depends on the engine's operating condition. Not all of the thrust is thus available for additional manoeuvring of the aircraft.

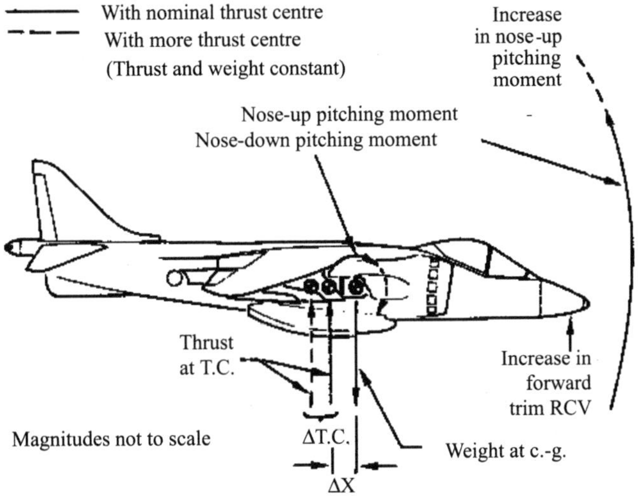

Fig. 11.9 Effects of Shift in Thrust Centre (Groll, 1989)

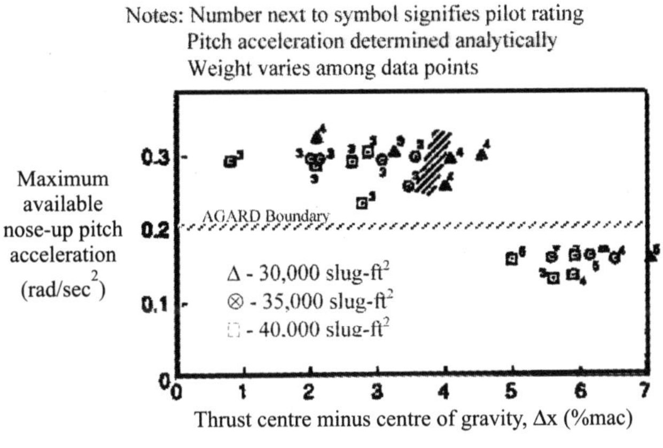

Fig. 11.10 Pilot Ratings for Pitch Acceleration (Groll, 1989)

Results of the simulation for pitch control power are shown in figure 11.10. The maximum nose-up pitch acceleration available for manoeuvring in a hover is plotted against distance between thrust centre and centre of gravity, or Δx in per cent mean aerodynamic chord (mac). Next to each point on the plot is the pilot's rating for pitch and power doublet tasks. The recommended boundary for pitch acceleration is 0.2 radian/second^2 (*ref*: Anderson et al., 1970), shown as a line with cross-hatching. Not all of the data above the recommended boundary has acceptable pilot ratings. A band is placed in the region where the pilots change their acceptable ratings to one calling for improvement. This indicates that the minimum acceptable pitch acceleration may depend on the Δx distance. There also may be an upper limit for acceptable pitch acceleration.

11.12 REVIEW PROBLEMS

Example problem # 11.1: An axial flow compressor is required to develop a pressure ratio of 4.75 and handle a mass flow m of 25 kg/second. Design point tip speed v_{tip} is to be around 325 m/second, axial flow velocity v_{axial1} at inlet about 170 m/second and root to tip ratio r_{root}/r_{tip} between 0.4 and 0.6. Provide a full design for the compressor.

Solution: Size of the annulus will be established first. Continuity requirement calls for:

$$m = \rho_1 A v_{axial1} = \rho_1 \pi (r^2_{tip} - r^2_{root}) v_{axial1} \quad ...(11.1)$$

where A is the annular area. Air temperature at sea level $T_a = T_{01} = 290$ K and pressure $p_a = 1.01$ bar. For 170 m/second axial velocity and $\gamma = 1.4$, the temperature is:

$$T_1 = T_a - (v_{axial1})^2/2p_a = 290 - (170^2)/(2 \times 1.01 \times 10^3) = 275.7 \text{ K}$$

$$p_1 = p_a(T_1/T_a)^{\gamma/(\gamma-1)} = 1.01 \times (275.7/290)^{\gamma/(\gamma-1)} = 0.846 \text{ bar}$$

$$\rho_1 = p_1/(RT_1) = (100 \times 0.846)/(0.287 \times 275.7) = 1.0693 \text{ kg/m}^3$$

$$r_{tip}^{\ 2} = 25/\{\pi \times 1.0692 \times 170 \times [1 - (r_{root}/r_{tip})^2]\} = .04378/[1 - (r_{root}/r_{tip})^2]$$

Tip speed $v_{tip} = 325 = 2\pi r_{tip}N$. The speed and tip radius are calculated over a range of root to tip ratios:

r_{root}/r_{tip}	r_{tip}, m	N, rps
0.40	0.2283	226.6
0.45	0.2343	220.8
0.50	0.2416	214.1
0.55	0.2505	206.5
0.60	0.2615	197.8

Using values calculated at the root to tip ratio of 0.5, tip radius of 0.2416 m and rotor speed of 214.1 rev/sec, or 12846 rpm, are selected. Corresponding tip speed is: $(2 \times \pi \times 0.2416 \times 12846/60) = 325.01$ m/sec. Speed of sound a and Mach number at the rotor tip M_{flow1} near the inlet should be checked. For constant axial velocity across the annulus:

$$v_{flow1}^{\ 2} = v_{tip}^{\ 2} + v_{axial1}^{\ 2} = 325.01^2 + 170^2, \text{ hence } v_{flow1} = 366.8 \text{ m/sec.}$$

$$a = (\gamma RT_1)^{1/2} = (1.4 \times 0.287 \times 1000 \times 275.7)^{1/2} = 332.8 \text{ m/sec.}$$

$$M_{flow1} = 366.8/332.8 = 1.102$$

Thus, the first compressor stage will be operating in the transonic range. However, the value is not considerable, and should not present a problem. Radius at the root is 0.1208 m, tip radius is 0.2416 m and mean radius is 0.1812 m. Compressor delivery pressure $p_{02} = 4.75 \times 1.01 = 4.80$ bar. For air temperature at exit, assume polytropic efficiency of compressor is 0.90. Then:

$$T_{02} = T_{01} \times (p_{02}/p_{01})^{(n-1)/n} \text{ where } (n-1)/n = 0.4/(1.4 \times .9) = 0.3175$$

Hence: $T_{02} = 290 \times (4.75)^{0.3175} = 475.6$ K. If it is assumed that air exiting the last stage of the compressor has no swirl and an exit velocity of 170 m/second, static temperature, pressure and density at exit are:

$$T_2 = 475.6 - 170^2/(2 \times 1.01 \times 1000) = 461.3 \text{ K}$$

$$p_2 = p_{02} \times (T_2/T_{02})^{\gamma/(\gamma-1)} = 4.80 \times (461.3/475.6)^{3.5} = 4.31 \text{ bar}$$

$$\rho_2 = p_2/(RT_2) = (100 \times 4.31)/(0.287 \times 461.3) = 3.256 \text{ kg/m}^3$$

Area of annulus at exit is: $A_2 = 25/(3.256 \times 170) = 0.04516 \text{ m}^2$. Since mean blade radius r_m is 0.1812 m, blade height at exit h is:

$$h = 0.04516/(2\pi r_m) = 0.0397 \text{ m}$$

Last stator exit radii may be calculated as follows:

$$r_{tip} = 0.1812 + 0.0397/2 = 0.2010 \text{ m}$$

$$r_{root} = 0.1812 - 0.0397/2 = 0.1614 \text{ m}$$

Thus, all pertinent data for the compressor has been established. It is recapped here:

Speed	N = 12,846 rpm	Inlet tip radius 0.2416 m
Tip Speed	325.01 m/sec	Inlet root radius 0.1208 m
Flow velocity	170 m/sec	Outlet tip radius 0.2010 m
Mean radius	0.1812 m	Outlet root radius 0.1614 m

Example problem # 11.2: In example problem # 11.1 determine the number of stages required to obtain the stated compression.

Solution: Rise in stagnation temperature in the compressor is 475.6 – 290 = 185.6 K. Increase in stage temperature rise may be estimated from the blade mean speed:

$$U = 2\pi r_{mean} N/60 = 2 \times \pi \times 0.1812 \times 12846/60 = 243.76 \text{ m/sec}$$

If the first stage sees only axial velocity (IGV's are absent), then inlet angle β_1 is:

$$\tan \beta_1 = U/v_{axial1} = 243.76/170 = 1.4339, \text{ or } \beta_1 = 55.11°$$

$$v_{flow1} = v_{axial1}/\cos \beta_1 = 170/\cos \beta_1 = 297.20 \text{ m/sec}$$

To determine maximum possible deflection, use the de Haller rule of $v_2/v_1 \le 0.72$. Then $v_2 = 297.2 \times 0.72 = 213.9$ m/sec and corresponding rotor blade angle at exit is:

$$\cos \beta_2 = 170/213.9 = 0.7945, \text{ or } \beta_2 = 37.39°$$

Then temperature rise per stage is:

$$\Delta T_{os} = \{243.76 \times 170 \times (\tan 55.11 - \tan 37.4)\}/1.010 \times 1000 = 27.47 \text{ K}$$

A temperature rise of 27.47 K per stage then suggests $(475.6 - 290)/\Delta T_{os} = 185.6/27.47 = 6.76$, or 7, stages are needed. Then the average temperature rise per stage will be 185.6/7 = 26.5 K. Since the first and last stages see a smaller increase, say 20 K, the remaining five stages may be expected to see a rise of about 27.6 K.

Example problem # 11.3: Prepare the velocity diagrams for the first two stages of the above problems. Also calculate pressure, temperature and related properties.

Solution: The expression relating the change in whirl velocity with temperature rise is:

$$\Delta v_{\text{whirl}} = \frac{c_p \Delta T_0}{\lambda U} = (1.010 \times 1000 \times 20)/(0.98 \times 243.76) = 84.56 \text{ m/sec.}$$

where the factor λ for work done is assumed to be 0.98 in a preliminary estimate. At stage 1 $v_{whirl1} = 0$, hence, $v_{whirl2} = 84.56$ m/sec.

Then: $\tan \beta_1 = 243.76/170 = 1.4339$, or $\beta_1 = 55.11°$.

$$\tan \beta_2 = (v_{tip} - v_{whirl2})/v_{axial1} = (243.76 - 84.56)/170 = 0.9364, \text{ or } \beta_2 = 43.12°.$$

$$\tan \alpha_2 = v_{whirl2}/v_{axial1} = 84.56/170 = 0.4974, \text{ or } \alpha_2 = 26.45°.$$

The velocity diagrams for the first stage are shown in figure 11.11. Note that deflection in the rotor blade is $\beta_1 - \beta_2 = 55.11 - 43.12 = 11.99°$, which is an acceptable value. Diffusion, temperature and pressure of the stage can be readily calculated.

$$v_2/v_1 = \cos \beta_1/\cos \beta_2 = 0.5720/0.7299 = 0.7837$$

$$(T_{03})_1 = 290 + 20 = 310 \text{ K}$$

$$(p_{03}/p_{01})_1 = \{1 + (0.9 \times 20/290)^{3.5} = 1.235$$

$$(p_{03})_1 = 1.010 \times 1.235 = 1.247 \text{ bar}$$

To select a value for the airflow angle at the exit from the stator, α_3, which is identical to α_1 into the second stage, consider the degree of reaction, Λ. An approximate value for Λ is given by $\Lambda \approx 1 - \{(v_{whirl2} + v_{whirl1})/(2 \times U)\} = 1 - \{84.56/(2 \times 243.76)\} = 0.827$.

For the second stage $\Delta T_{os} = \lambda U v_{axial1}(\tan \beta_1 - \tan \beta_2)/p_a = 27.6$ K and $\lambda = 0.93$, so:

$$27.6 = \{(0.93 \times 243.76 \times 170)/(1.010 \times 1000)\}(\tan \beta_1 - \tan \beta_2)$$

$$(\tan \beta_1 - \tan \beta_2) = 0.7232$$

From de Haller's criterion, $0.70 \approx (v_{axial1}/2U) \times (\tan \beta_1 + \tan \beta_2)$, so:

$$0.70 \approx \{170/(2 \times 243.76)\}(\tan \beta_1 + \tan \beta_2), \text{ so } (\tan \beta_1 + \tan \beta_2) = 2.007$$

Then: $\beta_1 = 53.78°$ and $\beta_2 = 32.70°$

Now $\tan \alpha_1 + \tan \beta_1 = U/v_{axial1} = 243.76/170 = 1.434$, hence $\alpha_1 = 3.92°$. Similarly, $\tan \alpha_2 + \tan \beta_2 = U/v_{axial2}$, so $\alpha_2 = 38.37°$.

Whirl velocities at inlet and outlet are:

$$v_{whirl1} = 170 \times \tan 3.92° = 11.65 \text{ m/sec}$$

$$v_{whirl2} = 170 \times \tan 38.37° = 134.60 \text{ m/sec}$$

Hence the change in whirl velocity is 122.95 m/sec. Velocity diagrams for the second stage is shown in figure 11.11. Temperature and pressure of the stage can be readily calculated.

$$(T_{03})_2 = 310 + 27.6 = 337.6 \text{ K}$$

$$(p_{03}/p_{01})_2 = \{1 + (0.9 \times 27.6/310)\}^{3.5} = 1.327$$

$$(p_{03})_2 = 1.247 \times 1.327 = 1.633 \text{ bar}$$

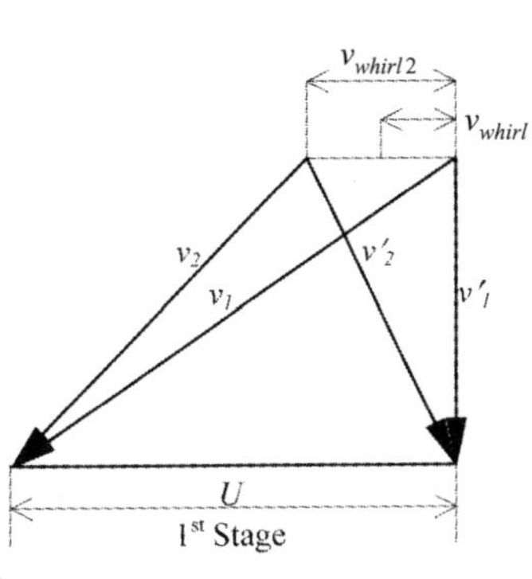

Fig. 11.11 1st and 2nd Stage Velocity Diagrams

Example problem # 11.4: Provide a construction method for the first stage blade assuming a circular arc camber-line.

Solution: The free vortex design is selected. At the mean radius of 0.1812 m angle $\beta_1 = 55.11°$ and $\beta_2 = 43.12°$, as determined earlier. Then nominal air deflection angle $\varepsilon = \beta_1 - \beta_2 = 11.99°$. From figure 11.12, $s/c = 1.1$, air inlet angle $\alpha_1 = \beta_2 = 43.12°$ and air exit angle $\alpha_2 = 38.37°$, $\beta_1 = 55.11°$. Chord length depends on the pitch spacing, which depends on the number of blades. The blade's aspect ratio (length/chord, or h/c) comes into play for controlling secondary losses in the determination of number of blades. Assuming a suitable value for this row of 3, blade height may be determined from the previous calculation of 0.1208 m, so the chord is $c = 0.1208/3 = 0.0403$ m.

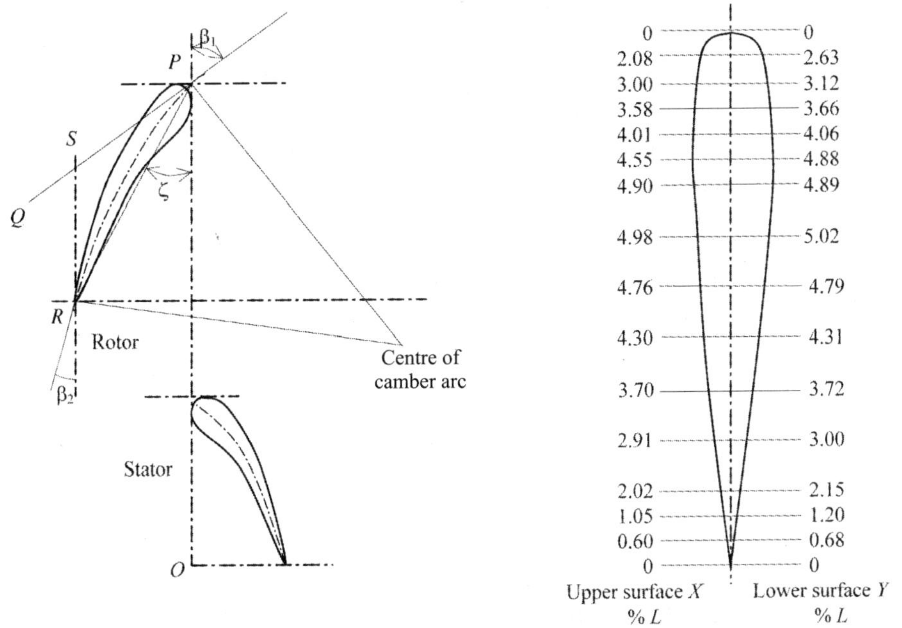

Fig. 11.12 First Stage Blade Design and Profile Development

Then pitch spacing $s = 1.1 \times 0.0403 = 0.04430$ m. The number of blades is given by $n = (2\pi \times 0.1812)/0.04433 = 25.70$. To ensure avoidance of common multiples for successive blade rows, and thus reduce the risk of resonant forcing frequencies, even number of stator vanes and a prime number for rotor blades is helpful. For this example 29 blades will be selected. Recalculation provides $s = 0.03926$ m, $c = 0.03569$ m and $h/c = 3.385$. The procedure for stator vanes is identical.

Developments in design methods have changed some views on the subject of using a prime number for rotating blades. If an even number of fan blades were selected, for example, balancing of the rotor is facilitated by replacing a pair of blades located 180° apart in case a single blade is damaged. Note also that the first stage blade's chord has been selected based on aerodynamic considerations. In practice, fan blade chord is decided on capability to withstand strike from a foreign object ingested into an aircraft engine, such as a bird.

Inlet and exit angles of a blade from experimental tests have shown a dependence on shape of the camber line of the blade section, as also the air angle at the outlet. An empirical rule for the variation is given by the term $\delta = m\theta\sqrt{(s/c)}$, where parameter $m = 0.23 \times (2a/c)^2 + 0.1 \times (\alpha_2/50)$, where a is the distance of point of maximum camber from the blade's leading edge and α_2 is in degrees. The value of $(2a/c) = 1$ is commonly used, hence:

$$\delta = \{0.23 \times 1^2 + 0.1 \times (38.37/50\} \times \theta \times \sqrt{(1.1)} = 0.3105 \times \theta$$

The procedure for laying the airfoil profile is as follows. Since $\theta = \alpha'_1 - \alpha'_2$ and $\alpha'_1 - \alpha_1 = \delta$, then $\theta = \alpha'_1 - \alpha_2 + 0.3105\theta$, or $0.6895\theta = \alpha'_1 - \alpha_2 = 55.11 - 38.37$. Hence, $\theta = 24.27°$ and $\alpha'_2 = 14.10°$. Angle of deviation is 5.67°. Position of the blade chord relative to the rotor axis is defined by stagger angle ζ, so $\zeta = \alpha'_1 - \theta/2 = 42.97°$.

In figure 11.12, chord PR is 0.03569 m long, oriented 42.97° to the axial direction. Lines PQ and RS at angles α'_1 and α'_2 are then added and a circular arc is drawn tangential to these lines, with PR as the chord. The arc represents the camber line of the blade around which the airfoil is to be developed. The method of defining ordinates at selected locations along the camber line is based on the NACA series of airfoil profiles.

Example problem # 11.5: A bar is subjected to a tensile preload of 12500 lbs and a varying tensile load from 0 to 21250 lbs. If a stress concentration factor of $k_t = 1.86$ is to be applied, determine a suitable diameter for an infinite life and a factor of safety of $k_s = 2.0$ from the following data: yield strength $S_y = 87.5$ kpsi, ultimate strength $S_{ut} = 112$ kpsi, endurance strength $S'_e = 0.5 \times S_{ut} = 56.0$ kpsi, surface finish modification factor $k_a = 0.77$, size factor $k_b = 0.65$, notch sensitivity $q = 0.85$.

Solution: Fatigue stress concentration factor $k_f = 1 + q(k_t - 1) = 1.731$. Hence endurance factor $k_e = 1/k_f = 1/1.731 = 0.578$, and endurance strength:

$$S_e = k_a\, k_b\, k_e\, S'_e = 0.77 \times 0.65 \times 0.577 \times 56.0 = 16.192 \text{ kpsi}$$

The static stress from the preload is $\sigma_{static} = 12500/\{\pi d^2/4\} = 15.916/d^2$ kpsi. The dynamic load stress range is $\sigma_{dynamic} = 21250/\{\pi d^2/4\} = 27.056/d^2$ kpsi. Alternating stress $\sigma_{alt} = 27.056/(2d^2) = 13.528/d^2$ kpsi. Mean stress $\sigma_m = \sigma_{static} + \sigma_{alt} = 29.444/d^2$ kpsi. Hence, $\sigma_{alt}/\sigma_m = 0.459$.

The fatigue diagram of figure 11.13 provides the relation between the stress and strength characteristics. The intersection of the modified Goodman line with another line at a slope of $\sigma_{alt}/\sigma_m = 0.459$ defines two values of strength. Alternating strength S_{alt} corresponds to stress σ_{alt} and strength S_m corresponds to stress σ_m. Using a factor of safety of 2.0, then $\sigma_{alt} \le S_{alt}/2.0$. Thus, $13.528/d^2 \le (29.44 \times 0.459/2.0)$, or $d = 1.414$ in. Choose $d = 1.5$ in.

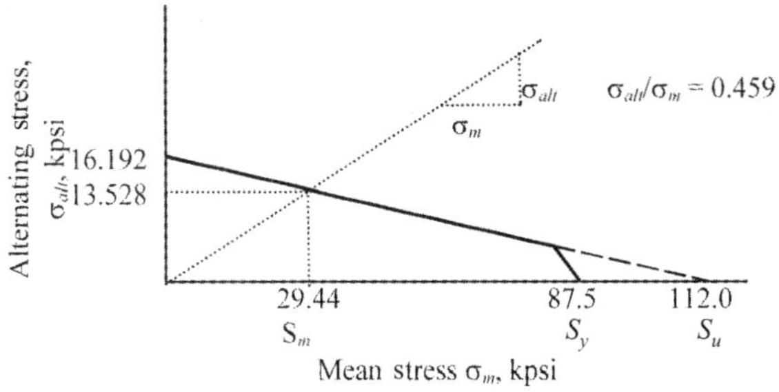

Fig. 11.13

Example problem # 11.6: The material for a part has S_y = 82 kpsi, ultimate strength S_{ut} = 105 kpsi, and endurance strength S_e = 27.0 kpsi with all corrections included. The mean and alternating stress components are $\sigma_m = 32/d^2$ kpsi and $\sigma_{alt} = 17/d^2$ kpsi. If the diameter of the part is 1.75 in., find the fact of safety using the Gerber theory.

Solution: The modified Goodman theory includes a number of conservatisms in order to account for unknown material characteristics. If the strengths are known accurately, the Gerber theory based on a parabolic relation is more extensively applied in the form:

$$(S_{alt}/S_e) + (S_m/S_{ut})^2 = 1$$

In this example, $\sigma_m = 32/1.75^2 = 10.45$ kpsi and $\sigma_{alt} = 17/1.75^2 = 5.55$ kpsi. Thus, $S_m = n\sigma_m = 10.45n$ and $S_{alt} = n\sigma_{alt} = 5.55n$. Substitution into the non-linear equation gives:

$$(5.55n/27) + (10.45n/105)^2 = 1$$

The factor of safety n is obtained by solving the equation, yielding $n = 4.067$.

Example problem # 11.7: In example problem # 11.6 find the diameter of the component if the factor of safety is reduced to 3.25.

Solution: $S_{alt} = n\sigma_{alt} = 3.25(17/d^2) = (55.25/d^2)$ kpsi and $S_m = n\sigma_m = 3.25(32/d^2) = (104/d^2)$ kpsi. Substitute into the Gerber relationship:

$$55.25/(27 \times d^2) + \{104/(105 \times d^2)\}^2 = 1$$

or:

$$d^4 - 2.046d^2 - 0.981 = 0$$

A root for d in the range $1 \le d \le 2$ is $d = 1.564$ in.

Example problem # 11.8: In example problem # 11.6 find the diameter of the component when multiple factors of safety are a requirement: $n_1 = 2.25$ for stress amplitudes, $n_2 = 1.55$ for mean stress, $n_3 = 1.29$ for the endurance limit, $n_4 = 1.22$ for the yield strength and $n_5 = 1.20$ for the ultimate strength.

Solution: The Soderberg theory will be used for this example, which takes the form:

$$(S_{alt}/S_e) + (S_m/S_y) = 1$$

Combining the safety factors to the strength characteristics gives: minimum $S_e = S_e/n_3 = 27/1.29 = 20.93$ kpsi, minimum $S_y = S_y/n_4 = 82/1.22 = 67.21$ kpsi, minimum $S_{ut} = S_{ut}/n_5 = 105/1.20 = 87.5$

kpsi. The permissible alternating stress $\sigma_{alt,p} = n_1\sigma_{alt} = 2.25(17/d^2) = 38.25/d^2$ kpsi and the permissible mean stress $\sigma_{m,p} = n_2\sigma_m = 1.55(32/d^2) = 49.6/d^2$ kpsi. Letting $S_{alt} = \sigma_{alt,p}$ and $S_m = \sigma_{m,p}$ and substituting into the Soderberg expression gives:

$$\{38.25/(20.93 \times d^2)\} + \{49.6/(67.21 \times d^2)\} = 1$$

This gives $d = 1.602$ in.

11.13 EXERCISES

11.1 An axial flow compressor capable of accommodating a mass flow of 70 lbs/second and developing a pressure ratio of 6.35 has inlet velocity of approximately 535 ft/second. Design point tip speed v_{tip} is to be about 975 ft/second and root to tip ratio r_{root}/r_{tip} between 0.5 and 0.75. Provide a full design for the compressor.

11.2 Determine the number of stages required to obtain the stated compression in exercise # 11.1.

11.3 Prepare the velocity diagrams for the first two stages of exercises 11.1 and 11.2. Also calculate pressure, temperature and related properties.

11.4 Provide a construction method for the first stage blade assuming a circular arc camber-line of the above exercises.

11.5 A bar is subjected to a steady tensile preload of 3275 lbs and a varying tensile load from −287 lbs to + 287 lbs. A stress concentration factor of $k_t = 1.79$ is to be applied. Calculate a suitable diameter for an infinite life and a factor of safety of $k_s = 2.0$ from the following data: yield strength $S_y = 92700$ lbs/in^2, ultimate strength $S_{ut} = 118700$ lbs/in^2, endurance strength $S'_e = 0.5 \times S_{ut}$, surface finish modification factor $k_a = 0.82$, size factor $k_b = 0.71$, notch sensitivity $q = 0.85$.

11.6 The material for a part has $S_y = 73500$ lbs/in^2, ultimate strength $S_{ut} = 97000$ lbs/in^2, and endurance strength $S_e = 29200$ lbs/in^2 with all corrections included. The mean and alternating stress components are $\sigma_m = 32000/d^2$ lbs/in^2 and $\sigma_{alt} = 16800/d^2$ lbs/in^2. If the diameter of the part is 1.875 in., find the fact of safety using the Gerber theory.

11.7 In exercise # 11.6 find the diameter of the component if the factor of safety is reduced to 2.50.

11.8 In exercise # 11.6 find the diameter of the component when multiple factors of safety are a requirement: $n_1 = 1.75$ for stress amplitudes, $n_2 = 1.55$ for mean stress, $n_3 = 1.32$ for the endurance limit, $n_4 = 1.20$ for the yield strength and $n_5 = 1.33$ for the ultimate strength.

References & Bibliography

Anderson, S.B. et al., "V/STOL Handling Qualities Criteria, Part I – Criteria and Discussion", NATO Advisory Group for Aerospace Research and Development, AGARD-R-577-70, 1970.

Emami-Naeni, A., "Application of Generalized Eigen-Structure Problem to Multi-Variable Systems and Robust Servo-Mechanism for a Plant Containing an Implicit Internal Model, Ph.D. Dissertation, Department of Electrical Engineering, Stanford University, Stanford, CA, 1981.

Groll, D.B. and Bossart, T.R., "Practical Design Considerations for Integrating Propulsion System with Aircraft for Jet-Borne Flight", ASME paper no. 89-GT-310, New York, 1989.

Lamport, L, Shotak, R. and Pease, M., "The Byzantine Generals Problem", ACM Transactions on Programming Languages and Systems, volume 4, no. 3, page 382–401.

McGlone, M.E., "Transition of Technology Base for Advanced Aircraft Gas Turbine Control Systems", ASME paper # 96-GT-107, New York, NY, 1996.

Ogata, K., "Modern Control Engineering", Prentice-Hall, Englewood Cliffs, NJ, 1970.

Polley, J.A., Adibhatia, S. and Hoffman, P.J., "Multi-Variable Turbofan Engine Control for Full Flight Envelop Operation", ASME Journal of Engineering for Gas Turbines and Power, volume 111, page 130–137.

Rosenblad, L.E., "Evaluation of Control Techniques for Aircraft Propulsion Systems", 34th International Gas Turbine and Aero-Engine Conference and Exhibition, Toronto, Canada, 1989.

Spang, H.A. and Corley, R.C., "Failure Detection and Correction for Turbofan Engines", presented at Joint Automatic Control Conference, San Francisco, June, 1977.

Tilman, K.D., "Integrated, Reliable, Fault-Tolerant Control for Large Engines", WRDC-TR-90- 2040, 1990.

Wright, W.E. and Hall, J.C., "Advanced Aircraft Gas Turbine Engine Controls", 34th International Gas Turbine and Aero-Engine Conference and Exhibition, Toronto, Canada, 1989.

12

Controller Design for Marine Gas Turbines

12.1 INTRODUCTION

The application of aero-derivative gas turbines to naval applications in the 1960's and 1970's initially relied on existing aircraft fuel controls. The controls were coupled to the ship's hydro-pneumatic machinery control system. The engines were adapted to naval requirements by including engine specific functions.

By 1980, major navies of the world had acquired experience with warships powered with marine gas turbines. The same year the Royal Navy of United Kingdom made the decision to replace steam turbines with gas turbine propulsion system for future warships after accumulating 20 years of working experience. The Royal Navy selected the Rolls Royce Olympus and Tyne engines as first generation marine gas turbines for propelling destroyers and frigates. The Rolls Royce Spey engines were later installed as intermediate power second-generation gas turbine for their new warships.

The advent of the gas turbine merely emphasized the need for increased automation on board the ships. For example, in comparison to steam plants the gas turbine responds more rapidly to changes in fuel flow and load, to the extent that the operator can no longer cope adequately with safe manual control and monitoring of the plant under continuous normal operation. Consequently, marine gas turbine control systems had to evolve.

Controller designs for marine gas turbine systems should consider three measures of performance: transient control, steady-state accuracy and disturbance rejection. Controller design is facilitated through the use of control tradeoff tools. For linear machinery they include rise time, overshoot, damping ratio and stability margin in gain and phase. But a marine propulsion system does not follow these rules because of its non-linear operational characteristics. Wide variations in non-linear classifications also tend to blur the distinctions in control theory. Hence, the control designer is forced to proceed with only two fundamental principles of control design: the system should display good command following (transient control and steady-state accuracy), and good disturbance rejection.

The need for centralized control and unmanned machinery spaces has been recognized for a number of reasons, ranging from improved plant performance, increasing complexity in plant machinery accompanied by faster response, personnel safety and the greater threat of nuclear, biological and chemical warfare. This has led to increased automation in remote control and remote surveillance.

12.2 ENGINE CONTROL REQUIREMENTS

Propulsion of a ship at some desired speed in its speed range is controlled by the thrust developed by its propellers. Since the direction of rotation of the output shaft of a gas turbine cannot be reversed, thrust reversal is generally achieved with the aid of a controllable pitch propeller (or CPP) instead of a reversing gearbox. The CPP is also useful in setting the thrust developed by varying the pitch and speed of the propeller. The pitch of the propeller is set by means of a pitch hydraulic actuating system, while speed is manipulated by adjusting the fuel flow to the gas turbine.

A dedicated engine control system is needed to regulate the fuel flow to a gas turbine, since a direct adjustment is not straightforward. For instance, the rotor speed and internal temperatures in the turbine must be restricted to certain limits from safety considerations. The fuel regulating mechanism interfaces with the propulsion control system of the ship.

The primary requirements for the control system of a marine gas turbine may be listed as follows:

- capability to start and accelerate the engine to idle through automatic execution of a sequence of operations by way of a start command or manually by control of individual components;
- capability to stop the engine normally through issuance of a stop command;
- capability to accelerate the engine to a given power level without encountering compressor stall or exceeding temperature limits;
- capability to govern the engine at any steady-state operating power level so as to maintain a constant gas generator speed in a stable manner for an undetermined period of time;
- capability to decelerate an engine from a given power level without causing the engine to flame out;
- capability to trip the engine upon sensing that key performance parameters are no longer within safe operating limits;
- capability to monitor the status of key engine and ancillary parameters to enable the operator to attempt corrective action prior to tripping the engine;
- capability to acquire data to monitor the health of the engine; and
- capability to perform all of the aforementioned functions reliably.

Secondary control requirements that should not be overlooked are:

- capability to adapt the system to other gas turbines;
- capability to make adjustments easily;
- capability to be repaired and maintained easily; and
- capability to be trouble-shoot easily.

12.3 FIRST GENERATION CONCEPTS

Control systems were implemented using mechanical, hydraulic, pneumatic and electrical or electronic medium or combinations thereof in the late 1960's and 1970's. Hybrid control systems using analog electrical/electronic and hydro-mechanical elements have been adopted for the first generation of aero-derivative marine gas turbines. The analog systems containing new logic interface with existing hydro-mechanical fuel control systems fitted on the engine.

Hydro-mechanical control systems fitted on engines are retained because they have evolved to become reliable as long as clean fuel is supplied, and are set up and maintained adequately (*ref*: Saville, 1977). The systems have undergone improvement over a period of two decades, with most improvements being made in operational procedures rather than in component design (*ref*: Lesson, 1974). Lack of viable alternatives resulted in fuel and metering functions remaining virtually unchanged.

Starting and stopping sequence logic and performance logic are specific to marine applications of the engine. In satisfying fundamental requirements to activate the starter, open the fuel valves and activate the ignitors to start and bring the engine to idle speed, the sequence depends considerably on the status of plant machinery, particularly the ancillaries of the gas turbine. The logic needs implementation to avoid overspeeding. In marine applications the power turbine replaces the exhaust nozzle originally fitted on the aero-engine. Overspeeding of the power turbine still remains a major concern since the engine could potentially self-destruct within one second of the event being detected. Hence, redundancy in the implementation of overspeed protection through different media and devices is sought after.

Electrical and electronic media for the implementation of control logic is perceived to be the solution to a number of problems plaguing many control systems. They are favoured because of their advantages over other control media (*ref*: Hunt, 1970):

- power supply requirements are modest, and available with high integrity;
- signal transmission is convenient and low in cost compared to other alternatives;
- system is clean and insensitive to contamination;
- simple and reliable electro-mechanical components can interface the system to plant machinery;
- fail-safe techniques and built-in tests are available;
- system is flexible and adaptable to other applications;
- electrical transducers to measure pressure, temperature, vibrations, flow, level, speed, position and time are available;
- complex control requirements can be implemented;
- rapid system response and precise and repeatable control are achievable; and
- perception that the medium is reliable in a marine environment.

12.4 PROPORTIONAL INTEGRAL CONTROLLER

The proportional integral (PI) regulator provides a safe and reliable method for controlling the speed of the rotating shaft. The PI controller is designed after the concept suggested by Rubis et al (1982, 1986), and is shown in figure 12.1. The regulator has an inner loop to control power developed in the gas generator and an outer loop to regulate the speed of the power shaft. The inner loop has all the non-linear safety limiting tests associated with shaft torque, acceleration rate and other related parameters. In this discussion, the non-linearities arising from safety limiting tests and control of pitch ratio are not taken into account. The structure of the controller is shown in

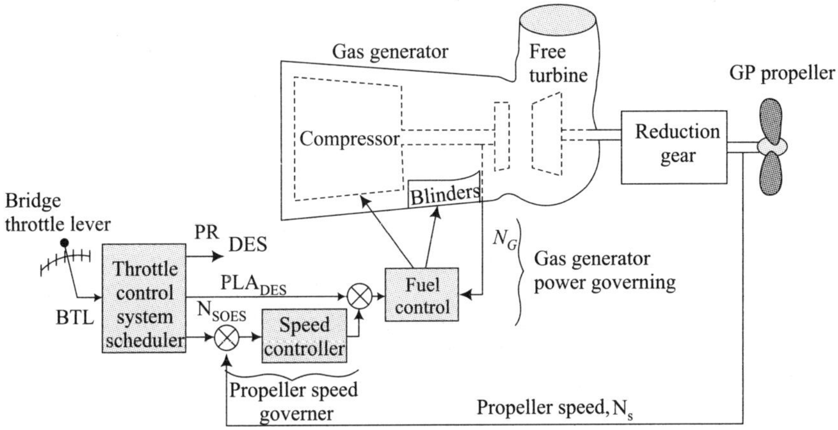

Fig. 12.1 Closed Loop Control for Marine Gas Turbine (Rubine, 1982,1962)

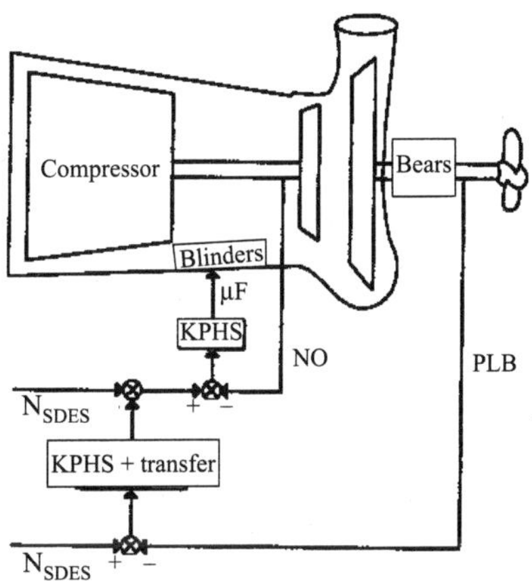

Fig. 12.2 Proportional-Integral Controller (Rubis, 1982, 1986)

figure 12.2, where the power lever actuator (PLA) reference signal is replaced with a control loop in the gas generator speed signal. Three gains need to be selected.

The gas turbine used in the evaluation, shown in figure 12.3, comprises of a 175 hp gas turbine coupled to a water brake dynamometer, although modeling and control of a fixed pitch propeller system under shaft speed control is attempted. The cause-and-effect model of the plant is shown in figure 12.4. The dynamometer is loaded at constant water volume, V, and operated in the same manner as a propeller loaded at a constant pitch. The single input to the plant is the fuel flow rate.

A cut-and-try approach is used for the design of the PI controller gains shown in figure 12.2. The goal here is to select three gains in order to obtain good command following. Thus, quick response time without any oscillation is to be achieved for all transients, with small or zero steady-state error in the target shaft speed. If the interest is in regulating large amplitude transients around the operating region, large changes in the plant's dynamics may be expected. A traditional method calls for opening both loops, design and fix the inner loop, and followed by designing the outer closed loop.

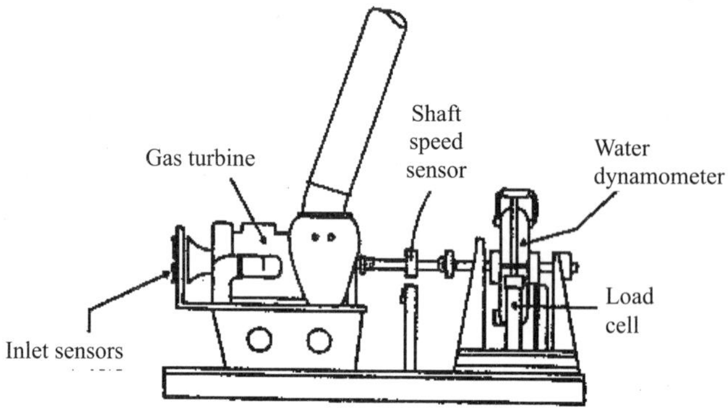

Fig. 12.3 Test Bed Installation

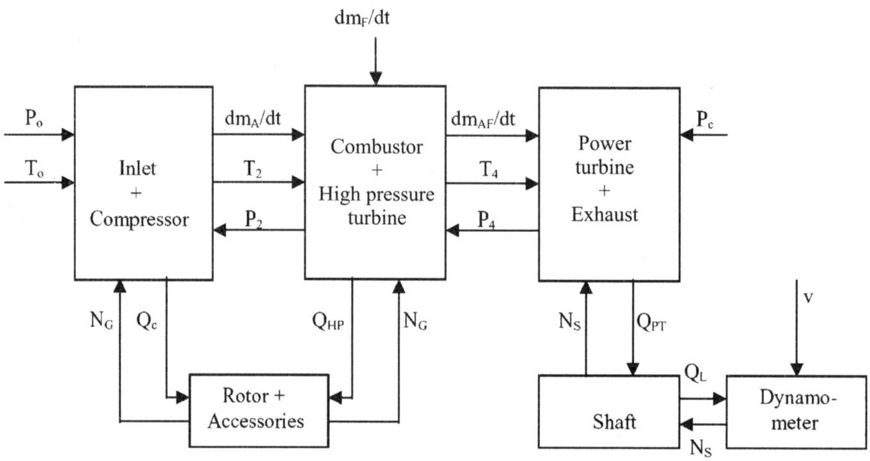

Fig. 12.4 Cause-and-Effect Model of Plant

The inner loop's design results are shown in figure 12.5, where the deceleration curves in the lower chart are less stable than the acceleration curves in the upper chart. The same gain value produces more oscillations during deceleration. Based on the goal of having no oscillations, the inner loop gain is set at KPNG = 0.001. Even though the chosen gain value leaves a large steady state error, it is not deemed to be a performance concern since the integral control in the outer loop is expected to dominate system performance.

The outer loop design is more difficult since there are two design degrees of freedom. Figure 12.6 shows the results. The outer loop's integral action does dominate the inner loop regulator. The gas generator moves to the required level to achieve the desired shaft speed regardless of the setting of its speed. This happens because of the summing junction that connects the two loops. In effect, the junction causes the set reference point for the gas generator to be constantly changed until the shaft speed integral regulator is satisfied.

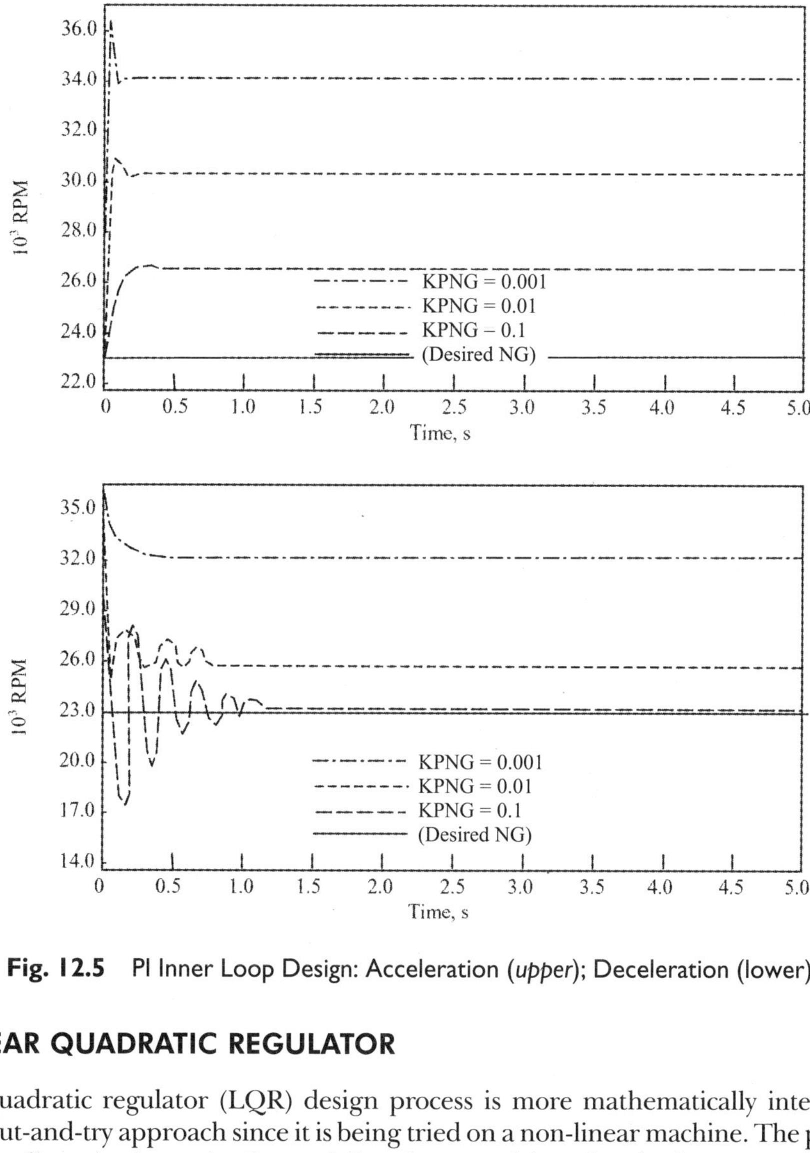

Fig. 12.5 PI Inner Loop Design: Acceleration (*upper*); Deceleration (lower)

12.5 LINEAR QUADRATIC REGULATOR

The linear quadratic regulator (LQR) design process is more mathematically intensive but still requires the cut-and-try approach since it is being tried on a non-linear machine. The process begins with selecting a linearization point for modeling the propulsion plant in the centre of the operating region, $N_G = 25{,}000$ rpm and $N_S = 1{,}500$ rpm. Propulsion dynamics is linearized in the form: **dx/dt = Ax + Bu**, where **u** is the fuel flow rate, **x** denotes the system state, **A** is state coefficient matrix, and **B** is input coefficient matrix. Assume (for gain determination only) that this is an adequate general model of the plant. The perturbed system state may be represented by $x(1) = \delta N_G$, $x(2) = \delta N_S$, and $x(3) = \delta E$, where E is the state error. Minimizing of the error is sought, and the solution to this problem for linear systems takes the form (*ref*: Ogata, 1970): $u = \mathbf{Ke}$, where **e** is the error vector and **K** depends on the plant model matrices **A** and **B** and the choices for the regulator Q and R. Q and R provide weighting for the error terms versus the input terms. Relatively large Q entries causes the

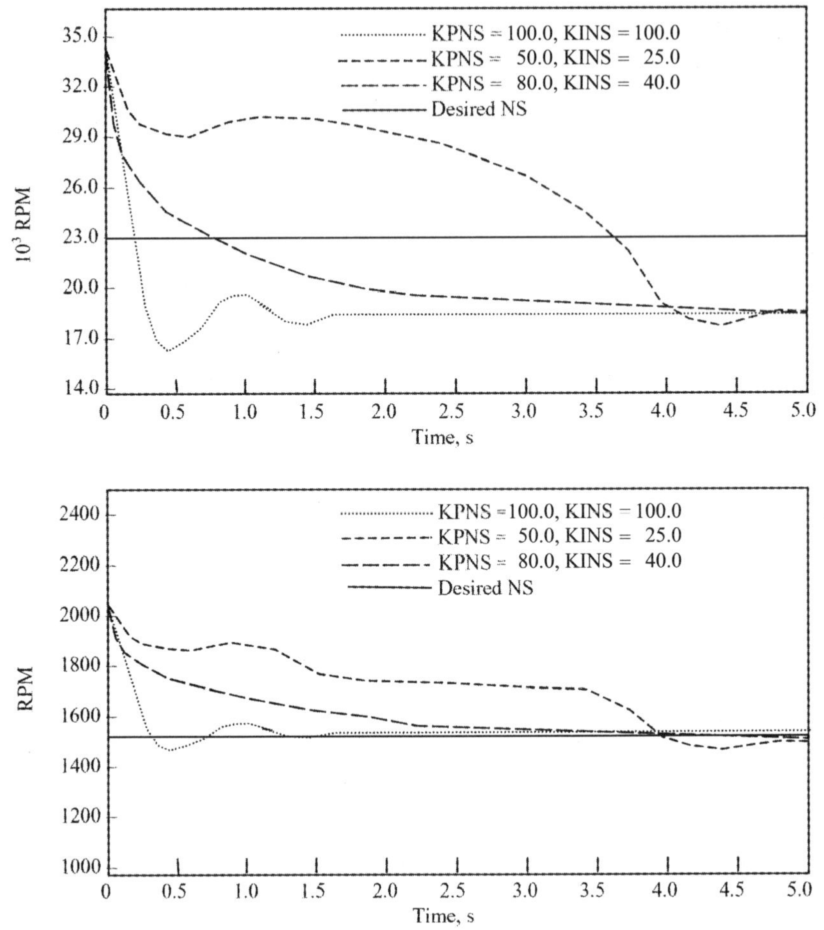

Fig. 12.6 PI Outer Loop Design: Gas Generator Speed (*upper*); Power Turbine Speed (*lower*)

system to respond more quickly in order to minimize the error terms at the expense of input magnitude, while the use of a large R causes the controller to act to minimize the input at the expense of error. For the selected values of Q and R, corresponding K entries and response of the non-linear system are obtained.

The two most important features of controlled system performance are command following and disturbance rejection. Good command following is selected as the criterion for good design in this discussion, which may be further split into acceptable transient control, or no oscillations and small steady state error.

Figure 12.7 shows a comparison of the simulated responses of the final designs of the PI and the LQR regulators for a non-linear plant model. The LQR regulator achieves its final steady state more quickly than the PI regulator. But while neither design displays any oscillations, the LQR model does indicate some steady state offset in shaft speed. This offset occurs because the system has no inherent integrating effect. The LQR regulator thus acts like a classical proportional controller that

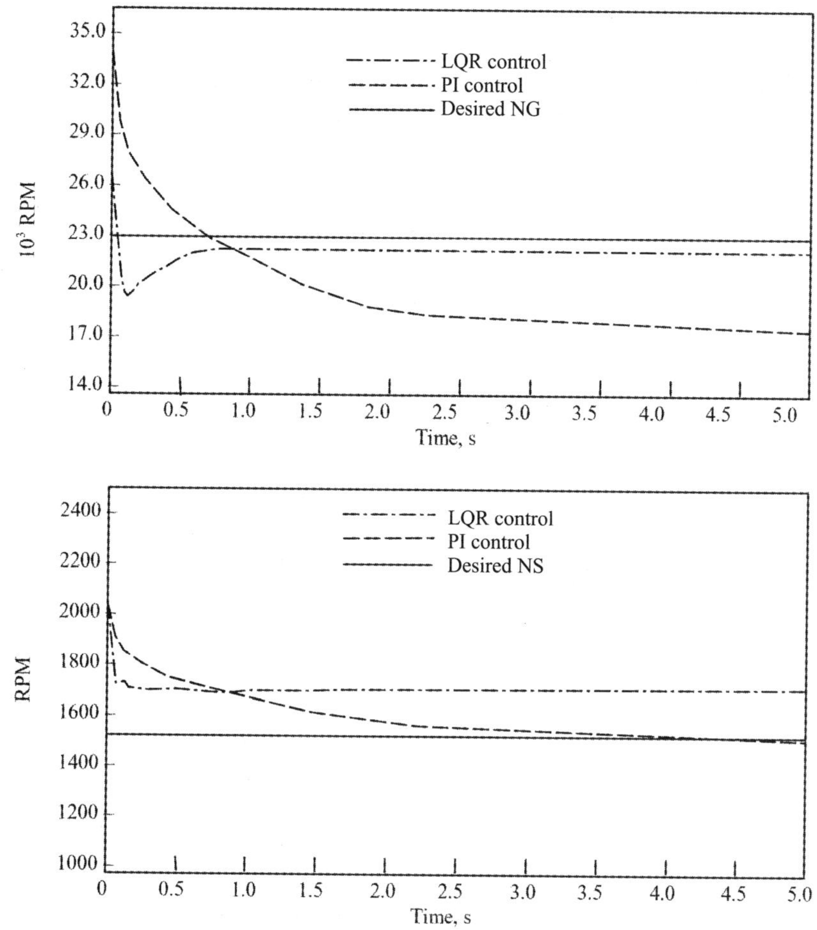

Fig. 12.7 Controller Comparison: Gas Generator Speed (upper); Power Turbine Speed (lower)

has steady state error. This offset could be reduced at the expense of a lower dip in gas generator speed. While the curves indicate a 10 per cent error in LQR steady state shaft speed for the deceleration manoeuvre, an acceleration manoeuvre between the same end states display only 1 per cent error, a clear result of the plant's non-linearity.

The two controller designs offer similar performance for this simplified case. The strength of LQR design lies in reducing gas generator oscillations and fuel consumed on the sea. The strength of PI design is in power turbine shaft speed regulation. The LQR regulator reaches final steady state more quickly, implying quicker ship response.

12.6 DDH 280 SYSTEM DESIGN

Originally developed as an all gas turbine warship by the Canadian Navy in 1966, the DDH 280 is a twin-screw vessel, with displacement of about 4200 long ton. Each shaft line is driven either by a

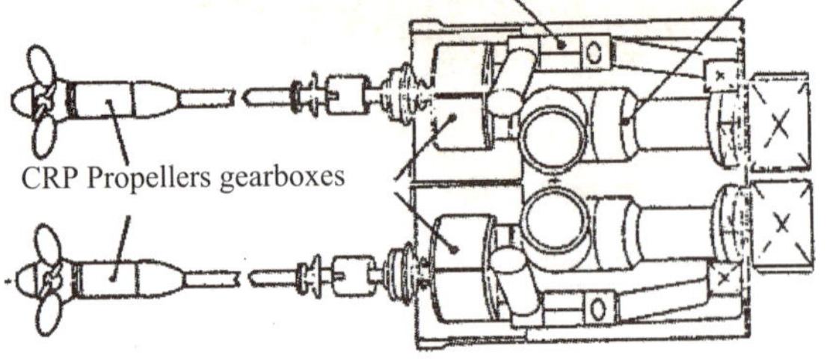

Fig. 12.8 DDH 280 Propulsion Machinery Arrangement

P&WA FT4A-2 boost engine or a P&WA FT12A-3 cruise engine in an arrangement as shown in figure 12.8. The power turbines of the engine drive a controllable pitch propeller through synchronizing self-shifting clutches and reduction gearing. Nearly twenty years later the DDH 280 evolved into Canadian Patrol Frigate (CPF).

The pneumatic propulsion control system for the ship is schematically shown in figure 12.9. The system performs several control and monitoring functions, including starting, stopping and protecting the engines and controlling power/pitch propulsion. The system makes use of different media for implementing control logic. With first-generation systems the surveillance and monitoring tasks are generally distinct from the control task. Digital technology successfully integrated the tasks.

The gas turbine control system is a hybrid system where the combined electrical/electronic/ pneumatic system interfaces with a hydro-mechanical fuel control unit. Besides providing for automatic engine start-up, shutdown and engine changeover, the analog electrical/electronic/ pneumatic system also trips the engine in the event of failure to reach ignition or idle speeds within a prescribed time limit and high vibrations (with battle override).

12.7 INTEGRATED MACHINERY CONTROL SYSTEM

The shipboard integrated machinery control system (IMCS) is a concept for automatic control and monitoring of the propulsion, ancillary and auxiliary machinery (*ref*: Khan et al., 1984, 1987). Developed by the Canadian Department of National Defence, the concept is implemented as a system of distributed digital computers interconnected by data highways.

A fundamental objective of the system is to improve the man-machine interface in warships. The impetus towards centralized control to reduce personnel, increase survivability and improve decision-making process is the only way to achieve integrated control and monitoring functions of the various components of the system. A digitally based IMCS offers a number of advantages, including standardization of equipment, adaptability, ease of adjustment and maintenance and spare capacity. The Canadian Navy decided to provide the IMCS for the patrol frigate. The frigate has two GE LM2500 gas turbines and a Pielstick Type 20 PA6-280 diesel engine in a Combined Diesel or Gas (CODOG) arrangement, as shown in figure 12.10. The gas turbines and diesel engine drive the controllable pitch propellers through a cross-connected reduction gearbox. For normal operation, the gearbox is cross-connected so that power can be supplied to both shafts by selecting two gas turbines, a single gas turbine (either port or starboard) or the diesel engine. The IMCS is

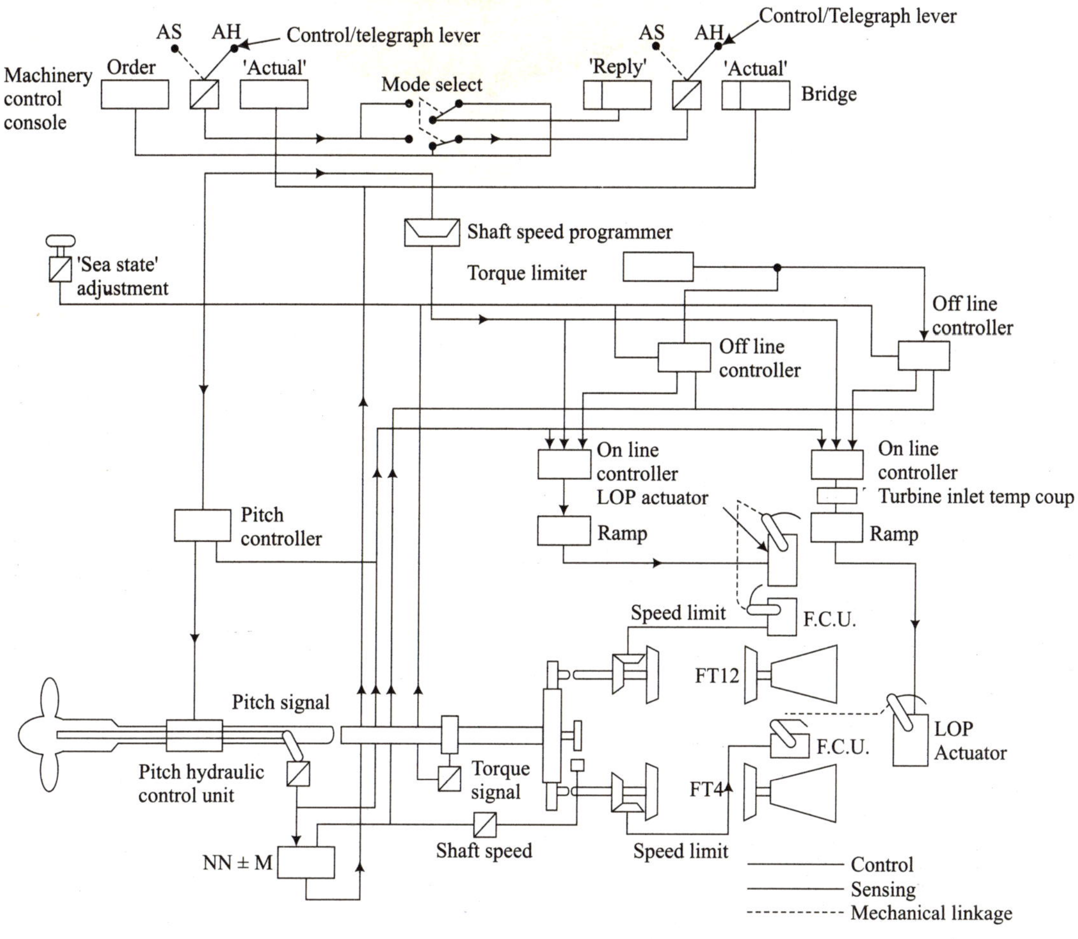

Fig. 12.9 DDH 280 Pneumatic Propulsion Control System

illustrated in block diagram form in figure 12.11. Microprocessors reside in Remote Terminal Units (RTU), which are distributed throughout the ship and located close to the machinery. A triplicate serial data bus provides communication between the RTU's and higher-level processors of the machinery control console, the maintainer/electrical console and the bridge console.

12.8 ELECTRONIC CONTROL MODULE

Since new components and shafting arrangements are continually selected for Naval vessels, propulsion systems may be configured in many different ways. In an attempt to standardize, reduce costs and improve reliability, manufacturers traditionally supply prime movers with dedicated control units. The electronic control module (ECM) for the patrol frigate is based on the same digital electronics technology selected for the IMCS.

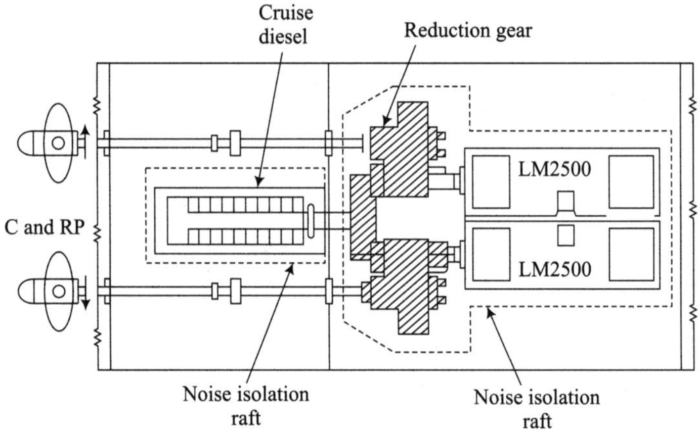

Fig. 12.10 Canadian Patrol Frigate Propulsion Machinery Arrangement

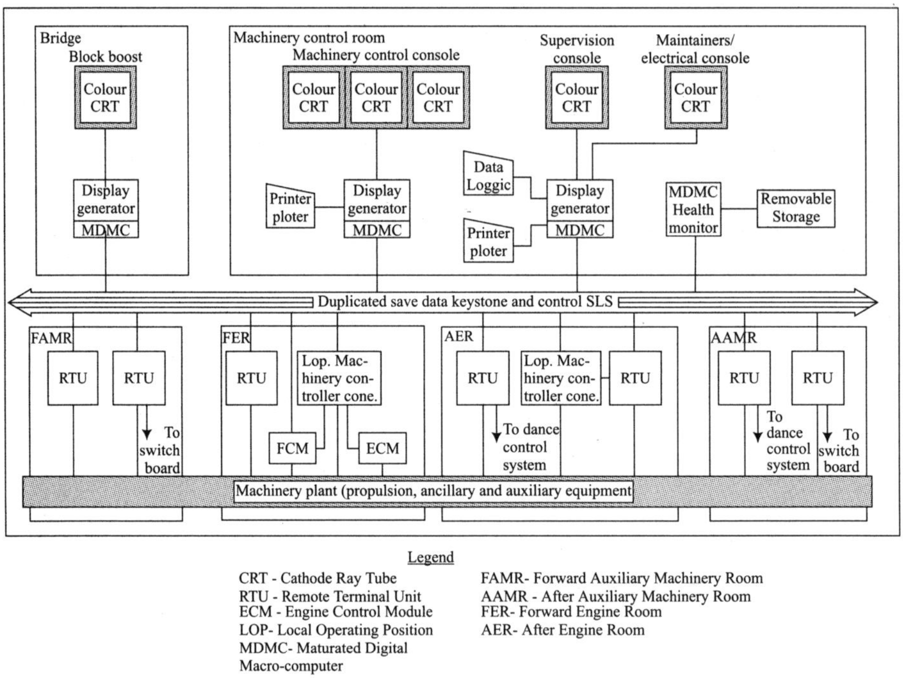

Fig. 12.11 Canadian Patrol Frigate Integrated Machinery Control System

For the patrol frigate, the LM2500 engines provide boost and intermediate power. The engine is a derivative of the commercial CF6 aero-engine, which in turn is a derivative of the military TF39 aero-engine. The manufacturer, General Electric, provides an electronic assembly for additional control logic specific to the application. The analog electronic system is the primary interface between the LM2500 engine and its hydro-mechanical fuel control unit. Besides manual or automatic engine start-up and one command shutdown, the electronic control module is capable of tripping the engine in the event of over-speed, over-temperature, loss of pressure in fuel manifold, low lube oil pressure or temperature, high engine vibrations or fire in the enclosure. The ECM offers closed-loop control for dc servomotor power lever actuator position, power turbine torque and acceleration limiting to protect the components, and surveillance with alarm annunciation.

12.9 DIGITAL FUEL CONTROL SYSTEM

The Canadian Navy's decision to adopt the digitally based IMCS for its warships spawned a number of activities aimed at enhancing the systems. One of these activities examined the feasibility to convert the fuel control units of the aero-derivative gas turbines from hydro-mechanical to digital systems. Eventual replacement of the hydro-mechanical system and integration of the fuel control functions in the ECM appeared as a logical step forward in the evolution of marine gas turbine controls.

Other factors also favoured the replacement of hydro-mechanical controls. The systems are reliable when supplied with clean fuel and maintained properly. Originally designed for the aircraft industry, stringent requirements are imposed to ensure the hydro-mechanical controls are indeed reliable. In a marine environment, however, controlling the quality of the fuel has been problematic, and the setup and maintenance of the hydro-mechanical control system has been less than ideal. This is in part due to the different emphasis on safety in the repair and maintenance philosophies of aero-engines as opposed to the gas turbine engines in marine applications. The inevitable result is that the hydro-mechanical fuel control system has not fared well in the marine environment.

Over the last fifteen years, technology advancements have made it possible to configure reliable and affordable microprocessor-based digital fuel control systems for industrial and marine applications. For instance, Woodward Governor Co. makes a full-authority microprocessor-based fuel control system for industrial gas turbines. The system can be designed to include sequencing, protection and fuel metering control functions. The system can be configured as a single microprocessor unit or as a triplex fault tolerant digital control system by integrating three control units.

Apart from the benefits derived from digital systems themselves, preliminary analyses employing machinery simulation suggest that simpler fuel control logic with reliance on fewer sensor measurements could be developed for speed governing and acceleration limiting of the marine gas turbine. Such potential improvements are desirable since they offer means for improving system reliability while reducing its complexity. With the emergence of reliable digital electronic control and display systems, navies of the world are in a position to apply this technology to the development of integrated machinery control systems, gas turbine engine control modules and full-authority digital marine gas turbine control systems for their warships. Figure 12.12 shows a typical digitally implemented fuel control concept.

12.10 REVIEW PROBLEMS

Example problem # 12.1: The output of a flow meter is 1250 ft^3/second and the meter proportionality constant is 0.737. The measuring device of the meter produces a signal proportional Q^2, where Q is the square of the volume flow rate of the gas. Calculate the flow rate through the flow meter.

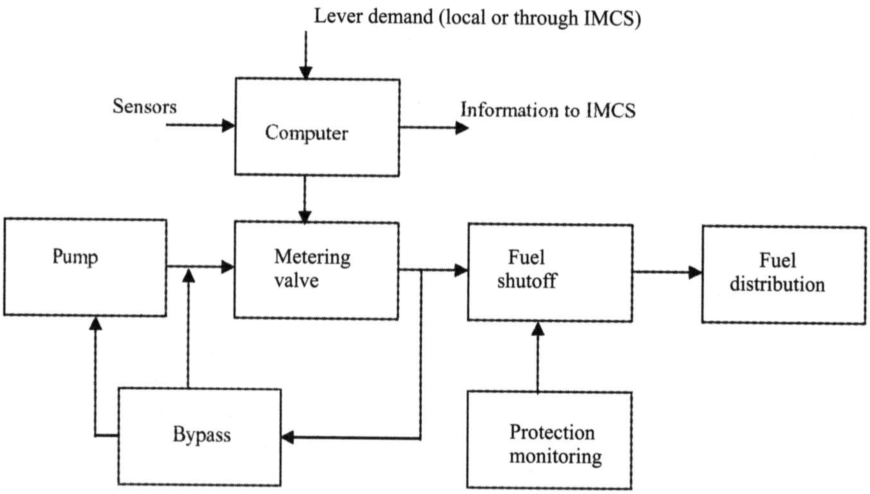

Fig. 12.12 Typical Digital Fuel Control Concept

Solution: If the output of the flow meter is denoted as D and meter proportionality constant is k_f, then the flow rate: $Q = k_f \times \sqrt{D}$

$$= 0.737 \times \sqrt{1250}$$

$$= 26.06 \text{ ft}^3\text{/second.}$$

Example problem # 12.2: The gram molecular weight M of the fuel gas for a marine gas turbine installation is measured to be 24, gas pressure p is 475000 pascal, and absolute temperature T is 520 K. The flow meter has a proportionality constant of .810 and its output is 339 m^3/sec. Find the flow rate by weight.

Solution: Metered flow rate $Q = k_f \times \sqrt{D}$

$$= 0.810 \times \sqrt{339}$$

$$= 14.91 \text{ m}^3\text{/second.}$$

Volume v, weight w, absolute pressure p, absolute temperature T, universal gas constant R = 8.314 J/K.mol and gram molecular weight M of the gas, are connected through the relation: $pv = wRT$, or $w = (pv/RT)$,

Or: $w = k \times \sqrt{D} \times (p/T)$,

where: $k = Mk_f/(10^3R) = 24 \times 0.810/(1000 \times 8.314) = 0.002338$

Then: $w = 0.002338 \times \sqrt{339} \times (475000/520) = 39.33$ kilogram/second.

Example problem # 12.3: A mechanical speed control system, also called governor, is used to regulate the speed of gasoline engines, gas turbines and steam engines. A drive shaft from engine rotates the fly ball and spring assembly at a speed proportional to the engine speed (figure 12.13). The spinning masses of the fly ball produce a centrifugal force that compress the spring and raises the valve stem. Thus the position of the valve stem depends on the engine speed. As the engine speed increases, the valve decreases the fuel flow, while reduced speed increases the flow of fuel. The engine will then settle out at a speed point that results in just enough flow of fuel that balances the load on the engine. If the weight of the fly balls is 2.3 lbs, stiffness of the spring is 12.85 lbs/in., and damping value is 1.27 lbs-sec/in, determine the transfer function for the mechanical vibrating system.

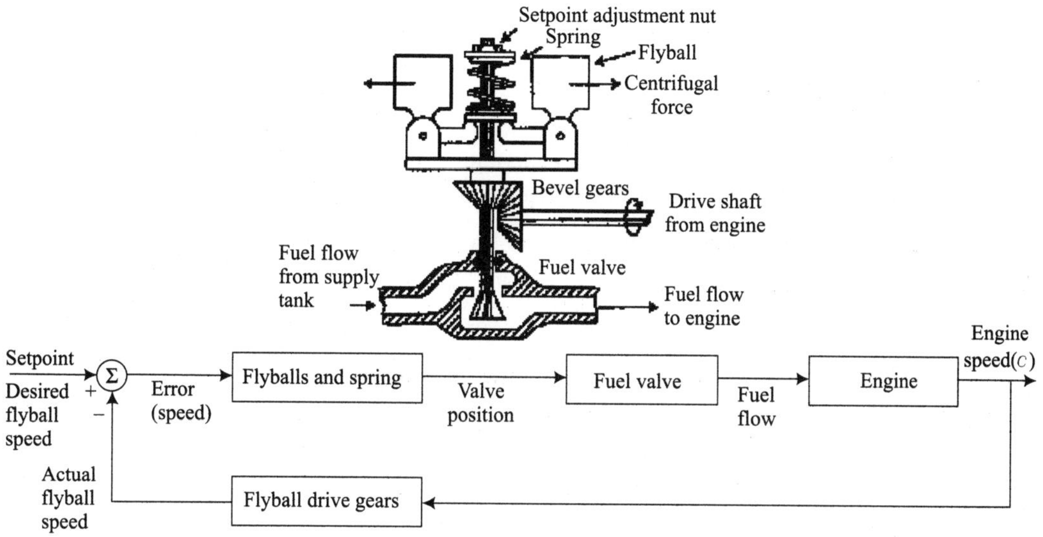

Fig. 12.13 Schematic Diagram of Speed Control Mechanism

Solution: From table 9.3: If h = output signal, f = input signal, time-domain equation for the mass, spring and damper system is: $MC_m \times d^2h/dt^2 + RC_m \times dh/dt + h = C_m f$

where:

$$MC_m = [(2.3 \text{ lbs})/(384 \text{ in/sec}^2)] \times (1/12.85 \text{ lbs/in}) = 0.0004661$$

$$RC_m = [1.27 \text{ lbs-sec/in}] \times (1/12.85 \text{ lbs/in}) = 0.09883$$

Time-domain equation: $0.0004661 \times d^2h/dt^2 + 0.09893 \times dh/dt + h = (1/12.85)f$

Transfer function: $H(s)/F(s) = 0.07782/(1 + 0.09883s + 0.0004661s^2)$

Example problem # 12.4: In example problem # 12.3 determine the resonant frequency.

Solution: Resonant frequency: $\omega = \sqrt{(1/MC_m)} = \sqrt{(1/0.0004661)} = 46.32$ radian/second.

Example problem # 12.5: In example problem # 12.3 determine the damping ratio.

Solution: Damping ratio: $\zeta = RC_m/[2 \times \sqrt{(MC_m)}] = 0.09883/(2 \times \sqrt{0.0004661}) = 2.2889$

Example problem # 12.6: The acceleration signal from a vibration transducer mounted on a turbine bearing running at 3000 rpm has peak reading of 237.5 in/sec^2. Express the vibration reading as displacement and velocity.

Solution: Operating speed $\omega = 2\pi N/60 = 2\pi \times 3000/60 = 314.16$ radian/second. From equation 2.7, the acceleration reading is:

$$a(t) = -X_0 \cdot \omega^2 \cdot \sin(\omega t) = -X_0 \times (314.16)^2 \times \sin(314.16t)$$

$$= -237.5 \times \sin(314.16t) \text{ in/second}^2, \text{ where } X_0 = \text{peak displacement amplitude.}$$

$X_0 \times (314.16)^2 = 237.5$, or $X_0 = 237.5/(314.16)^2 = 0.002406$ in.

$= 0.004812$ in. peak-to-peak.

From equation 2.6 the peak velocity reading is given by:

$$v(t) = X_0.\omega.\cos(\omega t) = X_0 \times 314.16 \times \cos(314.16t)$$

$$= 0.002406 \times 314.16 \times \cos(314.16t) \text{ in/second.}$$

Peak velocity reading $v_0 = 0.7560$ in/second.

Example problem # 12.7: A series electrical RLC circuit consists of inductor $L = 0.137$ H, capacitor $C = 27\mu$F, and resistor $R = 377\Omega$. Calculate the (*i*) time-domain equation, (*ii*) transfer function, (*iii*) resonant frequency, (*iv*) damping ratio, and (*v*) damping type.

Solution: From table 9.3: Time-domain equation for RLC circuit is:

$$A_2 \times d^2e_{out}/dt^2 + A_1 \times de_{out}/dt + e_{out} = e_{in}$$

where:

$$A_2 = LC = 0.137 \times 27 \times 10^{-6} = 3.70 \times 10^{-6}$$

$$A_1 = RC = 377 \times 27 \times 10^{-6} = 0.01018$$

Time-domain equation: $3.70 \times 10^{-6} \times d^2e_{out}/dt^2 + 0.01018 \times d\,e_{out}/dt + e_{out} = e_{in}$

Transfer function: $E_{out}(s)/E_{in}(s) = 1/(1 + 0.01018s + 3.70 \times 10^{-6}s^2)$

Resonant frequency: $\omega_0 = \sqrt{(1/A_2)} = \sqrt{(1/3.70 \times 10^{-6})} = 519.88$ radian/second.

Damping ratio: $\zeta = \alpha/\omega_0 = A_1/2\sqrt{A_2} = [0.01018/(2 \times \sqrt{3.70 \times 10^{-6}})] = 2.6463$

Damping ratio is greater than 1, so system is over damped.

Example problem # 12.8: A chemical process vessel has a diameter of 2.874 m and height of 4.92 m. A resistance of magnitude 5.0×10^5 Pa-s/m^3 is encountered at the outlet from the tank. The temperature of the contents in the vessel is 91°F, and density is 779 kg/m^3. Full-scale ranges are $FS_{in} = 0.00345$ m^3/s and $FS_{out} = 2.128$ m. Determine (*i*) the capacitance of the tank, (*ii*) the time constant of the process, (*iii*) the gain of the process, (*iv*) the time-domain equation, and (*v*) the transfer function.

Solution: From table 9.2: Vessel cross-sectional area:

$$A = \pi d^2/4 = \pi(2.874)^2/4 = 6.4873 \text{ m}^2$$

Capacitance: $C_L = A/\rho = 6.4873/779 = 83.277 \times 10^{-4}$ m^3/Pa

Time constant: $\tau = R_LC_L = (5.0 \times 10^5) \times (83.277 \times 10^{-4}) = 4164$ second.

Process gain: $G = (R_L/\rho) \times (FS_{in}/FS_{out})$

$$= (5.0 \times 10^5/779) \times (0.00345/2.128) = 1.0406$$

Time-domain equation: $\tau \times dh^*/dt + h^* = G \times q^*_{in}$

Or: $4164 \times dh^*/dt + h^* = 1.0406 \times q^*_{in}$

Transfer function: $H^*(s)/Q^*_{in}(s) = G/(1 + \tau s) = 1.0406/(1 + 4164s)$

Example problem # 12.9: An interacting two capacity electrical system has components $R_1 = 325\ \Omega$, $R_2 = 432\ \Omega$, $R_L = 550\ \Omega$, $C_1 = 33\ \mu F$, and $C_2 = 42\ \mu F$. Calculate the (*i*) time-domain equation and (*ii*) transfer function.

Solution: From table 9.3:

$$\tau_1 = R_1C_1 = 325 \times 33 \times 10^{-6} = 0.0107$$

$$\tau_2 = R_2C_2 = 432 \times 42 \times 10^{-6} = 0.01814$$

$$G = R_L/(R_1 + R_2 + R_L) = 550/(325 + 432 + 550) = 0.4208$$

$$A_1 = G\{(\tau_1 + \tau_2) + \tau_1R_2/R_L + \tau_2R_1/R_2\}$$

$$= 0.4208\{(0.0107 + 0.01814) + 0.0107 \times 432/550 + 0.01814 \times 325/432\}$$

$$= 0.02144$$

$$A_2 = G\tau_1\tau_2 = 0.4208 \times 0.0107 \times 0.01814 = 0.00008189$$

Time-domain equation: $A_2 \times d^2e_{out}/dt^2 + A_1 \times de_{out}/dt + e_{out} = G \times e_{in}$

Or: $0.00008189 \times d^2e_{out}/dt^2 + 0.02144 \times de_{out}/dt + e_{out} = 0.4208 \times e_{in}$

Transfer function: $E_{out}(s)/E_{in}(s) = G/(1 + A_1 \times s + A_2 \times s^2)$

Or: $E_{out}(s)/E_{in}(s) = 0.4208/(1 + 0.02144s + 0.00008189s^2)$

Example problem # 12.10: A bearing lubricant tank has a base 1.50 m × 1.50 m and height of 2.25 m. The inside film coefficient is 57 W/m², and outside film coefficient is 379 W/m². The steel (k = 48 W/m, heat capacity S_h = 2329 J/kg-K, density = 928 kg/m³) wall of the tank is 1.1 cm thick. Determine (*i*) the thermal resistance, (*ii*) thermal capacitance, (*iii*) time constant, (*iv*) time-domain equation, (*v*) transfer function.

Solution: Tank area: $A = b \times d + 4 \times (b \times h) = 1.50 \times 1.50 + 4 \times (1.5 \times 2.25) = 15.75\ m^2$

Thermal resistance: $R_T = (1/h_i + x/k + 1/h_o)/A$ kelvin/watt

$$= (1/57 + 0.011/48 + 1/379)/15.75 = 0.001296\ K/W$$

Thermal capacitance: $C_T = mS_h$

Mass of tank $m = \rho bdh = 928 \times 1.5 \times 1.5 \times 2.25 = 4698.0$ kg

Hence: $C_T = 4698.0 \times 2329 = 10.94 \times 10^6$ J/K

Time constant: $\tau = R_T C_T = (0.001296) \times (10.94 \times 10^6) = 14180$ seconds

Time-domain equation: $\tau \times d\theta_L/dt + \theta_L = \theta_j$, or: $14180 \times d\theta_L/dt + \theta_L = \theta_j$

Transfer function: $\theta_L(s)/\theta_j(s) = G/(1 + \tau s) = 1/(1 + 14180s)$, where gain $G = 1$.

Example problem # 12.11: A 470 foot long conveyor belt used in a chemical process plant travels at 29.5 ft/s. Determine: (*i*) dead-time lag, (*ii*) time-domain equation, (*iii*) transfer function, and (*iv*) the Bode diagram gain and phase line equations.

Solution: Dead-time lag: $t_d = D/v = 470.0/29.5 = 15.93$ seconds

Time-domain equation: $f_o(t) = f_i(t - 15.93)$

Transfer function: $F_o(s)/F_i(s) = e^{-15.93s}$

Bode diagram gain line is a constant 0 db, and the phase line equation is:

$$\beta = -57.3\omega t_d = -57.3 \times 15.93\omega = -912.92\omega$$

Example problem # 12.12: A DC motor has rotor mass polar moment of inertia $J = 15.93 \times 10^{-3}$ kg-m^2, damping resistance $B = 2.79 \times 10^{-3}$ N-m-s/radian, armature inductance $L = 0.038$ H, armature resistance $R = 2.70\ \Omega$, torque constant $K_t = 0.149$ N-m/A, EMF constant $K_e = 0.177$ volt-s/radian. Find the (*i*) mechanical time constant, (*ii*) electrical time constant, (*iii*) time-domain equation, and (*iv*) transfer function of the armature-controlled motor.

Solution: Mechanical time constant: $\tau_m = J/B = (15.93 \times 10^{-3})/(2.79 \times 10^{-3}) = 5.7097$ s.

Electrical time constant: $\tau_e = L/R = 0.038/2.70 = 0.01415$ s.

$$A_2 = \tau_m \tau_e RB/(K_e K_t + RB)$$

$$= (5.7097 \times 0.01415 \times 2.7 \times 2.79 \times 10^{-3})/(0.177 \times 0.149 + 2.7 \times 2.79 \times 10^{-3})$$

$$= 0.01795$$

$$A_1 = (\tau_m + \tau_e)RB/(K_e K_t + RB)$$

$$= (5.7097 + 0.01415) \times 2.7 \times 2.79 \times 10^{-3}/(0.177 \times 0.149 + 2.7 \times 2.79 \times 10^{-3})$$

$$= 1.2717$$

$$G = K_t/(K_e K_t + RB) = 0.149/(0.177 \times 0.149 + 2.7 \times 2.79 \times 10^{-3})$$

$$= 4.3945$$

Time-domain equation: $0.01795 \times d^2\omega/dt^2 + 1.2717 \times d\omega/dt + \omega = 4.3945e$ (where e = armature voltage)

Transfer function: $\Omega(s)/E(s) = 4.3945/(1 + 1.2717s + 0.01795s^2)$

Example problem # 12.13: A blending tank of dimensions 2.7 ft length, 2.4 ft wide and 1.5 ft height has a outflow rate of 0.63 ft^3/s. Calculate (*i*) the time constant, (*ii*) time-domain equation, and (*iii*) transfer function of the tank.

Solution: Volume of tank $V = lbh = 2.7 \times 2.4 \times 1.5 = 9.72$ ft^3

From table 9.2: Time constant: $\tau = V/q = 9.72/0.63 = 15.4286$ s

Time-domain equation: $15.4286 \times dc_o/dt + c_o = c_i$

Transfer function: $C_o(s)/C_i(s) = G/(1 + \tau s) = 1/(1 + 15.4286s)$, where $G = 1$.

12.11 EXERCISES

12.1 A speed control governor of a gas turbine engine's drive shaft turns the fly ball and spring assembly at a speed proportional to the engine speed. The weight of the fly balls is 1.85 lbs and damping value of the system is 0.72 lbs-sec/in. If the frequency of the mechanical vibrating system is 29.5 Hz, determine the stiffness of the associated spring.

12.2 The output of a flow meter is 1.89 m^3/second and flow rate through the meter is 1227 m^3/second. The measuring device of the meter produces a signal proportional Q^2, where Q is the volume flow rate of the gas. Calculate the proportionality constant of the meter.

12.3 The gram molecular weight M of the fuel gas for a marine gas turbine installation is 28 and gas pressure p is 472 lbs/in^2. Determine the absolute temperature T if the flow meter has a proportionality constant of 0.775, output is 1871 ft^3/second, and flow rate by weight is 112 lbs/second.

12.4 In exercise # 12.1 establish the time-domain equation and the transfer function.

12.5 In exercise # 12.1 determine the damping ratio.

12.6 A series electrical RLC circuit has inductance of 0.172 H, capacitance of 42×10^{-6} F and resistance of 428 Ω. Calculate the resonant frequency and damping ratio of the circuit.

12.7 Establish the time-domain equation and the transfer function in exercise 12.6.

12.8 The displacement signal from a vibration transducer mounted on a turbine bearing running at 5400 rpm has peak to peak reading of 2.17×10^{-3} in. Calculate the vibration reading in terms of velocity and acceleration.

12.9 A chemical process vessel has a diameter of 1.72 m diameter and height of 2.03 m. Flow resistance of magnitude 2.35×10^5 Pa-s/m^3 is encountered at the outlet from the tank. The temperature of the contents in the vessel is 91°C, and density is 779 kg/m^3. Full-scale ranges are FS_{in} = 0.00345 m^3/s and FS_{out} = 2.128 m. Calculate (*i*) the capacitance of the tank, (*ii*) the time constant of the process, (*iii*) the gain of the process, (*iv*) the time-domain equation, and (*v*) the transfer function.

12.10 A non-interacting two capacity electrical system has components R_1 = 27 Ω, R_2 = 39 Ω, R_L = 171 Ω, C_1 = 23.5 μF, and C_2 = 27.9 μF. Calculate the (*i*) time-domain equation and (*ii*) transfer function.

12.11 A 23.5 m long conveyor belt used in a plant travels at 7.45 m/s. Determine: (*i*) dead-time lag, (*ii*) time-domain equation, (*iii*) transfer function, and (*iv*) the Bode diagram gain and phase line equations.

12.12 A DC motor has rotor mass polar moment of inertia $J = 73.3789 \times 10^{-3}$ lb-ft^2, damping resistance $B = 33.452 \times 10^{-2}$ ft-lb-s/radian, armature inductance L = 0.067 H, armature resistance R = 21.9 Ω, torque constant K_t = 0.1726 lb-ft/A, EMF constant K_e = 0.239 volt-s/radian. Find the (*i*) mechanical time constant, (*ii*) electrical time constant, (*iii*) time-domain equation, and (*iv*) transfer function of the armature-controlled motor.

12.13 A blending tank of dimensions 1.4 m length, 1.97 m wide and 2.93 m height has an outflow rate of 0.176 m^3/s. Calculate (*i*) the time constant, (*ii*) time-domain equation, and (*iii*) transfer function of the tank.

References & Bibliography

Hunt, G.F., "An Introduction to the Use of Electronic Controls in Marine Automation", ASME paper # 70-GT-118, New York, 1970.

Khan, R., "Canadian Patrol Frigate Machinery Control System: An Update", Eighth Ship Control Systems Symposium, The Hague, the Netherlands, October 6–9, 1987.

Khan, R. and Eich, P., "Canadian Patrol Frigate Machinery Control System", Seventh Ship Control Systems Symposium, Bath, United Kingdom, September 24–27, 1984.

Lesson, P.J., "Gas Turbine Control Systems", AGARD Conference Proceedings No. 151, Power Plant Controls for Aero Gas Turbine Engines, September, 1974.

Ogata, K., "Modern Control Engineering", Prentice-Hall, Englewood Cliffs, NJ, 1970.

Rubis, C.J. and Harper, T. R., "Governing Ship Propulsion Gas Turbine Engines", SNAME Transactions, volume 94, page 283–308, 1986.

Rubis, C.J. and Harper, T. R., "The Naval Gas Turbine Ship Propulsion Dynamics and Control Systems Research and Development Program", SNAME Transactions, volume 90, page 321–338, 1982.

Saville, M. "Electronic Control Systems – Reliability Pitfalls and the Way Ahead", Second Symposium on Gas Turbine Operation and Maintenance, January, 1977.

Slagle, J.L., "A Triplex Fault Tolerant Digital Control", ASME paper # 87-GT-21, New York, 1987.

Smith, D.L. and Stammetti, V.A., "Comparative Controller Design for a Marine Gas Turbine Propulsion System", ASME paper # 89-GT-190, New York, 1989.

Sylvestre, R.A. and Dupuis, R.J., "The Evolution of Marine Gas Turbine Controls", ASME paper # 89-GT-204, New York, 1989.

13

Rotating System Vibrations

13.1 INTRODUCTION

Consider a disk of mass m mounted on a shaft of length l between two bearings, and spinning at a fixed speed ω. Assume the centre of gravity of the disk to be located at a radial distance e away from the centreline of the shaft. Then a rotating centrifugal force $m\omega^2 e$ acts on the disk. The rotating force vector may be divided into horizontal and vertical components, causing the disk to execute vibratory motions in the two directions. And if the spin speed ω coincides with the elastic shaft's natural frequency of vibration in the transverse direction, a state of resonance exists, and the shaft and disk assembly will experience considerable vibrations. Neglecting the effects of gravity, there are two forces acting on the disk, the elastic pull of the shaft tending to restore it to its original position and the centrifugal force on the centre of gravity of the disk, causing it to travel in a circle. The elastic force depends on the bending stiffness of the shaft, and is proportional to its deflection. This observation is also true of a shaft with many disks running in two or more bearings.

The sag or deflection of the rotating shaft bent by various concentrated (such as the centrifugal force of the eccentric disk) or distributed (such as the weight of the rotor) loads is calculated by integrating a differential form that expresses the local deformation of a small element of the beam. By considering the local deformations, bending moment M is given by:

$$M = EI\frac{d\varphi}{dx} \qquad ...(13.1)$$

where E is the Young's modulus of beam material, I is the area moment of inertia, or the second moment of the cross-section and $d\varphi$ is the change in the angle of slope of the deflection curve of a beam element of length dx. If the beam bends through only slightly, then approximately $\tan\varphi = \varphi = dy/dx$, and:

$$M = EI\frac{d^2y}{dx^2} \qquad ...(13.2)$$

This differential equation of flexure of a beam is written to express the varying bending moment M along the length of the beam in terms of the deflection y and the location coordinate x. For small slopes the term d^2y/dx^2 is the curvature of the beam, and the quantity EI is referred to as the bending or flexural stiffness of the beam. The corresponding bending stress in the beam is determined from $\sigma_b = (My/I)$.

The objective in a rotor dynamic study is to determine the deflections *y* due to dynamic forces. Failures in turbo-machinery shafts result from fatigue due to stress as it fluctuates between extreme values. In addition to torsion, fibres on the surface of a rotating shaft are subjected to bending loads, undergoing both tension and compression for each revolution of the shaft. If the shaft of a centrifugal compressor is coupled to the shaft of an electric motor operating at 1,740 rpm, the shaft is stressed in tension and in compression 1,740 times. If the shaft is also axially loaded (caused, for example, by a helical or worm gear), the shaft has an axial stress superimposed upon the bending component. These and other loads occurring in the shaft produce stresses often referred to as repeated, alternating or fluctuating stresses.

Bending in the shaft may arise from a number of operational factors, some of which arise internally in the machine and others from outside the machine. Dynamic balance of the machine's rotating assembly is an internal machine factor that causes bending in the shaft. As a result of manufacturing procedures and limitations, tolerances have to be provided on machined components. Thus, a minimum diameter shaft journal may be paired with a maximum diameter bearing, causing a larger than normal radial clearance in the bearing. In contrast with torsional vibrations, bearing and support stiffness plays a significant role in the shaft's lateral vibrations. Since the hydrodynamic stiffness of the bearing's lubricant film is reduced as a consequence of the larger bearing clearance, the shaft's natural frequency of vibration also is reduced, causing the machine to operate at or close to the natural frequency. Rotor unbalance arises from many sources attributable to manufacturing and field operation. With the appropriate amount and location of unbalance in the rotor, substantial bending moments are caused when operation takes place near a critical speed, resulting in flexural shaft vibrations along its lateral axes.

Shaft misalignment is another factor responsible for flexural vibrations. At the coupling between an internal combustion engine and an electric power generator thermal conditions may throw out the rotor from an aligned position during cold condition. Radial and angular misalignment of the rotor can radically alter the flexural vibration signature of a rotating assembly, with consequent bending moments in the shaft. This also has the adverse effect of putting unusual wear pattern in the bearings.

For flexural vibrations the most commonly encountered stimulus is rotor unbalance and shaft bow, resulting in whirling or whipping of the shaft. Unstable operation of the machine in the form of chatter (or stick-slip) and large deflections and loads are other undesirable consequences. To deal with such vibrations appropriate design strategies are required. For instance, placing the system's critical speeds out of the steady state operating range will minimize the degree of super-critical whirling and whipping phenomena. If for practical reasons it is not possible to raise the critical speed, or operation above the fundamental mode of vibration cannot be avoided, introduction of damping to reduce critical response peaks and to cancel the destabilizing forces in the operating range will be required. Minimizing the excitation through better balancing and aligning procedures will go a long way towards the stated objective of smoother operation of the machine.

An easy to use and accurate computational tool for prediction of the vibration characteristics of any equipment design is an indispensable requirement in the design process. The design process is too complex to permit automation in a direct sequence from specification to definition of the finished design. In spite of this drawback, a direct sequence can be approximated by a speedy iterative process that starts with posing design alternatives, then analytically evaluating them, followed by selecting and quantitatively refining the design that satisfies design requirements. Such analytical tools are made available in the form of computer codes, and are discussed in this and other chapters of this treatise. Analytical methods are useful for predicting the rotor dynamic behaviour of the preliminary design configuration, to assess its impact and to provide quantitative data for design refinement. The mathematical model developed provides valuable insight during testing of a new machine; when unexpected data from accelerometers during operations indicates trouble, analytical models are studied to provide understanding and guidance in remedial design, fault detection and correction. The analytical model may range from an idealized version of a single mode to refined and detailed finite element model of the total system to provide precise information in the form of natural frequencies, mode shapes, deflections and loads throughout the assembly.

Mathematical models for determining the vibration characteristics of a rotating machine simulate the kinematics of important degrees of freedom. They permit the determination of the response of the system due to the interaction of the geometric configuration of the rotor, bearings and support structure, with the dynamic loads acting on the system. The analytical model is used to analyse free vibrations, forced vibrations and self excited vibrations.

For each cycle or revolution of the shaft the most careful analysis may reveal that the maximum dynamic stresses are below the ultimate strength of the material, and quite frequently even below the yield strength; even then a fatigue failure is encountered. The primary distinguishing characteristic of fatigue failures is that the stresses have been repeated a very large number of times, and hence the failure is called a fatigue failure. A fatigue failure begins with a small crack, which may be small enough to prevent detection by the naked eye or even to locate in a Magnaflux or x-ray inspection. The crack will develop at a point of discontinuity in the material such as a keyway, a change in the cross section or a hole. Less obvious points at which fatigue failures are likely to start are inspection or stamp marks, internal cracks or irregularities caused by machining. Once a crack has developed, the stress concentration effect is increased and the crack propagation progresses at a faster pace. As the stressed area decreases in size, the stress increases in magnitude until the remaining area fails abruptly. Two distinct areas of failure, the first arising from progressive development of the crack and the second from sudden fracture, characterize the fatigue failure. This form of failure is in contrast to machine parts failing from a static loading; in this case the parts develop large deflection because the stress has exceeded the yield strength. Thus, many static failures are visible ones and give advance warning. Fatigue failures give no warning; they are sudden, total and hence dangerous. It is relatively easier to design against a static failure since load conditions are better defined and understood. Fatigue is a more complicated phenomenon, and requires intensive design engineering effort to come up with a successful product.

13.2 DYNAMICS OF SHAFT MOTION

Transverse or lateral direction, vibrations of a shaft are analytically different from torsional and axial vibrations in several respects:

- The deformation of the rotor is due to a combination of bending and shearing, and is inherently more complex than simple stretching or twisting.
- Shaft bending is not confined to a single plane, and motion must be looked at in two orthogonal planes. Within each plane it is necessary to characterize the motion of each rotor station by two independent parameters, displacement and angular rotation.

- Gyroscopic and other asymmetric effects cause the motion of the shaft in the horizontal and vertical directions to be coupled.
- The location and dynamic characteristics of the bearings play a substantial role in transverse direction vibrations. This calls for development of detailed models of the bearings and their support structure.

Despite these complications it is possible to develop equations of motion for transverse vibration of the rotor that are analogous to torsional vibrations. To establish some important concepts, consider a simplified model of a high-speed rotor that retains many of the essential characteristics of the complex system in its response to unbalance.

A disk is carried at the centre of a flexible but massless shaft supported in two bearings (figure 13.1). Two types of bearing supports will be considered, rigid and flexible. In the model with rigid bearings the only damping c_f will come from the fluid medium surrounding the disk. Two types of unbalance must be considered separately, one due to eccentricity in the disk where the centre of the disk is offset from the elastic axis of the shaft by a radial distance q, the other due to shaft bow, s.

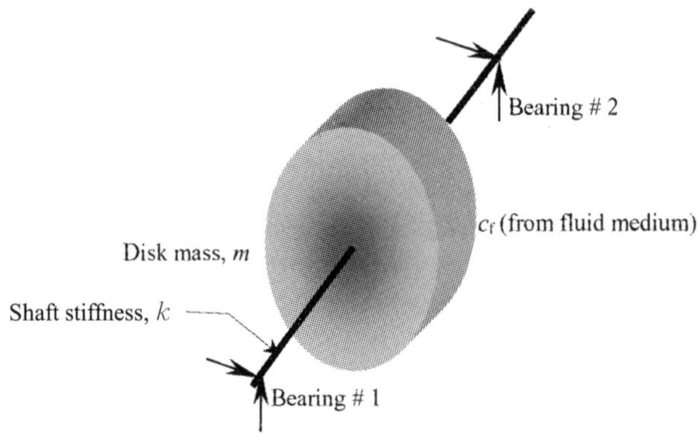

Fig. 13.1 Shaft Supported in Two Bearings

13.3 SUPPORT FLEXIBILITY AND CRITICAL SPEEDS

In the discussion so far the effect of mass distribution along the length of the rotor is absent, and so its usefulness for long rotors is limited. In addition to the fundamental, or first, critical speed, with distributed mass there will be an infinite number of critical speeds. Each critical speed has a characteristic mode shape. To understand this, consider a rotor of uniform cross section. Linn et al (1951) calculated the effect of bearing flexibility for the first four critical speeds of a uniform undamped rotor. The solution is derived from the differential equation of the elastic curve of the whirling shaft. The first critical speed on rigid bearings is given by:

$$\omega_1 = \pi^2 \times \sqrt{\frac{gEI}{WL^3}} \qquad ...(13.3)$$

where: g = acceleration due to gravity

E = shaft material modulus of elasticity

I = moment of inertia of shaft cross section

W = weight of rotor

L = length of rotor.

Figure 13.2 shows the impact of bearing flexibility on the first four critical speeds of a uniform rotor. A parameter representing support flexibility is plotted along the abscissa and the ratio of critical speed to first critical speed on rigid bearings along the ordinate. For a given value of rigid bearing, the first and second critical speeds continuously decrease with increasing bearing flexibility. This is accompanied by a gradual transformation of the mode shapes for these critical speeds. The numerical value between the first two critical speeds is equal to $\sqrt{3}$ for large values of bearing flexibility, the ratio obtained from theoretical considerations for rigid body modes. When bearing flexibility approaches infinity, these rigid body critical speeds approach zero. The third and fourth critical speed ratios initially decline rapidly with increasing bearing flexibility, then level out asymptotically to definite values. The third and fourth modes are referred to as free-free modes. This arises from the fact that conditions in the bearings at both ends have little impact on them.

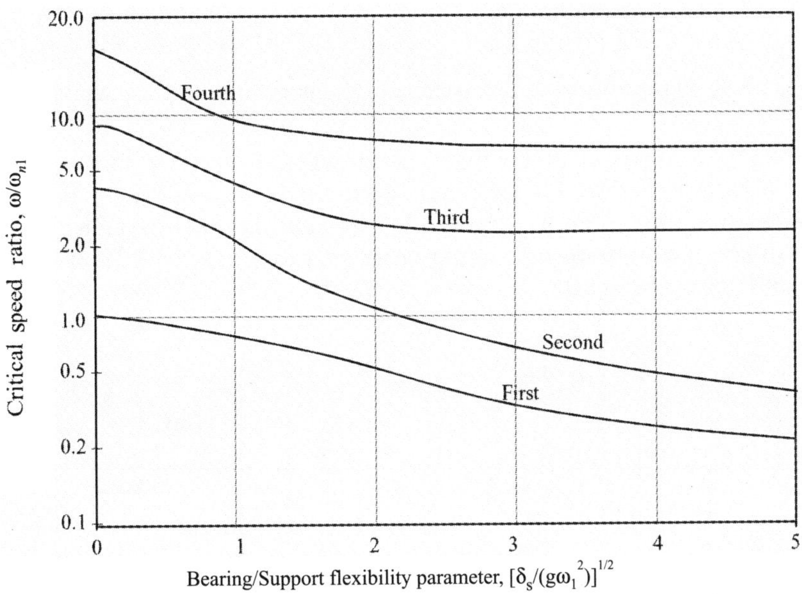

Fig. 13.2 Critical Speed as Function of Bearing Flexibility – Uniform Rotor

The location of the bearings along the length of a rotor will have a major effect on shaft response to unbalance. For instance, if the bearings were placed closer to each other and away from the ends to the nodal points of the first mode of the free-free rotor, as shown in figure 13.3, supporting structure damping will not be effective. Excessive response to unbalance will be experienced when operating near the first critical speed.

13.4 PROCEDURE FOR CRITICAL SPEEDS

The direct stiffness method or the transfer matrix method may be used for calculating free and forced response of a rotating system. Both procedures are developed using equations for force-

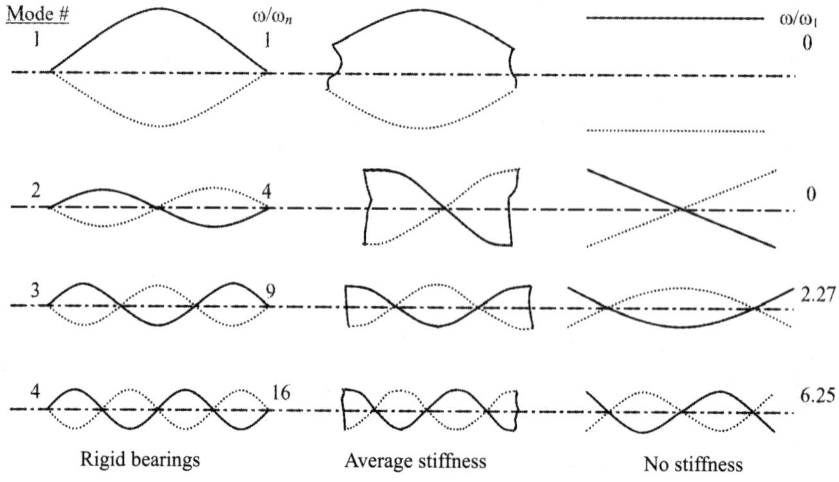

Fig. 13.3 Mode Shapes of Uniform Rotor as Function of Bearing Stiffness

displacement and force-velocity, together with momentum principles. The transfer matrix method, an extension of the Holzer method for torsional vibrations, is well suited for analysing rotating systems. Improvements to the system by Myklestad (1956) and Lund (1964) have made it suitable for computations of lateral vibrations of rotating beam structures. Figure 13.4 shows the dynamic model of a rotor for lateral vibrations.

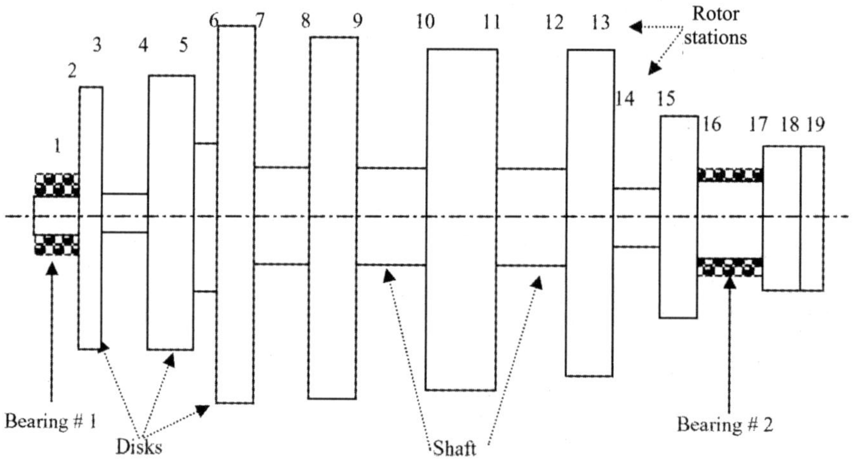

Fig. 13.4 Rotor Model for Lateral Vibrations

The method utilizes a sequential procedure, starting with boundary conditions at one end and proceeding along fixed locations in the structure to the other end. Elastic and inertia properties of the system are picked up as the calculation sequence proceeds for any shaft speed. Satisfaction of the boundary conditions at the other end for a specified speed value yields the natural frequency of vibration.

13.5 CRITICAL SPEED POSITIONING

Positioning of critical speeds in the operating speed range of a machine plays a major role in the design of rotors. A simple check of the undamped natural frequencies of vibration using proven support stiffness assumptions suffices in many cases. New machines under development and new applications call for a more detailed evaluation using appropriate computational methods. Evaluation of damped response to rotor unbalance is of special importance. Rotor instability as a source of excitation should also be explored.

The design of the rotor and bearing system must be carefully examined for forced synchronous vibrations resulting from residual unbalance. Residual unbalance may be due to a lack of coincidence of axis of the mass centres with the elastic axis of the shaft. It could also be the result of a bow in the elastic axis, or due to a combination of the two factors. Rotor bow often results from a temporary lack of symmetry in the temperature distribution in the rotor. Steam turbine rotors used in power generating plants have been known to end up with a permanent bow at start-up or shutdown times if the machine is not permitted to soak the temperatures at the proper speed level and time period. The soak cycle is required since the thickness of the shaft and length between bearing centrelines is large. A similar effect is encountered if splines or couplings are not properly fabricated and assembled. Although differences exist in the response curves away from critical speeds for the two types of unbalance, the importance of positioning of the critical speed in the operating speed range of the machine cannot be overstated.

The discussion will be illustrated for a single shaft two bearing gas turbine shown in figure 13.5. All major components of the system will be assumed to be axially symmetric. For proper positioning of critical speeds in the operating speed range of the machine, stiffness and inertia characteristics of both the rotor and the bearings and its supports at the right axial location must be known. The machine stator is generally more rigid than the rotor in large industrial gas turbines, and is not included.

The modes of vibration of the system may be divided into two parts, one where the rotor remains mostly rigid, and the other where the rotor undergoes flexure and whirls about the unbent axis.

There are two rigid body modes of the machine. In the first or fundamental mode the rotor supports move in phase with one another, and the shaft axis describes a cylindrical motion. In the second rigid body mode at a higher frequency the supports move out of phase with one another, with a node near the midpoint of the rotor. The rotor centreline describes two cones with their apex at the node. Usually these frequencies occur well below the frequency of the shaft flexural mode. If the stator mass is very large compared to the rotor mass, low rigid body mode frequencies will prevail, even with stiff mounts. When the rotor is operated at a speed at which these rigid body modes are excited, the rotor and stator will move in phase with little relative motion between them, assuming that the bearings are adequately stiff. With limited motion between rotor and stator, internal structural integrity of the machine will not be affected. There may, however, be undesirable transmission of vibrations to the surroundings.

From design considerations the rotor's flexural modes are of greater interest, and sometimes of concern. At each flexural mode the rotor whirls with a characteristic deflection shape, and the rotor and stator can vibrate out of phase. Thus, the potential for rotor rubbing against the stator exists when machine operation occurs at, or while passing through a rotor flexural mode. Rubbing of seals and other closely controlled clearance areas may take place, with a resultant loss in efficiency and functionality. In extreme cases major damage to the inside parts of the machine may occur, and in some cases to the supporting structure as well.

In order to define typical options available for the positioning of the rotor's flexural frequencies that are applicable to many different machine types, some examples will be discussed.

First, consider a single shaft, two bearing gas turbine where the operating speed range includes the first flexural rotor mode of vibration of the type shown in figure 13.5. The rotor span between the bearings when evaluated in rigid bearings has its lowest critical speed at 4,474 rpm, and the unit is required to operate up to a maximum speed of 6,200 rpm. The rotor is supported on journal bearings attached to a stiff and massive stator. The ratio of stator mass to rotor mass is 8. Machine stator is supported on a structural steel framework that is determined to have a natural frequency of 1,200 cycles/min. Damping in the steel framework is assumed to result in peak amplification factor ($1/2\xi$) of 10, implying a value of 5 per cent of critical damping. The rotor may be considered detached from driven machinery as far as lateral vibrations are concerned. The first step in the design process is to calculate the undamped critical speeds, and is determined to be 3,890 rpm for the first mode and 7,580 rpm for the second mode. Of special significance is the response to rotor unbalance, especially since machine operation takes place through the first critical speed. The shape of the modes dictates that unbalance at mid span portion of the rotor will be of interest for the first mode. Unbalance located at the ¼ and ¾ span length from either bearing will result in high deflections in the rotor when operation of the machine is near the second mode of vibration. Unbalance or misalignment at the flexible coupling will also be sensitive in the second mode. Coupling unbalance, however, will have little effect in the first mode. If operation were required through the second critical speed on this machine, then special attention would be needed to balance the coupling. Unbalance here could cause excessive whirl at the overhung coupling, together with high dynamic loads in the bearing at the higher operating speed. Rotor response at the location of unbalance is shown in figure 13.6.

Next, consider the case of a power generation steam turbine rotor with two spans and a total of four bearings, connected by a spacer type of coupling (figure 13.7). Individual spans will have their own stiffness and inertia properties, giving rise to a number of critical speeds, each critical speed characterized by a dominant amplitude in the span of the next higher stiffness. The journal bearings are supported in sturdy pedestals anchored to a massive concrete foundation. Due to the substantial characteristics of the mount, the effective spring and damping constants will be due to the journal bearing's lubricant film.

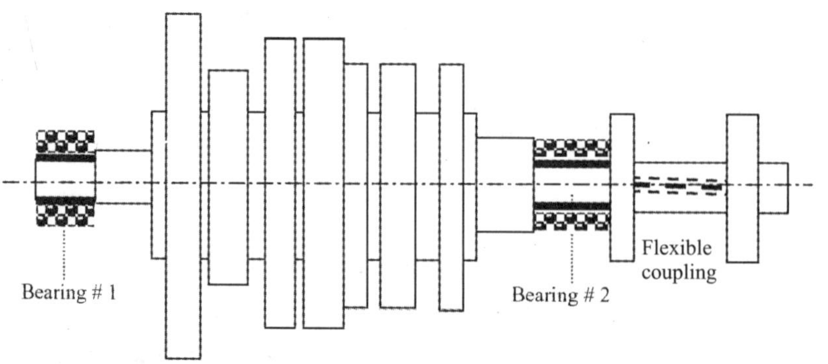

Fig. 13.5 Gas Turbine Rotor

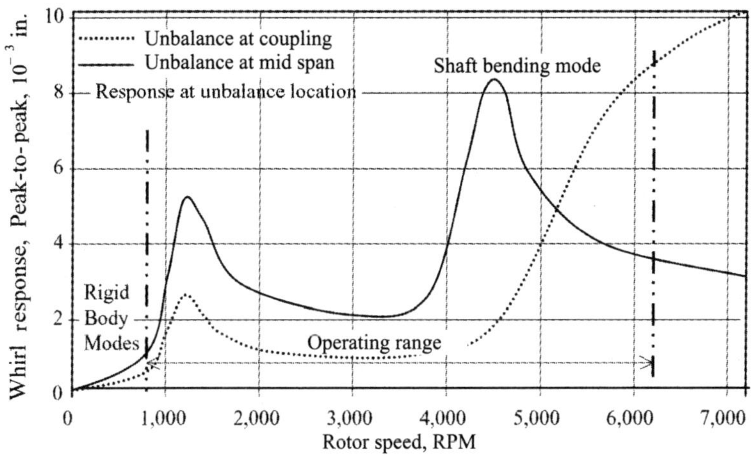

Fig. 13.6 Calculated Rotor Response due to Unbalance

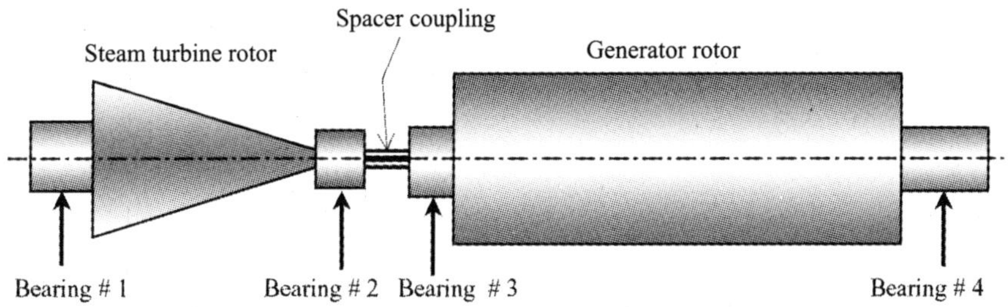

Fig. 13.7 Steam Turbine/Generator

The first three undamped critical speeds with bearing constant $K = 2.5 \times 10^6$ lbs/in are 1,550, 090 and 4,550 rpm. The spacing between the second and third critical speeds is large enough to commodate the constant running speed of 3,000 rpm needed for 50 Hz power generation, or of 600 rpm for 60 Hz electric power. Note that most electric power equipment manufacturers prefer margin of 10 per cent above and below the operating speed for vibration free operation. orresponding mode shapes are plotted in figure 13.8.

Rotor response due to unbalance will depend on the location and amount of unbalance. nsitivity to unbalance is related to the fact that the generator rotor has low stiffness relative to the pport stiffness provided by the journal bearings and the associated supporting structure. The cond critical speed is responsive to unbalance in the turbine span. To significantly excite the third tical speed requires a moment unbalance in the generator.

3.6 SYNCHRONOUS RESPONSE

nchronous response of a rotor at critical points that cannot be measured directly can be accurately ferred from measurements taken elsewhere in the machine. Accurate estimates of actual levels of oration at close clearance points on the shaft such as seals assists machine operators in

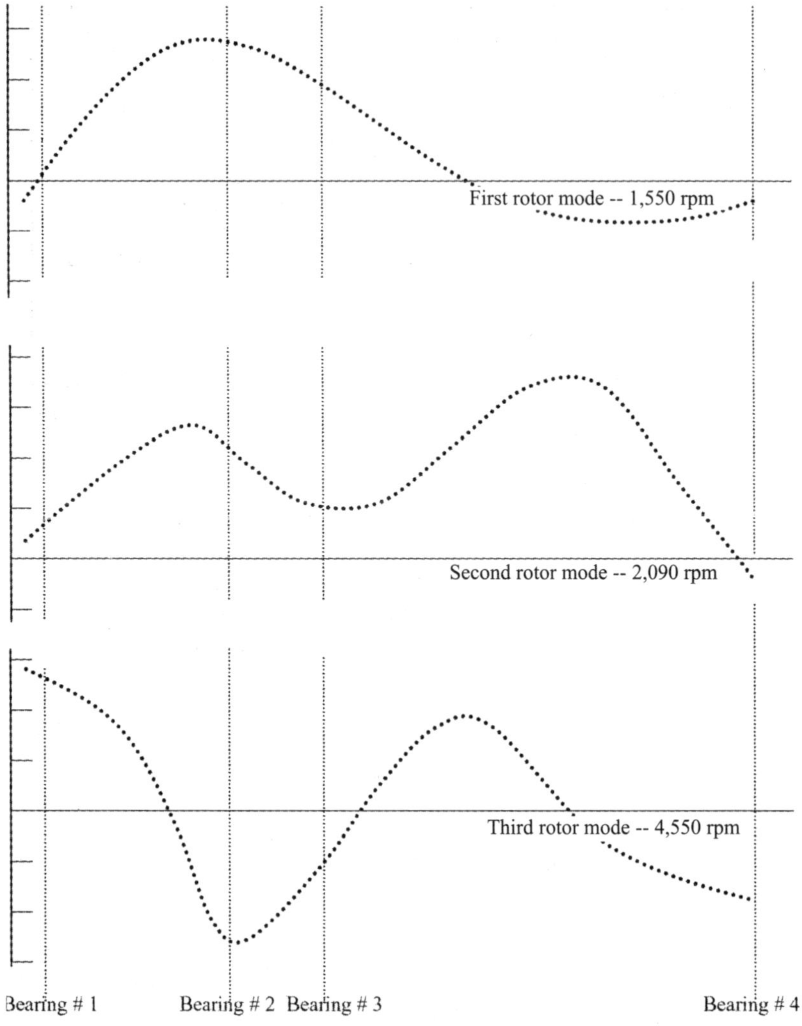

Fig. 13.8 Mode Shapes for Steam Turbine Rotor

troubleshooting and in protecting the equipment from expensive damage when the disturban forces acting upon the rotor are predominantly synchronous. A few sensors can be used to identify relatively large number of well-characterized synchronous forces, such as from mass unbalance if th measurements are taken at many different speeds (*ref*: Vance, 1988). Uncertainty in the rotor mod can substantially degrade this capability (*ref*: Sortore, 1999). But the utility of the estimation proce is also dependent on a reliable bound on the error in the estimate, and can be obtained if the size the model errors and forces acting on the rotor are available.

The response estimate is constructed using measurable response in conjunction with estimator gain matrix derived from a model of the transmissibility of the rotor system. There a two main components influencing the estimation, the nature of the dominant unknown excitin forces and the transmissibility between the forces and the measurable and non-measurable respon oints. Measured response y_m will typically be at discrete points determined by a proximity probe or n accelerometer, while non-measurable response y_u may be at selected points where, for example, aterial strain is required because of fatigue concerns. The frequency dependent transmissibility, or ansfer, functions between the responses y_m and y_u and the disturbing input forces P_m and P_u can be xpressed as:

$$\begin{Bmatrix} y_m(\omega) \\ y_u(\omega) \end{Bmatrix} = \begin{bmatrix} P_m(\omega) \\ P_u(\omega) \end{bmatrix} f(\omega) \qquad \text{...(13.4)}$$

The transmissibility coefficients P_m and P_u must be identified through a combination of odeling and system identification. The non-measurable displacements may then be computed as:

$$y_u(\omega) = P_u(\omega)[P_m(\omega)]^{-1} y_m(\omega) \qquad \text{...(13.5)}$$

ssuming $P_m(\omega)^{-1}$ exists. The unknown exciting force can also be identified as an intermediary step:

$$f(\omega) = [P_m(\omega)]^{-1} y_m(\omega) \qquad \text{...(13.6)}$$

In general, at a given speed only a subset of f can be estimated, because the dimension of f is reater than that of y_m, or only those forces lying in the domain of $P_m(\omega)$ produce the measurable y_m. orces that cannot be estimated through y_m can still produce a part of the response y_u, which portion f the response fundamentally cannot be estimated. By assuming the unknown forces to be egligible, however, a limit may then be imposed on the forces acting on the rotor, and a eaningful bound may also be placed on the y_u.

Missing in this formulation is the relative scales of the components of f and the relative nportance of errors in y_u. As a dimensioned measure the bound is hard to interpret, hence a better ay is to scale the vectors f and e to encapsulate physical limits on the elements of f and engineering equirements on residual error vector e. For instance, specifications for balancing a rotor typically mit the magnitude of allowed gram-millimetres of unbalance at each balance plane, and this may e converted to unbalance force at a given speed. In a similar manner, target limits on error in stimation will reflect how precisely the measurements need to be known to be useful. So if aximum rotor motion at a seal needs to be of the order of 25 μm, then the corresponding bound n the error might be set at 10 μm.

To illustrate the procedure, consider a large turbo-machine rotor used in a single stage entrifugal pipeline compressor shown in figure 13.9, indicating locations for vibration sensors and otentially close clearance points (*ref*: Maslen, 2000). The free-free natural frequencies of the rotor at zero speed) and modal parameter uncertainties arising from bearing stiffness and damping oefficients are shown in table 13.1. It is postulated that a strong localized model uncertainty and/ r a large localized excitation needs to be estimated at the balance piston seal, which may impose onsiderable destabilizing forces on the rotor. If unaccounted, this aspect can introduce a ubstantial difference between the actual and modeled dynamics of the rotor.

The results of the analysis of the rotor with a cross-coupled seal acting at the critical location are hown in figure 13.10. The maximum performance (e_{yu}) error for clearance # 2 is provided, but ctual errors are likely to be smaller. The upper bound in the error estimation ($e_{yu\mu}$), calculated using analysis if the model is not perfect, is also shown. The magnitude of the cross-coupled stiffness oefficient is increased linearly with frequency from 0 at standstill to 1.751×10^7 N/m at 7,000 rpm. The peaks in the plot arise from the uncertainty in the lightly damped higher frequency modes. Of

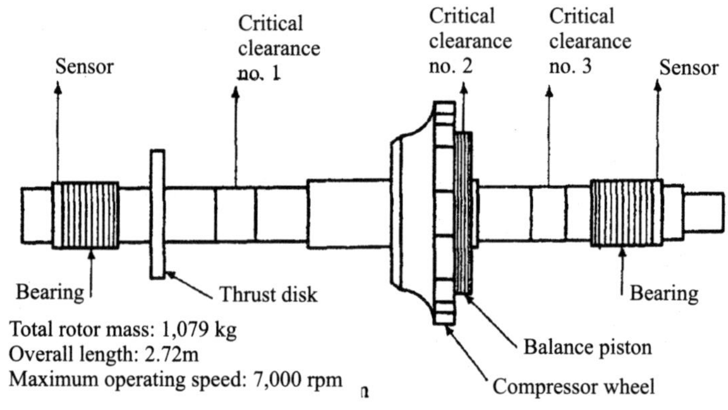

Fig. 13.9 Compressor Rotor (Maslen, 2000)

Table 13.1 Rotor Frequencies and Modal Uncertainties

Mode	*Natural Frequency, Hz*	*Error, per cent*		
		Frequency	*Damping*	*Mode Shape*
1	0	2	3	0.04
2	0	2	4	0.06
3	105.0	3	5	0.08
4	224.8	3	7	0.10
5	442.6	4	9	0.12
6	649.0	4	11	0.14

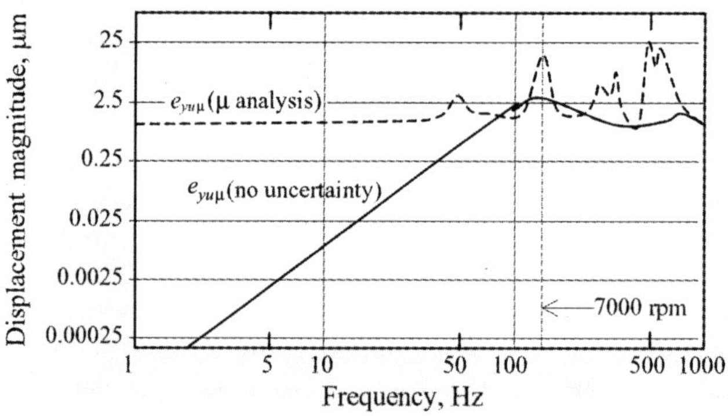

Fig. 13.10 Maximum Performance and Estimation Errors (Maslen, 2000)

particular interest is the visible split peaks in the error bound associated with the higher flexible modes. The peaks occur because of the presence of flexible modes on either side of the nominal value. Despite the combination of large potential unbalance and substantial model uncertainty, the estimator error bound is acceptable.

13.7 COMPONENT MODE SYNTHESIS OF LARGE SYSTEMS

When the total number of degrees of freedom of a dynamic system is large, analysts often resort to the modal method where the system is represented by its free vibration modes. The advantage of using the modal representation consisting of a few low frequency modes accrues from the reduction in the size of the associated problem, and is popular in problems such as bowed shaft analysis (*ref*: Choy, 1978) and transient behaviour of dual rotor aircraft engines (*ref*: Dennis et al., 1975). In an improved form of this method the dynamic structure is partitioned into a number of smaller sub-structures, with modal characteristics of each unit derived analytically or from vibration testing (*ref*: Klosterman, 1971). The total structure is then constructed from a reduced number of component modes from each sub-structure.

Component mode synthesis is used extensively in the aerospace industry to calculate undamped natural frequencies of large airframe structures. The sub-structures are required to share common degrees of freedom at the common interconnecting points to obtain a set of constraint equations for the reduced modal equations of motion. In rotating systems the interface between sub-structures consists of bearings and seals, and the supports between the engine casing and the attachment points are also elastic. If a common degree of freedom is not shared by the sub-structures, flexible connections may be introduced in the form of generalized damping, stiffness and force into the modal equations. Together with the rotor's free-free modes the undamped modes of the rotor and casing system are then obtained using the procedure. The system modes aid in developing the system modal equations with damping added.

Linearly elastic non-rotating structures yield a symmetric stiffness matrix, but the rotating subsystem stiffness matrix is asymmetric because of internal friction and asymmetric bearing stiffness coefficients. The symmetric and asymmetric parts may be separated, with only the symmetric portion used in the evaluation of undamped component modes. The asymmetric portion is added to the modal equations at a later stage. This approach provides a choice of evaluating only the free-free rotor modes containing the shaft stiffness or the shaft and bearing modes containing the shaft and principal bearing stiffness values. The shaft's flexural stiffness is then implicit in the mode shape and conveniently expressed by the eigenvalues.

The partitioned matrix equation has diagonal and square sub-matrices, with the dimension equal to the number of normal modes employed in the corresponding subsystem. The off-diagonal matrices represent the coupling of the subsystems from the interconnecting elements. Internal friction is represented by the dry friction loss angle and the viscous friction coefficient. The linearly elastic elements omitted in the undamped mode calculation contribute to the coupling stiffness and damping matrices, and consist of bearing stiffness and damping, gyroscopic moments and rotor acceleration. The linearly elastic elements may be generalized into four distinct linking categories, or their combination, with each link containing stiffness and damping coefficients:

- Direct links representing a linear element between the degrees of freedom in the subsystem and the stationary ground.
- Cross-coupled links between the coupling force at a degree of freedom in one subsystem and the motion at a degree of freedom in another subsystem.
- Direct/intermediate links consisting of an elastic element between the degrees of freedom in one subsystem and the degrees of freedom in another subsystem.
- Cross-coupled/intermediate links developing equal and opposite forces at various degrees of freedom between two subsystems due to the coupling effect of the relative motion. An

example of this is the cross-coupled dynamic coefficient of a journal bearing installed intermediately between two concentric rotating shafts.

The total effect from the linear linking of the elements is the sum of the individual contributions. Thus, a fluid film bearing mounted in a rigid housing may be represented by two direct links and two cross-coupled links between the shaft and the ground. Gyroscopic forces associated with a rotor represent additional modal contributions, and may be dealt with in a similar manner using modal contributions. Modal forces are related to actual forces from unbalance, disk skew or non-linear forces acting at nodes in accordance with the mode shapes.

Thus, the governing equations of a complex rotor system may be expressed in the generalized coordinates by adding up the contributions of all the flexible links to the basic uncoupled modal equation. After establishing this, critical speeds, stability, forced or transient response may be solved in the modal coordinates, and the final results then transformed back into real coordinates. In non-linear transient analysis it is necessary to transform back and forth between the real and modal coordinates during computation of the non-linear forces.

This scheme of computations is illustrated for a two-spool aircraft gas turbine engine (*ref*: Li, 1981), shown in figure 13.11. The engine consists of an inner, or power turbine, rotor supported by two main bearings placed close to the end extremities. Two differential bearings connect the gas generator and the power turbine rotors. In addition, the gas generator rotor has three primary bearings, with a squeeze film damper incorporated at the third bearing to reduce vibration amplitudes and bearing forces. Component modes of the rotors are obtained independently of each other by omitting the gyroscopic effects of the disk, squeeze film damper and differential bearings. A total of 12 undamped component modes containing two gas generator modes and four power turbine modes in the horizontal and vertical lateral planes below 30,000 rpm are used.

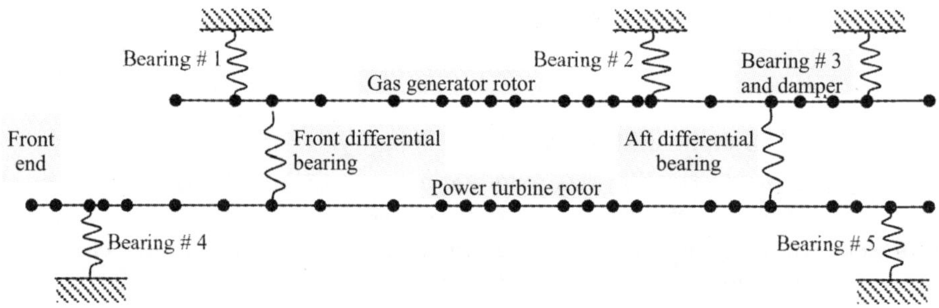

Fig. 13.11 Dynamic Representation of Two-Spool Aircraft Engine (Li, 1981)

The first portion of the computation calls for transient response of the system due to a suddenly applied unbalance, as may occur in the event of loss of part of the blades. Both rotors are assumed to be initially balanced when the unbalance distribution is applied, with the gas generator operating at 15,000 rpm and the power turbine at 17,000 rpm. The unbalance comprises of 25 g-cm in the 2nd stage turbine and an equal amount at the rotor mid-span 180° away. With the assumption of linear system behaviour, figure 13.12 provides the orbits at the 2nd stage gas turbine for the first 10 cycles of simulation. The steady state orbit is not fully developed even after 40 cycles, but the magnitude and phase angle of the vibration vector can be distinguished. A comparison with an independent steady state response calculation indicates close correlation. The maximum response amplitude at the start of the transient motion is about double the size of the steady state orbit.

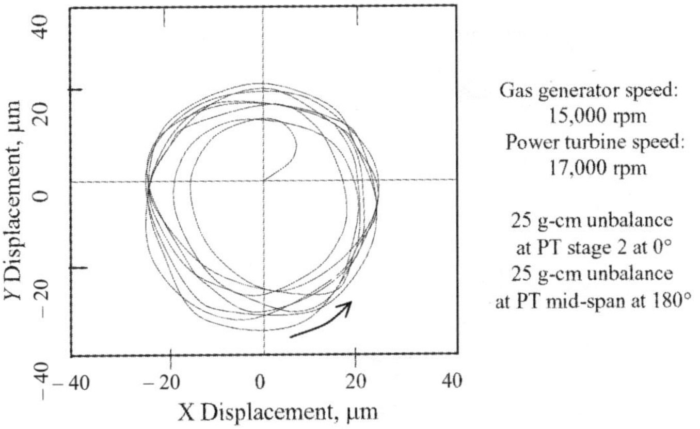

Fig. 13.12 Transient Orbit of Gas Generator Disk (Li, 1981)

The squeeze film damper employed at bearing # 3 is deficient in terms of design requirements, and a non-linear jump is observed during vibration tests of the engine. The damper is 11 mm long, 129 mm in diameter and has 0.1 mm radial clearance. A retainer spring of 2.15×10^7 N/mm is used to centre the outer raceway of the rolling element bearing. The diaphragm supporting the squeeze film damper and bearing is assumed to be rigid. Bearing force is calculated using the short bearing theory in the simulation. The second portion of the computation has the power turbine rotor operating at 14,000 rpm with an unbalance of 50 gm-cm at the turbine end. The gas generator rotor has the same amount of unbalance at the 2nd stage turbine. The speed of the gas generator rotor is increased from 9,000 rpm in steps of 2,000 rpm until the jump phenomenon occurs. At the start of the transient analysis the motion is set at zero displacement and velocity. Non-linear response is calculated for each speed for 10 gas generator rotor revolutions. Starting conditions at rotor speeds other than the initial speed are taken to be equal to the final displacement, and velocity is computed at the end of the simulation cycles for the previous gas generator speed. The transient response of the system is computed due to the suddenly applied unbalance in the rotor under steady speed operation.

Figure 13.13 provides the transient orbits of the gas generator rotor at the damper location as the speed is increased until the jump phenomenon is encountered. The initial transient motion at 9,000 rpm is shown in (*a*), with the orbits calculated for about 9 turns after release from the center of the bearing. The timing mark is synchronized with the unbalance vector, and is shown as an asterisk at the end of each revolution. Because the two rotors are operating at different speeds, the unbalance forces have two frequency components. Consequently, a response pattern repeating itself for every revolution is not observed even when the system reaches steady state operation. In (*b*) the non-linear motion at the damper location occurs at 13,000 rpm. Limiting orbits of large amplitude are formed at a rapid rate, in less than one revolution of the gas generator rotor. The damper is now operating at eccentricity ratios in excess of 0.9 of the radial clearance, indicating overloading of the damper bearing. The phase angle lag fluctuates around 45° throughout the simulation at this speed point. Finally, in (*c*) the non-linear jump occurs at the gas generator speed of 17,000 rpm. The orbit of the rotor is reduced to about a quarter of its initial size within 5 revolutions. The shift in the amplitude is accompanied by a shift of the timing mark, and approximately aligns itself with the

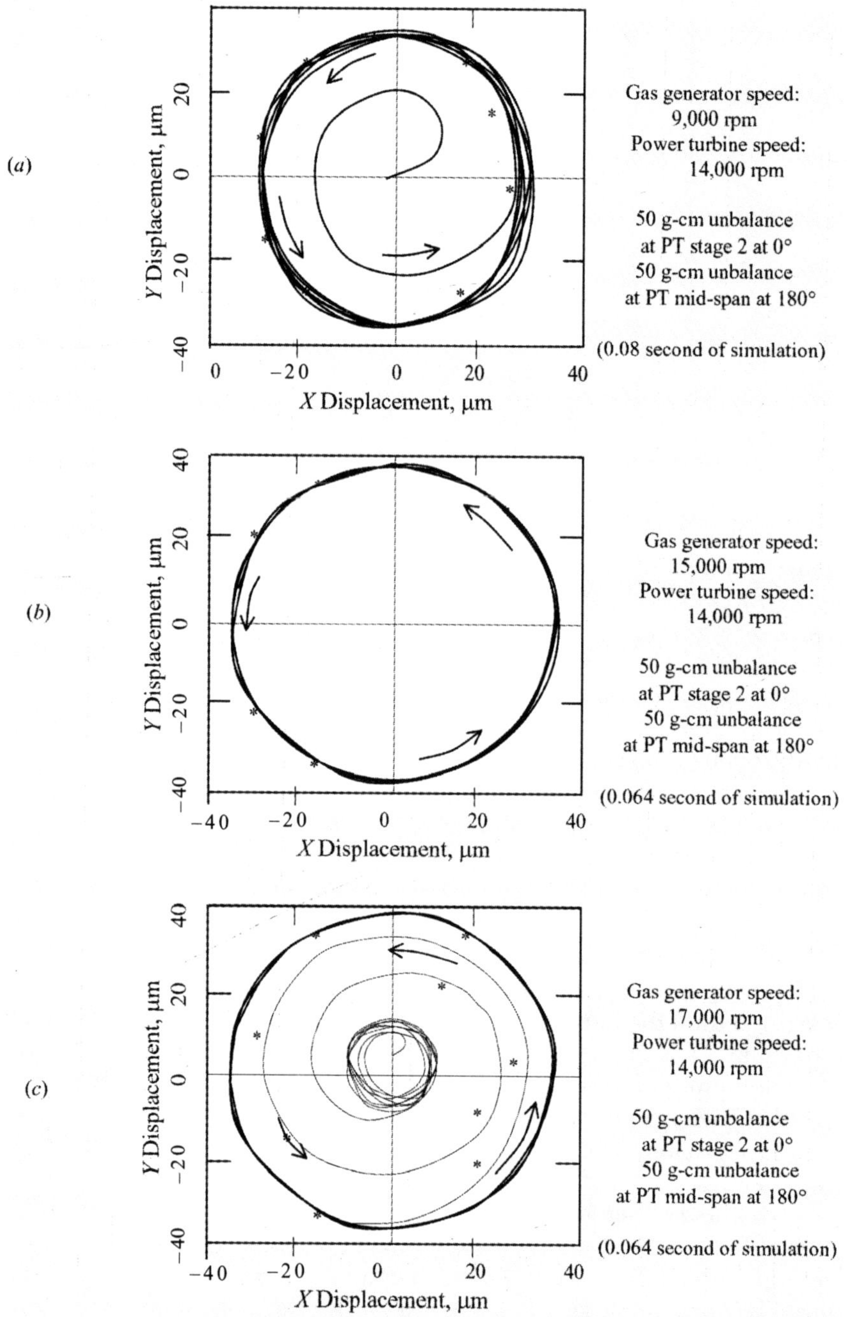

Fig. 13.13 Non-linear Rotor Orbits at Squeeze Film Damper Bearing (Li, 1981)

negative x-axis. The latter change is significant since it indicates that the gas generator rotor has rapidly passed through the critical speed of the gas generator rotor mode due to the sudden reduction in the effective damper stiffness and the consequent smaller orbit size.

13.8 ROTOR SUPPORTED IN FLEXIBLE CASING

In many turbo-machines the rotor operates above one or more critical speeds, hence the unbalance response at resonance is of major significance. In the early phase of the design, guidelines from parametric studies prove useful. The unbalance chart shown in figure 13.14 provides some useful results in this respect (*ref*: Balda, 1975), illustrating the interaction between critical speed ratio ω/ω_0 (ω_0 representing the frequency of a simply supported rotor), response factor a_d/e_d (amplitude/eccentricity) and bearing stiffness and damping ratios K/C and H/K. However, aircraft engine rotors are enclosed and supported in light and flexible casings, consequently the vibration pattern of the rotating system is affected by the stiffness and mass characteristics of the case and its supports. In analysing the dynamic aspects of the power plant it is necessary to include the casing and support structures in addition to the rotating assembly and the bearings. The first two critical speeds and response factors due to unbalance generally will vary widely with changes in casing mass and bearing and support stiffness, and the damper plays a useful role in reducing the vibratory response in the modes. But the third critical speed and response factor are little affected by the damper (*ref*: Miyachi, 1983).

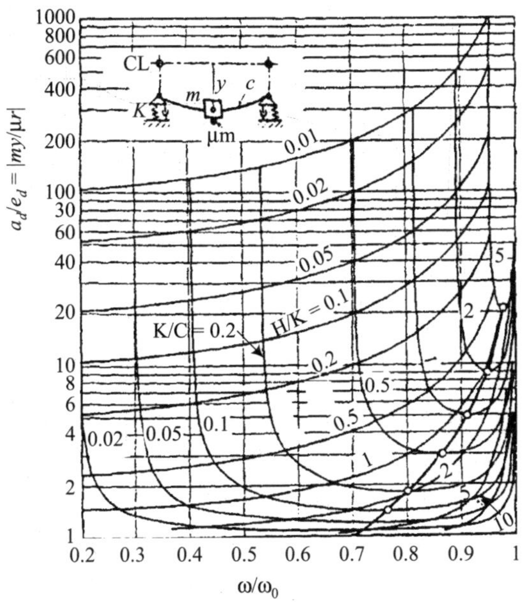

Fig. 13.14 Unbalance Response for Single Degree of Freedom Rotor (Balda, 1975)

To understand the impact of the flexible casing and supports, consider a uniform shaft with a disk of mass m_d and mass moment of inertia I_d mounted at the centre of the span, two rolling element type bearings at the ends, a casing and its supports. The ratio of the mass of the casing and the rotor is M_c, bearing to shaft stiffness ratio is K_b and support to shaft stiffness ratio is K_s. Since rolling element bearings offer poor vibration damping capacity in comparison to oil film supported sliding bearings, a squeeze film damper type of damping in the right bearing is considered in this evaluation.

The eccentricity of the disk e_d will be used to obtain the exciting force. The shaft is represented by m_j and I_j concentrated at its ends, m_j being one tenth of m_d, and I_j being one hundredth of I_d. Bending stiffness of the rotor EI_r is constant over the span. The bearings are supported in the casing by struts, while the casing is mounted to the supporting structure, connected to fixed locating points. The mass of the casing m_c, as also its mass moment of inertia I_c, are distributed half at the center and quarter at each of the two ends. Bending stiffness of the casing EI_c, assumed to be ten times that of the rotor, is constant over its length. To simulate the effects of light and heavy engines the mass ratio of casing to the rotor is varied from 0.5 to 20. Rejection of secondary effects of mass and bending stiffness of the casing is achieved by letting the bearings and casing mounts to be positioned at the same axial coordinates. Detailed description of properties such as axial distribution of the casing and rotor mass may affect the vibratory characteristics at higher speeds, but this evaluation is intended to focus on the lower modes.

The transfer matrix method for the forced vibration of the rotating beam is used, with the unbalance force arising from the eccentricity of the disk. Figure 13.15 provides the mode shapes for the model with and without a damper. Bearing stiffness ratio K_b is 0.5, casing support stiffness ratio is 2.0 and casing mass ratio M_c is 5.0. Without the damper the casing sees little participation in the first two modes. When the squeeze film damper is included in the right bearing the mode shapes tend to become asymmetric. Viscous damping coefficient of the squeeze film damper c_q is assumed to be optimum for ω_o, or equal to k_b/ω_o. With the damper, in the first mode the rotor and the casing move in phase, but in the second mode the two components move in reversed phase. In the third mode rotor extremities vibrate in inverse phase with the disk, and the amplitude of the rotor end and relative displacement between the rotating and stationary parts at the bearings is much larger

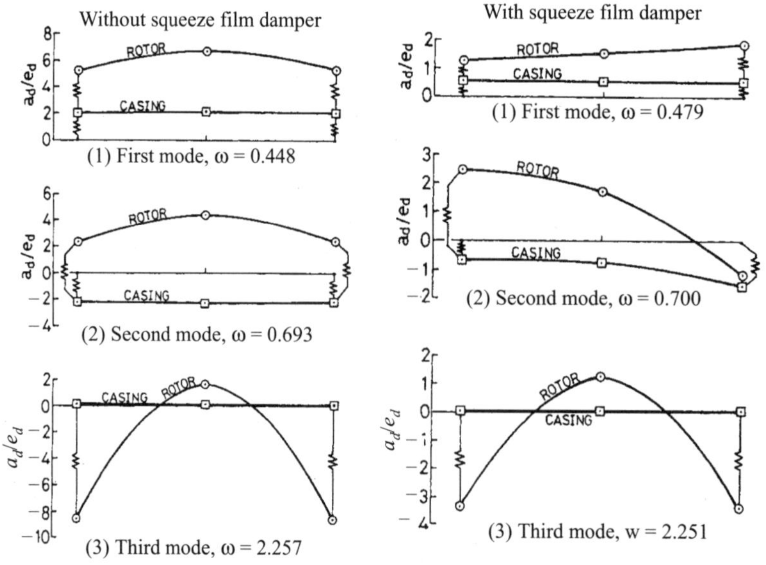

Fig. 13.15 Mode Shapes (Miyachi, 1983)

than the amplitudes of the center disk and the case. The dynamic load in the bearing may be expected to maximize in this mode, in spite of the low activity at the disk location. The effective stiffness of the bearing appears to be reduced by the damper. The phase angle of vibration cannot be clearly split into positive and negative, but the two dimensional display defines the intermediate phase angle simply as above or below the reference line. Thus, the relative displacement between the rotor and casing at the right bearing in the second mode appears small, but the difference in the phase angle at this location is 70°, making the actual relative displacement considerably larger.

Figure 13.16 provides the unbalance response without the damper, with speed ranging to 5 times the fundamental natural frequency. The disk experiences three peaks of vibratory motion, the higher first two being about equal in magnitude. In the absence of the damper the structural damping of the casing dissipates the dynamic energy, the supports and the shaft. Peak vibratory motions do not coincide at the same shaft speed for all the structural members.

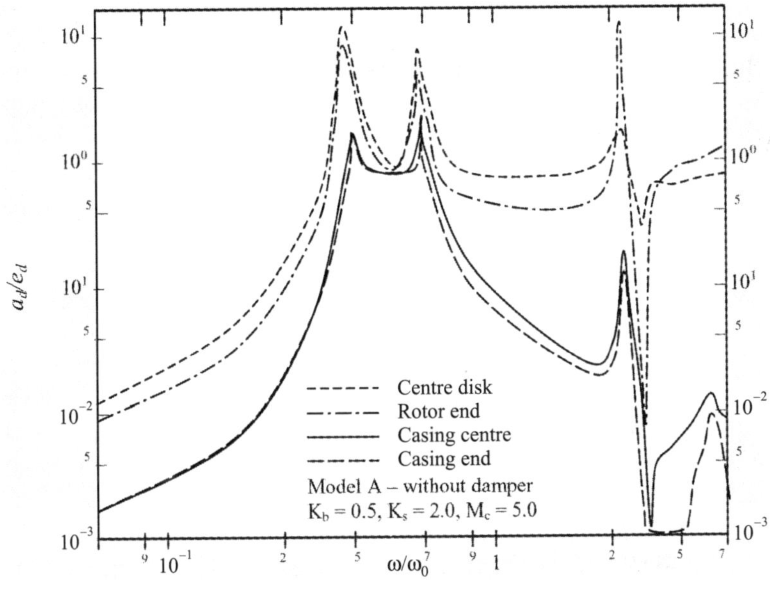

Fig. 13.16 Unbalance Response Without Squeeze Film Damper (Miyachi, 1983)

The squeeze film damper is effective in reducing the unbalance response at the critical speeds, especially in the region where response is high. Maximum amplitude is reduced by 50 to 80 per cent of those experienced without the damper. For another indication of the unbalance response, consider the dynamic load on the bearing, normalized by half the centrifugal force due to the eccentricity of the disk for all three modes. Table 13.2 provides a comparison of the bearing loads.

13.9 DRY FRICTION RUBBING BETWEEN ROTOR AND SEAL

Rubbing between the rotating and stationary parts of a rotating machine is a serious malfunction, holding the prospects of a catastrophic failure. The rub represents a secondary effect, caused by excessive lateral vibrations of the rotor or high displacement of the centreline of the rotor due to

Table 13.2 Peak Bearing Loads

Mode #	*Without Damper*		*Squeeze Film Damper*	
	Frequency Ratio	*Bearing Load*	*Frequency Ratio*	*Bearing Load*
1	0.448	17.3	0.479	5.1
2	0.693	8.5	0.700	4.7
3	2.257	31.2	2.251	9.8

misalignment or radial forces arising from the flow of the working fluid. The trend to increase operating efficiency, and also to improve stall margin in the case of a compressor, leads to minimization of the clearances at the blade tip and inner seals, which in turn increases the hazard of rubbing.

There are two types of rub. In the light and partial rub the contact forces are minimal, with the rotor occasionally touching the stationary elements during a fraction of its precessional period. A more serious phenomenon in its destructive abilities is a full annular rub that may occur in seals and bearings. The rotor maintains contact with the seal almost continuously. Self-excited reverse precessional full rubs, known as dry whip, are more severe than synchronous full annular rubs.

The synchronous rub regime is marked by interaction from the seal, with the latter acting as an additional bearing to yield a coupled rotor/seal system with increased stiffness and natural frequencies. The reduced friction force permits the shaft to slide along the seal's surface without rolling at a particular speed. Figure 13.17 shows a rotor's horizontal (X-direction) and vertical (Y-direction) response during a run up and down sequence with and without rubbing. Differences in the angular acceleration cause the run up and down curves to differ slightly without contacting the seal. The seal has 500 mm clearance, and 3.3 g-cm unbalance is attached to the disk. The resonance range of speeds is wider during the run up than during the run down. In the range of original resonance the response reduces with rub at the seal since it acts as a bearing, even though the shaft is not in continuous contact with the seal. The orbit (shown as insets in the figure) indicates the rotor is bouncing inside the seal, but it maintains the forward precession. As the rotor speed increases or decreases, the operating condition changes from non-contact to contact and then to non-contact again, and is accompanied by jumps in the amplitude. A sudden decrease in the 1 × amplitude, accompanied by an onset of other frequency components, is observed in the X-direction when the rotor just contacts the seal during run down to indicate a hammer impact form of external disturbance. A jump-up is observed in the Y-direction during the run down, and is caused by insertion of lubrication oil on the contact surface to avoid severe reverse precession rubbing. Unlike the case in the X-direction where the contact surface is not fully lubricated, the rotor's orbit in the Y-direction is smooth, suggesting the shaft's bounce pattern is considerably influenced by the friction coefficient.

Reverse precession full annular rubs can occur when the ratio of radius to radial clearance ranges between 10 and 40 without an external disturbance. In the neighbourhood of a resonance, synchronous forward precession rub turns into the reverse direction when friction at the rotor and seal surface is sufficiently high. Once generated, the reverse precession rub is sustained over the whole speed range as long as slippage is maintained at the contact region. A reverse rub can be disengaged at a non-resonant speed. Figure 13.18 provides the transition from the synchronous forward precession rub to the reverse precession type without an external disturbing force. As speed

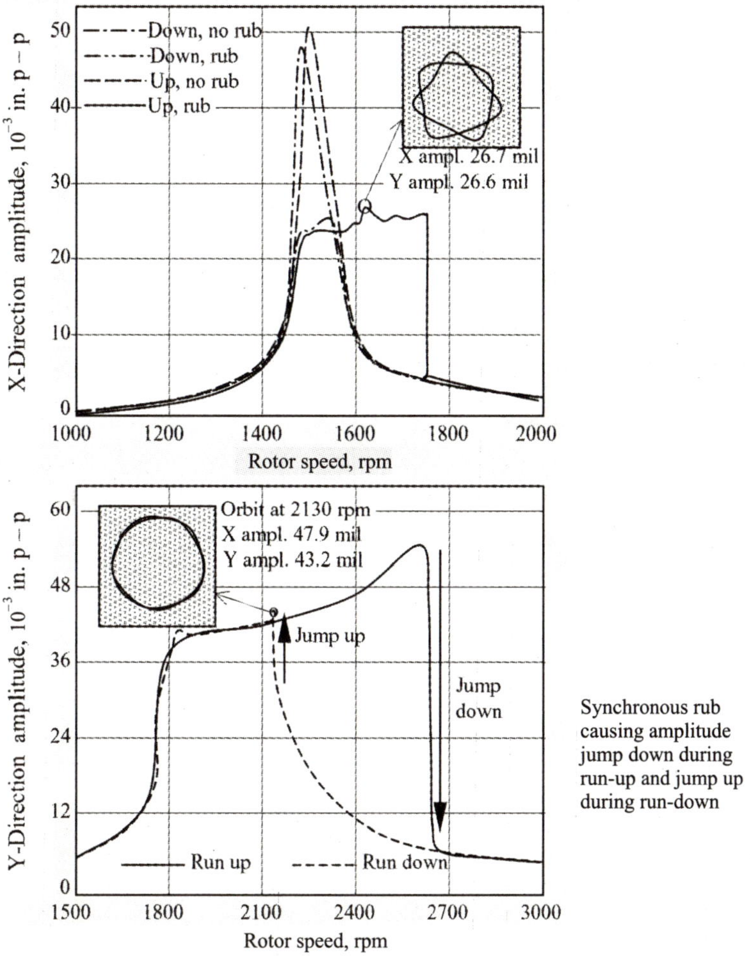

Fig. 13.17 Rotor Lateral Response With (*upper*), Without Rub at Seal (*lower*) (Yu, 2000)

approaches the first resonance, the rotor starts bouncing inside the seal as it develops elliptical orbits gradually rotating their major axes due to lateral stiffness anisotropy effects. Intermittent contact at multiple points also appears at this stage. Reverse-dominant precession initiates between 1550 and 1650 rpm, and around 1700 rpm suddenly the full annular rub appears at a frequency of 2300 cpm. Considerably higher vibration amplitudes occur in the reverse mode at –2300 cpm than in the forward frequency at 2300 cpm, also indicating an almost circular reverse orbit motion. Further increase in shaft speed does not affect the rubbing frequency, though the amplitude increases slightly due to thermal growth of the seal.

The reverse precession full annular rub can also trigger during the run down after the rotor contacts the seal close to the resonance speed, as shown in figure 13.19. The rub is maintained until shaft speed is low, and replaces the synchronous 1 × motion. When the applied unbalance is reduced to 1.2 g-cm for the same experimental setup, the reverse precession does not develop, even though the rotor bounces inside the seal at resonance speed. The reduced unbalance also

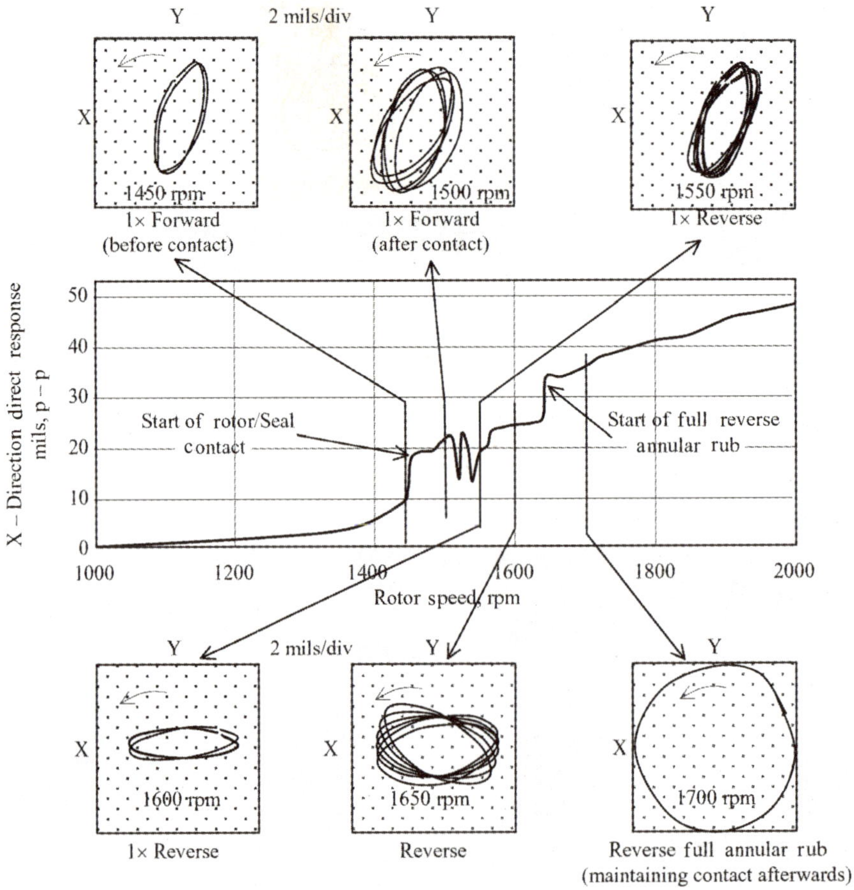

Fig. 13.18 Transition from Forward to Reverse Precession (Yu, 2000)

affects the starting point of the rub, as indicated in figure 13.20, where seal clearance is 250 μm. The first resonance speed of the original uncoupled rotating system is near 2000 rpm. The reverse rub develops after the resonance speed, and represents a switch from before resonance speed with the higher unbalance, with the same precession frequency of 3120 cpm. Although the rotor with higher unbalance initiates contact with the seal at about 1600 rpm, the reverse rub is not engaged quickly until around 1900 rpm, where an increasing contact force due to the synchronous 1 × rub motion is able to trigger it.

Considering the effects of shaft speed, once generated, the reverse precession full annular rub does not disengage by reducing the speed until it is very low, when slippage cannot be maintained. The ratio of reverse rub frequency to shaft speed ω/Ω is always less than the ratio of radius to radial clearance in the seal. Assuming the seal vibrates harmonically at the same frequency ω, the relative slip equals $(r\Omega - C_r\omega)$. The reverse rub disengages when this slippage velocity is near zero, so the friction force, also approaching zero, cannot balance the damping force. The reverse direction rubbing ceases when the slip velocity and the shaft speed are low.

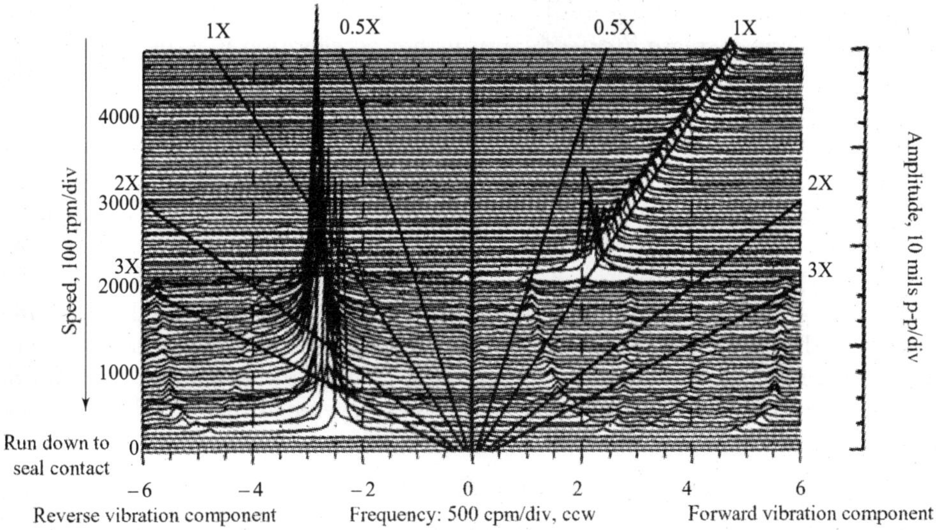

Fig. 13.19 Reverse Precession Rub Triggered During Run Down (Yu, 2000)

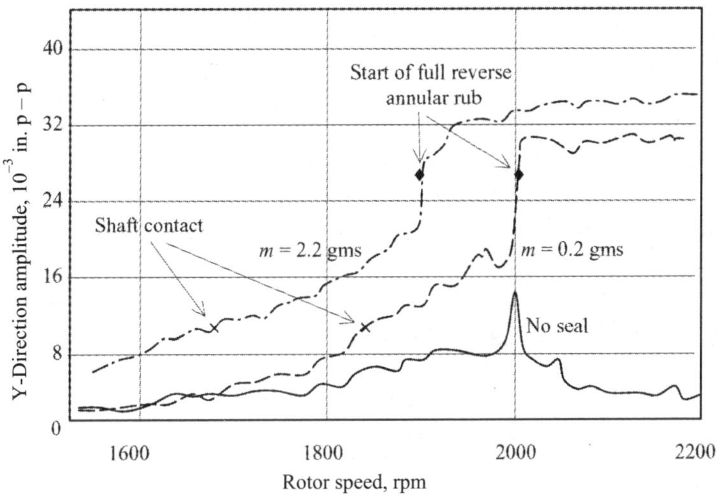

Fig. 13.20 Effect of Mass Unbalance on Start of Reverse Rub (Yu, 2000)

13.10 STABILITY CONSIDERATIONS

Rotating machines experience vibrations arising from a number of causes. Rotor unbalance is just one of them, although a more persistent one. Fluid flow forces, misalignment and manoeuvre loads in aircraft gas turbine engines are other important sources. There is also another class of vibration that mostly takes the form of flexural motion in the transverse direction at a rotor's natural frequency that is different from the running speed. The vibrations may be sub-synchronous, occurring at a frequency below the shaft speed, or super critical, when the frequency is above the shaft speed. A major feature of these non-synchronous modes of vibration is that the amplitude increases sharply with an increase in speed, and further acceleration may need to be aborted to avoid damage to the machine's components. Thus, such unstable vibrations are damaging and expensive, since they force delays in operation and production. Modification in engineering design may be required to overcome the problem, since they constitute a restraint on the capabilities of the rotating machine.

The mechanisms of instability may be classified by their characteristics, and by their effects on machine operation. Whirling and whipping of the shaft, stick-slip rubs and chatter are some commonly observed phenomena.

When a shaft experiences whirling and whipping, a tangential force normal to the direction of motion is developed. The magnitude of the force is proportional to the deflection, developing at some threshold speed when the force is large enough to offset the stabilizing external damping forces (*ref*: Ehrich, 1999). Figure 13.21 shows the forces developed in a whirling shaft. In polar coordinates the tangential force may be written as:

$$2m\omega(dr/dt) + c\omega r - F_\theta = 0 \quad ...(13.7)$$

where a constant rate of whirl is assumed. The whirling is based on the premise that some physical feature in the machine creates an exciting force that is normal to the radial deflection of the shaft and opposes the damping force, which tends to diminish the whirling motion. Often this disturbance force is proportional to the radial deflection.

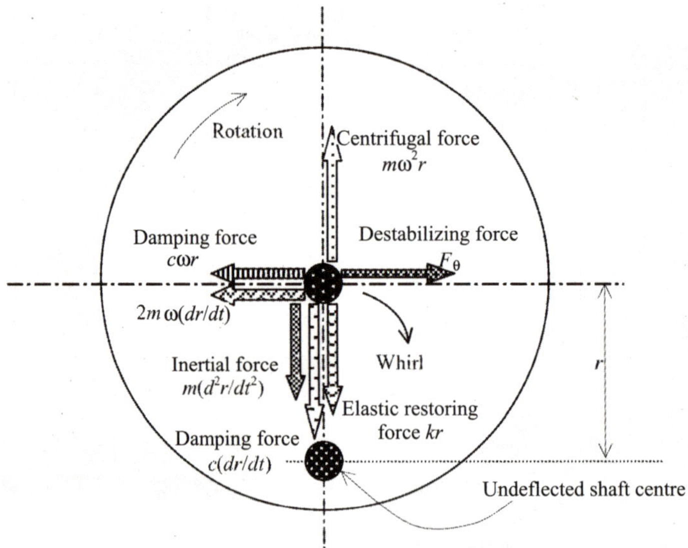

Fig. 13.21 Forces in a Whirling Shaft (Ehrich, 1992)

As the shaft speed increases to a threshold limit, the increased shaft radial deflection causes the disturbance force to reach a point where it balances the damping force to set up the condition for the onset of instability. Irrespective of the shaft spin speed, the whirling speed at the onset of instability equals the shaft's natural or critical frequency. The direction of the whirl may be in the same direction as the shaft rotation (called the forward whirl), or opposed to it (backward whirl), depending on the direction of the destabilizing force. In the unstable condition the shaft centre trajectory can be proved analytically to take the form of an exponential spiral up to a limit. Non-linear characteristics in the machine system will limit the growth of amplitude in the form of additional dissipating energy. Thus, the amplitude of whirl increases sharply with time before stabilizing at a steady state limit, and may be accompanied by damage to the machine's internal components.

Stability problems vary considerably in intensity. An operating unit may be marginally stable and operate for a long time without experiencing excessive vibrations. At many installations a fleet of identical machines may be operating satisfactorily from vibration considerations for many years, with the exception of a single machine. The machines may be operating basically on similar load cycles, handling the same fluid and work identical schedules. Under the circumstances it would be difficult to explain the dynamic experiences of the single machine with the high vibration signature. In such a unit, comparatively small changes often suffice to reduce the size of the problem. Changing the operating temperature of the fluid, going from one type of bearing to another or changing bearing preload may be enough to resolve the problem.

13.11 SELF-EXCITED VIBRATIONS

Self-excited vibrations in a system begin by themselves without any warning, initiating spontaneously. The amplitude increases until some non-linear effect limits any further increase. The excitation source for these vibrations is a uniform source of power associated with some inherent mechanism within the machine system. The destabilizing force sustaining the motion is derived from and controlled by the motion itself, and the force disappears when the motion stops.

Whirling and whipping instability arises from many sources:

- Internal rotor damping (hysteretic whirl), first reported by Newkirk (1924) and Kimball (1924), can be explained by using the force diagram of figure 13.22. Bending in the shaft due to some radial deflection sets up elastic restoring forces in a direction parallel and opposed to the direction of deflection. Internal friction in the shaft material also causes a phase shift in the developed stress, which results in a tangential force component. This sets up the condition for the shaft to whirl. The out-of-phase force from the bending stress increases as the shaft accelerates and reaches a point when the forces from the machine's damping system is exceeded to set up the condition for the onset of instability. Besides friction arising from relative motion between the molecules of the material, loose mating components also are responsible for this phenomenon. Bolted joints, shrink fitted assemblies and rabbets are prime examples. If the bolts are not tightened to the proper torque setting, or adequate interference fit does not exist, relative motion between the components will provide cause for hysteretic whirl in the system. Studies by Gunter (1966), Bolotin (1964) and Vance et al., (1973) have established that hysteretic whirl occurs at shaft speeds above the first critical speed. But once whirl sets in, the induced critical whirl frequency is about half the onset speed.

  Hydrodynamic journal bearings have been a major cause of unstable operation in rotating machines (Newkirk et al., 1925). Referring to figure 13.23, a shaft rotating in a journal bearing filled with a lubricant has a nominal radial deflection. Because of the shaft's

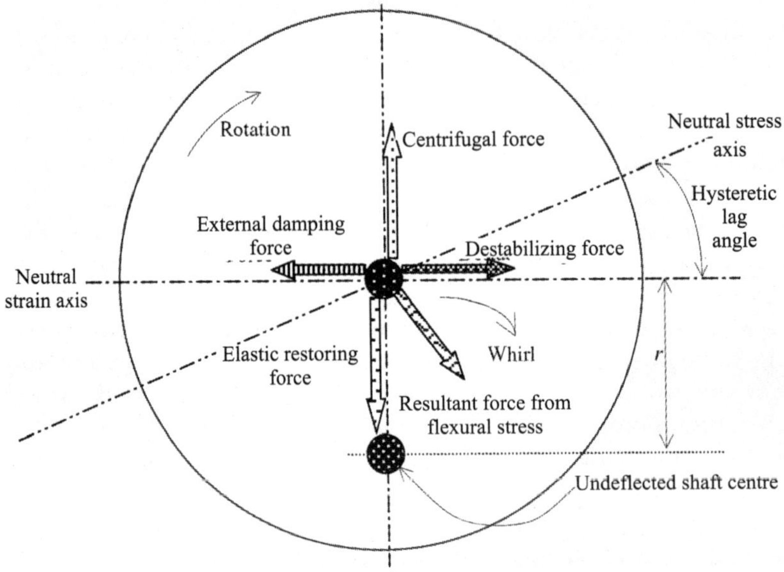

Fig. 13.22 Internal Rotor Damping Causing Hysteretic Whirl (Ehrich, 1987)

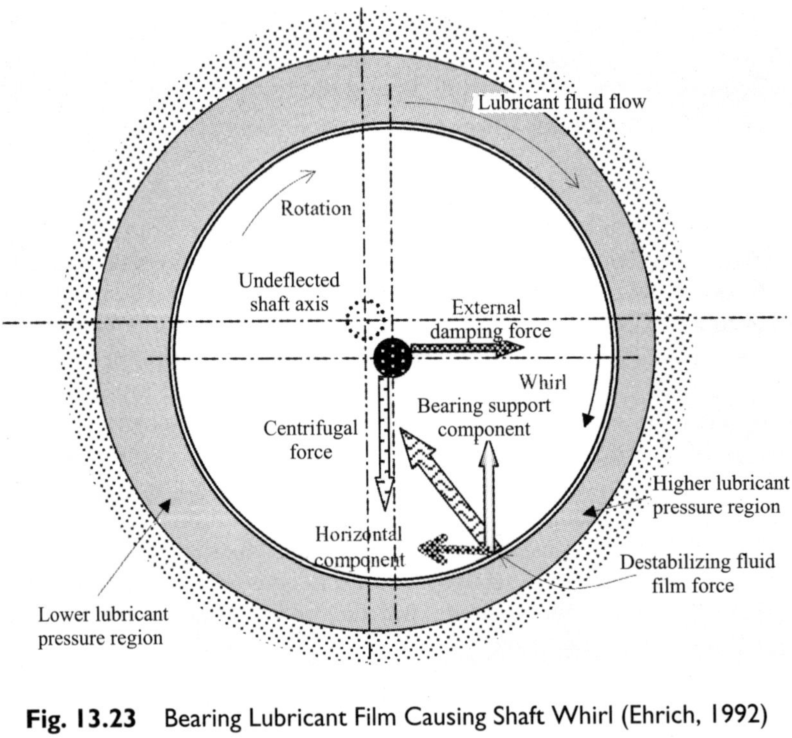

Fig. 13.23 Bearing Lubricant Film Causing Shaft Whirl (Ehrich, 1992)

eccentricity and the average fluid velocity (about half that of the shaft surface), the lubricant develops pressure, which is not symmetric about the line of radial deflection. Also, fluid pressure on the upstream side of the clearance region is higher than on the downstream side. This causes the resultant hydrodynamic force on the shaft to have a tangential component in the direction of rotation of the shaft, tending to whirl the shaft in the same direction. When this force exceeds the stabilizing damping forces, unstable operation will be the consequence. The stiffness and damping characteristics of the fluid film play a major role in determining the development of whirl. As the speed increases and the eccentricity ratio gradually becomes smaller, the cross coupling stiffness and damping terms become about equal. The speed ratio at which whirl will develop is given by $k_{r\theta}/c\omega$, where $k_{r\theta}$ is the film stiffness in the cross coupled radial and tangential directions. For plain journal bearings this ratio is about one-half. For a flexible rotor the threshold speed at which the instability occurs is approximated by $\omega_n/(k_{r\theta}/c\omega)$, where ω_n is the rotor critical speed. Thus, a flexible rotor supported in plain journal bearings usually experiences sub-synchronous unstable motion when shaft speed is greater than twice the lowest rotor critical speed.

- Steam and gas turbines require close control of radial clearance between the tip of the blades and the casing. The objective behind it is to prevent leakage of steam past the rotating blades without permitting the blades to rub against the case, and thus maximize operating efficiency. In many cases, however, it is difficult to maintain concentricity between the rotor and the stator, with the result that the tip clearance is not uniform along the full circumference. This can happen when the rotor lacks adequate stiffness, or due to larger than normal shaft span between the supporting bearings, causing the rotor to sag downwards. Overhung stages past a bearing support (aircraft jet engines are an example) also tend to exaggerate this situation. From aerodynamic considerations the result of the unequal space around the circumference is to set up a tangential force at the blade tip in the direction of rotation of the rotor. As the clearance distribution becomes more pronounced due to changes in speed or power setting, tangential forces may become large enough to overcome the energy dissipating forces of the damper, with the rotor whirling as a consequence. To overcome the problem, during the design stages bearing and shaft stiffness must be adequately provided for to avoid excessive radial deflection of the shaft. Figure 13.24 illustrates the forces acting on the rotor as a result of uneven tip clearance. Thomas (1958) offered an explanation for the destabilizing mechanism, with the expression for the reaction force $R_{r\theta}$ taking the form:

$$R_{r\theta} = T\beta/(DL) \qquad ...(13.8)$$

where: T = torque developed at stage

D = mean blade diameter

L = blade length

β = change in thermodynamic efficiency per unit change in clearance.

Subsequently, Alford (1965) offered similar explanations on stability of compressor rotors. Childs (1986) has done analysis of forces developed on a pump's impeller shroud.

Many examples of unstable rotor vibrations besides those given here can be cited. Aircraft engine propeller whirl has been recognized as a direct consequence of shaft whirl. When the plane of the propeller rotates because of the slope of the shaft's centreline, incremental axial velocities are induced (*ref*: Taylor et al., 1984; Houbold et al., 1981). At high air speeds destabilizing forces are set up, and can overcome damping forces to cause destructive whirling of the system.

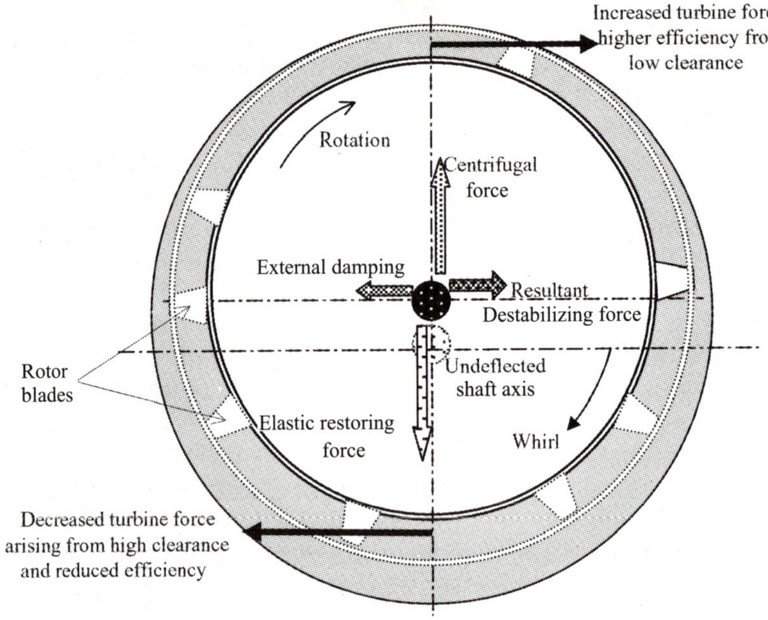

Fig. 13.24 Tip Clearance Variation Causing Whirl (Childs, 1987)

Fluid trapped in a hollow rotor can act as a high-speed centrifuge to set up destabilizing forces. Oil from bearing sumps and steam condensate may inadvertently collect and be trapped in the cavity of the rotor. When the shaft is operating with some nominal radial deflection, the fluid flies out, but does not remain in one fixed location. The surface of the shaft carries along some fluid with it in the direction of rotation. This advance sets up a force component in the tangential direction; the force is the source of instability because it induces forward whirl. The threshold speed for instability will be above the critical rotor speed.

Rubbing of a shaft against the stator case takes place in an environment where there is no oil or other fluid lubricant, so the rubbing is essentially dry. The dry rub sets up conditions for the shaft to whirl. Dry rub can also take place in a bearing that is starved of the lubricant, or with the loss of clearance in a journal bearing, or due to a loss of radial clearance in a labyrinth seal (see section 12.12), or between the blade tip and casing shroud of a turbo-machine. The mechanics of the rub and the consequent shaft whirl can be understood from figure 13.25. Coulomb friction due to rub between the rotating and static parts introduces a tangential force component equal to the product of the radial force and the coefficient of friction. The force induces the shaft into a whirling motion, and the associated centrifugal force produces greater radial deflection, which in turn causes an even greater radial force and whirl producing tangential friction force. Note that the friction force will be in the opposite direction of shaft rotation, thus causing a backward whirl of the shaft. Begg (1983) indicates that the frequency of whirl is generally less than the critical speed. A similar situation is encountered in an axial thrust bearing, where due to the slope of the shaft contact takes place between the rotating and stationary disks at a pivot point. Remedies to alleviate the shaft whirl due to rubbing will range from stiffening of the shaft to reduce radial deflection to improving assembly techniques for maintaining concentricity between rotor and stator. If the rub persists

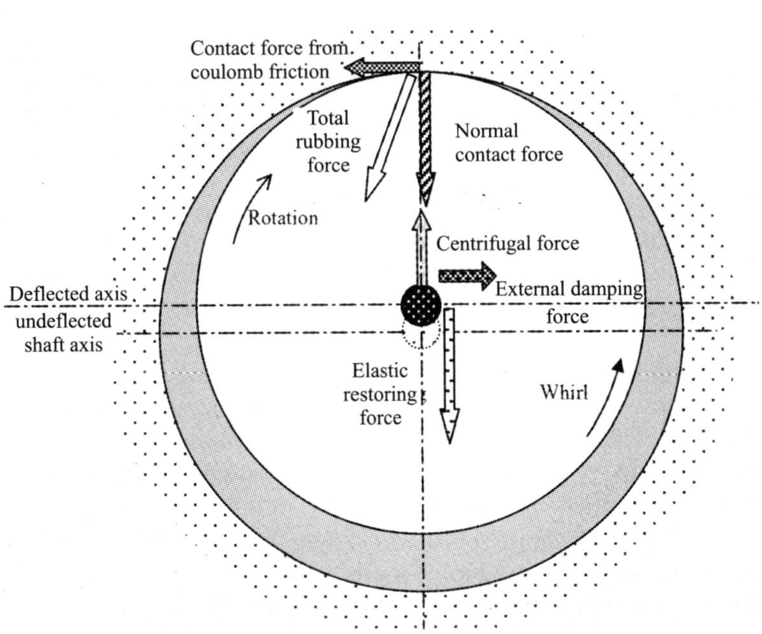

Fig. 13.25 Dry Friction Rub Causing Shaft Whirl (*ref.* Ehrich, 1987)

during operation, lubrication of the contact surfaces should alleviate the problem. If lubrication is not possible, another remedy would be to use an abradable lining material on the stator to permit disengagement between the interfering components.

Vance (1988) has suggested yet another source of excessive rotor vibration. Transmission of torque in high-speed rotors is accompanied by a rotational twist in the shaft. If that twist results in an angular deflection of the torque vector, then the torque will have a small radial deflection, which is nearly proportional to the radial deflection. The radial torque vector induces bending in the shaft and a deflection that is normal to the radial deflection. When this tangential component is large enough to overcome the energy dissipating forces in the system's damping mechanism, the threshold of instability is crossed and the shaft whirls.

13.12 PERIODIC SYSTEM CHARACTERISTICS

An analytical method to investigate the problem of dynamic instability in rotating machines is based on a system of linear differential equations with periodic coefficients, with the period in time t. A specialized form of this equation, called Mathieu's equation, can be used to represent many different physical problems in rotating machines:

$$\frac{d^2y}{dt^2} + [a - 2q \cos (2t)]y = 0 \qquad ...(13.9)$$

where a and q are constants. McLachlan (1947) has provided treatment of the theory behind this equation. Stability induced by periodic variation in a rotating machine's system parameters such as stiffness and inertia can be evaluated using this approach. In particular, asymmetric shaft cross-sections and bearing characteristics, pulsating torque and fluctuating axial compressive loads affect the lateral vibration stability of a rotating shaft.

When the stiffness characteristics of a shaft or stator are significantly different in two orthogonal planes, self-excited vibrations are likely to be a problem. An asymmetric shaft rotating at a speed away from whirl speed acts as a periodic change in stiffness in the plane of its radial deflection, causing instability in specific speed ranges determined by the degree of asymmetry. Unstable operation may be expected at one-third and one-half of critical speed, and at critical speed. But with even larger asymmetries in the shaft (or the support) stiffness, generalizations may not be possible.

Eshelman et al., (1980) carried out experiments to determine the effect of cyclically pulsating shaft torque and its effects on rotor stability. A varying torque was superimposed on a constant level torque in a rotating shaft. This caused the natural frequency of the shaft in the transverse direction to shift. The researchers investigated conditions under which the shaft whirled at the first critical speed, varying speed and pulsating torque. Certain combinations of the steady and pulsating torques caused instability in narrow speed bands at discrete fractions of the critical speed.

In a comparable manner, pulsating axial loads are also known to cause lateral instability in a rotating shaft. When the axial load approaches the critical buckling load for the rotating shaft, the natural frequency of vibration in the transverse direction reaches zero. Since the pulsating compressive load causes a loss in lateral stiffness in the shaft, a corresponding reduction in the frequency is the consequence, as noted by McLachlan (1947).

Dry friction was mentioned before, when the shaft rotating at a high velocity encounters instability due to contact between the rotating and static components of the machine. When the shaft is just beginning to turn, breakaway friction produces higher level of friction to cause another group of instability in shafts – chatter, or stick-slip. Periodic motion of the rotor's point of contact is self-excited as a result of negative damping, because motion is taking place in the direction of rotation, and increases the relative contact velocity. Rotor motion counter to contact velocity will reduce relative velocity and increase friction force, thus again reinforcing the periodic motion. In the absence of proper flow of cooling fluid at the cutting bit, dry friction in machine tools exhibit a similar characteristic, and is known as chatter. In both cases the vibration is more torsional in nature than lateral, but vibrations in the transverse direction can also ensue.

13.13 METHODS TO ALLEVIATE VIBRATION INSTABILITY

Rotating machines experience instability due to the presence of a tangential force set up by some physical characteristic related to the machine's design or operation. The tangential force arises due to a deflection of the rotor in the radial direction, thus providing a coupling between the two directions. If adequate steps are taken in the design stage of the machine, the tendency towards unstable operation will be diminished.

Stiffness of the shaft is an important consideration. Destabilizing forces generally become more pronounced when machine operation takes place above the first critical speed, since their sensitivity towards unstable operation increases with the shaft's flexibility. Thus, one possibility is to maintain the operating speed of the machine below the first critical. As the speed ratio ω/ω_n increases, the tendency for the shaft to whirl goes up. Keeping shaft span between the bearings low helps to increase the critical frequency of the rotor, and consequently the operating speed of the machine can be raised without encountering instability problems.

External damping plays a significant role in eliminating unstable operation in a rotating machine. Care must be taken to ensure that damping is external to the rotating system, so the energy dissipating forces in the damper depend on the relative motion between the rotor and the stator. Adequately designed bearing supports play a major role in this respect. In the preliminary design system parameters require adjustment in order to position the natural frequencies of whirl as required. Bearing and support stiffness plays a crucial role, together with support location and mass and stiffness distribution of the rotating components. For this purpose a plot of the natural frequencies of vibration against a particular design parameter provides useful information about the sensitivity of that parameter. Figure 13.26 illustrates a plot of critical speeds as a function of identical support stiffness at both ends. When a number of such plots are generated for variables such as distance between bearings and weight of rotor, a comparison of the design from shaft vibration considerations is possible. The comparison can also be made with test measurements during machine operation to improve the mathematical model with the help of critical speed maps. Use of energy maps for design evaluation has also been suggested (Ehrich, 1987). Potential (or strain) energy and kinetic energy distributions in major sub-assemblies in the machine for a particular vibration mode provide useful information. Since strain energy varies with the square of displacement and kinetic energy varies with the square of velocity, the energy distribution gives important data about the level of dynamic activity in different portions of the machine system. When a system design parameter is changed, its impact on the vibration characteristics can be studied from the energy distribution.

13.14 ROTOR DYNAMIC DESIGN OF HIGH PRESSURE COMPRESSOR

The reliability of centrifugal compressors is by far the most important aspect for many users in offshore petroleum platforms and pipeline pumping industries. With the high power density, aerodynamic forces arising in seals sometimes tend to endanger continuous operation. High

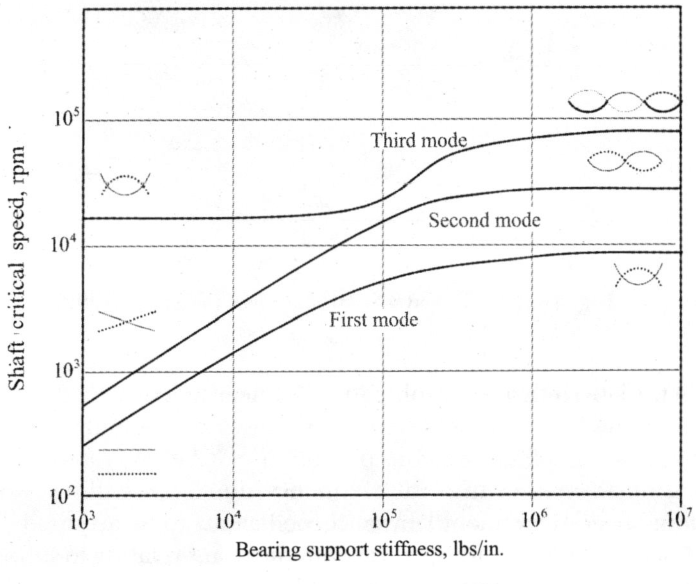

Fig. 13.26 Critical Speed

sub-synchronous vibrations make it necessary to perform alterations, incurring considerable cost and downtime. The problems indicate lack of adequate knowledge of the forces because computational fluid dynamics models are not able to deal with the complexity of flow through the seals.

A comprehensive experimental investigation for determination of stiffness and damping constants of the labyrinth seals must be a part of the design phase for simulation in the rotor dynamic analysis of the system (*ref*: Wagner, 1999). Thus, the test rig must be capable of producing and measuring significant gas forces encountered in the compressor. Active magnetic bearings are suitable in this respect because it is possible to set the static eccentricity independently of the rotor speed, and also to superimpose a circular motion of the shaft at adjustable frequency, amplitude and direction of precession. At the same time, force can also be measured from the bearings, permitting direct calculation of the transfer function without transducers. The dynamic seal coefficients may then be extracted from the transfer functions. Straight through, stepped and staggered labyrinth and honeycomb seals of precise geometry as used in the compressor can be used in the test rig shown in figure 13.27, which offers the advantage of eliminating extrapolation with reference to geometrical dimensions. Besides casing inlet and outlet pressures, static and total pressures are measured directly in front of the inlet to the labyrinth. The precise swirl at the labyrinth's inlet can then be determined, a primary influencing variable for the objectives of the

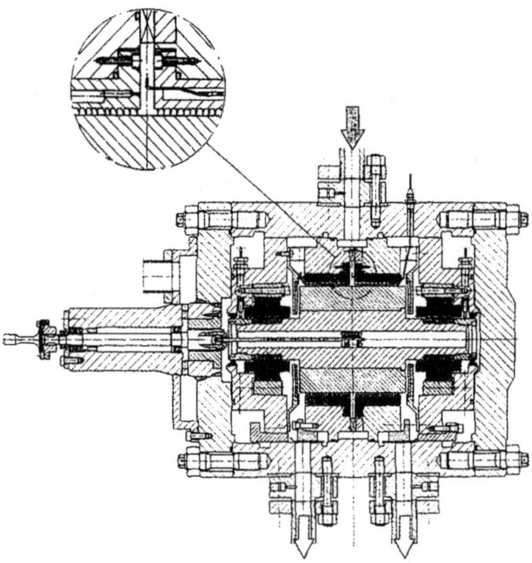

Fig. 13.27 Test Rig Sectional View (Wagner, 1999)

test.

Forces acting in the labyrinth are a combination of linear functions of displacement and velocity, and the stiffness and damping matrices become skew-symmetric only in the special case when eccentricity in the seal is zero. The equations of motion may be separated into real and imaginary components after transformation into the frequency domain, yielding two real equations for determining the four speed dependent labyrinth coefficients to be acquired. The identification is then reduced to a linear regression with speed ω, with the terms relating to stiffness obtained directly at the intersections and damping terms corresponding to the linear gradients. A number of seal types with different diameters, steps, rotating and stationary sealing tips, swirl brakes and abradable coatings on the rotor are examined in the test.

A linear relationship between labyrinth forces and displacement and velocity in the frequency domain is valid for labyrinth seals, but some restrictions are necessary for honeycomb seals. Thus, the labyrinth coefficients may be used as direct and cross-coupled variables for the restoring force $F_R = k_R + \omega d_Q$, but contradictions appear for the tangential force $F_Q = k_Q - \omega d_R$ in the form of sign changes. To overcome the problem, changing to the polar representation in the form $k = (k^2_R + k^2_Q)^{1/2}$ and $\tan \theta_k = k_Q/k_R$ alleviates the problem. The representation of swirl brakes is shown in figure 13.28. Swirl brakes reduce destabilization by reducing total stiffness and damping by turning the vector through a positive or negative angle. As opposed to stunt holes, swirl brakes have the advantage of not requiring additional bearing span and of having their effect precisely quantified during the test.

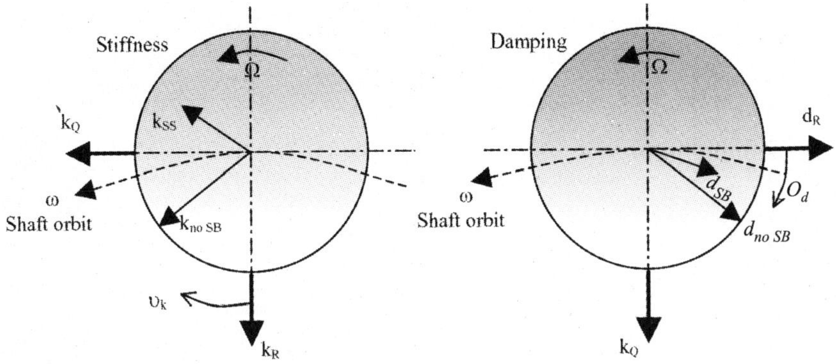

Fig. 13.28 Stiffness and Damping Vector With and Without Swirl Brake (Wagner, 1999)

Figure 13.29 provides a comparison of the dynamic coefficients for the different seal types. The seals are 200 mm in diameter, 80 mm long, have a radial gap of 0.2 mm and operate at inlet density of 200 kg/m^3 and pressure ratio of 0.45. Variations from operation at different shaft speeds and swirl ratios are indicated in the diagram. The staggered labyrinth produces the lowest stiffness and damping values. In comparison, forces arising in the honeycomb seal are greater by one order of magnitude. The restoring force in honeycomb seals is several times higher than the tangential force, and owing to the high cross-coupled damping, is heavily dependent on the oscillation frequency. The tangential force has a stabilizing effect in the examined honeycomb seals. The inlet swirl tends to have a larger impact on stiffness than on damping.

In rotor dynamic calculations honeycomb seals cause problems when they are employed at high densities because their dynamic coefficients then reach larger magnitudes than those of journal bearings. In the high-pressure range the lighter loading on the compressor rotor, critical speeds and logarithmic decrements rely more strongly on rotor eccentricity. Thus, it is difficult to get good correlation between calculated and measured results when honeycomb seals are employed.

The test results are applied to a re-injection compressor designed to boost pressure from 400 to 600 bar. The four stage unit has a power requirement of 8.5 MW, and density at discharge is 330 kg/m^3. Sealing is provided by a double acting mechanical contact seal with a closed high-pressure loop. The rotor (dimensions shown in figure 13.30) runs at 10,000 rpm and is supported in standardized four segment tilting pad bearings. The dimensions of the rotor are obtained for

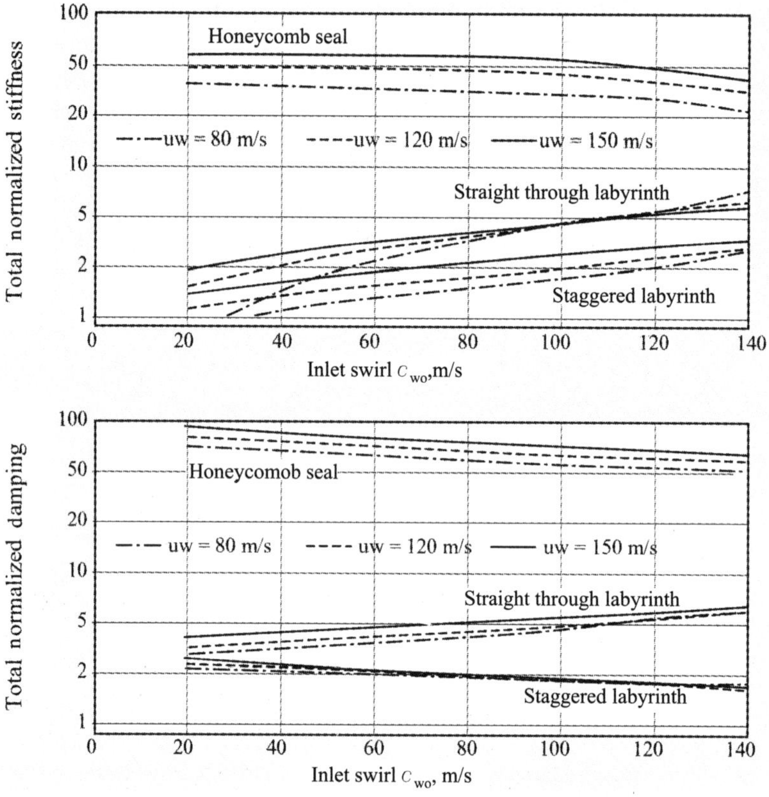

Fig. 13.29 Seal Dynamic Coefficients (Wagner, 1999)

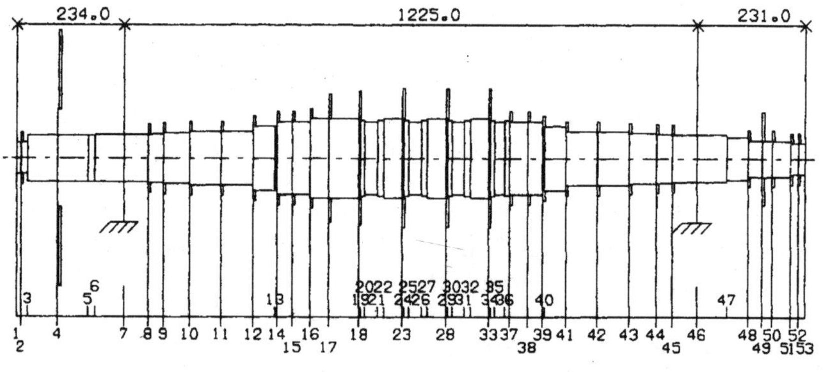

Fig. 13.30 Rotor Details (Wagner, 1999)

maximum stiffness from a proven concept, with deflection at mid-span limited to 16 times th bearing amplitude in the first mode. This achieves a high level of logarithmic decrement, an sensitivity to excitation from the labyrinth is lower than in an elastic rotor. A critical speed is in th vicinity of the operating speed, but in most applications operation of the compressor close t esonance is avoided, even though with adequate damping the rotor can operate near resonance ithout risk. Stability under all operating conditions is an essential requirement, since it may cause shutdown of the installation besides damaging the machine. Manufacturing tolerances of the ournal bearings play a major role for this relatively lightly loaded high-pressure compressor. ermissible variations in the lubricant oil for the bearings must also be taken into account.

Figure 13.31 shows variation of eigenvalues in the complex plane when density of the gas is ncreased from zero to double the design value. Calculations for minimum (0.05) and maximum).08) Sommerfeld (So) numbers indicate the impact of bearing tolerances and oil temperature. ower eigenvalues tend to react more sensitively to the Sommerfeld number than upper ones, and ne same is true of the effect of gas forces. In the case of the upper eigenvalues the gas forces cause ne natural frequencies to shift, yet the log decrement sees little reduction. The margin of separation equired in API 617 is observed even with double the design density. The curved branches in the igure are based on calculations using complete stiffness and damping matrices, and also with only he cross-coupled stiffness k_Q. The complete matrix solutions result in a lower degree of lestabilization, and also show the larger influence of gas forces on the natural frequencies. Dynamic oefficients for the seals at the impeller eye and at the balance piston labyrinth are calculated for arious clearance widths from available experimental data (*ref*: Zeidan et al., 1993). The mportance of examining the influence of clearance widths arises from observations that when the learance is enlarged through rubbing a compressor exhibits high sub-synchronous vibrations. The heory fails when the labyrinth excitation is calculated for large clearance widths.

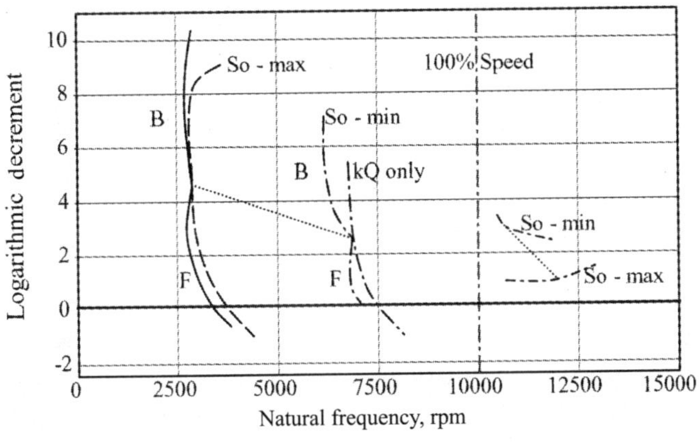

Fig. 13.31 Effects of Gas Density on Eigenvalues (Wagner, 1999)

The course of the minimum logarithmic decrement due to variations in the labyrinth gap clearance can be observed from figure 13.32, using the least favourable combination of bearing geometry and oil temperature. Adequate damping reserve is still present even after the labyrinth clearance is increased to 200 per cent of its nominal value.

One other factor to be considered is an asymmetric distribution of pressure development in the volute, causing the static force in the bearing to change. Figure 13.33 illustrates the influence of the static gas forces on the eigenvalues on the complex plane, using minimum damping. Here, too, the assumption of most unfavorable bearing geometry and oil temperature and of damaged labyrinth is used. Adequate damping reserve is still noticeable.

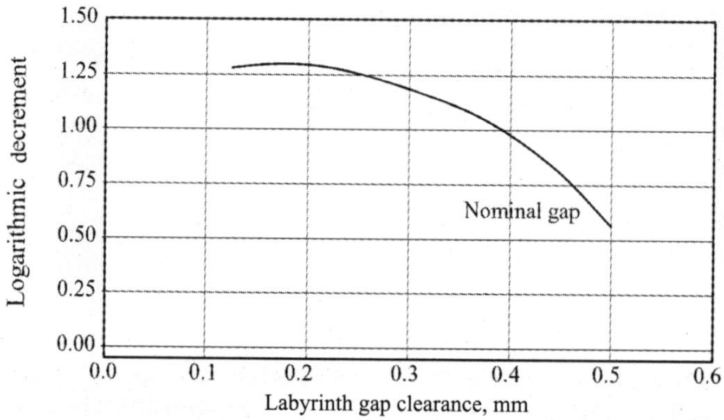

Fig. 13.32 Effect of Labyrinth Gap Clearance On Rotor Stability (Wagner, 1999)

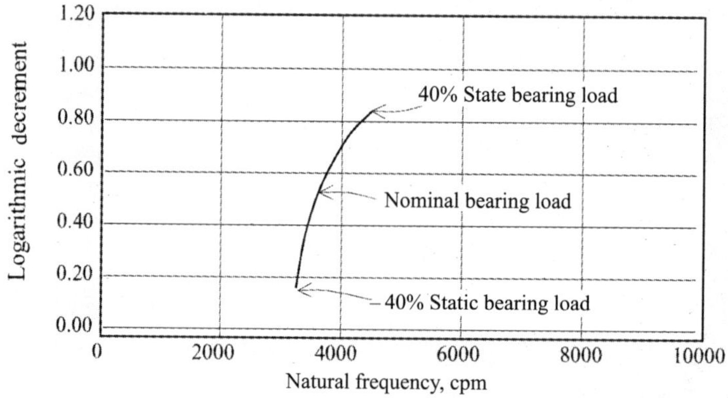

Fig. 13.33 Effect of Side Loads On Rotor Stability (Wagner, 1999)

Triple dry gas seals are not favourable from rotor dynamic considerations, mostly because the relatively large bearing span results in insufficient stability. One solution is to install swirl brakes on the shroud and the balance piston labyrinths. The swirl brakes have the effect of reducing the cross-coupled stiffness for the prescribed operating conditions by 70 per cent. The spring and damping coefficients of the labyrinths with and without the swirl brakes are obtained from test measurements. When a hydrocarbon mixture is envisaged for the compressing unit, the reliability of the stability calculations is of significance, and can be obtained from a back-to-back experimental verification using labyrinth seals manufactured and installed in the compressor with and without the swirl brakes.

Verification of the test and analytical results is obtained from measurements on an operating unit, with its design based on the lessons learned. With the swirl brakes installed, no noteworthy sub-synchronous vibrations were observed at any of the operating points. The logarithmic decrement is determined to be 0.3. For the second test run labyrinth seals without the swirl brakes are used. Instability is encountered at 100 per cent speed at discharge pressures between 240 and 320 bar, based on supply oil temperature. The frequency of vibration is around 6,000 cpm, also dependent on oil temperature. Based on the investigation, the necessary stability limit of 10 μm peak to peak sub-synchronous vibration limit is achieved by using a three-pronged approach: (*a*) increase machine speed at constant discharge pressure and oil temperature, (*b*) increase discharge pressure at constant speed and oil temperature, (*c*) increase oil temperature at constant speed and discharge pressure. The tests are also useful in ascertaining that the frequency of the sub-synchronous vibrations increases after the onset of instability. This effect is explained by the rub at the balance piston to effectively reduce the bearing span and raise the frequency of vibration as a direct consequence.

13.15 EFFECT OF SQUEEZE FILM DAMPER ON ROTOR RESPONSE AND STABILITY

The squeeze-film damper has the ability to attenuate vibration and stabilize a rotating assembly system by providing multi-directional damping. Broadly speaking, there are two different configurations of the squeeze-film damper. One is the arrangement where a centralizing retainer spring in the form of a squirrel cage is connected to the outer race of its rolling element bearing, and acts as a parallel element within the squeeze-film annulus. The second is a structure without a retainer spring, where, for a horizontal machine say, the outer race of the rolling-element bearing is assumed to remain at the bottom of the damper clearance circle until the unbalance forces become large enough to create lift.

One of the advantages of the configuration with a retainer spring is the ability to tune the critical speeds of the system away from the operating range to ensure smooth running. However, when different unbalances, static eccentricity ratios and supply pressures are applied, the system resonance will often migrate from their expected positions. In particular, if the damping is excessive, which can sometimes occur with increase in the oil supply pressure or in the static damper eccentricity ratio, resonant frequencies of the assembly may approach the corresponding frequencies for the model without a retainer spring. This is because the damper locks out the retainer spring, thus rendering it inoperative.

In an experimental investigation aimed at a direct comparison of the vibration performance of two configurations of a flexible rotor-bearing assembly incorporating a squeeze film damper, the subject is examined from the points of view of forced vibration and stability control (*ref*: Chu et al, 1996). The schematic diagram of the test rig is shown in figure 13.34. It consists of a flexible stepped shaft supported in two ball bearings, one of which is self-aligning; the other is flexibly mounted on four supporting bars, constituting a centralizing spring. A squeeze film damper annulus is placed around the outer race of the latter ball bearing to attenuate rotor vibration. Oil is pumped via three holes and a central groove into the annular clearance space of the damper. The shaft carries an overhung disk held by a taper-lock bush. The drive shaft is connected to the main shaft by a flexible rubber coupling to isolate the test shaft from any vibration influence from the motor. The squeeze film damper has a two-land geometry and is unsealed. Journal radius is 50 mm, radial clearance is 0.1 mm, land length is 10 mm and oil viscosity is 0.00375 N-s/m^2. The rig is first impulse tested to obtain its undamped natural frequencies. For the configuration with a retainer spring, in order to eliminate the influence of oil damping, the damper ring is temporarily detached from the rig. The test result at the damper ball bearing gives a first undamped natural frequency of 14 Hz and a second of 41 Hz. The retainer spring is then made inoperative by shimming the damper annulus within the damper ring. The corresponding impulse test result indicates the first two undamped natural frequencies of 31.2 Hz and 88.0 Hz.

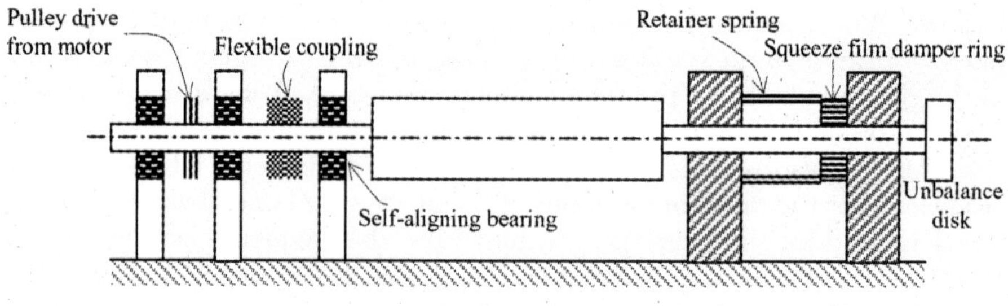

Fig. 13.34 Schematic Diagram of Experimental Rig (Chu, 1996)

The shaft and the overhung disk are first balanced before running, as also the two together. One of several unbalance masses can be screwed into the overhung disk. In order to observe the effects of different levels of unbalance, this is chosen such that the non-dimensional eccentricity u of the mass unbalance Q_c (= u/c) at the position of the overhung disk has values of 1.12, 0.649 and 0.148.

Radial and transverse squeeze film forces (P_r, P_θ) obtained by integrating the dynamic pressure distributions within the annulus are shown in figure 13.35, taking into consideration film rupture and the attitude angle α of the shaft. The squeeze film is rather difficult to specify, particularly with respect to the circumferential extent around the rolling element bearing. This is largely due to cavitation in the oil film, which arises when negative pressures are produced between rapidly separating surfaces during vibration and which cannot be compensated by sufficient supply pressure.

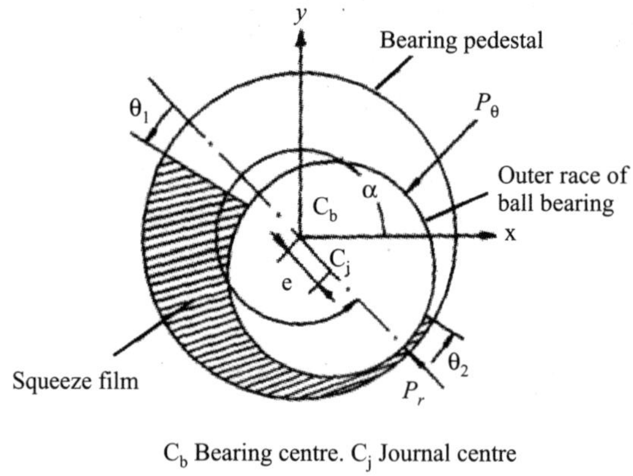

Fig. 13.35 Details of Squeeze Film Damper (Chu, 1996)

For a static eccentricity ratio of zero a higher supply pressure produces higher system damping, and such damping suppresses vibration. This is almost certainly due to maintaining the clearance full of oil. For concentric vibration ($\varepsilon_0 = 0$), as the supply pressure is increased from 0 Pa to 34.5 kPa, there is a shift in resonance position from 40.0 rev/s to around 34.0 rev/s. It is expected that if the supply pressure is further increased, the resonance will finally reach a value of 31.8 rev/s, the value without retainer spring. For eccentric vibration with a static damper eccentricity ratio of unity, increasing the supply pressure does not lead to change in the position of the resonance, remaining at around 31.8 rev/s. The reason for this is that the system damping is always high and is caused by the high static eccentricity ratio. As supply pressure and hence damping is increased, it can only reduce the peak value of the vibration and make no further shift in the resonant frequency.

For a high-speed rotating assembly it is important to ensure that the shaft is well balanced and runs in a quiet regime. Unfortunately, there will always be some residual unbalance, which, if large enough, will cause severe vibration. This can sometimes lead to fracture of the shaft and damage to the foundation. Also, even relatively low amounts of unbalance may change the position of the resonance for the configuration with retainer spring and make it unpredictable at the design stage. Figure 13.36 shows three response curves for concentric vibration at the squeeze film damper location for the configuration with and without the retainer spring. For low unbalance, say $Q_c = 0.148$, as shown by the lowest curve, the position of the resonance is at about 34 rev/s. If the unbalance is increased to $Q_c = 0.649$, a resonance is observed at around 33 rev/s and if unbalance is $Q_c = 1.12$, the position is at 32 rev/s. At the overhung disk location the shaft vibration with no supply pressure is nearly half that of the response when the supply pressure is 34.5 kPa.

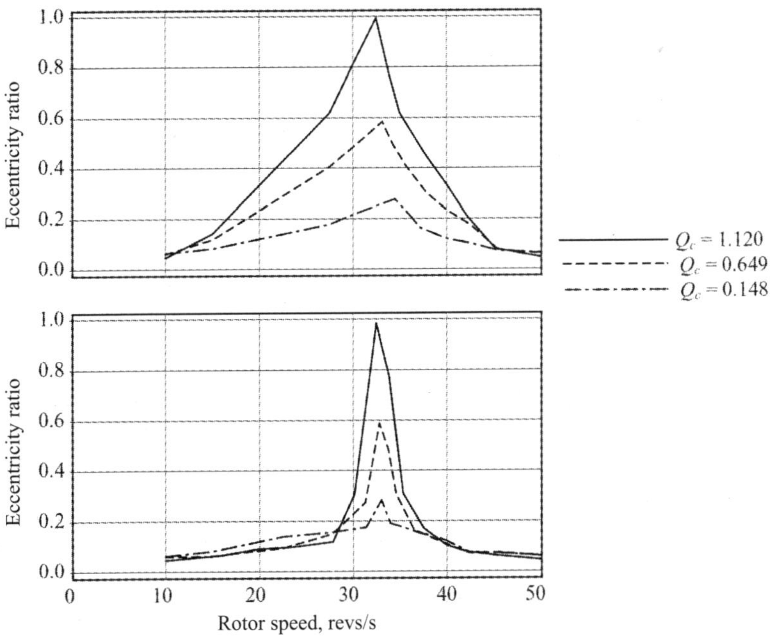

Fig. 13.36 Vibration Response With (*upper*) and Without (*lower*) Retainer Spring (Chu, 1996)

For a given operating condition, the system will be stable if all the eigenvalues λ that enable the eigen equation D(λ) = 0 to be satisfied are such that their real parts are negative. Even one root with a positive real part is sufficient to indicate unstable conditions. If the relation between the real part of the eigenvalue s and the rotating speed is obtained, operating stability of the system can be assessed. A numerical search for all the roots of the eigen equation can be very time-consuming, and tends to lead to a loss of physical insight. For the eigen-equation D(λ) = 0, the Leonhard locus (*ref*: Naslin, 1965) retains a feel for the underlying principles. Consider the complex λ plane and the complex D plane. The function D(λ) maps each point in the λ plane to a corresponding point in the D plane. The roots of the eigen equation can be regarded as those points in the λ plane that map to the origin of the D plane. A point traveling along the imaginary axis in the λ plane from $\omega = 0$ to $\omega = \infty$ will map to a curved locus in the D plane, and is known as the Leonhard locus. For a stable system all the roots will lie to the left-hand side of the locus in the λ plane. This implies that the Leonhard locus will be such that the origin is always to its left-hand side. Thus, for stability, the Leonhard locus must encircle the origin, in an anti-clockwise direction, as ω increases from zero. At the threshold of stability the Leonhard locus will pass exactly through the origin of the D plane. As an example of this theory, consider the instability problem for the chosen rotor dynamic system. The combined rotor-damper seal system is assumed to be supported by a linear damper with damping coefficient $c_o = 0.589 \times 10^3$ N-s/m say, to simulate the linear π model for the squeeze film damper. Figure 13.37 shows the two Leonhard loci for this case. It is seen that as the frequency ω increases the curve for $\Omega = 65.4$ rev/s encircles the origin in an anti-clockwise direction and the curve for $\Omega = 65.5$ rev/s does not. Therefore, 65.5 rev/s can be reasonably regarded as the onset speed for instability of the system.

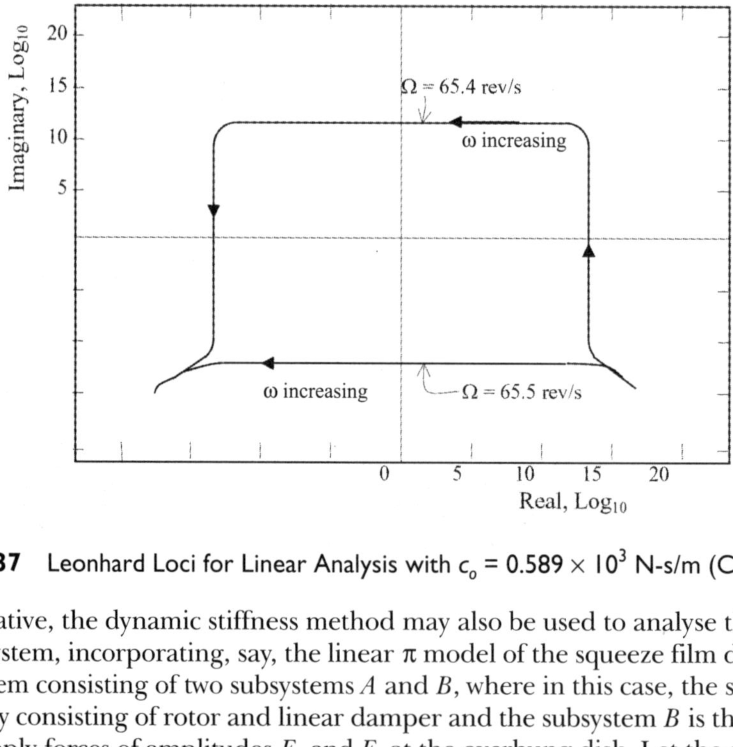

Fig. 13.37 Leonhard Loci for Linear Analysis with $c_o = 0.589 \times 10^3$ N-s/m (Chu, 1996)

As an alternative, the dynamic stiffness method may also be used to analyse the stability of the rotor dynamic system, incorporating, say, the linear π model of the squeeze film damper. Consider a composite system consisting of two subsystems *A* and *B*, where in this case, the subsystem *A* is the rotating assembly consisting of rotor and linear damper and the subsystem *B* is the seal. Subsystem *A* is isotropic. Apply forces of amplitudes F_x and F_y at the overhung disk. Let the amplitudes of the response be *X* and *Y*, also at the disk. Then the dynamic stiffness Z_x and Z_y for the shaft at the disk may be defined as follows. Since the shaft is circular and isotropic and the damper is also isotropic, these two dynamic stiffness coefficients will be complex and equal. They can be written as $Z_x = F_x/X$ and $Z_y = F_y/Y$. Subsystem *B* (the seal) has terms that include direct and cross-coupled force-deflection relationships, and some of the terms are functions of the rotating speed Ω. Now connect the two subsystems together at the overhung disk mass. At the borderline between stability and instability the system will vibrate at a natural frequency and no force will be required. This condition can be expressed to determine an expression for the borderline relation between stability and instability. The procedure calls for selecting trial values of rotor speed Ω, determining the real and complex portions of the two stiffness expressions and finding a value for the frequency ω such that the stiffness components come close to zero. A graphical approach provides the value of the instability onset frequency ω and the corresponding speed Ω at the intersection of the two curves. An added advantage of this procedure is that one subsystem, say the seal, can be maintained unaltered and the other, the rotator with the damper, revised as required to provide the requisite instability onset speed.

13.16 FULL LOAD TESTING FOR ROTOR DYNAMICS IMPROVEMENT

API 617 provides guidelines for rotor dynamics characteristics of compressors during shop tests. However, the requirements deal mainly with synchronous unbalanced response of the rotor, and do not relate to sub-synchronous vibrations induced by the flow of gas, or in other words, stability. In multi-stage centrifugal compressors when the gas crosses through the channel, distortions take place in the velocity distribution along the impeller, diffuser, channel return, inlet guide vanes and volute. This induces a complex pattern of dynamic forces on the rotor, resulting in unstable operation of the machine.

The aerodynamic forces may be split into two categories. The first type is forced excitation generated directly by the flow itself at low flow rates, and manifests in the form of a rotating stall. Prediction criteria for this event are not accurate when the flow coefficient is low in high-pressure centrifugal compressors. The second type of force arises from interaction between the fluid and the structure, and develops in regions where the fluid is confined. The displacement of the rotor creates local disturbances in the pressure and velocity, and appears as a special reaction. The interaction is of considerable significance when the compressor has several impellers, and hence a number of inter-stage and piston labyrinths, where the gas is confined.

The vibration problems are illustrated in the full load testing of two natural gas centrifugal compressors of the type shown in figure 13.38 (*ref*: Gelin et al, 1996). One unit is used for re-injection of natural gas driven by a variable speed electric motor, and the other is dedicated to a gas lift application powered by a fixed speed motor. Both machines are fitted with tilting pad bearings and dry gas seals. The units are split into two sections for intermediate cooling. The rotor for the re-injection application has higher stiffness since it handles gas at a greater pressure, and has a part of the gas path machined in the shaft at the impeller eye locations. The initial design called for vaneless diffusers.

Low-pressure performance tests were carried out according to ASME PTC 10 Class III requirements, and the units exhibited good vibratory characteristics. Sub-synchronous vibrations appear at 7 per cent speed, accompanied by unacceptable levels at the first natural frequency of the rotor, in the full load test of the re-injection unit. The rotating stall problem is overcome by using vaned diffusers in the second section and for the last impeller of the first section. The first natural frequency peak was diminished by installing a honeycomb seal in place of a toothed labyrinth on the balance piston, and by modifying the tilting pad bearings to provide increased stiffness.

Fig. 13.38 Compressor Rotor (*upper*) and Cross Section (*lower*) (Gelin, 1996)

The gas lift compressor exhibited high synchronous vibrations at the first bending mode while operating at 13,800 rpm, 17 bar suction pressure, 125 bar discharge pressure and 100 per cent flow in each section inlet. On dismantling the machine the thrust bearing pads, thrust collar and the dry gas seal were found to sustain extensive damage. Contact was noted on the impeller shrouds, but the radial bearings and the internal labyrinths were intact. The instability in the radial direction led to a major incident in the axial direction, with notable radial damage that was inexplicable. Investigating the rotor dynamics and thermal and aerodynamic excitation of the flow in the dry seals revealed two different and uncorrelated events: instability arising from the first bending mode under the prescribed operating conditions and a thermal problem with the gas seals. The problem was resolved by stiffening the shaft in a similar manner as on the other unit, installing a honeycomb seal on the balance piston, adjusting the pads on the radial bearings and redesigning the inlet and discharge volutes of the second section. Subsequent results form testing at full load were satisfactory, and permitted installation of the unit at the site.

Diagnosis of synchronous vibrations is obtained from a rotor dynamic analysis using the elastic beam representation of the shaft, each node having two radial displacement and two rotational degrees of freedom. Impellers, balance piston and gas seals are simulated as pure mass and inertia without stiffness. Radial tilting pad bearings are modeled by complete stiffness and damping matrices that include cross-coupling terms arising from the pad inertia. Fluid structure interaction due to the impellers and internal seals are simulated like the bearings, by stiffness and damping matrices of the form:

$$[K_f] = \begin{bmatrix} K & k \\ -k & K \end{bmatrix} \quad \text{and} \quad [C_f] = \begin{bmatrix} C & c \\ -c & C \end{bmatrix} \qquad ...(13.10)$$

Influence from the interaction between the fluid and the structure can be obtained from the difference ($k - C\omega$) because it represents the destabilizing cross-coupled force intensity. Dynamic characteristics of the radial tilting pad bearings come from the solution of the Reynolds equation with introduction of a perturbation (*ref*: Frene et al, 1990). Thermal equilibrium on each pad is verified to estimate the actual oil viscosity and clearance and preload in the bearing. The numerical resolution uses an iterative procedure based on a finite difference scheme. Natural frequencies and the associated damping in the form of logarithmic decrement are then computed by a parametric analysis from the rotor model at nominal speed. Figure 13.39 gives the corresponding chart for the first lateral mode in the forward direction for the re-injection compressor, where the diametral clearance is expressed as a ratio of bearing diameter and preload expressed conventionally. The diagram permits selection of the optimum bearing geometry to obtain a high natural frequency and good damping. Thus, paying special attention to clearance and preload definition achieves noticeable changes in the rotor dynamic characteristics. The Campbell diagram for the re-injection compressor is shown in figure 13.40.

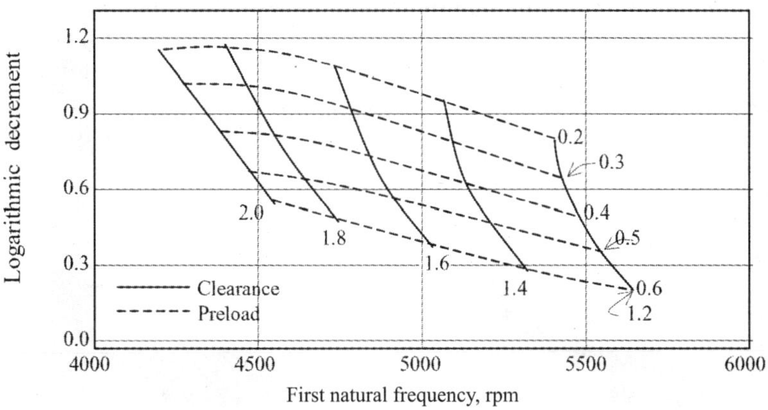

Fig. 13.39 Bearing Optimization Chart for Re-Injection Compressor (Gelin, 1996)

Even after optimization in the bearings both compressors exhibited sub-synchronous vibrations at the first natural frequency during full load tests. Attention was turned to the cross-coupled forces from the internal seals where the gas tends to confine. The seals on the impeller eyes are of teeth-on-stator labyrinth type, with only a shunt hole on the balance piston and a separation wall between the last impeller and the balance piston. With the higher flow rate the shunt hole became ineffective since pressure in the discharge volute was lower than the discharge pressure from the last impeller, and this caused the gas to enter the toothed labyrinth directly from this impeller's discharge. The undesirable condition was rectified by improving the feed into the shunt hole, installing a separation wall between the last impeller and the balance piston and using a honeycomb seal to improve the overall global damping.

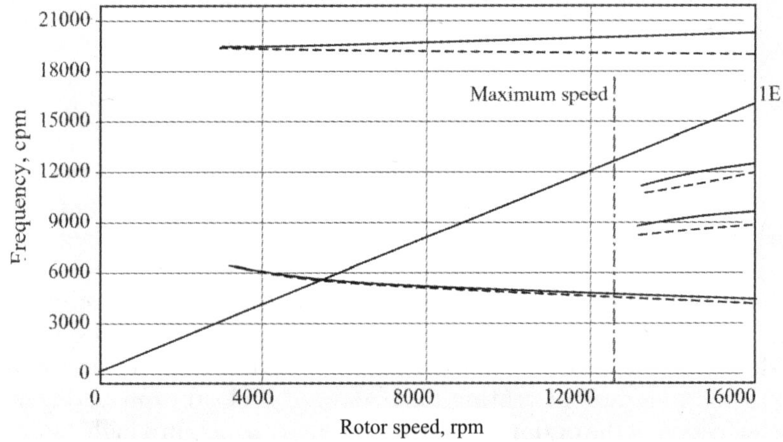

Fig. 13.40 Campbell Diagram for Re-Injection Compressor (Gelin, 1996)

The honeycomb seal conferred two interesting characteristics of increased natural frequencies. The first natural frequency increases mainly due to the mode becoming less sensitive to fluid-structure interaction. This increases the direct stiffness of the honeycomb seal. At the same time, a significant shift in the logarithmic decrement occurs due to the higher damping of the honeycomb seal. The corresponding mode shapes are shown in figure 13.41. The difference in shape is attributed to the high direct stiffness of the honeycomb seal relative to the toothed seal.

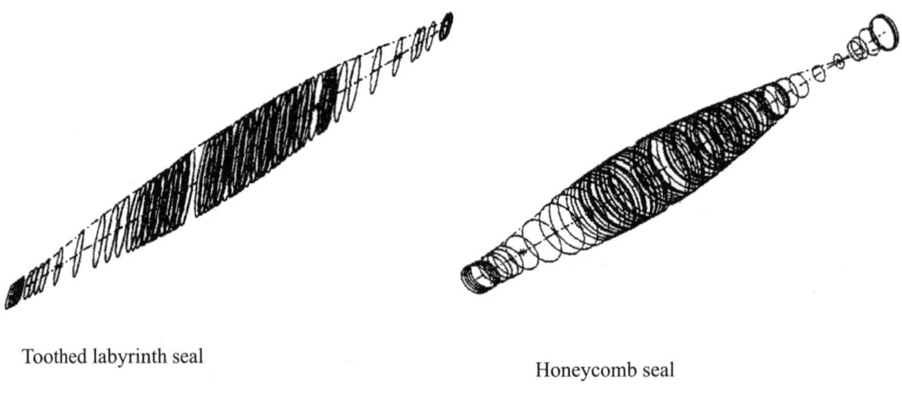

Fig. 13.41 Differences in First Mode due to Seal Type (Gelin, 1996)

13.17 LM2500 ENGINE VIBRATION CHARACTERISTICS

Among the many deleterious effects of vibrations, acceleration of wear in gas turbine engines is probably the most objectionable since it directly affects the durability characteristics of the components and the overall system. Critical speeds and associated mode shapes of the complete structure of a gas turbine engine must be understood to deal with vibration related problems. A multi-level, multi-rotor mathematical model that includes all rotating and stationary parts, including the support system, is developed to calculate the effects of rotating unbalance. Analytical predictions of natural frequencies of vibration and structural response must be correlated with actual engine vibration measurements.

Engines derived from aircraft power plant technology tend to have structurally flexible casing and support members because of thinner wall sections. Thus, many of the natural frequencies are associated with considerable dynamic activity in the stationary sections of the engine, driven by imbalance in the rotors and misalignment in the bearings. In the case of derivative engines used to power heavy military vehicles (see section 4.5), external excitation also plays a major role. For instance, when an army tank traverses a ploughed field at a speed of 20 miles an hour, the vehicle and the propulsion system are subjected to a spectrum of acceleration levels spread over a considerable frequency band. The casing and support system's geometric characteristics substantially affect vibration modes that are mostly associated with shaft bending. Understanding the analytical predictions for mode shapes provides useful insight in the matter of locating vibration measurement sensors and to determine vibration severity during engine testing.

To illustrate the process of analysing vibration characteristics of an engine system, General Electric Company's LM2500 marine gas turbine (shown in figure 13.42) will be reviewed (*ref*: Thompson et al, 1988). The US Navy has accumulated extensive fleet experience with this power propulsion engine for ships. Two gas generator rotor and two power turbine rotor critical speeds exist in the turbine's operating range. One mode, identified with the compressor rotor and observed between 8,100 and 8,500 rpm, exhibits significant dynamic activity at the compressor rear frame, a standard vibration monitoring location. Another mode attributed to the high-pressure turbine displays high activity in the 5,000 and 7,000 rpm range, and also is predominantly observable at the compressor rear frame. In the low pressure rotating system one of the natural frequencies occurs between 1,500 and 1,700 rpm and the other between 3,200 and 3,500 rpm, and the low-pressure turbine drives both modes. Interestingly, both modes are also seen primarily by the compressor rear frame accelerometer.

To understand the wear accelerating effects of vibrations in General Electric Company's LM2500 engine, the US Navy's engineers developed an analytical model for calculating the rotor dynamics and structural response (*ref*: Thompson et al., 1988). The mathematical procedure is a multi-level extension of the classical Myklestad-Prohl method (*ref*: Mykelstad, 1944; Prohl, 1945), with the high and low pressure rotating systems and the engine's casing and support structure included in the model. Previous efforts in evaluating the vibratory characteristics of the marine engine during the design phase were restricted to determining the critical speeds. The dynamic model is now adjusted to reflect the actual behaviour of the engine during operation. The gas turbine engine, shown in the upper half of figure 13.42, provides the main propulsion power for the Navy's frigates, destroyers and cruisers, and is considered a reliable workhorse of the fleet. But during certain conditions the engine exhibits vibrations that tend to accelerate wear in the components.

A number of parts cannot be modeled accurately with constant cross-section beam elements, so finite element techniques are necessitated to define deflection and rotation coefficients due to the action of shearing force and overturning bending moment in the beam. The development of expressions for the coefficients then permits their introduction into a transfer matrix form. The matrix multiplication process than proceeds normally across the hybrid elements. Transitional elements occur at joints between different levels of the structure and at points where an overhanging structural element is attached to a support point. Some examples of the latter are cantilevered

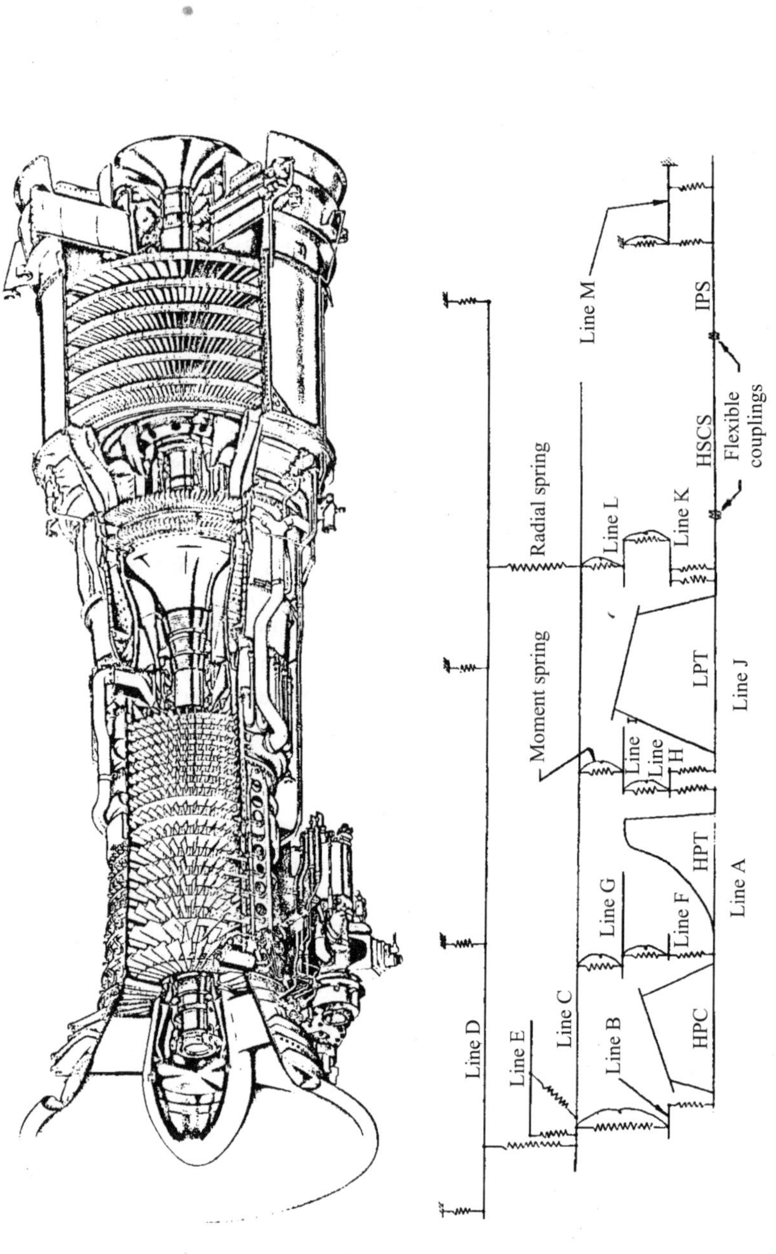

A Gas generator D Outer module G Combustor inner liner J Power turbine/HSCS/Pinion M Reduction gear case
B A – Sump E Accessory gearbox H C – Sump K D – Sump
C Outer casing F B – Sump I Turbine mid frame L Turbine rear frame

Fig. 13.42 LM 2500 Marine Gas Turbine Engine and Dynamic Analysis Model (Thomson, 1988)

segments on compressor and turbine rotors. Compatibility equations are then written for these elements to define the transfer of shear, bending moment, slope and deflection across the element or from one structural level to another.

Polar moment of inertia of the components tends to cause varying effects due to the difference in speed of the multiple rotors. This aspect is compensated by specifying the reference rotor to operate at synchronous speed, and by defining the other rotor speeds as a function of the reference rotor. In this respect the non-rotating components are assumed to be operating at the same speed as the reference rotor, so the results may be interpreted to say that the stationary parts execute dynamic motion that is synchronous with the reference rotor. The stiffness of flexible couplings, bearings and squeeze film damper bearing are defined in a similar manner. Offline calculations are required to develop the functional relationship between rotor speed and bearing load for the squeeze film damper bearing.

Model definition starts with sectioning the full structure into a series of connected elements, referred to as a 'Line'. Starting with the gas generator rotor, the lower half of figure 13.42 shows it as 'Line A'. Interconnections representing bearings and their supports, struts and mounts are defined by force and moment springs to structurally couple the line members. The springs may even be provided with viscous and damping properties for a forced response analysis. Boundary conditions for the rotating and static components with a free end take the form of zero moment and shear force at the end. Upon completion of the matrix multiplication process a square matrix is obtained, the determinant of which is zero for the natural frequencies.

The mathematical model is first used in the calculation of critical speeds and their associated mode shapes in the turbine's operating speed range. The high and low-pressure rotating systems operate in their own speed range, hence two executions are required for each rotor considered as the reference. And since engine mount system stiffness properties are determined for motion in the horizontal plane separately from motion in the vertical plane, a total of four sets of results are obtained. The four known critical speeds were predicted, and their mode shapes also correlated with the vibration survey measurements. The mode shapes provide added insight for further refinement of major structural members participating actively in the modes. The mode shapes also confirm the most sensitive locations on the outer engine case to measure the engine's response. Mode shapes for three of the critical speeds are shown in figure 13.43.

A number of the calculated modes are complementary pairs in the horizontal and vertical planes, and can be readily identified. But many other modes do not have a resembling complementary mode in the other direction, mostly because the effective mount stiffness values are substantially different, and this causes the pair's natural frequencies and mode shapes also to differ.

An investigative analysis of each of the critical speeds and associated mode shapes has been carried out (table 13.3). Many of these critical speeds turn out to be duplicates, representing a complementary pair of modes. The engine's internal components are practically symmetric about the centreline, but the mount system is asymmetric. The engine is supported by a system of links and brackets from yokes mounted to the base plate, with discrete attachment points between the engine and the yoke. The upshot of this mounting arrangement is that the stiffness property in the horizontal and vertical directions assumes differing values. In other words, a given horizontal force applied on the engine produces a horizontal deflection, but it does not match the vertical deflection if an identical force were applied in the vertical direction. But since most of the activity for a mode takes place in a selected portion of the engine, the analytical model calculates two separate modes of vibration, one of them corresponding to the vertically oriented mount stiffness and the other to the horizontal direction stiffness. To some extent this aspect of the mounting system of the engine is also responsible for the spread in the engine speed observed for any particular mode of vibration during engine operation. Note also that if a vibration mode does not show participation of the engine mount the variance in the horizontal and vertical critical speeds will diminish.

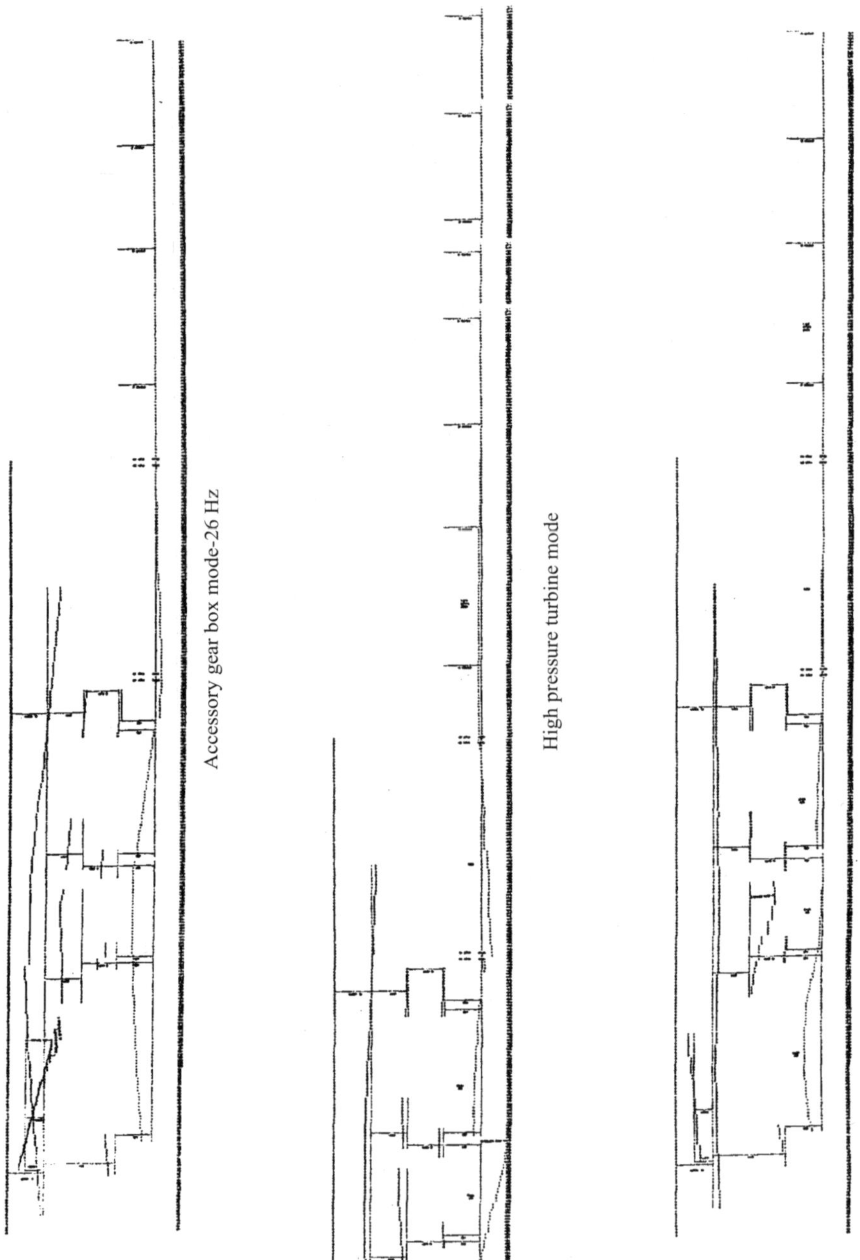

Fig. 13.43 Mode shapes for known critical speeds of LM 2500 engine (Thomson, 1988)h

Table 13.3 LM2500 Gas Turbine Engine Critical Speeds

Mode #	*Reference Rotor System*	*Speed RPM*	*Mount Stiffness Direction*	*Description of Associated Mode Shape*
1	PT	331	Vertical	System rocking Mode
2	PT	346	Vertical	System rocking Mode
3	PT	835	Vertical/Horizontal	1st stage HPT nozzle
4	PT	1090	Horizontal	Accessory gearbox
5	PT	1400	Vertical	Accessory gearbox
6	PT	1632	Horizontal	Gas turbine bending/Accessory gearbox
7	PT	1916	Vertical	Gas turbine bending/Accessory gearbox
8	PT	2565	Horizontal	PT rotor
9	PT	3520	Vertical/Horizontal	PT rotor/Case
10	GG	4400	Vertical/Horizontal	PT rotor/Diffuser
11	GG	5400–9000	Vertical/Horizontal	HPT rotor/Case
12	GG	5227	Vertical	PT rotor/CRF
13	GG	5615	Horizontal	Drive shaft/CRF
14	GG	5997	Vertical	PT rotor/CRF
15	GG	6226	Horizontal	Drive shaft/CRF
16	GG	6558	Horizontal	PT case/GG rotor/CRF
17	GG	7000	Vertical	PT case/GG rotor/CRF
18	GG	7500	Vertical	PT case/GG rotor/CRF
19	GG	8400/8600	Vertical/Horizontal	CRF/Compressor rotor
20	GG	9500/9750	Vertical/Horizontal	CRF/Compressor rotor

Notes: 1. Modes 11, 19 and 20 are observed critical speed, cannot be readily identified in analysis, response remains high in given speed range.

2. Modes 3, 9, 10, 11, 19 and 20 are not affected by mount stiffness.

3. See text for discussion on mount stiffness direction.

Legend: CRF - Compressor rear frame
GG - Gas generator rotor
PT - Power turbine rotor

For analysing dynamic characteristics of the mount system careful attention must be paid to the pair of horizontal and vertical vibration modes. Dynamic loads and stresses in the components will vary for the modes. Finite element evaluation is then required to establish and evaluate the suitability of the component's design traits.

Stiffness, or spring, values used in the analytical model needs some elaboration. For a component such as the compressor rear frame a three-dimensional finite element model is a prerequisite. Boundary conditions reflecting the physical realities of the component as it is assembled and operates in the engine system must be applied in the model. A static force is then applied, and deflections calculated at appropriate locations of the model for deriving the stiffness coefficients. Derivation of mass and inertia values, on the other hand, may not be as complex. As a first check the component may be pan weighed. Most structural mechanics computer programs generally provide location of the component's centre of gravity and mass inertia properties about the major axes as a matter of routine.

In the calculation sequence for system response, damping properties must be described. Aside from damping associated with the lubricant film in the bearings and material damping because of motion between the molecules in the metal, frictional damping in the structure plays a major role in the determination of response. Thus, if a bolted flange joint is not adequately tight as called for in drawing specifications, slippage between the mating parts will be the result. Damping values are generally obtained from bench test on components, and from observations during engine test.

Table 13.3 gives a listing of critical speeds for the LM2500 engine. The speed parameter applies to the high or low-pressure rotor to which the associated mode is ascribed. The ratio of the power turbine speed to the gas generator speed introduces important and substantial inertial effects, which is maintained at a constant level in the calculation sequence. However, this ratio assumes varying values in the full operating speed range, hence the power turbine speed can change both the frequency and the mode shape of the gas generator's critical speeds.

A series of vibration modes involving components in the back end of the engine, such as the turbine rear frame, power turbine case, diffuser and coupling shaft, may hold significance. Higher than usual wear patterns have been detected in these components, possibly because of the large dynamic movement. To investigate the situation, a sudden increase in activity was uncovered at the predicted modes. A detailed analysis of the vibration signature of the accelerometers mounted on the turbine rear frame and the power turbine case confirmed the mathematical predictions. Primary exciting force behind this mode was determined to arise from unbalance in the gas generator rotor. Additional trim balancing of this rotor was successful in reducing the response in the back end of the engine.

Evaluation of the previously unknown modes indicated many of them display considerable participation of components in the back end of the engine. The turbine rear frame, power turbine case, diffuser and the high speed coupling shaft play a major role in two of the modes, shown in figure 13.44. They provide some reasons for the wear pattern observed in the Navy's operating LM2500 engines, and are corroborated by a sudden increase in response at the turbine rear frame at speeds predicted in the analysis. For further confirmation, detailed analysis of vibration data from an engine experiencing a severe case of the wear out modes gave the right correlation with the mathematical predictions. The primary excitation behind this mode comes from unbalance in the gas generator rotor. In-place trim balancing of the high pressure compressor was successful in lowering the response at the compressor rear frame from 6.0×10^{-3} in. double amplitude at 8250 rpm and 4.5×10^{-3} in. double amplitude at the turbine rear frame between 7000 and 7500 rpm down to 2.7×10^{-3} in. at the compressor frame and 2.2×10^{-3} in. at the turbine frame.

The multi-level dynamic model also offers the potential for evaluating deflections in the engine casing and internal structural components and relative motion between the elements, referred to as a 'Line'. Starting with the gas generator rotor, the lower half of figure 13.42 shows it as 'Line A'. Interconnections representing bearings and their supports, struts and mounts are defined by force and moment springs to structurally couple the line members. The springs may even be provided with viscous and damping properties for a forced response analysis. Boundary conditions for the rotating and static components with a free end to take the form of zero moment and shear force at the end. Upon completion of the matrix multiplication process a square matrix is obtained, the determinant of which is zero for the natural frequencies.

The mathematical model is first used in the calculation of critical speeds and their rotor and the casing. Several engines removed for high vibrations were noted to have broken retaining pins for the combustor. High dynamic activity of the combustor in most of the high-speed modes was able to explain the cause of failure of the retaining pins.

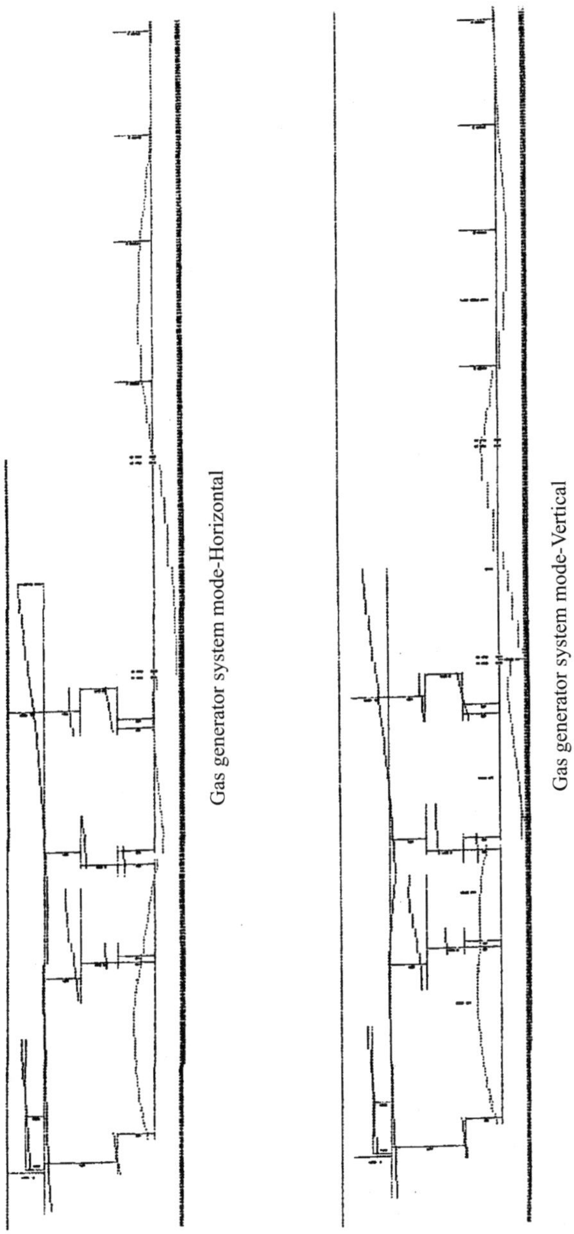

Fig. 13.44 Newly Identified Mode Shapes of LM 2500 Engine (Thomson, 1988)

Four of the modes in the power turbine's low speed range between 1000 and 2000 rpm are of special interest, where the accessory gearbox is the most active part on the engine. Two of the modes see the gearbox moving in phase with the engine casing, while the other two modes have the gearbox motion out of phase with the case. The analytical results confirm and add clarity to observations on the fleet engines: the mount links of the gearbox wear, become loose and permit the assembly to move relative to the engine. Since the gearbox supports the main fuel pump and its control, the lubricating oil pump, the starter, the air/oil separator and the transfer gearbox, redesign of the mounting system is under study.

Two other modes in the low speed range involve translation and rocking motion of the engine's enclosure module, mostly due to the stiffness values of the mount linkages, but engine operation does not take place at the speeds.

Vibration surveys on a number of the engines indicated that response of the high-speed modes in the gas generator running range was to some extent dependent on the power turbine speed, and occurred during pier side operation when the power turbine is unloaded. To understand this behaviour, note that the two rotating systems in the engine are not mechanically connected, but a number of structural members in the bearing housings, support system and casing are present to cause the two rotating systems to interact. The ratio of the power turbine speed to the gas generator speed introduces considerable inertial effects. A given power turbine speed defines the stiffness of the flexible diaphragms in the high speed coupling shaft, through which engine power is transmitted to the reduction gears. Thus, the frequency at which the mode displaying high response at the turbine rear frame due to compressor unbalance depends on the speed at which the power turbine runs. Figure 13.44 shows gas generator system modes.

The dynamic representation of the engine shown in figure 13.42 includes four input pinion shaft bearings for the reduction gear. The bearings are of the plain journal type, and their stiffness and damping coefficients rely on shaft speed and imposed load. The flexible diaphragms in the coupling do not cause the pinion shaft to affect the high-speed modes, but the effect from the bearings on the low speed modes can be pronounced. The coupling shaft tends to isolate the power turbine rotor from excessive loads and motion arising in the drive train. But if a condition arises where the pinion shaft bearings run with little load, their reduced effective stiffness and damping could cause damaging vibrations in the coupling.

13.18 PREVENTING SUB-SYNCHRONOUS VIBRATIONS IN LARGE STEAM TURBINES

Low frequency vibrations in turbine rotors are an important issue in developing higher capacity and advanced steam condition units. The non-synchronous vibrations set in as the load is increased, and are caused by steam whirl, flow disturbance in the control stage or rubbing between the rotor and the casing. Whirling of the shaft from the steam is a self-excited vibration of the first vibration mode at high loads, and is similar to, but not as severe as, oil whip. In addition, steam whirl can occur even when tilting pad bearings are used to control the oil whip. Exciting forces generated by the blades and seals have been investigated but cannot be accurately estimated during the design stage (*ref*: McClosky et al., 1992). On the other hand, vibrations caused by flow disturbance in the control stage and by rubbing are forced vibrations that result in exciting the first rotor mode (*ref*: Wu et al., 1993).

To solve the problem, an analytical and vibration testing of the turbine's rotor under actual operating conditions has been carried out by Mitsubishi Heavy Industries of Japan (*ref*: Kanki et al., 1997). The results indicate that the low frequency vibrations are due to a reduction in the available damping at high loads, and the falloff in the damping is caused by the presence of unstable forces in the control stage. As a counter-measure, a squeeze film damper has been designed and tested to verify its damping properties. The damper mechanism is then installed on the turbine unit and the test sequence performed at actual operating conditions. The field tests confirmed the increased damping in the system introduced from the squeeze film damper, and the low frequency vibrations have been eliminated.

The rotor in question comes from a 450 MW high-pressure steam turbine operating at 3000 rpm. Eight governing valves connected to the inlet adjust the flow of steam. The unit experienced a strong low frequency (27 Hz) vibration. Figure 13.45 provides the shaft vibration signature in bearing # 1 and low frequency trajectory in bearing # 2 at full power output during a typical run. The direction of the whirl is forward. The occurrence and threshold of the low frequency vibration change if the opening sequence of the eight valves is altered. The damping ratio of the first mode is measured using an inertia type of exciter placed on the bearing through a jig, and the natural frequency and damping ratio are obtained by measuring the frequency response. The stability of the rotating system is evaluated by using a swept frequency sine wave to excite the bearing pedestal, followed by measurement of response through an appointed natural frequency. The procedure is repeated for a number of speed and load points. A gradual reduction in the damping ratio of the rotor is noted with the increase in turbine output up to about 300 MW, when the damping coefficient suddenly decreases at points where the load exceeds 300 MW.

The test thus clarifies that the low frequency vibration is self-excited by the steam whirl when the damping is low. The destabilizing force is generally considered to arise from stage excitation, also known as the Alford force, and from the labyrinth seals. These forces may be represented by eight stiffness and damping coefficients. To determine the effects of these forces on the stability of the system, the dynamic coefficients are incorporated into the complete rotor model in the same manner as bearing coefficients. The analytical model for the full high and low-pressure turbine and generator rotors is shown in figure 13.46. The vibration analysis uses the transfer matrix method. The objective of this stability analysis is to determine the quantitative effects of partial admission of the control stage and alignment between the control stage and the nozzle.

Improvement in the damping ratio for the first rotor mode is achieved by replacing the # 2 bearing with a squeeze film damper bearing, shown on the left in figure 13.47. The inner ring of the bearing is fitted to the outer ring by 36 pins. The design procedure for the squeeze film damper aims at placing its stiffness and damping characteristics in between those for the bearing pedestal and the tilting pad bearing. The spring constant for the pins and the film's damping coefficient are selected by taking into consideration the need for maximum damping ratio and to avoid contact between the inner and outer rings under the combined action of the rotor weight and partial admission at the control stage. Setting the bearing clearance to 0.25 mm provides the optimum damping coefficient, even when the static displacement (calculated from the three-dimensional finite element model shown on the right in figure13.47) produced in the housing is taken into account and shaft eccentricity becomes zero.

Verification of the damping effect developed in the squeeze film damper bearing is obtained from shop tests. The required increment in the damping coefficient can be seen in figure 13.48 (*left*). The high-pressure rotor is then installed at the site and vibrations measured at different operating

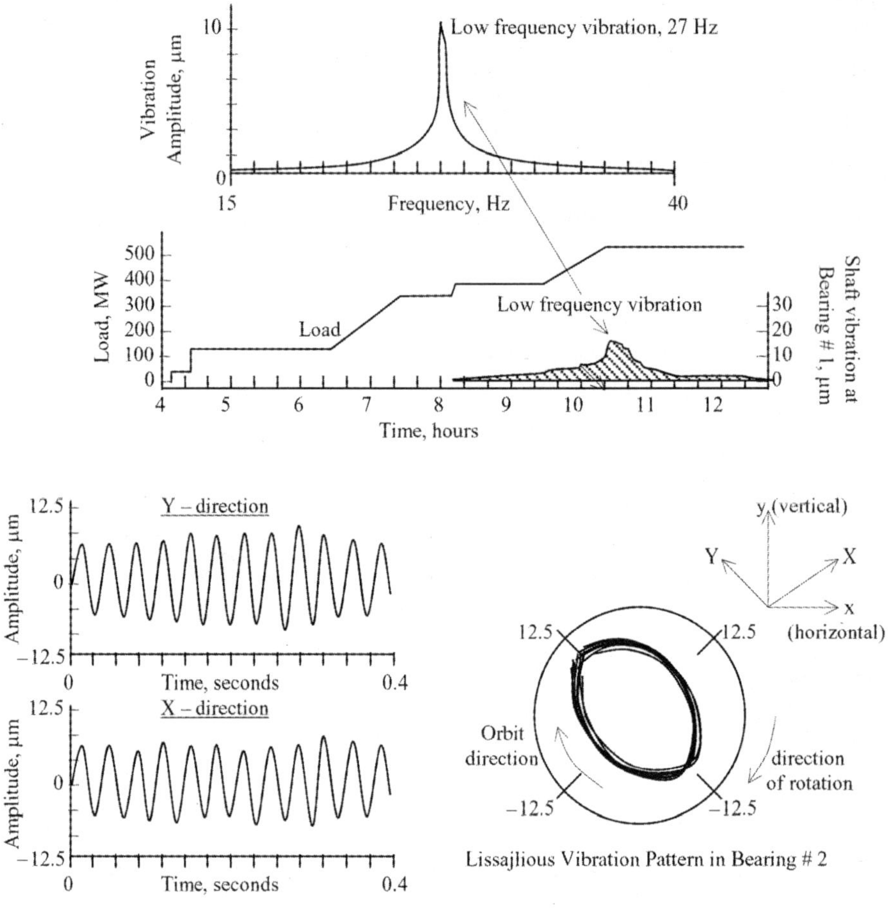

Fig. 13.45 Shaft Low Frequency Vibrations (*upper*) and Trajectory (*lower*) (Kanki, 1997)

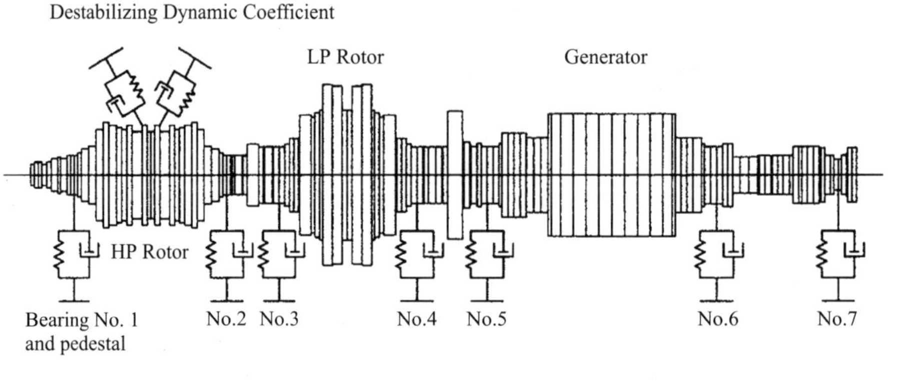

Fig. 13.46 Rotor Dynamics Model with Destabilizing Dynamic Coefficients (Kanki, 1997)

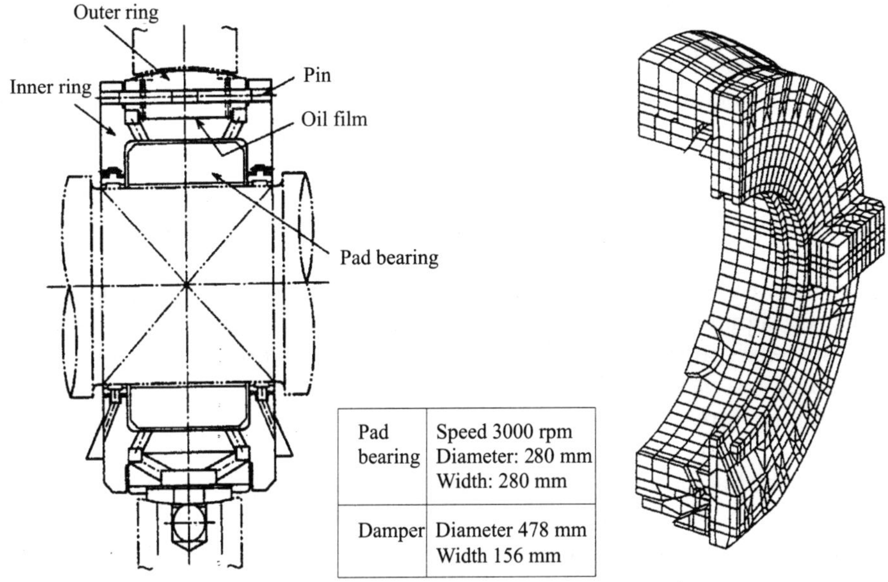

Fig. 13.47 Squeeze Film Damper Bearing: Design (*left*), Finite Element Model (*right*) (Kanki, 1997)

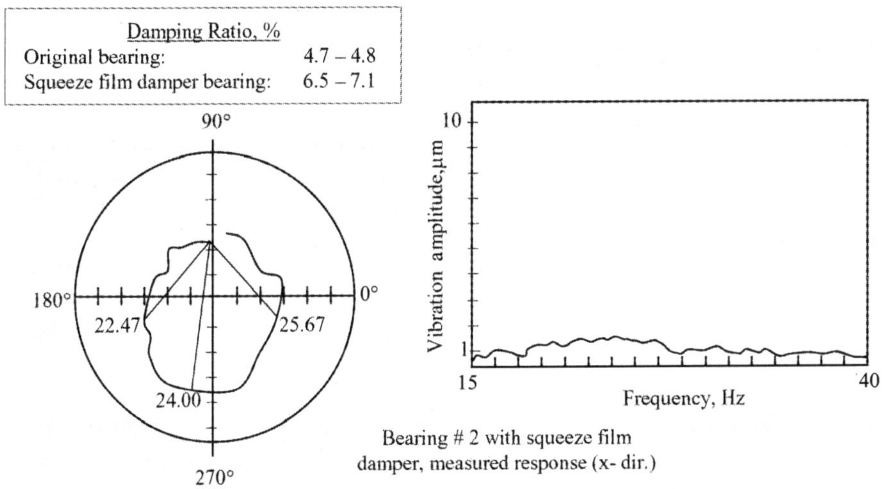

Fig. 13.48 Measured Damping Ratio (*left*), Shaft Vibrations in Bearing (*right*) (Kanki, 1997)

conditions. The first mode's damping ratio is also measured. Figure 13.48 (*right*) shows the spectrum of shaft vibration at rated load. The results provide evidence that the low frequency vibration is fully eliminated under all conditions as a consequence of the increased damping by using the squeeze film damper bearing. The turbine has been satisfactorily operating free of the low frequency vibrations for two years after the installation of the squeeze film damper.

13.19 WHIRL INDUCING FORCES IN AXIAL COMPRESSOR

An increase in the radial clearance often results from rubs between rotating and stationary aircraft engine components operating at tight levels of clearance. The most frequent causes are force induced by rotor unbalance, lateral deceleration during a hard landing, lateral forces induced by high acceleration and turning manoeuvres, thermal bowing and/or asymmetric ovalization of the casing, especially for fans. However, a potentially much more destructive mechanism for inducing rubs is whirl instability. Any radial deflection of the rotor relative to the stator creates circumferentially non-uniform clearances and unsteady aerodynamic forces on the rotor as each blade traverses the varying clearance gap. These unsteady forces are orthogonal to the deflection, and therefore are a significant driver of rotor whirl instabilities. The forces increase in magnitude as the deflection increases so that above the onset speed, where destabilizing forces overwhelm the stabilizing damping forces, the deflections are ultimately limited only by damage to the interacting parts or by damping forces. Consequently, an accurate determination of their magnitude and direction is of major importance in the design of safe, stable turbo-machinery components. Hysteretic whirl, whirl associated with fluids trapped within cylindrical rotor cavities and plain journal bearings (see section 13.12) are some typical examples.

Unsteady and destabilizing aerodynamic cross-axis stiffness forces that promote rotor whirl were first postulated by Thomas (1958) and Alford (1965) to explain rotor whirl instability mechanisms. In a deflected turbine rotor the airfoils in the closure zone are more highly loaded by aerodynamic forces than the airfoils in the open clearance zone because the former are operating more efficiently. This situation is shown schematically in figure 13.49(*a*) for a turbine rotor whose centreline has been displaced upward along the ordinate by an amount $+Y$. This gives minimum clearance at the top of the turbine and maximum clearance at the bottom. The forces at these two locations are the vector sum of the mean blade force F_m and the unsteady blade force resulting from the centreline offset F_u. Summing the forces perpendicular to the axis of displacement results in a net force $F_X = F_m + F_u$, due to the difference in airfoil loading. Since F_X acts normal to the axis of displacement, it is a cross-coupled stiffening force, with its direction acting to drive the rotor in an orbital whirling motion about the centreline in the same direction as rotor rotation. Thus, F_X promotes forward whirl for turbines, with the cross-coupled aerodynamic stiffness coefficient relating acting torque T and coefficient β by the expression (D_p = mean blade diameter, H = blade height):

$$K_{XY} = F_X/Y = T\beta/(D_pH) \qquad ...(13.11)$$

As the rotor whirls the circumferential variation in radial tip clearance causes a circumferential variation in efficiency, so that the blading with the smallest clearance is the most efficient. Alford hypothesized that the compressor would pump to a circumferentially uniform exit static pressure and therefore the more efficient blading at tight clearance would have a lower loading than the blading with larger clearance 180° away. This situation, illustrated in figure 13.49(*b*), shows that the net force, F_X, tends to cause forward rotor whirl. Thus, Alford concluded that compressors have positive β's so that whirl-inducing forces for both compressors and turbines are in the same direction.

Ehrich (1993) hypothesized differently from Alford. He reasoned that compressor airfoils with the smaller clearance would sustain a higher static pressure differential across their tips and would therefore be more highly loaded than the airfoils with larger clearance 180° away. As shown in

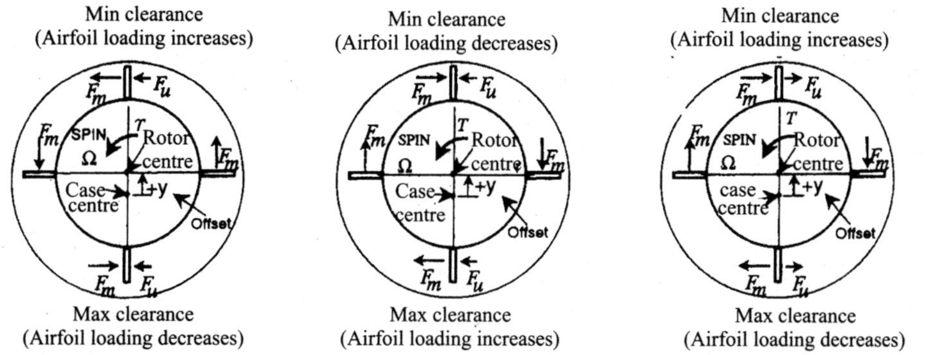

Accepted model for turbine drives forward whirl, β is positive.

(a)

Alford's model for compressors drives forward whirl, β is positive.

(b)

Ehrich's model for compressors drives backward whirl, β is negative.

(c)

Fig. 13.49 Mechanisms for Whirl Inducing Forces (Storace, 2000)

igure 13.49(*c*), this dictates that the net destabilizing force in compressors, F_X, tends to produce otor whirl counter to the direction of rotation. Thus, Ehrich concluded that compressors tend to ave negative β coefficients so that the direction of whirl-inducing forces for compressors would be pposite to those for turbines.

In view of the disparity between Alford's and Ehrich's conjectures and the absence of a decisive esolution of rotor whirl issues, General Electric's Aircraft Engines formulated a research program *ref*: Storace et al., 2000 and Ehrich et al., 2000). The test is conducted on GE's Low Speed Research Compressor to simulate the eccentricity of a whirling rotor and to measure the developed non-uniform and unsteady flow field. The test rig, which has a constant casing diameter of 1.524 m, is set up with four identical stages in order to simulate the repeating stage environment. The third stage s the test stage. The blading is representative of current design practice. Three different low-speed blading configurations are tested, being typical of modern designs, and have high hub/tip ratios of .85 with low aspect ratio, high solidity blading and shrouded stators. These compressors are low-peed, aerodynamic models of the middle and rear block of highly loaded, high-reaction 65 to 70 per cent HP compressors in commercial gas turbine engines currently in service. Another design, lso in commercial engine service, has cantilevered stators and blading with a lower hub/tip ratio of .70, lower reaction of 0.55, and higher aspect ratios.

Tests were performed for the configuration with two displacements of the casing centreline: a displacement of 0.1905 cm and a smaller one of 0.0965 cm. The rotor tip and the stator shroud seal re ground to allow both to run at the absolute minimum safe clearance of 0.051 cm, permitting the ehicle to enter a rotating stall. High-resolution pressure transducers, accurate to 0.01 per cent of the full-scale values of either 0.068 or 0.136 bar (1 or 2 lbs/in^2), are used to record steady state static nd total pressures for determining both overall compressor performance and the static pressures n the stator airfoil surfaces. High response Kulite pressure transducers, having a frequency esponse of 20 kHz, are imbedded inside the rotor airfoil surfaces to measure the unsteady static ressures acting on the suction and pressure surfaces. The signals from the pressure transducers are igitized and ensemble-averaged. The analog data are low-pass filtered at 1 kHz to avoid aliasing sing sufficient ensemble averages to reduce the effects of time-unresolved unsteadiness. A once-er-revolution pulse from an optical encoder in the casing senses the trigger airfoil on the rotor blade and initiates the sampling for the data windows for each rotor revolution. The raw pressure data for each transducer are processed using fast Fourier transform (FFT) methodology to give a filtered waveform consisting only of the first harmonic. The first harmonic is dominant. The 54th and 74th harmonics observed in the pressure patterns are associated with 54 rotors and 74 stators. Unsteady pressures measured at the discrete locations on the airfoil surfaces are bi-directionally curve-fit along the radial and chord wise directions to obtain continuous pressure distributions on the suction and pressure surfaces. By using the unsteady pressure distributions computed above and the geometric orientation of the airfoil surfaces in the compressor, integration over the airfoil surfaces obtains the unsteady forces acting on the various airfoils around the circumference. The unsteady forces are resolved into their tangential and radial components in the blade's coordinate system. The component of force acting perpendicular to the direction of the rotor offset (F_X, the Thomas/Alford force) is obtained by transforming the tangential and radial blade forces to the global coordinate system fixed to the bladed disk. The β coefficient is calculated from equation (2.27).

The effects of clearance variation and centreline offset on the overall compressor performance is observed as a four-stage average of pressure coefficient and efficiency plotted as a function of the flow coefficient, referred to as the pressure characteristic. Stall margin is computed for these low speed tests in terms of throttle margin. Variations in the loading levels from high flow (low-loading) to stall were achieved by varying mass flow rate through the compressor using a discharge throttle. Lines of constant throttle setting are shown in the figures 13.50 and 13.51 to indicate the different loading levels along the pressure characteristic. The tests are run at the design tip speed of 64.0 m/s requiring a rotational speed of 804 rpm. This gives a Reynolds number of 3.6×10^5, representative of engine conditions. The baseline performance of compressor A, shown as curves Al, is established with no centreline offset and with circumferentially uniform, nominal levels of rotor-tip clearance and stator shroud-seal clearance. The design point is shown in both figures. The negative slope of the baseline pressure characteristic over all of the flow range from high flow to near peak pressure provides stable operation over this range, after which it begins to roll over. Stall occurs at a flow coefficient of about 0.335, as indicated by the short vertical line at the low-flow end of the pressure characteristic. The baseline compressor A has high efficiency that peaks at 90.4 per cent. It also has a good throttle margin of 30.2 per cent, as indicated by the 17.9 per cent flow range from the design point to stall. This baseline is the performance standard against which all of the other configurations are to be compared.

The effects of varying axi-symmetric (circumferentially uniform) clearance on the performance of the compressor are shown relative to the baseline performance in figure 13.50. The clearance variation is obtained without centreline offset and thus is typical of what occurs in a uniform rotor tip rub. Both rotor tip clearance and stator shroud-seal clearance are varied independently to separate their effects. Detailed surveys of flow properties show that when rotor tip clearance alone is increased, the dominant effect is seen in the outer 25 per cent of span. Similarly, when the stator seal clearance alone is increased, the dominant effect is seen in the inner 25 per cent of the span.

The effect of variation in the centreline offset on the performance is observable from the pressure and efficiency characteristics for the two levels of casing centreline offset, and are compared to those for the baseline configuration. As expected, the offset configurations with their larger average and maximum clearances have lower efficiencies and lower peak pressure rise than those for the baseline case. There is an 8.3 per cent and 10.8 per cent loss in peak pressure and 1.5 point and 2.7 point loss in peak efficiency for the small and large offsets. Throttle margin is 11.7 per cent which is 39 per cent of the baseline value. Thus, stall margin has suffered considerably. Note that there is little change in loss of stalling flow range between small and large offsets.

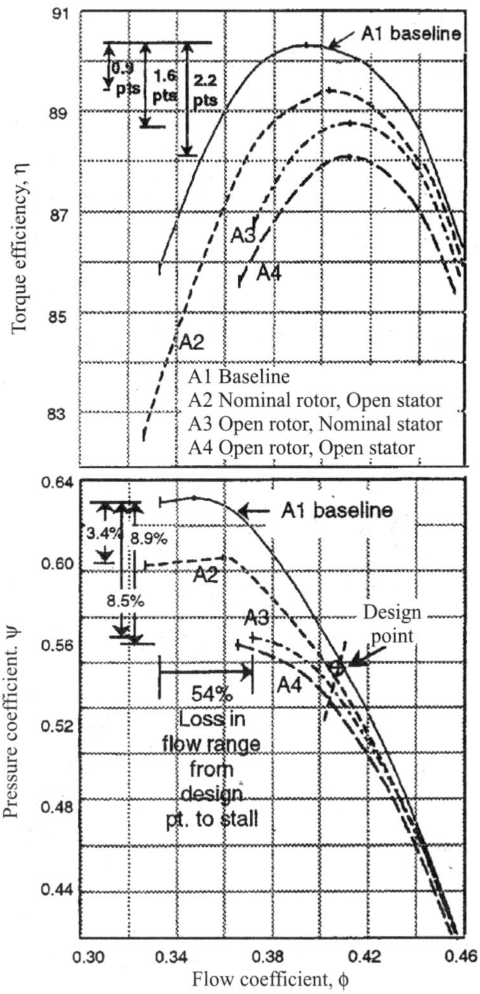

Fig. 13.50 Effects of Variation of Clearance on Performance of Compressor A (Storace, 2000)

Unsteady pressures on the rotor airfoils are measured as the rotor travels through varying levels of clearances caused by the centreline offset. Suction surface pressures are subtracted from pressure surface pressures to give unsteady pressure difference loading on the airfoil. Representative results showing this loading for three clearance levels around the circumference are presented in figure 13.53 as contours of differences in unsteady static pressure, with the rotor at near minimum, nominal and near maximum clearance. The measured steady state surface static pressure distribution at 80 per cent span for the rotor airfoil is shown in figure 13.54. Airfoil unsteady loading across the span increases in the region of minimum clearance and decreases in the region of

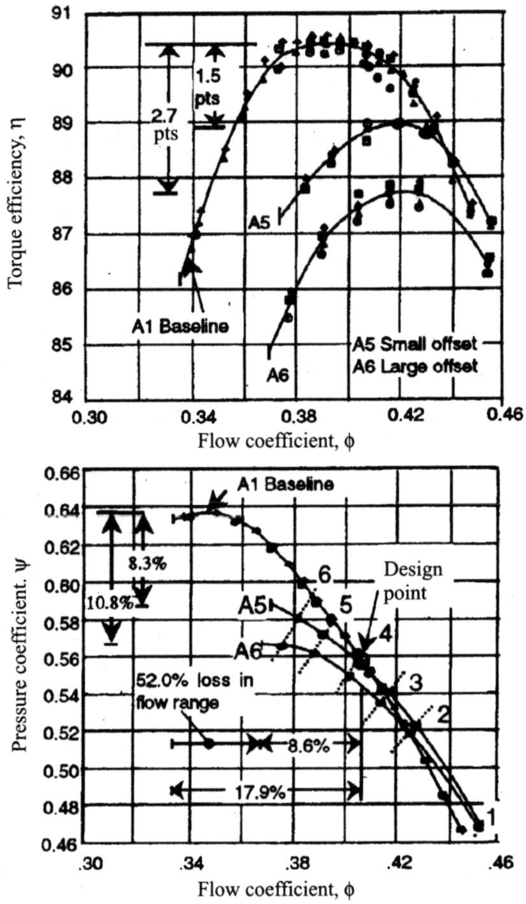

Fig. 13.51 Effects of Rotor Centreline Offset on Performance of Compressor A (Storace, 2000)

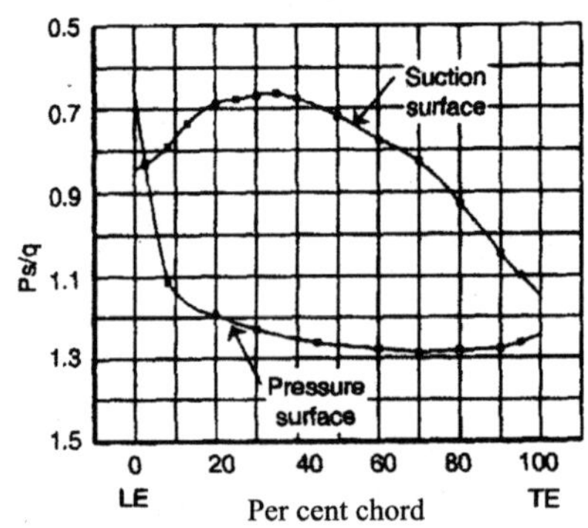

Fig. 13.52 Steady State Static Pressure Distribution on Airfoil (Storace, 2000)

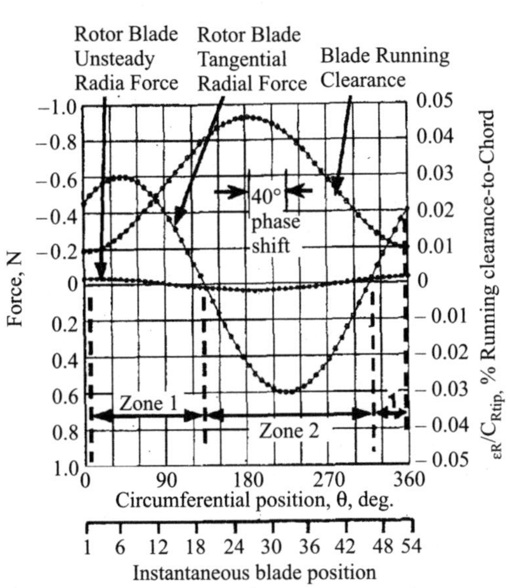

Fig. 13.53 Variation of Whirl Inducing Forces with Tip Clearance (Storace, 2000)

maximum clearance relative to the mean loading at nominal clearance. It is also clear that near the rotor tip in zones 1 and 2 of figure 13.53 the airfoil unsteady tangential loading increases at near minimum clearance and decreases at near maximum clearance. This finding confirms the correctness of Ehrich's hypothesis in figure 13.49(*c*) about the nature of rotor whirl-inducing forces in compressors.

The unsteady pressures are reduced to unsteady blade forces as described earlier. A representative result for the large offset configuration running at high compressor loading (point 5 in figure 13.51) is shown in figure 13.53. The tangential and radial components of the unsteady, whirl-inducing blade force are plotted as a function of circumferential position. The circumferential variation of rotor tip clearance is also plotted. The forces need to be understood relative to the direction of rotation and the direction of the driving torque, which are counterclockwise. The centreline is offset upward, which places the minimum clearance at the top dead centre. Two forces act on each airfoil, a mean force, F_m, and the tangential component of the unsteady force due to the offset, F_u. The magnitude and direction of the unsteady force will determine the direction of rotor whirl. The radial component of unsteady force is very small compared to the tangential component. The maximum tangential force occurs near the minimum clearance and the minimum tangential force occurs near the maximum clearance, again confirming the correctness of Ehrich's hypothesis. Careful use of the sign convention leads to the conclusion that in the region of negative forces shown as zone 1 in figure 13.53 the net forces acting on the airfoils increase because the unsteady force, F_u, adds vectorially to the mean force, F_m. This occurs around the minimum clearance. In the region of positive forces shown as zone 2 in figure 13.53 the net forces on the airfoils decrease as the unsteady force opposes the mean force. This occurs around maximum clearance near the bottom dead centre. The net effect of the force distribution tends to drive the offset rotor shaft counter to the direction of rotation to drive a backward whirl.

13.20 OPEN-LOOP CONTROLLER FOR AERODYNAMIC FORCE MEASUREMENT

Measurement of hydro and aerodynamic forces in rotating machinery is complicated by a number of features that contribute undesirable parasitic forces to the measured or test forces. The parasitic forces arise from seals, drive couplings and hydraulic or inertial unbalance. Accurate and sensitive measurement of the hydro and aerodynamic forces can be accomplished only by subtracting the parasitic forces.

Both the tests and parasitic forces are generally dependent on the machine's operating conditions, including the specific motion of the rotor. Hence, to properly remove the parasitic forces the vibration orbits and operating conditions must be identical in tests for determining the hydro and aerodynamic forces and in tests for determining the parasitic forces. This in turn requires a method by which the test rotor's motion can be accurately controlled to an arbitrarily defined trajectory.

The motion of a rotor can be obtained with the help of a multiple harmonic open-loop controller (*ref*: Baun et al., 2001). The technique is implemented on a laboratory pump with its rotor supported on magnetic bearings. The apparatus is intended for studying hydrodynamic impeller reaction forces, which are directly measured by the magnetic bearings. Forcing precise user defined and repeatable motion on the pump's rotor demonstrate the utility of the controller. The multiple harmonic open-loop controllers produces sinusoidal current signals at sub-harmonic, synchronous and super-harmonic frequencies of the rotor's spin speed.

The controller is used on a Plexiglas pump rotor shown in figure 13.54. The pump has a nominal speed of 620 rpm. The magnetic bearings have four lateral and one axial control degrees of freedom; open-loop control is used for the four lateral control axes. The open-loop controller, implemented through a commercial PC and a 100 kHz analog to digital converter card for vibration signal input and a digital to analog converter card for open-loop control signal output. Figure 13.55 provides the block diagram for the rotor-bearing system, stabilizing feedback controller, open-loop controller, and interrupt service routine. An encoder driven directly from the pump shaft generates a 256 times running speed phase-locked digital signal.

A pre-calculated schedule of control signals, $u_{fb}(t)$, is applied to a system to tailor its forced response in an open-loop control. The open-loop signals are superimposed on the stabilizing feedback control signals in this case. In the adaptive control system the elements in the system influence coefficient matrix are updated, or adapted, to the previous schedule of open-loop signals.

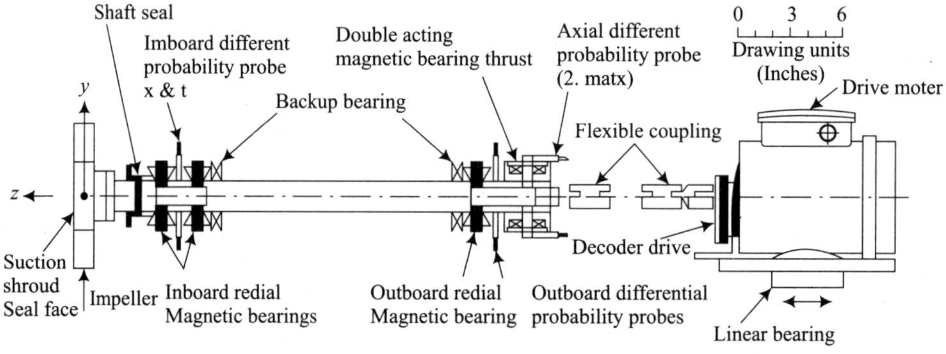

Fig. 13.54 Plexiglas Pump Rotor

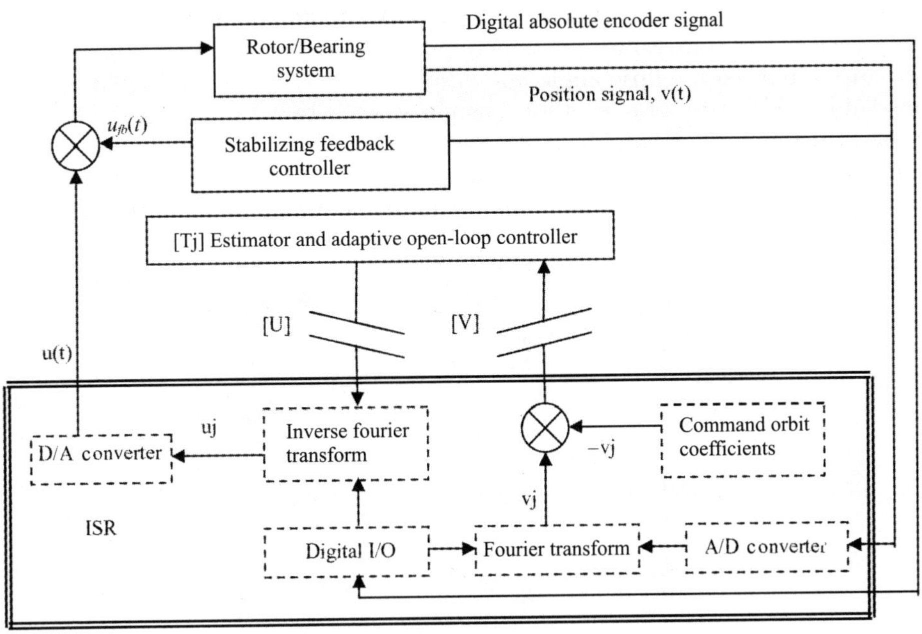

Fig. 13.55 System Block Diagram

This method is well suited to turbo-machinery rotors that are subject to synchronous and multiple super synchronous disturbance forces. These forces may also vary as the machine wears or when the operating speed changes.

The method requires prior knowledge of the system. To determine the amplitude and phase of the open-loop control signals, it is necessary to estimate a system influence coefficient matrix, which relates system response at a given frequency due to the open-loop control signals at the same frequency. This matrix may be estimated by the least-squares balancing technique where influence coefficients are determined by applying known unbalance forces at balancing planes along a shaft and measuring the rotor's response at position sensor locations. Since this rotor is supported in magnetic bearings, the bearings themselves may be used to apply rotating magnetic forces. The position sensors, required to close the feedback loop in a magnetic bearing system, provide the system response.

To demonstrate the utility of the procedure, the open-loop controller is used to manipulate the lateral (x-y plane) motion of the rotor. Total shaft run-out at each position sensor location is mapped in a slow roll test. Fourier coefficients of the run-out position signals are superimposed with the command orbit Fourier coefficients.

In the absence of open-loop control the predominant vibration at one control plane occurs at four times the running speed, 41.33 Hz, and is due to the vane passing frequency (4 vane impeller interacting with a single tongue in the volute). The vibration level, plotted on a diagram for one complete shaft revolution, is measured to be 38 μm double amplitude in the y direction and 8 μm double amplitude in the x-direction. Lesser spikes in vibration are noted at frequencies ranging from the second to the eighth multiple of shaft speed. With the application of the open-loop controller, the orbit size is reduced to a dot at the origin.

The open-loop controller is designed to accept a user requested command orbit at any of the control frequencies. The command orbit is entered as a vector whose components represent the semi-major and minor axes, ellipse angle and amplitude phase angle of an ellipse. Figure 13.56 shows the resulting rotor vibration orbit for a synchronous command orbit vector (0.004", 0.001", 25°, 45°}, or {101.6 μm, 25.4 μm, 25°, 45°}, while the second, third and fourth order vibration components are attenuated to zero. Within the range of open-loop control frequencies, only a synchronous vibration component is observed in the frequency spectra of the X and Y position signals.

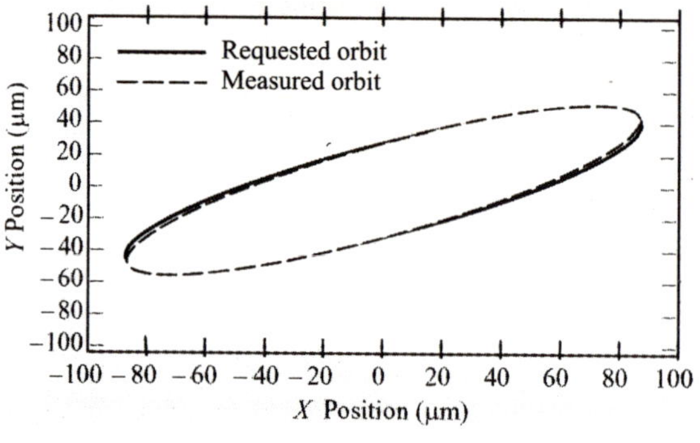

Fig. 13.56 Comparison Between Requested and Measured Orbits (1×)

The example illustrated here, while not exhaustive, demonstrates the effectiveness of the open-loop controller to simultaneously control multiple harmonics to a user prescribed condition. In the selected test cases, open-loop control is applied to four or five harmonics. While in principle any number of harmonics may be controlled at the same time, in practice the number will be limited by the controller processor speed and the sampling frequency of the input and output hardware. Poor signal to noise ratio, aliasing, and coupling between the harmonics can degrade the controller performance, particularly for higher harmonics.

13.21 REVIEW PROBLEMS

Example problem # 13.1: A 24.0 in. diameter pulley B weighs 145 lbs and receives 33.5 HP at 325 rpm from a shaft located at an angle of 45° below the pulley (figures 13.57 and 13.58). A 19.25 in. pitch diameter gear C weighing 57.5 lbs. delivers 35 per cent of the power horizontally at the top to gear G, and a 12.375 in. pitch diameter gear E weighing 28 lbs. delivers the remaining power to another gear H downward to the left at an angle of 30° below the horizontal. Both gear sets have 20° pressure angle. The system is required to have corrosion protection for long-term effects. The shaft is supported in bearings at A and D. Determine the torque transmitted by the gears and the pulley, and obtain the bending moment distribution in the shaft.

Solution: The torque acting on the shaft between B and C is:

$$T_B = 63025(\text{hp})/\text{rpm} = 63025 \times 33.5/325 = 6496 \text{ in-lbs.}$$

$$T_c = 63025 \times 0.35 \times 33.5/325 = 2274 \text{ in-lbs.}$$

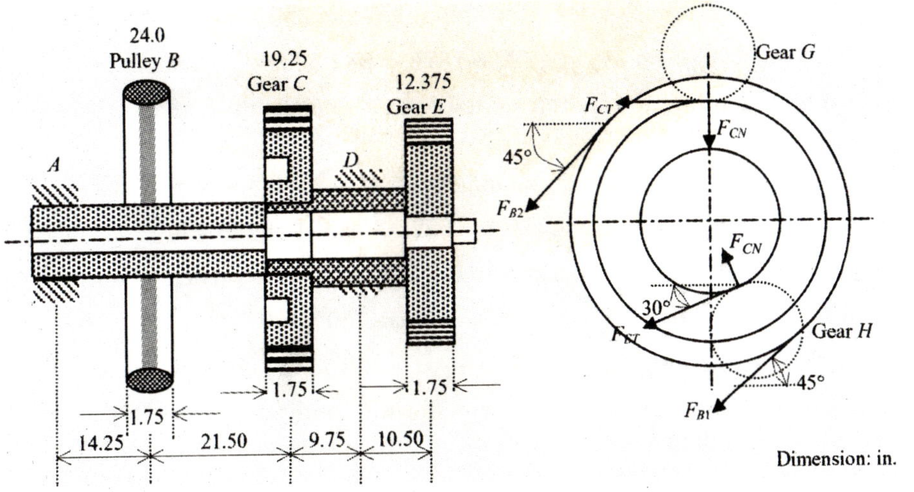

Fig. 13.57

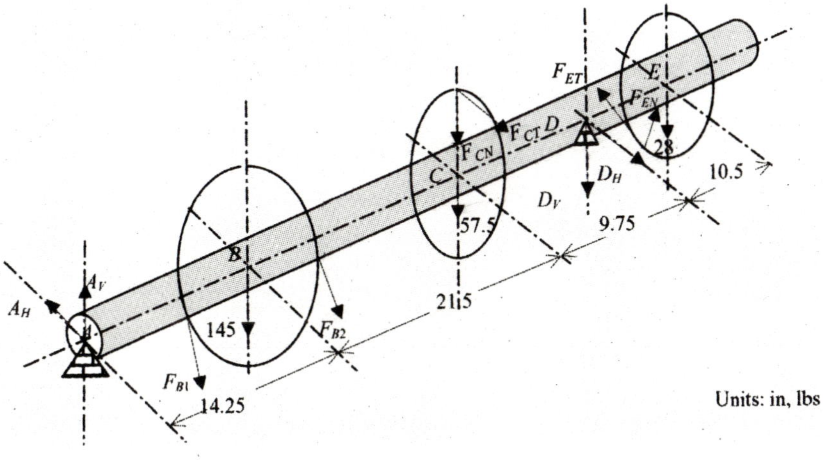

Fig. 13.58

The remaining power is delivered by gear E. Hence torque between C and E:

$$T_E = 63025 \times .65 \times 33.5/325 = 4223 \text{ in-lbs.}$$

The torque on the pulley is: $(F_1 - F_2) \times D_B/2 = 6496$ in-lbs.

Hence, $\qquad F_1 - F_2 = 6496/12 = 541.4$ lbs.

Assume belt tension ratio $F_1/F_2 = 3$ for a flat belt, with adequate initial tension.

Then: $F_1 = 812.1$ lbs., and $F_2 = 270.7$ lbs.

Total bending force at pulley B is: $F_B = F_1 + F_2 = 1082.7$ lbs. Then reactive tangential forces on gears C and E are:

$$F_{CH} = F_{CT} = T_C/r_C = 2274/9.625 = 236.2 \text{ lbs}$$

$$F_{ET} = T_E/r_E = 4222/6.1875 = 682.5 \text{ lbs}$$

The normal force tending to separate contact at the gear teeth in gears C and G and mating gears E and H are:

$$F_{CV1} = F_{CN} = F_{CT} \times \tan(20°) = 236.2 \times 0.364 = 86.0 \text{ lbs}$$

$$F_{EN} = F_{ET} \times \tan(20°) = 682.3 \times 0.364 = 248.4 \text{ lbs.}$$

Referring to figure 13.59, sum the forces in the horizontal and vertical planes:

$$A_H - F_B \sin(45°) - F_{CT} - D_H + F_{ET} \cos(30°) - F_{EN} \cos(60°) = 0$$

$$A_V - F_B \cos(45°) - 145 - F_{CN} - 57.5 - D_V + F_{EN} \cos(30°) + F_{ET} \cos(60°) - 28 = 0$$

Bearing reactions are obtained from static equilibrium conditions (sum of forces and moments is zero) in each plane. Bending moment and shear force diagrams can now be prepared for the horizontal and vertical planes, shown in figure 13.59.

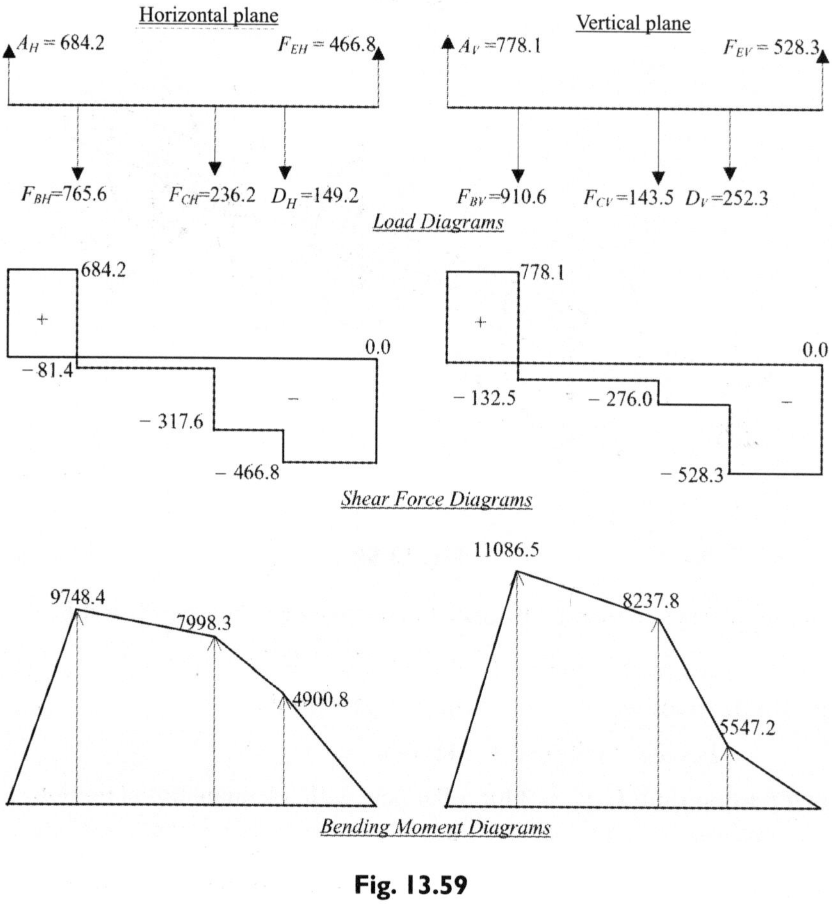

Fig. 13.59

Maximum bending moment occurs at pulley B. Total bending moment at B is:

$$M_B = [(9748.4)^2 + (11086.5)^2]^{1/2} = 14764.6 \text{ in-lbs}$$

Note that at pulley B the shaft also transmits a torque load of 6496 in-lbs.

Example problem # 13.2: In problem # 13.1 assume the shaft is to be designed for maximum loads. The shaft is to be machined from C1040 steel. Determine the shaft diameters. Assume the shaft is hollow, with $D_i/D_o = 0.5$.

Solution: For a hollow shaft the normal, shear and combined stresses are:

$$\sigma_x = \frac{32M}{\pi D_o^3 [1 - (D_i/D_o)^4]} \quad ...(13.12)$$

$$\tau = \frac{16T}{\pi D_o^3 [1 - (D_i/D_o)^4]} \quad ...(13.13)$$

$$\tau_{max} = \frac{0.5S_{yp}}{N} = \frac{16\sqrt{M^2 + T^2}}{\pi D_o^3 [1 - (D_i/D_o)^4]} \quad ...(13.14)$$

where N is a safety factor.

Normalized C1040 steel has the following material properties:

$$S_u = 85{,}000 \text{ lbs/in}^2 \qquad S_{yp} = 50{,}000 \text{ lbs/in}^2$$

Endurance limit for C1040 may be assumed to be: $S_n' = 0.5S_u = 42{,}500$ lbs/in^2. If the shaft has profiled keyways at B and C, K_f (bending) $= 1.6$ and K_{fs} (torsion) $= 1.3$. For the keyway at E assume K_f (bending) $= 1.3$ and K_{fs} (torsion) $= 1.3$. For operation in a non-corrosive environment and a survival rate of 95 per cent, the corrosion factor will be assumed at $C_R = 0.869$, and for a machined surface $C_F = 0.77$. Also, since the shaft will be larger than 0.50 in. diameter and smaller than 6.0 in. diameter, assume $C_S = 0.85$. The appropriate endurance limit is then obtained from:

$$S_e = (1/K_f)C_R C_S C_F S_n' = (1/1.6) \times 0.869 \times 0.85 \times 0.77 \times 42500$$

Or, $$S_e = 15{,}108 \text{ lbs/in}^2$$

The factor of safety N is assumed to be 3.75. Since the load is to be gradually applied, $K_{sb} = K_{st} = 1.0$. Stress concentration at the pulley is due to the profiled keyway and light press fit will be $K_t = 1.6$ for shafts subjected to fluctuating bending loads, and $K_f = 2.5$ for fretting corrosion. Thus, for the keyway and press fit combination,

$$K_f K_t M_B = 2.5 \times 1.6 \times 14762.9 = 59051 \text{ in-lbs.}$$

$$\tau_{max} = (0.5 \times 15108)/3.75 = 2014.4 = [(16/\pi D_o^3) \times (M^2 + T^2)^{1/2}]/[1 - (D_i/D_o)^4]$$

Substituting the values for moment and torque at location B, then $D_o^3 = 160.2$, outer diameter $D_o = 5.432$ in. Select $D_o = 5.50$ in. and inner diameter $D_i = 2.75$ in. Shaft segment between A and B experiences $M_{AB} = [(9748.4)^2 + (11086.5)^2]^{1/2} = 14764.6$ in-lbs and $(M^2 + T^2)^{1/2} = 16131$ in-lbs. using $K_f = 1.6$ for corrosion. Then $D_o^3 = 69.60$, outer diameter $D_o = 4.113$ in. Select $D_o = 4.125$ in. and inner diameter $D_i = 2.0625$ in.

Example problem # 13.3: Shaft deflections must be calculated in order to establish minimum permissible clearance between pulleys, gears and assembly enclosure. Explain the numerical integration method. How are shaft deflections related to the fundamental natural frequency of vibration?

Solution: Many methods are available for determining the deflection along the length of a shaft subject to a set of operating loads. The numerical integration approach is based on the following differential equations:

$$\text{Load equation}: \frac{d^4 y}{dx^4} = \frac{w}{EI} \quad \text{...(13.15)}$$

$$\text{Shear force}: \frac{d^3 y}{dx^3} = \frac{V}{EI} \quad \text{...(13.16)}$$

$$\text{Bending moment}: \frac{d^2 y}{dx^2} = \frac{d\theta}{dx} = \frac{M}{EI} \quad \text{...(13.17)}$$

$$\text{Slope}: \frac{dy}{dx} = \theta \quad \text{...(13.18)}$$

By considering finite distances along the shaft length (in the x-direction), the above equations can be integrated through numerical summation. As an example, $y = \Sigma \theta \Delta x$. The deflections are obtained by using the following step-by-step procedure, using a tabular format:

- Place a station number at each shear force location in the shaft, as also at each change in its cross-section. When long sections of the shaft with constant section are encountered, place stations at intermediate points. (1st column).
- Enter station numbers on alternate lines of the table, and force on the same line as the station number. Upward forces are positive, downward forces are negative. (2nd column).
- Calculate vertical shear at each station by summing forces in column # 2, and enter shear values in 3rd column.
- Enter the axial distance of the station from the previous station on the same line as station number. (4th column).
- Calculate bending moment at each station by multiplying shear force with axial distance. The product when added to the previous entry in 5th column gives bending moment at that station.
- In 6th column, on a line between stations enter area moment of inertia I at the section between the two stations.
- In 7th column enter product EI on the same line as I (E is material Young's modulus).
- In 8th column enter the value of (M/EI). The values are entered both at the station location as well as between stations.
- In 9th column enter the average value of M/EI on the line in between stations.
- Calculate slope at each station by entering running total of the product of 9th column and shaft segment length (4th column). Enter in 10th column on same line as station numbers.
- In 11th column enter arithmetic average of slope from 10th column, in between stations.

- Calculate deflection increment by multiplying 11th column with 4th column, and enter on the next lower line from average slope value. Enter in 12th column on same line as average slope.
- Total all deflection increments between bearing reactions, reverse the sign and divide by distance between the bearings. This gives the integration constant. Multiply this constant by shaft spans of 4th column and enter in 13th column on the same line as deflection increments.
- Add deflection increments of 12th column and the integration constant in 13th column. Enter the sum as part of a running total in 14th column on the same line as station number.

In this sequence of calculation the bearings are assumed to be rigid, hence the deflection at the bearing supports is zero. When shaft loads are separated into horizontal and vertical components, total shaft deflection is obtained by calculating the square root of the sum of the squares of the horizontal and vertical deflections at each station.

Practical difficulties in the manufacturing and assembly of a built-up rotor do not permit the centre of mass of a symmetric rotating system to coincide with the actual centre of rotation. Additionally, static deflections due to weight of the components and constant lateral loads cause a further shift of the mass centre from the axis of rotation. As the rotating shaft's speed increases, the kinetic energy of the attached masses also increases. When the kinetic energy reaches the potential energy of the deflected shaft, the vibrations in the shaft will be at a peak. The rotor speed at which this disturbance occurs is called the fundamental frequency of vibration, or critical speed, of the shaft. Equating the kinetic energy of the rotating masses to the potential energy of the deflected shaft results in an equation that will define the critical speed of the shaft. This equation is called Rayleigh's equation. Thus the lowest, or fundamental, frequency of vibration for a shaft simply supported at two points is given by Rayleigh's method:

$$f_n = 187.7\left[\frac{W_1y_1 + W_2y_2 + W_3y_3 + \ldots + W_ny_n}{W_1y_1^2 + W_2y_2^2 + W_3y_3^2 + \ldots W_ny_n^2}\right]^{1/2} \quad \ldots(13.19)$$

where W_n is the weight of the rotating mass in pounds and y_n is the static shaft deflection due to W_n. The constant factor = $(60/2\pi) \times \sqrt{386}$ = 187.7 and frequency is in cycles/min.

Example problem # 13.4: Shafts are used in many different ways in all types of mechanical equipment as power shafts, line shafts and camshafts, among others (figure 13.60). An axle is a stationary member used to support rotating wheels, idler gears and pulleys. A stub, or head, shaft is generally integral with an engine, motor or other prime mover, and has a projection to facilitate connection with other shafts. A jackshaft, or counter shaft, connects a prime mover with a line shaft or a machine. Flexible shafts may be used to transmit motion between two components where the rotating axes are at angle with each other.

Typically a shaft is subjected to fluctuating loads, so considerations for stress concentration and endurance strength play a major role in its design. Of equal importance, and sometimes more so, is the consideration of shaft rigidity, or stiffness.

Excessive lateral deflections can cause premature wear and failure of the bearings, while also reducing the critical speed, which can cause unacceptable shaft vibrations when operating at or near this speed.

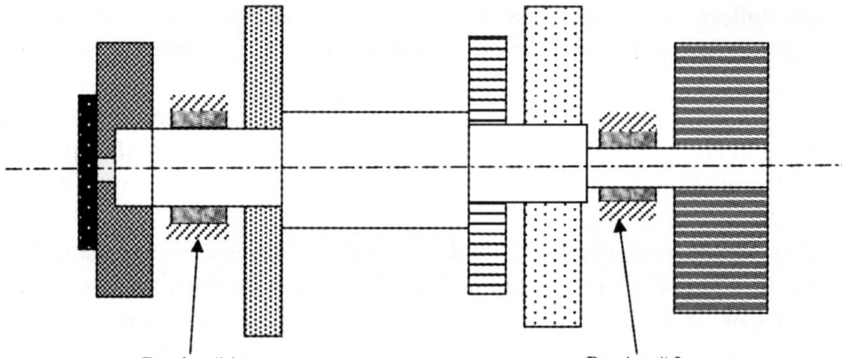

Fig. 13.60 Rotor-Bearing System Design Example

Generally, shafting of 3 to 3½ in. diameter is made of cold drawn carbon steel round bars. If toughness, shock resistance and strength are of prime concern, alloy steel bars (for example, 1347, 3140, 4150, 4340, 8650) may be more suitable. Heat treatment may also be specified to produce desirable properties. Surface wear resistance requires a carburized grade steel (1020, 1117, 4820, 8620, among others). Cost control can be obtained by using a plain carbon steel, if possible. Nitriding, cyaniding, flame hardening and other case hardening methods provide greater surface wear resistance.

Shafts larger than 3½ in. diameter call for machining from hot rolled carbon steel. Adequate stock must be removed in the machining process to eliminate decarburized scale caused by the hot rolling process. Larger diameter shafts are often times forged and machined to the required size from 0.45 per cent carbon steel.

Shaft diameter of a rotor system can be selected once the distribution of bending moments and torque along its axial length is available. The diameter of each section of the shaft will depend on stresses due to the combination of bending moment and torsion. The location of maximum stress in an application is sometimes not difficult to tell. In general, however, a systematic approach calls for drawing shear and bending moment diagrams to locate the points of maximum moments.

Example problem # 13.5: Table # 13.4 provides the geometric definition of a rotor and bearing system. A number of factors are to be taken into account in the computation of flexural (or transverse) natural frequencies and their corresponding modes of vibration of a rotating system. First, the deformation of the rotor is due to a combination of bending and shearing, which is inherently more complex than simple twisting or stretching. Bending is generally not confined to a single plane, so two orthogonal planes (for example, xz and yz) must be considered. Within each plane it is necessary to characterize the motion of each point i by two independent coordinates: a transverse displacement and an angular rotation. Finally, when the rotor is spinning, the motions in two orthogonal planes are coupled because of gyroscopic and other asymmetric effects. The location and nature of the bearings play a major role in flexural vibrations, and requires the development of more detailed models of bearings and support structures.

It will be assumed that all deformation takes place in the massless connecting shafts. The axially symmetric rigid disks have mass and axial and transverse moments of inertia, but are of negligible axial length. Each disk has four degrees of freedom; the centre may displace transversely through small displacements, and the disk may rotate through small angles about the two lateral axes.

Table 13.4 Rotor-Bearing System Design Data

Stn. #	Length in.	Diameter Outer in.	Diameter Inner in.	Shape Factor κ	Material Properties E 10^6 lbs/in^2	G 10^6 lbs/in^2	ρ lbs/in^3	Bearing Stiffness K_{xx} 10^6 lbs/in.	K_{yy} 10^6 lbs/in.	Added Weight Weight lbs	WR^2 lb-in^2
0	0	0	0	1	30	12	0.28	-	-	-	-
1	8.06	8.93	7.13	1	27.25	11.09	0.28	1.38	1.45	-	-
2	3.77	9.76	7.85	1.25	27.13	10.88	0.28	-	-	43.2	11057
3	12.44	11.12	9.38	1.38	26.89	10.83	0.28	-	-	111.7	45092
4	9.48	12.29	10.54	1.13	26.75	10.77	0.28	-	-	-	-
5	11.33	14.13	12.51	1.13	26.5	10.27	0.28	-	-	124.8	109000
6	9.48	12.89	10.91	1.13	25.5	9.85	0.28	-	-	-	-
7	12.44	11.74	9.9	1.13	26.7	10.32	0.28	-	-	179.3	187050
8	10.38	10.72	9.15	1	27.25	11.09	0.28	-	-	38.7	5852
9	8.87	9.13	7.49	1	27.5	11.13	0.28	1.43	1.35	-	-
10	9.24	8.53	6.56	1	27.5	11.13	0.28	-	-	84.2	23400
11	0	5	0.00	1	30	12	0.28	-	-	-	-

Assume the shaft spin speed is 3600 rpm. Station 0 is a dummy station, located at the left hand end of shaft. External masses in the form of disk, coupling or gear may be located here, as also a bearing. Shaft data – length, diameters, material properties – at station 1 correspond to shaft geometric properties between stations 0 and 1, other station data following the same format. Shear shape factor is used for shear compliance, as defined in Timoshenko's beam theory (1974). The x and y directions are located in a plane normal to shaft axis, and are mutually perpendicular to each other. Ignoring material and bearing damping, determine the first critical frequency and mode shape.

Solution: The dimensional data provided at the shaft stations must first be converted into geometric properties such as cross-section area, area and mass moments of inertia.

This data is shown in table 13.5. Equations (13.19) and (13.20) require parameters L, $L/(EI)$, $L^2/(2EI)$, $L^3/(6EI)$ and factor $\varepsilon = L^2/(\text{k}GA)$ for all elements, from which $[F_i]$, $[S]$, $[T]$ and $[W]$ matrices are formed. The required elements are shown in table 13.6. The assembled field matrix $[F_1]$ for the first shaft segment is provided below:

$$[F_1] = \begin{bmatrix} 1 & 0 & 0 & 8.057 & 6.40\times10^{-9} & 0 & 0 & 3.82\times10^{-8} \\ 0 & 1 & -8.057 & 0 & 0 & -6.4\times10^{-19} & 3.82\times10^{-8} & 0 \\ 0 & 0 & 1 & 0 & 0 & 1.6\times10^{-9} & 6\times10^{-9} & 0 \\ 0 & 0 & 0 & 1 & 1.6\times10^{-9} & 0 & 0 & -6\times10^{-9} \\ 0 & 0 & 0 & 0 & 1 & 0 & 0 & -8.057 \\ 0 & 0 & 0 & 0 & 0 & 1 & 8.057 & 0 \\ 0 & 0 & 0 & 0 & 0 & 0 & 1 & 0 \\ 0 & 0 & 0 & 0 & 0 & 0 & 0 & 1 \end{bmatrix}$$

Table 13.5 Geometric Properties of Rotor

Stn.	Cross sectional		Total mass	Total WR^2	
	Area, A	Inertia, I_{diam}	at station	Shaft axis	Diameter
#	in.	In	lbs.	lbs-in^2	lbs-in^2
0	0	0	25.803	421.2	246
1	22.704	186.298	39.826	696.2	387
2	26.417	259.018	106.352	12631.7	6478
3	28.015	370.568	202.749	47765.3	24100
4	31.379	514.09	95.055	3783.7	2110
5	33.895	754.502	228.38	113107.7	56800
6	37.011	659.677	104.286	3379.3	1960
7	31.273	460.955	269.983	189556.6	95000
8	24.501	304.184	101.319	7208.7	3730
9	21.407	186.589	57.179	906.8	551
10	23.348	168.971	114.602	23840.1	11974
11	19.635	30.68	0	0	0

Table 13.6 Shaft Element Field Matrix Parameters

Station #	L	L/(EI)	$L^2/(2EI)$	$L^3/(6EI)$	ε
0	0.00	0.00	0.00	0.00	0.00
1	8.06	1.60×10^{-9}	6.43×10^{-9}	1.73×10^{-8}	2.58×10^{-7}
2	3.77	5.36×10^{-10}	1.01×10^{-9}	1.27×10^{-9}	3.95×10^{-8}
3	12.44	1.25×10^{-9}	7.76×10^{-9}	3.22×10^{-8}	3.71×10^{-7}
4	9.48	6.89×10^{-10}	3.26×10^{-9}	1.03×10^{-8}	2.36×10^{-7}
5	11.33	5.67×10^{-10}	3.21×10^{-9}	1.21×10^{-8}	3.28×10^{-7}
6	9.48	5.63×10^{-10}	2.67×10^{-9}	8.43×10^{-9}	2.19×10^{-7}
7	12.44	1.01×10^{-9}	6.28×10^{-9}	2.61×10^{-8}	4.26×10^{-7}
8	10.38	1.25×10^{-9}	6.49×10^{-9}	2.25×10^{-8}	3.96×10^{-7}
9	8.87	1.73×10^{-9}	7.67×10^{-9}	2.27×10^{-8}	3.30×10^{-7}
10	9.24	1.99×10^{-9}	9.18×10^{-9}	2.82×10^{-8}	3.28×10^{-7}
11	0.00	0.00	0.00	0.00	0.00

The procedure for assembling field matrices for the other shaft segments is identical, and the remaining matrices for the shaft segments of the rotor are not listed here.

Similarly, the point matrix $[P_i]$ defined in equations (13.14) and (13.15) can be assembled from its elements. Material and bearing damping coefficients are disregarded, as also out of plane gyroscopic effects. Since damping and gyroscopic effects constitute the imaginary elements, the point matrix will only have real terms. The point matrix for the first station on the shaft is assembled from the $[B_i]$, $[I_i]$ and $[O_i]$ sub-matrices, where $[I_i]$ is the identity matrix and $[O_i]$ is the null matrix. The values shown are for the trial frequency of 5495.5 rpm, which is determined to be the first natural frequency of the rotor. The assembled point matrix is shown in table 13.7.

Table 13.7 Assembled Point Matrix

$$[p_1] = \begin{bmatrix} 1 & 0 & 0 & 0 & 0 & 0 & 0 & 0 \\ 0 & 1 & 0 & 0 & 0 & 0 & 0 & 0 \\ 0 & 0 & 1 & 0 & 0 & 0 & 0 & 0 \\ 0 & 0 & 0 & 1 & 0 & 0 & 0 & 0 \\ 0 & 0 & -5249 & -331827 & 1 & 0 & 0 & 0 \\ 0 & 0 & -331827 & 5249 & 0 & 1 & 0 & 0 \\ 0 & 1.42 \times 10^{-6} & 0 & 0 & 0 & 0 & 1 & 0 \\ 1.34 \times 10^{6} & 0 & 0 & 0 & 0 & 0 & 0 & 1 \end{bmatrix}$$

The system transfer matrix $[M]$ as defined in equation (13.22) is obtained with the help of the point and field transfer matrices, which sequentially transfer the rotor's state from the left end at station 1 to the right end at station n. Sequential multiplication of the state and field transfer matrices gives the matrix $[M]$ shown in table 13.8.

Table 13.8 System Transfer Matrix $[M]$

1	0	0	0	0	0	0	0
0	1	0	0	0	0	0	0
0	0	1	0	0	0	0	0
0	0	0	1	0	0	0	0
0.0014	0	−179750	−10262969	1	0	0	0
0	−0.0014	−10262969	179750	0	0	0	0
0	−98226	0	0	0	0	1	0
−98226	0	0	0	0	0	0	1

Boundary conditions at the left and right ends of the shaft are to be imposed next. The left end of the rotor is supported in a bearing, while the right end is free. For the pinned end the required conditions are $u = 0$ and $M_y = 0$, and at the free end the conditions are $V_x = 0$ and $M_y = 0$. Similar conditions are set up for the vertical plane. Applying the boundary conditions as described in equations (13.23) or (13.24), a non- trivial solution for $\{z^L\}$ exists only if the coefficient determinant vanishes. For the trial frequency of 5495.5 rpm, the coefficient determinant is calculated to have the minimum value of 1.266×10^{-24}. The corresponding relative deflections in the horizontal and vertical planes are shown in table 13.9. Since the bearing support stiffness K_{yy} at both ends of the shaft is substantially less than K_{xx}, the dynamic motion in the yz plane is also higher than in the xz plane, thus identifying this mode with the y-direction. The relative deflections in the y-direction are plotted in figure 13.61. Note that there is little flexural motion in the shaft since most of the activity is in the bearings, thus identifying this mode as a rigid body (or, rotor bounce) mode of vibration. Also, activity in the left end bearing is considerably more than in the right end bearing located 9.2 in. from the shaft's free end.

Example problem # 13.6: The life of a bearing is defined as the total number of revolutions, or the number of hours at a fixed speed, of bearing operation required for the failure criteria to develop. Under ideal conditions the first evidence of fatigue failure will consist of a spalling of the load carrying surfaces. The Timken Company goes by the failure criterion of pitting or spalling of

Table 13.9 Relative Shaft Deflections

Station #	*Deflection u, xz plane*	*Deflection v, yz plane*
0	27	-42.7
1	27.2	-42.7
2	24.7	-32.6
3	26.3	-32.5
4	36.6	-38.7
5	41.1	-41.0
6	44.1	-42.1
7	42.5	-39.3
8	35.4	-31.8
9	23.0	-20.1
10	10.1	-8.2
11	1.0	1.0

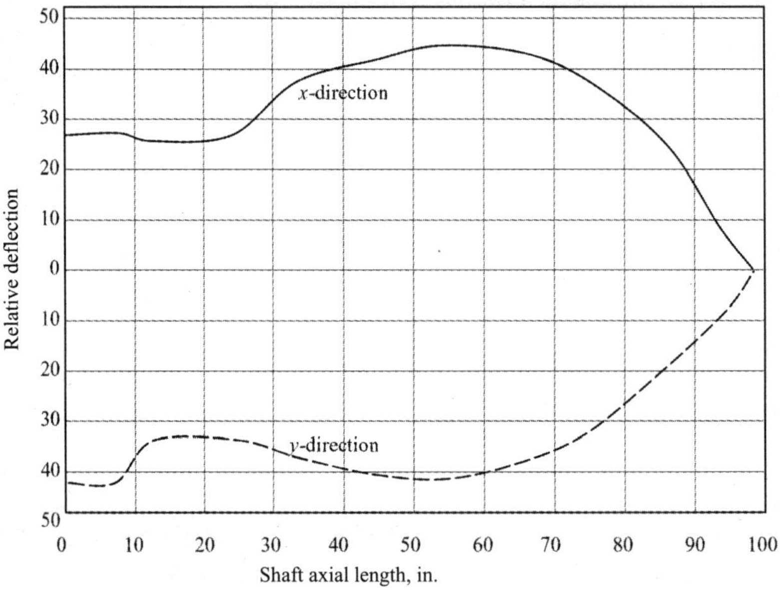

Fig. 13.61 Shaft Mode Shape at 5495.5 rpm Frequency

an area of 0.01 in^2. In testing a group of bearings the objective is to establish the median and the L_{10}, or rated, life. The L_{10} implies that 90 per cent of identical bearings operating at a constant speed and load will complete or exceed the test before the failure criterion develops. When a number of batches of bearings are under test, the median life is usually between four and five times the L_{10} life.

The concept of probable survival of a batch of bearings also needs examination. If a machine uses N bearings with each bearing having the same reliability R, then the reliability of all the bearings is $(R)^N$. The distribution of bearing failures can be approximated by the Weibull procedure. By making adjustments to the Weibull parameters (*ref*: Mischke, 1965), the distribution of bearing failures takes the form:

$$R = \exp\left[-\left(\frac{L}{6.84L_{10}}\right)^{1.17}\right] \qquad ...(13.20)$$

If a certain application requires a reliability of 98 per cent for the bearing to last for 2500 hours, determine the rated life.

Solution: Using $L = 2500$ and $R = 0.98$ in the above equation:

$$0.98 = \exp\left[-\left(\frac{2500}{6.84L_{10}}\right)^{1.17}\right]$$

Then the rated bearing life L_{10} is calculated to be 10262 hours.

Example problem # 13.7: Experiments indicate that identical bearings acting under different radial load F_1 and F_2 and operating at speeds n_1 and n_2 have lives L_1 and L_2 according to the relation:

$$F_2 = F_1\left[\frac{n_1L_1}{n_2L_2}\right]^{1/a} \qquad ...(13.21)$$

where $a = 3$ for ball bearings and 10/3 for roller bearings. A roller bearing can safely accept a load of 4.5 kN at 650 rpm for an L_{10} life of 1400 hours. Determine its life if the load is reduced to 3.75 kN and the speed is increased to 725 rpm.

Solution: In equation (10.42) $F_1 = 4.5$ kN, $F_2 = 3.75$ kN, $n_1 = 650$ rpm, $n_2 = 725$ rpm, $L_1 = 1400$ hours and $a = 10/3$. Then $L_2 = 2305$ hours.

Example problem # 13.8: Equations (13.20) and (13.21) may be combined to obtain a specified levels of load, speed, life and reliability factor. Then:

$$F_2 = F_1\left[\frac{n_1L_1}{n_2L_2(6.84)}\right]^{1/a}\frac{1}{(\ln(1/R))^{1/1.17a}} \qquad ...(13.22)$$

Determine the load rating F_2 if the bearing is to have a reliability level of 95 per cent.

Solution: Substituting the values gives $F_2 = 4.51$ kN.

Example problem # 13.9: Data for a journal bearing is as follows: viscosity $\mu = 3.95 \times 10^{-6}$ reyn, speed $N = 1800$ rpm, radial load $W = 525$ lbs, radius $R = 0.875$ in., clearance $c = 0.0014$ in. and length $l = 1.625$ in. Determine the bearing's characteristic number, minimum film thickness and its angular location.

Solution: Bearing $l/d = 0.929$ and pressure $P = (W/2Rl) = 184.6$ lbs/in^2. The bearing Sommerfeld, or characteristic, number is defined by the expression: $S = \mu N/P(R/c)^2$, and is calculated to be: $S = (0.875/0.0014)^2 \times \{3.95 \times 10^{-6} \times 1800/(184.6 \times 60)\} = 0.251$. From charts for minimum film thickness and eccentricity ratio (*ref*: Raimondi et al, 1958), the ratio of minimum film thickness and clearance $h_o/c = 0.54$ and eccentricity ratio $\varepsilon = 0.46$, and since clearance $c = 0.0014$, then minimum film thickness $h_o = 0.00076$ in. and eccentricity $e = 0.00064$ in. For no load, eccentricity $e = 0.0$ and film thickness $h = 0.0014$. As the load is increased the journal is forced downward. Figure 13.62 shows the distribution of hydrodynamic pressure in the lubricant film.

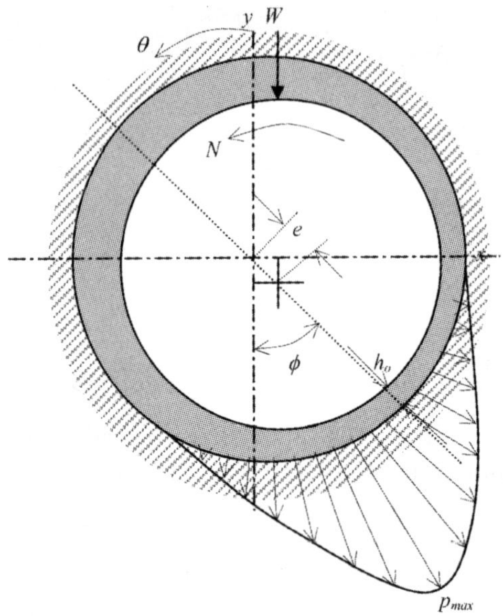

Fig. 13.62 Hydrodynamic Pressure Distribution in Fluid Film Journal Bearing (Raimondi, 1958)

Example problem # 13.10: In example problem # 13.9 determine the coefficient of friction, lubricant flow, film pressure and temperature rise.

Solution: From the chart for coefficient of friction (*ref*: Raimondi et al., 1958), the value is $(R/c)f$ = 5.2. Hence, the coefficient f = 0.00832. The torque required to overcome this frictional loss is $T = fWR$ = 3.82 in-lbs, which is equivalent to HP = $TN/63000$ = 0.1092 hp. From the chart for lubricant flow, $60Q/RcNl$ = 4.08 and Q = 0.244 in^3/s. Leakage at the two ends of the bearing is obtained from the charts, where Q_s/Q = 0.54, so Q_s = 0.132 in^3/s.

Heat generated by friction is dissipated by conduction, convection and radiation, and also carried away by the oil flow. A conservative assumption calls for all the heat to be extracted by the oil. The temperature rise in degrees Fahrenheit is given by the expression:

$$\Delta T_F = \{0.103 \times P \times (Rf/c)\}/\{[1 - 0.5(Q_s/Q)][60Q/RcNl]\}$$

$$= 0.103 \times 184.6 \times 5.2/[(1 - 0.54/2) \times 4.08]$$

$$= 33.2°\text{F}.$$

The maximum pressure developed in the film is obtained from the charts in the form of pressure ratio P/P_{max} = 0.47. Since P = 184.6, then P_{max} = 392.8 lbs/in^2.

13.22 EXERCISES

13.1 A 39.735 in. long uniform cantilever beam has a cross section of 1.735 in. width and 2.842 in. depth. Made of steel (E = 27.23 × 10^6 lbs/in^2, ρ = 0.275 lbs/in^3), determine the fundamental natural frequency of vibration of the beam.

13.2 The cantilever beam of exercise 13.1 carries a concentrated load of 425 lbs at mid-span and a uniformly distributed load of 22.75 lbs/in from the free end up to the mid-point, where the concentrated load is placed. Determine an expression for the shear force (*i*) from the fixed end to the mid-point and (*ii*) from the mid-point of the beam to the free end.

13.3 In exercise 13.2, derive an expression for the associated bending moment (*i*) from the fixed end of the beam to the mid-point and (*ii*) from the mid-point of the beam to the free end.

13.4 Calculate the maximum bending stress in the cantilever beam in exercise 13.3.

13.5 Derive an expression for deflections in exercises 13.1 to 13.3 (*i*) from the fixed end of the beam to the mid-point and (*ii*) from the mid-point of the beam to the free end.

13.6 If the concentrated load of 425 lbs in exercise 13.2 is shifted from the mid-span ($L/2$) to $3L/4$ from the fixed end of the cantilever, calculate the maximum deflection in the beam and its location from the fixed end.

13.7 In exercises 13.1 to 13.5, replace the uniformly distributed load of 22.75 lbs/in from the free end up to the mid-point with concentrated forces at 7 equidistant points. Calculate the defections at each of the load points along the length of the beam.

13.8 In exercise 13.7, use Rayleigh's method as defined by equation 13.19 to determine the fundamental natural frequency of vibration of the beam.

13.9 In exercises 13.1 and 13.2, assume that the depth of the beam is three times its width. If the permissible stress in the beam is 42500 lbs/in^2, calculate the dimensions of the cantilever beam.

13.10 In exercises 13.1 and 13.2, assume that the depth of the beam is three times its width. If the maximum deflection in the beam is limited to 0.0075 in., calculate the dimensions of the cantilever beam.

13.11 A journal bearing is designed as follows: viscosity $\mu = 4.23 \times 10^{-6}$ reyn, speed $N = 3600$ rpm, radial load $W = 15275$ lbs, radius $R = 3.725$ in., clearance $c = 0.0085$ in. and length $l = 7.25$ in. From charts for minimum film thickness and eccentricity ratio, the ratio of minimum film thickness and clearance $h_o/c = 0.61$ and eccentricity ratio $\varepsilon = 0.57$. Determine the bearing's characteristic number, minimum film thickness and its angular location.

13.12 In exercise # 13.11, assume the value of $(R/c)f = 5.2$, $60Q/RcNl = 3.98$ and $Q = 0.244$ in^3/s. Determine the coefficient of friction, lubricant flow, film pressure and temperature rise.

References & Bibliography

Alford, J., "Protecting Turbo-Machinery from Self-Excited Rotor Whirl", ASME Journal of Engineering for Power, 87(4):333–334, New York, 1965.

American Petroleum Institute, "Centrifugal Compressors for Petroleum, Chemical and Gas Service Industries", API Standard 617, 6th edition, 1995.

American Society of Mechanical Engineers, "Compressors and Exhausters", Power Test Code # 10, 1965 (reaffirmed 1979), ANSI PTC # 10, 1979.

Balda, M., "Dynamic Properties of Turbo-set Rotors", Symposium on Dynamics of Rotors, IUTAM, Lyngby, Denmark, 1975.

Barrett, L.E., Allaire, P.E. and Gunter, E.J., "Proceedings of the Conference on the Stability and Dynamic Response of Rotors with Squeeze Film Bearings", University of Virginia, Charlottesville, May 1979.

Barrett, L.E., Allaire, P.E. and Gunter, E.J., "Optimum Bearing and Support Damping for Unbalance and Response and Stability of Rotating Machinery", Journal of Engineering for Power, 100(1):89–94, 1978.

Begg, I.C., "Friction Induced Rotor Whirl – A Study in Stability", ASME, paper # 83-DET-10, New York, 1983.

Bentley, D.E., "The Re-Excitation of Balance Resonance Regions by Internal Friction", 82-PET-49, ASME, New York, September, 1982.

Bolotin, V.V., "Non-Conservative Problems of the Theory of Elastic Stability", Pergamon Press, New York, 1964.

Booker, J.F. and Huebner, K.N., "Application of Finite Element Methods to Lubrication, An Engineering Approach", Journal of Lubrication Technology, 94(4):313-323, October, 1972.

Booser, E.R., "Handbook of Lubrication", Vol. 2, CRC Press, Boca Raton, FL. 1973.

Brown, R.D. and Hart, J.A., "A Novel Form of Damper for Turbo-machinery", CP2443, NASA, Washington, DC, June, 1986.

Cataford, G.F. and Lancee, R.P., "Oil Free Compression on a Natural Gas Pipeline", 86-GT-293, ASME, New York, June, 1986.

Childs, D.W., "Force and Moment Dynamic Coefficients for Pump Impeller Shroud Surfaces", Proceedings of 4th Workshop on Rotor Dynamic Instability Problems in High Performance Turbo-machinery, NASA, CP, Texas A & M University, College Station, TX, June, 1986.

Childs, D.W., "Identification and Avoidance of Instabilities in High Performance Turbo-machinery", Von Karman Institute for Fluid Dynamics, lecture, Rhode-Saint Genese, Belgium, January, 1987.

Choi, Y. S. and Noah, S. T., "Non-Linear Steady State Response of a Rotor Support System", Journal of Vibrations, Acoustics, Stress and Reliability in Design, 169(3):255–261, July, 1987.

Choy, K.C., "Dynamic Analysis of Flexible Rotor Bearing Systems Using A Modal Approach", Ph.D. Thesis, University of Virginia, 1978.

Chu, F. and Holmes, R., "The Effect of Squeeze Film Damper Parameters on the Unbalance Response and Stability of a Flexible Rotor", ASME paper # 96-GT-377, New York, 1996.

Crede, C.E. and Ruzicka, J.E., "Theory of Vibration Isolation", Shock & Vibration Handbook, 3rd edition, Mc-Graw Hill, New York, 1988.

Den Hartog, J.P., "Mechanical Vibrations", Dover Publications, New York, 1985.

Dennis, A.J. and Erikson, R.H., Seitelman, L. H., "Transient Response Analysis of Damped Rotor Systems by the Normal Modes Method", ASME Paper # 75-GT-58, New York, 1975.

Deutschman, A.D., Michels, W.J. and Wilson, C.E. "Machine Design – Theory and Practice", MacMillan Publishing Co., New York, 1985.

Dimarogonas, A.D., "Newkirk Effect: Thermally Induced Dynamic Instability of High Speed Rotors", ASME paper # 83-GT-26, New York, April, 1983.

Ehrich, F.F. and O'Connor, J.J., "Stator Whirl with Rotors in Bearing Clearance", Journal of Engineering for Industry, 89(3):381 – 390, August, 1967.

Ehrich, F.F., "The Dynamic Stability of Rotor/Stator Radial Rubs in Rotating Machinery", Journal of Engineering Ind., page 1025 – 1028, 1969.

Ehrich, F.F., "Self-Excited Vibration", Shock & Vibration Handbook, 3rd edition, Mc-Graw Hill, section 5, 1987.

Ehrich, F.F., "The Role of Bearing Support Stiffness Anisotropy in Suppression of Rotor Dynamic Instability", DE-Vol. 18-1, ASME, New York, 1989.

Ehrich, F.F., "Handbook of Rotor Dynamics", Krieger Publishing Co., Malabar, FL, 1999.

Eshelman, R.L. and Eubanks, R. A., "Effects of Axial Torque on Rotor Response: An Experimental Investigation", ASME paper # 80-WA/DE-14, New York, 1980.

Fowlie, D.W. and Miles, D.D., "Vibration Problems with High Pressure Compressors", ASME paper # 85-Pet-28, New York, 1985.

Frene, J., Nicolas, D., Degueurce, B., Berthe, D. and Gaudet, M., "Lubrification Hydrodynamique", Eyrolles Edition, Lecreusot, France, 1990.

Gelin, A., Pugnet, J.M., Bolusset, D. and Friez, P., "Experience in Full Load Testing of Natural Gas Centrifugal Compressors for Rotor Dynamics Improvement", ASME paper # 96-GT-378, New York, 1996.

Gunter, E.J., "Dynamic Stability of Rotor Bearing Systems", SP-113, NASA, chapter 4, Washington, D. C., 1966.

Gunter, E.J., Barrett, L.E. and Allaire, P.E., "Design of Non-Linear Squeeze Film Dampers for Aircraft Engines", ASME Journal of Lubrication Technology, 99(1):57–64, January, 1977.

Hagg, A.C. and Sankey, G.O., "Some Dynamic Properties of Oil Film Journal Bearings with Reference to Unbalance Vibration of Rotors", Journal of Applied Mechanics, 23(2):302–306, June, 1986.

Hildebrand, F.B., "Advanced Calculus for Applications", Prentice-Hall, New Jersey, 1986.

Houbold, J.C. and Reed, W.H., "Propeller Nacelle Whirl Flutter", Institute of the Aerospace Sciences, 61 – 34, New York, January, 1981.

Humes, B. and Holmes, R., "The Role of Sub-atmospheric Film Pressure on the Vibration Performance of Squeeze Film Bearings", Journal of Mechanical Engineering Science, volume 20, no. 5, page 283 – 289, 1978.

ISO 1940/1, "Mechanical Vibrations — Balance Quality Requirements of Rigid Rotors - Part I, Determination of Permissible Residual Unbalance", International Organization for Standardization, Geneva, Switzerland, 1986.

Kanki, H., Kaneko, Y., Kurosawa, M. and Yamamoto, T., "Prevention of Low Frequency Vibration of High-Capacity Steam Turbine Units by Squeeze Film Damper", ASME paper # 97-GT-11, New York, 1997.

Kirk, R.G., "Labyrinth Seal Analysis for Centrifugal Compressor Design – Theory and Practice", Proceedings, International Federation for the Theory of Machines and Mechanisms 1986 International Conference on Rotor Dynamics, pages 589–596, Japan Society of Mechanical Engineers, Tokyo, September, 1986.

Kirk, R.G., Mondy, R.E. and Murphy, R.C., "Theory and Guidelines to Proper Coupling Design for Rotor Dynamics Consideration", ASME paper # 83-DET-93, New York, 1983.

Kimball, A.L., "Internal Friction Theory of Shaft Whirling", General Electric Review, 27(4): 244, April, 1924.

Kramer, E., "Instabilities of Rotating Shafts", Proceedings of the Conference on Vibration of Rotating Systems, Institution of Mechanical Engineers, London, page 230–246, 1982.

Klosterman, A.L., "On the Experimental Determination and Use of Modal Representations of Dynamic Characteristics", Ph.D. Thesis, University of Cincinnati, 1971.

Lawen, J.L. and Flowers, G.T., "Interaction Dynamics Between a Flexible Rotor and an Auxiliary Clearance Bearing", ASME Journal of Vibration & Acoustics, Vol. 121, page 183 – 189, New York, 1999.

Li, D.F. and Gunter, E.J., "Component Mode Synthesis of Large Rotor Systems", ASME paper # 81-GT-147, New York, 1981.

Li, D.F., Gunter, E.J., Barrett, L.E. and Allaire, P.E., "The Dynamic Analysis of Multi-Level Flexible Rotor Systems Using Transfer Matrices and Component Mode Synthesis", Final Report UVA/628145/MAE81/102, University of Virginia, Charlottesville, 1991.

Linn F.C. and Prohl, M.A., "The Effect of Flexibility of Support upon the Critical Speeds of High Speed Rotors", Transactions, Society of Naval Architects and Marine Engineers, 59:536–553, 1951.

Loomis, A.L., "Compressed Air & Gas Data", Ingersoll Rand Co., Woodcliff Lake, NJ, 1989.

Lund, J.W., "Spring and Damping Coefficients for the Tilting Pad Journal Bearing", Transactions American Society of Lubrication Engineers, 7(4):342–352, 1964.

Lund, J.W. and Thomsen, K.K., "A Calculation Method and Data for the Dynamic Coefficients of Oil Lubricated Journal Bearings", Topics in Fluid Film Bearing and Rotor Bearing System Design and Optimization, ASME, New York, pages 1 – 28, 1978.

Mabie, H.H. and Reinholtz, C.F., "Mechanisms and Dynamics of Machinery", John Wiley & Sons, New York, 1987.

Marmol, R.A. and Vance, J.M., "Squeeze Film Damper Characteristics for Gas Turbine Engines", ASME Journal of Mechanical Design, 100(1):139–146, January, 1982.

Maslen, E.H., Sortore, C.K., Vazquez, J.A. and Knopse, C.R., "Synchronous Response Estimation in Rotating Machinery", ASME paper # 00-GT-397, New York, 2000.

McLachlan, N.W., "Theory and Applications of Mathieu Functions", Oxford University Press, New York, 1947.

McCloskey, T.H. and Adams, M.L., "Troubleshooting Power Plant Rotating Machinery Vibration Problems Using Computational Techniques: Case Histories", Institute of Mechanical Engineers, # C432/137, pages 233 –244, London, 1992.

Miyachi, T., "A Parametric Study of the Unbalance Vibration of a Rotor Supported in a Light Flexible Casing", Gas Turbine Society of Japan paper # 83-TOKYO-IGTC-125, Tokyo, 1983.

Muller, P.C., Bajkowski, J. and Soffker, D., "Chaotic Motions and Fault Detection in a Cracked Rotor", Non-linear Dynamics, volume 5, page 223 – 254, 1994.

Muszynska, A., "Partial Lateral Rotor to Stator Rubs", paper # C281/284, Institution of Mechanical Engineers, London, 1984.

Muszynska, A., "Whirl and Whip – Rotor/Bearing Stability Problems", Journal of Sound & Vibration, 110(3):443–462, 1986.

Myklestad, N.O., "A New Method of Calculating Natural Modes of Uncoupled Bending Vibration of Airplane Wings and Other Types of Beams", Journal of Aeronautical Sciences, Vol. 12, page 153 –162, 1944.

Myklestad, N.O., "Fundamentals of Vibration Analysis", McGraw-Hill, New York, 1956.

Naslin, P., "The Dynamics of Linear and Non-linear Systems", Blackie, 1965.

Nayfeh, A.H. and Mook, D.T., "Non-linear Oscillations", John Wiley & Sons, New York, 1979.

Newkirk, B.L., "Shaft Whipping", General Electric Review, 27: 169 – 178, 1924.

Newkirk, B.L. and Taylor, H.D., "Shaft Whipping due to Oil Action in Journal Bearing", General Electric Review, 28(8): 559 – 568, 1925.

Orcutt, F.K., "The Steady State and Dynamic Characteristics of the Tilting Pad Journal Bearing in Laminar and Turbulent Flow Regimes", Journal of Lubrication Technology, 89(3):392 – 404, 1967.

Pincus, O. and Sternlicht, B., "Theory of Hydrodynamic Lubrication", McGraw-Hill, New York, 1984.

Prohl, M.A., "A General Method for Calculating the Critical Speeds of Flexible Rotors", Journal of Applied Mechanics, pages 142 – 147, 1945.

Raimondi, A.A. and Boyd, J., "A Solution for the Finite Journal Bearing and Its Application to Design and Analysis, Parts I, II and III", Transactions of ASLE, volume 1, number 1, pages 159 – 209 in Lubrication Science and Technology, Pergamon, New York, 1958.

Rangwala, A.S., "Turbo-machinery Dynamics: Design and Operation", McGraw-Hill, New York, 2005.

Roark, R.J. and Young, W.C., "Formulas for Stress and Strain", 5th edition, McGraw-Hill, New York, 1989.

San Andres, L.A. and Vance, J.M., "Effect of Fluid Inertia on Finite Length Squeeze Film Dampers", 86-TC-4D-1, American Society of Lubrication Engineers, Park Ridge, IL, October, 1986.

Sawyer, T., "Gas Turbines – Volumes I, II and III", International Gas Turbine Institute, Atlanta, ASME, 1984.

Schollhorn, V.K., Ebi, G. and Steigleder, K., "Frettinganrisse in Einen 936 MW Generator", VGB Kraftwerkstechnik, volume 73, no. 4, pages 340 – 344, 1993.

Sortore, C.K., "Observer Based Critical Response Estimation in Rotating Machinery", Ph.D. Dissertation, University of Virginia, Charlottsville, VA, 1999.

Taylor, C.F., "The Internal Combustion Engine in Theory and Practice", volume 2, The M.I.T. Press, Cambridge, MA., 1997.

Taylor, E.S. and Browne, K.A., "Vibration Isolation of Aircraft Power Plants", Journal of the Aeronautical Sciences, 6(2): 43 – 49, 1984.

Thomas, J.J., "Instabile Eigenschwingungen von Turbinenlaufern, Angefacht durch die Spaltstromungen Stopfbuschen un Beschaufelungen", AEG-Sonderdruck, 1958.

Thompson, B.D. and Badgley, R.H., "Application of an Advanced Hybrid Rotor Dynamics Model to the Complete Structure of a Marine Gas Turbine Engine", ASME paper # 88-GT-123, New York, 1988.

Timoshenko, S., Young, D.H. and Weaver, W., "Vibration Problems in Engineering", John Wiley & Sons, Inc., New York, 1974.

Vance, J.M. and Lee J., "Stability of High Speed Rotors with Internal Friction", ASME paper # 73-DET-127 , New York, 1973.

Vance, J.M. and Laudadido, F. J., "Experimental Measurement of Alford's Force in Axial Flow Turbo-machinery", 84-GT-140, ASME, New York, June, 1984.

Vance, J.M., "Torque Whirl – A Theory to Explain Non-Synchronous Whirling Failures of Rotors with High Torque Load", Journal of Engineering for Power, 100(2): 235–240, April, 1988.

Vance, J.M., "Rotor Dynamics of Turbo-machinery", John Wiley & Sons, New York, 1988*a*.

Wachel, J.C. and Tison, J.D., "Vibrations in Reciprocating Machinery and Piping Systems", Proceedings of the 23rd Turbo-machinery and Piping Systems, Texas A & M University, College Station, TX 1994.

Wachel, W.D., "Non-Synchronous Instability of Centrifugal Compressors", ASME paper # 82-Pet-22, New York, 1982.

Wagner, N.G., "Reliable Rotor Dynamic Design of High Pressure Compressors Based on Test Rig Data", ASME paper # 99-GT-150, New York, 1999.

Wang, Z. and Lund, J.W., "Calculations of Long Rotors with Many Bearings on a Flexible Foundation", C291/84, Proceedings of 3rd International Conference on Vibration in Rotating Machinery, Institute of Mechanical Engineers Conference Publication, York, England, 1984.

Wu, F. and Flowers, G. T., "A Study of the Influence of Rubbing on the Dynamics of a Flexible Disk Rotor System", ASME paper # 93-GT-35, New York, 1993.

Yu, J.J., Goldman, P., Bently, D.E. and Muzynska, A., "Rotor/Seal Experimental and Analytical Study on Full Annular Rub", ASME paper # 00-GT-389, New York, 2000.

Zeidan, F.Y., Perez, R.X. and Stephenson, E.M., "The Use of Honeycomb Seals in Stabilizing Two Centrifugal Compressors", Proceedings, 22nd Turbo-machinery Symposium, Texas A & M University, College Station, Texas, 1993.

Zhou Ren-Mu, "Instability of Multi-Stage Compressor K 1501", CP2443, NASA, Washington, DC, June, 1986.

14

Control of Rotor-Stator Clearances

14.1 INTRODUCTION

Substantial improvements in the specific fuel consumption of an engine can be obtained by decreasing the radial gap between the tip of the turbine blade and the outer casing and in the inner inter-stage seal areas between the mating stationary and rotating components. Sealing of the gas path has been a primary concern in all turbo-machines practically since their inception, and continues to draw considerable attention even in modern engines. The problem of controlling the clearances is significant enough to cause sealing play a leading role in the design and development of the compressor section and an even greater role in the hot turbine section. As the trend towards designing the turbine engine with fewer components, increasing the service life and enhancing the performance envelop continues, the need for advances in sealing becomes evident.

A number of concepts for controlling the clearance between the rotating airfoil tips and the stationary casing have been proposed. The approaches rely upon thermal, pneumatic and mechanical means, and depending on the method used, may be divided into passive and active means of control. Thermally activated designs rely on air bled from the compressor and fan sections to either heat and thus expand or cool and hence contract the stationary outer casing components. Positional and dimensional accuracy play a major role in a gas turbine engine's clearance control system, although manufacturing and assembly tolerances will be prime factors in sealing and issues related to the rotor dynamics of the system. Maximum displacement and minimum response time for actuation systems to be used in the clearance control systems are established from Federal Aviation Administration requirements.

Besides fuel efficiency, sealing of the gaps at the blade tips and in the inner seal areas have a major impact on operational aspects such as mission range capability, available payload and maintenance costs. Stability of operation and stall margin in the compressor are directly affected by the radial clearance between the rotating blade tips and the stationary outer casing. Excessive blade tip clearance is a root cause in accelerated degradation of hot section components. Increased clearances in the high-pressure (HP) turbine components cause the engine to operate at a harder pace, and hence at a higher temperature, to produce the same amount of power. Careful management of clearances in the HP turbine offers handsome dividends in the form of extended engine service life or time on the wings. In conjunction with other degradations in the components, the effects manifest in the form of excursions from the normal exhaust gas temperature (EGT) level. The EGT is a reliable indicator of the condition of an engine, and its limit is based on the upper temperature level at which the HP turbine can safely operate. The Federal Aviation Administration (FAA) certifies aircraft engines with a definite EGT limit. Crossing of the EGT limit requires removal of an engine from the wing for service maintenance (*ref*: Wiseman et al., 2001).

Increased clearance at the blade tip is a consequence of three factors: rubbing of the rotating blade tip against the casing, thermal fatigue and erosion. Effects of thermal fatigue can be mitigated to a considerable extent by ensuring adequate cooling of the blade tips and the casing shroud. Improved processing of the materials and coatings also play a vital role in diminishing the effects of thermal conditions. Cooling efficiency will deteriorate because of oxidation deposits fouling the passages inside the turbine airfoils, blockage and distortion of the passages. Erosion of the tips due to a number of factors also leads to increased clearance. Blade incursion is a continuous source throughout the useful life of an engine. Setting of clearances during the initial build of an engine are based on some amount of incursion at extreme operating conditions such as take-off and re-acceleration in order to hone the tip seal during the engine's preparatory run-in. The engine will then operate with line-to-line contact under repeated maximum power conditions. Erosion and thermal fatigue effects can also be exaggerated because of tip rubs wearing the protective coatings, such as thermal barrier coatings.

Definition and quantification of the rate of deterioration of gas turbine engines and associated performance losses has been the subject of a number of research programs undertaken by the government, military, manufacturers and users. For instance, flight loads and their effects on losses of a Pratt and Whitney JT9D engine due to increased running clearances between the blade and the outer air seal have been documented by Olsson and Martin (1982). Aerodynamic and inertial loads tend to impact performance losses to a greater extent than gyroscopic effects, affecting the fan clearance closures to a greater extent during take-off at maximum angle of attack. On the other hand, the HP turbine experiences adverse clearance closures under the combined effects of aerodynamic, thrust bending and thermal growth occurring during aircraft take-off. Larger than normal inertial loads, developed as an aircraft turns and manoeuvres, combine with high power operation during take-off to adversely affect clearance closures in the HP turbine.

Since aircraft engine casings are required to have small nominal thickness, they are flexible and lose their circular shape. In a study of a General Electric CF6-50 engine running on a test stand, performance characteristics were correlated with increased airfoil tip clearances and casing out-of-roundness effects (*ref*: Howard et al., 1982). HP turbine tip clearances and stator out-of-roundness were measured during steady state and transient operation of the engine. Variations in the airfoil tip clearance during throttle settings at take-off, deceleration, reburst and restart were correlated with performance degradation.

14.2 BLADE TIP SEALING IN COMPRESSOR AND TURBINE

Annular gas and liquid seals are used in steam and gas turbines, compressors and pumps to isolate the process fluid from the bearing lubricant and to prevent leakage to the environment. The geometry of a seal may be similar to that of a plain journal bearing, but there are major differences between the two on account of the high pressure drop across the seal and the consequent high turbulence in the flow. Thus, the Reynolds equation is not appropriate for annular seals. The forces developed by gas seals are roughly proportional to the pressure drop across the seal and the fluid density inside. Because of the density dependency, gas seals have a greater impact on steam turbines and high-pressure compressors than on gas turbines.

Labyrinth seals are used in several ways. In a multi-stage centrifugal compressor, for example, the shaft seal labyrinth may be used to restrict leakage flow along the shaft to the backside of the preceding impeller (figure 14.1). The eye-packing seal limits return flow leakage down the front of the impeller, and the shaft seal restricts leakage along the shaft to the preceding stage. To alleviate the axial force due to the pressure difference between the discharge and inlet pressure on an in-line machine, a balance drum is used to exert force in the other direction by directing the leakage flow through the drum to the inlet. Hence, the balance drum absorbs the full Δp of the compressor, and the fluid within the seal has an average density proportional to the inlet and discharge pressures. The eye-packing and shaft seal configurations may be considered to be half-labyrinth seals, while the one at the balance drum is an interlocking or full labyrinth (*ref*: Childs, 1993).

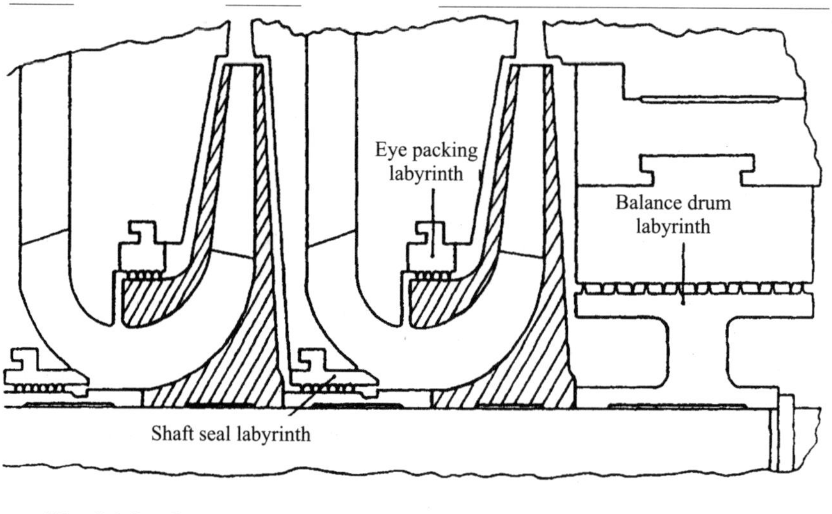

Fig. 14.1 Seal Arrangement in Multi-Stage Compressor (Kirk, 1987)

Generally the fluid forces do not interfere with the rotor's natural frequency or with the damping system. However, high fluid swirl activity due to lack of concentricity between the rotating and stationary parts has been known to create substantial destabilizing forces. Forces developed in a compressor seal labyrinth are one order of magnitude lower than its liquid seal counterpart. The direct stiffness term in a gas seal is typically negligible, and may even be negative. Seal leakage flow in a high-pressure steam turbine operating at constant speed has been studied by Greathead et al (1976) to evaluate stability of the rotor under varying loads. The unit operated up to 90 per cent of full power, but at higher power levels sub-synchronous whirl at the lowest natural frequency was experienced. This observation has also been noted in centrifugal compressors, and points to a linkage between power level and instability as opposed to speed related onset of instability in hydrodynamic bearings.

A honeycomb seal is shown in figure 14.2. The roughened stator helps reduce leakage, and also offers the benefit of reducing average circumferential velocity within the seal and the destabilizing

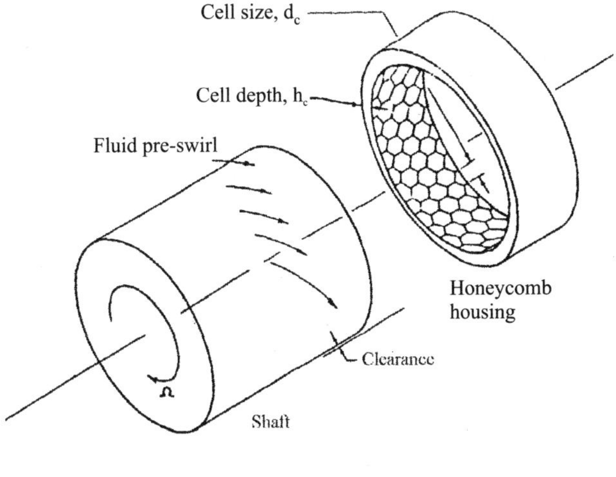

Fig. 14.2 Honeycomb Seal (Childs, 1993)

cross-coupled stiffness coefficient. This seal type has been successfully used in compressors for the balance drum and as a turbine inter-stage seal for the high-pressure oxygen turbo-pump of the space shuttle main engine.

Brush seals use a biased pattern of wires in contact with a ceramic coating on the shaft, and sharply reduce leakage when compared with a labyrinth or honeycomb seal. Based on tests (*ref*: Ferguson, 1988), the leakage rate of a single stage brush seal is a mere 5 per cent of a four-cavity tooth on a rotor labyrinth seal. And because of their compliant nature, damage due to transient impact between the rotor and stator components is held to a minimum.

Figure 14.3 depicts a floating seal construction. Seal segments consist of spring loaded inner and outer halves. The spring pre-load causes the ground and finished external faces of the segments

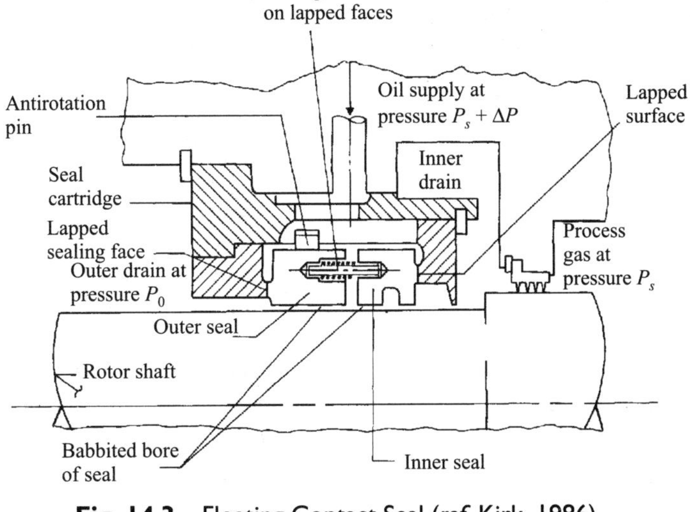

Fig. 14.3 Floating Contact Seal (*ref.* Kirk, 1986)

to be in contact with its housing, with a shear pin used to prevent rotation of the segments. Oil is supplied between the segments, then flows axially along the shaft. Most of the oil is recovered, but some of it may be lost to the process fluid stream. The seal is of interest from rotor dynamic considerations because the oil in the annulus provides a form of support for the shaft. But the seals lack load-carrying capability. Fluid forces developed from the relative position and motion in the seal is reacted by the friction at the contact face. A spiral-grooved face on the seal provides a refinement for primary gas sealing in applications calling for high pressures (*ref*: Sedy, 1979). The grooves create a separating force between the two mating faces, causing the seal to operate at tight clearances between the rotating and stationary faces, resulting in a low leakage rate.

Squeeze film dampers have the ability to attenuate vibrations and improve stability of a rotating machine assembly by providing multi-directional damping (*ref*: Rangwala, 2004). Aircraft jet engines use rolling element bearings, which are inherently free of damping as witnessed by their alternative name of anti-friction bearings. Squeeze film dampers are employed to introduce damping into the system to reduce synchronous response amplitudes, to enhance stability, and to provide a margin of safety for blade loss conditions. Squeeze film dampers have also been developed and successfully employed for protection against sub-harmonic rotor instabilities operating on tilting pad bearings. Some high-pressure compressor manufacturers use the dampers for this stated purpose (*ref*: Shemeld, 1986). In this application the squeeze film damper resembles a non-rotating plain journal bearing, providing damping to the lateral motion of the rotor as a consequence of pure squeezing motion of the damper's journal element.

The dampers may be broadly categorized into two different configurations. In one arrangement a centralized retainer spring in the form of a squirrel cage is connected to the outer race of a rolling element bearing, and acts as a parallel element within the squeeze film annulus. In another scheme, a structure without a retainer spring, where, say for a horizontal machine, the outer race of the rolling element bearing remains at the bottom of the damper's clearance circle, until the unbalance forces are large enough to create the lift. The advantage of the design with the retainer spring is the ability to tune the critical speeds of the system away from the operating range to ensure smooth running. But when differing unbalance, damper static eccentricity ratio, and oil supply pressures are applied, the system resonant speeds often migrate from their expected place. In particular, if the damping is excessive because of increased oil supply pressure or static eccentricity ratio, resonant frequencies of the assembly may approach frequencies without the retainer spring. The squeeze film damper then essentially locks out the retainer spring, rendering it inoperative (*ref*: Holmes et al., 1992).

Specially designed seals are required to prevent leakage of hot gases in the turbine in the confined space where a shaft meets a stationery component. Relative thermal growth between the components aggravates the problem. Labyrinth seals have been expressly developed to resolve the problem (figure 14.4). A number of grooves are machined into the rotor, separated from one another by a ridge with a knife-edged crest. The mating location on the stationary part has a flat cylindrical surface. Knife-edges cut into the stator's rub surface to provide the sealing. When hot gases leak past the gap, throttling reduces the pressure from one knife-edge to the next. At the last location the throttled pressure equals the pressure on the aft side, thus effectively reducing the leakage rate to zero. The method may be applied at the blade tips as well as at the inner inter stage locations. Blade shrouds are designed with the cutting tip to control leakage flow. Softer honeycomb materials are more suitable in the turbine casing.

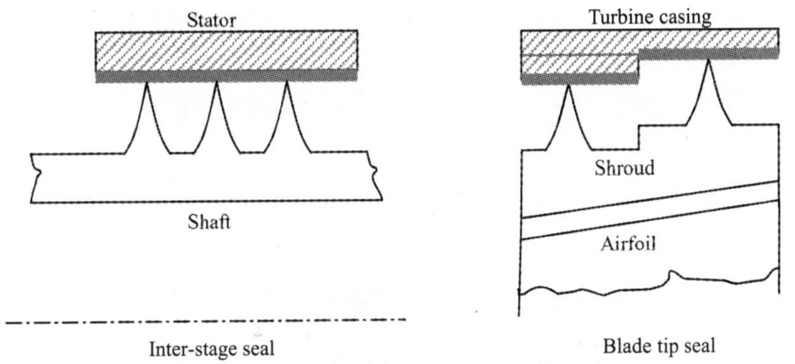

Fig. 14.4 Labyrinth Seal

Control of radial clearances between blade tips and the casing is achieved by blowing cooler air on the external surface of the case in order to shrink the casing. Air is directed from holes in tubes placed around the case in the plane of the rotor stages. In the active clearance control system, generally employed for the gas generator turbine, temperature sensors and valves are employed to control the flow of cooling air during periods of varying temperature operation. Passive cooling systems for the low-pressure turbines merely blow cooler air on the shell. The expense of installing sensors and valves is not required in the latter approach, but the cool air is being continually robbed from the system. In either case performance gains are achieved, because leakage of hot gases past the blade tips is minimized.

14.3 CAUSES OF TIP CLEARANCE VARIATION

The location of HP turbine blade tip seals in a modern gas turbine engine is shown in the combustor and turbine cross section of figure 14.5. The inside of the casing is lined with air seals to form a continuous shroud around the rotating blades. The arrangement provides for controlling the spillage of gas over the tip of the blades. The clearance over the blade tips varies according to the power level produced by the engine. The differences around the 360° arise from the distortion occurring in both the stationary and rotating components due to the combination of engine power-induced and flight loads acting on them. Centrifugal, thermal, internal pressure and thrust loads constitute engine loads, while external aerodynamic, inertial and gyroscopic forces contribute to flight loads. Linear and angular deflections in the components from the engine and flight loads may act in a specific direction or may be multi-directional. As a consequence, the deflections may be symmetric or asymmetric with the engine centreline.

14.4 BLADE RUB FROM MECHANICAL LOADS

Centrifugal and thermal loads are responsible for the maximum radial variations in tip clearance. Consider, for instance, a disk of mass m mounted on a shaft of length l between two bearings, and spinning at a fixed speed ω. Assume the centre of gravity of the disk to be located at a radial distance e away from the centreline of the shaft. Then a rotating centrifugal force $m\omega^2 e$ acts on the disk. The

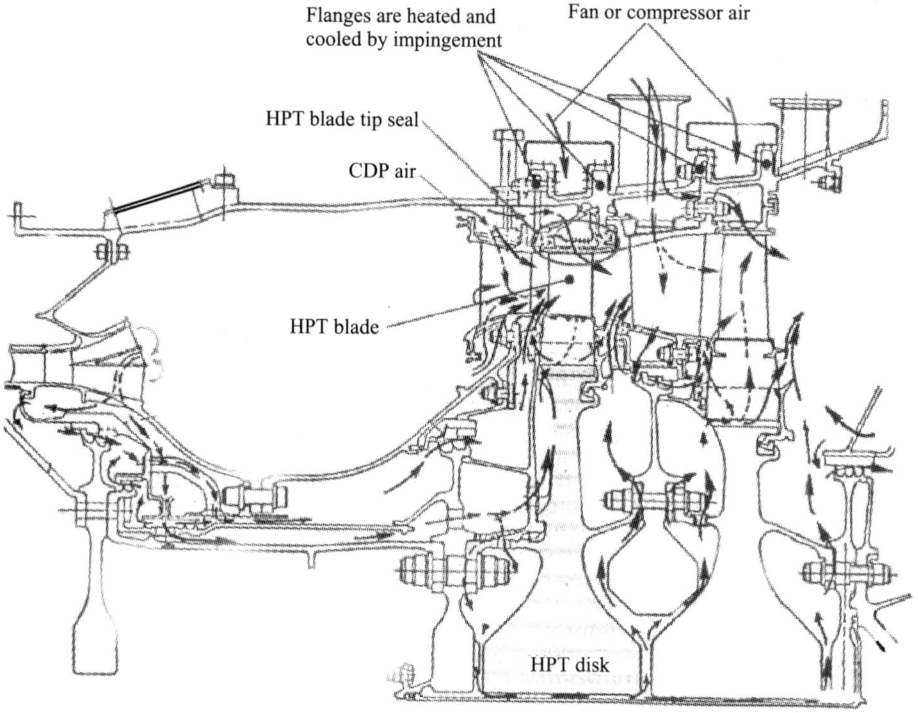

Fig. 14.5 Blade Tip Seal Locations in an HP Turbine

rotating force vector may be divided into horizontal and vertical components, causing the disk to execute vibratory motions in the two directions. And if the spin speed ω coincides with the elastic shaft's natural frequency of vibration in the transverse direction, a state of resonance exists, and the shaft and disk assembly will experience considerable vibrations. Neglecting the effects of gravity, there are two forces acting on the disk, the elastic pull of the shaft tending to restore it to its original position and the centrifugal force on the centre of gravity of the disk, causing it to travel in a circle. The elastic force depends on the bending stiffness of the shaft, and is proportional to its deflection. This observation is also true of a shaft with many disks running in two or more bearings.

The objective in a rotor dynamic study is to determine the transverse direction deflections due to dynamic forces. Failures in turbo-machinery shafts result from fatigue due to stress as it fluctuates between extreme values. In addition to torsion, fibres on the surface of a rotating shaft are subjected to bending loads, undergoing both tension and compression for each revolution of the shaft. If the shaft of a centrifugal compressor is coupled to the shaft of an electric motor operating at 1,740 rpm, the shaft is stressed in tension and in compression 1,740 times. If the shaft is also axially loaded (caused, for example, by a helical or worm gear), the shaft has an axial stress superimposed upon the bending component. These and other loads occurring in the shaft produce stresses often referred to as repeated, alternating or fluctuating stresses.

Bending in the shaft may arise from a number of operational factors, some of which arise internally in the machine and others from outside the machine. Dynamic balance of the machine's rotating assembly is an internal machine factor that causes bending in the shaft. As a result of manufacturing procedures and limitations, tolerances have to be provided on machined components. Thus, a minimum diameter shaft journal may be paired with a maximum diameter bearing, causing a larger than normal radial clearance in the bearing. In contrast with torsional vibrations, bearing and support stiffness plays a significant role in the shaft's lateral vibrations. Since the hydrodynamic stiffness of a journal bearing's lubricant film is reduced as a consequence of the larger bearing clearance, the shaft's natural frequency of vibration also is reduced, causing the machine to operate at or close to the natural frequency (figure 14.6). Rotor unbalance arises from many sources attributable to manufacturing and field operation. With the appropriate amount and location of unbalance in the rotor, substantial bending moments are caused when operation takes place near a critical speed, resulting in flexural shaft vibrations along its lateral axes.

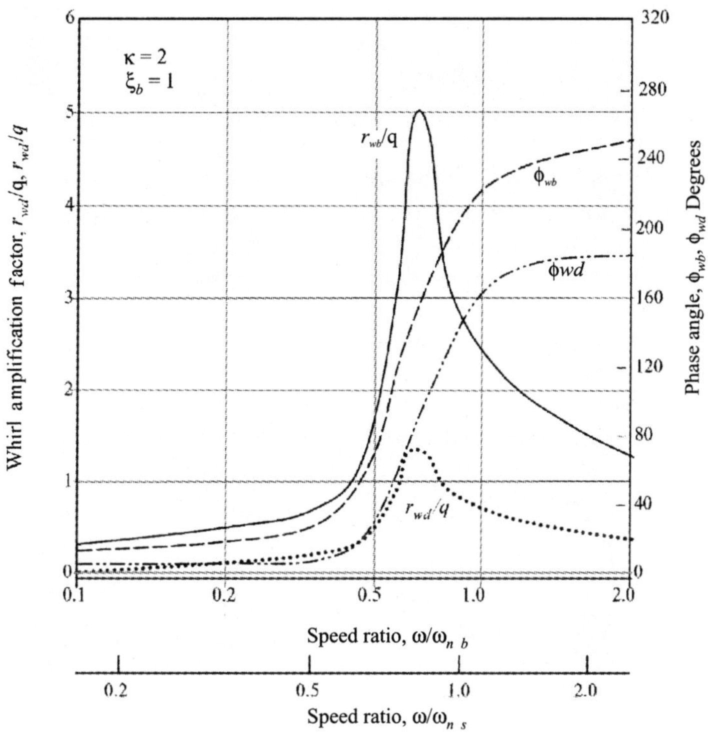

Fig. 14.6 Shaft Supported in Flexible Bearings – Response to Unbalance

Shaft misalignment is another factor responsible for flexural vibrations. At the coupling between an internal combustion engine and an electric power generator, thermal conditions may throw out the rotor from an aligned position during cold condition. Radial and angular misalignment of the rotor can radically alter the flexural vibration signature of a rotating assembly, with consequent bending moments in the shaft. This also has the adverse effect of putting unusual wear pattern in the bearings.

For flexural vibrations, the most commonly encountered stimulus is rotor unbalance and shaft bow, resulting in whirling or whipping of the shaft. Unstable operation of the machine in the form of chatter (or stick-slip) and large deflections and loads are other undesirable consequences. To deal with such vibrations, appropriate design strategies are required. For instance, placing the system's critical speeds out of the steady state operating range will minimize the degree of super-critical whirling and whipping phenomena. If for practical reasons it is not possible to raise the critical speed, or operation above the fundamental mode of vibration cannot be avoided, introduction of damping to reduce critical response peaks and to cancel the destabilizing forces in the operating range will be required.

14.5 SYMMETRIC AND ASYMMETRIC CLEARANCE SHIFTS

Uniformly acting loads such as centrifugal, internal pressure and thermal conditions will generally produce radial displacement and hence, axisymmetric clearance changes at the blade tips. Thermal loads produce axisymmetric clearance changes due to thermal growth and contraction of the stationary and rotating components. Asymmetric clearance changes may also develop if heating and cooling of the parts does not occur uniformly.

Large dimensions of the components (exceeding 36 in. diameter of HP turbine bladed disk and shrouds) combine with extreme variations in rotor speed (up to 11,000 rpm) and temperature range (over 2,500°F) to produce considerable deflections in the structures of a large aircraft engine.

A typical variation of tip clearance arising from axisymmetric loads and rotor speed as a function of time for a given mission (comprising take-off, cruise, deceleration and re-acceleration) is shown in figure 14.7. Cold build engine clearances and response of the rotor and the casing as the engine speed and temperature change are taken into consideration. Axisymmetric clearance changes to the extent of 0.030 in. are oftentimes encountered during take-off and re-acceleration in large commercial engines. Military engines place a greater emphasis on performance and lesser concern on efficiency, so clearance variations are even more exaggerated. The clearance available between the shroud and blade tips when the engine is started from the cold quickly decreases as the

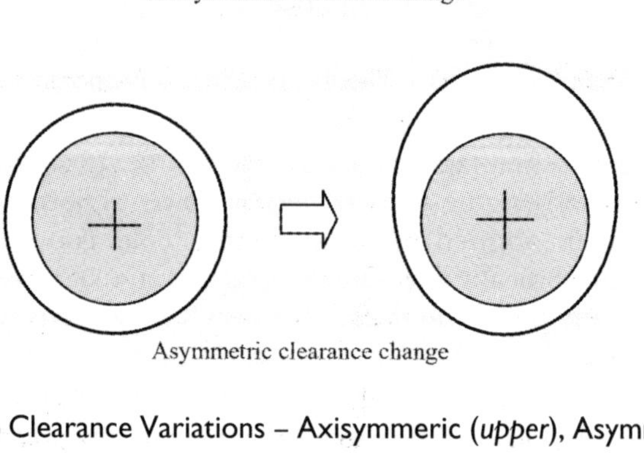

Fig. 14.7 Tip Clearance Variations – Axisymmeric (*upper*), Asymmetric (*lower*)

engine speed increases for take-off under the combined action of centrifugal forces on the rotor and heating of the airfoils. At the same time, growth in the outer casing proceeds at a slower pace. The consequent radial growth results in a minimum gap, commonly referred to as the 'pinch point'. As the casing gradually grows from the heating after the occurrence of the pinch point, the clearances increase. This is followed a short while later by casing growth, and the rotor continues to gain in temperature and grow, albeit at a slower pace due to its mass, and the clearance closes again. Approaching the cruise condition, both the casing and rotor reach thermal equilibrium and the clearances remain relatively steady. If during cruise a step shift in the altitude were to occur, for instance to avoid collision, tip clearances will be affected and must be accounted for when deciding design cruise clearance to prevent the blades from running into the case. At the time of landing the engine is decelerated, the clearance rapidly increases from the unloading of mechanical forces of the rotor and the blades, as also from thermal lag of the casing as the two components cool. If for some reason an evasive manoeuvre to abort landing is made and the aircraft were to climb during the deceleration, another pinch point can occur. Referred to as a re-acceleration or re-burst event, the centrifugal growth of the rotor in tandem with the thermal state of the rotor and the stator will lead to even lesser clearances when compared with the take-off point. The larger thermal mass of the rotor will delay cooling of the rotor as opposed to the casing. This situation will thus determine the cold build clearances.

Figure 14.8 helps in understanding the concept behind 'active clearance control', or ACC, to reduce operating clearance during cruising operation. The intent of the ACC system is to maintain minimum clearance and at the same time prevent rubs over the complete flight profile. Due to the length of the cruise portion of the flight, it holds the most potential for achieving gains in specific fuel consumption (SFC). At the same time it is essential to open the clearance at pinch points to prevent rubs, with minimal clearances maintained at take-off to obtain the necessary thrust while maintaining the EGT within limits.

Probably the most strenuous event in an aircraft flight takes place during an in- flight shutdown,

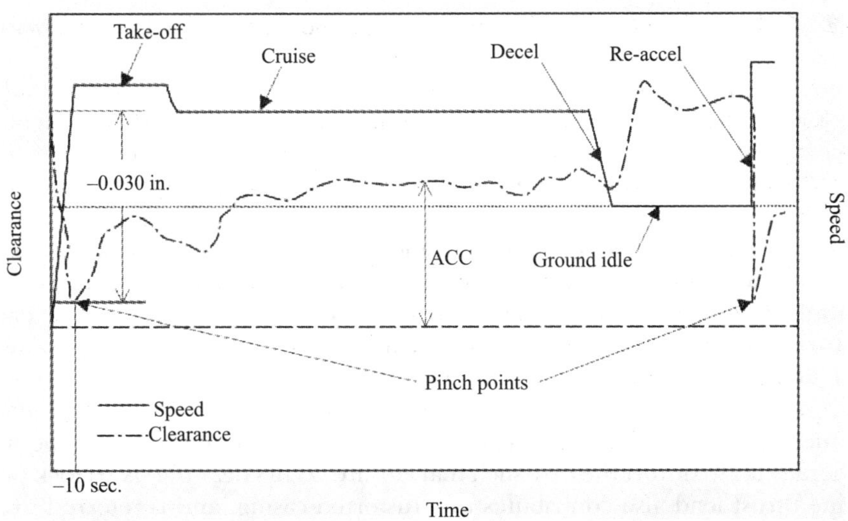

Fig. 14.8 HP Turbine Tip Clearance Variation over a Mission Cycle

or stopcock, followed by a relight in the combustor. As part of a new aircraft's acceptance testing, the engine is shutdown at altitude and followed by a dwell time during which the engine windmills due to the aircraft's forward momentum. The engine is then restarted and accelerated back to the same power setting at which the engine was shutdown. Since the engine is abruptly shutdown without going through deceleration, the casing cools even more than a normal shutdown, while the centrifugal forces are less because windmill speed is less than ground idle. Restart followed by acceleration leads to greater clearance closure than a take-off condition due to lower temperatures of both the rotor and the casing. Figure 14.9 provides clearance data from General Electric CF6-50 engine in a test stand operation during a stopcock event (*ref*: Howard et al., 1982).

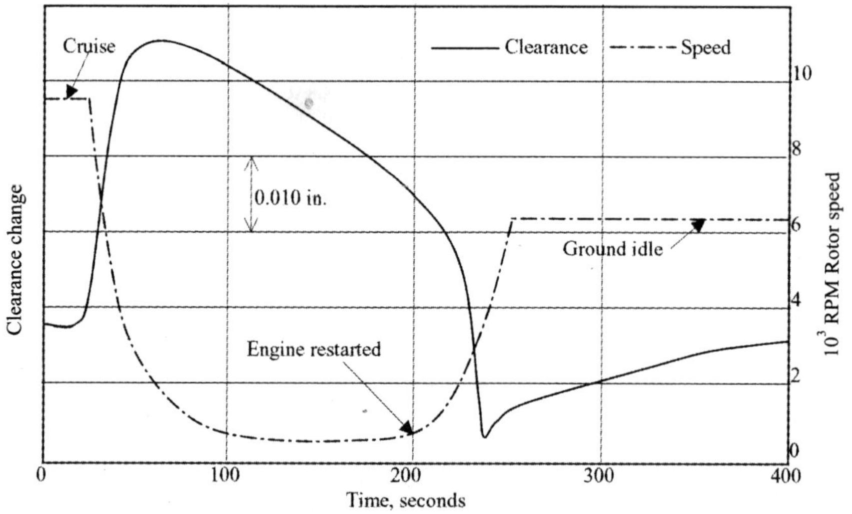

Fig. 14.9 CF6-50 HP Turbine Tip Clearance Variation during Stopcock Event (Howard, 1982)

Asymmetric clearance variations arise from non-uniform loads usually acting on the stationary structures of the engine. Thermal, thrust, inertial and aerodynamic loads produce non-uniform radial displacement (figure 14.8). Distortion of casing arising from non-uniform heating can cause the casing or the shroud to open clearances along a portion of the segment or segments, of the circumference while reducing them at other areas. This leads to an ovalized casing. Asymmetric distortion of the casing also arises from aerodynamic, thrust and manoeuvre loads. Since the engine is not attached to the wing of the aircraft at its centreline, aerodynamic and thrust load reactions produce an applied moment, leading to a bent structure relative to the rotor. Aerodynamic loads on the inlet cowl arise from bending of the air stream entering the fan inlet, and depend on fan airflow, inlet angle of attack, and aircraft speed; these produce shear forces and bending moments on the fan case that pass through the engine and deflect the casings throughout. Just after a take-off rotation the inlet angle of attack is extreme, hence the aerodynamic loads will also be at a peak. The loads will generally act to distort the case such that closure occurs near the six o'clock position of the engine. Engine thrust load also contributes to a distorted casing, and is referred to as backbone bending. Backbone bending typically causes the front flange of the fan case of a JT9D engine to deflect upward more than the front section of the low speed rotor, while the back end of the high speed core rotor deflects downward more than the rear section of the core casings (*ref*: Olsson et al.,

1982). Figure 14.10 shows the HP turbine clearance flight data for the same engine during take-off. Four clearance probes around the circumference are used for measuring the clearances. Initial closure is due to centrifugal effect as the engine accelerates. At take-off power the clearances open axi-symmetrically as the casing gains in temperature, but soon close along the bottom half and increase along the top half of the HP turbine from the increasing thrust load. Aerodynamic effects are most intense just after take-off rotation when the angle of attack is greatest, decreasing after low climb as this angle reduces. The most significant clearance closures occur under the combined effects of climb after take-off and a 2-G manoeuvre turn. Wind gusts and hard landings affect both the aerodynamic and inertial loads. Gusts, or turbulence, produce dynamic inertial loads due to movements of the structure.

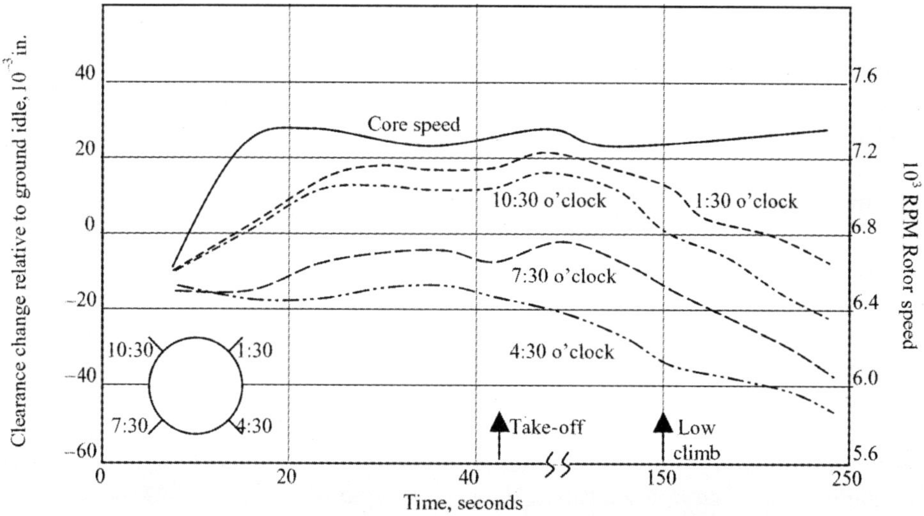

Fig. 14.10 JT9D HP Turbine Tip Clearance Variation due to Flight Loads (Olsson, 1982)

14.6 EFFECTS OF DETERIORATED CLEARANCE AT TIP

The clearance between blade tips and the surrounding casing, or shroud, varies primarily due to changes in thermal and mechanical loads on both the rotating and the stationary structures. Opening of the gap clearances in both the HP compressor and the HP turbine will dramatically alter the specific fuel consumption, compressor stall margin and engine efficiency. Payload and flight range capability will also be seriously affected.

Close control of the clearances, especially in the HP turbine, will improve engine service life, or time on the wing. Together with related degradations occurring in the components as more operating hours accumulate, the effects manifest in the form of increased exhaust gas temperature (EGT). The EGT is useful in gauging the overall condition of an engine, and its limit is based on the upper temperature level at which the HP turbine can safely operate. The Federal Aviation Administration (FAA) certifies aircraft engines with a definite EGT limit. Crossing of the EGT limit requires removal of an engine from the wing for service maintenance. The cost of maintenance for a major overhaul of large commercial engines is prohibitive, and will likely keep climbing.

14.7 COMPRESSOR STALL MARGIN AND TIP CLEARANCE

Air flowing through the annulus of an axial compressor performs the relatively more difficult task of going against an adverse pressure gradient. As the compression ratio increases, the danger of stall troubles rises correspondingly, which is all too prevalent in axial compressors. In individual airfoils, stalling is encountered when the difference between the flow direction and the blade's angle of incidence becomes excessively large. The radial clearance between the rotating blade tips and the stationary outer casing in the compressor directly affect the ability of operation and stall margin. The stability of the flow is jeopardized by the very fact that the movement is taking place in the same direction as that of the increase in pressure. Flow reversals occur under the right conditions of mass flow and rotor speed, and which vary from those for the blade's design.

Physical effects of a surge in an aircraft engine's compressor can be grave. A sudden stoppage in the airflow and accompanied by a loud noise emanating from the engine is common during such an event. Damage to the engine structure may be expected when repeated a number of times, particularly in the compressor and fan areas. Generally, the aircraft is capable of recovering from the stall and continue to proceed, albeit at a lower speed. If the aircraft is operating above Mach one, the consequences may be more severe. Unusually large pressures generated at the inlet due to the shock wave may result in distortion and damage to the structure. Loss of thrust experienced under such circumstances, although only for a short period of time, may cause the aircraft to go out of control under certain operating conditions.

Air is first imparted acceleration by the rotating blades, then decelerated in the stator passages, where the kinetic energy is converted to static pressure. A series of diffusions take place in both the rotor and stator passages. Absolute velocity of the working fluid is increased in the rotor, but the relative velocity will decrease. It is imperative for the flow area to diminish moderately in a diffusing flow, hence a single compressor stage can only deliver a small increment in the pressure, considerably less then what a turbine can advantageously use with its efficient pressure gradient, converging blade passages and accelerating flow. This explains why a few turbine stages can power many more stages of the compressor. As fluid density increases with progression of the flow, axial velocity does not change appreciably. During operation at lower speeds density in the latter stages deviates further from the design point, causing the flow velocity to reach a point where the blades may stall and the compressor to surge.

As the onslaught of instability progresses, mass flow in the unstable embedded cells reaches to a near zero, while in the non-stalled cells it is normal. A single stage fan or compressor may even experience unstable cells originating in a part of the blade span. With further throttling, the cell may propagate to cover the whole blade and spread to fill more of the annulus. Rotating stall cells cover the full blade in multi-stage compressors. Progression of stall along the blade row may be explained by considering the direction of flow. With a given passage partially blocked by the stall, flow is diverted to the neighbouring passages. This results in an increase in the incidence angle in the next blade in the direction of stagger and a decrease in incidence in the adjacent blade in the opposite direction. This causes the stall region to push in the direction of stagger, propagating at a speed of 40 to 60 per cent of the blade tangential velocity. The limit of stability in compressors may be defined by rotating stalls. Further pressure gains result in unsteady flow, causing considerable vibrations in the blades.

Inception of two forms of rotating stall has been observed in axial compressors. One of them, called modal stall, is characterized by waves with length scale on the order of the compressor's circumference, and propagates at a speed of one-fourth to one-half the speed of the rotor (*ref*: Haynes et al., 1994 and Tryfonidis et al., 1995). Another considerably different rotating stall is characterized by disturbances with a dominant length scale much shorter than the circumference, on the order of several blade pitches, and a propagation speed of 70 to 80 per cent of the rotor speed.

In contrast to inception of a modal stall, mechanistic description of this phenomenon is difficult due to the shorter length, and may best be referred to as spikes. Experiments indicate that this rotating stall inception possesses a radial structure (*ref*: Silkowski, 1995), and may be affected by clearance at the tip of the blades. The observations imply that a different and detailed approach is essential to capture the development of such short length inception. More specifically, it is necessary to include a description of the tip clearance flow structure within individual blade passages. It is hypothesized that the unstable motion of the tip clearance vortex forward of the compressor leading edge is a mechanism for the development of short scale length disturbances leading to a compressor stall.

14.8 ROTATING INSTABILITY FROM VORTEX AT BLADE TIP

Aerodynamic rotating instability is responsible for intensifying tip clearance related noise and for setting up excitation of blade vibrations. This flow characteristic may be due to a group of superimposed modes or a part span stall with fluctuating cell numbers. In pressure or velocity frequency spectra the instability appears in the form of a significant increase in amplitude within a frequency band that is well below the blade passing frequency, and is superimposed by lesser peaks with nearly constant spacing in the frequency. The pattern is marked in a group of consecutively numbered modes.

Primary parameters associated with the instabilities in axial compressors vary over a wide range of values. Mode orders of one-third up to twice the number of rotor blades have been reported. Correspondingly, circumferential wavelength of the disturbance is half to three times the blade pitch spacing. Rotational speed of the instabilities in different machines varies between 25 and 90 per cent of rotor speed. Tip clearance at the rotating blades may be considered a major influencing factor in the initiation of the instability. Pressure difference across the blade row and axial component of flow velocity also plays a substantial role.

Mailach et al., (2000) carried out an experimental investigation to determine the influence of blade tip vortex on the origination of rotating instabilities. A four-stage compressor preceded by inlet guide vanes with a hub diameter of 1260 mm and hub to tip ratio of 0.84 is used for the tests. Design point speed is 1000 rpm, mass flow is 25.35 kg/second and Mach number at inlet mid-span is 0.22. The instrumented row has 63 blades of 110 mm chord length at mid-span and 116 mm at tip, solidity of 1.55 at tip and stagger angle of 40.5°. Nominal clearance at the tip is 1.55 mm. The stator has 83 vanes with a chord length of 89 mm. Signs of instability for the nominal clearance were not observed, hence it was enlarged to 2.3 mm, 3.5 mm and 5.0 mm in steps. A two-dimensional laser Doppler anemometry system determines the axial and circumferential velocities of the flow field within the rotor blade passage. Kulite sensors acquire unsteady pressures on the suction and pressure surfaces of two adjacent blades.

Within the clearance volume, measurements can be made only for the biggest magnitude. Post processing of the data is by the ensemble-averaged method. An array of microphones distributed circumferentially in one axial plane determines higher order modes of the rotating instabilities. Simultaneously acquired signals permit correlation between various sensors (*ref*: Rangwala, 2004).

Figure 14.11 shows a typical frequency spectrum measured at the casing wall for an operating point near the stability limit; blade tip clearance is 5.0 mm, the machine is operating at design speed, and mass flow coefficient is 0.82. The largest disturbance amplitude occurs at 265 Hz, which is one-quarter of the blade passing frequency. Other notable spikes are seen at the blade passing frequency and at the sum of the frequencies of blade passing and the rotating instability. During the operation the first two modal waves are noted throughout the compressor, appearing as discrete peaks at considerably lower frequencies then those attributable to the rotating instabilities (1 per cent of blade passing frequency for the first mode, and 50 per cent of rotor speed for the second mode).

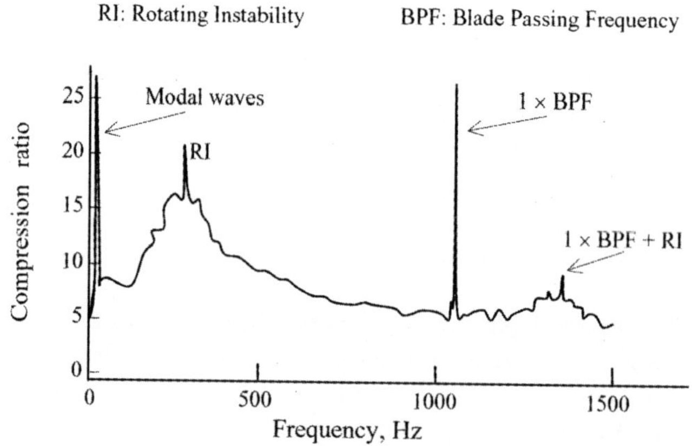

Fig. 14.11 Frequency Spectrum at Casing Wall (Mailach, 2000)

Rotating instabilities propagate in the blade's tip region in the form of a wave front. Since the dominating mode order is about half the number of blades, the process repeats itself at every other blade at the same time. Direction of the propagation in the tip region is determined by correlating the signal from the pressure transducers on the pressure and suction sides of adjacent blades. In figure 14.12 the + and – marks indicate maximum and minimum pressures. In the rotating frame of reference the propagation occurs against the direction of the shaft, and is directed downstream. The phase difference between the leading and the trailing edges of the profile is about 180°. Thus, the disturbances spread in a spiral form (figure 14.13).

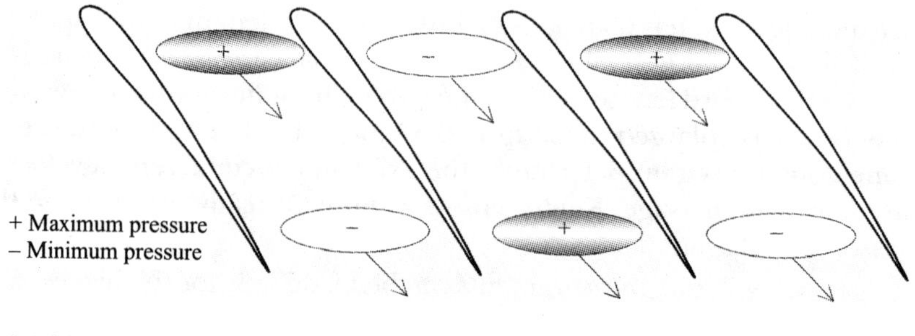

Fig. 14.12 Propagation of Rotating Instabilities in Blade Tip Region (Mailach, 2000)

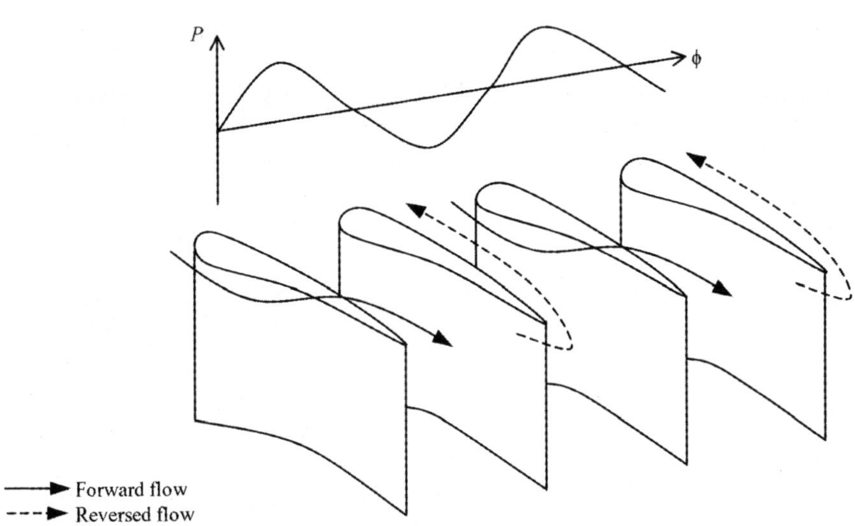

Fig. 14.13 Rotating Instability Propagation in Blade Tip Region (Mailach, 2000)

Pressure difference between the pressure and suction sides along the chord is nearly constant at the design point, hence the time averaged tip vortex intensity is not dependent on the axial position. But at an operating point near the stability limit the driving pressure difference at the leading edge initiates, indicating a propensity towards intensification of the vortex.

This trend is also verified by the anemometer signals. Large regions of reversed flow can be observed due to the axial velocity component and the relative flow angle rolling up of the blade tip vortex, shown in figure 14.14. The positive pressure gradient across the rotor blade row intensifies this reversed flow in the plane inside the tip clearance. In the front part of the blades a zone of low

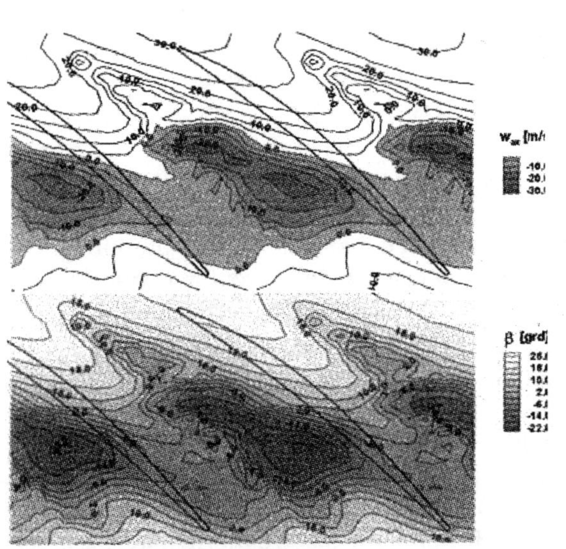

Fig. 14.14 Flow Characteristics in Blade Tip Clearance: Axial Velocity (*upper*); Relative Flow Angle (*lower*) (Mailach, 2000)

axial velocity suggesting an under-turning of the flow can be seen in the middle of the blade passage. Relative flow angle remains positive. The reversed flow region within the tip clearance is limited to the rear part of the blades and extends over the whole circumference. Minimum axial velocities and relative flow angles appear between the middle of the blade passage and the pressure side, mostly due to the large vortex at the tip.

14.9 TURBINE THERMAL STATE

Operating clearances in an engine are determined to a considerable extent by the operating temperature in an engine. For instance, in the generalized data for a cold engine start of figures 14.8 and 14.9, the take-off pinch point would be lower for a hot engine take-off since the rotating components will grow radially due to the larger thermal mass of the turbine rotor and blades as compared to the casing. An adequate period of cooling time must be provided before an aircraft can take-off with hot engines.

Besides the thermal growth, substantial thermal stresses in the airfoils and in the rotor are induced when the take-off is attempted with hot engines. Thermally induced growth is also responsible for another phenomenon known as 'bowed rotor'. Shortly after an engine is shutdown, heat build-up takes place in the upper portion of the engine from the heat retained in the components of the engine and lack of venting for this trapped heat. As a consequence, temperature gradients develop along the diameter of the high and low-pressure rotors and cause the shafts to deflect or bow. Depending on the prevailing wind in the vicinity of the engine and the consequent wind milling of the rotors, the maximum radial deflections from the chimney effect generally occur between 1 and 2 hours after the engine is shutdown. If the engine were to be restarted in this time period, the bowed rotor start will lead to considerably large amounts of unbalance in the two rotors, excessive vibrations in the engine, and cause rubbing between the blade tips and the casing. Bowed rotor starts are known to be severe enough to cause the low-pressure rotor of an aircraft engine to lockup due to the interference between the blade tips and the casing shroud.

Industrial and power generation steam and gas turbines display similar traits in regards to bowed rotor starts. As a rule, immediately following a shutdown, the rotor is then placed on turning gear to achieve better temperature distribution. Since the rotor is of a substantially heavier mass in this case, the cooling time required is also larger.

14.10 PASSIVE CONTROL METHODS

A number of methods have been devised for ensuring the optimum gap between the rotating airfoil tips and the stationary casing. The approaches, broadly divided into passive and active control methods, rely upon thermal, pneumatic and mechanical means.

Thermal systems rely on the stator's material characteristics, in particular the thermal coefficient of expansion, and operating condition of the engine to obtain the same level of growth in the rotor and the stator (*ref*: Carpenter et al., 1997). Frequently used on smaller engines, thermal expansion of the shroud at lower temperatures is controlled to better match the centrifugal growth of the rotor during take-off while the casing is still gaining temperature. During aircraft climb the casing is at a higher temperature and is required to grow at a slower pace, and the rotor is only growing from thermal effects. However, the passive thermal means lacks in accuracy and reliability in achieving optimum clearance for the minimum clearance condition necessary for take-off and for manoeuvres.

Passive pneumatic systems are actuated by gas pressures generated in the engine or by hydrodynamic effects. The designs are based on floating shroud segments (*ref*: Wardle et al., 1960) and blade tip cooling air discharge (*ref*: Huber et al., 1997). Development of force from a hydrodynamic system, however, suffers from the limited surface area of blade tips and high positioning and alignment tolerance requirements. Aggravating the situation are the problems of friction and secondary seal hang-up, and the concepts become unattractive. Pneumatic systems relying on flexible piping or bellows are subject to high cycle fatigue from the flexing motion, thin walls of the materials needed for operation, vibration, thermal and pressure loadings in the high-pressure turbine's environment.

14.11 ACTIVE CLEARANCE CONTROL REQUIREMENTS

Active systems remain the mainstay for controlling clearances in modern gas turbines. Active clearance control methods also depend on thermal, pneumatic and mechanical means, either singularly or in combination with electro-mechanical, magnetic and other devices for the functioning.

Thermal designs depend on compressor and fan air to respectively heat and expand or cool and shrink the supports for shroud seal segments to change the blade tip clearance. Figure 14.5 shows details of a thermal active clearance control system. The drawback of the system is slow thermal response, which does not provide for further gain from clearance closure during cruising conditions. Improvements in the system have been made through the use of more thermally efficient materials and geometric changes in the components. One method calls for attaching the seal segments to a thermally efficient channel structure that is cooled on the inside and decoupled from the substantial casing flange for obtaining a faster and more responsive control system (*ref*: Ciokajlo et al., 1992).

Active mechanical designs combine linkages and some form of actuating mechanism, which may be hydraulic, electro-mechanical or electro-magnetic for shifting the tip clearance. One system calls for a segmented shroud with the segments attached to an actuating ring through linkages and radial spokes. When the ring is turned in any direction, the segments move radially inward or outward in unison (*ref*: Tseng et al., 1991). Because of the elevated temperature level in the casing, actuation is generally required through the casing. However, the mechanically complex system adds weight, poses problems in secondary sealing and poses tolerance stack up issues arising from the number of linkages used. Many of these considerations may be tackled successfully, but the primary issue of one of positioning the shroud segments remains. Clearance measuring sensors must be able to withstand the operating temperatures and imposed vibrations at the seal locations while providing acceptable working life. High temperature sensing electronics is needed to resolve the problem.

Active pneumatic concepts for clearance control depend on the engine's own internal pressure, or pressure from other sources and bellows and valves. Load is directly applied on the flexible and sealed shroud segments to change the radial tip clearance. Precise balancing of the pressure loads is essential for proper operation, and high cycle fatigue issues may also arise. Since pressure requirements are likely to be substantial, the method may detract from efficiency considerations.

14.12 PROSPECTS FOR ACTIVE COMPRESSOR STALL CONTROL

In an effort to understand generic features of stall inception in aircraft engine compressors, Whittle Laboratory of Cambridge, England performed the task of theoretical and experimental work on four different engines (*ref*: Day et al., 1997). In addition to learning more about initiation of the phenomenon, the project aimed to evaluate data analysis techniques and to assess the range of incoming signals with which an active control system might be expected to cope. Earlier low speed work at the Laboratory has shown that stage matching and radial flow distribution are perhaps far more important than tip clearance in the stall inception process, especially the mechanism associated with modal oscillations of the entire flow field. In high-speed compressors the stage matching not only depends on design of the blades, but also on rotation speed and disposition of any variable stator vanes. This assignment for the Laboratory called for providing stall inception measurements covering the full speed range from ground idle to over speed.

Figure 14.15 shows the four compressor designs, each representative of aero-engine practice. MTU's three stage compressor develops intermediate pressure at first stage tip Mach number of 1.08, overall pressure ratio of 2.5, 20 per cent true chord inter blade row axial spacing and no variable vanes in the stator. The SNECMA compressor has four stages for high pressure design, relative first rotor tip Mach number of 0.798, overall pressure ratio of 2.0, variable inlet guide vanes and 50 per cent chord inter blade axial spacing. DRA's five-stage compressor is from a military core engine, first rotor tip Mach number of 1.001, overall pressure ratio 6.0, variable inlet guide vanes and 60 per cent chord inter blade axial spacing. Rolls Royce Viper engine compressor has eight stages, single shaft, first rotor tip Mach number of 0.929, overall pressure ratio 5.0, no variable guide vanes and 30 per cent chord inter blade axial spacing.

High frequency transducers are placed in circumferential arrays, and are placed equally within the limit of mechanical restraints (*ref*: Rangwala, 2005). Closer spacing is possible where blade rows are limited. In addition, speed, mass flow rate, stage-to-stage and overall pressure increment and operating temperatures are recorded using standard test-bed instrumentation. Guide vane scheduling, inlet distortion and reduced Reynolds number tests are set in accordance with test speeds.

Occurrence of modes and spikes in the compressors are considered first. Although their incidence is not particularly noteworthy, the role of stage matching in determining the pattern of stability in a given situation is of interest. Representative examples of modes and spikes are shown in figures 14.16 and 14.17. The distinction between the two is that in a modal type stall inception flow becomes unstable due to a large-scale circumferential oscillation, while in a spike style stalling flow breakdown originates from a distinctly localized disturbance. Physical features of the two processes may be described as follows. The modal perturbations rotate comparatively slowly at about 50 per cent of rotor speed, and appear as gentle waves, or undulations, in the pressure traces. Spike type disturbances are characterized by sharp peaks in the time traces, and generally rotate faster when first detected, at 70 to 80 per cent of the rotor speed, then slow down rapidly as they grow in size. In figure 14.18 the modal perturbations rotate at 33 per cent rotor speed, while spikes in figure 14.19 rotate at 70 per cent rotor speed, where a spike is visible at the first rotor nearly a full turn before detection in the second rotor.

Both modes and spikes can occur in the same compressor, and stage matching is one of the parameters that can change one stall pattern to another. In high-speed compressors, stage matching changes with operating speed. Compressibility effects shift the position of highest loading from the

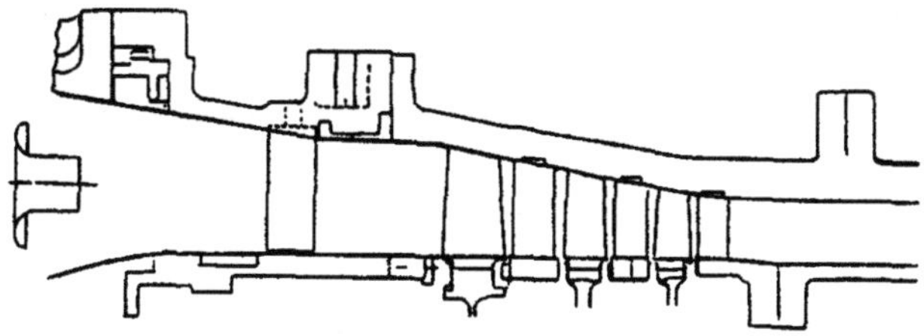

MTU Compressor

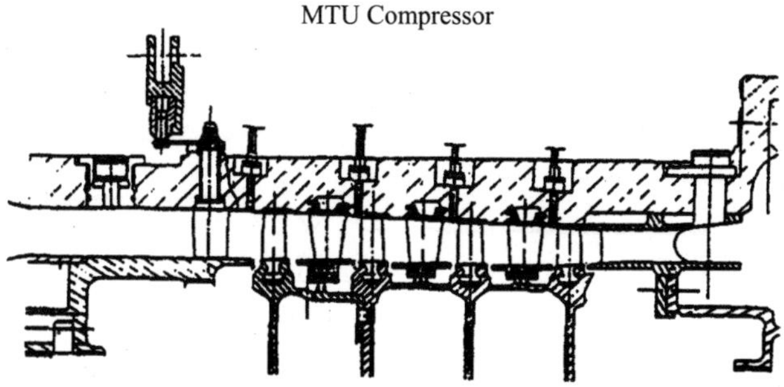

SNECMA Compressor

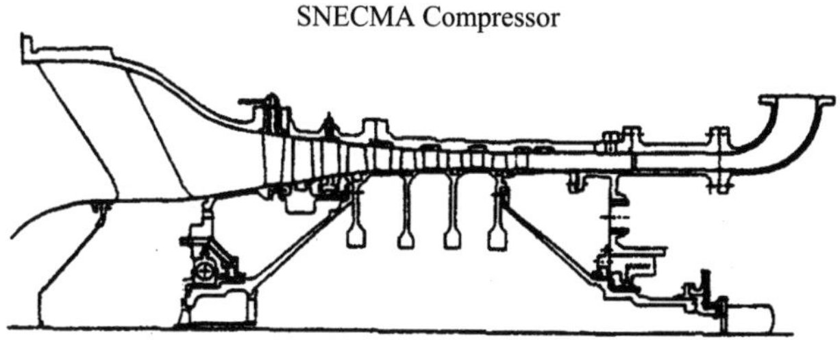

DRA Compressor

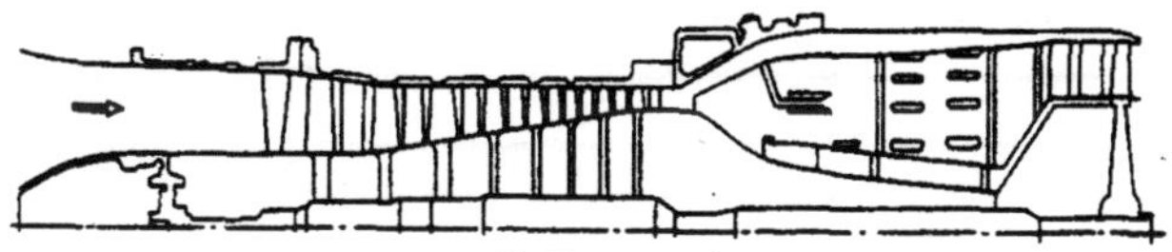

Rolls royce viper

Fig. 14.15 Compressor Designs (Day, 1997)

front to the rear of the compressor as the speed of rotation changes. Front stages are heavily loaded at low speeds, whereas rear stages are more likely to stall at high speeds. Even matching of all the stages near the stall point is achieved in the medium speed range.

The spike form of stalling is a localized disturbance, initiating as a small incident in just one row. Hence, it may be assumed to occur at all speeds of rotation where the individual stages at the front or rear of the compressor are more highly loaded than the rest. At middling speeds of rotation, where all stages are evenly matched, it might likewise be assumed that modal type stall inception would occur where the perturbations, being of larger proportions, can grow evenly throughout the machine. These concepts concerning the impact of compressor speed on stall inception pattern are supported by tests at Whittle Laboratory.

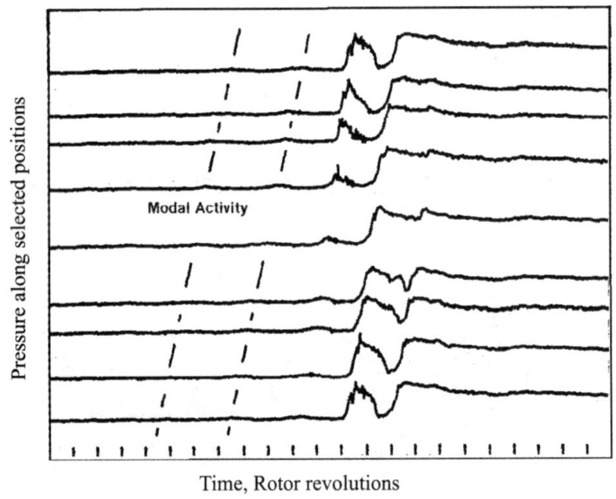

Fig. 14.16 Modal Style Stall in MTU Compressor (Day, 1997)

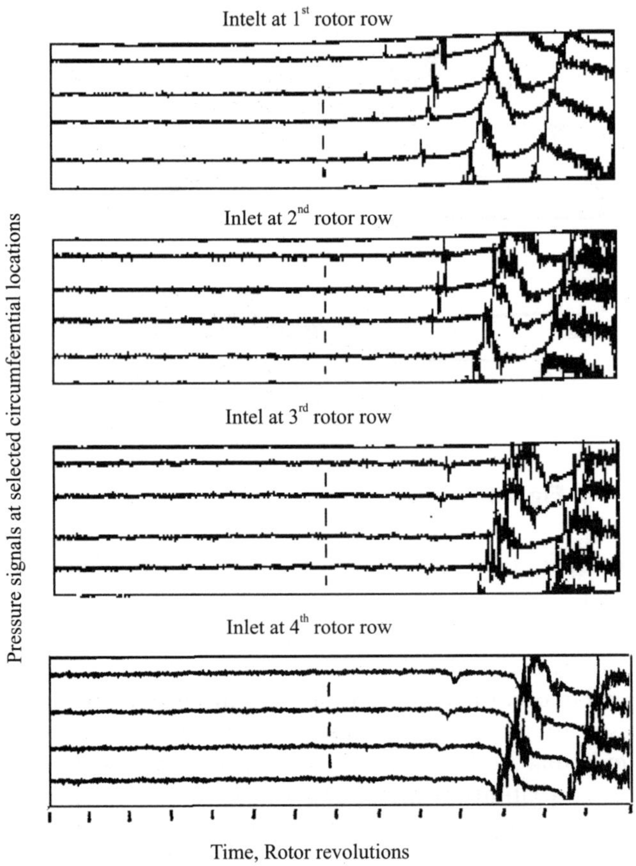

Fig. 14.17 Localized Spike Stall Inception in SNECMA Compressor (Day, 1997)

A clear example of the changing stalling patterns with a change in speed is seen in the MTU compressor (figure 14.18), where spike like stall occurs at low speed and modal type in the mid speed region. A similar trait is exhibited by the Rolls Royce Viper in the change from a spike to a mode as the speed increases. Spikes are observed in figure 14.19 at 70 per cent speed and the modal type activity at 85 per cent speed. At 100 per cent speed where spikes might be expected in this compressor, a rotating pattern does precede surge, but the structure of the pattern is ill-defined. The DRA compressor follows this trend, but again the stalling process is too complicated at full speed to permit distinguishing between the spikes and the modes. The SNECMA machine behaves differently from the others, where the stall is initiated by spikes throughout the speed range. This observation cannot be explained, except that this compressor has relatively large inter blade row axial spacing and a low overall pressure ratio.

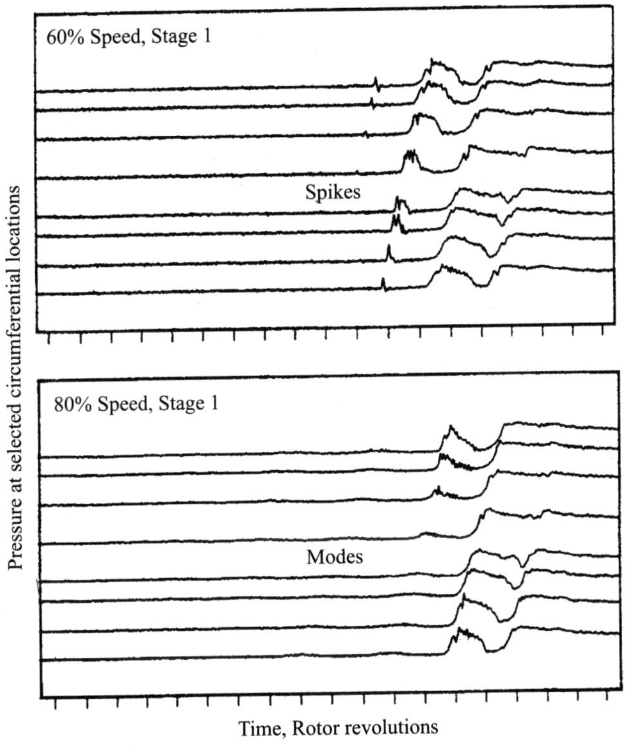

Fig. 14.18 MTU Compressor Displaying Spikes at Low Speed Changing to Modes at Middling Speed (Day, 1997)

To determine the impact of inlet guide vanes, the DRA and SNECMA compressors were tested throughout the speed range with different settings. Stall inception was not affected in the DRA compressor, despite large (9°) changes in both positive and negative directions. In contrast, the SNECMA compressor is very sensitive to closing and opening of the vanes. Modal activity is not exhibited at any speed, but the vane setting does play a role on the axial location of stall inception. At design setting of the vanes, spikes appear at the front of the compressor at low speeds and at the rear at high speeds. But a small change in IGV setting is sufficient to reverse this pattern. The reasons for the observations on the two machines cannot be readily explained.

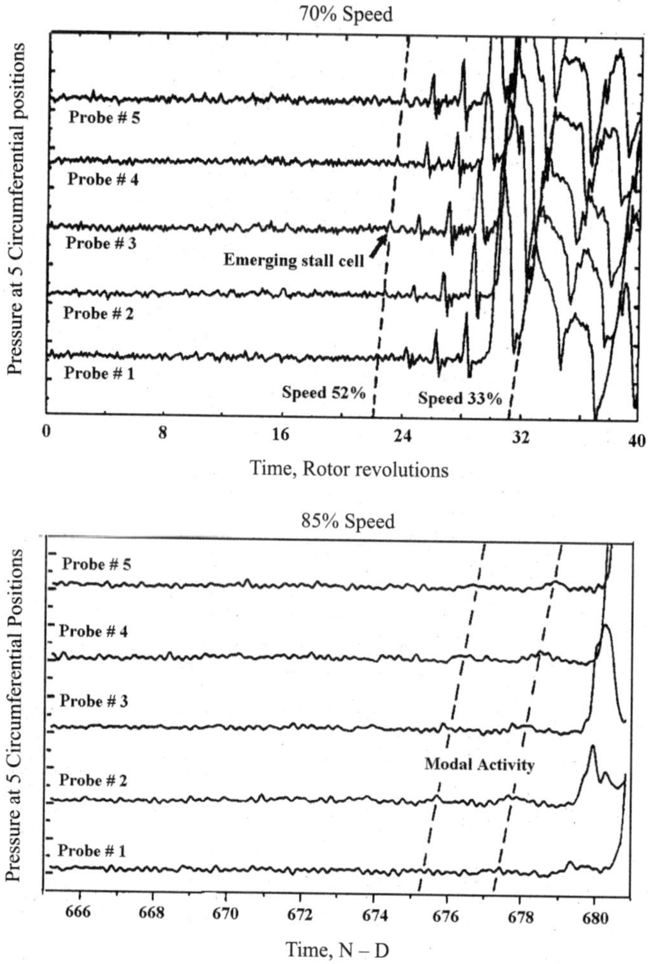

Fig. 14.19 Spike and Modal Stall Inception in Viper Engine (Day, 1997)

At low rotational speeds around 40 to 60 per cent most compressors tend to display stall like characteristics when the unit is started. The occurrence is natural, with operation taking place in the stable stall regime. This 'front-end' stalling comes about because the initial stages of a long compressor suffer from through flow velocity at low rotational speeds, and the blockage created by the stall cells is a self-correcting mechanism for reducing the frontal area of the compressor, thus improving the matching. The behaviour is confined to the first few rows, and the stall cells are nearly always of the part-span type. Characterized by multiple cells, in the event of throttling they organize themselves in a circumferential pattern, then coalesce through a merging process to give rise to a single stall block of larger proportions. Figure 14.20 from the Viper compressor reveals this process of three cells coalescing into one group. While the cells retain their multi-cell form, they are relatively benign, doing little harm to compressor or engine performance. The larger full-span cell, when it forms, has a bigger effect on performance because it spreads throughout the compressor and necessitates engine shutdown.

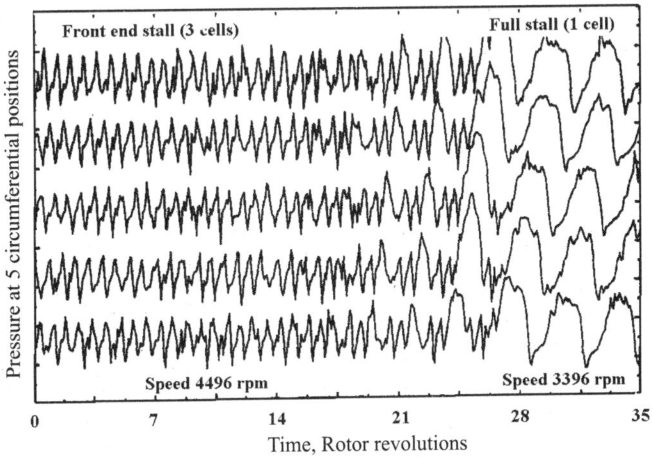

Fig. 14.20 Multi-Cell Front End Stall Developing into Full Span Stall, Viper Engine (Day, 1997)

At full speed it is difficult to make a clear statement of generic trends. All compressors surge at this speed; in most cases the surge event is preceded by a brief period of rotating stall. The stalling process is rapid, with less than three revolutions separating the appearance of the first stall cells and establishment of reversed flow. The number of revolutions mentioned here compares with about five at lower speeds. In the DRA compressor 10 milliseconds divides the forward flow from the reversed flow. Flow breakdown seems to initiate at a fixed circumferential location in the annulus. This is true of the SNECMA engine and to a lesser extent in other machines.

Circumferential bias is also a feature of another interesting observation in these tests. Sometimes the compressors display signs of protracted non-rotating pre-stall flow disruption. In the DRA, MTU and Rolls Royce compressors a gentle rise in pressure appears over three or four rotor turns prior to stall. The slow ramp occurs over a wide circumferential area. An example of this fixed location stall from the DRA compressor is shown in figure 14.21.

High frequency signals between 10 and 13 times rotor frequency are observed in the DRA and SNECMA compressors, the disturbances occurring at or near top speed.

The somewhat audible disturbances are not considered a consequence of aliasing in data acquisition. An example of the signal for the SNECMA compressor is shown in figure 14.22. The perturbations are the strongest at the rear, but detectable throughout the compressor. In a unit with a large number of blades in each row signals of higher frequency than rotor passing may be generated by an array of considerably small cells. As a generality, the smaller the stall cell, the faster it rotates, rotor speed being the upper limit.

Shaft order perturbations are present to a greater or lesser extent in all compressors, the effect magnifying at high speeds. Lack of concentricity of rotor and casing, tip clearance size and non-uniformity of the blade angles throughout the compressor area are contributing factors. The point of interest, however, is not the existence of the perturbations but whether or not the disturbance actually participates in the stalling process. It has been suggested (*ref*: Hendrick et al., 1997) that high-speed modal perturbations may lock into shaft order disturbances with the combined effect of triggering a stall, but these tests do not support the premise.

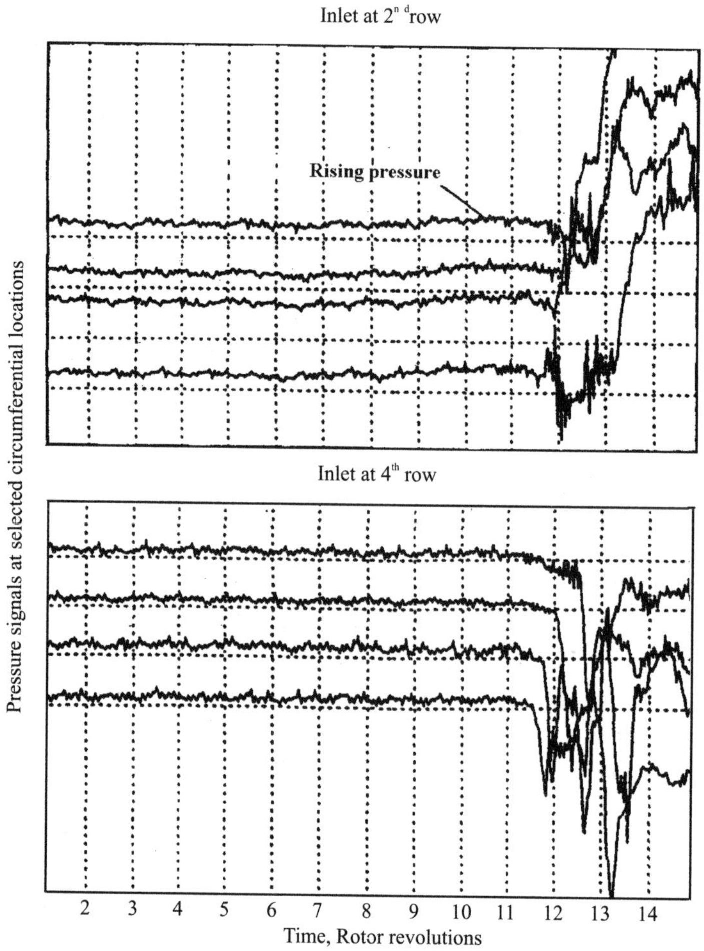

Fig. 14.21 Non-Rotating Pressure Ramping Prior to Stall at Full Speed in DRA Unit (Day, 1997)

From the conducted tests a pronounced link exists between the Reynolds number and compressor behaviour. A comparison of stalling traces do not indicate an appreciable difference when the average Reynolds number is reduced from 7×10^5 to 4×10^5, but the latter number is still above the threshold at which an effect on the blade row's performance may be detected. The impact of inlet distortion on specific changes in stalling behaviour is likewise difficult to determine, especially at high speeds where the stalling process is in any case difficult to categorize.

In the classical sense of the word, active control makes use of real time measurements in the compressor to detect stalling disturbances, and to take action to suppress them. System response is then tailored to the particular disturbance in the compressor at the time. The control system must then be capable of identifying and responding to all forms of stall inception. Misinterpretation of the incoming signals will lead to inappropriate instructions, which could possibly result in the exactly opposite direction of the compressor stalling prematurely.

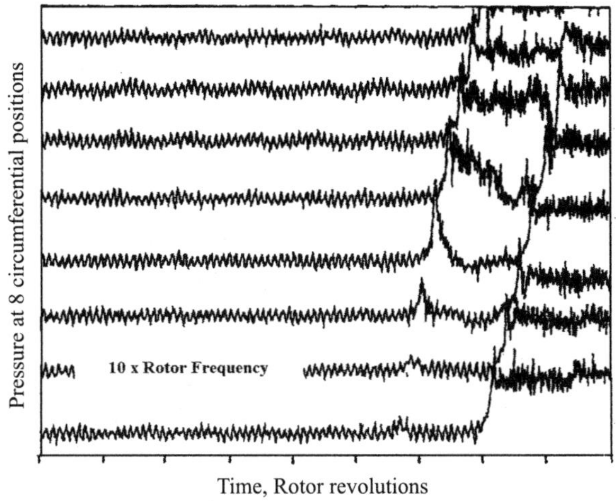

Fig. 14.22 High Frequency Stall in SNECMA Compressor at Full Speed (Day, 1997)

The structure of both modal and spike-like stall inception disturbances are reasonably well understood, and algorithms are available to detect their presence. Modal stalls are easier to detect because the whole compressor is affected at the same time, so all measuring probes will contribute to the identification. By contrast, the spike form of disturbance occurs mostly at low rotational speeds. Their localized appearance makes them more difficult to track than the modes. The disturbance originates in a small way from flow breakdown in a particular blade passage, and has no discernible impact on the rest of the circumference. It is even possible for a small disturbance to travel from one probe to another and remain undetected. Thus, early detection above the background noise is a difficult proposition at best.

Besides modes and spikes, other forms of stall that could have an influence on reliable application of active control are: front end start-up stall, high frequency stall, abrupt stall at full speed, fixed location stall, shaft order disturbances and distorted inlet flow. This list provides an indication of the variety of signals that may be observed during the operation of high-speed compressors. An individual compressor, of course, is not going to be susceptible to all of these problems, but it is useful to have foreknowledge of the possible range of signals that might be encountered.

An alternative approach to active control is to disregard the physical details of the flow generating the signals and to concentrate on deriving a stall risk index from the sum of all inputs. Reliability of the index would be essential for a scheme of this type, but the problem of identifying the necessary corrective action in the variable vanes or operating speed or both still remains. Thus, the goal of active control still seems to be some way off.

14.13 TIP CLEARANCE CONTROL MANAGEMENT

A successful clearance management for gas turbine engines of all military or commercial aircraft requires at the start technical requirements to be met. Operational conditions (temperature, pressure, vibrations, among others) and functional needs (radial clearance change, response rate, cycle life, positional accuracy, concentricity, ability to maintain circularity of the shroud segments, output force) must be identified at all mission points. In addition, design considerations of weight, space requirement, material cost, maintenance and reliability must be closely scrutinized.

Centrifugal growth of the rotor during aircraft take-off and re-burst conditions is the maximum contributor to HP turbine clearance variation. Symmetric and asymmetric loads on the rotor and stator components can cause up to 0.050 in. variation in the tip clearances. FAA specifications call for gas turbine engines to have the capability of reaching 95 per cent rated take-off power from flight idle (or from 15 per cent rated take-off power) in 5.0 seconds (*ref*: Electronic Code of Federal Regulations, 1971). This directive sets a limiting case of maximum displacement and minimum response time for an actuation system of an active clearance control mechanism. Thus, for the maximum displacement of 0.050 in. this will require the actuation system to provide radial clearance change of 0.010 in./second. This rate, however, will not accommodate the considerably faster clearance changes occurring every rotor revolution.

The other important considerations of positional and dimensional accuracy that affect sealing of the clearance gaps will rely heavily on manufacturing and assembly tolerances. The components of the engine and of the ACC system attempting to control tip clearances to within 0.010 in. will call for high precision. In addition, temperature excursions, vibrations, and fluid flow forces present major challenges in achieving target positions. Safety margin requirements generally do not permit thermal ACC systems to close cruise condition clearances below 0.015 to 0.020 in. This is because of their slower response in the event of throttle transients (for instance, step change in altitude) or manoeuvres at cruise.

The backside of the HP turbine shroud, or outer air seal, is usually cooled with compressor discharge temperature T_3 at 1200° to 1300° F. This permits the shrouds to survive rotor inlet gas temperatures of around 2500°F. The cooler air also purges the leading and trailing edges of the shroud segments and provides back-flow margin from the inlet flow into the rotor (figure 14.23). Air pressure around the shroud segment is expressed as a function of compressor discharge pressure (CDP) P_3. Pressure difference across the shroud's radial faces leads to a load that acts inwards towards the engine's centerline. The ACC system must be able to overcome this load, as also the overturning moment arising from the non-uniform pressures on the shroud's axial faces. If the mean pressure drop across a two-stage HP turbine shroud is 120 lbs/in^2, the shroud is of 30.0 in. inner diameter and 2.0 in. wide, this results in a radial force of 240.0 lbs/in. of arc length. If there are 16 shroud segments of 6.0 in. length, the pressure difference will produce a force of 1440.0 lbs on each segment. Accurate placement of the shrouds, within 0.005 in. tolerance for instance, under this load poses a difficult problem. Hence, the need for careful selection of the actuating mechanism for the ACC is easily seen. Elevated temperatures in the region will also limit the selection of materials for the components.

14.14 CLEARANCE SENSORS AND CONTROL ACTUATORS

Sensors for measuring blade vibrations and tip clearances rely on a number of technologies, including x-ray, inductive, capacitive, optical, eddy current, microwave and acoustic (*ref*: Flotow et al., 2000). Blade passage sensors typically are capable of measuring the clearance at the tip and the time of arrival. Accuracy level in the sensing devices should be of the order of 0.001 in. to avoid detraction from the ACC system's own tolerance stack up. Accuracy of measurement will also depend on the physical condition of the blade tip.

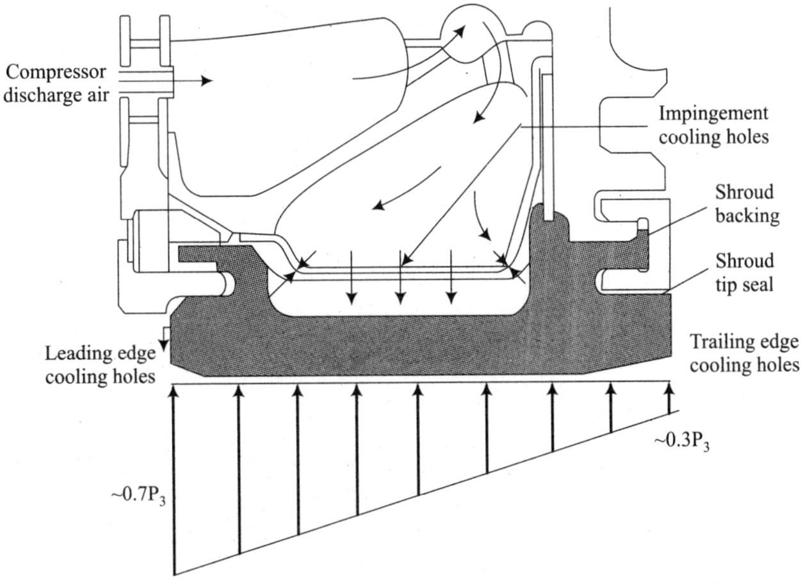

Fig. 14.23 Pressure Distribution Around HP Turbine Blade Tip Seal

Measurement of clearances at the blade tip may be imperative in an ACC system if the objective is to improve upon and replace the existing technology. To permit multiple measurement of clearance per blade of larger engines, sensor response should be of the order of 50 kHz. High-temperature capable sensors and the associated communication system may need cooling, and must also be durable to withstand vibrations and impact while providing long service life. The sensors must be tolerant to moisture, dirt and combustion gases.

Non-contacting measurement techniques such as blade tip timing measure the clearance gap between the rotating and stationary components and blade deflection at the tip using optical probes mounted in the assembly casing. When vibrations of the external casing structure are not substantial, the time for the tip of a probe, the so-called blade arrival time, depends on rotor speed alone. But when the blade is vibrating, blade arrival time also depends on the amplitude and frequency of vibration. In principle, blade motion can be characterized from such data. As with other transducers such as strain gauges and accelerometers, whether a particular mode can be captured by a given laser probe depends on the measurement location. The measurements are typically taken near the leading edge of the blade to maintain reasonable sensitivity to the motion of the first few bending and torsion modes.

A tip timing system consists of three main elements: (*i*) a number of probes placed in the casing to acquire raw blade arrival time, (*ii*) derivation of primary parameters (clearance gap, displacement, velocity or acceleration) from the measured data, (*iii*) processing of the parameters to describe the gap between the blade and the casing and the vibration pattern of the bladed disk assembly. Figure 14.24 provides details of the scheme (*ref*: Heath et al., 1997). The first two items are readily available using data acquisition and derivation technology. The third step plays a major role in determining the success of the concept. For asynchronous vibrations, the mean value of the measurements over a limited number of samples (or speed points) aids in deleting the steady part of

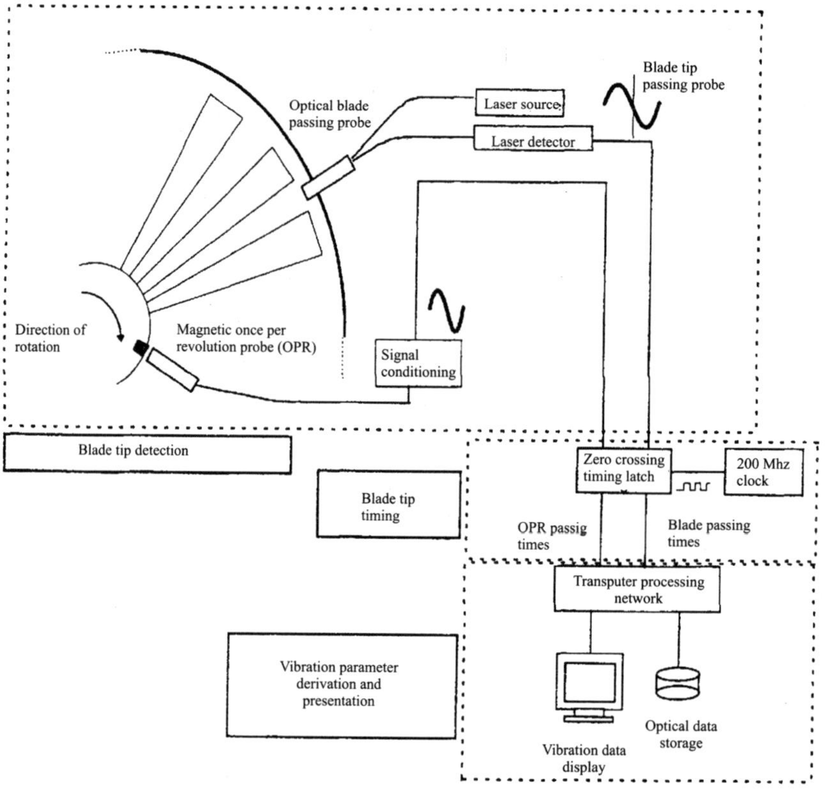

Fig. 14.24 Elements of Blade Tip Timing System (Heath, 1997)

the response. For synchronous vibration, the unsteady components rely on location of the blade relative to a non-vibrating rotating datum from a shaft encoder or a once per revolution probe (*ref*: McCarty et al., 1980).

Advanced actuator designs and smart materials such as piezo-electric actuators and sensors for optical positioning play a major role in many different control system applications. Shape memory alloys posses a number of design capabilities, but are not high-temperature resistant for high-pressure turbine application.

Actuation of high-temperature metals through thermal expansion and contraction provides a more precise and predictable system, but act at a slow pace. The response may be accelerated with a more efficient geometry or material, so long as structural integrity is not compromised.

Hydraulic actuation also finds application in a number of gas turbine engines. Fuel is presently used for valve actuation, and also for actuating fan and compressor inlet guide vanes (*ref*: Eder et al., 1997). Military engines with thrust vectoring also use high-pressure fuel (around 3500 lbs/in^2) to actuate and control exhaust nozzles.

On/off, model-based and feedback types of control systems are commonly used for the management of tip clearance systems. The on/off method provides two-position actuation, which may be optimized for a selected operating point. The method has advantages since during cruise the variation in clearances is limited. Military transport aircraft typically spend 89 per cent of the fuel and 93 per cent of the mission time at cruise. Fighter aircraft, on the other hand, spend 69 per cent of the fuel and 75 per cent of the time at cruise. Detrimental clearance shifts during flight transients, however, cannot be accommodated.

The model-based control method relies on a number of parameters such as engine speed, temperature, pressure, speed and power output measurements to estimate the clearance as a function of the operating condition. The engine manufacturer establishes the values of the operating parameters for different flight conditions together with blade tip clearance data from extensive engine tests and calculations, and stores the data within the onboard microprocessor. The model-based control system then determines from extrapolation of the stored data the amount of force to be applied through the actuating mechanism at a certain operating point. The method helps to hold clearances at some intermediate position for the flight operating condition.

The method of feedback control relies on continuous measurement of clearances throughout the flight. Based on the measured data, the actuating force is modified through the feedback mechanism to determine a new level of load for better adjustment of clearances between the rotating blade tips and the stationary casing. And since new measured data is continuously coming in from the feedback sensors, the actuating force also goes through adjustments without interruption.

14.15 RUB AVOIDANCE AND REGENERATION CONTROL

Prevention of rubs between the rotating blade tips and the stationary engine casing while maintaining an adequate level of clearance through a combination of feedback and model-based control of the ACC management system for all operating points is a goal that is worth achieving. The benefits of enhanced power output, stability in operation, improved fuel consumption and the associated reduced exhaust emissions provide a combination that cannot be overlooked.

Besides the passive and active methods of controlling the clearance, another control method may be identified that stands apart from the avoidance of rub approach. Restoration or regeneration of tip seals calls for using both passive and active control means to restore deteriorated tip seals that are worn from erosion and rubbing. Mechanically adjusted linkages that may be ground for tip seal restoration and seal materials that undergo permanent changes in volume illustrate this concept.

Mechanical restoration faces the same restrictions about tolerances as in the active mechanical designs. Also, the weight of additional hardware has to be justified, although deletion of the actuation driver may compensate to some extent. A more fundamental issue will be accessibility for routine ground maintenance.

Specially engineered materials that display permanent growth under the action of thermal, chemical and electrical processes are envisioned for regenerative seals for passive control. Oxidation is a leading example for such interaction when operating in the hot combustion gas environment of HP turbine. The potential for active control arising from chemical reactions that may be accelerated or decelerated through application of electric potential, additives, or heat exists (*ref*: Jorgensen, 1962). The issue of achieving uniform growth around the full circumference has to be adequately addressed. Also, the regenerative growth must take place in only one direction (radially) up to a certain limit to preclude blade wear and must have the right level of strength.

14.16 TIP CLEARANCE ACTUATION WITH MAGNETIC BEARINGS

Stable operation of axial flow compressors as they are employed in modern jet engines and gas turbines is often limited by two flow breakdown processes, known as surge and rotating stall. Surge is a circumferentially uniform pulsation of the mass flow through the machine, while rotating stall appears as a reduced flow region in part of the circumference, which travels around the compressor annulus at a fraction of the rotor speed. Theoretical and experimental investigations for the active control of rotating stall have been conducted at the NASA Glenn Research Centre on a single stage transonic core compressor inlet stage (Spakovszky et al., 2000). Active stabilization of rotating stall and surge using unsteady air injection was first presented by Weigl et al., (1988) in the NASA high-speed stage. The experiments showed a significant benefit in the stable operating range. Blade tip clearance in axial flow compressors is known to have a strong impact on compressor performance and stability; it also plays a major role in the interaction between the rotor dynamic shaft deflections and the aerodynamic behaviour of the compressor (*ref*: Rangwala, 2005).

Magnetic bearings are widely used as active suspension devices in rotating machinery, mainly for active vibration control purposes. The concept of active tip-clearance control suggests a new application of magnetic bearings as servo-actuators to stabilize rotating stall in axial compressors. The magnetic bearing servo-actuator is used to actively whirl the shaft, inducing an unsteady variation of the rotor blade tip-clearance distribution as shown in figure 14.25. The general sequence of steps required in the analysis and design of the magnetic bearing system is shown in figure 14.26. Starting with the compressor flow specifications, an unsteady compressor tip clearance and a rotor dynamic model of the compressor are implemented to determine the control authority and the detail design of the magnetic bearing's actuation system.

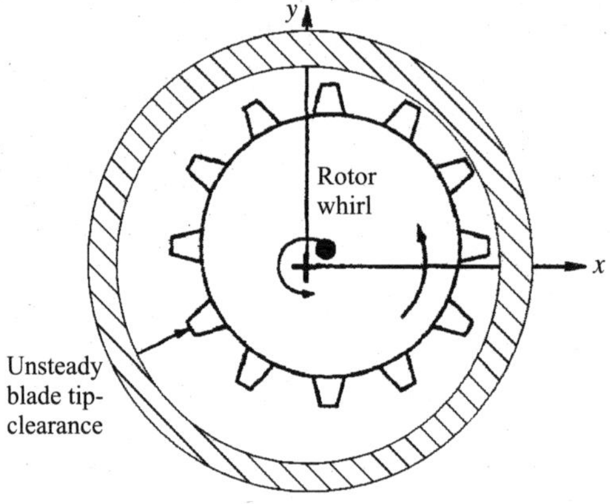

Fig. 14.25 Active Tip Clearance Control Concept (Spakovszky, 2000)

The NASA Stage 37 test compressor, originally designed as an inlet stage of an eight-stage 20:1 pressure ratio core compressor, has a total pressure ratio of 2.05, mass flow of 20.2 kg/s, rotor tip speed of 454 m/s, and rotation frequency of 286 Hz at design conditions. The rotor consists of 36 blades with an aspect ratio of 1.19, hub-to-tip radius ratio of 0.7, and blade tip diameter of approximately 50 cm. The mean-line rotor chord length is 56 mm. Atmospheric air is drawn into the test facility through an orifice plate and a plenum chamber upstream of the test section. Downstream of the compressor the flow is regulated with a sleeve type throttle valve. The compressor shaft is coupled through a drive train to a 2.2 MW DC drive motor. The shaft setup of the test compressor is an overhung rotor with radial fluid film bearings at the front (near the rotor disk) and at the back of the compressor (near the motor drive coupling), as well as a fluid film thrust bearing on the motor coupling side. A schematic of the test section and the compressor shaft is shown in figure 14.27.

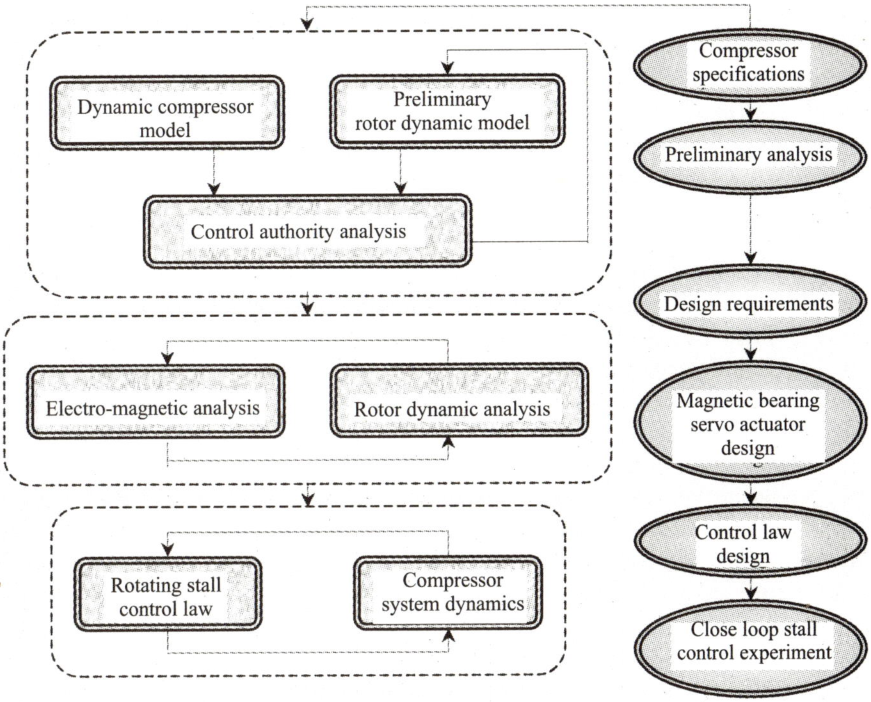

Fig. 14.26 Active Tip Clearance Control Concept (Spakovszky, 2000)

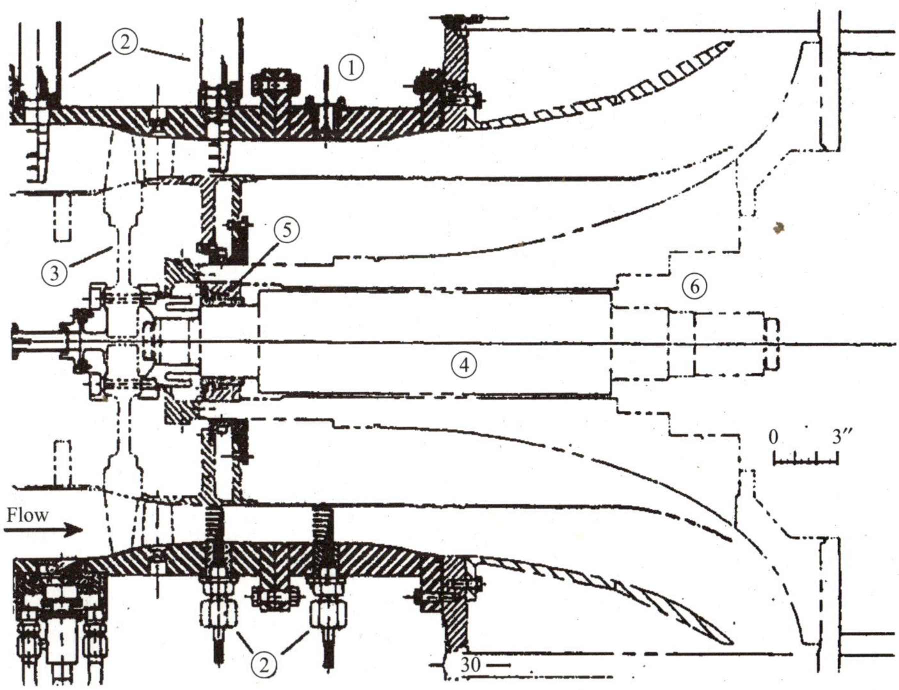

1. High frequency static pressure sensors 2. Steady/Unsteady survey probes 3. Rotor disk 4. Shaft 5. Front fluid journal bearing 6. Rear fluid journal bearing

Fig. 14.27 NASA High Speed Single Stage Compressor Rig (Spakovszky, 2000)

The effect of tip-clearance asymmetries due to shaft deflections on compressor performance and stability is addressed next. The objective of the preliminary analysis is to determine the magnetic bearing force bandwidth and the stall control authority required to conduct the rotating stall control with tip-clearance actuation. The specific question is: how much shaft motion and magnetic bearing force is required to stabilize a rotating stall? To answer this question a rotor dynamic design analysis and a unique stochastic estimation and control analysis are conducted. A design of the rotor with a magnetic bearing rotor is shown in figure 14.28. The solid shaft in figure 14.27 is replaced by a hollow shaft, and includes the magnetic bearing rotor laminations. The shaft is pinned at the rear journal and thrust fluid film bearings, and coupled to the motor drive train. Typical catcher bearing designs do not contact the shaft during magnetic bearing suspension. However, for the proposed stall experiments a fail-safe suspension system is mandatory. In particular, the compressor blades must be protected from possible rubs at the blade tips; destructive impacts must also be avoided in the case of a loss of magnetic levitation. A possible fail-safe solution is to use a spring-loaded catcher bearing that is always in contact with the shaft. This allows for the shaft deflections, but still yields a hard stop in case of an emergency. To ensure safe transient operation without large vibrations when critical frequencies are crossed during an emergency shut down, the damping must be added in parallel to the soft spring-loaded support. One possible compact solution is an integral squeeze film damper (ISFD) setup as reported by Santiago et al (1998). The ISFD's are comprised of accurate squeeze film pads rendering visco-elastic support and wire-electrical discharge machined webs acting like a squirrel cage. The open loop whirl speeds, natural frequencies and mode shapes are obtained from an eigenvalue problem resulting from the equations of motion. Assuming that the rotor is spinning at design speed (286 Hz) the first four eigenvalues are plotted in figure 14.29 and the mode shapes are reconstructed from the corresponding eigenvectors (figure 14.30). In order to determine the effective shaft motion (i.e., blade tip deflection) for the control of a rotating stall the closed loop system is employed. The compressor pre-stall dynamics are denoted by the transfer function $G(s)$. The outputs of $G(s)$ are the pre-stall pressure perturbations sensed upstream of the rotor $\delta p(i)$, which are fed back to the rotating stall feedback controller $K(s)$. The controller outputs are the six actuator commands $\delta c(t)$, which are modified by the magnetic bearing servo-actuator dynamics to yield the actual shaft position and the corresponding tip clearance distribution $\delta\varepsilon(t)$.

The open loop stable magnetic bearing servo-actuator dynamics parameters consist of the shaft rotor dynamics and the magnetic bearing servo control loop. The inputs to the compressor pre-stall transfer function $G(s)$ are the tip-clearance distribution $\delta\varepsilon(t)$ and background noise modeled by unsteady velocity fluctuations $\delta\omega(t)$.

The magnetic bearing servo-actuator design requirements include:

- Stable rotor dynamic operation over the entire compressor speed range (0 to 286 Hz).
- Maximum shaft deflection of 250 μm to avoid blade tip rubs between 0 and 143 Hz excitation frequency.
- Desired minimum whirl radius of 75 μm at the maximum excitation frequency of 286 Hz.
- Maximum bearing diameter of 0.356 m to fit into the compressor hub housing without modifying the compressor gas path.

Standard magnetic bearing suspension devices usually only yield large forces when the shaft levitates. The magnetic bearing servo-actuator for active stall control, however, must deliver a large dynamic load capacity, and high magnetic forces must be generated over an air gap substantially

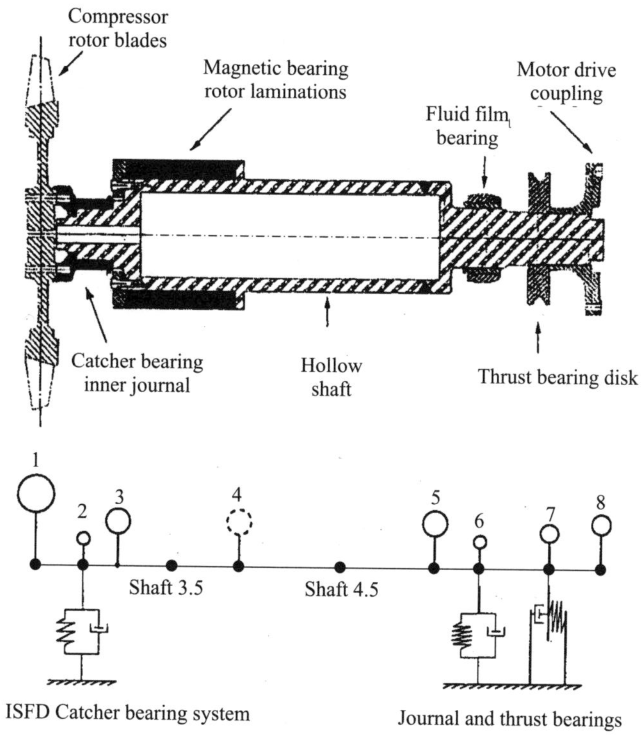

Fig. 14.28 Rotor Dynamic Model of Compressor with Magnetic Bearing (Spakovszky, 2000)

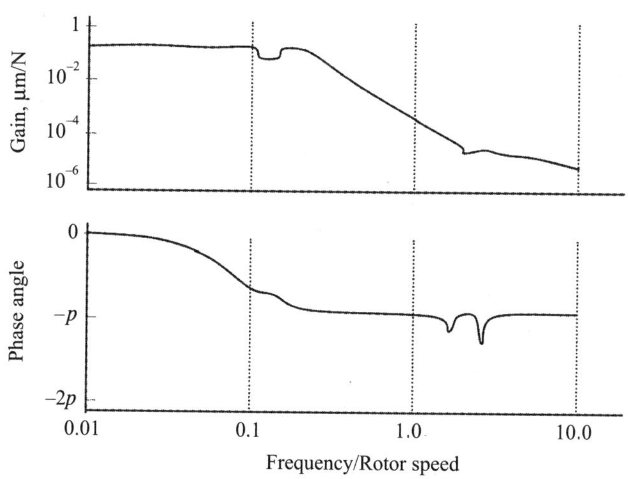

Fig. 14.29 Transfer Function Between Bearing Force and Tip Deflection (Spakovszky, 2000)

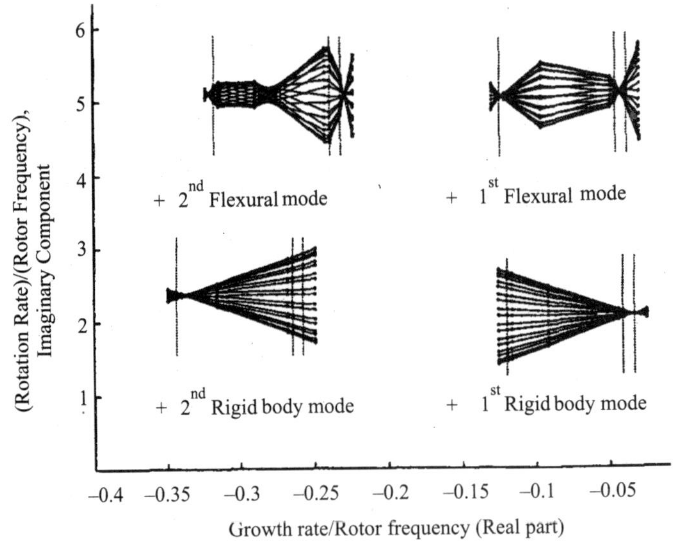

Fig. 14.30 Preliminary Rotor Model Eigenvalues and Mode Shapes (Spakovszky, 2000)

larger than the nominal gap since the rotor is whirling offset from its centreline. In addition, the outer bearing diameter and the space within the compressor hub housing are limited in this application. Hence, a special magnetic pole configuration and a new soft magnetic material with relatively small magnetic losses and very high saturation flux densities are considered. Apart from the magnetic bearing actuator itself, the above issues will also strongly influence the actuator electronics design. A 16-pole N-S-N-S configuration with rotor laminations, illustrated in figure 14.31, is selected to satisfy the requirements.

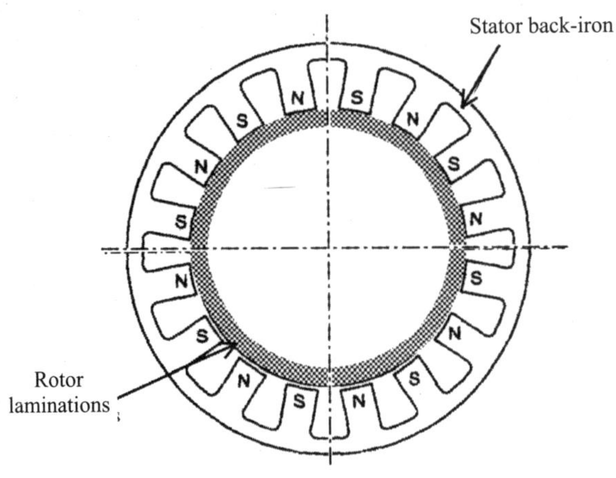

Fig. 14.31 Magnetic Bearing Servo-Actuator (Spakovszky, 2000)

Predicted results of the overall dynamic system are shown in the Campbell diagram of figure 14.32. The two rigid body forward and backward whirling modes are crossed by the one-per-rev line (1E) at about 30,180 and 200 Hz. Note that none of the flexural modes are crossed despite the strong gyroscopic effects, and the rigid body mode frequencies compare well to the preliminary results obtained from the simple lumped parameter model. The frequencies of the two rigid body modes mostly depend on the stiffness of the fluid film and the catcher bearings, the mass properties of the compressor shaft and the actuator feedback control. In order to limit the vibration level when the rigid body critical frequencies are crossed, either the rotor must be well balanced or the bearings must dissipate a substantial amount of energy, thus calling for sufficient external damping.

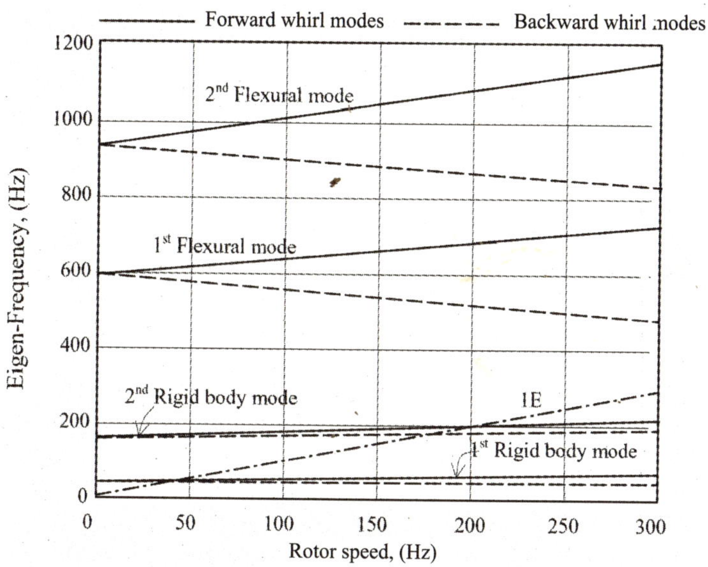

Fig. 14.32 Campbell Diagram (Spakovszky, 2000)

Non-linear effects are included in the voltage and current saturation characteristics of the power amplifier because of the influence of the actuator inductance and resistance and the back electromotive force term in the rotor. The resulting higher order system of the non-linear differential equations is solved by an iterative procedure. The results are shown in figure 14.33 for the shaft spinning at the design speed of 286 Hz. The plot depicts a family of non-linear current-force characteristics of the magnetic bearing servo-actuator and the locus of operating points with the excitation frequency ranging from 0 to 300 Hz at maximum available power (thick solid line). The effect of a varying air gap is included in the calculation. The achievable whirl radii at several shaft locations are plotted on the right hand side within the operational constraints of the amplifier and the whirl radius limitation at the magnetic bearing location (250 μm). Thus, the results show that the stated design requirements of 250 μm compressor blade tip deflection up to 185 Hz are fulfilled.

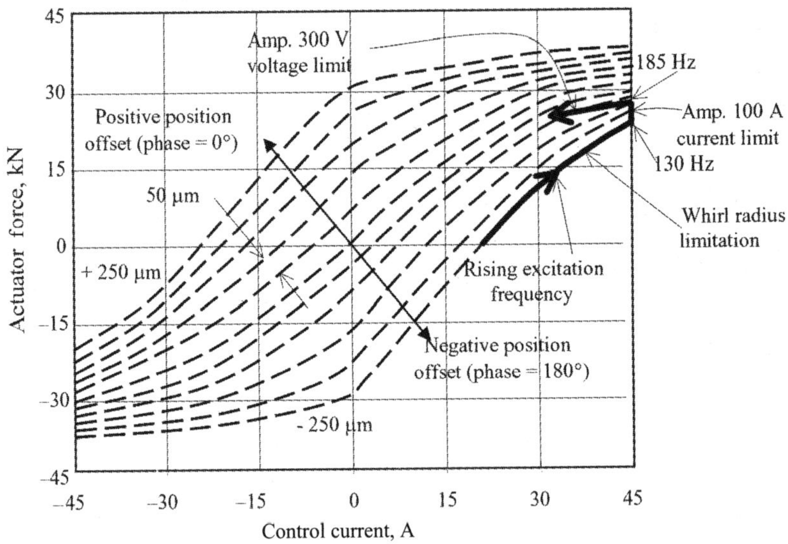

Fig. 14.33 Current-Force Characteristic of Magnetic Bearing Actuator (Spakovszky, 2000)

14.17 LABYRINTH AND HONEYCOMB SEAL EVALUATION

Labyrinth seals are extensively used in centrifugal compressors operating at high pressure and speed to control leakage. Field experience and tests indicate the labyrinth seals can generate unstable operation of the rotating system. In comparison, honeycomb seals of up to 50 mm in length have displayed better stability characteristics, and are also insensitive to pre-swirl at the inlet (*ref*: Childs et al., 1992).

Shunt injection is one approach to improve the performance of labyrinth seals. High-pressure gas, usually from the compressor's discharge, is injected into an intermediate labyrinth cavity at several circumferential locations in the radial direction or against the shaft rotation. The negative swirl flow injection is sufficient to reverse the flow direction in the seal, and the consequent reduction of circumferential velocity leads to a reduction in the cross-coupled stiffness k. The performance penalty arising from the diverted flow from the compressor discharge is the disadvantage of the technique.

An experimental study to compare the sub-synchronous vibrations and instability characteristics of a conventional labyrinth seal, labyrinth seal with shunt injection and a honeycomb seal has been conducted in a joint NASA-USAF funded program (*ref*: Soto et al., 1998). The test seal, 130 mm in diameter and 63 mm in length, is provided with feed holes for shunt injection, and choking of the holes is avoided by controlling the exit Mach number to less than 0.33. The seal has 20 teeth with a pitch of 3.2 mm and height of 3.17 mm. Shunt injection is facilitated by deleting the fourth tooth, and the flow occurs in both directions of the seal (figure 14.34). The rig is designed to measure axial pressure distribution, airflow rate, temperature, vibration amplitudes and seal reaction forces. Rotor dynamic coefficients are then calculated from transient measured reaction force and motion data. The test apparatus in the form of a rotor coupled to an electric motor has provision for controlling

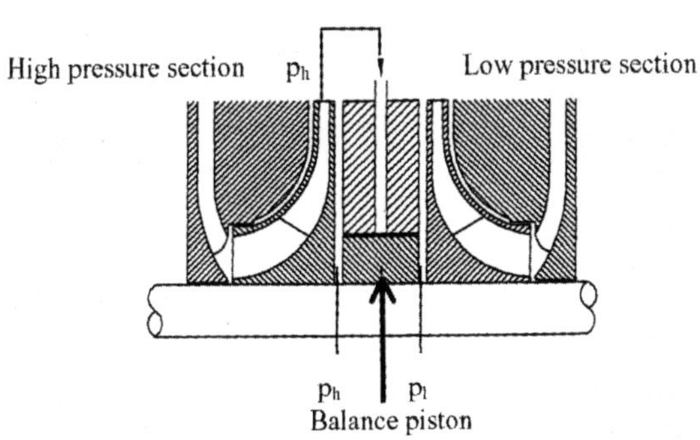

Fig. 14.34 Shunt Injection at Balance Piston Between High and Low Pressure Sections

the eccentricity in the seal. The rotor is suspended in the style of a pendulum from another rigidly mounted shaft at the top, and permits control of the motion of the rotor in the horizontal plane. A cam within the pivot shaft controls the system's vertical position. The equation of motion for the seal stator housing is:

$$\begin{Bmatrix} f_{sx} - M_s d^2X_s/dt^2 \\ f_{sy} - M_s d^2Y_s/dt^2 \end{Bmatrix} = \begin{bmatrix} K & k \\ -k & K \end{bmatrix} \begin{Bmatrix} X_s \\ Y_s \end{Bmatrix} + \begin{bmatrix} C & c \\ -c & C \end{bmatrix} \begin{Bmatrix} dX_s/dt \\ dY_s/dt \end{Bmatrix} \quad \text{...(14.1)}$$

where M_s is mass of stator housing, f_{sx} and f_{sy} are measured forces on the housing, K and k are the direct and cross-coupled stiffness and C and c are the direct and cross-coupled damping. Acceleration components of the housing, d^2X_s/dt^2 and d^2Y_s/dt^2 are measured by accelerometers. The coefficients are described in the frequency domain by using a swept sine wave in the 40 to 70 Hz range.

Rotor dynamic performance and seal leakage flow comparisons are carried out for the labyrinth seal with pre-swirl and with shunt injection and for the honeycomb seal. Shunt injection pressure ratios are set at 0.85, 0.90 and 0.95. In normal applications this pressure varies between 0.4 and 0.6. Rotor surface speeds corresponding to speed points between 30 and 110 m/s, against a normal operating range of 100 to 170 m/s.

Effective damping, defined by $C_{eff} = C \times (1 - k/C\omega)$, is used to evaluate overall seal stability characteristics. The parameter is useful in comparing annular seals since it combines the effects of direct damping and cross-coupled stiffness, both of which are associated with rotor stability. Since k is responsible for instability, it must have a low or negative value. Measured values for shunt injection at the maximum speed of 16,500 rpm and constant pressure ratio across the seal of 0.45 are shown in table 14.1. Injection against rotation provides the largest negative values of k, and increasing injection pressure also tends to increase k. Cross-coupled stiffness values due to radial shunt injection, on the other hand, remain practically the same for all cases.

Table 14.1 Cross-Couple Stiffness and Direct Damping Values

Shunt Injection Pressure Ratio	*Direction of Injection*	*Cross-Couple Stiffness k, kN/m*	*Direct Damping C, kN-s/m*
0.85	with rotation	–175	0.82
0.85	against rotation	–2025	0.83
0.90	with rotation	--	0.76
0.90	against rotation	–1810	0.77
0.95	with rotation	-	0.69
0.95	against rotation	–1475	0.72

A positive and large magnitude for direct damping in the annular seal improves rotor dynamic response, since it counteracts the destabilizing effects of cross-coupled stiffness in the rotor. Table 14.1 shows the coefficient magnitudes at the maximum speed of 16,500 rpm and constant pressure ratio across the seal of 0.45. The direct damping value increases with injection pressure, and the direction of injection has minimal impact.

The tangential force on a synchronous precessing (or forward whirl) seal is given by $f_\theta = (k - C\omega)A$, where A is the radius of precession. The whirl frequency ratio is expressed by $f_\omega = (k/C\omega)$. For the shunt injection seals with radial injection the whirl frequency is calculated to be negative and small, implying the seal could destabilize the seal into a backward whirl mode. However, backward whirl normally associated with rubbing is rarely experienced in the operation of turbo-machines, as opposed to the prevalent forward whirl instability condition.

Table 14.2 shows the comparison for all seals. The honeycomb seal has better effective damping than the labyrinth seal with radial injection, but shunt injection nearly doubles this value without injection. The highest positive cross-coupled stiffness (destabilizing) is obtained from the honeycomb seal, but the labyrinth seal has superior numbers with or without the injection. Whirl frequency ratio is negative for labyrinth seal with shunt injection, and is caused by the negative cross-coupled stiffness. In terms of leakage, the honeycomb seal performance is superior than that of the labyrinth with or without shunt injection.

Table 14.2 Comparison of Seal Performance

Seal/Configuration	*Effective Damping kN-s/m*	*Cross-Coupled Stiffness, kN/m*	*Whirl Frequency Ratio*	*Mass Flow Rate g/s*
Labyrinth/pre-swirl	0.3	20	0.15	110
Labyrinth/radial injection	0.8	–200	–0.20	165
Labyrinth/injection against rotation	1.5	–1500	–1.20	170
Honeycomb	1.6	1700	0.35	80

14.18 DAMPING SEAL DYNAMIC CHARACTERISTICS

Vibration problems in the turbo-machinery of the space shuttle's main engine have required closer evaluation of rotor dynamic stability margins. Annular seals are used in the machines for leakage control, but they also play a major role in controlling vibration response levels. Prior to entering the seal the fluid has a significant tangential velocity component, in contrast with the axial velocity down the length of the seal. A good example of the pre-swirl condition is the discharge from an impeller entering a seal, and is a significant source of destabilizing forces in the machine.

In an experiment conducted at NASA's Marshall Space Flight Centre, test measurements focused on the capacity of annular damping seals to minimize the whirl frequency ratio in the presence of a highly pre-rotated fluid at the inlet (*ref*: Darden et al., 1999). Dynamic coefficients for smooth seals tend to have a strong dependence on eccentricity over a wide range of speed, pressure and eccentricity ratios (*ref*: Marquette et al., 1997). Results from the smooth seal are used as a baseline for the investigation of damping seals. Details of the test rig are shown in figure 14.35. Designed to support testing of seal pressure differential of 2,000 lbs/in^2, the rig is powered by a 250 horsepower steam turbine and has a maximum speed of 20,000 rpm. De-ionized water between 70° and 90°F is used in the experiment. The test bearing is located in the seal carrier assembly at the mid-span of the shaft, and is supported axially by the upper and lower orifice compensated thrust bearings. This allows the test bearing to translate in response to an external load without recourse to additional load paths, while ensuring the seal assembly remains parallel to the shaft during the test. Pre-rotation of the fluid medium is achieved by using a shaft with 10 equally spaced radial fins placed just upstream of the entrance to the seal. Blade passing frequencies are well outside the frequency range of interest to avoid degradation of the frequency response functions.

The test seals are characterized at a combination of shaft speeds, seal pressure differentials and inlet pre-swirl conditions. The applied force is a band limited random excitation between 20 – 200 Hz. High-frequency data comprising of displacements, forces and pressures is recorded. Duration of each test is limited to 30 s. Relative displacement of the fluid film between the test seal and the shaft are measured by 4 inductive probes spaced equally around the periphery of the seal assembly. These measurements are fed back for inertial correction of the excitation force applied to the carrier assembly. Displacements are differentiated in the frequency domain to obtain accelerations. Acceleration levels of rig housing, measured by accelerometers at same locations as proximity probes, must also be taken into account. Seal reaction forces are:

$$-\begin{Bmatrix} f_y \\ f_z \end{Bmatrix} = \begin{bmatrix} K & k \\ -k & K \end{bmatrix} \begin{Bmatrix} \delta y \\ \delta z \end{Bmatrix} + \begin{bmatrix} C & c \\ -c & C \end{bmatrix} \begin{Bmatrix} d\delta y/dt \\ d\delta z/dt \end{Bmatrix} + \begin{bmatrix} M & m \\ -m & M \end{bmatrix} \begin{Bmatrix} d^2\delta y/dt^2 \\ d^2\delta z/dt^2 \end{Bmatrix} \quad ...(14.2)$$

where M and m are the direct and cross-coupled mass of the seal carrier assembly, f_y and f_z are measured fluid film reaction forces, K and k are the direct and cross-coupled stiffness and C and c are the direct and cross-coupled damping. Based on the symmetry of the centered seal state and flow conditions, all of the coefficients are assumed to be skew-symmetric. The measured parameters in the individual time history are transformed to the frequency domain, and the frequency response functions are curve fitted to yield the rotor dynamic coefficients.

The test matrix is considerably influenced by the high pre-swirl condition, and the stability of the seal is predicted to degrade with increasing shaft speed. Theoretical predictions are obtained from the code developed by Padavala et al., (1993) using Hir's friction factor and inlet pre-swirl values of 0.19 and 1.1. The results of the experimental investigation, with the predicted data, are shown in table 14.3. Non-dimensional values are obtained by multiplying the stiffness coefficients by $[c/(\pi dl\Delta P)]$ and the damping coefficients by $[c\omega_s/(\pi dl\Delta P)]$, where ω_s = shaft speed, c, d and l are seal radial clearance, length and diameter and ΔP = seal pressure differential.

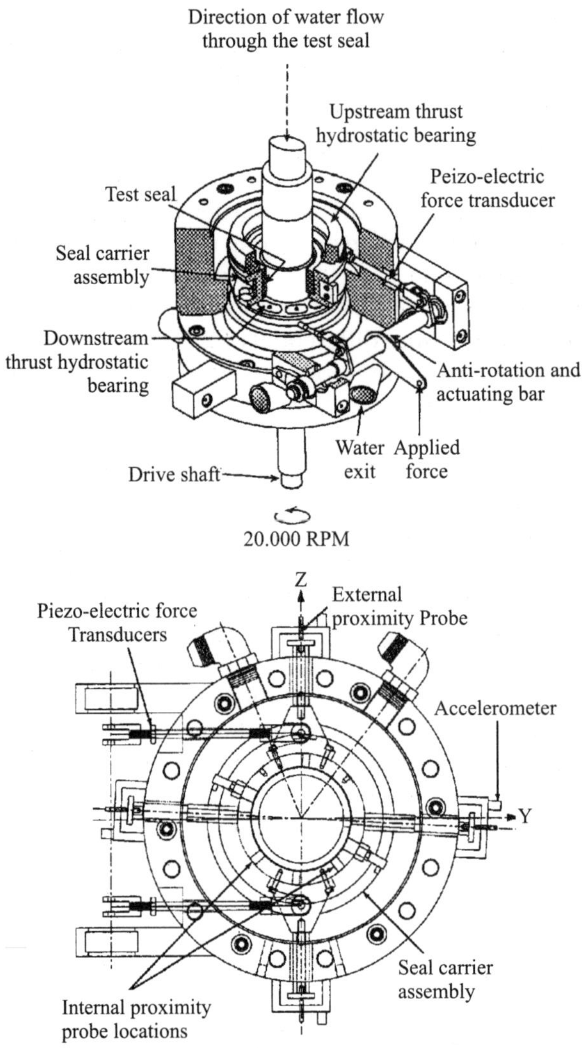

Fig. 14.35 Seal Test Rig: (*upper*) Cut-away View, (*lower*) Top View; (Darden, 1999)

The direct stiffness term is over predicted by the theory by 50 per cent for both the high and low pre-swirl conditions, but there is agreement in the overall upward trend with increasing seal pressure differential. The cross-coupled stiffness term is over predicted by the theory by 15 per cent for both the high and low pre-swirl conditions, and there is agreement in the upward trend with increasing shaft speed. Static tests were also conducted concurrently by applying a series of loads and recording the resulting displacements. Agreement between the static and dynamic tests provided verification of the lower than expected stiffness values in the operational tests. The damping coefficients under-predict the calculated values by 15 to 25 per cent for the low and high pre-swirl cases (*ref*: Rangwala, 2005).

Table 14.3 Damping Seal Rotor Dynamic Coefficients

Test Condition			Stiffness		Damping		Inertia		Leakage
Pre-swirl	Pressure kPa	Speed rpm	K N/m	k N/m	C N-s/m	c N-s/m	M Kg	m Kg	Rate ml/s
0.19	10300	5850	6.82E6	1.81E6	2.01E4	7.55E2	2.19	−0.60	14.8
0.19	10300	10500	7.00E6	3.19E6	2.09E4	8.90E2	2.87	−0.76	14.7
0.19	10300	15100	7.07E6	5.33E6	2.06E4	2.11E3	1.94	−0.32	14.7
0.19	13400	5780	8.85E6	1.98E6	2.31E4	1.02E3	2.61	−0.81	17.2
0.19	13300	10100	9.58E6	3.28E6	2.31E4	1.23E3	4.03	−0.91	17.1
0.19	12900	15800	9.18E6	6.04E6	2.37E4	1.72E3	2.38	−0.60	16.8
1.10	13100	5130	8.32E6	9.23E6	2.39E4	5.41E3	7.44	−3.84	16.7
1.10	12700	6140	8.37E6	1.04E7	2.44E4	6.54E3	7.85	−4.17	16.6
1.10	12300	4020	7.11E6	7.48E6	2.40E4	2.07E3	4.82	−2.24	16.3
1.10	16000	5280	1.01E7	1.10E7	2.73E4	3.57E3	5.62	−3.40	17.4
1.10	15700	6350	1.07E7	1.26E7	2.53E4	7.02E3	8.32	−3.08	17.1
1.10	14400	3980	8.89E6	8.15E6	2.52E4	2.17E3	5.20	−1.87	16.7

The whirl frequency ratio is given by $k/(C\omega_s)$, and is over predicted by 20 per cent for the low pre-swirl condition. The average whirl frequency ratio for the high pre-swirl case is 0.73, representing a large increase in the damping seal's stabilizing capacity.

14.19 REVIEW PROBLEMS

Example problem # 14.1: A uniform disk of 4.75 in. radius and 1.725 in. width is attached to a hollow shaft of inner radius 1.127 in. and thickness 0.173 in. The disk material has $E = 25.45 \times 10^6$ lbs/in^2 and density $\rho = 0.278$ lbs/in^3, and the shaft material has $E = 24.83 \times 10^6$ lbs/in^2 and density $\rho = 0.285$ lbs/in^3. The shaft is supported in two roller bearings that are placed 1.715 in. and 7.328 in. from the disk centreline (figure 14.36). Disregarding the weight of the hollow shaft, calculate the reactions at the bearings due to the weight of the disk.

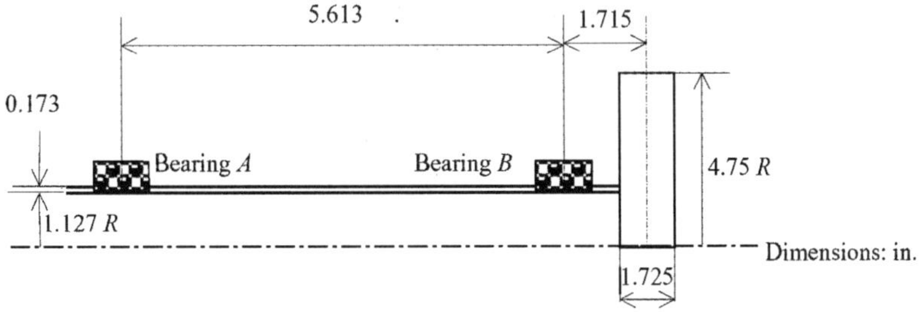

Fig. 14.36

Solution: Weight of the disk: W = Volume × density = $\pi R^2 l\rho$

$$= \pi \times 4.75^2 \times 1.725 \times 0.278 = 33.9915 \text{ lbs}$$

Assume x as the distance along the shaft, with $x_A = 0$, $x_B = 5.613$ in., and $x_{Disk} = 7.328$ in. For static equilibrium, $(R_A + R_B - 33.9915) = 0$, or $R_A = 33.9915 - R_B$.

Also: $\Sigma M = 0$. Taking moments about A: $33.9915 \times 7.328 - R_B \times 5.613 = 0$

Hence: $R_B = 44.3773$ lbs ↑, and $R_A = 33.9915 - 44.3773 = 10.3858$ lbs ↓

Example problem # 14.2: In example problem # 14.1, determine the expression for bending moment in the shaft due to the weight of the disk.

Solution: In the shaft segment between the bearings, consider a section located at a distance x from bearing A ($x < 5.613$).

Moment $M(x) = R_A \times x = 10.3858x$ in-lbs counterclockwise.

Hence, $M_A = 0$

$M_B = 10.3858 \times 5.613 = 58.2955$ in-lbs counterclockwise.

In the shaft segment between bearing B and the disk ($x > 5.613$):

$$M(x) = -R_B \times (x - 5.613)$$

$M_B = -[33.9915 \times 1.715] = 58.2955$ in-lbs counterclockwise.

$$M_{Disk} = [44.3773 \times (7.328 - 7.328)] = 0$$

Example problem # 14.3: In example problem # 14.1 and 14.2, assume the shaft is spinning at 2800 rpm, and the disk has an unbalance of 0.10 lbs at the rim. Determine the bending moment in the shaft due to the rotating unbalance.

Solution: The unbalance in the rotating disk creates a centrifugal force in the radial direction that also rotates with the disk. This force is given by: $F = (W/g) \times \omega^2 r$, where Wr provides the amount of unbalance in the form of weight W placed at a distance r from the axis of rotation, ω is the rotor speed, and g is the gravitational constant.

Centrifugal force $F = [0.10 \times (2 \times 3.14159 \times 2800/60)^2]/386.0$

$= 22.2733$ lbs

Corresponding reactions at the bearing supports are determined as follows.

For static equilibrium, $(R_A + R_B - 22.2733) = 0$, or $R_A = 22.2733 - R_B$.

Also: $\Sigma M = 0$. Taking moments about A: $22.2733 \times 7.328 - R_B \times 5.613 = 0$

Hence: $R_B = 29.0787$ lbs ↑, and $R_A = 22.2733 - 29.0787 = 6.8054$ lbs ↓

In the shaft segment between the bearings ($x < 5.613$), consider a section located at a distance x from bearing A.

Moment $M(x) = R_A \times x = 6.8054x$ in-lbs counterclockwise ($x < 5.613$).

Hence, $M_A = 0$

And $M_B = 6.8054 \times 5.613 = 38.1988$ in-lbs counterclockwise.

In the shaft segment between bearing B and the disk ($x > 5.613$):

$$M(x) = -R_B \times (x - 5.613) = -29.0787x + 163.2190 \text{ (for } x > 5.613)$$

$$M_B = -[29.0787 \times (7.328 - 5.613)] = 49.8700 \text{ in-lbs counterclockwise.}$$

$$M_{Disk} = -[29.0787 \times (7.328 - 7.328)] = 0$$

Example problem # 14.4: In example problems # 14.1, 14.2 and 14.3, determine the deflection in the shaft due to the rotating unbalance.

Solution: Bending moment in a beam is also given by the expression:

$$M(x) = EI \times (d^2y/dx^2), \text{ where } y = \text{deflection in the downwards direction.}$$

Hence, between the bearings ($x < 5.613$):

Moment $M(x) = R_A \times x = 6.8054x$ in-lbs counterclockwise.

$$M(x) = EI \times (d^2y/dx^2) = 6.8054x$$

Shaft material $E = 24.83 \times 10^6 \text{ lbs/in}^2$

And shaft cross-section $I = (\pi/4) \times (R_o^{\,4} - R_i^{\,4}) = (\pi/4) \times (1.127^4 - 0.954^4) = 1.0006 \text{ in}^4$

Thus: $(d^2y/dx^2) = (1/EI)(6.8054x)$

$$= [1/(24.83 \times 10^6 \times 1.0006)] \times (6.8054x)$$

$$= 2.7391 \times 10^{-7}x$$

Integrating: $dy/dx = 1.3696 \times 10^{-7}x^2 + C_1$

$$y = 4.5652 \times 10^{-8}x^3 + C_1x + C_2$$

At bearing A ($x = 0$) the deflection y is zero. Hence, $C_2 = 0$.
Also, at bearing B ($x = 5.613$) the deflection y is zero.

Hence: $0 = 4.5652 \times 10^{-8} \times (5.613)^3 + 5.613C_1$, or $C_1 = -1.4383 \times 10^{-6}$

Or deflection: $y = 4.5652 \times 10^{-8}x^3 - 1.4383 \times 10^{-6}x$ (for $0 < x < 5.613$)

Between bearing B and the disk ($x > 5.613$):

$$(d^2y/dx^2) = (1/EI)[-29.0787x + 163.2190]$$

$$= [1/(24.83 \times 10^6 \times 1.0006)] \times [-29.0787x + 163.2190]$$

Thus: $(d^2y/dx^2) = -1.1704 \times 10^{-6}x + 6.5694 \times 10^{-6}$

Integrating: $dy/dx = -5.8520 \times 10^{-7}x^2 + 6.5694 \times 10^{-6}x + C_1$

$$y = -1.9507 \times 10^{-7}x^3 + 3.2847 \times 10^{-6}x^2 + C_1x + C_2$$

At bearing B ($x = 5.613$) the slope dy/dx is zero.

Hence: $0 = -5.8520 \times 10^{-7}x^2 + 6.5694 \times 10^{-6}x + C_1$

$$0 = -5.8520 \times 10^{-7} \times (5.613)^2 + 6.5694 \times 10^{-6} \times (5.613) + C_1$$

$$C_1 = -1.8440 \times 10^{-5}$$

At bearing B ($x = 5.613$) the deflection y is zero.

Hence: $$0 = -1.9507 \times 10^{-7} \times (5.613)^3 + 3.2847 \times 10^{-6} \times (5.613)^2 + (-1.8440 \times 10^{-5}) \times 5.613 + C_2$$

Or: $$C_2 = 3.4496 \times 10^{-5}$$

Thus, deflection $$y = -1.9507 \times 10^{-7}x^3 + 3.2847 \times 10^{-6}x^2 - 1.8440 \times 10^{-5}x + 3.4496 \times 10^{-5}$$

(for $5.613 < x < 7.328$).

Example problem # 14.5: The rotor of example problems # 14.1 to 14.4 is placed inside a stator casing. Determine the diameter of the casing if the minimum radial clearance between the rotor and the stator is required to be 0.0010 in. on the radius when the rotor is spinning at the rated speed.

Solution: Deflection of the rotor where the disk is located in the radial direction when the rotor is spinning at 2800 rpm for the specified unbalance is to be determined. At the disk, where $x = 7.328$ in., the deflection due to dead weight is:

$$M(x) = -R_B \times (x - 5.613) = -44.3773 \times (x - 5.613) \text{ where } 5.613 < x < 7.328.$$
$$= -44.3773x + 249.0898$$

Or: $$(d^2y/dx^2) = (1/EI)[-44.3773x + 249.0898]$$
$$= [1/(24.83 \times 10^6 \times 1.0006)] \times [-44.3773x + 249.0898]$$

Thus: $$(d^2y/dx^2) = -1.7826 \times 10^{-6}x + 9.9855 \times 10^{-6}$$

Integrating: $$dy/dx = -8.9309 \times 10^{-7}x^2 + 9.9855 \times 10^{-6}x + C_1$$
$$y = -2.9770 \times 10^{-7}x^3 + 4.9928 \times 10^{-6}x^2 + C_1x + C_2$$

At bearing B ($x = 5.613$) the slope dy/dx is zero.

Hence: $$0 = -8.9309 \times 10^{-7} \times (5.613)^2 + 9.9855 \times 10^{-6} \times (5.613) + C_1$$
$$C_1 = -2.7912 \times 10^{-5}$$

At bearing B ($x = 5.613$) the deflection y is zero.

Hence: $$0 = -2.9770 \times 10^{-7} \times (5.613)^3 + 4.9928 \times 10^{-6} \times (5.613)^2 + 5.613 \times (-2.7912 \times 10^{-5}) + C_2$$

Or: $$C_2 = -0.0002613$$

Thus, deflection $$y = -2.9770 \times 10^{-7}x^3 + 4.9928 \times 10^{-6}x^2 - 2.7912 \times 10^{-5}x - 0.0002613$$

At the disk, $x = 7.328$ in.

Hence: $$y = -2.9770 \times 10^{-7} \times (7.328)^3 + 4.9928 \times 10^{-6} \times (7.328)^2 - 2.7912 \times 10^{-5} \times (7.328) - 0.0002613$$
$$y = -0.0003149 \text{ due to dead weight.}$$

Deflection from rotating unbalance is:

$$y = -1.9507 \times 10^{-7}x^3 + 3.2847 \times 10^{-6}x^2 - 1.8440 \times 10^{-5}x + 3.4496 \times 10^{-5}$$

$= -0.0002080$ in. due to rotating unbalance.

Hence, total deflection due to rotating unbalance and dead weight is:

$$y = -0.0003149 - 0.0002080 = -0.0005229$$

Minimum radial clearance = 0.0010 in.

Hence radius of the casing:

$$\begin{aligned} R_{casing} &= \text{(disk radius)} + \text{(minimum radial clearance)} + \text{(radial deflection)} \\ &= 4.750 + 0.0010 + 0.0005229 \\ &= 4.751523 \text{ in.} \end{aligned}$$

Example problem # 14.6: The coefficient of expansion of the material of the rotor of example problems # 14.1 to 14.4 is $\alpha = 5.23 \times 10^{-6}$ in/in/°F, and for the casing material is a $= 8.97 \times 10^{-6}$ in/in/°F. If the running temperature of the unit is 1625°F, determine the difference in the thermal growth of the rotor and the stator from the cold condition of 60°F.

Solution: Thermal growth in the diameter $= \alpha D \Delta T$, where D is the diameter and ΔT = temperature differential.

$$\begin{aligned} \text{Thermal growth in rotor} &= 5.23 \times 10^{-6} \times 4.75 \times 2 \times (1625 - 60) \\ &= 0.07776 \text{ in.} \\ \text{Thermal growth in casing} &= 8.97 \times 10^{-6} \times 4.75 \times 2 \times (1625 - 60) \\ &= 0.13336 \text{ in.} \end{aligned}$$

Difference in thermal growth = 0.13336 – 0.07776 = 0.05560 in.

Example problem # 14.7: Combine the diameter of the disk with the minimum radial clearance and radial deflection calculated in example problem # 14.5 with the relative thermal growth in example problem # 14.6 to determine the diameter of the casing at cold condition if rubbing is to be avoided.

Solution: Diameter of casing from example problem # 14.5:

$$D_{casing} = 2 \times 4.751529 = 9.50305 \text{ in.}$$

Difference in thermal growth between rotor and casing = 0.05560 in.

Hence, to avoid rubbing, casing diameter at cold condition is:

$$D = 9.50305 - 0.05560 = 9.44744 \text{ in.}$$

But the disk diameter is 9.50 in. and minimum radial clearance required is 0.0020 in on the diameter. As the unit runs up to full speed, the deflection from centrifugal force and thermal growth will occur. If deflection from the weight of the disk is not taken into consideration, then the cold casing diameter = 9.500 + 0.0020 = 9.502 in.

Example problem # 14.8: In example problems # 14.6 and 14.7 calculate the running clearance when the unit is operating at the rated speed.

Solution: Cold casing radius = 4.751 in.

Radial deflection due to rotating unbalance and dead weight = 0.0005229 in.

Thermal growth in casing = 0.13336 in.

Thermal growth in rotor = 0.07776 in.

Combined disk radius, deflection from rotating unbalance and dead weight and thermal growth of disk = 4.750 + 0.0005229 + 0.07776 = 4.828283 in.

Combined casing radius at cold and thermal growth = 4.750 + 0.001 + 0.13336

= 4.88436 in.

Hence running clearance = 4.88436 – 4.828283 = 0.056077 in.

14.20 EXERCISES

14.1 A uniform disk of 5.25 in. radius and 1.745 in. width is attached to a hollow shaft of inner radius 1.139 in. and thickness 0.187 in. The disk material has $E = 22.75 \times 10^6$ lbs/in^2 and density $\rho = 0.283$ lbs/in^3, and the shaft material has $E = 23.73 \times 10^6$ lbs/in^2 and density $\rho = 0.289$ lbs/in^3. The shaft is supported in two roller bearings that are placed 28.785 in. apart, and the disk centreline is located equally between the bearings (figure 14.37). Ignoring the weight of the hollow shaft, calculate the reactions at the bearings due to the weight of the disk.

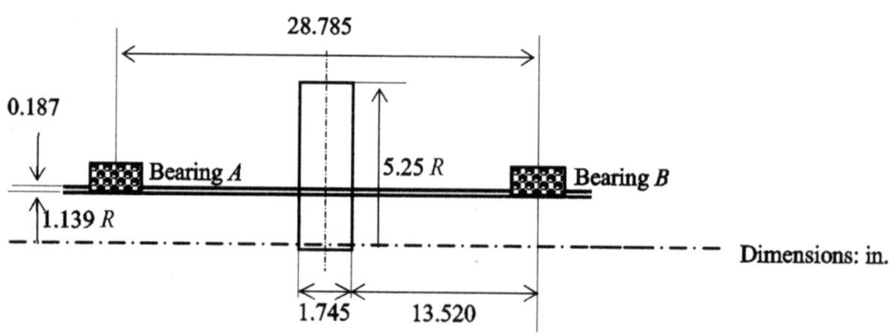

Fig. 14.37

14.2 In exercise # 14.1, determine the expression for bending moment in the shaft due to the weight of the disk.

14.3 In exercises # 14.1 and 14.2, assume the shaft is spinning at 3600 rpm, and the disk has an unbalance of 0.10 lbs at the rim. Determine the bending moment in the shaft due to the rotating unbalance.

14.4 In exercises # 14.1, 14.2 and 14.3, determine the deflection in the shaft due to the rotating unbalance.

14.5 The rotor of exercises # 14.1 to 14.4 is placed inside a stator casing. Determine the diameter of the casing if the minimum radial clearance between the rotor and the stator is required to be 0.0010 in. on the radius when the rotor is spinning at the rated speed.

14.6 The coefficient of expansion of the material of the rotor of exercises # 14.1 to 14.4 is $\alpha = 6.72 \times 10^{-6}$ in/in/°F, and for the casing material is $\alpha = 7.89 \times 10^{-6}$ in/in/°F. If the running temperature of the unit is 1625°F, determine the difference in the thermal growth of the rotor and the stator from the cold condition of 60°F.

14.7 Combine the diameter of the disk with the minimum radial clearance and radial deflection calculated in exercise # 14.5 with the relative thermal growth in exercise # 14.6 to determine the diameter of the casing at cold condition if rubbing is to be avoided.

14.8 In exercises # 14.6 and 14.7 calculate the running clearance when the unit is operating at the rated speed.

References & Bibliography

Bureau of Transportation Statistics, 400 7th Street, SW, Room 3103, Washington, DC, 20590.

Carpenter, K.D., Wiedmer, J.D. and Smith, P.A., "Rotor Blade Outer Tip Seal Apparatus", US Patent 5,639,210, 1997.

Catlow, R., "Blade Tip Clearance Control Apparatus", US Patent 5,211,534, 1993.

Cawley, J.D., "Oxidizing Seal for Turbine Tip Gas Path", US Patent 4,540,336, 1985.

Childs, D., "Turbo-machinery Rotor Dynamics", John Wiley & Sons, New York, 1993.

Childs, D. and Kleynhans, G., "Experimental Rotor Dynamic and Leakage Results for Short (L/D = 1/6) Honeycomb and Smooth Annular Pressure Seals", Proceedings, Institute of Mechanical Engineers, 5th International Conference On Vibrations in Rotating Machinery, London, 1992.

Ciokajlo, J.J., "Blade Tip Clearance Control Apparatus Using Shroud Segment Annular Support Ring Thermal Expansion", US Patent 5,116,199, 1992.

Darden, J.M., Earhart, E.M. and Flowers, G.T., "Comparison of Dynamic Characteristics of Smooth Annular Seals and Damping Seals", ASME paper # 99-GT-177, New York, 1999.

Day, I.J., Breur, T., Escuret, J., Cherret, M. and Wilson, A., "Stall Inception and the Prospects for Active Control in Four High Speed Compressors", ASME paper # 97-GT-281, New York, 1997.

Eder, M., "Engine Fuel System Design Issues", Aviation Fuels with Improved Fire Safety: A Proceedings, page 61–64, National Academy Press, Washington, DC, 1997.

Electronic Code of Federal Regulations, Title 14, Aeronautics and Space, Chapter 1, Federal Aviation Administration, Department of Transportation, Part 33, Airworthiness Standards: Aircraft Engines, 33.73 Power or Thrust Response, Amdt. 33-1, 36 FR 5493, March 24, 1971.

European Regions Airline Association (ERA) "ERA Fast Facts", September, 2001.

Fasching, W.A., "CF6 Jet Engine Performance Improvement Summary Report", NASA CR-165612, 1982.

Ferguson, J., "Brushes as High Performance Gas Turbine Seals", ASME paper # 88-GT-182, New York, 1988.

Flotow, A., Mercadal, M. and Tappert, P., "Health Monitoring and Prognostics of Blades and Disks with Blade Tip Sensors", Aerospace Conference Proceedings, IEEE, volume 6, page 433–440, 2000.

Gaffin, W.O., "JT9D-70/59 Improved High Pressure Turbine Active Clearance Control System", NASA CR-159661, 1979.

Greathead, S. and Bostow, P., "Investigations into Load Dependent Vibrations of High Pressure Rotor on Large Turbo-Generators", Proceedings, Institute of Mechanical Engineers, Conference on Vibrations in Rotating Machinery, Cambridge, page 279 – 286, 1976.

Halila, E.E., Lenahan, D. T. and Thomas, T. T., "Energy Efficient Engine, High Pressure Turbine Test Hardware Detailed Design Report", NASA CR-167955, 1982.

Heath, S. and Imregun, M., "A Survey of Blade Tip Timing Measurement Techniques for Turbo-machinery Vibration", ASME paper # 97-GT-213, New York, 1997.

Haynes, J.M., Hendricks, G.J. and Epstein, A.H., "Active Stabilization of Rotating Stall in a Three Stage Axial Compressor", ASME Journal of Turbo-machinery, volume 116, page 226 – 239, New York, 1994.

Hendrick, G.J., Dabinis, J.S. and Feulner, M.R., "Analysis of Instability Inception in High-Speed Multi-Stage Axial Flow Compressors", ASME Journal of Turbo-machinery, volume 119, page 714 – 722, New York, 1997.

Holmes, R. and Box, S., "On the Use of Squeeze Film Dampers in Rotor Support Structures", Machine Vibration, volume 1, page 71 – 92, 1992.

Howard, W.D. and Fasching, W. A., "CF6 Jet Engine Diagnostics Program: High Pressure Turbine Roundness/Clearance Investigation", NASA CR-165581, 1982.

Huber, F.W. and Dietrich, D.J., "Clearance Control for the Turbine of a Gas Turbine Engine", US Patent 5,667,359, 1997.

Jorgensen, P.J., "Effect of an Electric Field on Silicon Oxidation", The Journal of Chemical Physics, volume 37, no. 4, 1962.

Kawecki, E.J., "Thermal Response Turbine Shroud Study", Air Force Aero Propulsion Laboratory Technical Report AFAPL-TR-79-2087, 1979.

Kirk, R.G., "Oil Seal Dynamics: Considerations for Analysis of Centrifugal Compressors", Proceedings, 15th Turbo-machinery Symposium, Texas A & M University, College Station, Texas, 1986.

Kolthoff, P., "Device for Controlling Clearance Between Rotor and Shroud of a Turbine", US Patent 3,039,737, 1962.

Marquette, O.R., Childs, D.W. and San Andres, L., "Eccentricity Effects on Rotor Dynamic Coefficients of Plain Annular Seals: Theory versus Experiment", ASME, Journal of Tribology, volume 119, page 443 – 448, New York, 1997.

Martin, R.L., "Nacelle Aerodynamic and Inertial Loads (NAIL) Project, Test Report", NASA CR-165760, 1981.

McCarty, P.E. and Thomson, J.W., "Performance Evaluation of a Prototype Non-Interference Technique for Measurement of Turbine Compressor Blade Stress", Report # AEDC-TR-80-5, Arnold Air Force Station, TN, 1980.

Olsson, W.J. and Martin, R.L., "B747/JT9Dflight Loads and Their Effect on Engine Running Clearances and Performance Deterioration: Nacelle Aerodynamic and Inertial Loads (NAIL)/JT9D Jet Engine Diagnostics Programs", NASA CR-165573, 1982.

Rangwala, A.S., "Turbo-machinery Dynamics: Design and Operation", McGraw-Hill, New York, 2005.

Rocky Mountain Institute (RMI), "Climate, Air Travel Emissions", 2002.

Santiago, O., San Andres, L. and Oliver, J., "Imbalance Response of a Rotor Supported on Open End Integral Squeeze Film Dampers", ASME paper # 98-GT-6, New York, 1998.

Sedy, J., "Improved Performance of Film-Riding Gas Seals Through Enhancement of Hydrodynamic Effects", ASLE Transactions, volume 23 (1), page 35 – 44, 1979.

Shemeld, D., "A History of Development in Rotor Dynamics – A Manufacturers Viewpoint", Rotor Dynamic Instability problems in High Performance Turbo-machinery, NASA Report # CP 2443, page 1 – 18, 1986.

Silkowsky, P. D., "Measurements of Rotating Stall in a Matched and a Mismatched Multi-stage Compressor", GTL Report # 221, Gas Turbine Laboratory, Massachusetts Institute of Technology, Cambridge, MA, 1995.

Soto, E.A. and Childs, D.W., "Experimental Rotor Dynamic Coefficient Results for (a) a Labyrinth Seal With and Without Shunt Injection and (b) a Honeycomb Seal", ASME paper # 98-GT-8, New York, 1998.

Spakovszky, Z.S., Paduano, J.D., Larsonneur, R., Traxler, A. and Bright, M.M., "Tip Clearance Actuation with Magnetic Bearings for High-Speed Compressor Stall Control", ASME paper # 2000-GT-528, New York, 2000.

Tseng, W. and Hauser, A.A., "Blade Tip Clearance Control Apparatus with Shroud Segment Position Adjustment by Unison Ring Movement", US Patent 5,035,573, 1991.

Tryfonidis, M., Etchevers, O., Paduano, J.D., Hendricks, G.F. and Epstein, A.H., "Pre-Stall Behaviour of Several High Speed Compressors", ASME Journal of Turbo-machinery, Volume 117, page 62 – 80, New York, 1995.

Wardle, R.L., "Floating Blade Shrouds", US Patent 2,927,724, 1960.

Weigl, H., Paduano, J., Frechette, L., Epstein, A., Greitzer, E., Bright, M. and Strazisar, A., "Active Stabilization of Rotating Stall and Surge in a Transonic Single Stage Axial Compressor", ASME Journal of Turbomachinery, volume 120, page 625 – 636, New York, 1998.

Wiseman, M.W. and Guo, T., "An Investigation of Life Extending Control Techniques for Gas Turbine Engines", Proceedings of the American Control Conference, Arlington, VA, June 25–27, 2001.

15

Combustion and Emissions Control

15.1 INTRODUCTION

Combustion in gas turbines describes the exothermic reaction of a gaseous or liquid fuel and oxygen in the air. A flame propagating through the unburned charge of air and fuel, and defining a rapid chemical change occurring in a thin layer, accompanies the combustion process. The deflagration regime of combustion, requiring 1×10^{-3} s to complete 80 per cent of the task, is marked by a luminescent flame front that may be viewed as an interface between the burned gases and the unburned mixture. The process is characterized by steep temperature gradients and species concentration. Relative to the fresh air and fuel mixture, the burned gases are far higher in volume and temperature and lower in density, with the waves traveling at under 1 m/second. Instead of the flame (or combustion wave) spreading through a static gas mixture, it is usual to stabilize the flame to a steady condition by supplying it with a continuous flow of combustible mixture. In the detonation part of the combustion process, a shock wave connected with and supported by the chemical reaction zone propagates at velocities ranging between 1 and 4 km/second.

Both physical and chemical aspects are embraced during combustion. This subject of physics includes mass and heat transfer, thermodynamics, gas and fluid dynamics, while chemistry influences pollutant emission among the products of combustion, heat release rate and radiation properties of the flame at high temperatures. In aviation applications, the chemical process also impacts lean light-off and flameout limits at high altitudes. Flames may be categorized as pre-mixed type when the fuel and air are mixed before combustion and as diffusion type when the two components are diffused within the flame zone. The two flame types may also be described as laminar or turbulent, depending on flow velocity. When burning liquid fuels, complete vaporization may not take place before entering the flame zone, resulting in a diffusion flame burning of fuel droplets superimposed on a pre-mixed turbulent flame zone.

Engine specifications and efficient use of available space will influence the type and layout of the combustion chamber. Larger engines generally call for the air to flow nearly parallel to the axis of the combustor, but in smaller engines the flow reverses direction in the annular system to provide compactness and closer connection between the compressor and the turbine. A tubular can form of construction calls for a cylindrical liner mounted concentrically inside a cylindrical case, with between 6 and 16 cans arranged inside an engine. The larger length and weight of the resultant assembly restricts their usage to industrial turbines, where relative simplicity and accessibility are of prime significance. Annular combustors have the liner inside an annular casing to give a compact design and a clean aerodynamic flow path with little loss in pressure. But the absence of radial load carrying members generates buckling problems in the outer liner. Multi-can configurations with a group of between 6 and 10 tubular liners arranged inside a single annular casing combine the compactness of an annular chamber with the mechanical strength of the tubular type. The liner functions to contain the combustion process and to permit circulation of cooling air in various zones in set amounts. Besides the pressure differential on either side, the liner must have thermal resistance to withstand continuous and cyclic operation at elevated temperatures.

With conventional combustors, any modifications to alleviate generation of smoke and oxides of nitrogen (NO_x, most notably nitrogen oxide and nitrogen dioxide) will invariably result in increased levels of emission of carbon monoxide (CO) and unburned hydrocarbons, with the reverse also being true. Regulating the amount of air entering the primary combustion zone through a variable geometry mechanism has been successful to a considerable extent in overcoming this problem. Larger quantities of air are admitted at higher pressures to obtain more complete combustion and to minimize the incidence of soot and NO_x formation. When the pressure is low the primary airflow is restricted to raise the air-fuel ratio and to reduce the velocity, thereby improving combustion efficiency and lowering emission of CO and unburned hydrocarbons. This feature also is helpful in initiating the combustion process during light up. The variable geometry arrangement for controlling the flow of air calls for complex control and feedback mechanism, adding to the cost and weight while raising reliability questions, but may be justifiable for larger industrial gas turbines.

The concept of staged combustion attempts to achieve the same objectives by using two or more separate zones, each designed specifically to optimize certain features of combustion. The process may take the form of axial or radial staging. In the lightly loaded primary first zone, high combustion efficiency and minimized CO and unburned hydrocarbon production are achieved by operating at a relatively high equivalence ratio ϕ of around 0.8, where equivalence ratio is defined by the actual fuel ratio divided by the stoichiometric fuel ratio. The primary zone, placed at the upstream end along the central axis of the chamber provides the temperature increase needed at low power settings up to idle speed. At higher power requirement, it acts as a pilot heat source for the main combustion region located downstream in the axial staging version or at different radial distance from the pilot in the radial staging style. The second and subsequent zones are supplied with the fuel-air mixture of suitably adjusted equivalence ratio to all zones at varying levels of power setting. Staged combustion is extensively used in industrial turbines using gas fuels, achieving acceptable levels of pollutant emission.

Pressure rise in axial flow compressors heavily relies on flow velocity, so a high velocity is required to minimize the number of stages. In many aircraft engines compressor discharge velocity may exceed 150 m/second, and it is not possible to burn the fuel in such a high velocity stream. Prior to combustion, flow velocity is reduced to about a fifth of the compressor discharge velocity by placing a diffuser between the compressor exit and the inlet to the liner. Essentially a diverging passage in which the airflow decelerates and velocity head converts to increase the static pressure, the length of the diffuser is of significance in order to attain the twin objectives of maximum efficiency of the conversion process and of limiting the overall engine length.

Aerodynamic processes play a major role in a combustion system. A primary objective of a good combustion system is to achieve satisfactory mixing within the liner and a stable flow pattern throughout, with no parasitic losses and minimal pressure loss. Inside the combustion liner, considerable flow re-circulation is essential to stabilize the flame, and maximum benefit must be derived from the cooling air along the walls. Mixing in the combustion and dilution areas is of significance. Proper mixing in the primary zone enhances the rate of burning, while also minimizing soot and nitric oxide formation. A satisfactory temperature distribution in the exiting gases relies on the degree of mixing of the air and the products of combustion in the dilution zone.

Current emission regulations call for combustion efficiency in excess of 99 per cent. Failure to achieve this level is objectionable partly because fuel is wasted, but more so because it is manifested in the form of excessive pollutants in the environment. For aircraft engine combustors, a design hallmark calls for it to be large enough to ensure an adequate level of combustion efficiency during engine restart at the highest altitude at which re-light is required. After a flameout during flight, combustion efficiency in the 75 to 80 per cent range is justified, because with the engine windmilling the temperature and pressure of ambient air are low enough to affect the stability of the flame. The engine control system attempts to compensate by supplying more fuel to the combustor, preventing the flame to stabilize due to the overly rich air-fuel mix. As a consequence, the combustion efficiency deteriorates.

Mechanical stresses in the combustor liner are relatively low when compared with other engine components, however, it is required to withstand high temperatures and considerable temperature gradients. Commonly used materials for the liner such as Nimonic 75, Hastelloy X and HS 188 are restricted to maximum temperature levels of 1100 K, beyond which mechanical strength of the nickel and cobalt based alloys rapidly deteriorates. Thus, innovative means of removing the heat from the liner walls must be provided. A sizeable part of the heat transferred from the hot gases contained within the liner to the liner walls is by radiation. Cooling air, injected in regions that are primarily heated through internal radiation, provides an effective barrier between the hot gas and the liner wall. Internal convection is difficult to estimate, because the hot gases are undergoing rapid physical and chemical changes, and the difficulty is compounded by the presence of considerable pressure and temperature gradients. A realistic model defining the airflow pattern, boundary layer development and effective gas temperature thus faces substantial uncertainties. Among the many schemes employed to extract heat from a liner, film-cooling calls for a number of annular slots placed at 40 to 80 mm intervals along the length, through which the coolant is axially injected along the inner wall of the liner. Other devices in the form of wiggle strips and rings of various profiles are used in film cooling to suit the conditions of pressure and temperature, and for manufacturing ease.

15.2 FUELS FOR VARIOUS APPLICATIONS

Gas turbine liquid fuels may be split into two main groups. Lighter weight high performance aircraft engines require higher grades, while heavy duty industrial engines operate on a wide range of heavier and more easily obtainable fuels. Natural gas is now the preferred fuel for power plant and many other industrial gas turbines. Derivative aircraft engines for marine and industrial applications use modified combustion systems to enable them to burn commercial distillate fuels.

JP-4 and JP-5 are used widely by the US military for turbo-jet operation. JP-4 is a naphtha fuel with vapour pressure of about 2.5 psi and aromatic content under 25%. JP-5 is a blended kerosene fuel, has a flash point of 140°F and freezing point of –51° F. Airplanes operating at supersonic speed up to 3.5 Mach require thermally stable fuels. JP-7 and JP-8 provide varying degrees of flash and freezing points, thermal stability, aromatic content and flame luminosity. Commercial airlines rely on ASTM jet aviation turbine fuel. This is also a kerosene-based fuel with a flash point of 110°F and freezing point of –36°F.

Atomization of liquid fuels into droplets with extensive surface area is necessary for ignition and combustion of liquid fuels, since they are not sufficiently volatile to produce vapours. The rate of evaporation is enhanced by reducing the size of the droplets. Forcing the fuel under pressure through an orifice aids in atomization. In a simplex atomizer a swirl chamber is placed before the orifice, injecting into a spray cone angle of about 90°. Dual orifice atomizers have two swirl chambers, with a pilot injector located concentrically inside the main. At low fuel flow the pilot delivers all the fuel at adequate pressure. At a pre-determined pressure a valve activates to let fuel flow through the main atomizer, permitting acceptable atomization over a larger range of fuel flow. Pressure-swirl atomizers can sustain combustion at weak fuel and air mixture strength, but may get plugged by contaminants in the fuel, and have a tendency to form soot at higher pressures. Another concept calls for the fuel to flow at lower pressure over a lip in a high velocity air stream, causing atomization by the air as it enters the combustion zone. When the liquid sheet at the atomizing stream is exposed to the high velocity air on both sides, drop size is minimized to provide maximum contact between the air and the fuel. In this form of air-blast atomizers the airflow pattern controls the distribution of the fuel to reduce soot and exhaust smoke formation. However, the system has limited stability limits, and at start-up lack of air velocity leads to poor atomization. The problem may be controlled by combining the features of a pilot pressure-swirl atomizer to obtain easy light-up and improved stability limits.

Gaseous fuels such as natural gas are comparatively easier to burn when the calorific value is higher. In multi-fuel engines, lower heat content fuels tend to take up more of the combustor mass flow and combustion zone volume to cause a mismatch between the compressor and the turbine. Gas fuel injection may be accomplished by using orifices, swirlers, slots and venturi nozzles.

Lean pre-mixed combustion is the preferred method for controlling the pollutants in natural gas fired turbines, but as the system operates close to the edge of combustion stability to reduce emissions of NO_x, modest upsets in operating conditions or variations in fuel composition tend to have repercussions in the overall power generating capability of the unit. Constituent concentration for natural gas available in United States tend to average 93.9 per cent methane, with ethane, propane and higher hydrocarbons rounding out the remaining composition at 3.2, 0.7 and 0.1 per cent. But methane, ethane and propane composition can approach values of 74.5, 13.3 and 23.7 per cent, respectively. Such large swings pose a considerable challenge for the combustor to maintain optimal performance. As a consequence, delineation of the key phenomena is necessary to render the combustor insensitive to the variations.

The effect of fuel composition on engine performance has been evaluated in a project conducted at University of California at Irvine (*ref*: Flores et al., 2000). A model combustor using a flexible injection system to provide radial jets, shown in figure 15.1, mimics key features of those found in practice. Fuel is injected at multiple points to control flow split between three independent injection circuits. Radial and axial injection from a centrepiece (labeled CB and pilot in the figure) and wall injection from equally spaced holes (WJ) into the swirling air stream are provided. Firing rate for the system is 15 kW at 0.0093 kg/second, and inlet air is maintained at 700 K. Exhaust emissions are analysed by an integrated sampling and data acquisition system. The objective of the experiment is to obtain performance maps based on CO, NO_x and lean blow off as a function of fuel characteristics.

A comprehensive fuel blending system combines natural gas, ethane and propane in desired proportions (figure 15.2). Since methane constitutes the largest component (nearly 97 per cent) in the natural gas used for the study, adding 15 per cent ethane or 20 per cent propane by volume has little effect of the same constituents present in the base natural gas. Four fuel compositions and their associated properties used in the experiment are shown in table 15.1.

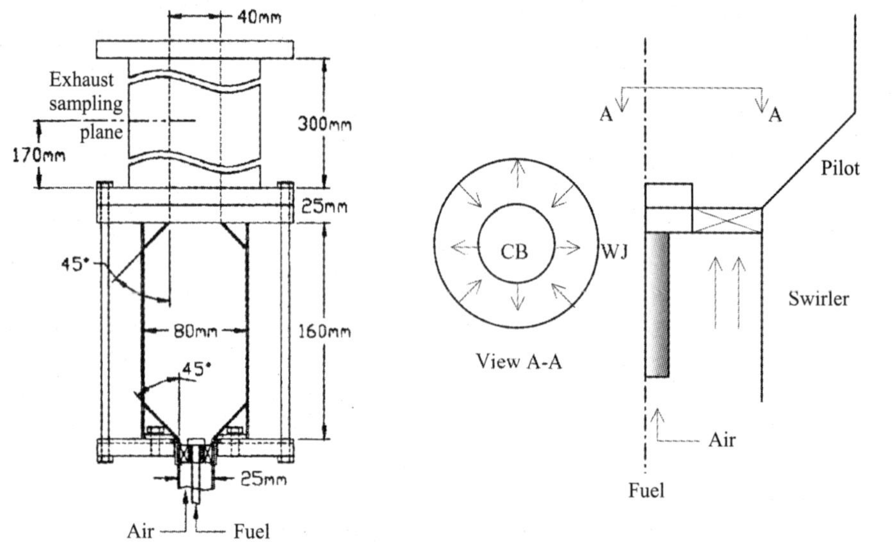

Fig. 15.1 Model Combustor (*left*), Fuel Injection Details (*right*) (Flores, 2000)

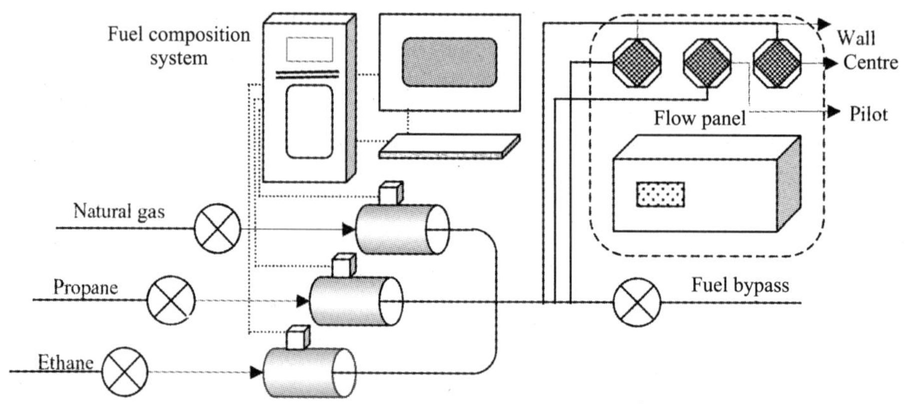

Fig. 15.2 Fuel Blending System (Flores, 2000)

Table 15.1 Gas Composition and Properties

Blend	Wobbe Index* MJ/m^3	Specific Gravity Relative to Air	LHV MJ/m^3
Natural gas (NG)	44.3	0.576	33.6
85% NG, 15% ethane	46.8	0.645	37.6
80% NG, 20% propane	50.5	0.765	44.2
Propane	70.1	1.523	86.5

* Wobbe index = LHV/√(SG)

Characteristics of the emitted gases indicate that fuel split between the centrepiece and wall injectors has little impact on NO_x emission, but the presence of CO reveals increased levels under extremely lean conditions and at equivalence ratios exceeding 0.5. The pilot fuel contributes higher levels of CO and NO_x, although lean blow-offs are not equally affected. Since the pilot nozzle injects fuel axially into the central re-circulation zone, a rich high-temperature core with limited mixing results to create the situation. A distinct improvement in performance is observed for cases without the pilot, with the response appearing in the form of ridges for a broad range of equivalence ratios, peak values being a function of the fuel split. Table 15.2 illustrates features of this behaviour, together with reaction temperatures. Average pre-mixed temperatures are well below the 1900 K threshold for NO_x formation, suggesting local regions with higher temperatures. Peak temperatures also rise when the natural gas is mixed with other components, indicating the reason for higher levels of pollutants. But the results are not indicative of relative contribution of non-thermal mechanisms (for example, N_2O path) suggested by Nicol et al., (1995).

Table 15.2 Relative Emissions and Gas Composition

Blend	CO/CO_{NG}	$NO_x/NO_{x\,NG}$	$T_{adiabatic,\,NG}$, K
85% NG, 15% ethane	0.54	1.25	1616
80% NG, 20% propane	0.86	1.25	1620
100% Propane	0.55	1.87	1637

Emission provided as ratio for baseline 100% natural gas (NG)

Pure methane gas has the richest lean blow-off limit, while pure propane offers the widest range of lean blow-off. With pilot fuel injection this range is broadened by enriching the re-circulation zone, but overall performance deteriorates when compared with no fuel at the pilot. Without the pilot fuel, composition impacts the lean blow-off limit consistent with the reaction rate, indicating the presence of a kinematic mechanism to stabilize the reaction.

The effects of chemistry from mixing are isolated by injecting the fuel well upstream of the inlet to the combustor to obtain fully pre-mixed conditions. The resulting NO_x and CO emissions are shown in figures 15.3 in terms of the adiabatic flame temperature. Lower NO_x levels are achievable with improved mixing, but requirements for turndown, avoidance of flashback and auto-ignition and robustness to fuel composition variability make a completely pre-mixed system less attractive than the controlled lean and rapid mix injection strategy.

Gas turbines offer the flexibility of burning any petroleum-derived fuel, including unrefined crude and residual oil. But turbine manufacturers place restrictions on excessive contaminants such

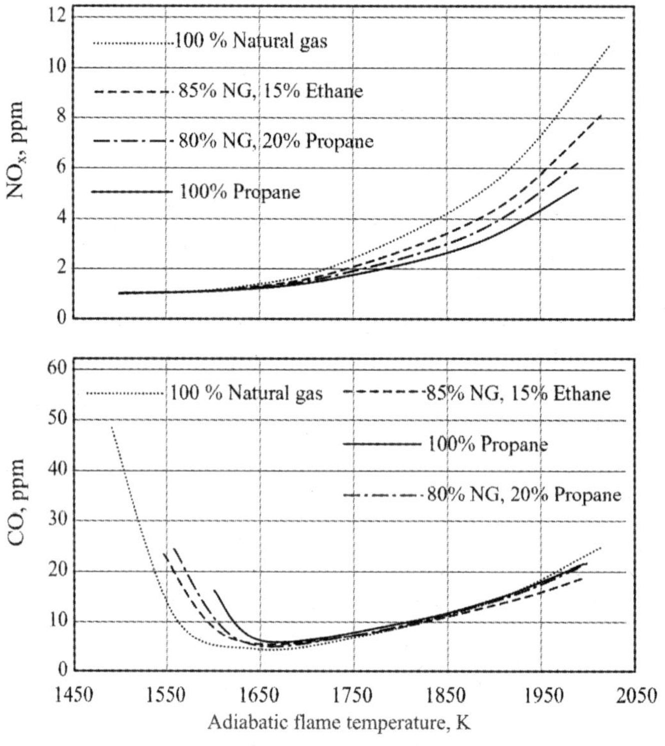

Fig. 15.3 Concentration of Emissions for Fully Pre-Mixed Case (Flores, 2000)

as sodium, potassium, calcium, sulphur and other elements that contribute to deposits and corrosion in the hot gas path. Deposits formed on turbine airfoils by contaminated fuels can seriously impair performance of a turbine through fouling and high temperature corrosion. Oil soluble and particulate elements may occur naturally in the crude oil, or may be introduced during storage, refining and transportation. Solid matter such as sand, rust particles and foreign objects can be effectively filtered out to 5 to 50 μm size. Smaller sand particles, although chemically benign, may cause unacceptable erosion, and can be removed by centrifuging.

Test data shows that both corrosion and deposits can be successfully suppressed by additives through change in the morphological nature of oil ashes. The ashes have the characteristic of being relatively non-adherent or easily removable. But the additives have the potential for high temperature corrosion, and do not provide equal protective effects for all variations.

Since many of the contaminants are water soluble, the water washing process is used to develop an emulsion containing most of these compounds after intimate mixing with the fuel. The emulsion must be broken from the fuel to separate the fuel from the water. Fuel washing systems generally remove about 90 per cent of the entrained inorganic salts.

Cost considerations play a major role in fuel selection for industrial gas turbines. Natural gas, petroleum distillates, refinery and chemical plant gases and liquids and residual fuel oil are some of the most common fuels. Natural gas, propane and butane have distinct advantages of burning readily in a small space with minimum smoke, possess low flame luminosity and generate negligible corrosive residues. Petroleum distillate fuels, available in negotiated specifications for gas turbines, are widely used for electric utility peaking plants. Distillate fuel combustion characteristics are good and mechanical atomization may be used without preheating the fuel (*ref*: Sawyer, 1982).

Turbine installations in steel mills burn blast furnace gas, a mixture of CO and N_2 that burns slowly and has bulk. Consequently, combustor design must have special features to account for the gas characteristics.

Storage, handling, metering, atomization and combustion are affected by a fuel's properties, and are discussed at length in ASTM specifications D-2880 and D-1655. Flash point provides an indication of the maximum temperature at which a fuel may be safely stored and handled without risking a fire hazard. Regional government agencies have a say in permissible limits. Pour point determines the lowest temperature at which oil may be stored and still flow under gravity. Water and sediments in oil fouls the fuel handling equipment, and is also detrimental to the turbine's fuel system. Water is also responsible for corrosion of steel storage tanks and equipment, and may cause emulsion in residual fuels. Presence of carbon residue in a fuel approximates formation of carbon deposits in the combustor. Viscosity of a fluid determines how easily it flows and atomizes in the nozzles. When viscosity is too low the fuel cannot be pumped satisfactorily; excessive pressure losses in piping and poor atomization are consequences of high viscosity. Specific gravity does not directly affect combustion of a fuel oil, but plays a role in weight-volume relation and heating value calculations.

The hydrogen content of a fuel has a strong linkage with combustion in a gas turbine. Reduced hydrogen in a fuel enhances its aromatic content, resulting in soot formation and flame radiation in the combustor. Increased metal temperature and reduced life of the components are a direct consequence. Smoke also is created. Figure 15.4 shows the relation between the fuel's hydrogen content and a temperature related parameter for the liner.

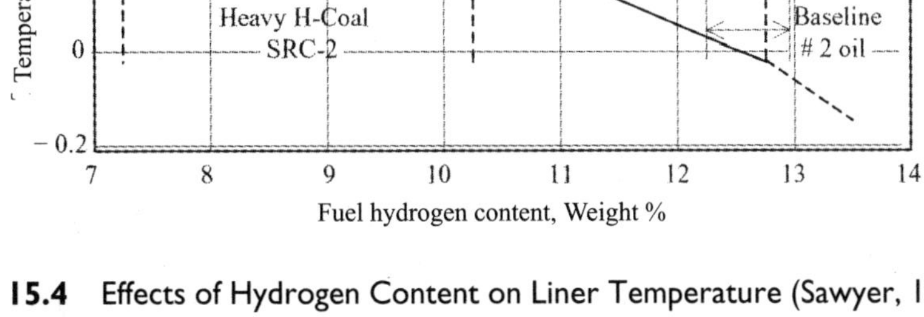

Fig. 15.4 Effects of Hydrogen Content on Liner Temperature (Sawyer, 1982)

When fuels with excessive water, water-soluble dissolved alkali salts or oil soluble compounds of metals are encountered, equipment and processes are available to alleviate the problem. Water washing, chemical injection, heating and chemical analysing methods are widely used to protect the investment for gas turbines, running into several million dollars. Centrifuges present a well-known industrial process for separating, clarifying and purifying liquids of different specific gravity and removing solids. Electro-static systems are also available for water washing, calling for water emulsion between charged plates. Demulsifiers injected into a fuel stream is also well recognized in treating water and oil emulsions seen in the petroleum industry; flocculants are used to further remove oils from the effluent water.

15.3 COMBUSTION PRINCIPLES

The rate at which a combustion wave travels as a plane flame front in a direction normal to its surface through the flammable mixture is controlled by flame radiation (and hence temperature), local gas properties such as viscosity and diffusion coefficient and the imposed variables of pressure, temperature and air-fuel ratio (*ref*: Lefebvre, 1999). This basic property of the combustible mixture establishes the rate of heat release and stability limits of the flame. A reproducible constant value for any fuel can then be used when the imposed variables are fixed from the burning velocity, and approaches 0.43 m/second for stoichiometric mixtures of many fuels at normal atmospheric pressure and temperature. Chemical reaction rate determines the laminar flame speed in pre-mixed systems, which is governed by the equivalence ratio, temperature and pressure. Turbulence appreciably increases the flame speed, but the manner and extent of its influence is not well understood. Increase in the burn rate may be attributed to wrinkling of the flame front by the turbulence by enlarging the specific surface area (*ref*: Damkohler, 1947).

Heterogeneous mixtures of fuel drops, fuel vapour and air experience increased flame speed with reduced mean drop size until a critical value of about 60 – 70 μm for kerosene type fuels is reached, when the curve flattens. In finely atomized sprays the flame speed is less dependent on evaporation rate, and is governed mostly by the chemical reaction rate. Flame speed also gains from the beneficial effects of the accompanying increase in flow velocity due to increase in the turbulence intensity. Subsequent evaporation of the droplets requires simultaneous heat and mass transfer from the surrounding air or gas before diffusion back into the gas stream.

Ignition of combustible mixtures relies on a transient ignition source, usually an electric spark, to supply sufficient energy to create a volume of hot gas that just satisfies the necessary and sufficient condition for propagation, so that the rate of heat generation exceeds the rate of heat loss. For an incipient spark kernel to survive and propagate unaided through a gaseous mixture, its minimum dimension should exceed a certain quenching distance. A minimum level of ignition energy is needed to heat the smallest volume of gas to its adiabatic flame temperature, whose minimum distance equals the quenching distance. The ignition energy depends on pressure, velocity, turbulence and mixture composition. When ignition is successful, combustion of the fuel vapour continues to produce heat, which diffuses outward from the kernel to raise the temperature and initiate combustion in the surrounding unburned mixture. The flame then spreads rapidly to all regions where the air and fuel mixture are in combustible proportion.

Spontaneous, or auto-ignition, can occur when a combustible mixture undergoes a chemical reaction that leads to rapid evolution of heat in the absence of a flame or a spark. In pre-mixed combustors and other arrangements relying on mixing of air and fuel prior to combustion, spontaneous ignition may result in damage to the components while increasing pollutant emissions. The time interval between the creation of a combustible mixture and the onset of flame depends on fuel composition, method of fuel injection and pressure and temperature of the mixture. The continuing trend toward engines of higher pressure ratio requires increased attention to this phenomenon. Pre-mixed combustion systems also experience flashback when the flame travels upstream from the combustion zone into the pre-mixing part of the combustor. If the flame speed exceeds the approach flow velocity, flashback can occur in the free stream or in the slower flowing boundary layer along the walls of the mixing section. Flow reversal in the bulk flow through the combustor as a consequence of compressor surge and instability is the prime cause of flashback in the free stream. Lean combustion tends to reduce flame speed, but other factors of an engine cycle such as elevated temperature, pressure and turbulence level can increase the flame speed.

The adiabatic temperature of the flame, attained when the energy liberated by the chemical reaction is fully used to heat the products, plays a substantial role in the rate of reaction. In practice, however, some heat is lost by radiation and convection. At flame temperatures around 1800 K dissociation of combustion products accelerates to absorb much heat. At low temperatures combustion of a lean mixture produces only CO_2 and H_2O, but at higher temperatures the products destabilize to partially revert to simpler molecular and atomic species and radicals, principally CO, H_2, O, H and OH. Considerable energy is needed to dissociate the components, thereby reducing the flame temperature.

Reduction of NO_x emissions to meet regulatory levels may be obtained by lowering reaction temperature, which in turn calls for operating the combustor under fuel-lean conditions. The lean pre-mixed, pre-vaporized concept introduces a uniformly lean mixture of fuel vapour and air into the dome region, burning the fuel in its vapour phase. The process, however, is prone to auto-ignition and flashback. In the lean direct injection method gaseous fuel is injected upstream of the reaction zone to reduce these drawbacks.

Atomizing and vaporizing a liquid fuel for low NO_x can be achieved with the lean burn injector (*ref*: Leong et al, 2000). The injector, consisting of a fuel tube centrepiece, swirler and venturi mixing section, prepares a vaporized fuel-air mixture by injecting spray jets of fuel radially into a swirling cross flow. The jets are formed by air-blast atomization to permit high fuel turndown ratios, then ejected into the primary combustor dome. The swirling component of the mixture flow induces a re-circulation zone to anchor the combustion process downstream of the venturi throat, which also aids in eliminating flashback. This concept may be applied to aero and industrial engines. An experimental evaluation has been conducted to evaluate robustness of the design process, primarily at low power conditions when low preheat temperatures and pressures may not fully atomize and vaporize the droplets before the mixture exits the mixing section. Figure 15.5 shows a schematic diagram of the injector assembly.

Table 15.3 Measured Average Emission from Lean Burn Injector Combustor

CO_2 ppm		*O_2 ppm*		*NO_x ppm*		*CO ppm*		*UHC ppm*	
Max	*Min*	*Max*	*Min*	*Max*	*Min*	*Max*	*Min*	*Max*	*Min*
10	5	12	8	117	24	153	7	2	0

Emission samples are obtained downstream of the throat where bulk residence time of the combustion products is 8×10^{-3} seconds. Pressure difference within the assembly and the exhaust is used to drive the sample through a multi-port water-cooled probe to yield area-weighted emissions across the combustor can. The gas sample then passes through ice water for a short duration in preparation for the analysers to measure pollutant concentrations. Analog signals from the emission

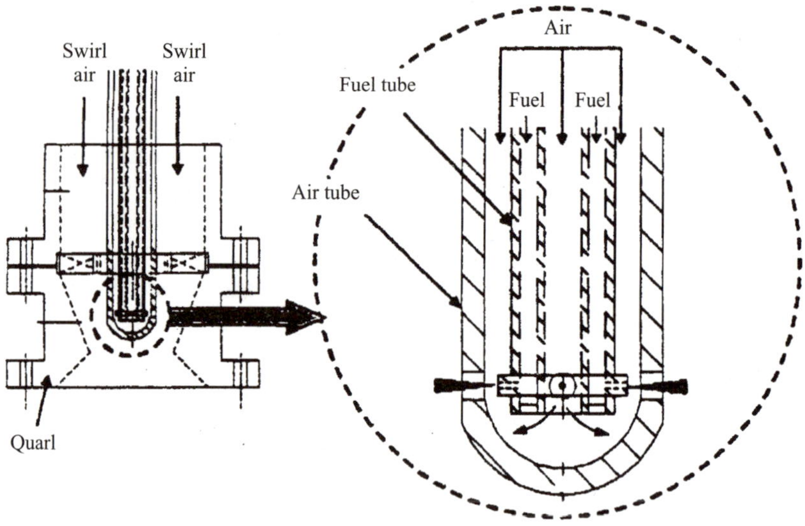

Fig. 15.5 Lean Burner Injection System Assembly (Leong, 2000)

analyser are then collected and processed to obtain a stationary temporal distribution. Gas flow rates in the test are comparable to the low power regime of ground idle and subsonic cruise conditions. While overall air mass flow rate is held steady, the fuel-air equivalence ratio (between 0.55 and 0.65) is changed by varying the liquid fuel flow rate. Two ambient pressure conditions of 4.0 and 6.6 atmospheres and two air pressure drops typical of aero-engine combustors (4 – 7 per cent) are used in the experiment.

Table 15.3 provides results from the test. Minimum values for the pollutants occur for the lower level of 0.45 for the air-fuel equivalence ratio, and maximum values are reached when the ratio is 0.65. The radial profile of the emissions then yields an area-weighted average, which is compared with calculated values at the corresponding equivalence ratio. Agreement is within 2 per cent of predicted values for the 4.0 atmosphere case, varying up to 15 per cent for the 6.6 atmosphere condition. Lean blow out levels are 0.415 for the 4.0 atmosphere case and 0.39 for 6.6 atmospheres, with pressure drops between 3 and 7 per cent.

Combustion efficiency and production of pollutants are prime indicators of the performance of the combustor. Efficiency is determined from the level of CO and unburned hydrocarbon formation. The lean burn injector combustion system displays efficiency levels above 99.90 per cent for all conditions, as seen in figure 15.6. Higher efficiency is obtained for the 7 – 8 per cent air-blast pressure drop condition than for the 3 per cent case. NO_x emissions are held at a low level because the reaction temperatures are held under 1800 K, so the exponential growth in NO_x formation associated with the thermal mechanism is avoided.

15.4 COMBUSTOR DESIGNS AND SELECTION

Reduced pressure losses because of the relatively straight flow path of the gases and compactness in design make the annular combustor suitable for aviation engines. The annular configuration used

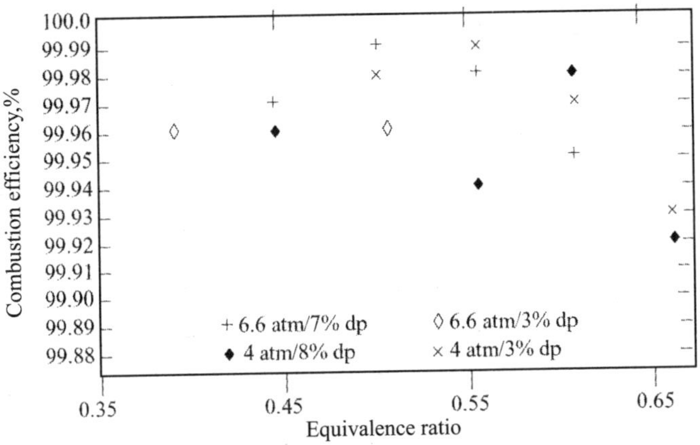

Fig. 15.6 Lean Burner Combustion Efficiency (Leong, 2000)

on General Electric's CF6 and GE90, Pratt and Whitney's JT9D and Rolls Royce RB211 engines has achieved high degrees of reliability in spite of increased operating pressures. Figure 15.7 shows an annular combustor for the CF6-50 engine.

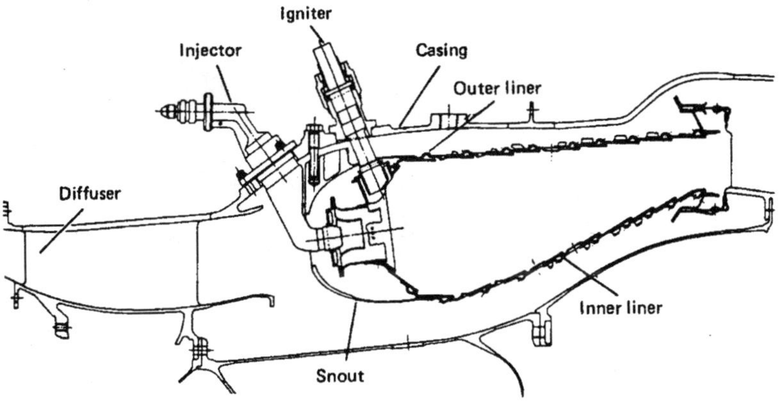

Fig. 15.7 Annular Combustor Design for CF6-50 Engine (*Courtesy*: General Electric Co.)

Can combustor designs call for a cylindrical liner placed concentrically inside a cylinder. From 6 to 16 cans are placed equally spaced around a central core casing in many early jet engines. The construction offers extra rigidity, but the extra length adds to engine weight. Industrial turbines extensively use this combustor type because of accessibility and ease of maintenance. The MS3002 and MS5002 gas turbines are manufactured by General Electric Company for pipeline pumping applications (*ref*: Feitelburg et al., 2000). The two-shaft engines have a high-pressure turbine to power the air compressor and a low-pressure turbine to drive the load. In the regenerative cycle configuration the hot turbine exhaust gas heats the compressor discharge air to nearly 900°F prior to combustion. A lean head end combustor is employed on the engines to control the emissions. A single fuel nozzle centered at the inlet of cylindrical cans of 27 cm diameter produces a swirl stabilized diffusion flame. The walls of the can are louvered for cooling, and contain an array of mixing and dilution holes to provide the air required for completing the combustion and dilute the burned gas to the desired turbine inlet temperature. Tighter spacing of the louvers helps to cool the regenerative cycle liners because the air temperature is higher. Six cans are provided on the smaller model of the engine and twelve on the larger capacity engine.

The combustor cans retain key features of conventional designs, the arrangement of the cooling and dilution holes being the primary difference (figure 15.8). The number, diameter and location of the holes is optimized to bring down NO_x emissions. An annular vortex generator is mounted in the combustor cap around the fuel nozzle. A flexible seal at the downstream end of the combustor liner fits into the transition piece. Crossfire tubes at the upstream end are used to equalize pressure between the chambers and to propagate the flame upon ignition. By placing the mixing and dilution holes closer to the fuel nozzle, air is redistributed from the downstream end towards the upstream side to produce a more fully aerated diffusion flame. The technique achieves a reduced turbulent flame length while decreasing the amount of time spent in the stoichiometric flame zone, thus lowering NO_x emissions.

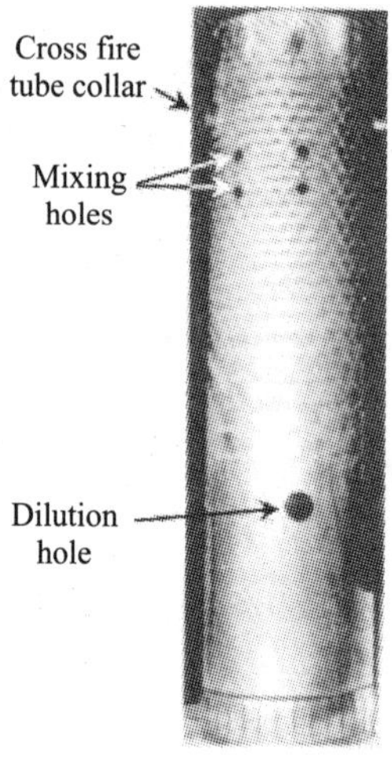

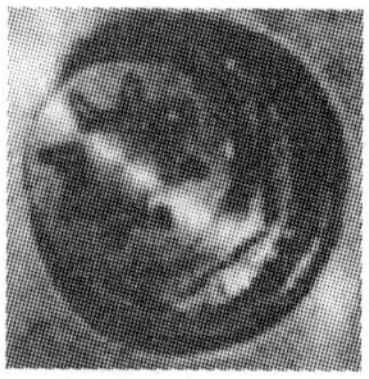

Fig. 15.8 Cylindrical Combustor Liner (*left*), Gas Fuel Nozzle (*right*) (Feitelburg, 2000)

The laboratory test stand used in the development of the lean head end combustor is shown in figure 15.9. The inner liner is placed in a pressure vessel that is supplied with pre-heated, non-vitiated, oil-free air entering through a distribution plenum around the circumference. Air flows over a simulated transition piece and continues upstream towards the fuel nozzle, mimicking the reverse flow configuration of the turbines. The burned gas is sampled with an uncooled stainless steel probe to analyse CO, CO_2, O_2, NO_x and unburned hydrocarbons with continuous emission analysers. Inside the combustor pressure is established with an orifice plate, with the backpressure cooled by a water jet sprayed into the exhaust gas. The sampling probe extends across the diameter of the duct, and has inlet holes equally spaced across its length to collect samples representing an integrated average across the duct.

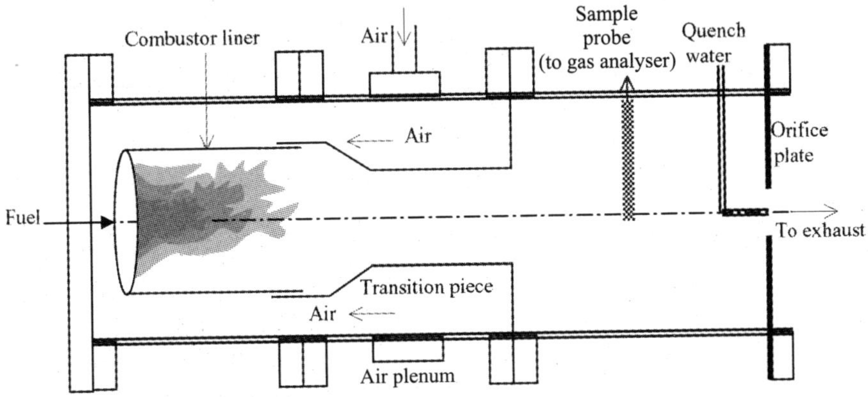

Fig. 15.9 Test Stand for Combustion Evaluation (Feitelburg, 2000)

Results from the tests of the standard and the lead head end combustors for the regenerative cycle MS5002 engine are shown in figure 15.10. NO_x emissions from the regenerative cycle are considerably greater than for the simple cycle turbine due to the higher combustion air temperature and the subsequent stoichiometric flame temperature. The latter is a dominant factor controlling NO_x emissions in diffusion flame combustors. Regression analysis shows that NO_x emissions from the standard liner are proportional to $P^{0.55}$ and for the new design are $P^{0.58}$, the difference being dependent on pressure P, and is within experimental uncertainty of the data. Scaling the data to a single pressure indicates that at base load conditions (combustor discharge temperature = 1006°C), NO_x emissions from the lean head end design are 50 per cent lower than for the standard version. CO emissions nearly double from 10 ppmv (dry) to 20 ppmv (dry) at base load, and for the diffusion flame combustors are insensitive to the total airflow rate and pressure.

A number of steam injection techniques have been employed in gas turbine operation to increase power output and to reduce NO_x emissions. Steam may be injected into the turbine wrapper, which, due to the reverse flow configuration of this machine, results in steam entering both the head end and the dilution zone of the combustor. Other options call for steam injection through the fuel nozzle directly into the head end of the combustor, or into the combustor casing with the jet directed towards the head end of the liner. However, customer acceptance for pipeline pumping applications is limited because of lack of required steam availability. The lean head end combustor liner design was developed by the manufacturer in order to reduce the NO_x emissions without resorting to dilution injection.

A catalytic combustion system has good potential for reducing NO_x emissions levels, below conventional systems such as diffusion or premixed combustion. It also has features that enable burning of a variety of different fuel without regard to specifications such as octane number or cetane index. In a project to develop a 100 kW ceramic gas turbine for passenger automobiles, a catalytic combustor that permits raising outlet temperature to 1473 K has been investigated by Kaya (1997). When this catalytic combustor is installed on an engine the required performance must be maintained over a range of conditions, from cold starting to steady state operation. During the initial cold start the catalyst is not activated, and requires heating by other means.

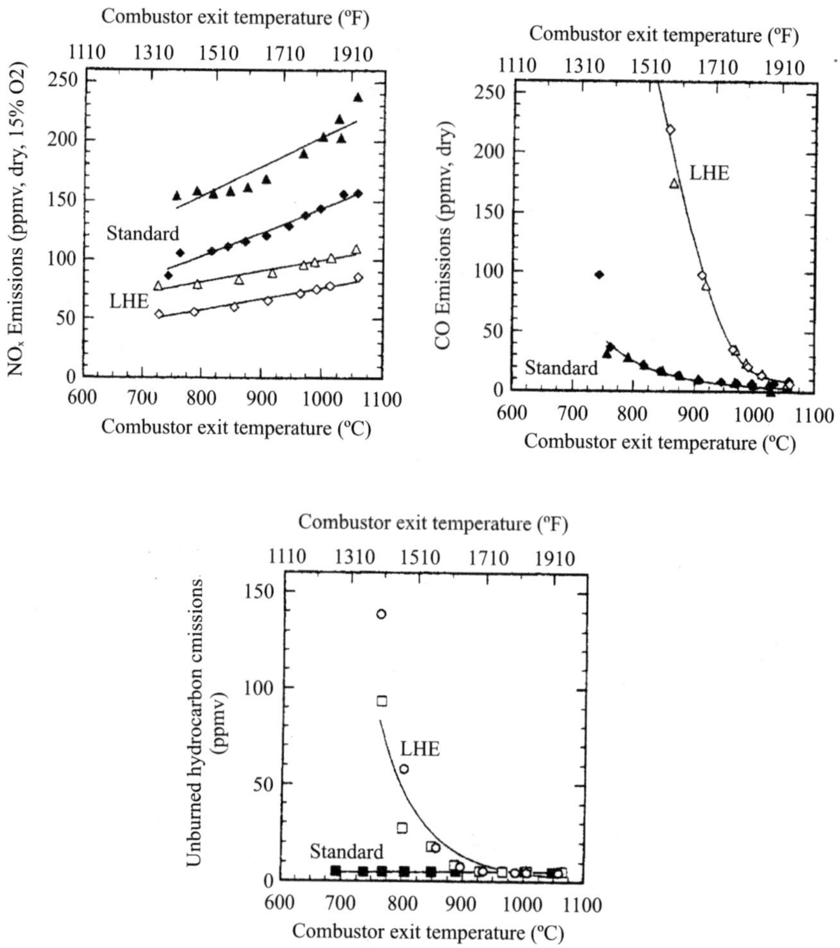

Fig. 15.10 Comparison of Emissions Measured in Combustor Tests (Feitelburg, 2000)

A concept for direct heating in order to downsize the combustor and reduce warm-up time during a cold start has been proposed by Yoshida et al., (2000). A vaporizing tube to mix the gas and function as a starting burner is employed, with its effectiveness experimentally verified. Furthermore, hydrocarbons, NO_x and CO emissions during engine start are reduced to achieve ultra low limits.

Figure 15.11 illustrates the operating principle during cold start, transition and steady state cruise condition. A variable swirler, air-assist fuel nozzle, vaporizing tube, ignition plug and three-staged catalyst constitute the major components of the catalytic combustor. The first stage catalyst offers high activity at low temperatures, but can withstand elevated levels. The second and third stage catalysts offer even greater resistance to heat. Inflow ports are provided in the liner between

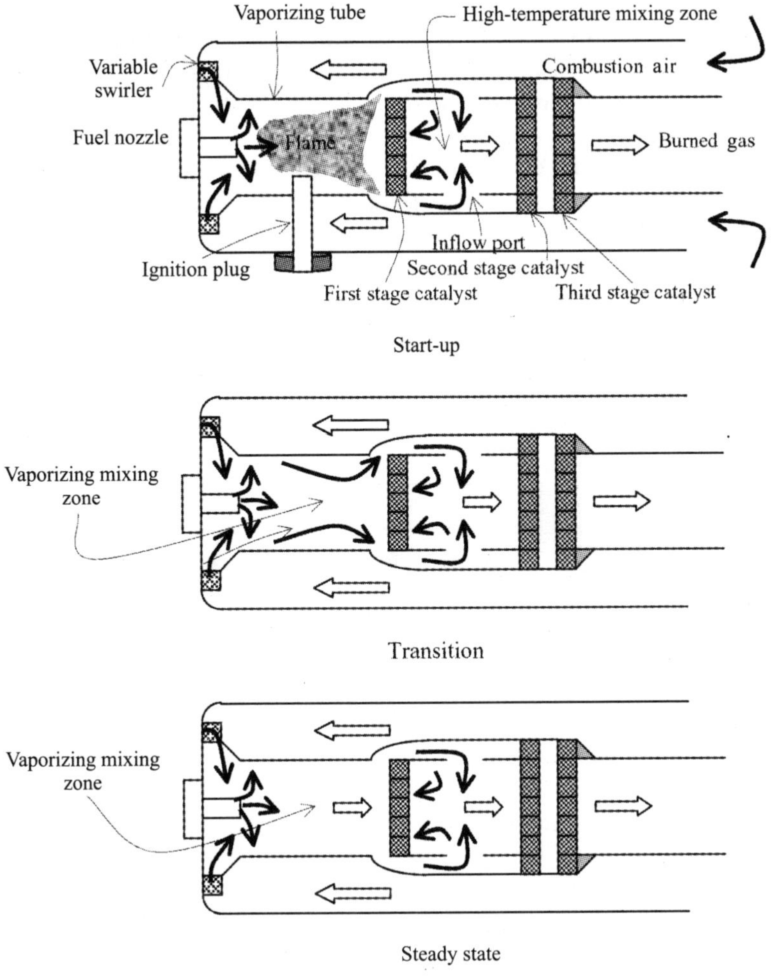

Fig. 15.11 Catalytic Combustor with Starting Burner (Yoshida, 2000)

the first two catalyst stages to allow gas flow into the high-temperature mixing zone in the event of excessively high swirl strength. Combustion gas is then forced to move towards the outer direction by its own momentum. A large amount of gas flows through the slit section, with the circulation passing through the first stage catalyst and the vaporizing tube. Note that the high temperature flame does not directly heat the first stage catalyst.

During start-up cold air is swirled by the variable swirler, then enters the vaporizing tube. Spray from the fuel nozzle is ignited by the ignition plug to form a swirling flame in the starting burner tube. First stage catalyst is heated by the circulation of burning gas flowing in the mixing zone. Burning gases then flow through the subsequent catalyst stages, where the unburned hydrocarbon and CO are burned through catalytic reactions.

During the transition from start to steady state the temperature of the catalyst rises rapidly and becomes active at ignition temperature. First, the catalyst temperature increases from heat radiated by the second stage catalyst. A heat exchanger may be available to recover exhaust waste heat. The starting burner's flame, supported near the inner wall of the vaporizing tube, extinguishes by the increased atomizing airflow. The catalytic combustion is continued. The incoming vaporizing gas is heated by the reversed reacting gas from the first stage catalyst, thus maintaining circulation in the vaporizing tube. Switching from the transition stage to a steady operating state is accomplished by adjusting the swirl strength of the combustion air.

A conventional honeycomb catalyst made of Pd-Al_2O_3-cordierite with heat resistance up to 1273 K and high activity at low temperatures is used for the first stage. The second stage uses a honeycomb body made of aluminium titanate, capable of resisting 1773 K. Kerosene is used for fuel. Figure 15.12 provides details of NO_x emission index between start-up and transition.

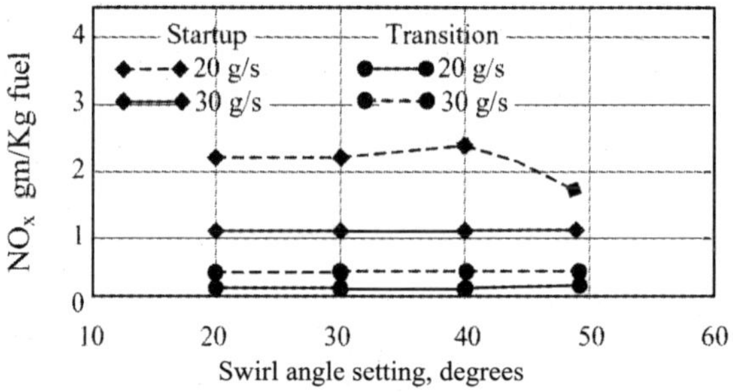

Fig. 15.12 NO_x Emissions Index (Yoshida, 2000)

Directionally solidified and single crystal super alloys have reached the point of application in engines for supersonic aircraft and industrial gas turbines where further gains in raising combustion outlet temperatures are not possible. In response to the drive to achieve combustion at temperatures that exceed the limits of durability of metallic materials, new materials such as inter-metallic compounds and oxide dispersed super alloys are being developed.

Japan's Research Institute of Advanced Material Gas-Generators has conducted a development program to apply ceramic matrix composite liners in the combustor (*ref*: Nishio et al., 1998). Silicon carbide fibre reinforced by silicon carbide fibre is selected as the ceramic material for its excellent resistance to high temperatures and oxidation. Conceptual configuration of the combustor and liner, made using the filament winding method, is shown in figure 15.13. The winding technique permits adjustment of the orientation of the fibre, making it easier to form a near net shape. The fibre has a carbon surface layer that becomes the interface laminating material, and is composited by the polymer impregnation and pyrolisis method. Polycarbosilane is used as the matrix precursor polymer. In a tensile strength at room temperature the fibre orientation angle of the test piece is varied incrementally in the range of 0° to ± 82.5°. Combustor liner forming angle is between ± 22.5°. The tensile strength at 20° is determined to be 250 MPa. In separate tests reaching 1473 K the material's strength is found to be virtually unaffected by the temperature. The gas swirlers are made of TiAl.

Thermal stresses are calculated with the aid of an axi-symmetric finite element model. A spring connection is used to tie the liner with the swirler in order to relieve thermal stresses arising from the difference in thermal expansion rates between the two materials. Temperature distribution in the

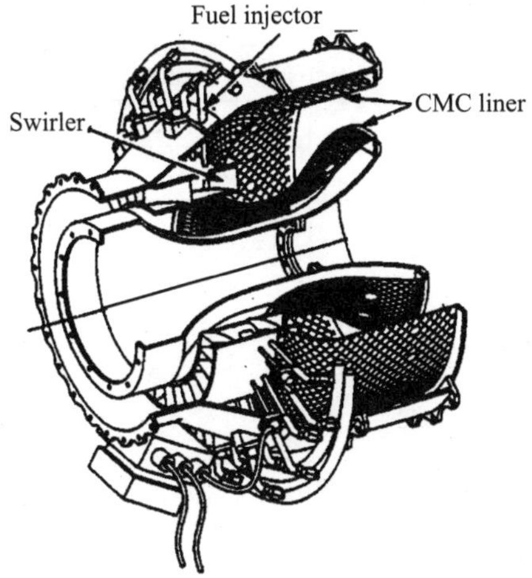

Fig. 15.13 Ceramic Matrix Composite Liner Design (Nishio, 1998)

liner's axial direction and through the material thickness is established from a heat transfer analysis using heat boundary conditions estimated from the results of combustion tests of a metal liner. Material characteristics data used for the analysis is shown in table 15.4. Swirler temperature is set at a constant 973 K. The finite element model is shown in figure 15.14.

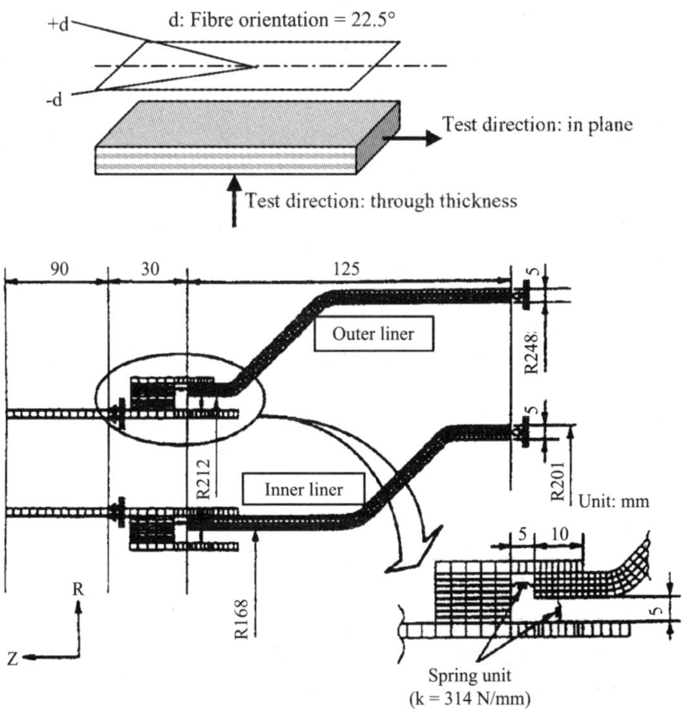

Fig. 15.14 Thermal Stress Analysis Model (Nishio, 1998)

Table 15.4 Material Characteristics

Property	Ceramic Matrix Composite			TiAl	
	Temperature K	Orientation		Temperature K	Magnitude
		In-plane	Normal		
Young's Modulus, GPa	1273	100	50	973	150
Poisson's Ratio	1273	0.16	-	973	0.3
Thermal Expansion Coeff., 10^{-6}/K	1273	4.30	6.9	973	11
Thermal Expansion Coeff., 10^{-6}/K	1573	4.20	5.0	-	-
Thermal Conductivity, W/m-K	1273	-	1.59	-	-
Thermal Conductivity, W/m-K	1573	-	2.34	-	-
Specific Heat, J/kg-K	1273	1.47	1.47	-	-
Specific Heat, J/kg-K	1573	1.62	1.62	-	-

Steady state maximum thermal stress is a little under 200 MPa, and occurs in the liner near the swirler tie-in, but the peak stress is dependent on the spring constant, decreasing with a softer spring. In the outer liner the thermal stress peaks at 170 MPa and 146 MPa in the inner liner at material temperature of 1325 K. A shift in the combustion state from engine idle to design point and back to idle is used for the transient evaluation, with the heat boundary conditions varied in stages. The results of the heat transfer and thermal stresses are collated to focus on the centre point on the region where the combustion temperature reaches a maximum and at the swirler tie-in. Peak circumferential stress occurs 9 seconds from the idle point during the shift to the design point, and is calculated at 108 MPa in the outside liner. Temperature differential between the liner's exterior and interior surfaces is responsible for the rapid build up in the stress level, although the maximum steady state stress is not exceeded.

To evaluate the applicability of the design a prototype liner made of the ceramic matrix composite material is subjected to a combustion test. Changes in the basic characteristics of the liner after the combustion test are observed for external appearance, dimensions and weight. The shape and dimensions of the part preclude the use of non-destructive techniques such as scanning by X-ray. A check of material strength from cut samples is also ruled out since further tests for damage assessment are needed. The combustion tests are carried out using 0.5 kg/s fuel flow, inlet atmospheric pressure and gas temperature of 798 K, air-fuel ratio of 30 and exhaust gas temperature of 1873 K. The liner's internal surface temperature measured 1473 K. Combustion is repeated at 10 minute intervals, and the combustion and pause cycles are repeated 18 times, giving the high temperature retention time of 3 hours.

Inspection of the external appearance do not indicate abnormalities in the form of cracking or delamination. On the combustion side the surface formation of an oxide layer is imparted with a bluish tinge. The liner's diameter at the inlet and outlet ends, thickness and overall length are measured at eight points along the circumference and compared with the corresponding values before the test. Figure 15.15 shows the comparison. Some degree of change is noted in practically all measurements, and is attributed to undulations caused by the reinforcement fibre flux appearing on the surface. This causes variations in the circumferential direction. The undulations affect measuring precision by ±1 mm, and lack of coincidence in the measurement locations before and after the test is considered to be double of this accuracy level to ± 2 mm. If the precision figures are included, the variations fall within the measurement tolerance range. Hence, the deformation experienced by the liner during the combustion test is practically negligible. The weight of the liner experienced a loss of 10 g, but here again most of the loss is attributable to the thermo-paint coating. From this short-term test using methane gas as the fuel to gauge the performance, the ceramic matrix composite material is determined to be satisfactory.

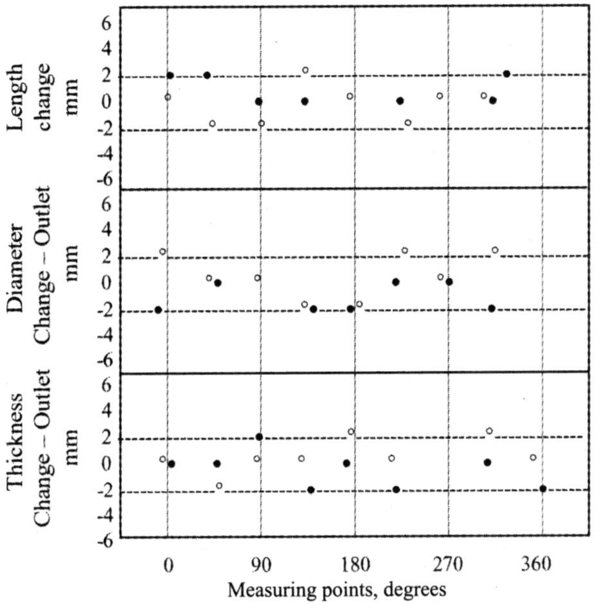

Fig. 15.15 Dimensional Changes from Combustion Test (Nishio, 1998)

15.5 CONTROL OF POLLUTANTS

All combustion processes generate exhaust gases composed of carbon monoxide (CO), carbon dioxide (CO_2), water vapour (H_2O), oxides of nitrogen (NO_x), oxygen, nitrogen, unburned hydrocarbons (UHC) and particulate matter in the form of carbon. Because of their impact on people's health and the environment, government agencies have passed aggressive rules to control emission of pollutants from turbines. Carbon monoxide is poisonous even in small quantities, unburned hydrocarbons are objectionable because of toxicity, NO_x produces acid rain, and smoke and soot lead to visibility concerns. Of considerable interest is the formation of ozone in the troposphere above the earth by the pollutants. Recent research indicates a number of respiratory problems are a direct consequence of excessive pollutants in the air.

Carbon monoxide is formed when combustion of fuel occurs in the absence of sufficient oxygen. Combustion of a lean fuel mix may cause dissociation of CO_2 to form CO. Inadequate residence time in the primary zone, improper mixing of air and fuel and chilling of the products of combustion by cooling air near the liner walls are also responsible for incomplete combustion of the fuel and high concentrations of CO. Unburned fuel particles sometimes leave the combustor for the same reasons, when the parent fuel is degraded into other components of different molecular weight. Soot in the form of a fine powder is produced when re-circulating burned gases move back towards the flame close to the spray. Local regions of the fuel vapour then are surrounded by oxygen deficient gases at elevated temperature, and emerge from the combustor as smoke (*ref*: Lefebvre,

1999). Soot and smoke formation becomes more pronounced at higher pressures. Chemical reaction rate tends to rise with increase pressure, causing combustion to initiate earlier and a greater portion of the fuel to burn in the region near the spray. Fuel properties such as viscosity and volatility can also affect smoke and soot production because drop size, penetration and evaporation are affected. A decrease in the hydrogen content of the fuel is known to raise smoking tendencies.

Production of NO_x occurs mostly in flame zones where the temperature exceeds 1850 K. Oxygen molecules in the air split up to combine with nitrogen rather than with the fuel to form NO, specially when the charge is fuel lean. Production of NO reduces at an exponential pace as the flame temperature reduces. Liquid fuels tend to generate higher levels of NO_x than gaseous fuels for the same flame temperature, but at higher temperatures this difference is practically absent. Figure 15.16 illustrates the difference based on experimental tests (*ref*: Snyder et al, 1994). Air temperature at the inlet to the combustor plays a role, since it affects the flame temperature. Larger residence time within the combustion chamber tends to increase NO_x emissions, except when the rate of formation is low due to considerably lean mixtures. Figure 15.17 provides data from results determined by Anderson (1975) for premixed air-fuel combustion.

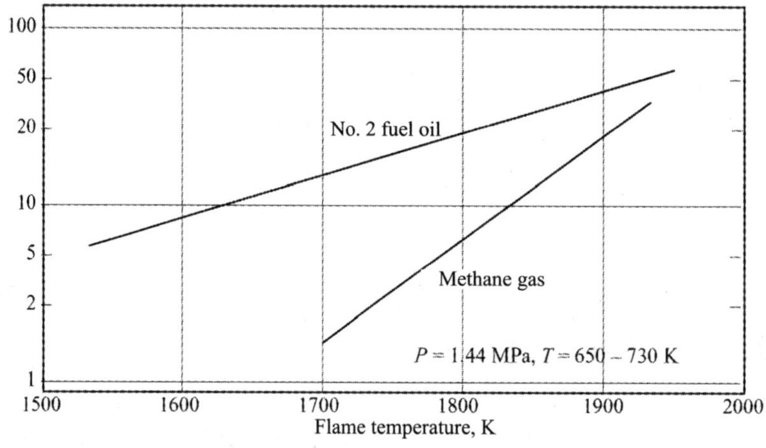

Fig. 15.16 NO_x Variation with Flame Temperature (Snyder, 1994)

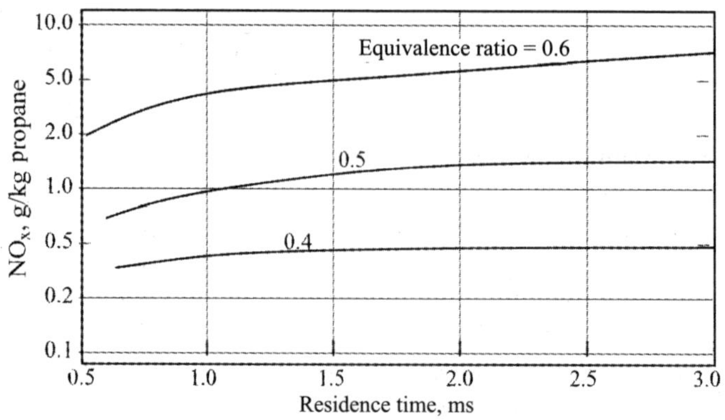

Fig. 15.17 NO_x Variation with Residence Time (Anderson, 1975)

Gas pressure (*P*) increase tends to raise the formation of NO_x in conventional combustor designs, being proportional to P^n, where n varies between 0.5 and 0.8 (*ref*: Maughan et al., 1994). The relation is derived from test measurements of turbulent premixed methane-air flames at pressures from 1 to 10 atmospheres and equivalence ratios from 0.5 to 0.9. Oxides of nitrogen are also affected by the size of fuel droplet, depending also on equivalence ratio (*ref*: Rink et al., 1989). Emission of NO increases with the size of the droplets at low equivalence ratio, mostly because local burn regions of high temperature develop around the larger drops.

Reduction in the levels of CO and UHC can be achieved through more complete combustion of the fuel, which also has the advantage of raising overall efficiency of the unit. Re-circulation of the airflow to obtain better mixing of fuel and air with an equivalence ratio of 0.8 in the primary zone helps towards this objective. Restricting the amount of air for cooling of the liner surface in the primary zone also can be effective in reducing CO and UHC. Improved fuel atomization and increasing residence time in the primary zone are also factors for enhancing combustion efficiency. Soot and its accompanying smoke can be controlled by inserting more air in the primary zone to obtain better mixing. Exhaust smoke will then diminish by preventing formation of fuel-rich regions in the flame.

One method to diminish temperature of the flame and consequent NO_x formation in stationary engines is by injecting water to extract combustion heat. Water or steam may be directed into the flame in the form of a spray from nozzles located at the upstream end of the combustor or through holes combined with the fuel nozzle. The water can also be introduced upstream of the combustion liner into the air stream prior to entry through the swirlers. If steam is to be used for controlling NO_x, it may be introduced at the compressor discharge, in the upstream region of the airflow or into the combustion zone. The effectiveness of this procedure has been demonstrated for gas and liquid fuels on the General Electric MS7001F gas turbine (*ref*: Claeys et al., 1993). Water injection provides better results when NO_x emission rates are at a peak due to high operating condition pressures and temperatures. A larger quantity of steam is generally required than water to achieve the same level of reduction in NO_x emission.

But this procedure has disadvantages in the form of higher fuel consumption of about 2 to 3 per cent, and also adds to the initial cost of the system by \$10 –15 per kilowatt (*ref*: White et al., 1982). CO and UHC emissions may also be expected to rise as a consequence of reduced combustor performance. The purity of the water introduced into the combustor must also be taken into account to inhibit corrosion of the turbine blades. Oscillations in combustion chamber pressure have also been known to increase with the injection of water.

NO_x molecules can also be chemically reduced to nitrogen and water vapour by introducing ammonia on a catalytic bed, and this is the underlying principle behind the selective catalytic reduction (SCR) method. Either alone or in combination with other techniques, SCR is a popular choice for NO_x reduction. In a combined cycle unit the ammonia injection system and catalyst bed may reside between heat exchanger circuits at a point where temperature is within the window from 560 to 670 K for maximum efficiency. The method needs a control system to deliver the proper quantity of ammonia.

The SCR procedure can introduce some potentially complicating factors. If the NO_x – NH_3 reaction does not proceed to total completion, a small amount of NH_3 slips through the catalyst untouched, eventually escaping to the atmosphere as a hazardous air pollutant. When natural gas is the sole fuel, ammonium sulphate [$(NH_4)_2SO_4$] and bisulphate [$(NH_4)HSO_4$] formation is not of concern, but problems can arise when dual fuel units are switched from gas to fuel oil due to the presence of sulphur. Reaction of ammonia with sulphur tri-oxide (SO_3) in the flue gas produces the sulphate and bisulphate compounds, which can foul and corrode turbine blades and other downstream components in heat exchangers. Problems of storage of ammonia in concentrations above certain levels at the site are also experienced due to toxicity issues. In spite of the demerits, the method finds favour among many turbine manufacturers.

Since flame temperature plays an overwhelming role in the production of pollutants, most emissions control mechanisms focus on constraining it within set limits. As indicated in figure 15.18, CO levels are low when combustion chamber temperatures exceed 1650 K, while NO_x production rate becomes objectionable beyond 1900 K. Hence, the range of 1650 to 1900 K must be achieved for all power settings of the engine. Thus, a larger amount of air is required at the entrance to the combustor liner at full power to reduce the flame temperature. As engine load decreases, more of this air must be directed towards the downstream end to dilute the burned gases and maintain the temperature above the low limit. Variation in the distribution of flow can be accomplished by employing variable area swirlers to change flow rate into the combustion zone (*ref*: Micklow et al., 1993) and variable air openings into the dilution zone (*ref*: Sasaki et al., 1991). However, the associated mechanism tends to be complex due to the need for feedback.

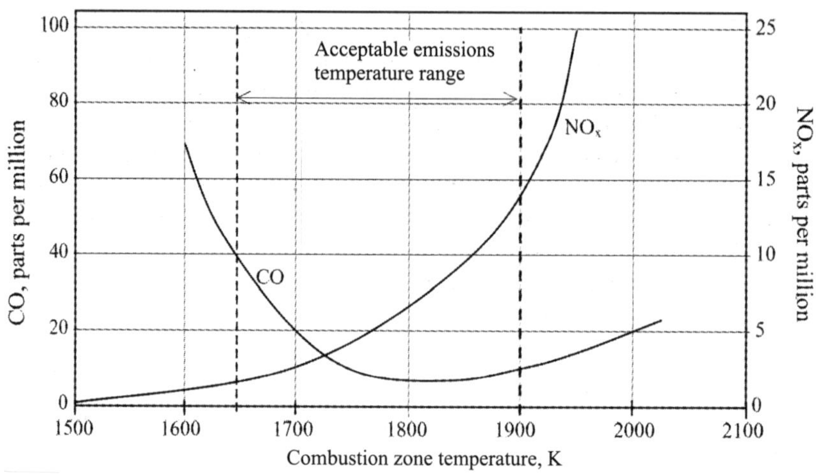

Fig. 15.18 Combustion Zone Temperature and CO, NO_x (Lefebvre, 1999)

Staged combustion is another technique towards the objective of holding combustion temperatures within the narrow limits. Instead of altering airflow distribution, fuel flow is switched from one zone to another (*ref*: Bahr, 1987). Fuel is provided in stages to selected combinations of injectors at various engine operation points. For example, in an aircraft engine combustor, every alternate injector may not be active at light off and at idle condition, but the full complement participates during cruise and take-off. The equivalence ratio is then increased to raise combustion temperatures in local regions when power output is low. Lean blow out limits are also enhanced in addition to controlling CO and UHC pollutants by this modulating technique. But at the fringes of individual combustion regions incomplete combustion has been noted to occur, reducing combustion efficiency and increasing pollutant formation. Also, the staged combustion process yields uneven temperature distribution of the gases around the circumference, tending to reduce turbine efficiency. Some manufacturers have opted to sacrifice some of the objectives, mostly concentrating on primary zones to achieve the required temperature rise and CO and UHC emissions at low power operation. At higher power output, the primary zone acts as a pilot to anchor the main combustion zone receiving the properly mixed air and fuel mixture. The equivalence ratio in the primary zone at low power is kept at 0.8, and 0.6 in both regions at high power to reduce NO_x and smoke.

Staging may be accomplished by placing the injectors radially or axially. Radial staging achieves the performance targets in a larger diameter combustor without affecting the overall length of the engine, a consideration in weight and dynamics aspects of the rotating system. But the arrangement of fuel tubing and injectors becomes complex, lowers the lean blow out limit due to the asymmetry in the airflow, shifts the radial temperature profile and requires more cooling of the outer liner. Performance characteristics at intermediate power setting are also found to be below optimum requirements. Figure 15.19 illustrates the radial staging concept used in a General Electric Company's CFM56-5B aircraft engine.

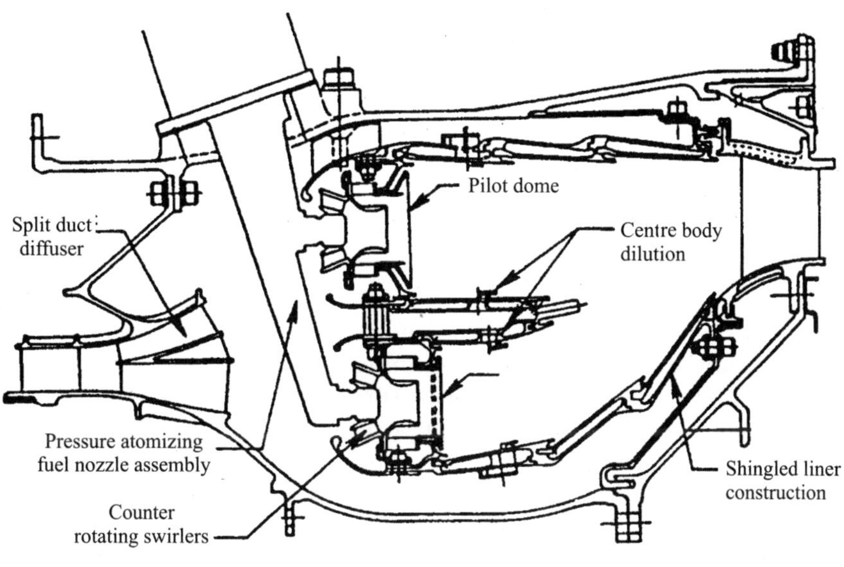

Fig. 15.19 Radial Staged General Electric Combustion System (Bahr, 1987)

Axial staging of the fuel system calls for injection of some of the fuel in the primary combustion region, and the remainder, usually premixed, in the main combustion zone downstream. The primary injectors operate at low power settings, and aid in rapid combustion in the main zone at higher loads and speed. Axial staging offers more reliable and faster ignition, and combustion is more complete even at low equivalence ratios. A uniform radial temperature profile at combustor exit is available using dilution holes (*ref*: Segalman et al., 1993). But this in-line arrangement makes the liner and the engine longer. Figure 15.20 provides details of an axial staged combustor design from Pratt and Whitney Company for V2500-AS engine (*ref*: Koff, 1993).

Alternative methods to control the NO_x are coming on the scene. One interesting process calls for injection of ozone (O_3) directly into the gas stream (*ref*: Buecker, 2002). Because of its powerful

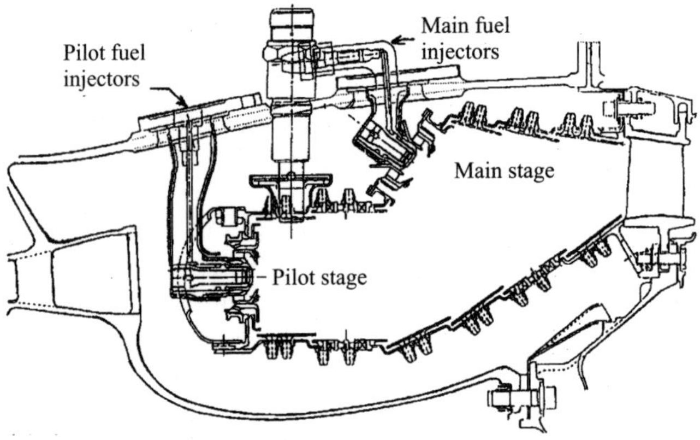

Fig. 15.20 Axial Staged Pratt and Whitney Combustion System (Koff, 1993)

oxidizing characteristics, ozone converts NO and NO_2 into readily soluble nitrogen penta-oxide (N_2O_5). This compound reacts with water to form dilute nitric acid (HNO_3), which subsequently reacts with alkaline scrubbers such as limestone to produce $Ca(NO_3)_2$. A distinct advantage of this procedure is the absence of a catalyst, thus eliminating substantial initial and periodic replacement cost. The process can also serve as a polishing device on units with other NO_x control devices such as low NO_x combustors.

15.6 NO_X FORMATION

A combustion model is required to describe the interaction between turbulence and chemistry to understand the formation of nitrogen oxides. Beginning with fast chemistry without the influence of turbulence, eddy currents may be modeled with simplified kinetics and then followed by the complex joint probability density function approach. A model of the flamelet for combustion simulation is advantageous because it offers a physically clear basis while offering easy implementation with a multi-dimensional computational fluid dynamics code. In the flamelet model the only parameter responsible for turbulence-chemistry interaction is the scalar dissipation, N, and the model equations can be solved separately for use in the CFD code for estimation of flame properties. Chemical calculations are detached from the hydrodynamics, and limits are not placed on the kinetic scheme.

Based on Kuznetsov's theory (1982), combustion in diffusion flames occurs in thin layers in the vicinity of nearly stoichiometric zones, where the gas temperature and the concentration of radicals are mainly high. For non-premixed flames the system of reacting species transport equation with mixture fraction Z as an independent variable is given by the expression:

$$N_s = \frac{d^2C_i}{dZ^2} + R_i = 0 \qquad ...(15.1)$$

where $N_s = D(\nabla Z)^2$ is the scalar dissipation on the stoichiometric surface $Z = Z_s$, C_i is the mass fraction of the ith component, R_i is the corresponding chemical source term and D is the molecular diffusivity. Boundary conditions may be posed at $Z = 0$ (pure oxidizer) and $Z = 1$ (pure fuel), and averaging is needed for evaluation of mean values in the turbulent case. The equation can be solved for various values of N_s as a parameter to determine the dependence of NO_x formation rate on mixture fraction and scalar dissipation, $R(Z, N_s)$. The data is created for different combinations of pressure, composition and temperature in the range of operation.

Flow field calculations using incompressible variable density are carried out using Favre averaged Navier-Stokes equations for several characteristics (mixture fraction, kinetic energy). The resulting data is then used for modeling of formation of nitrogen oxides in the post-processing mode. Turbulent stresses are approximated using appropriate viscosity values. A fast (equilibrium) chemistry assumption together with a bi-modal probability density function for the mixture fraction is used to determine heat release. Mean NO_x formation is then estimated at every computational cell using values already determined:

$$\text{Mean } \rho R_{NO_x} = \int \rho R_{NO_x}(Z, N) p(Z, N) dZ dN \qquad \text{...(15.2)}$$

where ρ is density, $p(Z, N)$ is joint probability density function of the mixture fraction and scalar dissipation. NO_x formation rate R depends on two opposing effects: temperature reduction and radical concentration increase with growth of scalar dissipation. Hence, fluctuations in scalar dissipation are not needed. Uniform velocity profiles are assumed in all inflow holes and slots, and scoop holes are treated as plain holes with a discharge coefficient of 1.0.

Experiments have been performed on an installation fitted with natural gas fueled can combustor of the type shown in figure 15.21 (*ref*: Volkov et al., 2000), and modifications to lower NO_x emissions tested. Air distribution along the liner is modified to generally make the head end leaner, and the data is used to validate the numerical procedure. Calculations are performed for a standard and three modified operating regime parameters (1.06 MPa pressure and 616 K inlet air temperature). Table 15.5 provides comparison of the calculated and measured values of NO_x formation. Absolute values of calculated NO_x concentrations are lower than test values, but the overall trend of change in the level is proper, so the relative reduction in NO_x is reasonable. Inaccuracies in the prediction arise from limitations of the detailed chemistry and flamelet models. Available grid cells are not sufficient for a detailed description of the flow near the walls of the liner with air holes and cooling films, so the turbulent flow's complexity inside the combustor needs refinement.

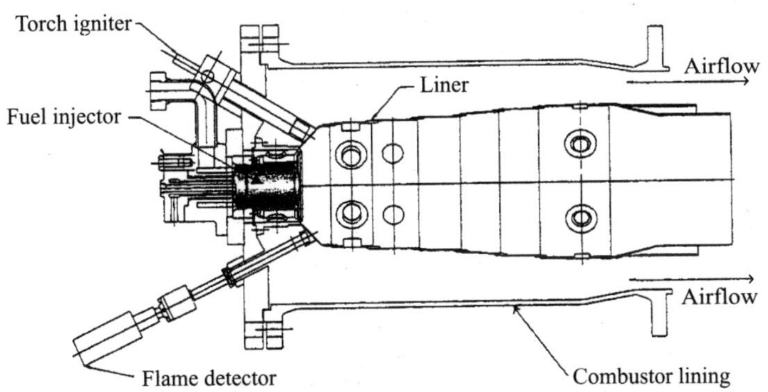

Fig. 15.21 Test Combustor Layout (Volkov, 2000)

Table 15.5 Comparison of Predicted and Measured NO_x Levels

Configuration	*Air at Head %*	*Measured Reduction, %*	*Calculated Reduction, %*	*Measured ppmvd*	*Calculated ppmvd*
Standard	29.5	0	0	129	95.6
Modification # 1	53.7	22.5	31.0	100	66
Modification # 2	59.0	24.8	35.1	97	62
Modification # 3	62.0	36.4	39.0	82	58.3

15.7 EFFECTS OF SWIRL

Lean pre-mixed combustion offers the advantage of lower NO_x emissions by reducing peak flame temperature at the fuel-air interface in traditional diffusion flame type of combustors. Fuel and air are mixed upstream of an arrangement to provide swirl to the flow for stabilizing the flame. Stabilization is achieved by transporting hot and chemically active combustion species from the downstream region to the root of the flame, creating a thermal non-uniformity between the re-circulating hot gases and the cooler gases flowing from upstream of the swirler and the flame zone. The extent of non-uniformity is difficult to characterize in premixed flames, but can be expected to depend on combustor configuration, the degree and distribution of swirl and operational parameters. The non-uniformity influences the efficiency of the combustor and the NO_x emissions level.

Significantly different combustion characteristics can be obtained by altering the radial distribution of the swirl in a burner (*ref*: Qi et al., 1997). In a research project conducted at the University of Maryland, swirl flow direction in the outer annulus of a double concentric burner was observed while maintaining the inner swirl direction fixed. Each annulus may be given a desired degree of swirl, so the flame can have either co-swirl in both annuli or co- and counter-swirl between the inner and outer tubes. Pre-mixed air fuel mixture in the desired ratio may be admitted into any annulus or central nozzle of the burner (*ref*: Gupta et al., 2000). A 30° swirler for annulus # 1 and + 50° and –50° swirlers for annulus # 2 are used to investigate flames produced from the change of swirl direction in the annuli. The arrangement with both swirlers having positive angles is referred to as producing co-swirl flame, while angles of opposite directions have a counter-swirl flame. Figure 15.22 provides a diagram of the burner outlet where the flame stabilization zone occurs, and figure 15.23 shows details of the swirling flow field and the regions of a swirl stabilized pre-mixed flame (*ref*: Marshall, 1996).

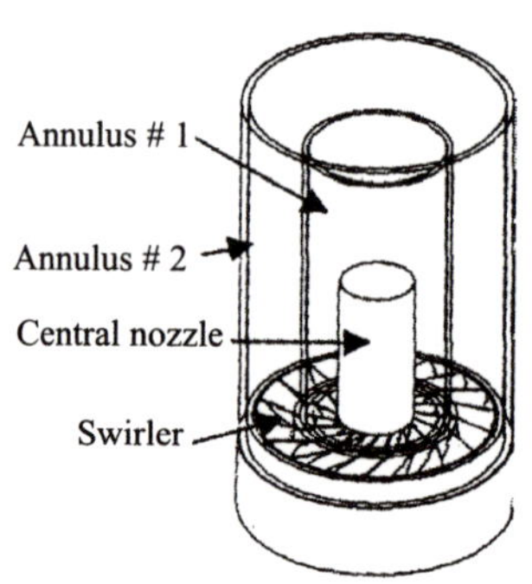

Fig. 15.22 Double Concentric Burner Outlet Region (Gupta, 2000)

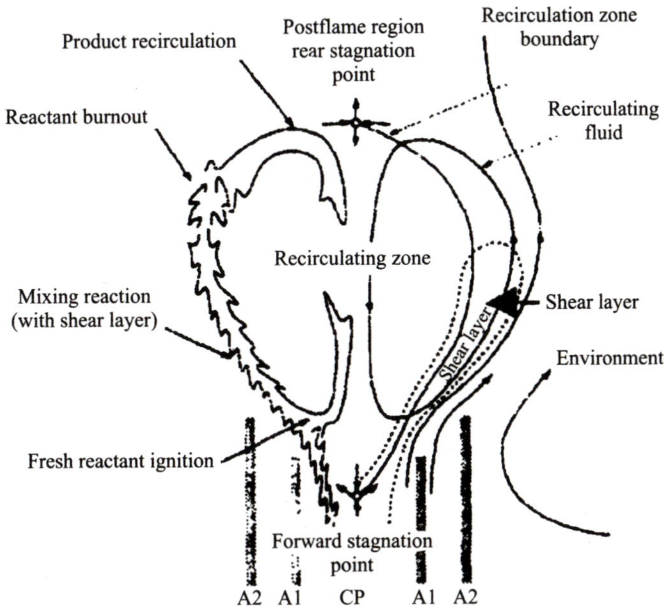

Fig. 15.23 Swirling Flow Field of Pre-Mixed Flame (Marshall, 1996)

High frequency temperature measurements are taken with a micro-thermocouple probe, with a wire diameter small enough not to cause interference on the flame's structure while providing rigidity for the probe. At every location in the flame the signal is amplified and digitized for a sampling time of 30 seconds to allow averaging over low frequency temperature measurements and to assure a good statistical representation of the thermal field. Large variations in the temperature are present at any location in the flame. The sampling frequency used is 10 kHz, which is high enough to resolve small thermal time scales in the flame. Direct flame photographs taken during the tests provide data about the overall features of the flame and its stability. Negative images of the photographs determine the size of the flame in proportion to the burner.

Raw temperature data has to be compensated for radiation losses and thermal inertia effects of the thermocouple. Radiation losses can be significant, particularly at high temperatures. Similarly, the level of fluctuations obtained without compensating the thermocouple output can be considerable. Fluctuating temperatures are lower by as much as 250°C at some locations in the flame without compensation. Qi et al., (1997) provide a method for making corrections. The compensated mean temperature maps, shown in figure 15.24, display substantial differences between the left and right sides of the counter-swirling flame, with a flat hot shear layer present at the left side where temperatures exceed 1700 K. The shear layer on the right is steeper but comparatively cooler at about 1500 K. The non-symmetric behaviour of the counter-swirling flame is observed by comparing it with the co-swirling map. In the post-flame region large differences exist in the mean temperatures on both sides. The co-swirling map tends to be wider with a long area of reduced temperature fluctuations. A thin but intense reaction zone in the counter-swirl case causes non-symmetrical fluctuations in a smaller area of the flame. But overall differences in mean temperatures between the two cases are not large.

The temperature maps make it possible to locate the combustion area, the re-circulation zone and the post-flame region. Outside the shear layers the flame tends to show higher fluctuating

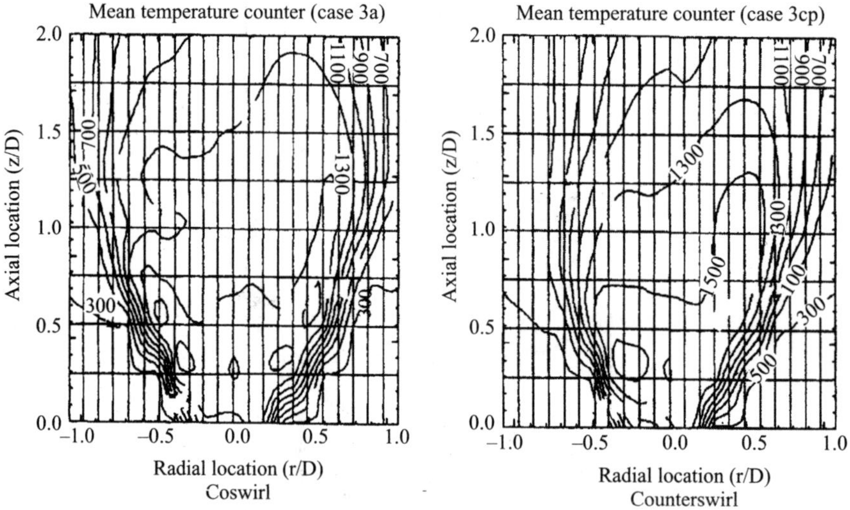

Fig. 15.24 Compensated Mean Temperature Maps (Gupta, 2001)

temperatures than in the recirculation zone. The regions of low fluctuations are caused by continuous combustion, and represent pockets of burned gases within the recirculation zone. High temperature fluctuations outside the combustion zone are caused from mixing between the hot reaction products and the surrounding air, and suggest a stream of ambient air entrained towards the flame caused by the recirculation zone of the swirling flow field. Thus, the regions of high temperature fluctuations are outside the hot regions. Examination of the effects of swirl on the flame shape, mean and fluctuating temperatures can be useful in evaluating eddies present in the flame, which subsequently affect formation of NO_x.

15.8 DRY LOW NO_X COMBUSTION SYSTEM

Cogeneration systems using a gas turbine as the prime mover offer high total thermal efficiency. They are subject to strict NO_x regulations since air pollution in major population centres shows no sign of improvement. Water or steam injection or selective catalytic reduction is widely used in gas turbines to reduce the emissions, but the methods tend to increase operating cost. Lean pre-mixed combustion offers a convenient method to reduce NO_x emissions with low initial and running cost.

Tokyo Gas has focused on the development of dry low NO_x combustors for a cogeneration system in the 1 to 4 MW output range (*ref*: Sato et al., 1996). Engine output is controlled by varying the fuel gas flow, thus eliminating the need for complex variable geometries for air flow control. The double swirler staged combustor uses tertiary premix nozzles located around the liner. Multi-staged combustion offers the benefit of sustaining stable combustion with flame temperature in a range under 1650 K.

Figure 15.25 shows the flow of air and gas in the double swirler combustor concept. The primary and secondary pre-mixing nozzles are placed co-axially with the radial swirlers. A pilot nozzle installed at the centre of the premixing nozzles generates a diffusion flame rather than a perfect premixed flame, hence it stabilizes the flames at the other nozzles adequately. The swirlers generate swirling flows in the same direction. Four separate fuel lines lead to the pilot, primary, secondary and tertiary nozzles. Figure 15.26 provides the fuel supply schedule to the nozzles. The schedule is designed to provide constant air excess ratios for the pilot and primary nozzles over the whole engine operating load regime to generate stable combustion with low NO_x emission. In the 0 to 30 per cent load mode the schedule eliminates fuel supply to the tertiary nozzle to generate a lean fuel-air mixture with an excess air ratio of about 2.0 in the secondary nozzle, thereby igniting and oxidizing while directly contacting the stable combustion products of the primary and pilot nozzles. The flame temperature is low because of the excess air, producing practically no NO_x. In the high engine load mode up to 100 per cent fuel is supplied to the tertiary nozzle, fuel flow to the other nozzles remaining constant at maximum levels. Excess air ratio in the tertiary nozzle is also high to reduce NO_x formation substantially.

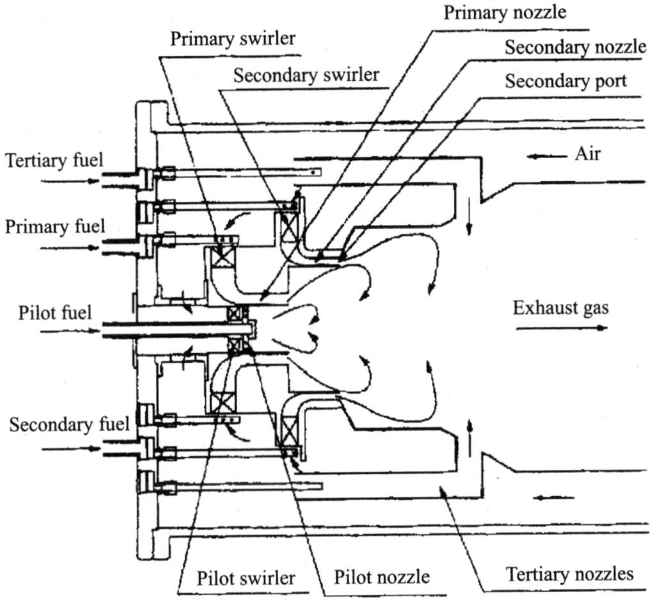

Fig. 15.25 Double Swirler Staged Combustor Arrangement (Sato, 1996)

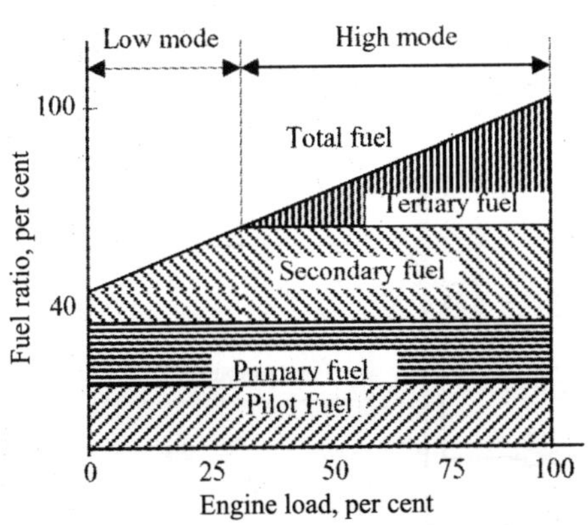

Fig. 15.26 Fuel Gas Supply Schedule (Sato, 1996)

Operating conditions and target performance of the combustor are shown in table 15.6. Target NO_x level is 9 ppm at engine load between 50 and 100 per cent, the normal operating range of gas turbines for cogeneration. At less than 50 per cent load the target is 25 ppm. These target levels convert to 3.0 and 8.3 ppm under atmospheric pressure, assuming the general relationship that NO_x emission is proportional to the square root of operating pressure.

Table 15.6 Operating Conditions and Target Performance

Inlet air pressure	0.91 MPa
Inlet air temperature	640 K
Full load outlet gas temperature	1473 K
Full load excess air ratio	2.7
Fuel, LHV	Natural gas, 41.6 MJ/Nm^3
NO_x target: 0 – 50% load	< 25 ppm (15% O_2)
50 – 100% load	< 9 ppm (15% O_2)
Combustion efficiency target: 0 – 50% load	> 95%, 50 – 100% load > 99%
Pressure drop	3%

A schematic of the test facility is shown in figure 15.27. Air is preheated and rectified before introduction into the test combustor. The natural gas used in the test is 89 per cent methane, with ethane, propane and other hydrocarbons forming the rest. Combustion exhaust gas is sampled with a five point water-cooled probe positioned downstream of the exhaust, then introduced into the gas analyser. O_2 is analysed by a magnetic analyser, CO and CO_2 with a non-dispersive infrared analyser, NO_x with a chemi-luminescence analyser and unburned hydrocarbons with a flame ionization detector. Temperature distribution at the combustor outlet is measured in the same plane as the gas sampling probe at 24 positions to obtain an acceptable pattern factor. Total flow rate of the process air is calculated from exhaust gas composition and measured fuel flow, and flow rate to each nozzle assumes a split proportional to the open area of the respective air nozzle.

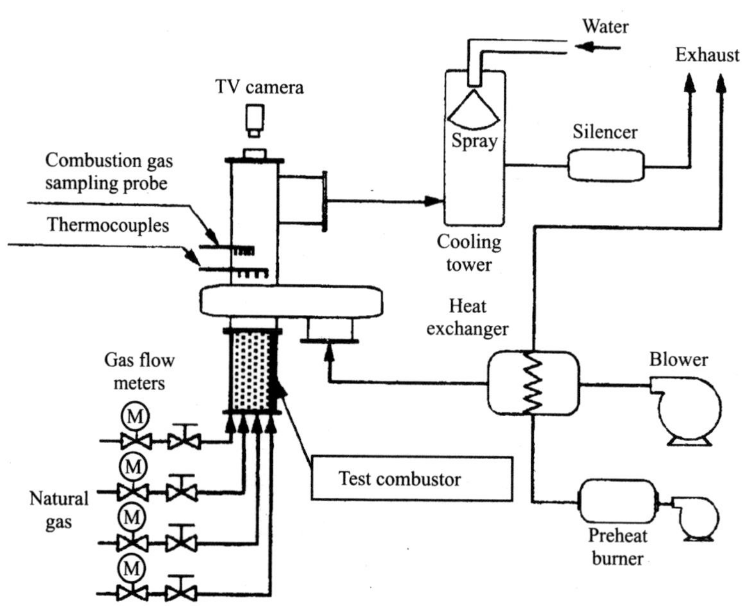

Fig. 15.27 Schematic Diagram of Test Facility (Sato, Mori, and Nakamura, 1996)

The charts of figure 15.28 provide performance characteristics of the combustion system, showing the effects of excess air ratio and corresponding engine load on NO_x, CO, unburned hydrocarbons and combustion efficiency. Engine loads of 100, 50 and 0 per cent are associated with 2.7, 3.9 and 6.8 of excess air ratio λ_{tot}. NO_x level is considerably influenced by the high/low engine load. In the low mode, when λ_{tot} is between 5.0 and 7.0, the NO_x holds steady at 5 ppm. The secondary flame produces virtually no NO_x in this range. But a sharp increase is observed when λ_{tot} decreases from 5.0 to 4.0, with NO_x level reaching 8 ppm. In the high mode the additional tertiary fuel with its sufficiently high excess air produces lesser NO_x, dipping under 2 ppm when λ_{tot} is 3.3. Thus, a higher excess air ratio helps to curtail thermal NO_x production.

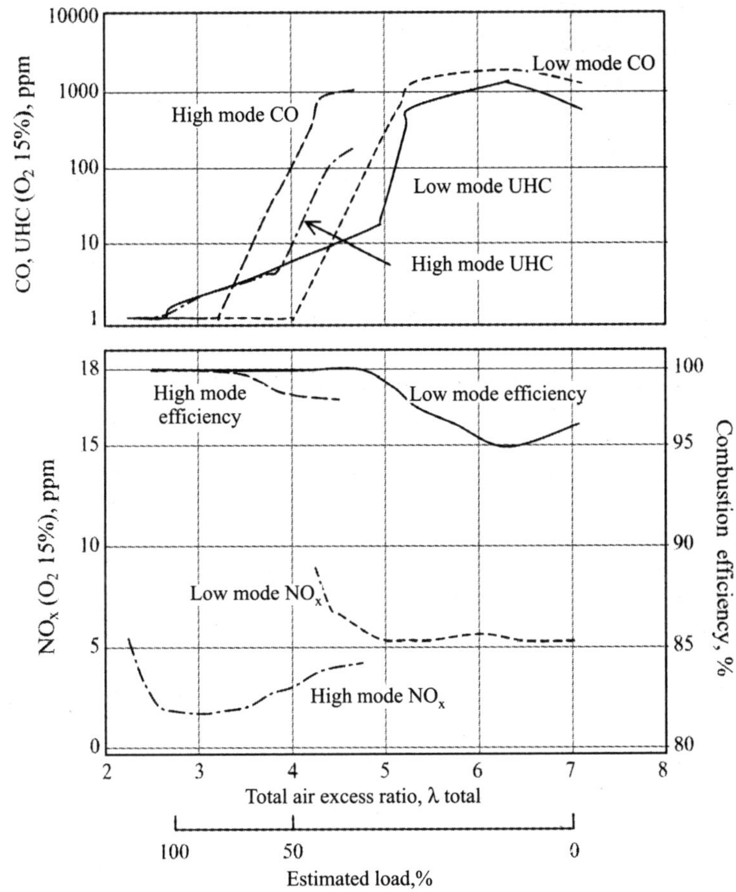

Fig. 15.28 Combustor Performance Characteristics (Sato, 1996)

Combustion efficiency reaches a low value of 95 per cent during the low mode engine operation for $\lambda_{tot} = 6.2$, with a corresponding increase in UHC formation. The efficiency curve recovers as excess air ratio reaches a maximum. Combustion efficiency and CO and UHC emissions at maximum λ_{tot} are affected by the combination of pilot and primary flames. A similar drop in combustion efficiency occurs in the high operating engine mode when λ_{tot} increases from 4.0 to 4.6, where the tertiary flame plays the same role as the secondary flame does in the low mode. Considering that the pilot burner is designed for a stable diffusion flame, the majority of CO and UHC is originated in the primary flame. Unlike the pilot, the primary flame mixes directly with the secondary airflow, so the swirling primary flame has a limited residence time and results in high emissions of CO and UHC. The relatively simplified geometry of the double swirler staged combustor designed to operate at standard atmospheric pressure is shown in figure 15.29.

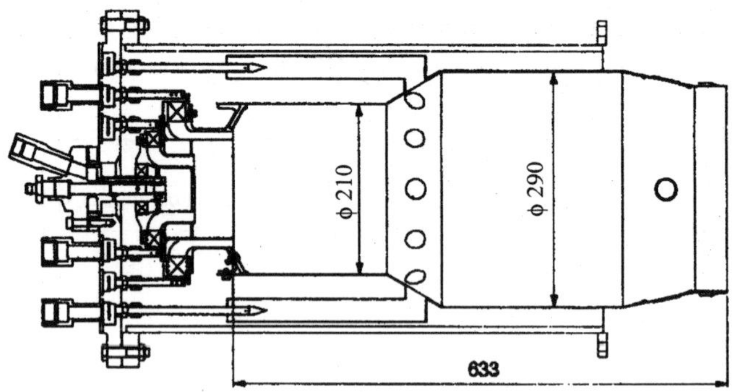

Fig. 15.29 Double Swirler Staged Combustor Design (Sato, 1996)

15.9 CATALYTIC COMBUSTOR FOR UTILITY TURBINE

The selective catalytic reduction method is useful in chemically reducing NO_x into nitrogen and water vapour, however, the costs associated with heat rate deterioration due to diluent injection and the capital and operating costs for the required systems make it financially unattractive for application in combined cycle and cogeneration power plants incorporating gas turbines. Direct catalytic combustion offers good potential for reducing formation of NO_x, CO and UHC in tests carried out at General Electric for model MS9001E gas turbine (*ref*: Dalla Betta et al., 1996; Schlatter et al., 1997). The design calls for partial reaction of fuel-air mixture within the catalytic reactor to generate a gas temperature of 800 – 1000°C at reactor exit. At this temperature in the reactor the catalyst can include precious metals, and the substrate may be cordierite or metal.

The combustion system design (figure 15.30) requires a pre-burner, fuel and air preparation system, catalytic reactor and a combustion liner downstream of the reactor. The pre-burner carries machine load at conditions when temperature levels do not allow satisfactory catalytic combustion, and also preheats to achieve catalytic reactor ignition at high loads. Catalytic staging initiates at turbine inlet temperature of 700°C when the main fuel injector activates. The fuel and air preparation system provides the components and pre-burner products to the reactor bed at a uniform strength, pressure, velocity and temperature.

The catalytic reactor promotes oxidation of hydrocarbons and CO for lean mixtures at adiabatic flame temperature below the threshold for thermal NO_x formation. Combustion initiated by the catalyst is then completed by homogeneous burning in the post catalyst region where high temperatures are obtained. Catalytic reactor technology developed by the manufacturers gives a bed for full fuel and airflow required for maximum power, while avoiding exposure of the catalyst to high temperatures that may damage the supporting substrate. Use of ceramic catalytic materials

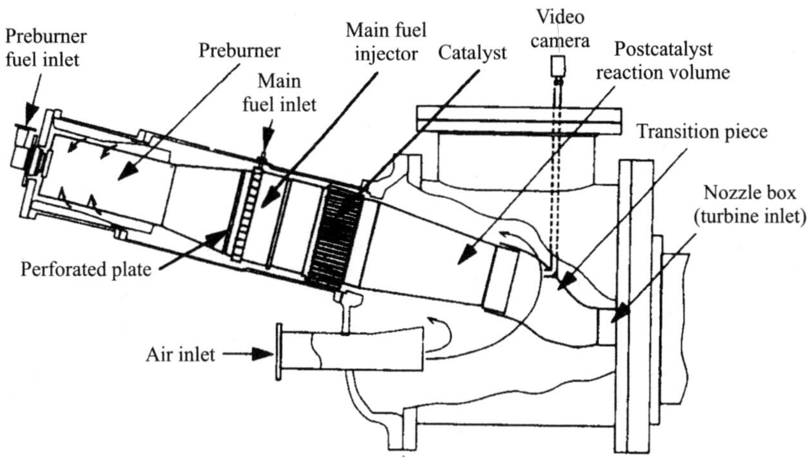

Fig. 15.30 Catalytic Combustion Test Rig (Schlatter, 1996)

maintains the catalyst surface below the adiabatic combustion temperature. Advantage is taken of the palladium oxide in catalysing methane oxidation, while metallic palladium is appreciably less active (*ref*: McCarty, 1994). Palladium has the unique thermodynamic characteristic of oxidizing and reducing. Depending on pressure, the oxide decomposes to the metal between 780°C and 920°C. The reactor consists of three separate catalyst stages, with the stages formed by corrugating and foiling metal foil to constitute a channeled monolithic structure. Active ceramic material is coated on the foil. The stages are supported in a reactor container by large cell honeycomb structures made of Hastelloy X.

Experimental data is obtained over a range of conditions from full speed without load to base load of the engine. Combustor discharge temperature ranges from 543°C at no load to 1195°C at base load. Reactor operation is started by heating the system with the pre-burner, then turning on the main fuel flow to provide a smooth light-off of the reactor with a uniform temperature profile across the face.

Figure 15.31 shows measured pollutant emissions data corresponding to ISO ambient with 15 per cent oxygen concentration for average combustor temperatures at the nozzle box. The peak NO_x value of 55 ppm in the 519°C and 626°C range results from the diffusion flame in the pre-burner when no fuel is delivered to the main burner and the catalyst. As the combustor is taken to higher exit temperatures, fuel is shifted to the main fuel injector. NO_x levels drop to a less than desirable 11 ppm, mostly due to the need for higher temperature in the pre-burner to keep the catalytic reactor fully active. Introduction of steam into the pre-burner zone lowered the NO_x to 3 ppm. This data is consistent with NO_x suppression data for diffusion combustors using natural gas combined with steam injection. At base load the catalytic reactor fuel is about 80 per cent of the total, indicating essentially no NO_x production by the reactor.

CO emission shows a similar trend, peaking to 3200 ppm at 930°C during pre-burner only operation, when catalyst staging is in a transient condition between no load and base load. Combustion temperature rise subsequently transfers to the catalytic combustor. At base load the reaction temperature is noted at 1196°C, when CO emission falls to a minimum value of 10 ppm. Pre-burner exit temperature must be maintained at a high enough level to keep the catalyst fully

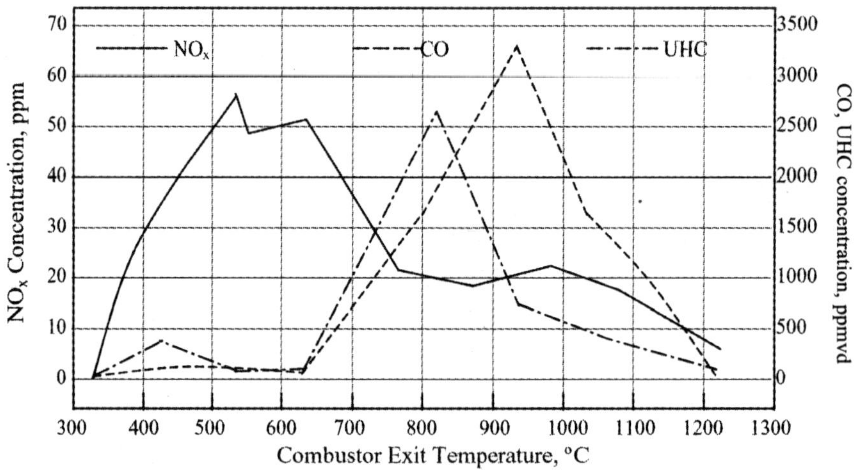

Fig. 15.31 NO$_x$, CO, UHC Emissions (Dalla Betta, 1996)

active. Considerable scatter is noted at the base point, mostly due to sensitivity to the pre-burner exit temperature. UHC emissions show a major peak at 800°C during pre-burner fuel operation, reducing to negligible levels at 1200°C. At the simulated base load operating point with pre-burner exit temperature at 563°C the overall combustion of fuel to equilibrium combustion products is greater than 99.99 per cent.

Dynamic pressure measurements indicate the catalytic combustor system experiences oscillations lower than in conventional combustors. The maximum discrete peak has a magnitude of 0.00173 MPa at a frequency of 252 Hz, and occurs during steam injection. Maximum overall rms noise level of 0.00836 MPa also occurs at the same time. Without steam injection the dynamic pressure measures about 20 per cent less. Combustor exit temperature distribution factor, defined by the ratio of maximum variation from the mean to the overall combustor temperature, is 0.138. Pre-burner exit temperature non-uniformity contributes substantially to this variation. The tests indicate improvement in the structural integrity of reactor, with the diameter experiencing minimal distortion after several hours of operation.

15.10 ACOUSTIC RESONANCE

Combustion of the air and fuel mixture is accompanied by noise directly as a consequence of the process and indirectly due to the flow of burned gases through the turbine and exhaust nozzle. Combustion noise can become detrimental when instabilities arising in the burning process couple with acoustic modes inside the chamber. The natural frequencies of the combustor can be excited by resonant pressure waves in the main gas flow along the axial and radial directions, as also by lateral modes in the tangential direction (*ref*: Paxson et al., 1995; Ohtsuka et al., 1998; Rangwala, 2004). Sustained oscillating phenomena due to a higher level of mixing of the fuel and air prior to combustion lead to engine noise and vibration problems.

Premixed combustion in gas turbines helps produce low levels of NO$_x$ emissions, but practical application of this concept is limited by self-excited combustion oscillations. When operation in a lean, premix combustor is close to the flammability limit slight changes in operating conditions can lead to sudden flame extinction or to excessive CO emissions. In addition to static stability, lean premix combustors must achieve dynamic stability, meaning the combustion must not oscillate. Oscillation must be eliminated in a combustor design because the associated pressure oscillations tend to have life shortening consequences (*ref*: Richards et al., 1997). Figure 15.32 shows cracks experienced in a transition piece due to excessive acoustic oscillations. Operation near the lean limit is especially prone to oscillation problems, where minor variations in fuel/air ratio lead to appreciable variations in combustion reaction rate. When these variations in the reaction rate couple with the acoustic modes, significant pressure oscillations occur, with frequencies ranging from hundreds to a few thousand hertz.

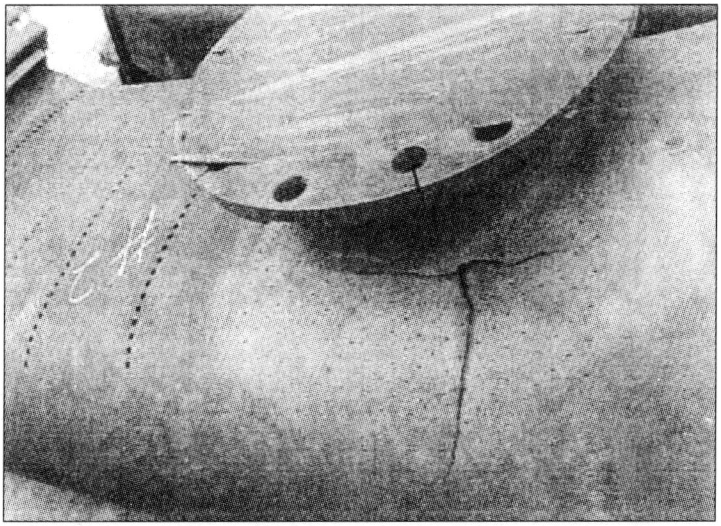

Fig. 15.32 Acoustic Oscillations Damaged Transition Piece (Lieuwen, 2002)

The task of studying and eliminating combustion oscillations in a gas turbine is complicated by the specific acoustic response of a combustor's design. The combustion process interacts with the acoustic field, leading to instabilities. Rapid changes in air and fuel supply and aerodynamic disturbances may lead to the instability because of a sequence of extinction and re-ignition of the flame in parts of the combustor. If the heat release rate does not take place uniformly and periodic spikes occur, acoustic waves of the same frequency may be expected in the combustion zone. Reflection from the liner causes pressure waves to be returned to the combustion zone after a time delay, and the waves are reinforced when the heat release and pressure wave peaks coincide. As defined by Lord Rayleigh's criterion, oscillations set in when changes in heat release are in phase with acoustic pressure disturbances. Conversely, oscillations are dampened when heat release fluctuations are out of phase with pressure fluctuations. This criterion serves as the cornerstone for the development of combustion oscillation analysis. Variation in heat release result from changes in flame structure produced by acoustic pressure disturbances. Time delay between pressure disturbance and heat release variation determines the phase and, consequently, the stability of the system.

Based on these observations, lean premix combustors can be characterized by a simple time lag approach. Figure 15.33 shows for a specific case a schematic of important processes, where a sinusoidal pressure disturbance produces a sinusoidal variation in airflow 180° out of phase with the pressure. Time lag τ is estimated from the distance between the point of fuel injection and the flame front divided by average axial velocity, or:

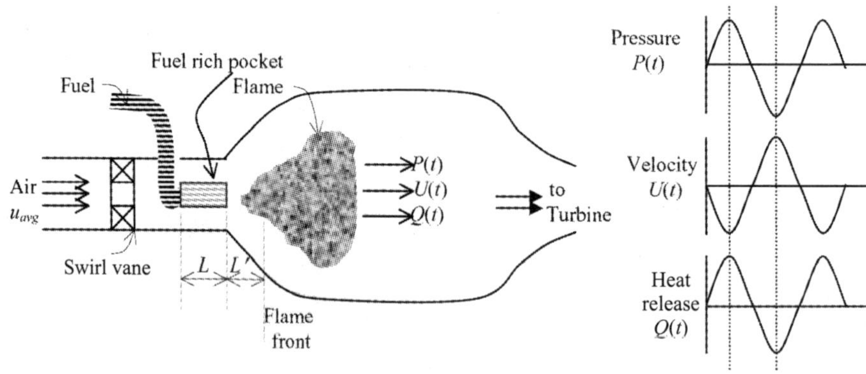

Fig. 15.33 Flow Characteristics During Acoustic Oscillation (Richards, 1997)

$$\tau = (L + L')/U_{avg} \qquad ...(15.3)$$

where L is the distance from fuel injection point to nozzle tip, L' is distance of nozzle tip to flame front and U_{avg} is average velocity of air-fuel mixture in the nozzle.

A positive pressure fluctuation in the combustor produces a momentary decrease in airflow. If the fuel supply is choked, fuel flow rate will not change with pressure variation. Thus, the reduced airflow will receive a proportionally higher amount of fuel, creating a fuel rich pocket. This richer pocket arrives at the flame front with a time lag, indicated by the equation given above. If the additional fuel produces an immediate increase in heat release, oscillations will be most likely when the pressure fluctuation peak is in phase with the increased heat release; that is, when time lag $(t_2 - t_1)$ is an integer multiple of acoustic period. This criterion for oscillations may be stated as: (time lag)/(acoustic period) = 1, 2, 3, ... Since acoustic period is the reciprocal of frequency f, then: (time lag) × (frequency) = 1, 2, 3 ... or, $f(L + L')/U_{avg}$ = 1, 2, 3 ... This is a restatement of Rayleigh's criterion. In practice, heat release and pressure do not necessarily need to be exactly in phase to drive oscillations. Heat release fluctuations leading or lagging pressure by as much as ¼ of the acoustic cycle will also cause some oscillations, although driving is greatest for integer values where pressure and heat release are exactly in phase.

The discussion above is specific to the example where positive pressure produces an immediate decrease on airflow, and assumes that the fuel rich pocket produces an immediate increase in reaction rate when arriving at the flame front. Other mechanisms for variable heat release can complicate the criterion for oscillations such that the expression may have values other than 1, 2, 3, ... Similar criteria can be developed to account for fuel system impedance, or to describe oscillations linked to tangential velocity component in the fuel nozzle swirl vane. Geometry of flame front has also been shown to produce a numeric series.

Radiated sound may have frequencies ranging from 100 to 2000 Hz. Sound pressure frequencies mostly do not depend on engine power or flame temperature, but radiated noise level tends to vary with these factors. In the presence of combustion instability, a rumbling or growling form of noise is audible in the low frequency 50 to 180 Hz range when the engine may be in the sub-idle operating condition. The growl is objectionable because it increases the time to start an engine, while also reducing the stall margin in the compressor. At higher frequencies corresponding to take-off condition (200 to 500 Hz) the generated noise takes a more distinct howling or humming pattern. Unstable operation in the compressor tends to play a role, and may even act to trigger the noise. Increase in air temperature to combustor inlet has been noted to decrease the rate of occurrence and intensity of growling noise, while raising combustor pressure has the opposite effect. Fluctuations in fuel pressure may also induce high frequency noise.

Thermo-acoustic response of a gas turbine engine combustor for two different fuel injectors has been investigated in a study conducted by the US Air Force (*ref*: Arana et al., 2000), with the intention of identifying design features that cause increase in the acoustic pressure. A hybrid air blast injector presently in use with inner and outer flow passages is selected as a baseline design. To lower the smoke production level the investigation focused on using higher swirl flow in the proximity of the spray point, while lean blowout and dynamic stability can be obtained with lower swirl in the zone. Simultaneously achieving the apparently conflicting requirements for high and low swirls near the spray point led to the development of a new design concept with variances. Figure 15.34 provides details of the baseline and new fuel injector designs.

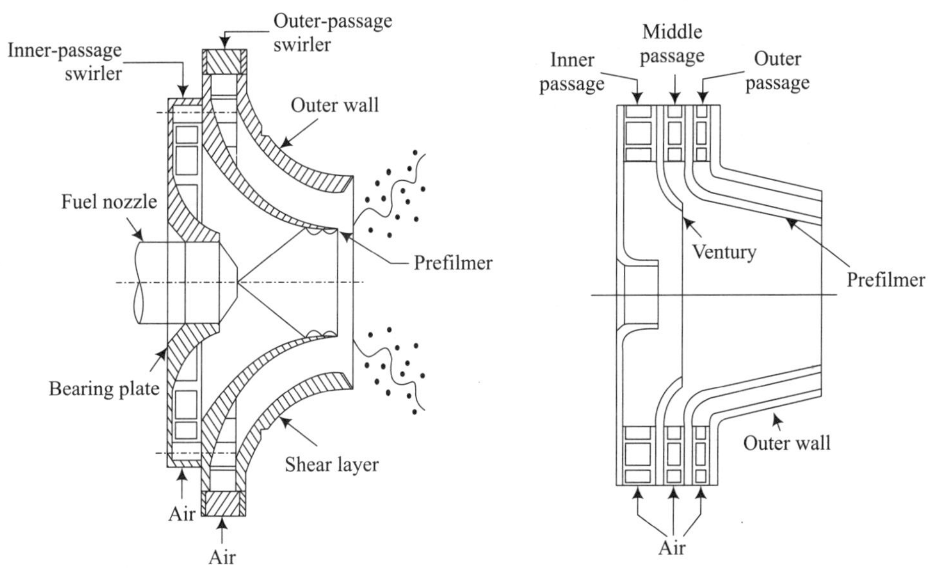

Fig. 15.34 Baseline (*left*) and New (*right*) Fuel Injector Designs (Arana, 2000)

The new injector design differs from the baseline in the configuration of the venturi, the counter-rotating swirlers of the venturi and the middle passage and the ratio of vane and discharge areas. The last parameter is 50 per cent larger than in the baseline design, suggesting that the new swirler exhibits less resistance to dynamic changes in pressure at higher frequencies. Use of velocity on the downstream side may be expected to maintain a higher level of the transfer function.

Co-rotating and counter-rotating swirlers are characterized for different passages. The injectors are initially tested at atmospheric and high-pressure conditions in an ignition rig, then assembled in the combustor of a development engine demonstrator employing 24 injectors around the circumference of the bulkhead. Air is fed to the combustor through a stepped diffuser.

The radial swirlers are the primary conduits of air between the external combustor shrouds and the internal combustion chamber. If coupling and amplification between the chambers is the root cause of the instability, then swirler response to a forcing function needs to be checked by measuring the impedance of the conduit. Impedance defines the total resistance and reactance opposition exerted by the swirlers to the forced, or pulsed, airflow of a given frequency, and is determined by measuring the transfer function of the swirlers. Pressure is measured as the upstream parameter and velocity on the downstream side for a number of frequencies.

Figure 15.35 shows the measured transfer functions and corresponding phase angles for the two designs. The new design swirlers exhibit a higher value of the transfer function in the 400 to 500 Hz frequency range, where the natural frequency of the annular combustor occurs. The phase angle relation between the pressure and velocity oscillations also points to this aspect, and is indicative of a dynamic response as opposed to a static one. The implication is that if the frequency of the acoustic chamber of less than 400 Hz is obtained, the new fuel injector design provides better attenuation and less acoustic response.

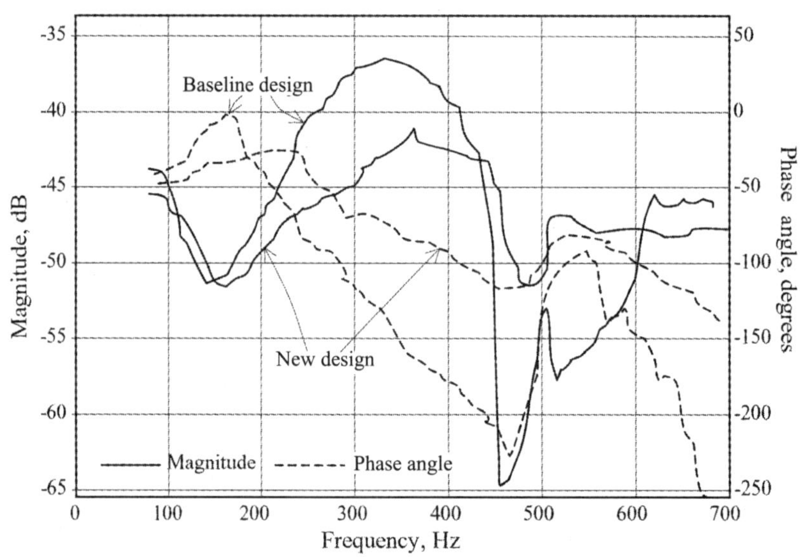

Fig. 15.35 Transfer Function of Upstream Velocity and Downstream Pressure (Arana, 2000)

15.11 ACTIVE COMBUSTION INSTABILITY CONTROL

Combustion instabilities are difficult to predict analytically in the design phase for all operating conditions due to the complex geometry of the system. Noticeable humming caused by self-excited vibrations can occur during shop tests in the premixed mode operation of the turbine. Pressure oscillations may exceed unacceptable levels, and a quick and flexible response in the form of control mechanism may be necessary while the combustor design is optimized.

A similar situation has been experienced on a Siemens model V84.3 gas turbine equipped with a new ring (or annular) combustor design (*ref*: Seume et al., 1997). Dynamic pressure and heat release rate are measured at different locations in the combustor, and dominant signals are recognized at 217 Hz and 433 Hz due to oscillations in the combustor. Several cross power density spectra and transfer functions are derived from two dynamic pressure signals in different areas of the combustor. Modal analysis indicated the oscillations excite standing sound waves in the structure. The standing waves consist of alternating regions of high and low sound pressure amplitudes, related to each other by a characteristic difference in phase. Azimuthal modes in the form of waves are distributed along the circumferential coordinate. With a mean diameter of d = 2.5 m and speed of sound c = 844 m/s at a mean temperature of 1500°C, the frequency of vibration $f_n = nc/\pi d$ yields 215 Hz and 430 Hz for the second and fourth harmonics, indicating good agreement with the measurements. Significant amplitudes are not observed for the first and third harmonics of 108 and 326 Hz. Figure 15.36 provides details of the second and fourth modes.

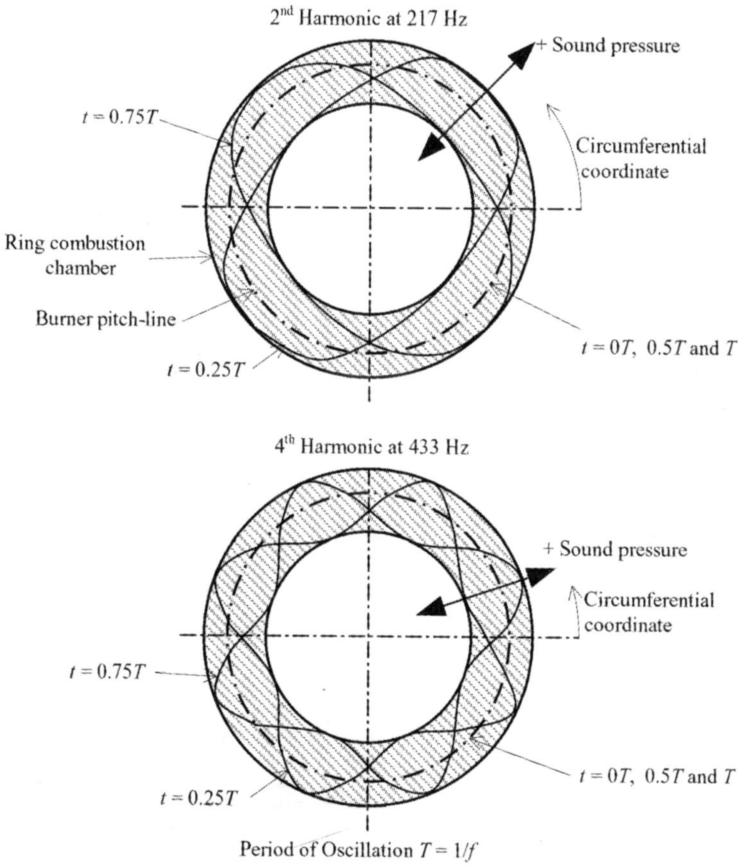

Fig. 15.36 Excited Modes in Combustion Chamber (Seume, 1997)

Passive methods rely on making changes in operating parameters (such as equivalence ratio) or geometry of the combustion system to hinder the self-exciting mechanism. The sound pressure amplitude can also be decreased to a tolerable level by dissipative baffles or mufflers (*ref*: Culick, 1988). By contrast, active methods use a feedback control loop. Heat release or pressure in the combustor is processed by a controller and used as an input signal for an actuator to influence the oscillating combustion that counteracts the self-excitation process (*ref*: Candel, 1992; McManus et al., 1993). Fluid stream inside the combustion chamber can be modulated to reduce pressure fluctuations by introducing inversed sound pressure oscillations, as in a loud speaker. The method is impractical for bigger turbines because of large amounts of air and exhaust gas to be handled. Combustion oscillations can be suppressed if the rate of fuel reaching the flame is anti-cyclical to the oscillations of the heat release rate. In either case, modulation of gases or fuel must take place at the frequency of the self-excited vibrations. Since these can often reach 1000 Hz, suitable actuators must be used to meet the requirement.

The active stability control system for the V84.3 gas turbine uses pressure transducer measurements in the chamber, and the signals are sent to a control unit to derive an input signal for the actuator to modulate the fuel flow rate. A few basic problems needed to be solved for operation on a ring combustor. In the premixed mode and at base load the engine uses 9 kg/s of gas. Thus, the actuator has to handle this mass flow rate at the observed frequencies. Active control is secured through additional diffusion flames that contribute about 10 per cent of the total power of the burner to stabilize the main premixed flame. Fuel flow rate to the pilot flame is modulated to influence the heat release in the main flame accordingly. A special high-speed direct-drive valve serves to actuate the pilot gas flow. The success of the control mechanism heavily depends on the pressure amplitudes in the pilot gas pipes, with higher amplitudes increasing the heat release in the flame. Hence, the length of the pipe to the pilot must be acoustically tuned to the frequency to be controlled. When the pilot's gas piping layout is complex and introduces damping, a suitable device may be placed upstream of the actuator to acoustically decouple it. Figure 15.37 shows a schematic of the control mechanism.

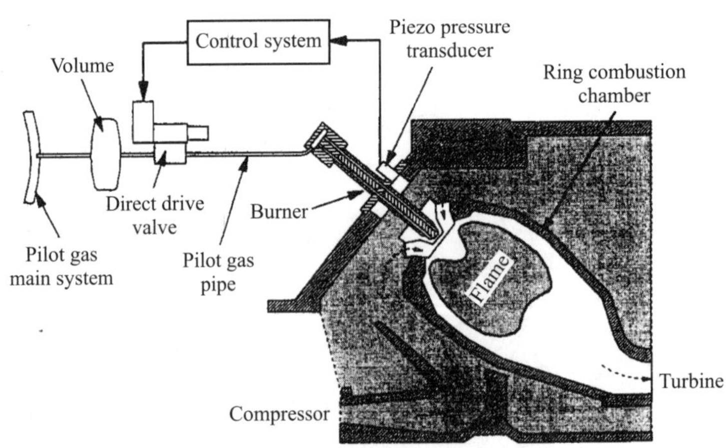

Fig. 15.37 Active Instability Control System for V84.3 Gas Turbine (Seume, et al. 1997)

The second problem is associated with the azimuthal modes of the instabilities. Since several control systems are placed in different locations along the circumference, the control devices are also situated in different regions of the excited acoustic field, with prevailing oscillating parameters strongly differing in amplitude and phase. The symmetry of the azimuthal modes is marked by a characteristic distribution of nodes and anti-nodes, with regions of high and low amplitudes related to each other by a constant phase shift. Consequently, it is possible to use a signal measured at a certain circumferential location on the ring combustor to calculate not only the actuator signal for the particular location but also for other defined locations. One control unit can then be used for all the actuators in the system. Figure 15.38 depicts this principle for the second harmonic to provide the input signal for four actuators located 90° from each other.

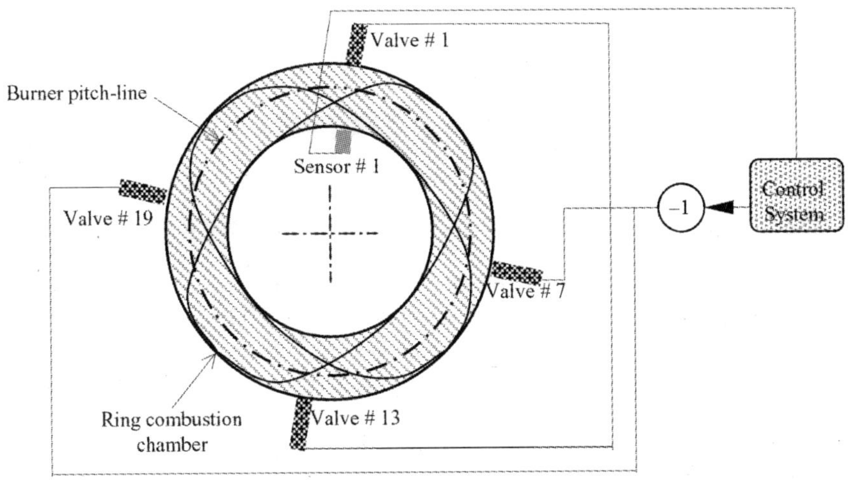

Fig. 15.38 Sensor and Signal Input Controller for Second Harmonic Mode (Seume, 1997)

Performance of the active control system can be gauged from shop test measurements of the 170 MW gas turbine's ring combustor (figure 15.39). The mechanism reduced the oscillations at the dominant frequency of 433 Hz by up to 17 dB. With active control turned off, measured sound pressure amplitudes rose to 210 mbar (corresponding to sound pressure level of 177 dB), falling to about 30 mbar with the control system turned on (*ref*: Rangwala, 2004).

Modulation of fuel flow rate is commonly achieved by using reciprocating flow devices where instability occurrence is at about 200 Hz or when the level of modulation required is small. In instances where instability frequencies are in the 200 to 500 Hz range and attenuation requires modulation of large fractions of engine fuel flow rate of hundreds of pounds per hour, a spinning drum valve has proved more useful (*ref*: Barooah et al., 2002). The spinning valve design is based on a rotary concept to generate maximum frequency response. A rotating drum with a selected number of holes equally spaced around the circumference is used, with the holes aligned in the surrounding enclosure to pass the liquid fuel flow. By minimizing the clearance between the drum and the enclosure, leakage is reduced when the holes in the rotating and stationary components come in line. The holes in the enclosure are radially opposed to balance the pressure and to minimize the traverse loads. Unlike a reciprocating device, the upper frequency limit is not affected by the inertia of the spool or the low power requirement to accelerate it.

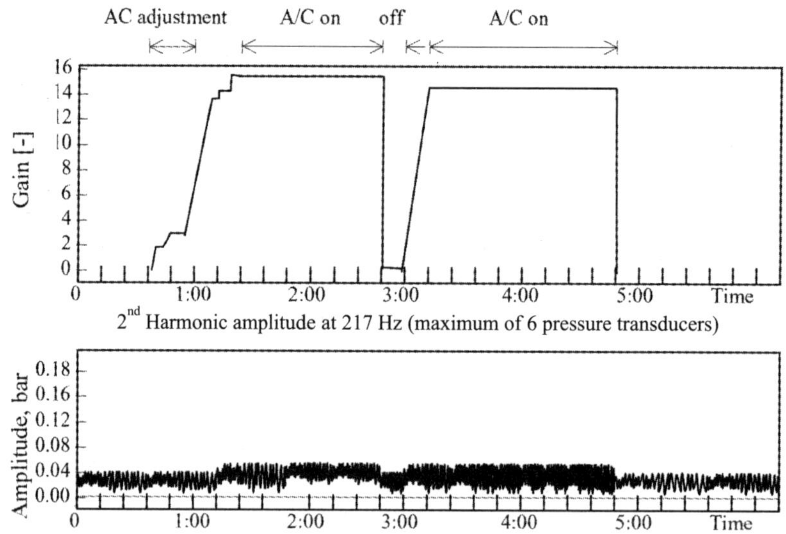

Fig. 15.39 Control System Test Operation at Base Load (Seume, 1997)

Liquid fueled low NO_x combustors can mitigate combustion instability at realistic operating conditions by modifying the fuel nozzle (*ref*: Cohen et al., 1998). The fuel is injected through axial tubes with spray tips protruding from the nozzle centrepiece. A pilot injector is placed a short distance downstream of the fuel ejection plane. After passing through a venturi the airflow is split between the fuel nozzle and a bypass segment. Airflow from the bypass is injected at the downstream end combustor, and represents dilution air. To obtain control of the acoustic oscillations, one tube delivers the fuel to a metering system and a solenoid valve. The tubing between the valve and the injection point is minimized to reduce attenuation and time lag due to capacitive effects. The main fuel flow takes place through the other injection tubes. The solenoid valve is driven at varying frequencies independent of the combustor behaviour using a signal generator, with an on/off duty cycle of 50 per cent. Combustor pressure and heat release rate are measured. When the control sensor signal crosses a predetermined threshold level, a command sent to the solenoid valve turns it on or off. Time delay between the instant of crossing and valve command is also taken into account. The threshold level and the time delay are manipulated through a user interface to the control algorithm. A proper choice of the two parameters yields 15 dB attenuation of the objectionable oscillating mode. The control system is also effective in holding the NO_x emission relatively constant across the range of equivalence ratios.

15.12 THERMAL PROTECTION OF COMBUSTOR LINER

The functional nature of a combustion liner imposes on it high temperature levels and steep thermal gradients. Mechanical strength of nickel and cobalt based alloys used for the liner deteriorates considerably when temperatures exceed 1100 K, hence means must be designed to relieve the heat build up. With gas turbines employing higher operating pressures and temperatures to improve performance and power, the need for cooling the combustion liner becomes even more acute. At the same time the liner is required to possess a minimum number of operating hours (*ref*: Lefebvre, 1999).

The temperature of the liner increases due to the combination of heating by radiation and convection from the internal hot gas flow and cooling due to radiation to the outer case and convection to the air in the annulus. Depending on the volume, pressure, temperature and chemical composition of the gases flowing through it, as also the dimensions and shape of the component, a considerable amount of heat is radiated on the liner. The size and number of hot and glowing soot particles formed during the combustion also control the intensity of radiation. Geometric shape factor between the liner and the casing and surface areas of the liner and outer case will govern the heat radiated by the liner to the casing. Internal heat convected to the liner walls from the gases is complicated by the rapidly changing physical and chemical characteristics, as also the temperature, of the gases. Steep gradients in flow velocities and pressures add even more uncertainties, because the state of boundary layer development makes it difficult to prepare an adequate model. In a can combustor reversal of flow designed in the primary zone permits only a portion of the flow to be modeled using the pipe analogy. Using a Reynolds number based index consistent with observed parameters for extreme turbulence, an expression using the hydraulic diameter (proportional to the ratio of cross sectional flow area and wetted perimeter) may be used. When a swirler is used, local gas velocity at the wall increases by the factor $1/(\cos \beta)$ relative to the downstream velocity, where β is the angle between the velocity vector and axis of the combustor. Bulk gas temperature used for internal convection in the primary zone may also need to be modified by reducing the corresponding radiation temperature by about 15 per cent. External convection from the cylindrical surface requires the Reynolds number to be calculated using the hydraulic mean diameter of the annular air space.

Film cooling of the inner surface of the liner is useful in achieving additional extraction of heat, and is accomplished by injecting air along the wall axially through slots and holes machined in the liner. The small diameter, closely spaced holes may be provided by laser drilling or by the EDM method. Since the turbulent hot gases gradually eliminate this film, the hole and slot pattern is repeated at specific intervals along the length. Cooling air is supplied through rings rolled or machined in the liner, through stacked rings with holes sized to deliver adequate amounts of cooling air and through corrugated spacers attached between overlapping segments of the liner. Depending on the design and method of attachment, the rings also serve to provide stiffness to the liner. Cooling efficiency in the liner is enhanced by improving the heat transfer coefficient on the coolant side (*ref*: Nealy, 1980) and by increasing surface roughness of the heat transferring areas.

Cooling effectiveness can be increased by impinging the coolant flow against the wall, as shown in figure 15.40. The double-walled passage is closed at the upstream end, and the outer wall is provided with holes. The impingement jets may be located at selected high temperature locations. Provision must be made for the difference in thermal growth, however, and the consequent increased thermal stresses in the region. Convective heat transfer on the external surface of the liner can be improved by providing fins, ribs or other protrusions to add to the surface area for heat exchange by convection. The ribs may run longitudinally, and have been used on industrial turbine combustors. Rolls Royce has elected to use pedestals in the dome region of RB211 combustor liners.

Another method for obtaining a relatively uniform temperature distribution is a liner with a large number of small holes perforated in it to assure the impinging jets spread the flow close to the wall. The flow must be controlled to prevent rapid mixing with main hot gas flow and to deter the cooling film from gradually rising in temperature by the surrounding combustion gases. The process, called effusion cooling, uses a larger amount of cooling air, but is effective in suppressing local hot spots. Drilling the holes (approximately 0.4 mm in diameter) at a shallow angle of 20°

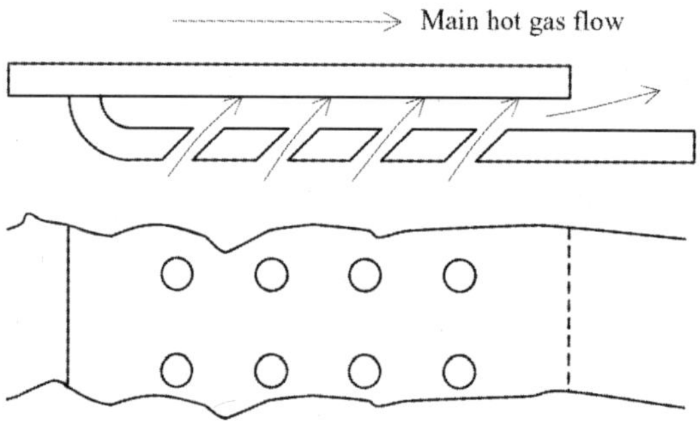

Fig. 15.40 Film Cooling of Liner with Impingement Jets (Lefebvre, 1999)

offers the twin benefits of increased surface area and reduced penetration of the exiting jet for a better film along the wall surface (*ref*: Dodds et al., 1989). Wall thickness needs to be increased to compensate for the holes and for protection against buckling. The angled effusion cooling hole concept is used on General Electric GE90 engine combustor. Manufacturing cost of drilling so many holes at precise locations is a factor to be taken into account in the use of this technique.

Many industrial turbine combustors are lined with refractory bricks to decrease heat flux into the supporting liner. The bricks are large in weight, but lighter metal tiles cast from turbine blade alloys with good resistance to high temperatures have been used on aero engines. Since the tiles are exposed to the hot gases, relatively lower temperatures and thermal stresses are experienced by the supporting shell, which may be made of a cheaper alloy. The cooler temperatures help to limit thermal growth in the shell. The tiles are equipped with pedestals on the rear surface, and cooling air flows around the pedestals to eject at the ends to form a layer of cooler air. The cost of repairs is reduced since only the tile experiencing distress needs to be replaced instead of repairing the liner.

Sheets of cobalt or nickel based alloys are commonly used to produce combustor liners exposed to severe temperatures and pressures. Resistance to oxidation and corrosion is a primary consideration in the selection of material and manufacturing process, but low coefficient of expansion and Young's modulus, resistance to thermal fatigue and high thermal conductivity also play a major role. Steep thermal gradients are encountered around the edges of cooling holes and in isolated hot spots. The pace of oxidation is noted to appreciably increase when metal temperature approaches 1300 K. A thin layer of a thermal barrier coating (TBC) may be applied on the inner surface of the liner for extra protection. The refractory material of low thermal conductivity has the capacity to reflect much of the radiated heat from combustion while offering increased resistance to heat flow to reduce the temperature of the metallic liner. To adequately provide the protection the thermal barrier coat must be chemically inert, be resilient to thermal shocks and have good erosion and wear characteristics. Also, its coefficient of thermal expansion must match that of the underlying substrate. A base metallic coat of Ni-Cr-Al-Y is overlaid with one or two coats of ceramics to constitute many TBC's. Plasma flame spraying is found effective in obtaining durability, consistency and thickness uniformity of the coats.

15.13 STRUCTURAL DESIGN FOR DYNAMIC PRESSURE

The liner and transition piece of a dry low NO_x combustor contain and direct the high temperature combustion gases. To allow for thermal growth the components are restrained at the fewest locations. The combustion process often creates pressure oscillations at discrete frequencies, especially during premixed burning. This stimulus on the flexibly attached structure can lead to substantial vibrations, develop fatigue cracks and eventual failure.

A trade-off between adequate flexibility to minimize thermal stress and stiffness to avoid vibration problems is thus required. Once a specific temperature distribution is available, thermal stresses may be computed with relative ease. Vibratory stresses, however, cannot be so readily defined. Complexities in geometry, loading pattern and contact behaviour are some reasons. Strain measurements may also be of limited usefulness because critical locations are not generally known, often are inaccessible and have high levels of metal temperature (*ref*: Rangwala, 2004).

Field experience gained from years of operation may permit setting limits on combustion dynamic pressure fluctuations to ensure structural integrity. Direct application of this experience to new applications, however, is questionable. Analytical prediction of component life and development of a method to include effects of dynamic pressure loading may thus be essential. Non-linear transient finite element analytical procedures are useful in predicting the dynamic behaviour, and can compare measured strains and accelerations with reasonable accuracy (*ref*: Barnes, 1996). Component stiffness and mass characteristics, as also distributed pressure loading, can then be accurately modeled. Contact behaviour, including sliding friction at supports and seals, is required to compute response, since they add considerable damping. Non-linear gap conditions capable of maintaining or breaking physical contact in accordance with relative displacements between components are also essential. Verification of natural frequencies and mode shapes computed from modal analysis may be made through laboratory modal testing, and to determine proximity to resonant conditions.

Figure 15.41 shows the finite element model. Mesh density requires special attention be paid to supports, fillets and areas of known stress concentration. This level of detail captures stress concentration in a forced response analysis, but may not be required for a modal analysis. Linear eight-noded brick elements help to maintain the number of elements within limits in a large model, especially in a computationally intensive solution. Effects of non-symmetric loading and support features generally do not permit taking advantage of geometric symmetry. Seals on forward and aft ends of the liner may be modeled as a combination of a normal spring and a tangential friction element. Seal stiffness may be computed independently from its own finite element model, while friction coefficient must be measured from seal fretting tests. Contact force at the seals will depend on relative thermal expansion of the components, hence the sequence of calculations may need repetition for various operating conditions.

Crossfire tubes between adjacent combustion liners are flexible and generally will not impact the calculations; hence they may be deleted in the analysis. Responding to pressure variation with unknown phase difference due to the crossfire tubes between adjacent combustors is at best a difficult proposition. Note that liner supports are considerably less rigid than turbine cylinders to which they are attached, and hence liner supports may be grounded at the other end.

The pressure loading represents test laboratory measurements simulating the machine's operating conditions. In this example peak-to-peak pressure variation of 21 kPa at 160 Hz defines the dynamics of combustion. Figure 15.42 provides computed oscillating displacements at the forward end of the transition piece. Strain gauges approximately centered in this region measures deformations that correlate very well with the analytical predictions. Radial motion at the aft end of the inner panel is affected by the damping provided by the seals, and the effectiveness of the damper is controlled by wear and by alignment of the mating components.

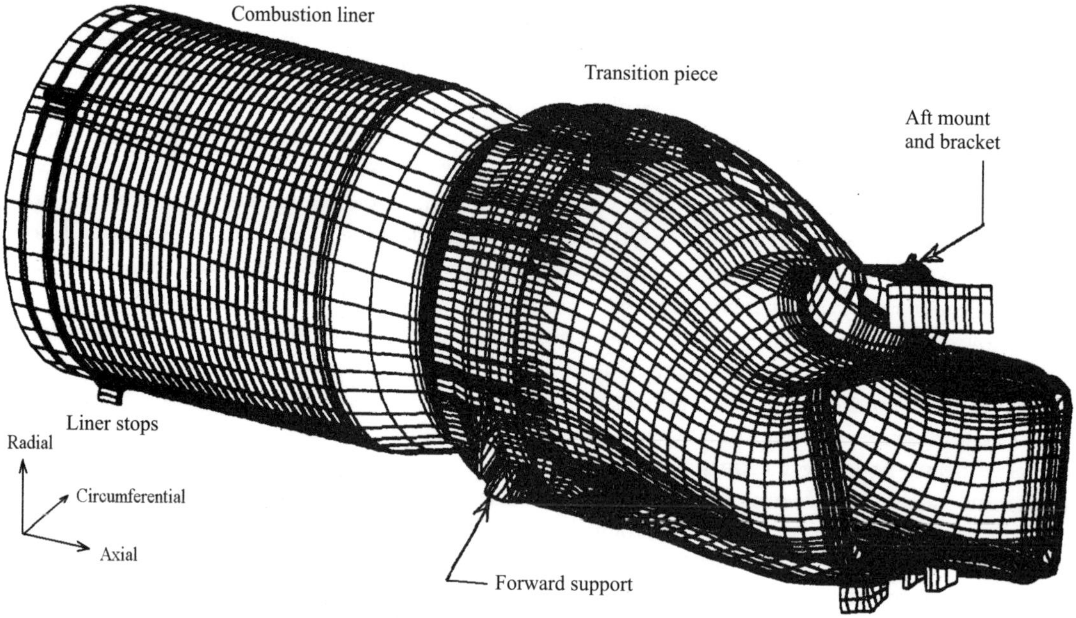

Fig. 15.41 Combustion Liner and Transition Piece Finite Element Model (Barnes, 1996)

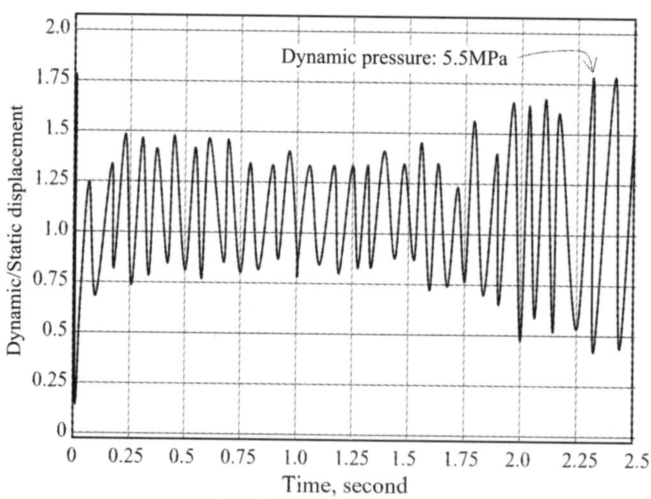

Fig. 15.42 Dynamic Axial Displacements at Forward End (Barnes, 1996)

High cycle fatigue capability of the components depends on vibratory and mean stresses, metal temperature and material characteristics. Critical locations are determined by computing the dynamic strain range, metal temperature and mean stress in the elements. Figure 15.43 provides a plot of equivalent vibratory strain divided by allowable strain at a critical location between the aft mount and the aft frame at the centreline of the transition piece. Measurement of strain at this location poses practical problems.

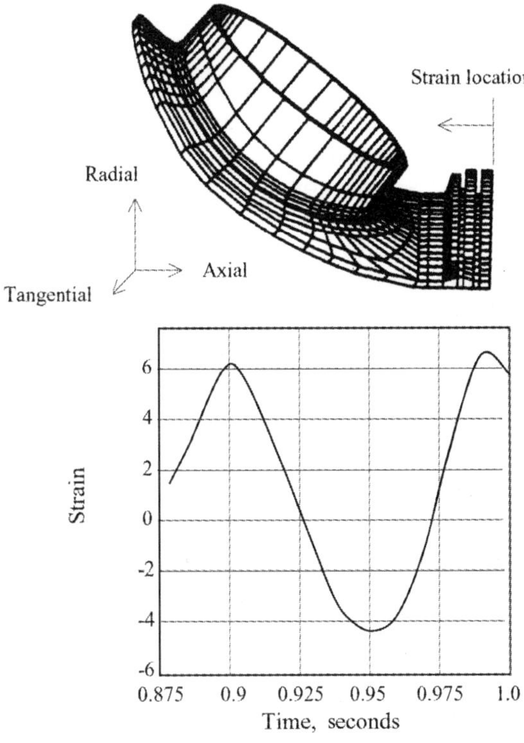

Fig. 15.43 Equivalent Strain History (Barnes, 1996)

Determination of response to initial impact of the dynamic pressure requires non-linear transient dynamic analysis, assuming oscillation to initiate at full amplitude. Dynamic amplification peaks at the forward support of the transition piece, mostly because a gap at the location causes it to alternate between contacting and free modes of vibration. This results in the reaction force at impact to multiply by a factor of 12 over the static reacting force value. Thus, the seat for the support must be designed to reflect this situation. To assure structural dynamic response is not significant due to small variations in stimulus frequency, a wider frequency range can be expeditiously checked through modal analyses of individual components in a freely supported condition. Analytical prediction of natural frequencies must agree with measurements, as also the corresponding mode shapes. Variations due to differing mass and stiffness properties from manufacturing tolerances and measurement inaccuracies must be maintained within acceptable limits. Where the natural frequencies are deemed too close to those of the combustion dynamic pressure, the structural response for these modes must be captured with the non-linear dynamic analysis process.

15.14 MACH NUMBER CONTROL WITH FUEL SYSTEM

Control theory related to compressor exit Mach number can be effectively used towards improved regulation of transient compressor working point trajectories of a turbine engine. The intent is to raise the steady working line towards the surge boundary, with the added advantage of increased thrust and decreased specific fuel consumption. Mach number limiters (or working line controls) can be designed to be independent of fuel metering accuracy or variable geometry position control accuracy.

The new control method can replace the rigid acceleration and deceleration limiters of the fuel in the fuel control loop, thus achieving superior avoidance of problems related to surge and flameout (*ref*: Dadd et al., 2000). The benefits are considerable. For instance, thrust response times are repeatable, and transient response trajectories on the compressor time are consistent in spite of variations in temperature and fuel characteristics. On the debit side, the rate of change of high-pressure compressor spool speed, dN_{HP}/dt, proves to be obstructive in that the occurrence of surge invokes an increase in the fuel flow in order to maintain the dN_{HP}/dt. This problem is conventionally countered by open-loop fuel flow limits, which may be sensitive to high-pressure compressor (HPC) delivery pressure P_3, to ensure over-fuelling does not occur. Hence, accurate metering of fuel flow remains a prime requirement for the fuel system, and the much sought after simplifications in the mechanical setup have yet to be achieved.

If the surge margin and the fuel limit are set using the worst power off-take scenario over the whole flight envelope, engine performance is seriously affected. The issue is particularly relevant when significant power take-off from the engine is anticipated (*ref*: Toyne et al., 1983; Cronin, 1983; Chan, 1998). When a backup limiter in terms of compressor exit Mach number (*ref*: General Electric Company, 1967) intercedes dN_{HP}/dt acceleration control loop in this situation, it removes the need for the less accurate fixed schedule fuel limiting. Transient performance is substantially improved for all levels of operation, leading to reduced reliance on fuel scheduling and enabling simplification of the fuel system hardware.

The control concept has been investigated and successfully demonstrated in a twin-spool military engine. The simplified fuel system uses a variable capacity fuel pump, has no flow feedback, and its design is considerably less complex than its predecessors. Figure 15.44 shows a generalized view of the control configuration adopted. Fuel control is achieved by manipulating a selected parameter within an inner closed loop to set the area of the fuel valve. The selected inner loop parameter must exhibit high sensitivity in the event of a surge, so its action can be used to reduce fuel flow. Under steady state and transient engine conditions the primary loop will then act to reduce fuel flow at the first signs of a surge.

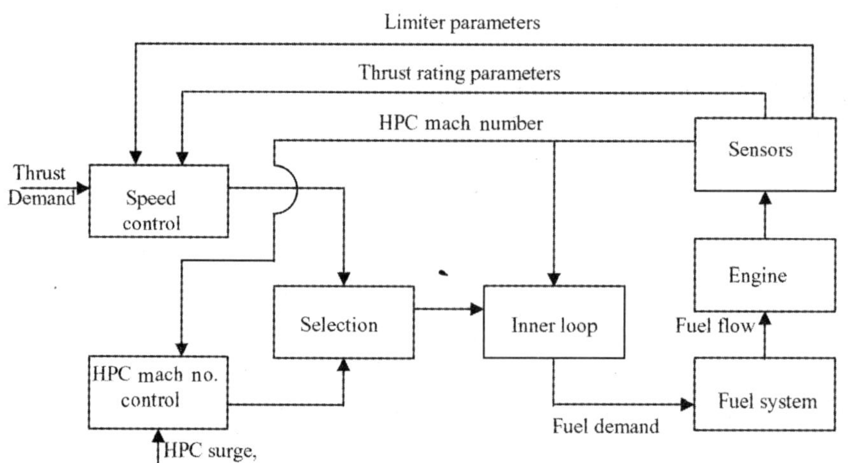

Fig. 15.44 Mach Number Control Configuration

The control structure is shown in greater detail in figure 15.45, where the engine control law uses the Mach number error (MNER) as the selected parameter. The HPC exit Mach number, HPCMN, provides a direct measure of the compressor's operating point. The compressor pressure ratio maps into the exit Mach number, which is an invariant if the geometry of the compressor is held fixed. Also, the exit Mach number displays nearly two to three times the sensitivity to displacement from the operating line than does the compressor pressure ratio. Also, HPCMN exhibits constructive behaviour in the presence of surge, thus permitting more successful recovery from surge. The Mach number limiting control blocks 'Acc. Control' and 'Dec. control' blocks in the control structure diagram are primarily proportional plus integral ($P + I$) controllers that compare the MNER feedback with scheduled limits and deliver the demands for selection. With the right set of gains, one of these demands will become dominant when the scheduled limit is reached. A Hanus feedback block is also incorporated to maintain an appropriate integrator output when the loop is not selected.

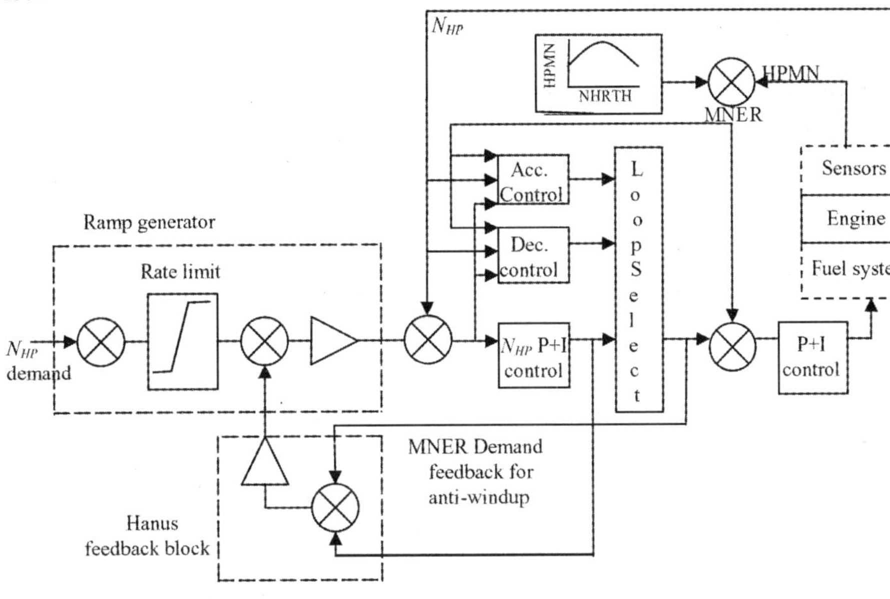

Fig. 15.45 Control Law Using Mach Number Error

The speed of the compressor spool dN_{HP}/dt may also be used to control the exit Mach number. Similar to the scheme for control using Mach number error, the difference lies in the use of speed as the selected parameter and the deletion of the inner loop feedback path. The loop in control of the engine then becomes dN_{HP}/dt demand instead of the Mach number error demand. This method eliminates the necessity for an integrator in the main speed control loop with its inherent need for resetting during a transient.

15.15 COMBUSTOR NOISE CONTROL

Westinghouse Electric Corporation manufactures the W501D5 combustion turbines for electric power generation. Rated at 104 MW, the turbines are designed to burn natural gas, and with field modifications, retrofitted to operate on synthetic fuel gas or a mixture of the two, together with steam injection for NO_x control and power augmentation (*ref*: Scalzo et al., 1981).

Field conversion to burn medium Btu fuel from a coal gasification plant of one of the turbines at a customer location resulted in excessive airborne sound and a corresponding non-synchronous engine vibration when burning natural gas with steam injection (*ref*: Scalzo et al., 1989). The synthetic fuel gas has a lower heating value than the natural gas. The retrofit program included replacement of fuel nozzles and combustor baskets and modification of control functions. A severe non-synchronous 105 Hz vibration in the rotor and the casing accompanied by noise are linked to engine load and steam flow rate. The threshold for the vibration signal is 85 MW and 10,000 lbs per hour of stem injection. A similar unit at the same location that has not been modified did not exhibit similar noise and vibration problems.

Experience with other combustion turbines indicated a discrete 209 Hz frequency signal, but the 105 Hz signal was absent. Physical dimensions of the combustor baskets, gas flow velocities and operating temperatures are nearly the same. Hence, the changes adopted in the new synthetic gas nozzles and the baskets appeared to hold the potential to alter the combustor acoustics and be solely responsible for the noise and vibration problems.

The lower heating value of the synthetic fuel increases the mass flow to the turbine. An upper limit on the synthetic fuel and steam flow based on compressor surge is required, as shown in figure 15.46. Rotor vibration data for the converted unit when operating at 100 MW on natural gas with 10,000 lbs/hour are: unfiltered and filtered seismic is 5.0 mils and 2.5 mils, respectively, and absolute rotor vibration is 5.0 mils double amplitude at the exhaust end. The 105 Hz vibration component has a significant magnitude when compared with the 60 Hz signal. Accelerometer readings conformed that a major portion of the rotor vibration is the non-synchronous 105 Hz signal is present at the exhaust end of the engine. A noise meter measured an acoustic reading of 112 dB in the 100Hz third octave. The evidence indicates a preponderance of vibrations and noise arising in the installed synthetic gas nozzles and baskets in the combustion system.

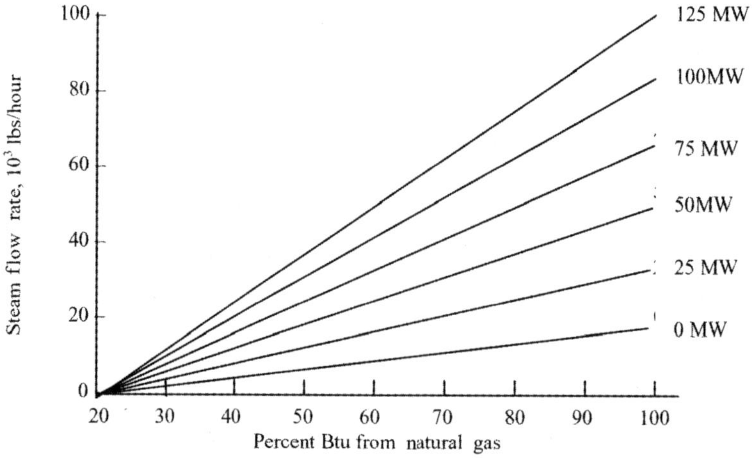

Fig. 15.46 Permissible Steam Injection as Function of Load and Gas Btu

A short-term plan to enable the unit to operate while a long-term fix was developed called for balancing of the rotor, noise measurements and tests to determine the effects of inlet guide vanes (IGV) closure on the 105 Hz noise. One balance plug applied at the compressor balance plane produced a substantial degree of decoupling from the response due to unbalance and the non-synchronous response due to combustion noise. Total vibration in the rotor decreased from 6 mils before balancing to less than 3 mils at the turbine end. Moreover, the 105 Hz shaft vibration reduced to only 1 mil double amplitude at minimum steam flow. Noise level tests focused on developing diagnostic trend data to determine suitable operating conditions.

The recirculation stabilization flow system provided by the primary scoop air flow within the combustor basket is shown in figure 15.47, where the original nozzle provides steam through the centre orifice and natural gas on the outer orifices. The included angles of the fuel and steam vectors provide good mixing, without directly opposing the recirculating flow. Raw fuel from the natural gas orifices is mixed with the recirculating air, and is carried into the six-lobed pattern provided by the six scoops to be ignited and to flow back to ignite new fuel. Steam mixes to reduce the flame temperature, thereby reducing NO_x formation.

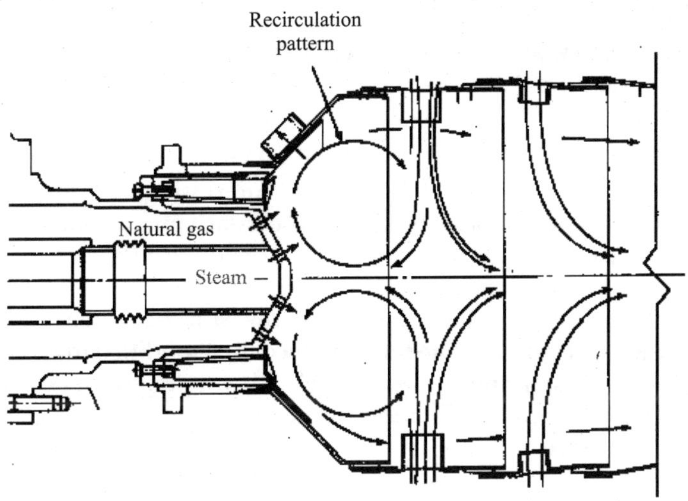

Fig. 15.47 Steam and Natural Gas Vectors with Standard Natural Gas Nozzle

The synthetic gas nozzle is shown in figure 15.48, together with the recirculating primary airflow. The centre hole fuel vector directly opposes the recirculating air pattern. The reduction in the angle for the synthetic gas and steam orifices provides better mixing, but it also opposes the re-circulating airflow. Achieving a balance between mixing and supporting the recirculating primary flow has the tendency to maintain stability in the combustion process and to control noise.

Verification of this influence of momentum vectors in the gas flow on the 105 Hz third-octave noise is obtained by plugging the centre hole of the natural gas orifice and measurement of noise inside the turbine enclosure at the exhaust end. Noise level is reduced from 103 dB to about 95 dB.

Additional cooling of the dome in the combustor basket with a compensating decrease in the primary airflow has also been investigated to maintain the original pressure drop in the combustor. The influence of the primary air recirculating flow is checked by reinstalling the original baskets with

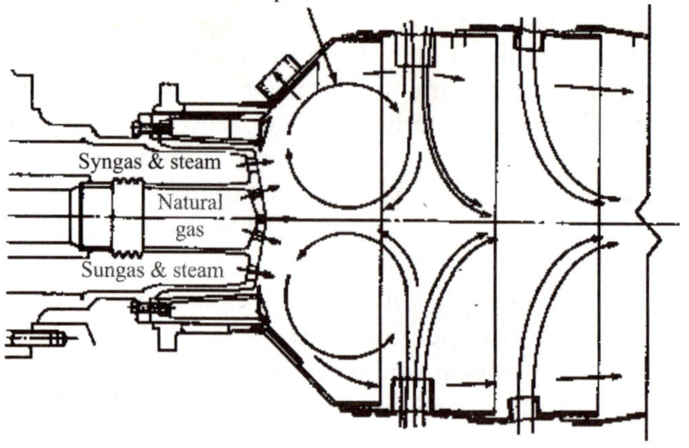

Fig. 15.48 Steam and Natural Gas Vectors with Synthetic Gas Nozzle

their 57 per cent larger airflow and conducting a noise test. The results indicated a reduction of 15 dB at 20,000 lbs/hour of steam.

The relative and seismic journal vibration signals were monitored at both ends of the turbine during acceleration and loading. No evidence of non-synchronous vibration was noticed.

Interpretation of all the test data led to some interesting conclusions. The undesirable combustion noise and non-synchronous vibrations arose from a pressure driven closed-end standing wave of one-quarter wavelength. The momentum vectors due to the flow of steam and fuel that oppose the primary air scoop recirculation pattern result in the closed-end pattern. On the other hand, the same momentum vectors that are in sympathy with the primary air scoop recirculation pattern result in an open-end pattern. The noise pattern then modified to an acceptable level with a velocity-driven open-ended standing wave of one-half wavelength. The modified combustion system reduced the 100 Hz third-octave noises from 115 dB to 97 dB, representing an intensity reduction of 64 to 1.

15.16 REVIEW PROBLEMS

Example problem # 15.1: Find the primary zone temperature of an uncooled liner for a tubular combustor from the following data:

Combustor inlet pressure = P = 2750 kPa

Combustor inlet temperature = T = 920 K

Case diameter = D_C = 0.2 m

Liner outer diameter = D_L = 0.15 m

Liner thickness = t_L = 0.001 m

Casing emissivity = ε_c = 0.4

Liner emissivity = ε_L = 0.65

Liner thermal conductivity = k_L = 25 W/(m-K)

Mass flow rate through combustor = dm_a/dt = 7.0 kg/s
Mass flow rate through primary zone = dm_p/dt = 2.5 kg/s
Fuel/air mass ratio = q = 0.06
Luminosity factor = L = 1.75
Primary zone combustion efficiency = η_c = 0.85

Solution: Temperature of gas in the primary zone is obtained from temperature increase due to combustion and added to the inlet temperature. Using temperature rise curves for a mixture of air and kerosene at 2750 kPa, 920 K and assumed 85 per cent combustion efficiency, the effective $q = 0.06 \times 0.85 = 0.051$ and $\Delta T = 1415$ K. Assume a decrease of 50 K for heat loss in the evaporation of fuel and raising its temperature to that of the surrounding gases. Then primary zone gas temperature T_g is 920 + 1415 – 50 = 2285 K.

Example problem # 15.2: Determine the radiation heat flux from the combustion gas to the liner for this combustor.

Solution: The rate of heat transfer R_1 by non-luminous radiation from a gas to its enclosure depends on the size and shape of the container and the mean, or bulk, conditions of the gas. The gas emits only a few narrow bands of wavelengths, and absorbs only those wavelengths included in its emission bands. Also, the surface exposed to the flame has an effective absorption rate under unity, and can be estimated by the factor $0.5(1 + \varepsilon_l)$. Then the heat transfer rate is given by:

$$R_1 = 0.5\sigma(1 + \varepsilon_L)(\varepsilon_g T_g^4 - \alpha_g T_{w1}^4) \quad ...(15.4)$$

where σ = Stefan-Boltzmann constant = 5.67×10^{-8} W/(m^2-K^4), ε_g = gas emissivity, T_g = gas temperature, T_{wl} = liner temperature on flame side and α_g = gas absorption rate at T_{wl}. α_g may be approximated from $\alpha_g = \varepsilon_g(T_g/T_{w1})^{1.5}$. At gas temperature of 2285 K gas emissivity ε_g is assumed to be 0.6. For tubular can type of combustors beam length is given by the expression (*ref*: Fishenden, 1950): l_b = 3.4 × (volume)/(surface area), working out to between 0.6 and 0.9 times the liner diameter, or $0.6 \times D_L = 0.09$ m. So:

$$R_1 = 0.5 \times 5.67 \times 10^{-8} \times (1 + 0.65) \times 0.6 \times 2285^{1.5} \times (2285^{2.5} - T_{w1}^{2.5})$$
$$= 765126 - 0.003066 \times T_{w1}^{2.5} \text{ W/m}^2$$

Example problem # 15.3: Determine the radiation heat flux from the liner to the casing for this combustor.

Solution: Radiation heat flux from the liner to the casing, R_2, is expressed by:

$$R_2 = \frac{\sigma\varepsilon_L\varepsilon_c}{\varepsilon_c + \varepsilon_l(1-\varepsilon_c)D_L/D_{\text{ref}}}(T_{w2}^4 - T^4) \quad ...(15.5)$$

where $\varepsilon_L = 0.65$, $\varepsilon_c = 0.4$, ratio of hydraulic mean diameters for a tubular combustor D_L/D_{ref} = 0.15/0.192 = 0.78 and T_{w2} is liner temperature on coolant side.

Then: $$R_2 = \frac{5.67 \times 10^{-8} \times 0.65 \times 0.4}{0.4 + 0.65 \times (1 - 0.4) \times 0.78}(T_{w2}^4 - 920^4)$$
$$= 2.0920 \times 10^{-8} \times T_{w2}^4 - 14987 \text{ W/m}^2$$

Example problem # 15.4: Determine the liner temperatures on the casing (T_{w1}) and the coolant (T_{w2}) sides in example problems 15.1, 15.2 and 15.3.

Solution: Gas properties at primary zone temperature of 2285 K, pressure of 2750 kPa and liner data are:

$$\text{Thermal conductivity } k_g = 0.153 \text{ W/(m-K)}$$
$$\text{Dynamic viscosity } \mu_g = 7.02 \times 10^{-5} \text{ kg/(m-s)}$$
$$\text{Mass flow rate through primary zone} = dm_p/dt = 2.5 \text{ kg/s}$$
$$\text{Liner outer diameter} = D_L = 0.15 \text{ m}$$
$$\text{Cross sectional flow area of liner} = A_L = \pi \times (0.15)^2/4 = 0.01767 \text{ m}^2$$

Since internal convection is difficult to estimate precisely because of rapid physical and chemical changes during combustion, some form of classical heat transfer expression for straight pipes may be assumed if the Reynolds number is consistent with the stated conditions of extreme turbulence. The expressions for casing (C_1) and coolant (C_2) sides are:

$$C_1 = 0.017\frac{k_g}{D_L^{0.2}}\left(\frac{dm_p/dt}{A_L\mu_G}\right)^{0.8}(T_g - T_{w1}) \quad \text{...(15.6)}$$

Hence:

$$C_1 = 0.017\frac{0.153}{0.15^{0.2}}\left(\frac{2.5}{0.01767 \times 7.02 \times 10^{-5}}\right)^{0.8}(2285 - T_{w1})$$

$$= 960004 - 420.1 \times T_{w1}$$

A similar expression: $$C_2 = 0.020\frac{k_a}{D_a^{0.2}}\left(\frac{dm_a/dt}{A_a\mu_a}\right)^{0.8}(T_{w2} - T_3) \quad \text{...(15.7)}$$

where:

$$\text{Thermal conductivity } k_a = 0.0553 \text{ W/(m-K)}$$
$$\text{Dynamic viscosity } \mu_a = 3.85 \times 10^{-5} \text{ kg/(m-s)}$$
$$\text{Mass flow rate through primary zone} = dm_a/dt = 7.0 \text{ kg/s}$$
$$\text{Annular width} = D_C - D_L = D_a = 0.2 - 0.15 = 0.05 \text{ m}$$
$$\text{Liner cross sectional flow area} = A_a = [\pi \times (0.2^2 - 0.15^2)]/4 = 0.01374 \text{ m}^2$$
$$\text{Combustor inlet temperature} = T = T_3 = 920 \text{ K}$$

Then:

$$C_2 = 0.020 \times \frac{0.0553}{0.05^{0.2}}\left(\frac{7.0}{0.01374 \times 3.85 \times 10^{-5}}\right)^{0.8}(T_{w2} - 920)$$

$$= 1003 \times T_{w2} - 922480 \text{ W/m}^2$$

Conduction heat transfer through a solid liner wall due to a temperature gradient in the wall K_{1-2} is:

$$K_{1-2} = \frac{k_L}{t_L}(T_{w1} - T_{w2}) \quad \text{...(15.8)}$$

Then: $$K_{1-2} = \frac{25}{0.001}(T_{w1} - T_{w2}) = 25000(T_{w1} - T_{w2})$$

For equilibrium, $R_1 + C_1 = R_2 + C_2 = K_{1-2}$ ...(15.9)

Substituting for R_1, C_1, R_2, C_2 and K_{1-2} gives the equations:

$$-0.003066 \times T_{w1}^{2.5} - 25420.2 \times T_{w1} + 1725130 = -25000 \times T_{w2}$$

$$2.0920 \times 10^{-8} \times T_{w2}^{4} + 26003 \times T_{w2} - 937466 = 25000 \times T_{w1}$$

Then: $T_{w1} = 1589$ K, and $T_{w2} = 1559$ K.

Example problem # 15.5: In the first four example problems cooling of the liner has not been considered, but the results do not necessarily indicate maximum temperatures in the liner. If in a region annulus flow velocity is substantially lower than the mean value on one side, and if the cooling flow is restricted in the same zone on the other side of the wall, the liner may reach exceptionally high temperatures in the form of hot spots.

Calculate the liner temperatures at a distance x downstream of a cooling slot of height s = 0.0020 m and thickness t = 0.0008 m. Ratio x/s = 18 where the coolant flow rate is 0.3 kg/s.

Solution: $s = 0.002$ m, $x/s = 18$, $x = 0.036$ m and surface area of slot:

$$A_s = \pi \times D_L \times s = 0.0009425 \text{ m}^2$$

Flow rate of coolant through the slot: $dm_a/dt = \rho_a U_a A_s = 0.3$ kg/s, hence:

$$\rho_a U_a = 318.3 \text{ kg/(m}^2\text{-s)}$$

At 2750 kPa and 920 K, properties of air are:

Dynamic viscosity $\mu_a = 3.85 \times 10^{-5}$ kg/(m-s)

Thermal conductivity $k_a = 0.0548$ W/(m-K)

Then flow Reynolds number in slot $RE_s = \dfrac{\rho_a U_a s}{\mu_a} = 16536$

And flow Reynolds number along liner $RE_x = \dfrac{\rho_a U_a x}{\mu_a} = 297640$

For flow along the liner, $A_L = 0.01767$ m^2, $T_g = 2285$ K, $\mu_g = 7.02 \times 10^{-5}$ kg/m-s, fuel-air ratio in the primary zone $q_p = 0.051$, $(dm_g/dt) = \rho_g U_g A_L = 2.5$ kg/s, $k_g = 0.153$ W/(m-K).

Then, $\rho_g U_g = 2.5/.01767 = 141.5$ kg/(m^2-s).

Also, $m = \rho_a U_a / \rho_g U_g = 318.3/141.5 = 2.25$. From an analysis of experimental data on the influence of slot thickness t on cooling effectiveness, a correction factor is available (*ref*: Ballal et al., 1973):

$$\eta = 1.28 \times \left(\frac{\mu_a}{\mu_g}\right)^{0.15}\left(\frac{xt}{s^2}\right)^{-0.2} = 1.28 \times \left(\frac{3.85}{7.02}\right)^{0.15}(18 \times 0.4)^{-0.2} = 0.788$$

$$\text{Cooling efficiency } \eta = \frac{T_g - T_{w,ad}}{T_g - T_a} = \frac{2285 - T_{w,ad}}{2285 - 920} = 0.788 \text{ and } T_{w,ad} = 1209 \text{ K.}$$

where T_a = combustor inlet temperature = 920 K.

In the heat transfer calculations with a film-cooled liner, constants R_1, R_2 and C_2 remain the same as for a non-cooled liner. Constant C_1 is altered by changes in coolant flow velocity and hot gas temperature near the wall, and is expressed by:

$$C_1 = 0.10 \times \left(\frac{k_a}{x}\right)(RE_x)^{0.8}\left(\frac{x}{s}\right)^{-0.36}(T_{w,ad} - T_{w1}) \qquad ...(15.10)$$

Hence:

$$C_1 = 0.10 \times \left(\frac{0.153}{0.036}\right) \times (297640)^{0.8} \times \left(\frac{0.036}{0.002}\right)^{-0.36}(1209 - T_{w1})$$

$$= -1287 \times T_{w1} + 1556038$$

From example problems 15.1 through 15.4,

$$R_1 = 765126 - 0.003066 \times T_{w1}^{2.5} \text{ W/m}^2$$

$$R_2 = 2.0934 \times 10^{-8} \times T_{w2}^4 - 14997 \text{ W/m}^2$$

$$C_2 = 1003 \times T_{w2} - 922719 \text{ W/m}^2$$

$$K_{1-2} = 25000 \times (T_{w1} - T_{w2})$$

Substituting for R_1, C_1, R_2, C_2 and $K_{1\text{-}2}$ in (15.9) gives the equations:

$$-0.003066 \times T_{w1}^{2.5} - 26287 \times T_{w1} + 2321164 = -25000 \times T_{w2}$$

$$2.0920 \times 10^{-8} \times T_{w2}^4 + 26003 \times T_{w2} - 937466 = 25000 \times T_{w1}$$

Then $T_{w1} = 1320$ K and $T_{w2} = 1303$ K

15.17 EXERCISES

15.1 Find the primary zone temperature of an uncooled liner for a tubular combustor from the following data:

Combustor inlet pressure = P = 2877 kPa
Combustor inlet temperature = T = 885 K
Case diameter = D_C = 0.227 m
Liner outer diameter = D_L = 0.143 m
Liner thickness = t_L = 0.00125 m
Casing emissivity = ε_c = 0.4
Liner emissivity = ε_L = 0.65
Liner thermal conductivity = k_L = 31 W/(m-K)
Mass flow rate through combustor = dm_a/dt = 8.37 kg/s
Mass flow rate through primary zone = dm_p/dt = 2.57 kg/s

Fuel/air mass ratio = q = 0.0575

Luminosity factor = L = 1.70

Primary zone combustion efficiency = η_c = 0.875

15.2 In example 15.1, determine the radiation heat flux from the combustion gas to the liner for this combustor.

15.3 In examples 15.1 and 15.2, determine the radiation heat flux from the liner to the casing for this combustor.

15.4 In examples 15.1 and 15.3, determine the liner temperatures on the casing (T_{w1}) and the coolant (T_{w2}) sides.

15.5 In the four examples 15.1 to 15.4 cooling of the liner has not been considered, but the results do not necessarily indicate maximum temperatures in the liner. If in a region annulus flow velocity is substantially lower than the mean value on one side, and if the cooling flow is restricted in the same zone on the other side of the wall, the liner may reach exceptionally high temperatures in the form of hot spots.

Calculate the liner temperatures at a distance x downstream of a cooling slot of height s = 0.0020 m and thickness t = 0.0008 m. Ratio x/s = 18 where the coolant flow rate is 0.3 kg/s.

References & Bibliography

Anderson, D.N., "Effects of Equivalence Ratio and Dwell Time on Exhaust Emissions from an Experimental Premixing Pre-Vaporizing Burner", ASME paper # 75-GT-69, New York, 1975.

Arana, C.A., Sekar, B., Mawid, M.A. and Graves, C.B., "Determination of Thermo-Acoustic Response in a Demonstrator Gas Turbine Engine", ASME paper # 00-GT-091, New York, 2000.

Bahr, D.W., "Technology for the Design of High Temperature Rise Combustors", Journal of Propulsion and Power, volume 3, number 2, page 179 – 186, 1987.

Ballal, D.R. and Lefebvre, A.H., "Film Cooling Effectiveness in the Near Slot Region", Journal of Heat Transfer, page 265 – 266, 1973.

Barnes, J.E., "Structural Integrity of a Gas Turbine Combustion System Subjected to Increased Dynamic Pressure", ASME paper # 96-GT-473, New York, 1996.

Barooah, P., Anderson, T.J. and Cohen, J.M., "Active Combustion Instability Control with Spinning Valve Actuator", Proceedings, ASME Turbo Expo, paper # GT-2002-30042, Amsterdam, June, 2002.

Buecker, B., "Emissions: SCR Design", Power Engineering, page 24 – 28, August, 2002.

Candel, S.M., "Combustion Instability Coupled by Pressure Waves and Their Active Control", 25th International Symposium On Combustion, Sydney, Australia, 1992.

Chan, K.K., "More Electric Engine Controls Technology Report", DERA Technical Report, Defense Research and Evaluation Agency, UK., 1998.

Claeys, J.P., Edward, K.M., Mick, W.J. and Symonds, R.A., "Combustion System Performance and Field Test Results of MS7001 F Gas Turbine", ASME Journal of Engineering Gas Turbines & Power, volume 115, page 537 – 546, New York, 1993.

Cohen, J.M., Rey, N.M., Jacobson, C.A. and Anderson, T.J., "Active Control of Combustion Instability in a Liquid Fueled Low NO_x Combustor", ASME paper # 98-GT-267, New York, 1998.

Cronin, M.J., "All Electric vs. Conventional Aircraft: The Production/Operational Aspects", J. Aircraft, volume 20, No. 6, 1983.

Culick, F.E.C., "Combustion Instabilities in Liquid Fueled Propulsion Systems – An Overview", AGARD Conference on Combustion Instabilities in Liquid Fueled Propulsion Systems, Bath, AGARD-CP-450, page 1-1 – 1-73, 1988.

Dadd, G.J. and Chan, K.K., "Gas Turbine Mach Number Control with simplified Fuel System", ASME paper # 00-GT-044, New York, 2000.

Dalla Betta, R.A., Schlatter, J.C., Nickolas, S.G., Cutrone, M.B., Beebe, K.W., Furuse, Y. and Tsuchiya, T., "Development of a Catalytic Combustor for a Heavy Duty Utility Gas Turbine", ASME paper # 96-GT-485, New York, 1996.

Damkohler, G., NACA TM # 1112, 1947.

Darling, D., Radhakrishnan, K., Oyediran, A. and Cowan, E., "Combustion-Acoustic Stability Analysis for Premixed Gas Turbine Combustors", NASA TM 107024, 1995.

Dodds, W.J. and Ekstedt, E.E., "Broad Specification Fuel Combustion Technology Program", Phase II, Final Report, 1989.

Feitelburg, A.S., Tangirala, V.E., Elliott, R.A., Pavri, R.E. and Schiefer, R.B., "Reduced NO_x Diffusion Flame Combustors for Industrial Gas Turbines", ASME paper # 00-GT-085, New York, 2000.

Fishenden, M. and Saunders O., "An Introduction to Heat Transfer", Oxford University Press, New York, 1950.

Flores, R.M., Miyasato, M.M., McDonell, V.G. and Samuelsen, G.S., "Response of a Model Gas Turbine Combustor to Variation in Gaseous Fuel Composition", ASME paper # 00-GT-141, New York, 2000.

General Electric Company, "Improvements in Control Systems for Axial Flow Compressors, Particularly Adapted for Use in Gas Turbine Engines", British Patent 1184369, 1967.

Gupta, A.K., Lewis, M.J. and Daurer, M., "Swirl Effects on Combustion Characteristics of Pre-Mixed Flames", ASME Journal of Engineering Gas Turbines & Power, volume 123, page 619 – 626, New York, 2000.

Kaya, H., "Catalytic Combustion Technologies", Journal of Gas Turbine Society of Japan, volume 25, number 98, page 48 – 51, Tokyo, 1997.

Koff, B.L., "Aircraft Gas Turbine Emissions Challenge", ASME paper # 93-GT-422, New York, 1993.

Kuznetsov, V.R., "Influence of Turbulence on High Non-Equilibrium Concentrations of Atoms and Free Radicals in Diffusion Flames", Fluid Dynamics, volume 17, page 815 – 820, 1982.

Lefebvre, A.H., "Gas Turbine Combustion", Taylor and Francis, Philadelphia, 1999.

Leong, M.Y., Smugeresky, C.S., McDonell, V.G. and Samuelsen, G.S., "Rapid Liquid Fuel Mixing for Lean Burning Combustors: Low Power Performance", ASME paper # 00-GT-116, New York, 2000.

Lieuwen, T. and McManus, K., "That Elusive Hum", Mechanical Engineering, page 53 – 55, June, 2002.

Marshall, A.W., Ph.D. Thesis, University of Maryland, College Park, MD, 1996.

Maughan, J.R., Lutts A. and Bautista, P.J., "A Dry Low NO_x Combustor for the MS3002 Regenerative Gas Turbine", ASME paper # 94-GT-252, New York, 1994.

McCarty, J.G., "Kinetics of PdO Combustion Catalysis", Proceedings, International Workshop on Catalytic Combustion, H. Arai, editor, page 108, 1994.

McManus, K.R., Poinsot, T. and Candel, S.M., "A Review of Active Control of Combustion Instabilities", Prog. Energy Combustion Science, volume 19, page 1 – 29, 1993.

Micklow, G.J., Roychoudhry, S., Nguyen, H. and Cline, M.C., "Emissions Reduction by Varying Swirler Airflow Split in Advanced Gas Turbine Combustors", ASME Journal of Gas Turbines & Power, volume 115, page 563 – 569, New York, 1993.

Nealy, D.A., "Combustor Cooling – Old Problems and New Approaches", Gas Turbine Combustor Design Problems, A H. Lefebvre editor, page 151 – 185, Hemisphere, Washington, DC, 1980.

Nicol, D.G., Steele, R.C., Marinov, N.M. and Malte, P.C., "The Importance of Nitrous Oxide Pathway to NO in Lean Pre-Mixed Combustion", ASME Journal of Gas Turbines & Power, volume 117, page 100 – 111, New York, 1995.

Nishio, K., Igashira, K.I., Take, K. and Suemitsu, T., "Development of a Combustor Liner Composed of Ceramic Matrix Composite", ASME paper # 98-GT-104, New York, 1998.

Ohtsuka, M., Yoshida, S., Inage, S. and Kobayashi, N., "Combustion Oscillation Analysis of Premixed Flames at Elevated Pressures", ASME paper # 98-GT-581, New York, 1998.

Paxson, D.E., "A Comparison Between Numerically Modeled and Experimentally Measured Loss Mechanisms in Wave Rotors", AIAA Journal of Propulsion Power, volume 11, number 5, page 908 – 914, 1995 (also NASA TM 106279).

Qi, S., Gupta, A.K. and Lewis, M.J., "Effect of Swirl on Temperature Distribution in Pre-Mixed Flames", 35th AIAA Aerospace Sciences Meeting, paper # 97-0373, 1997.

Rangwala, A.S., "Dynamic Analysis of Turbo-machinery", McGraw-Hill, New York, 2004.

Richards, G.A. and Janus, M.C., "Characterization of Oscillations During Premix Gas Turbine Combustion", ASME paper # 97-GT-244, New York, 1997.

Rink, K.K. and Lefebvre, A.H., "Influence of Fuel Drop Size and Combustor Operating Conditions on Pollutant Emissions", International Journal of Turbo and Jet Engines, volume 6, number 2, page 113 – 122, 1989.

Sasaki, M., Kumakura, H. and Suzuki, D., "Low NO_x Combustor for Automotive Ceramic Gas Turbine – Conceptual Design", ASME paper # 91-GT-369, New York, 1991.

Sato, H., Mori, M. and Nakamura, T., "Development of a Dry Low NO_x Double Swirler Stage Gas Turbine Combustor", ASME paper # 96-GT-134, New York, 1996.

Sawyer, T., "Gas Turbines – Volumes I, II and III", International Gas Turbine Institute, Atlanta, ASME, 1982.

Scalzo, A.J., Sharkey, W.T. and Emmerling, W.C., "Solution of Combustion Noise in Coal Gasification Cogeneration Application of 100-MW Class Combustion Turbines", ASME paper # 89-GT-217, New York, 1989.

Scalzo, A.J., Holden, P.C. and Howard, G.S., "The Westinghouse 501D Combustion Turbine Engine", ASME paper # 81-GT-32, New York, 1981.

Schlatter, J.C., Dalla Betta, R.A., Nickolas, S.G., Cutrone, M.B. and Beebe, K.W., "Single Digit Emissions in a Full Scale Catalytic Combustor", ASME paper # 97-GT-57, New York, 1997.

Segalman, I., McKinney, R.G., Sturgess, G.J. and Huang, L.M., "Reduction of NO_x by Fuel Staging in Gas Turbine Engines – A Commitment to the Future", AGARD Conference Proceedings 536, page 29/1 – 17, 1993.

Seume, J.R., Vortmeyer, N., Krause, W., Hermann, J., Hantschk, C.C., Zangl, P., Vortmeyer, D. and Orthmann A., "Application of Active Combustion Instability Control to a Heavy Duty Gas Turbine", ASME paper # 97-GT-119, New York, 1997.

Snyder, T.S., Rosfjord, T.J., McVey, J.B. and Chiappetta, L.M., "Comparison of Liquid Fuel/Air Mixing and NO_x Emissions for a Tangential Entry Nozzle", ASME paper # 94-GT-283, New York, 1994.

Touchton, G.L., "Influence of Gas Turbine Combustor Design and Operating Parameters on Effectiveness of NO_x Suppression by Injected Steam or Water", ASME paper # 84-GT-IPGC-3, New York, 1984.

Toyne, G.B. and Hodges, G.S., "All-Electric Accessory Drive System Implications in Engine Design and Performance", AGARD Conference Proceedings No. 382, 1983.

Volkov, D.V., Belokin, A.A., Lyubimov, D.A., Zakharov, V.M. and Opdyke, G., "Flamelet Model of NO_x in A Diffusion Flame Combustor", ASME paper # 00-GT-099, New York, 2000.

White, D.J., Batakis, A., Le Cren, R.T. and Yacabucci, H.G., "Low NO_x Combustion Systems for Burning Heavy Residual Fuels and High Fuel-Bound Nitrogen Fuels", ASME Journal of Engineering Gas Turbines & Power, volume 104, page 377 – 385, New York, 1982.

Yoshida, Y., Oyakawa, K., Aizawa, Y. and Kaya, H., "A High Temperature Catalytic Combustor with Starting Burner", ASME paper # 00-GT-087, New York, 2000.

16

Performance Enhancement

16.1 INTRODUCTION

Performance guarantees from the manufacturers of gas turbines play a major role in a customer's selection decision of one unit over another. Fuel, operating and other ownership costs are directly linked to the specific fuel consumption, power generating capacity and durability of the unit. Factors affecting the performance of a gas turbine are of considerable interest, both for the manufacturers and for their customers. Losses as a consequence of a number of operating circumstances, hardware configuration and combustion characteristics of the fuel deserve close scrutiny.

Sealing of the gap between the rotating blade tips and the stationary engine casing in the high-pressure compressor and turbine sections of a gas turbine plays a major role in the specific fuel consumption. Effective sealing of the gas path can also provide dramatic improvements in compressor stall margin, increased power and reduced maintenance costs. In the case of aircraft propulsion plants, controlling the radial gap contributes to longer time on the wing, increased payload and mission range capabilities. Removal of an engine from service is primarily because of spent exhaust gas temperature margin, which is caused by deterioration of the high-pressure turbine clearances.

Ambient conditions are another factor that affect the performance of a gas turbine. Pressure, temperature and relative humidity of the air in the vicinity of the gas turbine affect the mass flow of the air entering the inlet, and will have a substantial impact on the performance of the gas turbine. Generally referred to as International Standards Organization (ISO) Standards, 59°F (14°C) temperature, 14.7 lbs/in^2 (1.013 bar) pressure, and 60 per cent relative humidity are widely used as standard conditions in the power generation industry.

The compression ratio of a simple-cycle gas turbine is in the range of 12 and 45:1. The increase in pressure of the ambient air is accompanied by an increase in the temperature. The air temperature at the exit from the compressor is between 650°F and 900°F (345°C and 480°C). Entering the combustors, except for friction losses the pressure remains virtually unaffected during the combustion process. But the temperature in the combustors climbs to between 2200°F and 3000°F (1200° and 1650°C). As the hot and pressurized gases flow past the stationary vanes and the rotating blades of the turbine, the thermal energy of the hot gases is converted into mechanical work. Following the first law of thermodynamics, the thermal energy of the expanding hot gases is converted into kinetic energy as the flow velocity increases in the stationary passages between the vanes. The rotating blades of the turbine then convert the kinetic energy to work. The work developed in the turbine is adequate to not only drive the compressor but also the generator load in power generation turbines or develop thrust in aircraft power plants. Depending on the size of the gas turbine, the compressor may consume anywhere between 55 and 67 per cent of the total work developed by the turbine.

Single-shaft gas turbines have one continuous shaft, so the compressor and the turbine operate at one speed. The generator may be located at either the compressor or the turbine end. This arrangement is suitable for power generation applications where the operating speed is held constant at 3600 rpm for 60 Hz electric power or at 3000 rpm for 50 Hz power. Some gas turbines operate at 5400 rpm for 60 Hz power. The turbine shaft is then connected through a 2:3 speed reducer to the electrical generator.

A performance test is normally conducted after the gas turbine is installed, with parameters such as power, fuel and heat consumption duly recorded. The objective is to permit adjustment of the relevant parameters to the condition of the guarantee. The ASME Performance Test Code PTC-22-1985, "Gas Turbine Plants," establishes the testing procedures and calculation methods. All the instruments used for data collection must be inspected and calibrated before the test.

16.2 THERMODYNAMIC ANALYSIS

The gas turbine operates on the Brayton cycle. The Brayton principle consists of two isobaric and two isentropic processes, as shown in figure 16.1. The former takes place in the gas turbine's combustor and in the steam generator's gas side, and the latter represents the compression of air and expansion of gases in the turbine. Gas turbine cycle efficiency depends on the compression ratio and turbine firing temperature, increasing with both parameters according to the relation shown in equation 16.1. It is assumed in the derivation that the pressure ratios in the compressor and the turbine remains unaltered, fuel flow rate is considerably smaller than that for air, specific heat of the gases stays constant and that all components operate without incurring any losses:

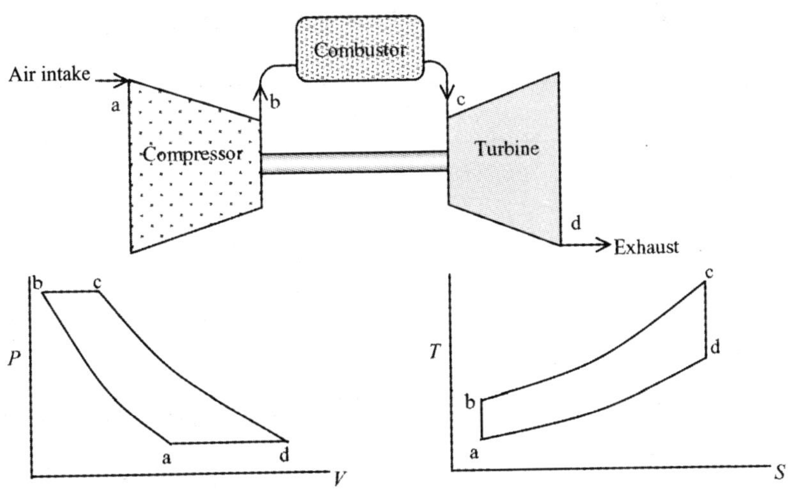

Fig. 16.1 Standard Brayton Cycle for Gas Turbine

$$\eta = 1 - \left(\frac{1}{r^{\frac{\gamma-1}{\gamma}}} \right) \qquad ...(16.1)$$

where η is ideal cycle efficiency, r represents compression ratio and γ is ratio of specific heats at constant pressure and constant volume. Turbine and compressor efficiencies will have a moderating effect on the overall efficiency. Figures 16.1 and 16.2 demonstrate the impact of the two parameters. Work performed per pound of air occurs at a lower pressure ratio than the point of maximum efficiency for a given firing temperature. Overall cycle efficiency is improved with increased pressure ratio, cooler compressor inlet temperature and higher turbine inlet temperature (figure 16.3). Evaporative coolers, direct water fogging and refrigerated cooling systems are some methods used to cool the air at the inlet. The impact of efficiencies in the compressor, combustor and turbine, as also system related pressure losses in a simple cycle turbine, is shown in figure 16.4.

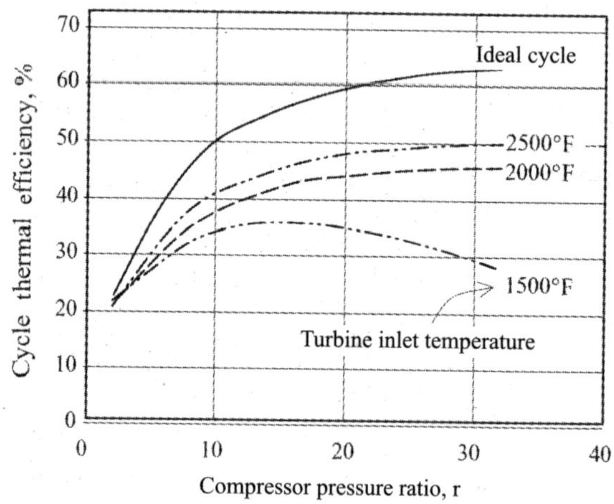

Fig. 16.2 Compression Ratio and Thermal Efficiency

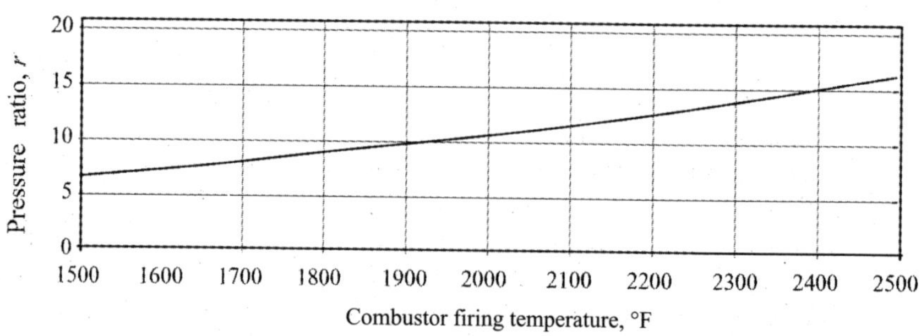

Fig. 16.3 Optimized Pressure and Temperature for Maximum Thermal Efficiency

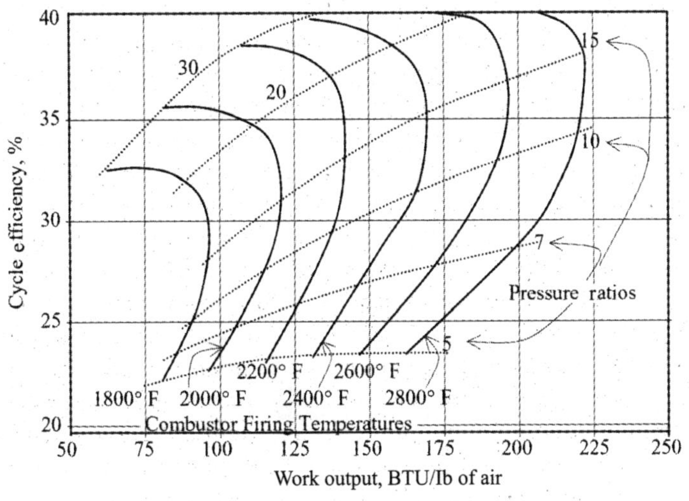

Fig. 16.4 Simple Cycle Gas Turbine Performance Map

Since gas temperature at the turbine exit is higher than that exiting from the compressor, insertion of a regenerator to preheat the air between the compressor and the combustor with the turbine's exhaust gases will reduce fuel requirement. If T_1, T_2, T_3 and T_4 represent inlet temperatures to compressor, regenerator, combustor and turbine and T_5 is the exit temperature from the turbine, ideal regenerator efficiency (assuming no pressure loss) may be expressed in the following form:

$$\eta_{\text{Reg}} = \frac{T_3 - T_2}{T_5 - T_2} \qquad ...(16.2)$$

System cycle efficiency then takes the form:

$$\eta_{\text{Reg-Cycle}} = \frac{(T_4 - T_5) - (T_2 - T_1)}{(T_4 - T_3)} \qquad ...(16.3)$$

Regenerator efficiency will depend on available surface area for heat transfer, but increased area will call for higher cost, pressure loss and space. Heat exchangers may be of regenerative or recuperative type. Regenerative systems call for another medium to effect transfer of heat between turbine exhaust gases and compressed air. Thus, heat flows into and out of the intermediate fluid. Recuperative heat exchangers have heat-transferring elements at a constant temperature, with the air and gas paths arranged in counter flowing directions (*ref*: Boyce, 2002).

Power output from a gas turbine can be increased in a number of ways. Inter-cooling of air between the stages in a compressor may be used to reduce work done to pressurize the air. Temperature reduction of the partially compressed air causes its volume to shrink, consequently less work is required to compress it to the next pressure level. Thermal efficiency of a simple cycle is decreased by the addition of an inter-cooler, but addition of an inter-cooler to a regenerative gas turbine cycle increases thermal efficiency and power output. The explanation is that a larger portion of heat required for regeneration now comes from the turbine exhaust instead of additional fuel consumption.

When turbine expansion is split into two or more steps, with constant pressure heating taking place before each expansion, the process is referred to as reheat cycle. As in inter-cooling the thermal efficiency of the simple cycle is reduced by the reheat, while work output rises. In combination with a regenerator, reheat can be made to increase thermal efficiency. Performance curves for a simple cycle gas turbine that includes provision for inter-cooling, regeneration and reheat are shown in figure 16.5. Figure 16.6 provides details of a single shaft gas turbine supported on two bearings.

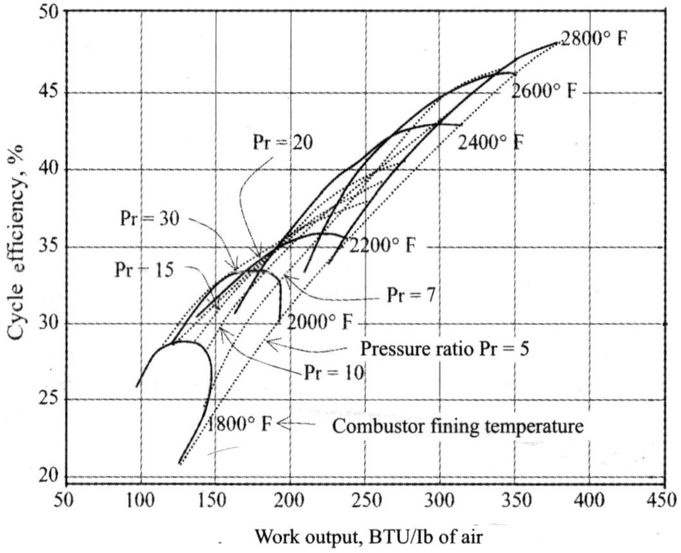

Fig. 16.5 Simple Cycle Gas Turbine with Inter-cooling, Regeneration and Reheat – Performance Map

The effects on cycle efficiency due to changes in the specific output (output power per pound of airflow) at different firing temperatures (in the combustors) and pressure ratios can be considerable. An increase in specific output will result in a smaller gas turbine for the same output power. Simple-cycle gas turbines increase in efficiency at a given firing temperature when the pressure ratio increases. Also, the increase in firing temperature results in increase in specific output for a given pressure ratio. The pressure ratio has less effect on efficiency in combined cycles. The specific output decreases when the pressure ratio increases. The thermal efficiency increases with increasing firing temperature.

The parameters giving optimum performance are different between simple and combined cycles. Increasing the pressure ratio increases the efficiency in simple cycles. Having a relatively modest pressure ratio and higher firing temperature increases the efficiency in combined cycles. For example, General Electric's MS-7001-FA design parameters are a pressure ratio of 14:1 and a firing temperature of 2350°F (1288°C). The combined-cycle efficiency of this machine is optimized. However, its simple-cycle efficiency is not. On the other hand, the pressure ratio of the LM-6000 is 24:1 and its simple-cycle efficiency is 40 per cent.

Fig. 16.6 W501F Compressor and Turbine Components on a Single Shaft (*Courtesy*: Siemens Westinghouse)

16.3 PERFORMANCE FACTORS

Pressure losses are encountered in a gas turbine system when air filtration, silencing, evaporative coolers, chillers in the inlet, or exhaust heat recovery devices are inserted. Effects vary with the unit. A pressure drop of 4 in (10 mbar) of water at the inlet may be accompanied by a decrease in the output power and efficiency of around 1.5 and 0.5 per cent, respectively, may be expected. The same pressure drop at the exhaust of a gas turbine will reduce the output power and efficiency by around 0.4 per cent.

The type of fuel used also has an impact on the performance of the gas turbine. In addition to being a clean fuel that leaves behind little or no residue after combustion, natural gas also offers the advantage of producing more power output then distillate oil. Since a higher concentration of water vapour is present in the products of combustion of natural gas due to a higher hydrogen-carbon ratio, the resulting higher specific heat is responsible for a difference of as much as 2 per cent more power production.

Gaseous fuels having lower heating values than natural gas can significantly alter the performance of a gas turbine. When the heating value falls, the fuel flow must increase to provide additional heat. However, the compressor does not pressurize the additional mass flow. Since the compressor power is not affected by this change, the turbine and the output power of the machine are increased. Some other side effects may also be expected, including:

- An increase in mass flow through the turbine increases the power developed by the turbine. The compressor absorbs some of this increased power, but the net result is an increase in the pressure ratio across the compressor. The compressor then operates closer to the surge limit.
- Because the increase in turbine power has the potential to take the turbine and the equipment in the power train above their 100 per cent rating, durability of the equipment rated at higher limits may be required in some cases.
- A higher volume of low or medium heating value (Btu) coal gases is usually required at elevated temperatures, which have a lower density. In tandem with the increased fuel flow, the consequent increase in the volumetric flow will require increments in the size and cost of the piping, fittings and valves.
- Gases having low heating values (Btu) have a propensity to be saturated with moisture before delivery to the turbine. This results in an increase in the heat transfer coefficients of the combustion products, leading to an increase in the substrate temperature in the turbine.
- The amount of air required for burning the fuel increases as the heating value decreases. Gas turbines having high firing temperatures may not be able to operate using low-heating-value fuel.

Each model of a gas turbine will require a set of guidelines for any application. The list includes specification of flows, temperatures and output power to ensure performance, efficiency and life of the machine. In most applications involving lower-heating-value fuel, it is assumed that the efficiency and power output will be equal to or higher than the ones obtained using natural gas. On the other hand, in applications where higher heating value fuels such as refinery gases are contemplated, the efficiency and output power will be less than or approach those obtained using natural gas.

Water and steam injection have been used during the last few decades to reduce NO_x emissions. This technique involves injecting water or steam in the cap area, or the head end, of the combustor liner. The output power and efficiency will increase due to the additional mass flow. However, each machine has limits on the amount of water or steam injected. These are imposed to protect the combustor and turbine section. Steam injection can increase the output power and efficiency by 20 and 10 per cent, respectively. Water injection can increase the output power by 10 per cent. However, it has very little effect on efficiency because more fuel is needed to raise the water to the combustor's temperature level.

Many gas turbines require air to be bled from the compressor. For example, aircrafts use the air from the auxiliary power unit for starting the main engines. Air from the main engines is used for items as diverse as cabin pressurization and cooling the combustor liner and the blades and vanes in the high-pressure turbine section. In general, up to 5 per cent of the flow can be extracted from the discharge casing of the compressor. This can be done without modification to the casings or on-base piping. Higher flow (from 16 to 20 per cent) can be extracted from the compressor. However, this requires modifications to the casings, piping and controls. Extracting the air from the higher-pressure stages of the compressor has a significant effect on the performance of the machine. The rule of thumb is that every 1 per cent of air extraction reduces the power output by 2 per cent. But when the intent is to lower the metal temperatures, the higher temperature at the turbine inlet aids in increasing the thermal efficiency of the unit. This provides ample justification for extracting the air from the compressor.

The performance values for a machine are normally given for base load ratings. The American

National Standards Institute (ANSI) B 133.6 Ratings and Performance guideline defines operation at base load at the rate of 8000 hours in a year with 800 hours per start, and peak load operation at the rate of 1,250 hours per year with 5 hours per engine start. Because of their ease in starting, gas turbines are the preferred option for meeting peak load requirements.

Since peak-load operating hours are shorter, increasing the firing temperature can increase the power output. Shorter operating periods will necessitate more frequent inspections of the gas turbine, but this disadvantage is justified since operation at peak levels is generally more cost-effective. The additional power is then generated in periods of higher power cost for the consumer. Generators also have peak ratings. These are obtained by operating at a higher power factor or temperature increase. The firing temperature can be adjusted between the base and the peak conditions. They are chosen to maximize the power output while remaining within the limits of the repair interval of the turbine hot section.

Since the mass flow of air entering the gas turbine determines the amount of fuel for proper combustion, ambient air conditions will determine the amount of power produced. Figure 16.7 illustrates the case of ambient air temperature affecting the output power, heat rate (reciprocal of thermal efficiency), heat consumption and the exhaust flow for a typical single-shaft heavy-duty gas turbine. Another example is the altitude of the location where the gas turbine is operating. Airflow and power output of a gas turbine will decline with increasing altitude due to the decrease in the barometric pressure. The reduction in these parameters is proportional to the decrease in the air density. Typically, the decrease in airflow and output power of a gas turbine is 1 per cent for every 100-metre increase in the altitude, although the heat rate and other cycle parameters will not experience any changes.

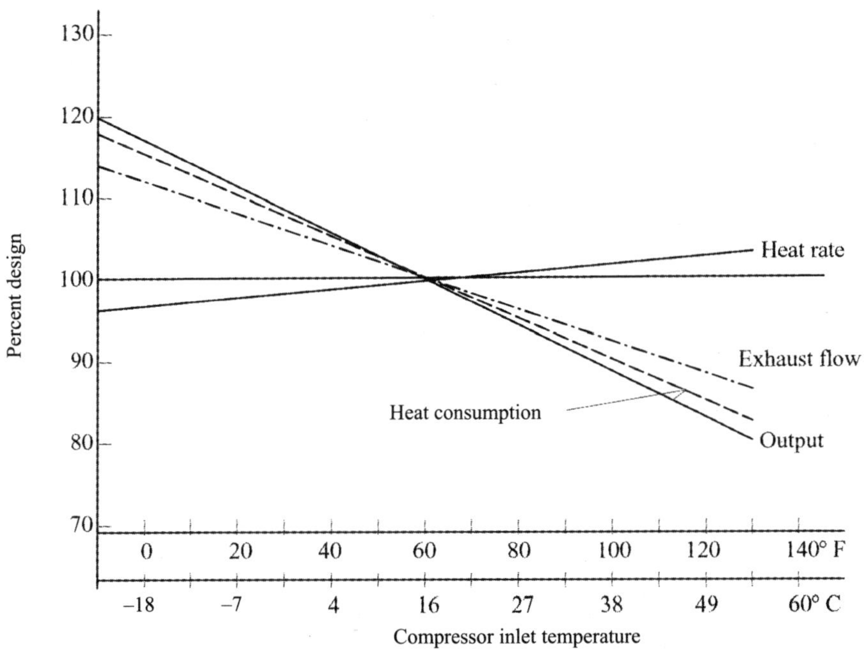

Fig. 16.7 Effect of Ambient Temperature (*Courtesy*: General Electric)

Because density of humid air is lower than that of dry air, an increase in ambient humidity will reduce the power output and efficiency of a gas turbine. An increase in specific humidity of 0.01 kg of water vapour/kg dry air will typically reduce the power output and efficiency by 0.0015 and 0.0035 per cent, respectively. Although sometimes considered negligible, water vapour content is of practical importance when water or steam injection is essential for controlling the level of nitric oxide (NO_x) in the exhaust gases. This humidity effect is mainly caused by the control system approximation of the firing temperature. Some gas turbine control systems reduce the power when ambient humidity increases. However, on some multi-spooled aero-derivative engines the power generated is increased because the discharge temperature from the gas generator is used to regulate the fuel flow. The increased level of fuel flow raises the temperature of the moist air to the set-point temperature. The increase in the fuel flow will increase the speed of the gas generator rotor. The operating speed of the gas generator is independent of the speed of the power turbine rotor. Thus, the increase in fuel flow rate compensates for the decrease in density of the moisture-laden air.

The performance of all turbo-machinery generally deteriorates as they accumulate operating time. Some of the losses encountered from the degradation are recoverable. Fouling of the compressor flow path occurs because of deposition of materials on the surfaces, which generally reduces the flow area for the air. The deposits may be partially removed through washing of the compressor with water. When the turbine is opened for maintenance, mechanical cleaning of the compressor blades and vanes will remove all of the deposits. On the other hand, increased tip clearances at the blade tip in the turbine and compressor will lead to non-recoverable losses. Changes in the surface finish due to pitting of airfoil contours will also cause non-recoverable losses. When sulphur is present in liquid fuels, it will result in the formation of chrome sulphides, which will lead to corrosion and unevenness of the surface of the airfoils. These losses can only be recovered if the affected parts are replaced. The normal recommended interval for inspection of the hot gas path is 24,000 hours of operation. The total performance degradation that may be expected in the time period lies between 1 and 1.5 per cent. Frequent off-line water washing is known to reduce the amount of recoverable and non-recoverable losses. In general, machines that operate in hot, dry climates degrade less than those in humid climates.

16.4 ADAPTIVE PERFORMANCE SIMULATION

Estimation of performance parameters and cycle details for any operating condition encountered in the field is of fundamental importance. Engine performance computer models that are based on division of the engine into its components according to the thermodynamic process occurring in each of them are required for the purpose.

A common feature of all simulation techniques is the need for component maps. The reliability of the predictions depends on the accuracy of the maps. But there is one aspect that cannot be covered by the maps. Assembly and manufacturing tolerances in different engines of a particular series exhibit small differences in their performance. The differences can be important during condition monitoring, since deviations from the baseline constitute the fundamental point of the procedure. Disassembly and rebuilding of an engine can also cause small shifts in its performance (*ref*: McLeod et al., 1988).

A performance model of a gas turbine views the unit as assembly of different components or modules. Each component is identified according to the thermodynamic process it carries out. The engine cycle at any operating point is defined by the values of the thermodynamic properties of the working fluid at stations placed at the inlet and exit of the component. Each station is characterized by a number of thermodynamic parameters of the fluid; namely, total and static pressures and temperatures, mass flow rate W, Mach number M, fuel/air ratio f, and specific enthalpy h.

For performance predictions, the flow is assumed to be uniform radially and circumferentially, thus yielding a one-dimensional model. The working fluid is assumed to be a perfect gas, with properties changing with the temperature. The parameters are interrelated through the compressible flow relations, which indicate that only four of them are independent. Determination of the independent parameters is sufficient for calculating the rest of them (*ref*: Stamatis et al., 1989).

If $\mathbf{Y}_{IN}$ is the vector of independent variables at the inlet and $\mathbf{Y}_{OUT}$ is the corresponding vector at the outlet, for each module there exists a relation of the form $f(\mathbf{Y}_{IN}, \mathbf{Y}_{OUT}) = 0$. This relation is derived from mass, energy and momentum conservation laws, as also from experience in the operation of the component. $\mathbf{Y}_{IN}$, $\mathbf{Y}_{OUT}$ do not necessarily include the same variables. It can be an analytic relation, possibly including empirical constants (for example, duct pressure loss) or a set of curves (for example, compressor map).

Compatibility of component functioning imposes matching conditions, as for example, power balance between turbine and compressor. A set of simultaneous equations that have to be satisfied by the fluid parameters is thus formed. Referred to as the straight engine model in figure 16.8, solution of the system for one operating point yields the full cycle details (*ref*: Saravananamuttoo et al., 1983). Being highly non-linear, the system of equations is solved numerically. The approach followed first calls for suitably choosing values for variables v_i, then solving the equations explicitly to obtain the full set of parameters. Error terms are then formed from the differences to get: $e_i = |PI_1 - PI_2|$, where PI_1 and PI_2 are values of a parameter Pi calculated by two different equations. For instance, turbine power can be calculated from balance with the compressor and from turbine flow, pressure ratio and efficiency. The guesses are then repeated until the error terms are reduced to zero. The procedure may be formulated as a minimization problem, following the least squares logic. Input data in the model is usually in the form of performance maps. Selection of a suitable set of independent parameters will define one steady-state operating point. All cycle details and performance parameters are then uniquely defined.

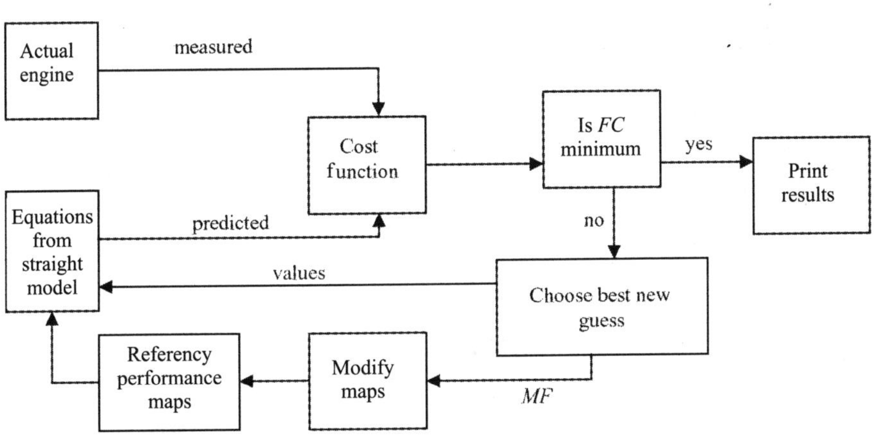

Fig. 16.8 Flow Chart for Adaptive Simulation Model

In an adaptive simulation model, the performance maps of each engine component are expressed by functional relations between performance parameters of the type $f(\mathbf{Y}_{IN}, \mathbf{Y}_{OUT}) = 0$. If a particular parameter has a value X_{ref} on the reference map and a value X_{act} on the actual map, then the correspondence between the two is expressed by a modification factor MF of the form $MF = X_{ref}/X_{act}$. Knowledge of the reference performance map and the values of MF leads to the possibility of reproducing the actual maps. The factors must be consistent with existing representation of the maps and the set of equations used. For instance, in the case of the compressor the rotor speed and pressure ratio may be used to define the factors MF. Thus, $P_{act} = MF \times P_{ref}$.

Having chosen a set of values MF_i, it is possible to define an optimization problem that allows determination of their values required for adapting the maps. The flow chart of the procedure is shown in figure 16.8. A cost function FC is defined by the equation: $FC = \Sigma a_i e_i^2 + \Sigma b_i (Q_{ci} - Q_{mi})^2$ where a_i, b_i are weight functions and Q_{ci} and Q_{mi} are calculated and measured mass flow parameters. Index i denotes axial stations as shown in figure 16.9, and is the index for the two summation terms in the definition for FC. If modified maps are introduced into the model through a set of values for MF_i, the value of FC obtained will be a function of the original guess for v_i's and MF_i's. Covering the entire operating range of the engine will yield the full set of MF_i's. The non-linear minimization problem is then solved using a generalized minimum residual method till convergence is attained.

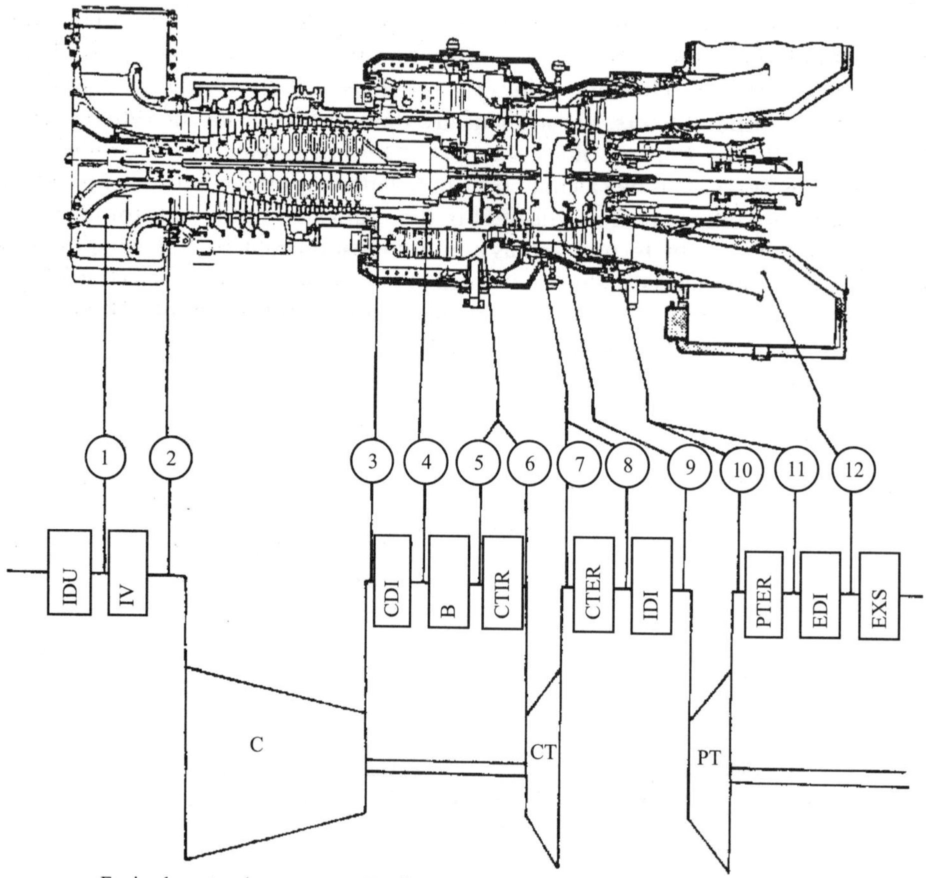

Engine layout and components; B = burner; C = compressor; CDI = compressor diffuser; CTIR = compressor turbine inlet remix; CT = compressor turbine; CTER = compressor turbine exit remix; EDI = exhaust diffuser; EXS = exhaust stack; IDI interduct diffuser; PT = power turbine; IDU = inlet duct

Fig. 16.9 Gas Turbine Engine Layout and Station Identification

The method is applied to the case of the single shaft version of the 'Tornado' gas turbine, manufactured by Ruston Gas Turbines (*ref*: Timperley et al., 1983). The core turbine and the power turbine can be readily decoupled from each other for the two-shaft version. The adaptive model offers the advantage of tracking deviations from the average engine through the use of a test bed data. The test data is obtained from an engine, which, after 13,000 hours of operation, has tip clearances that are considerably larger than normal values.

Component data maps obtained from the manufacturer are compared with test data acquired from a wide range of compressor operating conditions. For each equilibrium operating point, appropriate modifying factors based on actual and reference mass flow rate, pressure, temperature and polytropic efficiency are applied for the different components.

Predicted results from the straight engine model and the component maps model differ from measured values for different operating points. The comparisons for the straight engine model are shown for two rotor speeds in figures 16.10 and 16.11. Depending on the speed and load, predicted values differ from the measured ones in various degrees, and the predictions are nor generally accurate.

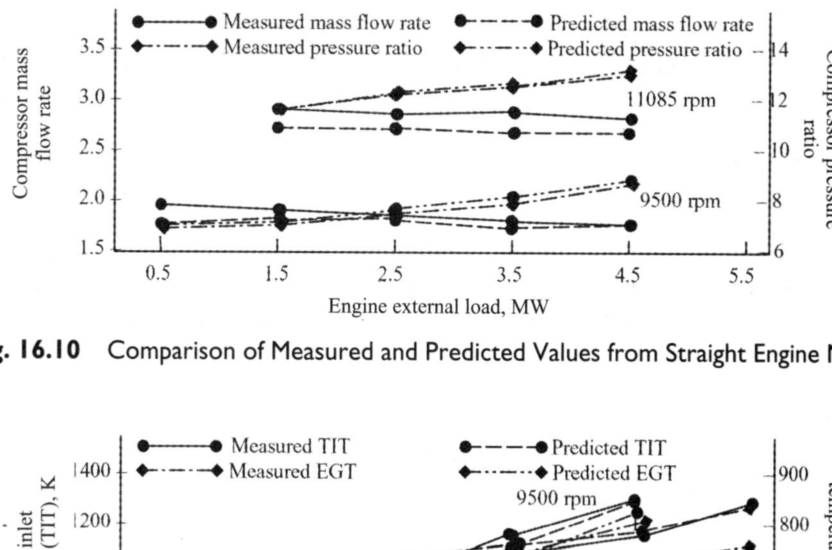

Fig. 16.10 Comparison of Measured and Predicted Values from Straight Engine Model

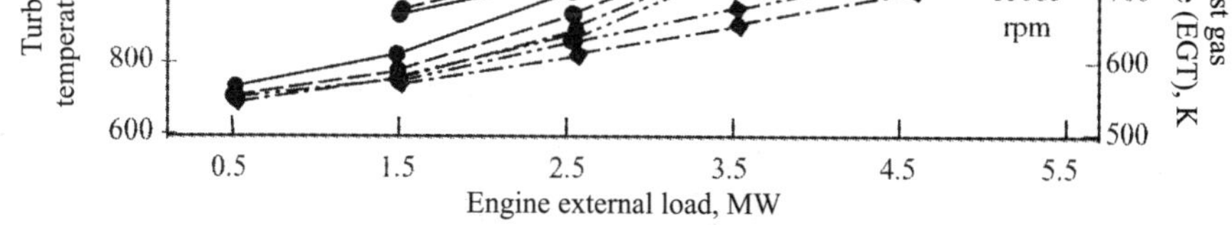

Fig. 16.11 Comparison of Measured and Predicted Values From Straight Engine Model

With the introduction of measured values into the adaptive model, a set of modification factors are produced for adapting the component maps. The modification factors for the range of operating points under consideration result in mass flow corrections of the order of between 3 and 5 per cent. This agrees with the manufacturer's experience about differences in flow capacity of different Tornado units. The degree of dispersion in the cluster of points indicates the scatter that may be expected for the modification factors.

Comparing the measured and predicted values of the total pressure and temperature, the temperature is much better predicted than in the straight engine model. Prediction of pressure does not display the same level of accuracy, but the error has been reduced from 8 per cent to 1.6 per cent.

Determination of the modification factors can also be used for detecting faults, either in the engine components or in the sensors. If, for instance, the efficiency of a component drops, it will show up as a change in the corresponding modification factor and may thus be detected. If on the other hand a sensor fails, its reading will influence the calculated value of the modification factor.

16.5 SEQUENTIAL LINEARIZATION FOR REAL-TIME SIMULATION

A real-time simulation of a gas turbine offers an essential basis for advanced propulsion control designs. Accuracy and speed are of the essence because the turbines respond rapidly to changes in the input, with time constants of less than 1 second. This implies that the process must be computationally simple.

A large body of evidence exists that indicates that the performance of a marine gas turbine system is governed by a non-linear set of dynamic relations (*ref*: Kidd et al., 1985; Rubis et al., 1986). One approach is to use sequential state space linearization approximation to the non-linear dynamics, which are simple enough to be computed in real time. The method of developing the model is applied to the gas turbine plant apparatus shown in figure 16.12. The equipment comprises of a Boeing 502-6A 175 hp gas turbine connected to a Clayton 17-300 water brake dynamometer. The multi-port diagram of figure 16.13 is a graphical means of representing the essential components of the gas turbine, the interaction among its components, and the identification of the variables in the interactions. The method could then be outlined in four steps: (*i*) determine the non-linear analytical relationships between the inputs and the outputs of each component, (*ii*) linearize the component models, (*iii*) combine these linear component models to obtain the overall linear system model, and (*iv*) simulate performance of the system to investigate speed and accuracy of the model.

The shaft, rotor and fuel components display significant time lag, and are modeled with rate

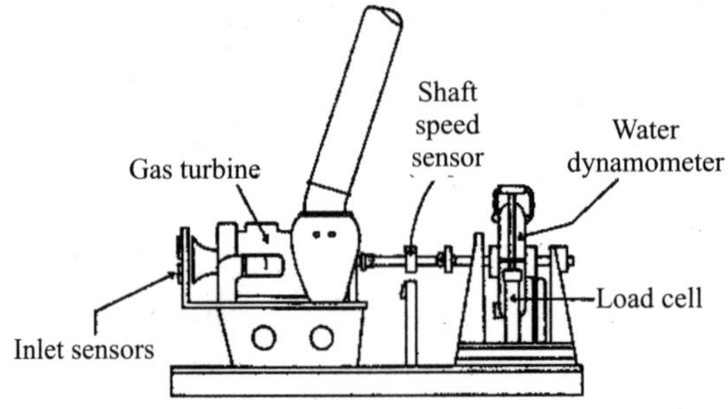

Fig. 16.12 Test Bed Installation

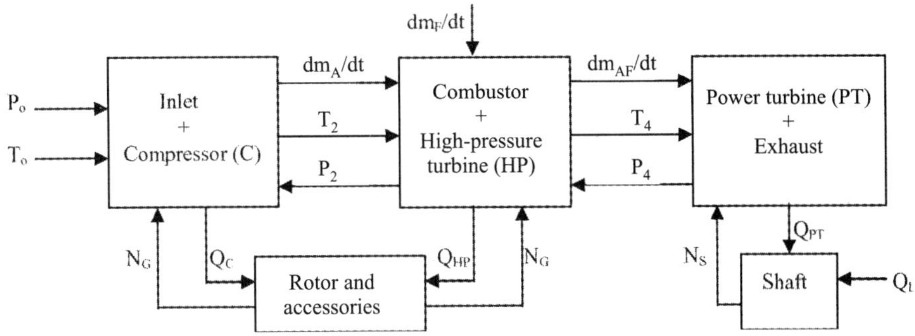

Fig. 16.13 Cause and Effect Plant Model

equations of the type: $dN_G/dt = (Q_{HP} - Q_C)/J_G$. The other components are modeled with simple algebraic equations. For example, the compressor torque is modeled as a quadratic function in terms of all its component input variables in the form: $Q_C = aN_G^2 + bP_2^2 + cN_GP_2 + dN_G + eP_2 + f$.

Each component output equation is linearized through perturbation analysis. Thus, the foregoing equations take the following form: $\delta dN_G/dt = (\delta Q_{HP} - \delta Q_C)/J_G$ and $\delta Q_C = 2aN_G\delta N_G + 2bP_2\delta P_2 + cN_G\delta P_2 + cP_2\delta N_G + d\delta N_G + e\delta P_2$, where a typical perturbation means: $\delta N_G = N_G - N_{Gi}$.

System linearization is found by combining the linearized component equations into a space form. The states of the gas turbine are those variables that comprise the minimum list necessary to predict the dynamic and steady state performance. The independent state variables describe the entire machine; the rest of them are dependent variables.

For controller design purposes, it is necessary to know the non-linear dynamic relationship between the states and the inputs, and may be expressed as a set of differential equations of the type: $d\mathbf{X}/dt = \mathbf{f}(\mathbf{X}, \mathbf{U})$. The overall plant dynamic behaviour may be approximated with a series of perturbational linearizations: $d\mathbf{x}/dt = \mathbf{Ax} + \mathbf{Bu}$, where perturbed state $x = \delta X$ and perturbed input $u = \delta U$. After expansion in terms of the more familiar variables, the problem then reduces to expressing the linear component output equations in the form of a matrix equation. Entries for the state coefficient matrix **A** and the input coefficient matrix **B** then become functions of only two independent variables, N_G and N_S, gas generator and power shaft speeds. The final step in the modeling process is to curve fit a function in N_G and N_S to each of the coefficient entries.

During simulation, each matrix entry is computed on each time step in order to preserve the nature of the underlying small perturbation analysis. Typical simulation results are shown in figures 16.14 and 16.15. A slight change in the fuel flow rate is accompanied by a large change in load torque. For reference, a simulation run is based upon a single linearization about the run start point in figure 16.15. The sequential linearization method (SLM) is seen to predict a better response than the single linearization. Also, the qualitative agreement between the actual data and SLM curves is quite good for the large excursions in the operating conditions. The issue about computational time is easily resolved through use of a larger processor.

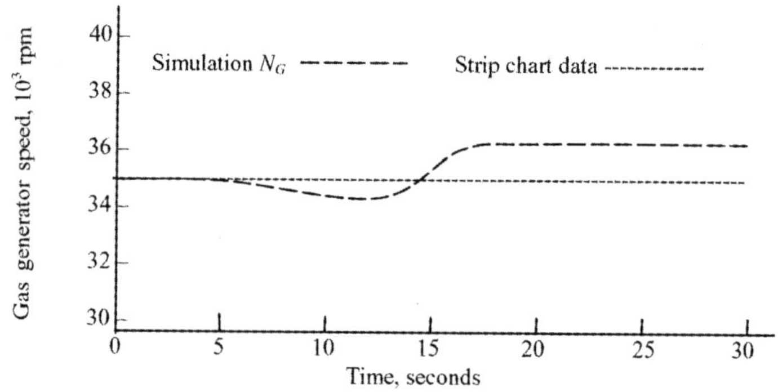

Fig. 16.14 Sequential Linearization for Gas Generator Simulation

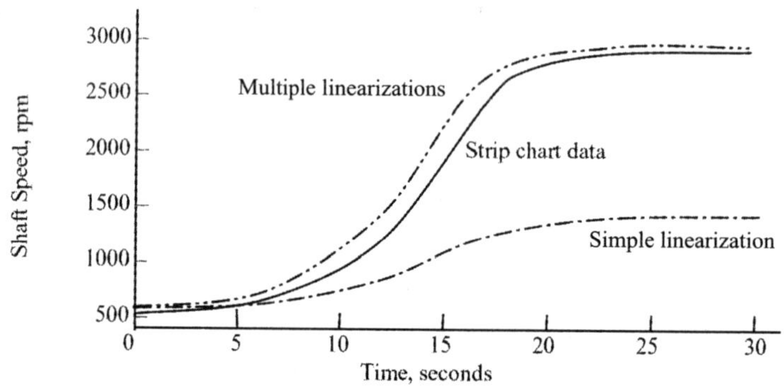

Fig. 16.15 Simulation of Shaft Speed Response

16.6 CYCLE ANALYSIS FOR HELICOPTER ENGINE

Combat and medium lift helicopters typically require a 1000 kW engine. The major cycle parameter used in the analysis is pressure ratio, with the technology level defined by the allowable material temperatures and the cooling effectiveness for the gas generator first stage turbine rotor. This translates into an achievable cycle temperature at the stator outlet (or, SOT) that depends considerably on the temperature of the cooling air bled from the compressor outlet, and hence also depends on the pressure ratio. Compressor and turbine efficiency trends are assumed to be similar to thos indicated by Niedzwiecki et al, 1987.

In a thermodynamic analysis the values of specific power and specific fuel consumption (sfc) are calculated over a range of pressure ratios. It is possible from these data to deduce the design options to obtain the minimum engine weight. The minimum engine plus fuel weight is likely to be the choice of the helicopter designer, who is mostly interested in designing for minimum take-off gross weight (TOGW). Minimum engine weight, when allied with a low component count, should provide the minimum first cost option.

In most applications where a certain power level is specified, technology improvements such as increased operating temperatures that lead to significant increases in specific power will also lead to smaller engines. But they will tend to have inferior component performance due to the effects of Reynold's number, clearances, surface roughness, manufacturing constraints and the limitations of design methods. The benefits of the higher temperature in the cycle have historically outweighed this disadvantage, and each new generation of engines has had lower sfc as well as smaller size. But at very high cycle temperatures the effects of dissociation will eventually negate that trend, even if component efficiencies are maintained (*ref*: Bewley, 1989).

Advanced technology engines have compressor efficiency that is 2 to 5 per cent higher than present engines in the power range of interest. The improvements result from the benefits of advanced fluid computational dynamics, reduced clearances and other factors. Smaller engines may be expected to benefit more from three-dimensional fully viscous codes because boundary effects and blade thickness are more critical. Similar trends are also expected for turbine stages.

The cost and maintenance implications of an additional stage in the compressor are modest. The gas generator turbine is more critical. Two expensive cooled stages are required to achieve peak efficiency. On the other hand, a single-stage turbine has attractive features such as lower first cost, reduced maintenance cost and a smaller cooling air requirement.

For modern engines where the components are cooled, blade metal temperatures represent more practical design parameters than gas temperatures. The highly stressed gas generator turbine blade is often the most critical component. An advanced design can accept a 50 K material temperature increase over a conventional metal turbine because of improved rotor cooling; a thermal barrier coating applied to the cooled and enhanced material a further increase of 100 K in surface temperature; a ceramic turbine has 1800 K material capability. Cooling performance has been demonstrated in larger engines, but there remains a significant challenge in the manufacture of complex cooling geometries for smaller blades.

All engines experience pressure losses that are associated with the intake, combustor, inter-ducts, exhaust duct and nozzle. The assumed losses as a percentage of the available pressure $\Delta P/P$ are: 3.5 in the combustor, 3.0 in the inter-duct, 1.6 in exhaust duct and 2.0 in the nozzle.

At the 1,000 kW power level, three curves define the envelope of best attainable performance as the pressure ratio is varied. In figure 16.16, the sfc is plotted against specific power for a gas turbine operating on simple cycle. At a low pressure ratio of 10:1, an engine with a single stage gas generator turbine offers the highest attainable specific power. As the pressure ratio is increased, sfc is improved, but this improvement is modest because of the effect of decreasing turbine efficiency. Engines with a two stage gas generator turbine give improved thermal efficiency at a given pressure ratio. Above 12:1 pressure ratio, the two-stage gas generator turbine option gives a lower sfc for the attainable specific power. At pressure ratios above 18.5:1 the gas temperature entering the power turbine rotor has dropped to a sufficiently low level for it to operate uncooled, with a consequent step change in efficiency.

Since TOGW is critical for helicopters, engines that require minimum fuel will be favoured. But fuel burn depends on the mission, and typically over half the fuel consumed occurs below half power. Fuel consumption at half power may therefore be taken as a reasonable average for a typical mission of duration of say 2.5 hours. Engine weight may also be generalized to a first approximation as proportional to core flow rate. With these simple assumptions, lines of constant engine plus fuel weight can be drawn on the sfc versus design specific plot in figure 16.17. Here the pressure ratio is 18.5:1, which is the lowest pressure ratio at which the more efficient and uncooled power turbine can be used.

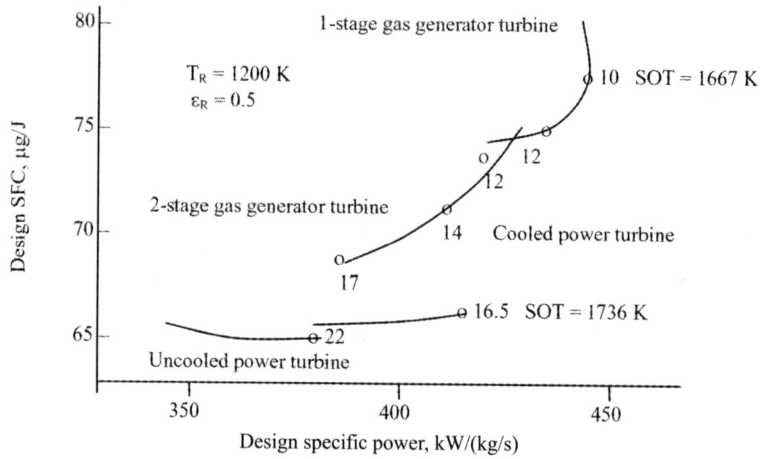

Fig. 16.16 Design performance of Simple Cycle 1,000 kW Gas Turbine

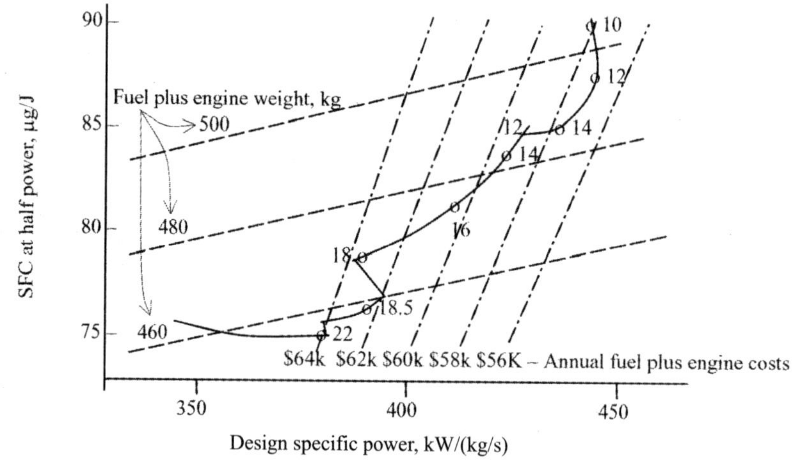

Fig. 16.17 Implications of Cost and Weight in Simple Cycle 1,000 kW Gas Turbine

The other important consideration is cost. The minimum cost of ownership of the whole helicopter would generally be the preferred optimizing parameter. If changes in the vehicle cost between different engines are not large, then changes in engine plus fuel cost will dominate. Fuel costs may be easily deduced, with fuel at \$0.55/kg and assuming 200 flight hours per year at half power. Engine first cost can be taken to a first approximation as proportional to the core flow to be about \$3200/kg. Annual engine costs, including maintenance, depreciation, investment and spare holding may be assumed to be 20 per cent of first cost. On this basis, annual engine costs are about 3.5 times the fuel cost. With these assumptions, lines of constant cost may be drawn as shown in figure 16.17. These indicate that minimum fuel plus engine cost would occur at close to maximum specific power. But any increase in fuel price or higher utilization would reduce the slope of the cost lines.

16.7 DYNAMIC SIMULATION OF COMBINED CYCLE PLANT

Dynamic simulation of the energy systems of a combined cycle plant can be used as a design tool for selecting control strategies and establishing operating procedures. A primary objective of the simulation system is to determine if the transient behavior of the system that is predicted by the model for various upset conditions is physically reasonable and in line with operating constraints imposed on the system.

The procedure is illustrated for the case of a gas fired combined cycle plant, with a gas turbine, steam turbine and alternator turning on a common shaft. A waste heat boiler produces steam at two pressures using heat from the gas turbine's flue gas, which is then expanded in a steam turbine. The plant is designed to operate under some unique conditions. A sizeable amount of the power, above that consumed in the local area, is exported to the regional grid, with the local and regional grids connected in parallel to the plant. During parallel operation, the plant experiences substantial changes in the load pattern. Another mode of operation occurs in the event of abrupt disconnection of the plant from the national grid, leading to rapid load shedding to match the power requirements of the local area alone. To permit safe switchover in this event, some unique control characteristics are necessitated in addition to the routine control features.

Important equipment and control characteristics are digitally modeled and tested. Pertinent equipment and controls are represented by differential equations. Steady state simulation is based on an iterative process using a numerical integration scheme.

Figure 16.18 provides a schematic of the combined cycle plant (*ref*: Ahluwalia, 1989). Flow of air through the compressor of the gas turbine is a function of machine speed and air temperature. Exhaust gas from the gas turbine passes through a waste heat boiler for steam production, and used in the high-pressure (HP) and low-pressure (LP) sections of the steam turbine. Also, hot air from the compressor produces LP steam in the waste heat exchanger, while exhaust air cools the gas turbine blades. At full load the gas and steam turbines generate 37 and 23 MW.

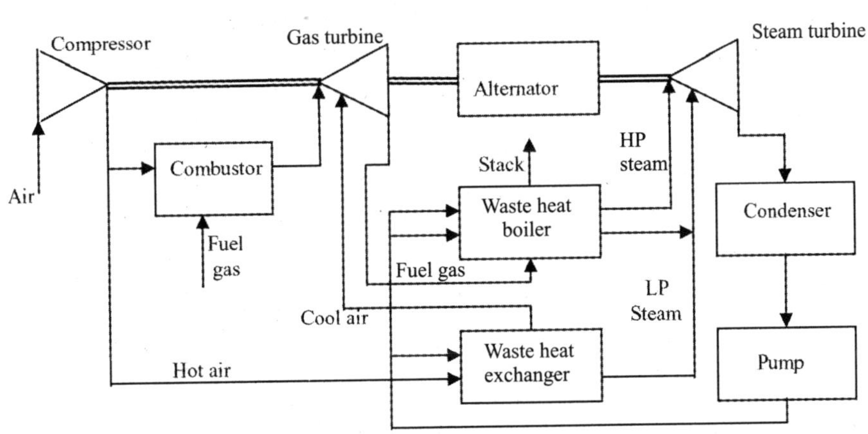

Fig. 16.18 Schematic Representation of Combined Cycle Plant

In parallel operation with the national grid, the combined cycle operates either under power control or under droop control on speed. A change in set point is executed through a ramp in order to slow variations in HP steam superheat temperatures and reduce consequent thermal stresses.

In the power-control case, the measured variable is the total power produced; in the droop-control case, the measured variable is the machine speed as set by the regional grid. The power-control method uses a Proportional-Integral-Derivative (PID); the droop-control method is through gain that is optionally corrected by a lead/lag compensator without integral action. The final action on the control valve is rate limited.

When the combined cycle plant is connected to the local grid, three control modes are possible for the gas turbine: speed droop control using proportional and lead-lag compensator, speed control with a PID controller and bypassed power control. In the last mode the power set point is biased to force the gas turbine toward a larger share of the total load, and the extra steam produced bypasses the steam turbine. The steam turbine is droop controlled through the HP steam inlet valve. The steam system has control loops based on level, pressure and temperature. The HP and LP level loops ensure that water level remains within certain limits for all operating conditions by regulating the feed water flow. The temperature loop controls the temperature of the HP superheated steam entering the turbine. Water from the HP feed water pump is sprayed to regulate the temperature of the superheated steam with the aid of a PI controller. Possible overriding of the HP steam valve position by speed control is also available.

Under parallel operation the machine speed is set by the regional grid, so only the pressure loop is active. Pressure varies according to the operating load on the steam turbine. But the HP drum pressure is allowed to float only above a minimum value; this value is maintained at lower loads by throttling the inlet control valve. A bypass line upstream of the HP steam throttle valve provides protection from over-pressure. Pressure control for the LP system is similar.

Steady state simulation at several load points represents initial conditions for the dynamic events, and is performed with three logical blocks for the gas turbine, waste heat boiler and steam turbine. In the dynamic model, differential equations describe the thermal balances. Various equipment time constants are determined at every integration step on the basis of variable relevance. Time constants of valves, obtained from the manufacturers, and time lags from piping elements between the elements are included in the model. A series of linear differential equations are obtained, with time as the independent variable and thermodynamic characteristics for the components as dependent variables. Additional differential and algebraic equations for proportional, integral, derivative and lead/lag compensation controls are superimposed on the differential equations.

Numerical integration is performed by Euler's method. The time increment in each step is estimated in the integration algorithm. For parallel operation simulation the load is reduced from 100 to 50 per cent, followed by 600 seconds of stabilization and then recovery of load to 80 per cent under power control. Major results from the analysis may be recapitulated as follows:

- The gas turbine attempts to compensate for the slower response of the steam turbine by overshooting its own steady state load level.
- Lifting of the HP and LP steam inlet valves reflects the control strategy that permits fluctuations in the circuits only above minimum steam pressures.
- The thermal inertia of the superheater tubes causes the steam temperature to be higher than the flue gas temperature during the transient.

During a switch from the parallel operation to the local area mode of operation, with a rejection of 16.5 MW of power under speed control, the gas turbine absorbs the largest portion of power variation. The quick closing and opening of the fuel gas and HP and LP inlet valves do not cause appreciable variations in steam drum pressures because of the relatively large volume of steam available in the system.

16.8 POWER AUGMENTATION

The power produced in a simple-cycle or a combined-cycle plant may be increased by injecting water or steam, including supplemental firing in the heat recovery steam generator (HRSG) and peak firing.

Injecting steam or water in the combustor of a gas turbine for nitric oxide (NO_x) control increases the mass flow of the air, and thus also results in increased power output. The amount of steam or water injected is normally limited by the amount required to control NO_x. This is done to minimize the operating and maintenance costs and impact on inspection intervals. The steam injected is normally mixed with the fuel entering the combustors. It can also be injected into the compressor discharge casing of the gas turbine. In combined-cycle applications, the heat rate increases with steam or water injection. The reason for this change is as follows. For water injection, a considerable amount of heat is required to vaporize the water. Steam is taken from the bottoming cycle (HRSG/steam turbine) for steam to be injected in the gas turbine. This reduces the efficiency of the steam cycle.

Most machines allow up to 5 per cent of the compressor airflow for steam injection. The steam must have at least 50°F (28°C) superheat and be at a similar pressure to the fuel gas. Most control systems allow only the steam flow required until the unit is fully loaded. At this stage, additional steam or water is admitted for further increase in power (*ref*: Kiameh, 2002).

Since only a small percentage of the air entering the gas turbine participates in the combustion process, the oxygen concentration in the discharge of the gas turbine allows supplementary firing in the HRSG. The definition of a supplementary-fired unit is an HRSG fired to an average temperature of not exceeding about 1800°F (982°C). Supplementary-fired HRSG's are installed in new units. It is not practical to retrofit them on existing installations due to the space requirements for duct burners and other significant material changes.

Some gas turbines can be operated at a higher firing temperature than their base rating. Referred to as 'peak firing', the output of the simple-cycle and combined-cycle plants will increase. This mode of operation results in a shorter inspection interval and increased maintenance. Despite this penalty, operating at higher firing temperatures for short periods is cost-effective due to the increase in power output.

Several output enhancement methods are shown in table 16.1, with the effect on performance for each method on a day when the temperature is 90°F (32.2°C), with 30 per cent relative humidity. The capability of each piece of equipment in the plant, including gas turbine, steam turbine and generator, must be reviewed to ensure that the design limits will not be exceeded. For example, the capability of the generator may be limited on hot days due to inadequate cooling capability.

Low-grade heat can be used to increase the temperature of gaseous fuels. Plant efficiency is improved by reducing the amount of fuel consumed when the fuel temperature is raised to combustion temperature. This method has minimal impact on the output of gas turbines. However, it results in a limited reduction in the output of combined cycles due to use of energy to heat the fuel rather than for steam production. The temperature of the fuel can be increased up to 700°F (370°C) if the fuel constituents are permissible. Further increase in the temperature will cause carbon deposits to form on heat transfer surfaces and the remainder of the fuel delivery system. Fuel temperatures of between 300° F and 450° F (150° and 230° C) are considered economically optimal for combined-cycle application. A typical gain in efficiency for a large combined-cycle plant is around 0.3. It is important to prevent the fuel from entering the steam system because the temperature of the steam is normally higher than the auto-ignition temperature of gas fuels. This can be accomplished by implementing the following modifications:

Table 16.1 Effect on Performance of Power Enhancement Option on Combined Cycles Compared with the Base Case

Change in Power Enhancement Option	*Change in Power Output*	*Efficiency*
Evaporative cooling of inlet air	+5%	-
Chilling of inlet air	+11%	−1.5%
Peak load of gas turbine	+5%	-
Steam injection (5% of air flow through the gas turbine)	+3.5%	−4.3%
Water injection in gas turbine	+6%	−5%
Supplementary-fired HRSG	+30%	−10%

- Maintaining the water pressure above the fuel pressure in direct fuel-to-steam heat exchangers ensures that no leakage will occur into the fuel system.
- Design and operational requirements must be checked to prevent the fuel from entering the steam system when the steam pressure is low.
- Use an intermediate heat transport fluid so that any leak in the fuel heat exchanger will not affect the steam system.

16.9 INLET AIR COOLING

The installation of an evaporative cooler or inlet chiller in the inlet ducting (downstream of the inlet filters) will lower the inlet temperature to the compressor. There is an improvement in power output and heat rate when the inlet temperature to the compressor is reduced. However, inadequate operation of this equipment can result in condensation or carryover of water into the compressor. This increases compressor fouling and degrades the performance. Moisture separators, or coalescing pads, are generally installed to reduce the possibility of moisture carryover (*ref*: Kiameh, 2002).

Inlet chillers are also used on aero-derivative engines to enhance the power output. The incoming air is cooled in this method to increase the output. It is usually desirable to chill the incoming air to the maximum power output. This occurs at an inlet temperature around 45° to 50°F (7° to 10°C). An absorption refrigeration system is normally used for this application. The steam is delivered to the absorption system from the low-pressure section of the heat recovery steam generator.

Chillers do not have the same characteristics as evaporative coolers. The wet-bulb temperature does not limit them. The temperature achieved is limited by the capacity of the chiller.

16.10 EVAPORATIVE INTER-COOLING

Hot and low-humidity climates gain the most from evaporative cooling by boosting the power output and heat rate. Evaporative cooling is generally limited to an ambient temperature higher than 59°F (15°C). The reason is concerned about potential formation of ice on the compressor blades. The information presented in figure 16.7 is based on the evaporative cooler having an effectiveness of 85 per cent. The effectiveness is measured by how close the cooler exit temperature is to the ambient wet-bulb temperature. For most applications, a cooler effectiveness of between 85 and 90 per cent provides the most economic benefit (*ref*: Kiameh, 2002).

Evaporative coolers are used to increase the output of gas turbines operating at high ambient temperature and low humidity. This system evaporates water into the inlet airflow upstream of the compressor, resulting in a decrease in the compressor inlet temperature. This method is economically justifiable for both base- and peak-load applications. The output of heavy-duty machines can increase by up to 9 per cent when the ambient temperature is 90°F (32°C) at relative humidity of 20 per cent Aero-derivative machines using an evaporative cooler operating at 85 per cent effectiveness will increase the output by about 22 per cent under the same ambient temperature and relative humidity conditions.

16.11 STEAM AND WATER INJECTION

The injection of steam or water into the combustor to reduce NO_x emissions results in increasing the mass flow. Therefore, the power output will increase. The amount of steam or water injected is limited to the amount required to meet the NO_x requirement. It is around 1.1 kg of steam/kg of fuel or 1kg of water/kg of fuel (*ref*: Kiameh, 2002).

Steam injection was used for power augmentation for more than 30 years. The steam is normally injected into the compressor discharge casing and combustor. It can increase the power output by up to 20 per cent and the efficiency by 10 per cent. Most machines are designed to allow up to 5 per cent of the compressor airflow for steam injection. The steam must have around 50°F (28°C) superheat. It is normally premixed with the fuel before being injected in the combustor.

16.12 OFF-PEAK THERMAL ENERGY STORAGE

Off-peak thermal energy storage is used where the cost of electricity during daytime peak periods is very high. Ice or cold water is produced during off-peak hours and weekends by mechanical chillers and stored in large holding tanks.

The increased power demand generally lasts for a few hours each day. The occurrence is restricted to periods when the ambient air temperature is very high during the summer months, mostly because of increased demand for air conditioning in homes, schools, hospitals and business places. The power output from the gas turbine is raised by chilling the intake air during the periods of peak power demand with the aid of the melted ice or cold water. The gas turbine inlet air temperature is reduced to between 50°F and 60°F by this system. Extensive refrigerated storage space is required for the ice or the cold water.

The procedure proves beneficial in multiple ways. Essentially the gas turbine is required for meeting peak demands in electricity for a limited number of hours each day, when the power will be developed at or near full-load capacity. But the unit can keep running even during periods of slack demand for running the refrigeration system that produces chilled water or ice, possibly at partial load. As a consequence, power generation will occur for longer periods in the day, and the duty cycle for the gas turbine will be more evenly spaced. Also, the thermal efficiency of the gas turbine will benefit from the cooler air coming in at the intake as opposed to the uncooled ambient air entering the inlet.

16.13 GLANDS AND SEALS

Glands and seals of steam turbines perform the vital function of preventing or reducing the leakage of steam between the rotating and stationary components of the turbine when the operating steam pressure is higher than atmospheric. The last few stages in the low-pressure (LP) turbine are normally under vacuum, and the steam pressure is less than atmospheric. The seals then prevent or reduce air ingress between the rotating and stationary components of the turbines.

Leakage of steam or air occurs where the shaft protrudes through the end walls of the turbine and extends to the atmosphere. The consequent loss of power associated with steam leakage or air ingress requires the design of glands and seals to be optimized to decrease any leakage. Modem steam turbines use labyrinth glands to restrict steam and air leakage, although the carbon ring gland is prevalent on older turbines.

The labyrinth gland can withstand higher steam conditions than the carbon ring gland. Figure 16.19*a* illustrates a simple form of a labyrinth seal. It consists of a ring having a series of machined fins. The fins form a number of fine annular restrictions. An expansion chamber follows each restriction. When the steam enters a restriction, the velocity increases and the pressure decreases (conversion of pressure energy into kinetic energy – the first law of thermodynamics). When the steam enters the expansion chamber between the fins, the kinetic energy goes through another conversion by turbulence into heat. At this point the pressure is not affected. As the flow progresses through successive restricted zones, the pressure goes through further reduction. The finned ring and the shaft are usually stepped to facilitate the conversion of energy (Fig. 16.19*b*). However, this design of the labyrinth is used only if the differential thermal growth in the axial direction between the rotor and the casing is of limited proportions. Figure 16.19*c* illustrates an alternative arrangement of stepped labyrinth glands. Where differential expansion in the axial direction presents a problem, the vernier gland shown in figure 16.19*d* proves advantageous. Here the shaft and the seal ring are both provided with fins. The arrangement results in a greater restriction because the fins on the two sides are generally pointing against each other.

Where leakage protection is required on axial and radial planes, the labyrinth gland design shown in figure 16.20*a* proves beneficial. The number of restrictions available in a given gland length also increases as a consequence.

Heavy rubbing between the stationary and rotating mating parts in case of an accidental 'rub' of the gland will produce a significant amount of heat. This could result in distorting the shaft in the local area and cause it to go out of balance. The thickness of the gland tips must be kept to a minimum if heat the generated is to be minimized.

The rate of leakage of steam is proportional to the annular flow area, and will vary with the radial clearance in the labyrinth gland. Maintaining the clearance gap to a minimum will ensure minimized leakage across the gland. The minimum radial clearance used in modern turbines is of the order of 0.5 mm. Rubbing in the close-clearance glands can be controlled by spring-loading the glands, as shown in figure 16.20*b*. The gland rings are normally made of four or more segments.

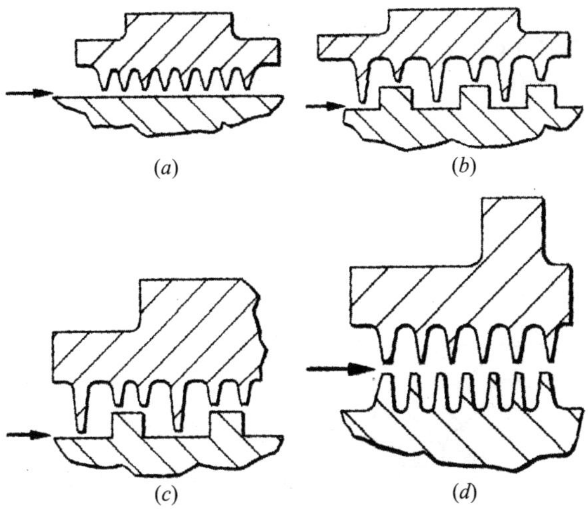

Fig. 16.19 Labyrinth Seals (*a*) Plain, (*b*) Stepped, (*c*) Double-stepped, (*d*) Vernier

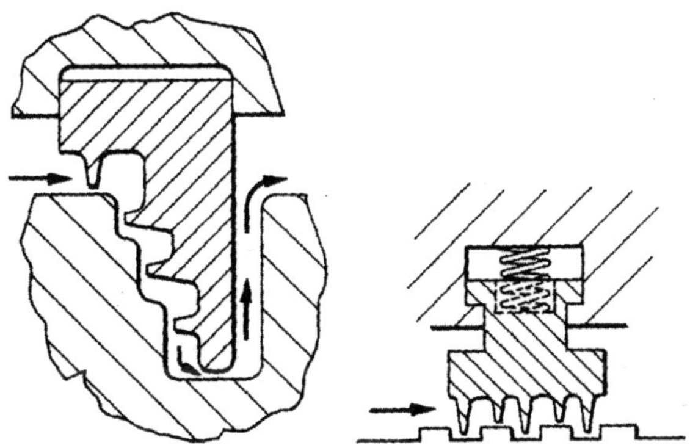

Fig. 16.20 Labyrinth Glands: Axial/Radial (*left*), Spring-Backed (*right*)

The gland sealing system supplies steam to seal the turbine shaft glands under all operating conditions while extracting leak-off steam from the glands. Typically split into two halves, one part supplies steam to the glands of the high-pressure (HP) and intermediate-pressure (IP) turbines, and the second supplies steam to the glands of the low-pressure (LP) turbine. The intent is to accommodate the range of temperatures experienced throughout the turbine.

The gland sealing system operates on dual modes. In the first mode, 'live steam' at the outlet conditions of the super-heater is used during start-up, shutdown, and when the unit is operating at low loads. The second mode of operation involves extracting steam from the HP and LP turbine and using it to seal the glands of the LP turbine during normal power operation. The use of steam from the HP and IP turbines rather than live steam results in increased efficiency. The changeover from one source of steam to the other is automatic.

The glands are normally divided into sections. After each section, the steam is fed back to an appropriate stage in the turbine or to a feed heater. Thus, energy is returned to the cycle to improve the efficiency. Figure 16.21 (*upper*) illustrates the final section of the glands in an HP turbine.

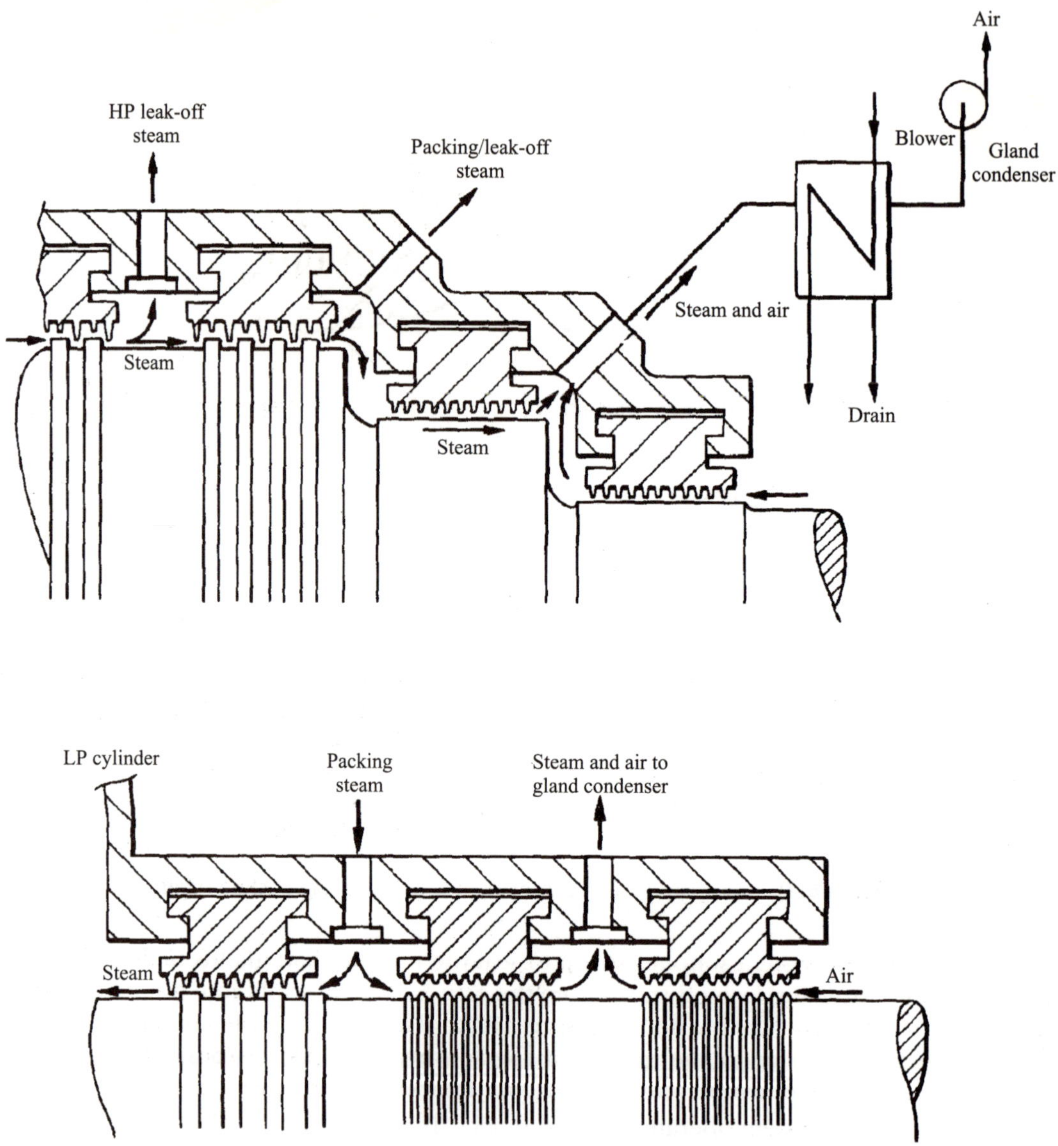

Fig. 16.21 Glands: HP Final Glands (*upper*), LP Glands (*lower*)

Steam leaked from the HP turbine is normally connected to the IP turbine, with the pressure maintained at the LP exhaust pressure. The steam pressure at the packing leak-off point is normally maintained slightly above atmosphere. The steam taken from the leak-off is normally used to seal the gland of the LP turbine (Fig. 16.21 *lower*). Since the steam is moving outward in the gland, it prevents air ingress into the turbine and condenser. The last line in all of the glands is connected to the gland steam condenser. The pressure in this line is maintained slightly below atmospheric to prevent steam leaks from the turbine. There is also a continuous air flow inward through the outboard section of the gland due to the sub-atmospheric pressure in the line.

At low loads, live steam enters the HP desuperheater through a pressure-reducing valve. The steam is cooled in the HP desuperheater to an acceptable temperature by the HP/IP glands. The steam then enters the HP/LP glands on its way to the LP desuperheater, where the steam is then cooled to an acceptable temperature by the LP glands.

At higher loads, the glands of the HP/IP turbines are self-sealing. Excess steam from these glands enters the LP desuperheater to seal the glands of the LP turbines. Two strainers are used to prevent impurities from entering the glands. The first is used for the HP/IP system. The second is used for the LP system. These strainers are installed after the desuperheaters.

Two desuperheaters are used to control the steam supplies to the glands. The first is for the HPIIP system and the second is for the LP system. The desuperheaters are normally of the water spray type.

During start-up and shutdown, a pressure-regulating valve is used to control the pressure of the live steam reaching the glands. The steam flow to the glands of the HP/IP turbines is reduced as the load increases. The steam flow to the glands eventually reverses direction when leak-off steam becomes available. As the pressure in the glands increases, the pressure-regulating valve of the live steam closes progressively (the function of the pressure-regulating valve is to maintain constant pressure at the glands). The pressure-regulating valve eventually closes completely when the pressure at the glands reaches a predetermined value. At this stage, the steam leaking from the glands of the HP and IP turbines is used to seal the glands of the LP turbines. A leak-off valve controls the pressure at the glands by dumping steam to an LP heater. This configuration ensures that the pressure at the glands is controlled by one regulating valve at any one time and that there is an automatic changeover from live steam to leak-off steam. The pressure of the steam in the sealing line is displayed locally and in the control room. Alarms annunciate low-pressure conditions.

16.14 REVIEW PROBLEMS

Example problem # 16.1: The discharge pressure from a compressor is 245.30 lbs/in^2 and the intake pressure is 14.69 lbs/in^2. If the ratio of specific heats at constant pressure and constant volume is 1.385, calculate the ideal cycle efficiency.

Solution: The compression ration $r = 245.30/14.69 = 16.70$

Equation 16.1 provides the ideal cycle efficiency:

$$\eta = 1 - \{1/r^{(1-1/\gamma)}\}, \text{ where } r = \text{compression ratio}$$

and γ = ratio of specific heats at constant pressure and constant volume = 1.385

Hence:

$$\eta = 1 - \{1/16.70^{(1-1/1.385)}\}$$

$$= 1 - \{1/(16.70^{0.2780})\} = 1 - (0.4572) = 0.5428$$

Example problem # 16.2: Calculate the compression ratio required in a gas turbine unit if the specific heat at constant pressure is 1.735 and at constant volume is 1.250 and the ideal cycle efficiency is required to be 0.575.

Solution: The ratio of specific heats $\gamma = 1.735/1.250 = 1.388$

Ideal cycle efficiency $\eta = 1 - \{1/r^{(1-1/\gamma)}\}$, where r = compression ratio

$$(\gamma - 1)/\gamma = 0.2795$$

$$1 - \eta = 1 - 0.575 = 0.425 = 1/r^{0.2795}$$

Hence: $r^{0.2795} = 2.3529$, and the required compression ratio $r = 21.3487$

Example problem # 16.3: Calculate the specific heat at constant pressure in a compression process where the specific heat at constant volume is 1.315, compression ratio is 17.5 and the ideal cycle efficiency is required to be 0.5592.

Solution: Ideal cycle efficiency $\eta = 1 - \{1/r^{(1-1/\gamma)}\}$, where γ = ratio of specific heats.

$$1 - \eta = 1 - 0.5592 = 0.4408 = 1/r^{(1-1/\gamma)}$$

$$1/[17.5^{(1-1/\gamma)}] = 0.4408, \text{ and } 17.5^{(1-1/\gamma)} = 2.2686$$

Hence: $(1 - 1/\gamma) = 0.2862$, $\gamma = 1.401$

And: Specific heat at constant pressure $c_p = 1.401 \times 1.315 = 1.8423$

Example problem # 16.4: A regenerator is used to preheat compressed air before entering the combustor, using heat from the turbine's exhaust gases. If the inlet temperature to the compressor is 73°F, inlet temperature to the regenerator is 674°F, inlet temperature to the combustor is 839°F, inlet temperature to the turbine is 1783°F, and exit temperature from the turbine is 1112°F, find the regenerator efficiency if pressure losses in the components are to be ignored.

Solution: Inlet temperature to the compressor is $T_1 = 73°F$

Inlet temperature to the regenerator is $T_2 = 674°F$

Inlet temperature to the combustor is $T_3 = 839°F$

Inlet temperature to the turbine is $T_4 = 1783°F$

Exit temperature from the turbine is $T_5 = 1112°F$

Regenerator efficiency from equation (16.2), if pressure losses are ignored:

$$\eta = (T_3 - T_2)/(T_5 - T_2)$$
$$= (839 - 674)/(1112 - 674)$$
$$= 0.3767$$

Example problem # 16.5: In example problem # 16.4 calculate the system cycle efficiency.

Solution: System cycle efficiency from (16.4): $\eta = [(T_4 - T_5) - (T_2 - T_1)]/(T_4 - T_3)$

Hence: $\eta = [(1783 - 1112) - (674 - 73)]/(1783 - 839)$

$$= 0.07415$$

16.15 EXERCISES

16.1 The intake pressure of a compressor is 15.25 lbs/in^2 and the discharge pressure is 229.75 lbs/in^2. If the ratio of specific heats at constant pressure and constant volume is 1.395, calculate the ideal cycle efficiency.

16.2 Calculate the specific heat at constant pressure in a compression process where the specific heat at constant volume is 1.287, compression ratio is 17.3 and the ideal cycle efficiency is required to be 0.5592.

16.3 Calculate the compression ratio required in a gas turbine unit if the specific heat at constant pressure is 1.489 and at constant volume is 1.157 and the ideal cycle efficiency is required to be 0.537.

16.4 A regenerator is used to preheat compressed air before entering the combustor, using heat from the turbine's exhaust gases. If the inlet temperature to the compressor is 121°F, inlet temperature to the regenerator is 564°F, inlet temperature to the combustor is 179°F, inlet temperature to the turbine is 1653°F, and exit temperature from the turbine is 1129°F, find the regenerator efficiency if pressure losses in the components are to be ignored.

16.5 In exercise # 16.4 calculate the system cycle efficiency.

References & Bibliography

Ahluwalia, K.S. and Domenichini, R., "Dynamic Modeling of a Combined Cycle Plant", ASME paper # 89-GT-133, New York, 1989.

American National Standards Institute (ANSI), "B133.6 Ratings and Performance," ANSI, Washington, D.C.

American Society of Mechanical Engineers, "Gas Turbine Plants, Performance Test Code PTC-22-1985", ASME International, New York, 1985.

Baun, D.O. Maslen, E.H., Knospe, C.R. and Flack, R.D., "A Multiple Harmonic Open-Loop Controller for Hydro/Aerodynamic Force Measurements in Rotating Machinery Using Magnetic Bearings", ASME paper # 01-GT-023, New York, 2001.

Bewley, A.D., "Cycle Analysis for Helicopter Gas Turbine Engines", ASME paper # 89-GT-328, New York, 1989.

Boyce, M., "Gas Turbine Engineering Handbook", Gulf Publishing Company, Houston, Tex., 1995.

British Electricity International, "Modern Power Station Practice", 3rd ed., Pergamon Press, Oxford, United Kingdom, 1991.

Brooks, F.J., "GE Gas Turbine Performance Characteristics, GER-3567H, GE Power Systems", Schenectady, N.Y., 2000.

Kiameh, P., "Power Generation Handbook", McGraw-Hill, New York, 2002.

Kidd, P.T., Munro, N., Thiruarooran, C. and Winterhouse, D.E., "Linear Modeling of Ship Propulsion Plants", Proceedings of the Institution of Mechanical Engineers, volume 199, no. A1, pages 53–58, London, 1985.

McLeod, J.C. and Laflamme, J.C.G., "The Effect of a Compressor Rebuild on Gas Turbine Engine Performance", AGARD CP-488, paper 26, June 1988.

Niedzwiecki, R.W. and Meitner, P.L., "Small Gas Turbine Engine Technology", NASA/Army Rotorcraft Technology Conference, Ames Research Center, Moffett Field, CA, 1987.

Rubis, C.J. and Harper, T.R., "The Naval Gas Turbine Ship Propulsion Dynamics and Control Systems Research and Development Program", SNAME Transactions, volume 90, pages 321–338, 1982.

Saravananamuttoo, H.I.H. and McIssac, B.D., "Thermodynamic Models for Pipeline Gas Turbine Diagnostics", ASME paper # 83-GT-235, New York, 1983.

Stamatis, A., Matioudakis, K. and Papailiou, K.D., "Adaptive Simulation of Gas Turbine Performance", ASME paper # 89-GT-205, New York, 1989.

Timperley, S., "A Data Logging and Performance Analysis System for Application to Industrial Gas Turbines", ASME paper # 83-GT-104, New York, 1983.

17

Engine Diagnostics and Trouble Shooting

17.1 INTRODUCTION

Turbo-machines are subject to many different abnormal modes of operation. Factors as diverse as loss of lubricating oil in the bearings, low vacuum in the condenser, and failure of the servomechanism in speed governor have the potential to degrade the performance of the system, and sometimes even lead to emergency shutdown. The start of a fire can create damage to equipment and pose safety problems for plant personnel. In aero engines, ingestion of a foreign object can force the aircraft to make emergency landing. Prompt remedial action is essential to ensure integrity of the equipment is not compromised.

Tripping events that are capable of foiling a sequence of events are known as 'category B trips'. In steam turbines, for instance, when the rate of steam flow drops below the one required to overcome friction losses in the bearings and rubbing of air or hydrogen against the generator rotor, the generator starts to function as a motor and pulls current from the grid to continue running at the rated speed. The circuit breaker now opens on reverse power. This set of events prevents the possibility of the steam turbine from over-speeding. The unit shuts down as a consequence of the closing of the valves, and the generator gets disconnected from the grid. However, if the generator remains on line when the steam valves close, the turbine will not tend to go beyond its rated speed. The unit can be shut down safely only by closing the boiler stop valves. Other events that can lead to a category B trip include failures in the governor mechanism and the lubricating oil mechanism, manual emergency stop, ingression of water and over speeding of the turbine rotor.

The bearings of a turbine require a steady flow of lubricant to permit them to perform the intended function of supporting the rotor of the turbine. Thus, the pressure, flow rate and temperature of the incoming oil into the bearings must be within specified limits. In the event of increased temperatures, for example, the lubricant loses its viscosity; the bearing essentially cannot support the rotor, and the shaft and the bearing surfaces will be scored because of direct contact between the metals in the stationary and rotating parts. Inadequate flow rate of lubricant and pressure will also lead to a similar sequence of events. The oil returning from the sump must also be treated, first, to remove any suspended metallic or other extraneous materials, and second, cooled with the aid of a heat exchanger to lower its temperature before recirculating it to the bearings.

A good maintenance program is vital to maximizing availability and operation of the equipment at peak performance. Advance planning prior to shutdown for maintenance must be in place to reduce downtime. The parts that require the most careful attention in a gas turbine are the combustors and the section exposed to the hot gases that are discharged from the combustors. Known as the 'hot-gas-path' parts, they include combustion liners, crossfire tubes, transition pieces, turbine nozzles, turbine casing shrouds and rotating blades.

Radial clearance between the rotating blade tips and the stationary shrouds on the casing must be carefully checked. A prime cause of deterioration in the exhaust gas temperature is because the radial clearances in the high-pressure turbine are out of bounds. In aircraft engines, a majority of the engines removed from the wing are primarily because the upper limit of the exhaust gas temperatures has been reached. In the compressor section also, loss of performance may be expected when the radial tip clearance cannot be maintained within their upper and lower bounds.

The main contributors of downtime in gas turbines are normally controls and accessories. Combustor, turbine, generator and the balance of the plant also require close scrutiny. The outages caused by controls and accessories normally are of a shorter duration, but they tend to be more frequent. The other systems usually cause longer-duration outages.

Closely following the maintenance and instruction manual procedures for turbo-machinery equipment is important to ensure that tasks are not overlooked. What might appear at first sight to be mundane and of little significance can result in problems that are difficult to resolve. Some manufacturers also provide a system of technical information letters that update the information that is included in the maintenance and instruction manuals. This ensures optimum installation, operation and maintenance of the unit.

17.2 INSTRUMENTATION

Turbine instrumentation may be split into categories based on the type of functions performed: supervisory, efficiency, auxiliary system, condition monitoring, protection and control equipment, and recording of post-incidents. Supervisory and efficiency instrumentation may be considered significant due to the role they play in monitoring plant safety and power production.

Supervisory instruments continuously determine the condition of the main rotating and stationary components. Their functional objectives are to ensure safe operation within acceptable limits, provide advanced warning of deterioration in the performance of the turbine generator, and maintain or temporarily restrict the operating mode. The measured parameters include axial position of the rotor in order to check relative axial movements of the rotor, and ensure that clearance margins are maintained under all operating conditions. These measurements are normally taken on each cylinder of the machine. Cylinder expansion is measured to control relative radial movement between the rotors and the stators. They are used to ensure that radial clearance margins are maintained under all operating conditions.

Vibrations at the pedestal of each of the bearings are continuously monitored to indicate the dynamic behaviour of the machine. The radial excursion of the rotor relative to the stationary parts is measured (peak-to-peak) on each rotor. This is done to indicate abnormal or unsafe conditions of eccentricity of the shaft.

The running speed of the shaft is measured independently of the turbine governor. The measurement is used mostly for operator reference, and is mainly used during run-up in power generation steam and gas turbines.

The position of each steam valve is measured. These measurements are used as a general reference for the operator, and are used to determine if the load can be increased or for diagnostic purposes.

The temperature of the components of the turbine is measured during normal operation and transient states. The instruments are located in the high-pressure (HP) and interceptor steam valve chests, and in the cylinders of the HP and intermediate-pressure (IP) turbines.

Measurements are taken to ensure that wear of the pads in the thrust bearing is within acceptable limits. If the wear is higher than the acceptable limit, the turbine generator rotor will move with respect to the stator. This could have disastrous consequences on the machine.

Measuring instruments are used to determine or infer the thermal efficiency of the plant. The information is stored to determine long-term trends. The temperature and pressure of steam and water are measured at different locations throughout the plant. These measurements are taken to ensure that equipment is operating efficiently. For example, the conditions of steam and water at the inlet and outlet to the feed water heaters are measured. Unsatisfactory operation will not likely result in shutdown of the plant; however, it will result in decreased efficiency.

17.3 CONTROL METHODS

Modern gas turbine instrumentation and control systems provide advanced monitoring and diagnostics that are designed to prevent damage to the unit and to enable it to operate at its peak performance. The following sections describe the various measurements and instrumentation used in gas turbines.

Microprocessor-based distributed digital control systems are used for control and monitoring of modern gas turbines and combined cycle power plants. The instrumentations are normally triplicated with 2 out of 3 voting logic. The start-up and shutdown of the gas turbine is fully automatic. The supervisory system of the gas turbine includes monitoring of speed, vibration, temperature and flame as well as an operating data counter. The human system interface (HIS) is provided locally (near the gas turbine) and remotely in the control room. It consists of operator station(s) with monitor, keyboard, mouse, printer and sequence-of-event recorders. Operator intervention is not required, except for selecting fuel type and for setting of the load. The controls are done in a sequential mode. All steps are monitored for execution. An incomplete step can prevent the program from advancing further. The cause of the incomplete step is indicated on the monitor. The active step of the program is also indicated.

The task of the closed-loop controller is to continuously monitor and adjust process variables (e.g., temperatures, pressures, flow) in the gas turbine to match their set points. They are used to control many parameters in the gas turbine and include start-up and shutdown speed, fuel flow, frequency and load, exhaust temperature and position of the guide vanes.

The control system of a gas turbine provides speed and temperature control in the machine, control the unit during normal operation, provide protection to the gas turbine, and perform start-up and shutdown sequence of events.

Magnetic transducers measure the speed of the shaft at a toothed wheel mounted on the shaft. The transducers provide an output in the form of AC voltage having a frequency proportional to the rotational speed of the shaft. A frequency-to-voltage converter is used to provide a voltage proportional to the speed. This measured value of the speed is then compared to the desired value of the speed (speed set-point). The difference between these two values is called the 'error'. If there is an error, the control system will adjust the opening of the fuel valve to eliminate it. It relies on a proportional-integral-derivative (PID) algorithm to eliminate the error within a minimal time period and without instabilities, such as oscillations in the speed.

A series of thermocouples mounted around the periphery at the exhaust of the turbine provides an input to the control system. They are normally made from iron-constantan or chromel-alumel, and are fully enclosed in magnesium oxide sheaths to prevent erosion. Two thermocouples are frequently mounted for each combustion can. The redundancy improves the reliability of the control system. The output of the thermocouples is generally averaged. The control system compares this measured value of the turbine exhaust temperature with the desired set point value. The difference between these values is called the temperature error. If the temperature error is different from zero, the control system will adjust the opening of the fuel valve to eliminate it. It relies also on an algorithm to eliminate the error within minimal time and without instabilities.

The data trending technique involves monitoring for a change in the slope of a curve derived from the received data. The slope of the curve is normally calculated for both a long-term trend (7 days, or 168 h) and a short-term trend based on the last 24 h. If the difference between the short-term slope and long-term slope is more than a predetermined value, this is an indication that the rate of deterioration has changed. The maintenance schedule will be affected because of this change. Figure 17.1 illustrates the difference between the slope of a short-term and a long-term trend.

The trended data may be used to predict the scheduled maintenance. For example, figure 17.1 is used to predict when the compressor cleaning is required. The data presented in this figure are obtained by recording the compressor exit temperature and pressure each day. These points are then joined and a line is projected to predict when cleaning will be required.

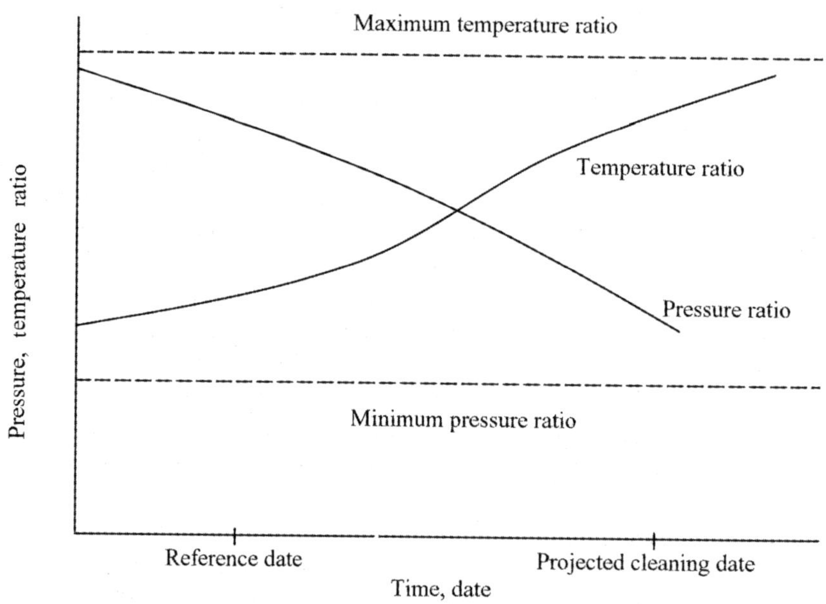

Fig. 17.1 Data Trend to Predict Maintenance Schedule

It should be noted that as the pressure at the exit of the compressor decreases, the temperature at the exit increases. The reason for this is that as the pressure at the exit of the compressor drops due to fouling buildup on the compressor blades, the efficiency of the compressor drops. This results in increased turbulence. Thus, a higher portion of the input mechanical energy from the turbine will be converted to temperature increase rather than pressure increase at the exit of the compressor.

17.4 EMERGENCY AND PROTECTIVE DEVICES

Protective systems in gas turbines are designed to provide protection in the event of over-speeding of the rotor, temperature rising beyond stated limits, excessive vibrations, loss of flame in the combustor, or loss of lubricating oil.

The over-speed protection relies on a transducer mounted on the accessory gearbox or on the shaft. The speed sensor will trip the unit at around 110 per cent of the operating speed. The over-temperature protection system relies on thermocouples similar to the ones used for temperature control. The flame detection system consists of at least two ultraviolet flame detectors that provide the status of the flame in the combustion cans.

Where multiple-combustor cans are used, the flame detectors are mounted in cans that are not equipped with igniters in order to ease propagation of the flame between cans during the ignition phase. During normal operation, a detector indicating a loss of flame in one can will only annunciate an alarm in the control room. At least two detectors must indicate a loss of flame to trip the machine.

The vibration protection system normally relies on velocity transducers to provide a constant trip set point throughout the complete speed range. Two transducers are normally installed on the gas turbine with additional transducers on the driven equipment, such as the generator. Vibration monitors provide an alarm at a specified level and a trip at a higher level. Most control systems provide a warning in the event of an open-circuit, ground, or short-circuit fault.

Although the instrumentation and control system in gas turbines vary considerably between different machines, most units are protected from a number of abnormal operational modes. In the event of low lube oil pressure downstream at the oil pump or in the bearing header, the turbine trips. Excessive lube oil temperature indication from redundant resistive thermal detectors (RTD) embedded in the babbitt (white metal) of the bearings to monitor the temperature will trip the unit. The reason for the action is that the oil viscosity drops with increased temperature, resulting in higher friction in the bearings. The elevated temperature inside the bearings will lead to bearing damage.

The shaft is prevented from moving axially in either direction at the thrust bearing. Tilted pads are installed at both sides of the shaft collar to prevent axial movement of the shaft. An oil film is established between the shaft collar and the pads at both sides of the collar. When the pads wear out, the shaft starts to move axially. This causes higher friction inside the bearing, leading to increase in the oil temperature. The RTD's trip the turbine on high thrust pad temperature. An alternative method relies on proximity probes to trip the turbine following axial movement.

Dual radial sensors monitor the vibration and trip the unit on high vibrations. A triple-redundant system protects the gas turbine against shaft over-speed. Magnetic pickup sensors are mounted near the tooth-wheel on the rotor inside the inlet bearing cavity (near the thrust and journal bearings) to monitor the speed. The sensors and the associated control components will trip the unit on over-speed. Over-speed bolts are also used to trip the unit when the speed reaches around 112.5 per cent.

The exhaust temperature of the air leaving the gas turbine is monitored with the aid of thermocouples. They have the capability to trip the unit on high exhaust temperature. The blade path temperatures are monitored by 16 thermocouples installed at about 1.3 cm (0.5 in) downstream of the last-stage blades of the turbine. The thermocouples are installed around the circumference of the turbine for monitoring the temperature difference between the combustor baskets. These thermocouples cannot be installed at the discharge of the combustors due to the high temperature, around 2500°F (1371°C). However, they can still monitor the difference in temperature between the combustor baskets due to a relationship between the location of the baskets and the location of the thermocouples. For instance, the air stream leaving a combustor flows through the turbine and is discharged from it at a 40° angle away from the combustor. Therefore, the temperature difference between two neighbouring thermocouples would be representative of the temperature difference between two combustors. The concern about having a high difference in blade path temperatures is thermal cycling and fatigue. An alarm is received when the blade path temperature difference is 90°F (50°C); the unit is unloaded when the difference is 120° F (67°C); and the unit is tripped when the temperature difference is 130°F (72°C).

Acceleration detectors trip the turbine on high shaft turning speed acceleration. They are used during start-up to prevent surge in the compressor. They are also used during normal operation to prevent over-speed following load rejection, when the circuit breaker connecting the generator to the grid opens suddenly following a fault in the grid.

The pressure at the inlet to the gas turbine is below atmospheric (under vacuum). A pressure switch monitors it. The condition of the intake filters is responsible to a considerable extent for the level of vacuum. The control system trips the unit when the vacuum drops, caused by a damaged filter, or when the vacuum increases because of a clogged filter.

The unit is tripped on high turbine exhaust pressure. This trip is needed in case there is a restriction in the flow of gases downstream of the turbine, for instance in the heat recovery steam generator.

17.5 STEAM TURBINE VALVES

Steam exiting from the superheater passes through emergency stop valves (ESVs) and associated governing valves (GVs) before entering the high-pressure (HP) turbine. The ESVs and GVs are housed inside a steam chest that is made from closed-die alloy-steel forgings welded together, or from alloy-steel castings such as 0.5-CrMoV steel. The simple geometry of the chest helps control buildup of thermal stresses. The reheat emergency stop valve (RESV) and interceptor valve (IV) also are placed inside a similar steam chest. The chests are located between the reheater and the intermediate-pressure (IP) turbines. The HP steam chests are heavier due to the higher steam pressure.

The removable plate type seats of the valves are screwed in place. The mating faces of the valves and their seats are made of Stellite in order to resist wear caused by steam erosion. The erosion wear occurs mainly when the valve is cracked open. A jet is propelled at high velocity due to the large pressure differential across the narrow port opening. Impact damage of the valve and seating faces occurs from frequent high-speed test closures. Cushioning devices placed inside the relay system or testing with a slower motion helps in alleviating the damage. Specially treated alloy steel sleeves mounted in the valve-cover support and guide the valve spindles.

A cylindrical strainer is placed inside each ESV. A number of 2 to 5 mm diameter holes in the strainer sleeve prevent solid particles of foreign matter from being carried with the flow of steam entering the turbine. The particles travel at a high speed, and have the potential to cause considerable damage to the turbine blades.

ESV's are intended to interrupt the flow of steam promptly during an emergency trip to cut off the steam supply when the unit is shut down. The valves need to be tested regularly to ensure they will close during an emergency.

Generally a single-seated unbalanced plug type of valve, the ESV has an internal small pilot valve that opens first that can open against main steam pressure. The pilot valve may be used during run-up when the steam flow is from 1 to 2 per cent of the full-load steam flow. The action reduces the force required to actuate the valve. When the machine reaches operating speed, the GV, which has been open, closes in. At this stage, it is possible to open the ESV because the upstream and downstream pressure of the valve has been equalized.

The governor valve controls the rate of steam flow entering the turbine. Since the turbine drives the generator, the governor valves also are responsible for controlling the generator load when the machine is synchronized to the grid. The governor valve throttles the steam flow during turbine run-up to speed. However, earlier designs rely on the pilot valve in the ESV in conjunction with the governor valves during the run-up of the system.

17.6 FUEL SUPPLY

Certain liquid fuels such as naphtha possess low lubricity, and may result in greater than normal wear problems in the pumping equipment. It, however, has clean combustion characteristics and leaves little residue. Other liquid fuels are thicker and have high viscosity over a wide range. They require heating before being pumped and used. When the fuel is of high density, it tends to collect heavy-metal components from the crude oil, and requires treatment before use. Chemical treatment counteracts the corrosive effects of the combustion products containing these metallic elements, especially sodium and vanadium. Similar problems are encountered with crude oil.

Diesel fuels also require heating to prevent blockage of the filter by waxes at the specified viscosity. Aero-derivative gas turbines typically use number 1 or 2 diesel-grade fuel oil. Number 3 fuel oil is used in some engines under special circumstances.

The ratings used for liquid fuels are viscosity or distillation ranges. There are specifications for diesel fuel, heating fuel oils and gas turbine fuels, which have similar viscosity and distillation ranges.

The metallic ion content of a fuel is affected by its mode of transportation from the refinery to the location of use. Aeronautical fuels are transported in dedicated tankers or tankers that have been rigorously cleaned after being used for a different fuel, but transportation of commercial fuels does not have the same restrictions. This fuel can become contaminated during sea transport if the vessel carried residual grades of fuel, heavy crude oil, or salt water (used as ballast) on a previous trip. Note that oftentimes specifications of the fuel at site may be different from the ones at the refinery.

The presence of water and sediment in the fuel leads to fouling of the fuel-handling system. Sediment tends to accumulate in storage tanks and on filter screens, leading to obstruction of the fuel flow. Water can also cause corrosion and emulsions. Sodium, potassium and calcium can be present in water. These salts are not soluble in the fuel. Water washing is effective in removing these salts. Vanadium salts are soluble in oil but not in water. Thus, they cannot be removed by water washing.

The presence of carbon is an indicator of carbonaceous material left in a fuel after vaporizing all the volatile components in the absence of air. The parameter indicates susceptibility of a gas turbine for carbon residue or varnish to form in the nozzles. Combustion systems that use lighter grades of fuel may require a limit on the carbon residue.

Soluble and insoluble contaminants may be present in a fuel, with contamination during sea transport often being a culprit. Vanadium is the most common constituent, and its presence can lead to formation of low-melting compounds such as vanadium pentoxide, which melts at 1275°F (691°C). It can severely corrode turbine blades, which are made of high-temperature resistant super alloys. If the fuel contains vanadium, the weight ratio between magnesium and vanadium must be higher than 3 to control corrosion. The presence of lead can cause corrosion and spoil the effect of magnesium. However, the presence of lead is rare in crude oil, being usually caused by contamination. Sodium and potassium combine with vanadium to form eutectics that melt around 1050°F (566°C). They can also combine with sulphur in the fuel to produce sulphates ($-SO_4$) that melt during normal operation. The levels of sodium and potassium must be controlled. Calcium does not have harmful corrosion effects, and in reality reduces the effect of vanadium. However, it can produce hard, bonded deposits, which cannot be removed by water washing. These deposits degrade the performance in a gas turbine but do not cause material damage. Water washing of the fuel will reduce the calcium content. Sulphur gets oxidized into sulphur dioxide when burned, and also partially to form sulphur trioxide. The latter can combine with sodium and potassium to form substances that can react with the oxide layer on the turbine's hot path components. This will leach away the layer and allow the nickel in the high-temperature alloys to be attacked, resulting in sodium sulphate or 'green rot' corrosion.

In general, it is impractical to prevent corrosion by limiting the level of sulphur alone in the fuel. Limiting the levels of sodium and potassium controls the rate of corrosion. Some stages of the rotating blades and stationary vanes normally have a protective coating to reduce the impact of corrosion.

Sulphuric acid can form in the exhaust of the heat recovery systems when the exhaust gas contains sulphur dioxide or trioxide. Limestone injection is used to reduce the impact on the metals.

Fully treated pipeline natural gas is the most common gaseous fuel. Some units operate on landfill or sewage digester gas. These fuels have heating values of 25 to 60 per cent of the heating value of natural gas. Gas produced from coal gasification is also used. It normally has 10 per cent of the heating value of natural gas. Gaseous fuels possess heating value between 300 and 500 Btu/ft^3, solid contaminants under 30 ppm, flammability limit from 2.2 to 1, sulphur, sodium, potassium, and lithium composition under 5 ppm, and H_2O by weight of under 25 per cent.

The presence of droplets of liquid fuel cannot be tolerated in the gas, since they possess heating values that are 20 to 70 times higher than that of gas fuel. Some fuel nozzles are designed for gas only. They are not atomizing nozzles as the ones used for liquids. Liquid droplets passing through these nozzles tend to burn on the surface of combustors and turbine blades. They cause extremely high thermal stresses, metal melting and component damage.

Compression is needed to raise the pressure of gas entering the combustors. The compressors are normally subjected to high wear. Pulsation dampeners are required for reciprocating compressors. Oil separators are required for oiled screw compressors. Gas coolers are required for most of them.

There are two basic sources of energy used for starting a gas turbine. One of them is stored energy from batteries, compressed air in bottles, compressed gas from gas pipelines and hydraulic oil from an accumulator. The other source includes electricity from the grid for either the starting motor or directly to the generator that will act as a motor during start-up, or an internal combustion engine to start the gas turbine directly.

The advantage of stored-energy systems is the ability to provide black-start capability. However, they have practical size limitations. Most black-start systems require the capability for three starts in case initial attempts are unsuccessful. Small units (up to 6 MW) rely on electric batteries for start-up. Larger units use air or hydraulic starter motors. Compressed gas from pipelines is used with units up to 25 MW. However, this method is becoming less popular due to environmental impact and cost concerns associated with exhausting large amounts of natural gas to the atmosphere. An electric motor powered from the grid can be used to drive a hydraulic pump or air compressor to provide high-pressure oil or air to the starting device. An internal combustion engine can also be used to start the unit. This provides black-start capability.

Engine starting systems are used to rotate the engine for washing the compressor. Some units use the starter to cool the engine by air following an emergency trip before a restart can be attempted.

A small DC motor is normally used to drive heavy-frame machines at very low speeds through a speed reduction gearbox as the engine cools down. This is done to prevent sagging of the rotor between the bearings due to uneven thermal growth as the engine cools. A typical barring speed is 6 r/min, which is maintained for a period of time. It is essential to maintain lubrication of the bearings to remove the heat that is generated during the barring phase.

Many gas turbines have a liquid fuel supply in addition to the gas fuel supply. The following are its main protective features: (*i*) Trip the fuel pump in the event of low suction pressure to prevent cavitation damage in the pump. (*ii*) The pressure difference between the inlet to the liquid fuel manifold and the compressor discharge is monitored. It is used to confirm that flow of fuel as it enters the combustors is sufficient. The unit is tripped when this differential pressure becomes very high (over-fueling) or very low (under-fueling). (*iii*) A pressure switch in the liquid fuel line monitors the transfer from gas to liquid fuel. The unit is tripped upon a fuel transfer failure.

17.7 LUBRICATION SYSTEM

The turbine's rotor shaft must be lifted off the bearing sleeve prior to starting, and the jacking oil system performs this function. Lubricating the bearing provides a hydrodynamic oil wedge between the bearing and the shaft, thus increasing the hydrodynamic pressure in the oil film to lift the rotor. The lubrication system provides the oil flow to maintain the lining of the bearing below 230°F (110°C). The temperature of the bearing increases during operation from thermal conduction, friction between the oil film, the journal (portion of the shaft inside the bearing) and the bearing's material and turbulence within the oil itself.

The temperature of the lubricant oil leaving the bearing is normally limited to 160°F (71°C). A directly driven centrifugal pump delivers the oil at 1.1 MPa. The oil from this pump goes through an oil turbine to reduce the pressure to 0.3 MPa. The oil turbine drives a booster pump that supplies oil from the main tank to the suction of the centrifugal oil pump. The system is protected against over-pressurization by a relief valve mounted on the oil tank. The relief valve is connected to the bearing oil supply line. During normal operation, the directly driven main oil pump suffices to provide a reliable source of lubrication oil. The alternating-current (AC) auxiliary oil pump provides lubrication during start-up and shutdown. The direct-current (DC) auxiliary pump provides lubrication during emergency shutdown and consequent loss of AC power or when the AC pump fails to start. Lubrication oil is also supplied to the generator hydrogen seals from this system. However, most modern units have a separate seal oil system to prevent contaminating the main oil with hydrogen.

Although some units use the same oil for lubrication and turbine control and protection, fire-resistant fluid (FRF) at a pressure of 7 to 17.5 MPa is preferable for the turbine control system.

A single lubricating system is usually used for heavy-frame gas turbines and driven equipment using mineral oil. Synthetic lubricating oils have the advantage of resisting fires. A viscosity of 32 centistokes (cSt) is recommended, although higher-viscosity oils can be used in high ambient-temperature areas. Heavy-frame and power turbines use oil-film bearings.

Aero-derivative gas turbines have two lubricating systems. The first is for the gas generator where the rotors are carried on ball-and-roller antifriction bearings, where synthetic oil that is used in the parent aero engine is essential. Most aeronautical oils used today meet or are very similar to two military specifications, MIL-7805 and MIL-23699. The latter oil is preferred due to its high-temperature capability. The second lubricating system is used for power turbines and driven equipment using oil similar to that used in heavy-frame machines. The aero-derivative gas generator requires an oil cooler to reject the heat removed from the engine to the atmosphere or to a glycol and water cooling loop, which rejects the heat to the atmosphere. Some liquid-fueled installations cool the lubricating oil in a shell and tube heat exchanger with the aid of the incoming fuel.

A separate lubricating system tank with heaters to maintain the required temperature is necessary for heavier gas turbines. Since most installations experience very low temperatures in the winter, a DC pump is used to keep the system reasonably full and warm, ready for an emergency start. This pump is used during start-up (black-start capability). When the generator reaches the rated frequency and voltage, the AC pump is started and the DC pump is stopped.

New oils normally have an acidity that is caused by the fact that oil oxidizes to organic acids. The level of acidity in the oil is a good indicator about the condition of the oil. The need for purification and conditioning of the oil will be based on the level of acidity. Water ingress into the turbine oil is a common problem. Water presence could lead to bacterial and fungal contamination in the oil systems. Table 17.1 provides a list of preferred characteristics for lubricant oils.

Heavy-frame machines, power turbines and driven equipment of most aero-derivative installations have a separate lubricating system tank with heaters to maintain required temperature. The main pump is driven by an alternating current (AC) motor or by the shaft, while an AC motor usually drives the auxiliary pump. A DC motor drives a smaller emergency pump. Smaller units use an oil-to-air, oil-to-glycol/water mixture that is cooled by air, or oil-to-cooling-water heat exchanger. An oil-to-cooling-water heat exchanger is normally used for large units. The water is usually cooled in a cooling tower.

Table 17.1 Turbine Lubricating Oil Requirements

Lubricant Characteristic	Type prefix ISO VG grade designation (BS4231) TO 32	TO 46	TO 68	TO 100
Viscosity, kinematic at 40°C, mm^2/s (cSt)	Min: 28.8	41.4	61.2	90
	Max: 35.2	50.6	74.8	110
Viscosity index	Min: 70	70	70	70
Flashpoint, closed, °C, Pensky-Martens	Min: 168	168	168	168
Pour point, °C	Max: −6	−6	−6	−6
Total acidity, mg KOH/g	Max: 0.20	0.20	0.20	0.20
Copper corrosion, classification (3 at 100°C)	Max: 2	2	2	2
Rust-preventing characteristics	Rusting absent	Rusting absent	Rusting absent	Rusting absent
Demulsification number,s	Max: 300	300	360	360
Foaming tendency	Max: 450	450	600	600
Foam, mL at 24°C at 93.5°C	Max: 50	50	100	100
at 24°C after test at 93.5°C	Max: 450	450	600	600
Foam stability after 300 s				
Foam, mL at 24°C	Max: Nil	Nil	100	100
at 93.5°C	Max: Nil	Nil	25	25
at 24°C after test at 93.5°C	Max: Nil	Nil	100	100
Air release properties,				
minutes to 0.2% air content at 50°C	5	6	7	10
Oxidation characteristics: Total oxidation products (TOP) with sludge limited to 40% of the determined TOP				
No catalyst, %	Max: 0.1	0.1	0.1	0.1
Solid copper catalyst, % duration of	Max: 1.0	1.0	1.0	1.0
Soluble metal catalysts, % tests 164 hours	Max: 1.0	1.0	1.0	1.0

Since most installations experience very low temperatures in the winter, the DC pump is used to keep the system reasonably full and warm, ready for an emergency start. This pump is used during start-up for black-start capability. When the generator reaches the rated frequency and voltage, the AC pump is started and the DC pump is stopped. The oil makeup should be monitored, and routine sampling and analysis of the oil should be carried out to ensure proper operation. During long outages, the tank should be drained and cleaned to remove the sludge and sediments at the bottom, and the electrical heater elements should be checked for cleanliness before the tank is refilled.

17.8 PARAMETER MEASUREMENT

Pressure transducers consist of a diaphragm and strain gauges. When pressure is applied, the deflection of the diaphragm is measured by strain gauges. The output signal varies linearly with the change in pressure over the operating range. Transducers having an operating temperature below 350°F (177°C) are located outside the machine due to temperature constraints. In these cases, a probe is installed inside the machine to direct the air to the transducer. Most modern gas turbines provide probes to measure the compressor inlet and outlet pressure and the turbine exhaust pressure. These probes are normally installed along the shroud of the unit. Some gas turbines have probes installed in each bleed chamber in the compressor and on each side of the inlet air filter.

Figure 17.2 illustrates the locations of pressure and temperature probes in a typical gas turbine.

Resistive thermal detectors (RTDs) determine the temperature by measuring the change in resistance of an element due to a change in temperature. Platinum is normally used in RTDs due to its mechanical and electrical stability. Platinum RTDs are used for measuring temperatures from –454 to 1832°F (–270 to 1000°C). The RTD requires an electrical current source to operate. Its accuracy is within ± 0.02°F (± 0.01°C).

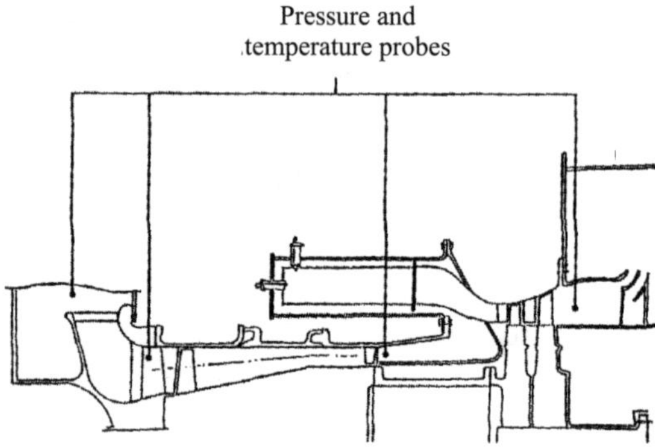

Fig. 17.2 Pressure and Temperature Probe Location on Typical Gas Turbine

Temperature detectors such as thermocouples and RTDs are used for high-temperature measurement, and are installed at two locations in a typical gas turbine. One set is at the turbine exhaust. Thermocouples are installed around the periphery at the exhaust of the turbine. Two thermocouples are installed at each location to improve the reliability of the measurements. The first set of thermocouples (e.g., 16) is installed about 0.5 in (1.3 cm) downstream of the last-stage blades of the turbine. These thermocouples measure the blade path temperature of the turbine. Differences between the readings obtained between thermocouples located at different locations around the periphery indicate differences in gas temperature leaving the combustors. This differential temperature between the combustors creates significant thermal stresses on the blades of the turbine. Most control systems reduce the load or trip the gas turbine when the differential temperature readings exceed predetermined values. The other set of thermocouples are installed at the exhaust of the gas turbine, a few metres downstream of the blade path thermocouples. They measure the air temperature leaving the machine. In heat recovery applications such as cogeneration and combined cycle plants, this is the air temperature entering the heat recovery steam generator (HRSG). This temperature is monitored to prevent overheating of the turbine components. The temperature inside the combustors, or the firing temperature, is not normally monitored for cost and safety reasons. Thermocouples that are able to detect a temperature around 2400° to 2600°F (1315° to 1427°C) are very expensive, and can cause damage to the turbine blades if a thermocouple were to break and pass through the turbine.

Redundant RTDs are embedded in the babbitt (white metal) of the bearing to monitor the oil temperature in the bearings. The unit is tripped on high lube oil temperature. It is also prevented from starting on a low lube oil temperature. Similarly, the compressor inlet and discharge temperatures are measured to evaluate the compressor performance.

Thermocouples are used for measuring temperatures from –330° to 5000°F (–201° to 2760°C). They operate by producing a voltage proportional to the temperature difference between two junctions of dissimilar metals. Thermocouples measure this voltage to determine the temperature difference. The temperature at one of the junctions is known. Thus, the temperature at the other junction can be determined. Since they produce a voltage, there is no need for an external power supply.

17.9 VIBRATION PROBLEMS

Table 17.2 shows a chart used for vibration diagnosis. It consists of general guidelines used for diagnosing mechanical problems. The collected vibration data is normally stored in the computer system. Previous vibration data is recalled and compared with recent data to identify problems in the machine.

Table 17.2 Vibration Diagnostics

*Usual Predominant Frequency**	*Cause of Vibration*
Running frequency at 0 – 40%	Loose assembly of bearing liner, bearing casing, or casing and support
	Loose rotor shrink fits
	Friction-induced whirl
	Thrust bearing damage
Running frequency at 40 – 50%	Bearing support excitation
	Loose assembly of bearing liner, bearing case, or casing and support
	Oil whirl
	Resonant whirl
	Clearance induced vibration
Running frequency	Initial unbalance
	Rotor bow
	Lost rotor parts
	Casing distortion
	Foundation distortion
	Misalignment
	Piping forces
	Journal and bearing eccentricity
	Bearing damage
	Rotor bearing system critical
	Coupling critical
	Structural resonances
	Thrust bearing damage
Odd frequency	Loose casing and support
	Pressure pulsations
	Vibration transmission
	Gear inaccuracy
	Valve vibration
Very high frequency	Dry whirl
	Blade passage

*Occurs in most cases predominately at this frequency, harmonics may or may not exist

Machine vibration is monitored using displacement probes, velocity pickup detectors and accelerometers. Displacement probes are used to measure the movement of the shaft in the vicinity of the probe. They cannot measure the bending of the shaft away from the probe. Displacement probes indicate problems such as unbalance, misalignment and oil whirl. Velocity pickup detectors have a flat response of amplitude as a function of frequency. This means that their alarm setting remains unchanged regardless of the speed of the unit. Their diagnostic role is somewhat limited. The velocity pickup detectors are very directional. They provide different values for the same force when placed in different directions. Accelerometers are normally mounted on the casing of the machine. They pick up the spectrum of vibration problems transmitted from the shaft to the casing. Accelerometers are used to identify problems having high frequency response such as blade flutter, dry frictional whirl, surge and gear teeth wear.

Figure 17.3 illustrates a chart used to convert from one type of measurement to another. The vibration limits are also shown on this chart. It demonstrates that the velocity measurement is independent of the frequency, except at very low frequencies where the displacement amplitude is constant.

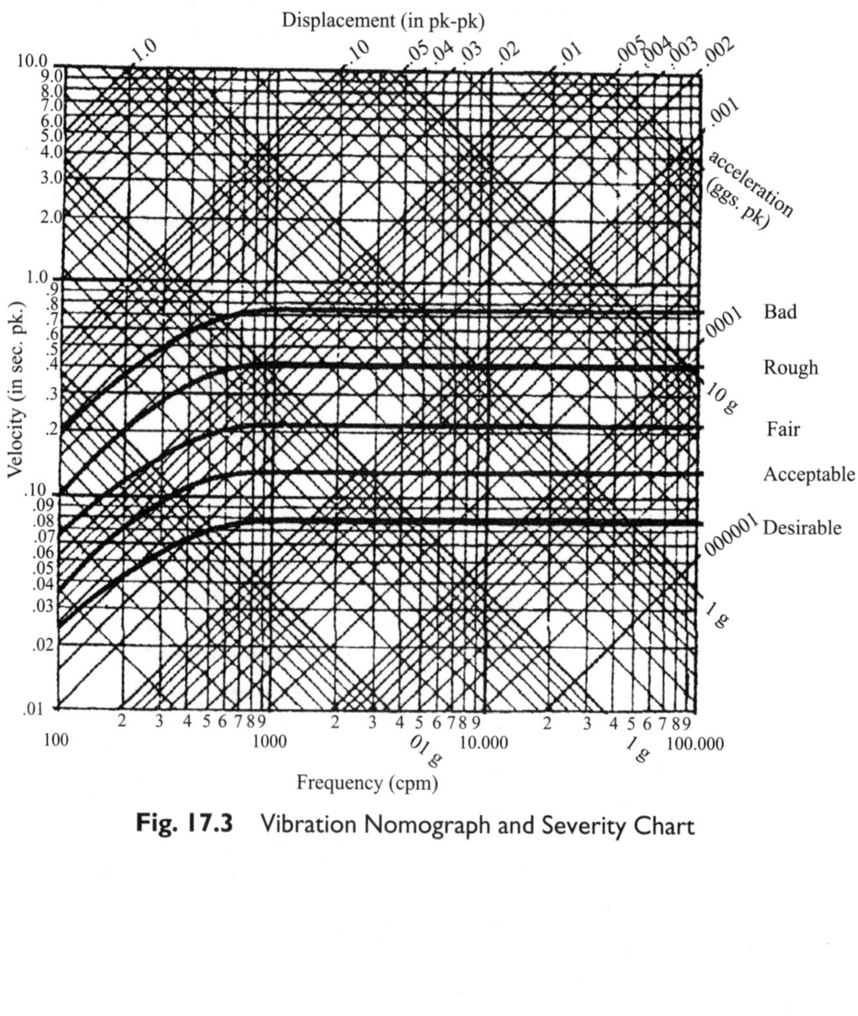

Fig. 17.3 Vibration Nomograph and Severity Chart

17.10 COMPRESSOR SURGE

Performance characteristics for centrifugal and axial compressors typically depict a pattern of constant aerodynamic speed lines and constant efficiency lines. The pressure ratio changes with the flow and speed. Compressors operate normally on a line known as the 'operating line' of the compressor, which is separated by a safety margin from the surge line. Compressor surge is a situation of unstable operation that results in flow reversal, high vibrations, overheating and possible damage to the compressor. Therefore, it should be avoided during operation of the unit.

Parameters that are monitored to perform a compressor analysis include inlet and exit pressures and temperatures, ambient pressure, vibration at each bearing and pressure and temperature of the lubricating system.

Table 17.3 shows how some problems affect the various parameters of the compressor. Monitoring these parameters allows the identification of some serious problems, such as a clogged air filter. A clogged air filter is normally indicated by an increase in the pressure drop across the filter. A surging compressor is another indication. Compressor surge is normally identified by a rapid increase in shaft vibration and instability in the discharge pressure. The pressure in the bleed air chamber will also fluctuate. A fouled compressor is normally indicated by a decrease in pressure and an increase in temperature at the discharge of the compressor. If the buildup of deposits on the blades is excessive, the vibration level will increase. A failed bearing is normally indicated by a loss of lubrication pressure in the bearing, an increase in temperature difference across the bearing, and an increase in vibration level.

Impurities in the incoming air build up on the internal components of the engine and diminish the flow area. This changes the compressor characteristics, and can lead to increased chances of surge. The efficiency of the compressor may also be reduced. Coarse dirt in the inlet air erodes the coatings of the components. Improper filtration in the intake housing can result in blocking of the cooling passages inside the turbine's rotor blades.

Table 17.3 Compressor Diagnostics

	Efficiency	*Pressure Ratio*	*Temperature Ratio*	*Mass Flow*	*Vibration*	*ΔT Bearing*	*Bearing Pressure*	*Bleed Chamber Pressure*
Clogged filter	↓		↓					
Surge	↑	Variable	↓		High fluctuations	↑	↑	High fluctuations
Fouling	↓	↓	↑	↓	↑			
Damaged blade	↓	↓	↑	↓	↑			
Bearing failure				↑	↑	↑	↓	High fluctuations

Inertial filtration is normally used to remove large particles. This is followed by self-cleaning filters, which detect an increase in pressure drop across the filter and release air pulses to remove the dirt. Most filtration systems have a 'blow-in' door located downstream of the filter. It opens automatically when the differential pressure between the area downstream of the filter and the outside exceeds a preset value. This door prevents excessive and damaging differential pressure across the filter. However, when the door opens, unfiltered air enters the engine.

Ambient air conditions should be evaluated carefully to determine the particle size and concentration in the area before specifying the type of filtration required. The filtration system should be specified to remove a complete range of particles that are likely to be encountered.

The flexible sealing bands between sections of the intake should be checked routinely for cracks. Unfiltered air enters the engine through such cracks causing damage to the engine.

Dirt accumulation on the compressor blades changes the characteristics of a compressor and reduces the output power. The exhaust temperature increases when the output power decreases, as may be seen from the conservation of energy theory. The compressor can be cleaned using these five methods:

- Disassemble the compressor partially to clean the blades of the rotor. This method gives excellent results, but it is very time-consuming.
- Some manufacturers recommend cleaning the compressor by using ground shell injected into the inlet by a high-velocity air stream. They remove buildup by an abrasive fashion. This method cannot be used with engines having a coating on the blades, because they will be eroded. This method is normally done while the compressor is being rotated with the starter mechanism. Many users have difficulty with this method because the ground shell enters the hydraulic fluid of the governing system and the pressurizing air for seals and bearings, thus blocking passages. This method cannot be used with units having air-cooled rotating or stationary turbine blades.
- Liquid wash while the compressor rotor is on the starter. Demineralized water mixed with a detergent is injected to wash the contaminants off the blades. This is followed by a demineralized water rinse and an air-dry cycle while the machine is on starter. Another technique, referred to as the soak-wash cycle, calls for injecting demineralized water while the engine is stopped. The water is allowed to soak to loosen the dirt accumulation before injecting a mixture of demineralized water and detergent into the engine.
- A 'crank cleaning' method involves a soak, followed by an abrasive shell cleaning, a rinse and a drying cycle.
- A recent method was developed for on-wing cleaning of aircraft engines (on-line or fired washing). It requires washing the gas generator by spraying a special cleaning liquid into the compressor inlet while the engine is running. The manufacturer specifies the speed of rotation during the cycle and the cleaning liquid to be employed.

Methods that include injection of liquids use a set of liquid spray nozzles at several positions around the inlet to obtain a uniform spray pattern into the compressor. Trans Canada Pipelines recommends a compressor cleaning at intervals of 1500 to 2500 hours using a soak wash cycle to recover the full performance of the engine. Some heavy-frame machines use a turbine wash cycle while the engine is shut down to remove deposits off the hot-end components.

Besides a vibration baseline spectrum, a gas turbine requires a combined aero-thermal performance baseline that represents its normal operating aero-thermal characteristic. Any deviation from the aero-thermal performance baseline beyond a predetermined value should trigger an alarm. When a compressor operates close to the surge line, an alarm should annunciate. Figure 17.4 illustrates a typical compressor aero-thermal characteristic. Other monitoring and operating outputs of a compressor include loss in compressor flow, loss in pressure ratio, and

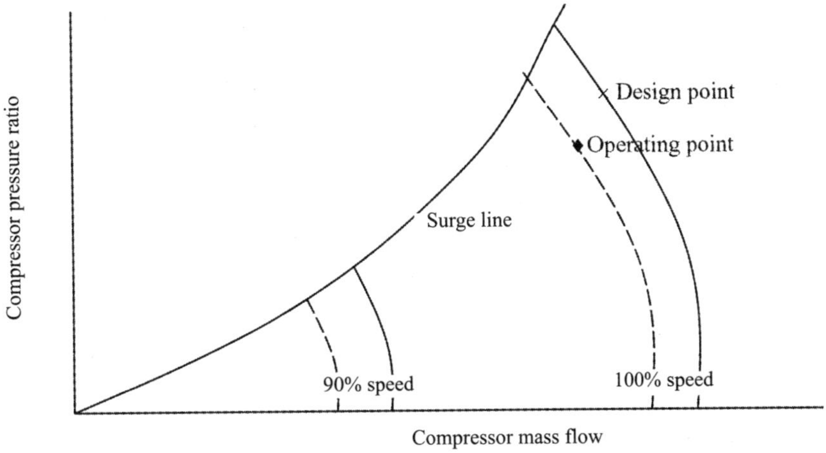

DATA INPUT
Ambient Pressure
Compressor Inlet Pressure
Compressor Discharge Pressure
Compressor Inlet Temperature
Compressor Discharge Temperature
Compressor Speed
Compressor Inlet or Discharge Flow if Available

DIAGNOSTICS OUTPUTS
Compressor Efficiency Lower than Design
Compressor Approaching Surge Conditions
Compressor Approaching Choke Conditions
Dirty Compressor

CONDITION MONITORING OUTPUT
Loss in Compressor Flow Through Put
Loss in Compressor Pressure Ratio
Fuel Cost Penalty
Projected Increase in Fuel Cost After One Month Operation
Surge Point Deterioration Trend and Anticipated Outage Del.

Fig. 17.4 Aero-Thermal Condition Monitoring for Compressors

increase in operating fuel consumption if operation occurs at off-design conditions or with excessive dirt buildup on the compressor.

The aero-thermal characteristic of compressors and turbines is sensitive to variations in temperature and pressure at the inlet. Thus, the aero-thermal performance parameters (e.g., flow, speed, horsepower) should be normalized to the standard-day condition. If these corrections to standard conditions are not used, the performance of the unit may appear to have degraded, even if the performance changed because of a change in ambient pressure or temperature.

17.11 COMBUSTOR EVALUATION

The measured parameters in the combustor are pressure of the fuel and evenness of combustion noise. The inlet temperature to the turbine is not normally measured due to the very high temperatures in the combustors. Table 17.4 shows how various problems affect the combustor parameters. Measurement of these parameters permits identification of a plugged nozzle, or a cracked or detached liner. A plugged nozzle is identified by an increase in the fuel pressure and unevenness of combustion noise. This problem is common when the unit burns residual fuels. A cracked or detached liner is identified by an increase in the reading of the acoustic meter and a large difference in the exhaust temperature of the combustors.

The decision to inspect or overhaul a combustor liner is based on the equivalent engine hours, which depends on the number of starts, fuel type and temperature inside the combustors. Figure 17.5 illustrates the effect of these parameters on the life of the unit. The strong effect of the fuel type and number of starts on the life of the engine should be noted.

Table 17.4 Combustor Diagnostics

	Fuel Pressure	*Uneven Combustion Sound*	*Exhaust Temperature Spread*	*Exhaust Temperature*
Clogged filter	↑	↑	↑	↑
Combustor fouling	↑ or ↓	↑	↑	↓
Crossover tube failure	↑ or ↓	–	↑	–
Detached or cracked liner	↑ or ↓	↑	↑	–

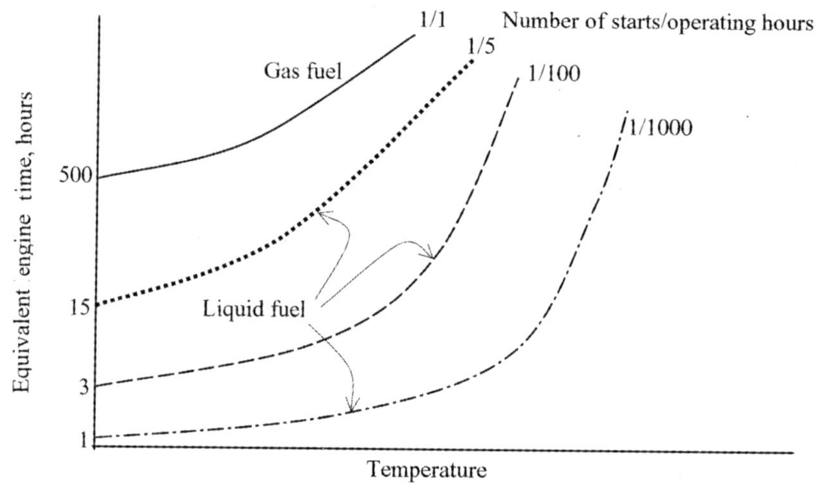

Fig. 17.5 Equivalent Engine Time in Combustor Section

17.12 TURBINE ANALYSIS

Analysing a turbine calls for monitoring the pressures and temperatures across the turbine, shaft vibration, and temperature and pressure in the lubricating system. Table 17.5 shows the effects of various problems on the turbine parameters. Monitoring these parameters permits identification of problems such as fouling, damaged blades, bowed nozzle, failed bearing and failure in the cooling system.

Fouling in the turbine is indicated by an increase in the exhaust temperature of the turbine. The vibration amplitude will also increase if the fouling is excessive and causes a rotor unbalance.

Damaged turbine blades are indicated by a large increase in vibration amplitude and an increase in the exhaust temperature of the turbine. A bowed nozzle results in an increase in the exhaust temperature and possibly an increase in turbine vibration. Turbine bearing problems have the same symptoms as in the compressor bearing.

Table 17.5 Turbine Diagnostics

	Efficiency	*Pressure Ratio*	*Temperature Ratio*	*Vibration*	*ΔT Bearing*	*Cooling Air Pressure*	*Wheel Space Temp.*	*Bearing Pressure*
Fouling	↓		↓	↑			↑	
Damaged blade	↓		↓	↑				
Bowed nozzle	↓	↓	↓	↑			↑	
Bearing failure				↑	↑			↓
Cooling air failure					↑	↓	↑	

A failure in the cooling system will lead to problems with cooling of the blades, and is normally detected by an increased pressure drop in the cooling line.

The 'equivalent engine time' is widely used to determine the scheduled maintenance of the turbine. The period between scheduled maintenances is a function of temperature, type of fuel and number of starts. Figure 17.6 illustrates the effect of these parameters on the life of the unit, with fuel type and number of starts playing a dominant role on the life of the unit.

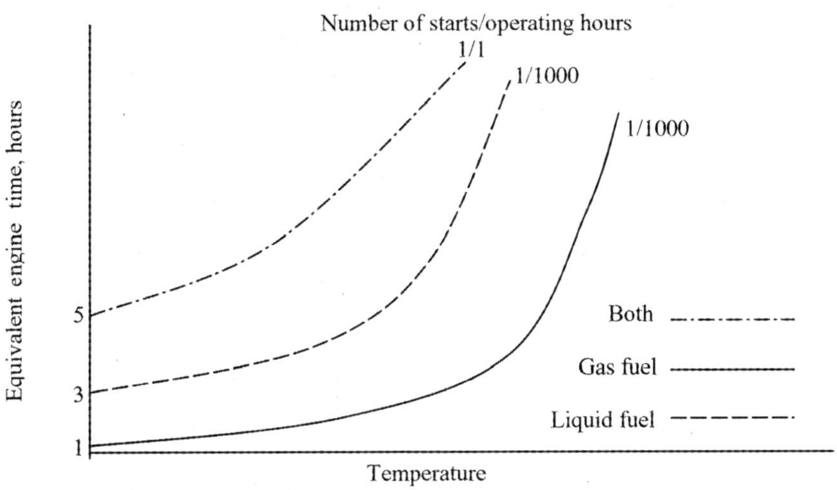

Fig. 17.6 Equivalent Engine Time in Turbine Section

Monitoring the efficiency of the gas turbine equipment and correcting for operational problems can achieve significant fuel savings. Some of these problems are easy to correct, such as cleaning of the compressor blades. Others may require a more elaborate solution to maximize the overall efficiency of the plant equipment. Figure 17.7 illustrates the impact on profits gained by operating at a slightly higher efficiency.

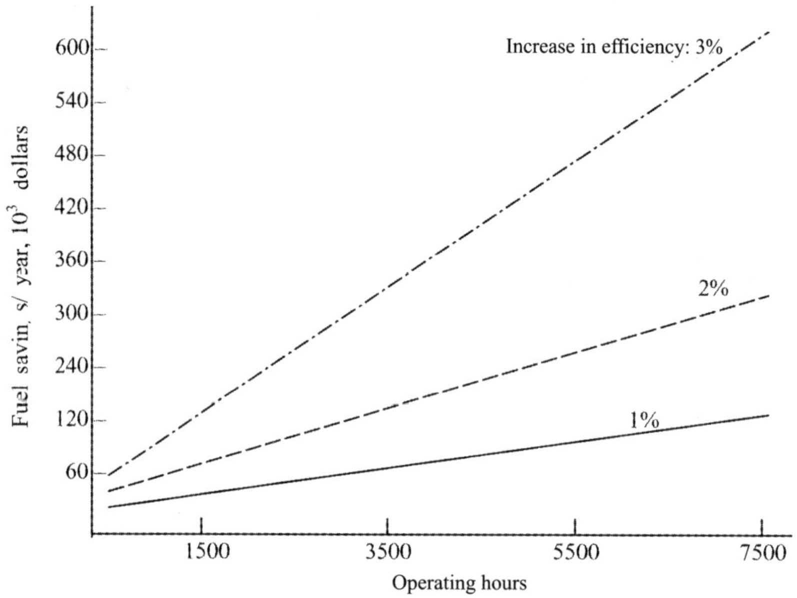

Fig. 17.7 Savings as Function of Efficiency for 15-MW gas turbine (Based on $1/gal, unit consuming 280x10^6 BTU/hr, $2.7 million)

17.13 INTERLOCKS

Interlocks, or permissives, are conditions that must be satisfied so that the control system can permit the continuation of the start-up sequence or continued operation in a gas turbine. Some commonly used interlocks are discussed below.

Ultraviolet detectors, or flame scanners, located in the upper section of the combustor monitor the combustion process. The sensors and the accompanying control mechanism interrupt the fuel to the turbine if combustion is not confirmed a few seconds after the fuel is injected into the combustors.

A pressure switch monitors the discharge pressure of the compressor. When there is sufficient airflow to support combustion, the compressor discharge pressure will be higher than a predetermined value, for example, 0.6 lbs/in^2 (4 kPa). At this stage, the control system opens the fuel valve. The condition where the compressor discharge pressure is higher than a predetermined value is referred to as a permissive or interlock. If the permissible value is not satisfied, the control system will stop the start-up sequence.

The start-up sequence is discontinued if the temperature of lube oil is low. The reason for this is the difficulty encountered in pumping the oil to the bearings due to increased viscosity at low temperature.

The turning gear is prevented from operating when the bearing oil pressure is low. This is done to prevent damage to the unit due to excessive friction inside the bearings.

The start of a gas turbine is aborted when the gas supply pressure is high or low. The objective is to prevent excessive or diminished gas flow into the combustors.

17.14 FAILURE DIAGNOSTICS

Gas turbine failures can be diagnosed. The following items discuss how some of the problems are diagnosed.

Increased vibration amplitudes are the consequence in the event of damage sustained by turbine blades, accompanied with an increase in the exhaust temperature of the turbine.

An increase in the exhaust temperature results in bowed nozzles, and possibly an increase in turbine vibration.

Problems in the turbine bearing result in symptoms that are similar to problems in the compressor bearing.

A failure in the hot path cooling system is normally detected by an increase in the pressure drop in the cooling line.

The 'equivalent engine time' is used to establish the turbine maintenance schedule. It is a function of temperature, type of fuel and number of starts. Fuel type and number of starts on the life of the unit have a strong effect.

17.15 PROBLEMS DURING START-UP AND SHUTDOWN OF GAS TURBINE

The control system of the gas turbine performs the start-up sequence. The objective is to ensure that all subsystems of the gas turbine perform satisfactorily, and the turbine rotor temperature does not increase too rapidly or overheat during start-up. The control system is designed to start the unit remotely, accelerate it to operating speed, synchronize it automatically with the grid and increase the load to the desired setting. The start-up sequence for a typical large gas turbine includes the following items.

Prior to start-up, all control and service breakers must be closed. The computer breaker must be closed if it has been de-energized, and the time of day entered. Under normal conditions, the computer operates continuously. All alarms must be acknowledged, confirm that all lockout relays are reset, and place the 'Remote-Local' switch to the desired position.

When all the preparations to start the unit are complete and the unit is ready to go through the start-up process, the 'Ready to Start' lamp will energize. At this stage, the operation of the start-up push button will initiate the start-up sequence. The initial steps in the start-up sequence call for energizing the auxiliary lubrication oil pump and the instrument air solenoid valve.

When the pressure downstream of the auxiliary lubricating oil pump reaches a predetermined value, the turbine turning gear is started. If the pressure downstream of the auxiliary lubricating oil pump does not reach the predetermined value within 30 s, the unit is shut down. When the signal indicating adequate operation of the turning gear is received, the start-up sequence continues.

At this stage, the starting device (e.g., starting motor) is activated if the lubricating oil pressure is sufficient (above the predetermined value). The turning gear motor is de-energized at around 15 per cent of the operating speed. When the turbine reaches the firing speed (when ignition should start), the turbine over-speed trip solenoid and vent solenoid are energized to reset. When the oil pressure is sufficient, the over-speed trip bolts will be reset. These bolts are used to trip the unit at around 12.5 per cent over-speed. They initiate the trip when the governing system fails to limit the over-speed to a lower value. When the over-speed trip bolts are reset, the ignition circuit is energized. It will initiate or energize the ignition transformers and timer, the fuel system, atomizing air and timer to de-energize the igniters at the proper time.

At around 50 per cent of the operating speed, the starting device is stopped. This is the 'self-sustaining speed' of the gas turbine, when the turbine is generating enough power to drive the compressor and continue the increase in speed. The bleed valves, which bleed air from the compressor during start-up to prevent surge, close around 92 per cent of the operating speed. Following fuel injection and confirmation of ignition, the speed reference no-load speed set point is increased. The fuel valve will open further to increase the speed of the unit. The shaft accelerates at a desired rate that is limited by the maximum permissible blade path and exhaust temperatures. The unit is tripped if the desired acceleration is high since chances of compressor surge increase. Also, the rotor could overheat at a much higher rate than the stator. The rotor blades would then expand at a higher rate than the stator blades, leading to rubbing between the blades, and considerable damage to the turbine.

When the unit reaches operating speed, it can be synchronized manually or automatically. Following synchronization, the speed reference becomes a load reference. In other words, since the speed of the unit cannot increase while the unit is synchronized, an increase in speed reference will result in an increase in the load. The speed/load reference is increased at a predetermined rate. This leads to further opening in the fuel valve until the desired load is reached. The computer will store the number of starts and operating hours at various loads. This information is used for maintenance scheduling.

Following a local or remote shutdown request, the fuel is reduced at a predetermined rate until zero-load is reached. The main circuit breaker connecting the unit to the grid and the circuit breaker connecting it to its own auxiliary loads (field circuit breaker) are opened and the fuel valves are tripped at this stage. During an emergency shutdown (e.g., a load rejection following a fault on the grid), the circuit breakers and fuel valves are tripped immediately without waiting for the load to be reduced. The turbine speed and the oil pressure from the motor-driven pump will drop. The DC auxiliary lubricating oil pump will start. At around 15 per cent of the operating speed, the turning-gear motor will be restarted. When the unit reaches the turning-gear speed (around 5 r/min), the turning-gear overrunning clutch will engage the shaft to rotate the rotor slowly.

The unit must be purged completely of any fuel before it can be restarted, and is done by moving air through it. The airflow must be greater than five times the volume of the turbine. The unit must be left on turning gear for up to 60 h, at which point sagging and hogging are no longer a concern due to low rotor temperature. The turning gear and auxiliary lube oil pump are stopped and the shutdown sequence is complete. The computer stores all the contact status and analog values. They can be displayed if required.

References & Bibliography

Boyce, M.P., "Gas Turbine Engineering Handbook", Gulf Publishing Company, Houston, TX, 1982, reprinted 1995.

British Electricity International, "Modern Power Station Practice", 3rd edition, Pergamon Press, Oxford, United Kingdom, 1991.

Kiameh, P., "Power Generation Handbook", McGraw-Hill, New York, 2002.

18

Manufacturing Operations

18.1 INTRODUCTION

The evolution of modern aircraft and industrial gas turbine engines has coincided with the evolution of super alloys and means to cast, cut, machine and join them into finished products. The toughness required of the materials for withstanding exceptionally high temperatures at stress levels approaching their elastic limits translates into difficult conditions for cutting, forming, machining and joining the parts. Dimensional instabilities arise from residual stresses and metallurgical alterations introduced by the manufacturing processes. The primary and secondary procedures lead to changes in the surface layer such as plastic deformation, which affect the surface integrity and stability of the dimensions.

A nearly net shape can be obtained from many different casting methods of super alloys. Alloying of wrought super alloys must be restricted to preserve their hot workability characteristies. Cast super alloy compositions are not so confined, and alloys with much greater strengths consistent with restraints are possible. Mechanical properties such as creep and rupture are maximized by the casting and heat-treating processes. Ductility and fatigue properties in the castings are generally not as good as their wrought counterparts, but refining the grain size alleviates many of the defects associated with casting. Production of hollow airfoils for turbine blades and vanes with an intricate system for cooling passages is aided by the 'lost-wax' or the investment casting process. The shapes are developed when the mold slurry flows around the wax pattern defining the part shape. Use of preformed ceramic cores adds to the capability to produce hollow airfoils.

High performance aerospace components are also produced economically using the powder metallurgy technology. Powder based alloys are used when cast or wrought components cannot meet the requirements of the application. Failure in conventionally cast and wrought parts often arises from segregation, resulting in inconsistent and reduced thermo-mechanical response. The powder-based process is then employed when cast or wrought components are not suitable. Some intrinsic attributes of powder materials make them appropriate for turbine components. A high rate of solidification results in smaller inter-metallic particles and reduced spacing between the particles, features that cannot be duplicated in casting. High levels of mechanical properties can be realized with the forgeable microstructures. Their unique structure has the ability to provide for special environments, such as strengthening from oxide dispersion. Broader application of powder metallurgy is hampered by performance limitations and cost. The alloys are sensitive to contamination, especially in highly stressed parts that may be critical from fracture and fatigue considerations. Thermo-mechanical processing is essential for neutralizing the ensuing defects. The powders are required to have a very fine size.

The magnitude of the stress and metallurgical alterations during machining depend on machining parameters such as feed, speed, depth of cut, cutting tool material, part geometry and cutting fluid. Normal industry process calls for relieving the thermal stresses to reduce the effects on dimensional instability at the machining stage. But there is no stress relieving cycle, for example for the Inconel 718 alloy, other than solutioning method that can change the mechanical properties and hardness. Also, only a limited aging process is permitted for this material to avoid degradation. This metallurgical restriction compels control of the residual stresses by substantially altering the machining parameters.

Welding plays a major role in the fabrication of parts for aircraft engines and industrial gas turbines. The procedure permits economical fabrication of sub-components without adding weight or cost. Welded joints do not suffer from problems related to deteriorated service capabilities. Cracks and fissures can develop in the welds, and represent a major drawback of the process. The welding may also cause reduction in the material properties in the region, mostly in the form of reduced ductility because the structure of the solidified weld material is segregated and less ductile than an equivalent wrought structure. This aspect is also responsible for deteriorated oxidation resistance. When elements with a high vacancy of electrons segregate, σ and other embrittling phases may precipitate during welding and even after placement in operational service. Thus, the alloys must be individually checked for degradation in properties as a consequence of welding. Reinforcement of the weld in the form of over and under-beads must be eliminated when fatigue is identified as the primary mode of failure.

Curvic couplings provide a precise method for connecting, centering and improving load carrying capacity of turbine shafts and disks. Their design calls for a high level of accuracy in the positioning and indexing of the teeth on the disk face. The concave and convex tooth profiles are generated by using the outside and inside surfaces of a cylindrical grinder, with a conical machining surface of the grinder.

Diffusion and overlay protective coatings are used for gas turbine components to enable them to withstand a severe environment. The corrosion and oxidation resistance provided by the coatings extend the component's operating life. Diffusion coats provide a surface enriched with aluminium, chromium or silicon. In earlier production the electron beam physical vapour deposition method was used, but because of the high capital cost in setting up a commercial plant, plasma-spraying systems are preferred. Air plasma spraying is a widely accepted procedure, in particular the argon shrouded and the vacuum methods. An inert gas or a vacuum allows application of the low oxygen content coating. Another recent innovation is the high velocity oxygen fuel system in open air, and has been established for production of coatings with a low oxide content, low porosity and high bonding strength. MCrAlY coatings may be applied with this procedure in open air and still achieve near chamber quality, mostly because of the higher particle velocity when compared with other thermal spraying systems. Thicker coats can also be applied with this technique for improving the residual stress. Cracking and delamination of the sprayed coat is mostly a consequence of the residual stresses.

18.2 CENTRIFUGALLY SPUN ALLOY STEEL CASTING

As the name implies, the essential features of the centrifugal casting process consist of subjecting molten metal to centrifugal pressure created in a rapidly rotating mold in such a manner that the metal is directed to assume the shape of the mold. All extraneous non-metallic material, being less dense than steel, is retained at the surface of the bore together with the micro-porosity formed by the directional solidification. The inner surface is then removed by machining to provide a sound and more homogeneous casting.

The cleanliness and convenience of this process is attractive for aircraft engine and industrial gas turbine manufacturers (*ref*: Nixon, 1987). Production of vertical and horizontal die-castings ranging from 6 in. to 60 in. diameter has been achieved. Over the past several years, centrifugal castings cast in refractory molds of up to 120 in. in diameter and 10,000 lbs weight have been perfected.

Two forms of centrifugal casting machines are employed, one with the axis of rotation oriented horizontally, the other being vertical (figure 18.1). The suitability of a preferred direction is based on the shape of the casting. When the diameter exceeds the length the direction is generally vertical, and is horizontal if the length is greater than the diameter. In the former process the die spins at a low rate when compared with the horizontal method, and the metal flows through an open top lid into an inclined injector, which transfers the melt on to the wall of the die at a tangent in the direction of rotation. In the horizontal method the die rotates on twin rolls, and the molten metal is introduced through an open end plate at one end of the die and moves along the die to give a casting of uniform thickness along its length.

The values for a number of parameters must be carefully selected in order to obtain the dimensions required for the finished part. In the horizontal method of production the outside diameter controls the metal die. The outside diameter is dictated by the finished diameter, with an allowance for surface roughness contraction and the thickness of the refractory coating inside the die. The inner diameter requires an allowance for the unsound material to be removed in the bore, a peculiarity of this process. From these dimensions the quantity of metal required to produce the casting is computed. Accurate digital scales located above the ladle ensure the exact amount of metal is poured into the die.

Production by the vertical method can be obtained with metal die or refractory shaped molds. Metallic molds are simpler in configuration, but to produce a shaped outer diameter calls for a corresponding cavity in the die. As an example, stiffening rings may be located on the outer surface. But this adds to the cost of the tooling, and can be warranted only when the production run is large. Conventional sand castings, on the other hand, incur the cost of making patterns, molding and drying. This process may be conveniently used for large castings with flanges at the top and bottom ends and on the split line. The process also has the added advantage of giving a good finish on the outer face. Typical split outer casings are shown in figures 18.2 to 18.5.

The shaped outer diameter on a centrifugal casting can save considerable amount of time during machining and grinding. Centrifugally spun castings also offer some distinct advantages over the non-rotating, or static, process. Progressive solidification occurs when the die is spinning. Commencing at the inner wall, whether metallic or refractory, the solidification progresses radially from the outer edge towards the axis of rotation. Foreign bodies and lighter inclusions are directed toward the bore initially by the centrifugal force, and then by progressive solidification of the metal. Columnar grains grow continually from the outer wall to the bore to give an unacceptable layer in the inner narrow zone, which is subsequently machined away. With the conventional sand casting

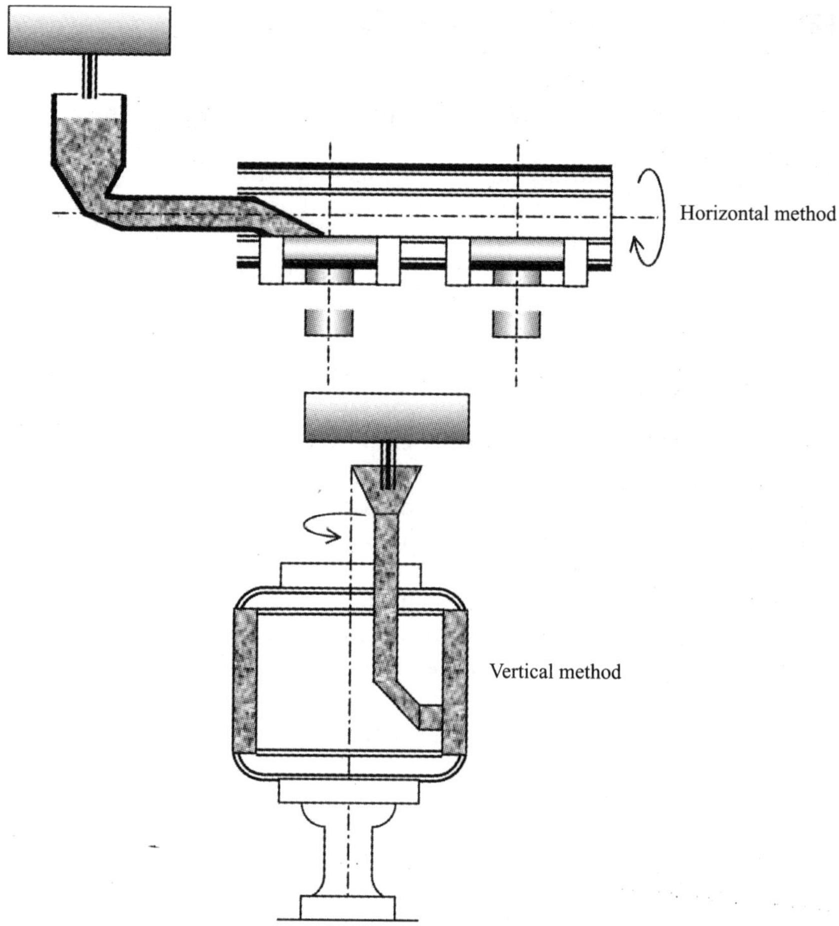

Fig. 18.1 Centrifugal Spinning Operation for Castings

Fig. 18.2 Measurements on a Split Outer Casing (Nixon, 1987)

Fig. 18.3 Centrifugal Spun Casing with Shaped Outer Cast Profile (Nixon, 1987)

Fig. 18.4 Centrifugal Spun Casting After Final Machining (Nixon, 1987)

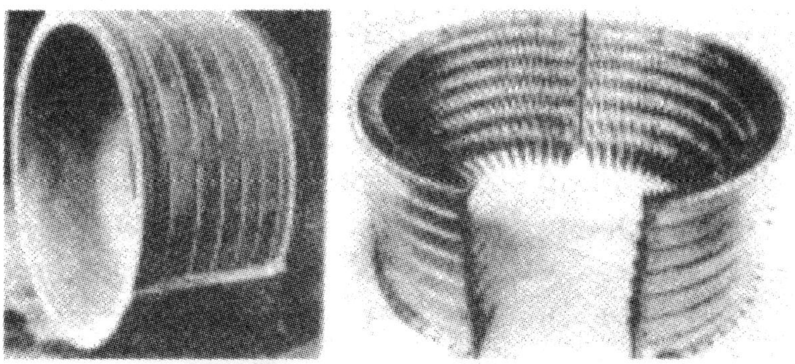

Fig. 18.5 Compressor Casing: As Cast (*left*), Machined, Assembled, Bladed (*right*) (Nixon, 87)

method cooling proceeds from two faces, accompanied by the formation of columnar grains meeting toward the centre of the wall of the casting. All of the unsoundness arising from shrinkage, tears and inclusions remain in the region, and is carried into the finished component. In general, a metal die centrifugal casting meets ASTM E446 Class I radiography criteria, and a refractory shape centrifugal casting falls in Class II, but in many circumstances the latter can qualify to be in Class I.

As with all fabrication procedures, the centrifugal casting requires good control on the shop floor. Control starts with the metal to be melted. The charge for the melt is selectively controlled with both quality and economy in mind. The castings generally undergo extensive machining operation, which at the end of the operation must yield high percentage returns in turnings. After chipping and degreasing the turnings are returned to the furnace for re-use in another melt. Prior to this it is essential to segregate the materials to avoid contamination. In the preparation of the die a mandatory requirement is to avoid lapped surfaces, cracks and hot tears. This calls for the dies to be sprayed with a slurry made from a refractory after an initial preheat to 400°C in a stove. The slurry aids in the flow of the metal along the surface of the die and in preventing fusion with the die. Refractory molds also similarly need the right mix of refractories and accurate control of the pattern and during molding. Pouring of precise quantity of the metal into the casting is important. If the quantity exceeds the requirements the excess amount must be machined, to the detriment of the cost of production. Insufficient metal into the die, on the other hand, produces an unsound casing, since there is not enough material at the bore to satisfy the feeding requirements.

Once the casting is extracted from the die it is stamped to identify the cast number, material and machine number on which it is cast. The identifying numbers remain on the part throughout its life and into the customer's records. An accompanying card with the casting details information on heat treatment, cast and machined size and weight, dies used, spinning speed and other related details. Emphasis on the correct spinning speed is placed because the castings tend to be unsound when the centrifugal force is insufficient, causing deleterious inclusions to accumulate at the bore.

Inspection segregates scrap castings before any work is performed on them. Final inspection of gas turbine engine components is routinely done by the red dye procedure, although the fluorescent dye line is more sensitive. The dye line is in the shape of a horseshoe and split into seven stations, and can accommodate castings up to 120 in. diameter. Defects are noted and compared with the requirements laid down by the industry, which in the case of aircraft engines are near perfection. A high proportion of castings are subjected to radiographic examination, particularly for aircraft engines.

18.3 INVESTMENT CASTINGS

The investment casting process has retained most of its features over the centuries in the making of items of jewelry, with innovations introduced in applications for gas turbine components made of super alloys. A precise replica of the part is first produced in wax or a plastic polymer. Compensation due to shrinkage of the component dimensions during the sequence of processing must be provided (*ref*: Sims et al., 1987).

Where cooling channels are part of the design, a preformed ceramic core of the same configuration is placed inside the cavity around which pattern material is injected. For smaller components, the pattern may be duplicated and assembled into a cluster, with the cavities held together and connected by a system of ducts through which liquid metal flows into individual cavities. The assembly of connected patterns is then dipped in a slurry of aqueous ceramic material. The fragile shell pattern is coated with a granular form of ceramic material to provide a semblance of rigidity, and the steps may be repeated for this purpose. The mold is then dried to eliminate all moisture content, followed by melting of the wax. The mold is fired in a furnace for further strengthening. Prior to casting the configuration is wrapped in an insulation blanket to reduce loss of heat and to obtain a controlled rate of solidification. After the casting is cooled the shell and the core are separated from the metal parts mechanically and chemically, and also split from the cluster at the connecting ducts. The parts then go through inspections, heat treatment or densification by the hot isostatic press (HIP) method.

The pattern for the preparation of the mold must precisely duplicate all the finer features of the component to be cast; hence a complex configuration such as a turbine airfoil will require a special pattern. The pattern must also have stable dimensions, have a smooth surface and permit its easy removal from the ceramic shell. Urea based compounds polystyrene and synthetic wax combined with various resins are generally used for making patterns. Strength, compatibility with ceramic shells and cores, little expansion or contraction and economy are some factors in the selection of the material. The low temperature liquid injection process may be used to make patterns from wax, while plastic patterns are made by the injection molding method. The patterns thus obtained often need to be reformed because of thin component walls by making small adjustments.

Internal cooling passages and other features in hollow components are created with cores made of silica based or alumina ceramics. The core is fixed inside the wax pattern during the injection according to engineering requirements. Room for relative thermal growth between the shell and the core restricts attachment between the two at a single point. Molten metal is poured into the cavity formed after the wax is melted around the core. The outer shell and the inner core are then separated from the casting.

The core must have adequate strength to withstand the pressures of the flowing wax and molten metal, be chemically resistant to the melt at a high temperature and be refractory to retain its configuration during the full cycle. Larger and less complex ceramic cores are made by the injection molding method. The ceramic is mixed with a thermoplastic material, heated in the barrel of the injection-molding machine and forced into the core-shaped cavity of the die. The binder material hardens over a short time period, the machine ram retracts and the part is removed from the die. The binder separates during baking at a low temperature, and the core is sintered. The transfer molding method uses a thermosetting form of binding material and yields a more durable green core, so more complex shapes can be handled by this procedure.

The core is subjected to a thermal shock, and must not crack or deform during the casting. The core is removed from the casting either mechanically or chemically. Cores made of materials with high silica content enjoy the advantage of leaching by bases such as sodium hydroxide and potassium hydroxide. Acids may also be used if the alloy is not affected. To avoid the bases from attacking the inter-granular region, the container material, caustic chemistry and process parameters must be strictly controlled. Larger cores may be separated by air blasting with sand or glass beads.

With adequate precautions to ensure the core does not shift or warp, cast parts can have wall thickness as low as 0.015 in. and holes of 0.020 in. diameter. The length of the core may exceed 15 in. (figure 18.6).

The shell of the mold is exposed to mechanical and thermal loads during the casting operation,

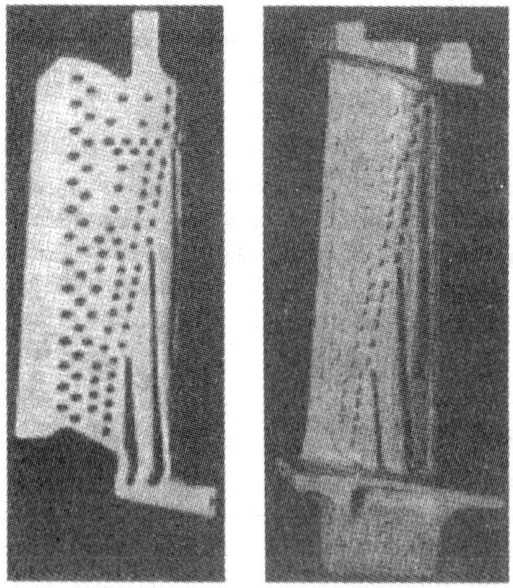

Fig. 18.6 Ceramic Core and Cut Section of Turbine Blade with Cooling Passages (Sims, 1987)

but must not be excessively durable to make it difficult to separate the shell and cause fracture of the casting as the metal freezes and shrinks. Mismatch in the thermal growth between the metal and the ceramic shell during solidification and cooling can affect the quality. The mold shell is subjected to elevated temperatures in processes calling for directional solidification of the metal for a period of time, and must avoid consequent distortion.

A slurry made from a finely grained ceramic refractory material and a binder of silica and a dry refractory grain are used to make the mold shell. The slurry and the dry grain are alternately applied to the wax pattern. The assembly is first dipped in the slurry, than immersed in a fluidized bed of the particles. The binder then cures by chemical reaction for further applications. The rate at which drying occurs must be controlled to avoid distortion. A typical shell mold may be coated 5 to 10 times to develop thickness and strength. The ceramic slurry is made of a finer grain then the subsequent particles.

Beside grain configuration, micro-structural features and freedom from inclusions, a sound casting requires controlled solidification, with sufficient time to permit the molten metal to flow into the geometry. Increased melt and mold temperatures reduce the rate of solidification, and this in turn improves the quality of the casting. Localized hot spots where the metal impinges on the shell walls result in porosity on the surface, and must be avoided by redirecting the flow. The melt enters the turbine airfoil castings through the root attachment. Blades with a tip shroud have a separate gating to achieve the right form at the junction of the airfoil and the shroud. Larger turbine blades may also need a gate for local feeding, but may not be advisable due to its impact on the surface finish and possible alteration of the airfoil's dimensions. Micro-shrinkage of the castings cannot be avoided as the material solidifies, but if it can be restricted to the region around the centreline of the part, then the hot isostatic press process may be able to eliminate it.

A vacuum avoids the risk of oxidation of the reactive elements during the casting for most super alloys. Some cobalt base alloys are cast in induction or indirect arc furnaces in the presence of air. Silica lined zirconia crucibles are generally used, with the pre-weighed charge introduced through a one-way device. The temperature measured during the melting is above that of the liquidus, and is critical in obtaining the proper grain size and orientation. The melt is then poured at a controlled rate into the preheated mold and transferred to the evacuated main furnace.

Directionally solidified single or poly-crystal castings are produced in special equipment, where the mold is kept at a higher temperature than the liquidus of the alloy to be cast. The mold, open at the bottom, is placed on a chilling plate to obtain the proper thermal gradient, and is then retracted from the heater at a controlled rate. The procedure results in the formation of a continuous grain.

A finer size of the grain produced by a relatively rapid rate of solidification results in improvement in the tensile, fatigue and creep properties at medium temperatures. On the other hand, high temperature rupture performance calls for slower solidification, and cooling rates aid in coarsening the grain size. In turbine blades the airfoil operating at high temperatures require a coarse grain, but the heavier dovetail section are less rupture dependent and need a fine-grain microstructure. The problem may be tackled by adding a gate at the edges of the airfoil for the melt to enter, and deliberately set up hot spots in the region to slow the solidification rate. This practice may also be followed to delete the formation of columnar grains at the edges.

Casting defects appear in the form of inclusions, hot tears and peculiar micro-structural features, and are of special interest if they cannot be identified by non-destructive methods (figure 18.7). Non-metallic inclusions are easier to identify on thinner sections, and may be caused during the manufacture of the alloy or during casting. Inadequate level of vacuum during re-melting and casting and improper operating practice at the furnace can cause dross to adhere to the crucible wall prior to pouring. Certain alloys such as INCO 718 require filtering through reticulated ceramic foam to alleviate the problem. Hot tears appear when high temperatures in the casting lead to plastic strains, causing the just solidified material to split in regions where the wall thickness changes suddenly. The tears are more likely to appear below the surface, making their detection by non-destructive methods difficult. Two flowing fronts of liquid metal may not be able to combine because of a thin film of oxide at the interface, and this can result in a cold shut. The problem can only be avoided by changing the casting procedure.

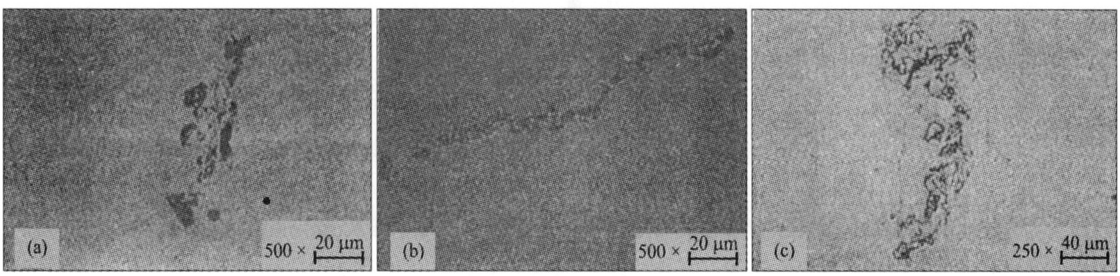

Fig. 18.7 Examples of Inclusions in Alloy Castings (Koul, 1985)

18.4 POWDER METALLURGY PROCESS

Compressor and turbine disks operating in the 1000° – 1400°F temperature range benefit from the powder metallurgy technique. The components are made from pre-alloyed atomizing powder, consolidated by the hot isostatic pressing or by hot extrusion and then forged. The process is also used for stator vanes.

The reactive nature of alloying elements in super alloy powders requires the process to be performed in an inert atmosphere or in a vacuum to obtain proper bonding between the particles (*ref*: Reichman et al., 1970). The powders usually are spherical, and filtered through a fine mesh to reduce contamination. Methods such as atomization by inert and soluble gas, the rotating electrode process and centrifugal atomization are found to be practical for commercial production of the parts. In the gas atomization method the vacuum refined alloy is re-melted under an inert gas and poured and metered, with a nozzle delivering a continuous and rotating stream of gas at a high pressure. This causes the liquid metal to break into spherical particles under the action of the cyclone gas stream. After a controlled rate of cooling the powder is collected in an atomization chamber (*ref*: Sims et al., 1987).

In the solution gas method the atomization is directed upwards from a crucible heated by induction, as shown in figure 18.8. A pressurized atmosphere of air and a gas such as hydrogen (soluble in the alloy) is used in the vessel containing the crucible. A ceramic tube is immersed in the liquid metal, with the other end attached to the evacuated atomization chamber located above. The molten metal flows through the tube to the upper chamber, where the abrupt change in pressure causes the metal to atomize. The powder is then cooled at a controlled rate.

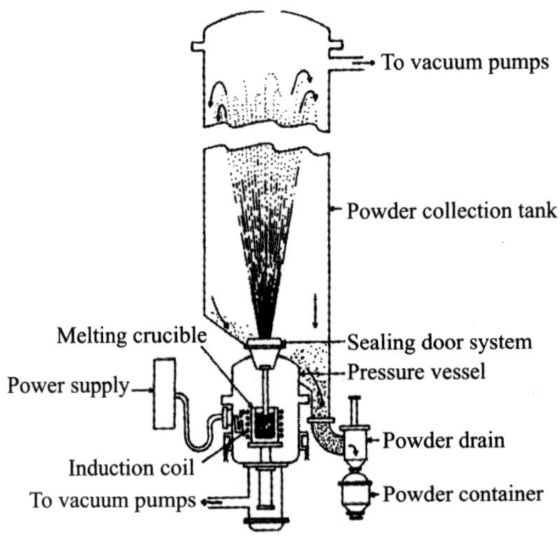

Fig. 18.8 Production of Super Alloy Powder by Soluble Gas Atomization (Sims, 1987)

A non-consumable electrode rotates at a high speed in an inert chamber in the rotating electrode method, striking an arc against the alloy surface. The melting metal at the surface is skimmed by the centrifugal action of the rotating electrode, and is split into atomized particles in the chamber. A plasma arc may also be used instead of the electrode. The common feature of all the methods is to provide for atomization of the melted alloy into a powder form free of contaminants.

The size of the average particle obtained from the rotating method tends to be larger than from the atomizing method, mostly due to differences in the physical aspects of the procedures. The presence of oxidized particles developed during the atomization has repercussions on physical properties such as ductility and low cycle fatigue life. Leakage of air into the system is primarily responsible, and even low levels of incidence tend to reduce the minimum properties. Hollow particles may also develop mostly from trapped gas or from shrinkage during solidification. Consolidation of the powder may help to close the pores, but the entrapped gas may expand during high-temperature operations. Change in density after exposure to elevated temperatures helps in determining the level of porosity, and the density change must be limited to a certain ratio.

Contamination of the alloy powder, mostly from oxidation and from organic elements can be quantified by mixing a sample of the alloy powder in de-ionized water. Lower density particles can then be separated, characterized and counted, with acceptability determined from the type and amount of the contaminants. Contamination from other alloys produced in the same facility also may be responsible in the deterioration of mechanical properties, especially if the cross contamination is from a lower hardness alloy.

Compaction of the powder alloy into a consolidated form relies on hot isostatic pressing (HIP) to its nearly full density, followed by hot extrusion. Increasing the density of the material calls for packing the powder into a metallic container or a closed die after it is evacuated of air, which is then sealed. All forms of extraneous materials must be kept away from the powder. During the HIP procedure the powder in the container is heated, either to a temperature above the γ' solvus or below it, depending on the required size of the grain in the final product. External pressure to the tune of 15,000 psi is applied, and the combined temperature and pressure consolidates the powder to a fully dense product. Time, in addition to adequate temperature and pressure levels, also plays a role in determining the success of the procedure (*ref*: Tien et al., 1984). The mechanism for compaction is influenced by plastic deformation and by creep, and time, temperature and pressure are the controlling factors. Many components can be used after the HIP and the subsequent heat treatment process.

Extrusion following the hot compaction procedure is a more predominant method for producing powder metallurgy components. After the HIP process the consolidated powder has achieved about 95 per cent of its rated density. Hot extrusion obtains a fully dense billet, and possesses a re-crystallized structure with nearly complete elimination of the left over boundaries and dendritic formations (*ref*: Shamblen et al., 1975). The extruded bar may be used after heat treatment or further worked by forging. The final step may be in the form of directional re-crystallization that yields exceptionally stable grains. The grains may be several inches in length, and impart substantial strength to the material by deleting the transverse grain boundaries (*ref*: Cockell et al., 1985).

Thermo-mechanical processing aims to neutralize the intrinsic and extrinsic defects arising from contamination. Intrinsic defects occur during the powder making process from the argon pores and oxidized particles. Extrinsic effects may occur from handling of the powder and during fabrication. Improvement in the properties of the consolidated alloy powders is hard to come by through the thermo-mechanical processing, since the properties are already high. Forging is mostly conducted at isothermal conditions to break up defect structures through the flow of metal, with the tooling and the metal at the same temperature. Rene 95 alloy powder, for example, may be consolidated to 2 – 5 μm range grain size, and exhibits super-plastic behaviour. Thus, after reaching peak stress as a consequence of deformation, the flow softens with increasing strain at a constant rate of strain.

Monotonic properties of powder metallurgy materials such as tensile, creep and stress rupture are governed by the chemical composition and grain structure. Data from the Metal Powder Report (1980) is provided in table 18.1.

Table 18.1 Mechanical Properties of Selected Powder Metallurgy Super Alloys

Alloy	Tensile					Stress Rupture		
	Temperature °F	UTS ksi	0.2% YS ksi	Elongation %	R/A %	Temperature °F	Stress ksi	Life h
U-700[a]	1400	149	148	20	28	1400	85	25
IN-100[b]	1300	184	154.5	20	21	1400	95	35
Rene 95[c]	1200	218	165	13.5	14.9	1200	150	54
Astroloy[d]	1200	192	142	25.6	25.9	1400	92	89

Abbreviations: UTS - ultimate tensile strength; YS - yield strength; RA - reduction of area

[a]HIP + heat treat

[b]Minus 100 mesh, extruded, gatorized

[c]Minus 150 mesh, 2050°F, HIP, heat treat

[d]HIP at 1150° F, 4 h/1975°F oil quench + 24 h/1200°F air cool + 8h/1400°F air cool.

18.5 WELDING METHODS

Welding may be conducted by many different methods, and some commonly used techniques for super alloys such as shielded metal arc, gas-tungsten arc, gas-metal arc, resistance and electron beam are shown schematically in figure 18.9 (*ref*: Sims et al., 1987; Kearns, 1984). Welding requires generation of heat and melting in a local region where two metals meet, and in the process cause them to bond. The main feature distinguishing the various processes is the generation of heat in the selected area. In aircraft engine components wall thickness of the components is small, so the objective is to minimize the amount of external heat provided to eliminate warping and cracking.

The shielded-metal arc process is based on the formation of an electric arc between an electrode and the metal component. Essentially a metal rod of selected diameter coated with a flux material, the electrode's metal core fuses to fill the crevice between the mating components, with the coating also melting to cover the weld beads and prevent oxidation. Being mostly a manual procedure, it does not find much usage for welding of turbo-machinery components because of the difficulty in scraping the flux and the need for precision of the path and the size of the weld.

Gas tungsten arc welding method relies on an arc between a tungsten electrode and the component, with an inert gas (argon or helium) shielding the welding region. If a filler material is required in the joint, it is fed from a different source during the process. A relatively clean procedure, the method is adaptable to joining thin metal sections. A plasma needle arc (*ref*: Union Carbide Corporation) may be substituted for a lesser electric current for welding thin sections (0.010 in. thick) in airfoils. Filler wires may be made from nickel, cobalt or iron based materials. The procedure can be automated to enhance precision. Except for the substitution of the consumable metal electrode instead of the tungsten, the gas-metal arc method is similar to the shielded metal arc process.

Resistance welding relies on generation of heat at the joint by a discharge of a high-density current at a low voltage through the interface. Application of force prior to, during and following

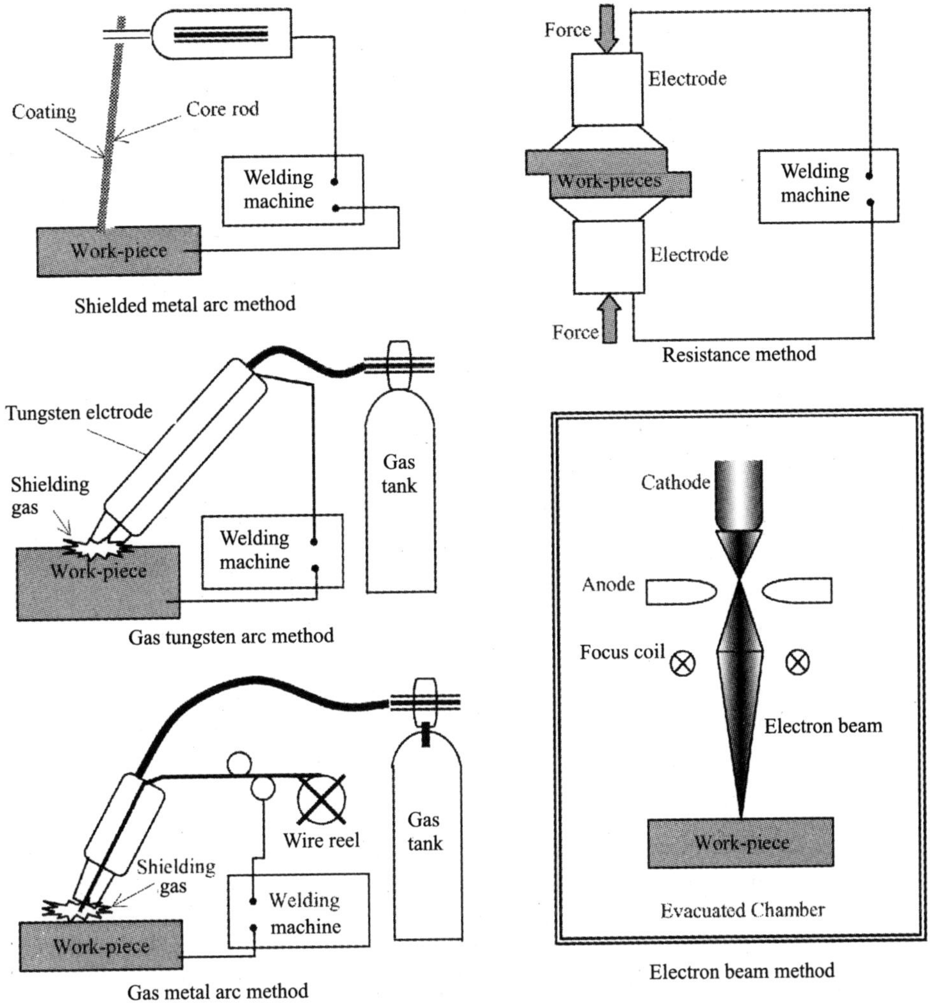

18.9 Welding Methods

the process to maintain a continuing circuit achieves temperatures above the melting point of the metal, and is essential for obtaining the bonding. In dispersion-strengthened alloys, however, agglomeration of the dispersion may be encountered at the high temperature.

If contamination is a problem, an evacuated chamber provides the solution. An electron beam generating and focusing equipment and the job pieces are placed in the chamber. The electron beam welding method is known for its superior quality, especially where the joint is narrow (~ 0.060 in.) and deep penetration (0.050 in.) is a requirement. Dissimilarity in the thickness and the material composition of the two components do not pose a problem in this approach. The capital equipment requirement is high, but the payoff comes from the ability to weld complex structures in relatively inaccessible locations, low levels of thermal distortion and potential for automation.

A welding has two distinct parts, the metal fusion region and the zone affected by the process heat (*ref*: Savage, 1969). In the former, stirring of the filler and the base materials results in modification of the chemical composition. An unmixed layer persists in the end region of the weld metal, where the base metals solidify without being affected by the filler. A solidified sub-structure exists at the surface around the region where complete melting occurs. Partially melted base metal ranging from 0 to 100 per cent occurs just outside the weld surface. Changes in the microstructure in the solid condition as a consequence of the welding take place in the heat affected zone (figure 18.10).

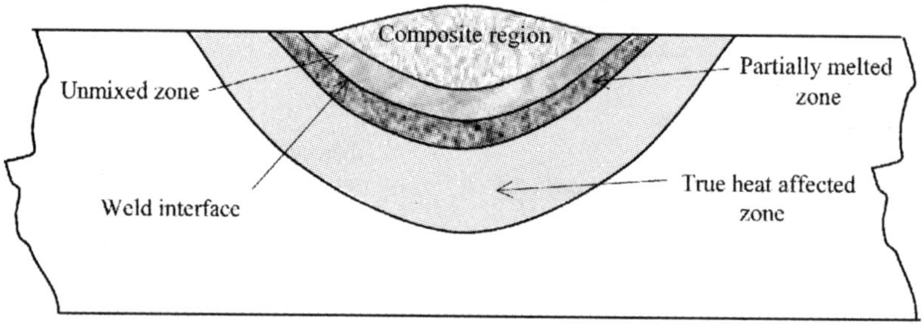

Fig. 18.10 Welding Regions (Savage, 1969)

The occurrence of cracks in a welding may be cause for scrapping of the part. In production environments the welding and its inspection are done prior to expensive machining and finishing processes. Sensitivity to cracking may be obtained on a mockup of the part where the critical features are duplicated. A number of cracking test procedures are available for developing stress during the welding (*ref*: Vagi et al., 1968). The stress level at which the cracks initiate then provides the guidelines for restraining the component during the welding. Strains developed during the procedure may then be changed in incremental amounts. Metallurgical changes during the welding can be observed through hot ductility tests in the region adjacent to an arc weld. The work specimen is held in a high-speed tensile testing machine. Tensile strength and ductility are measured at several temperature points during the simulated weld thermal cycle. Zero ductility and strength define the corresponding temperature points when the two parameters assume zero value. The test parameters can considerably affect the hot-ductility response and hence the sensitivity to cracking. As an example, variations in the peak temperature in the weld heat region of Hastelloy X are shown in figure 18.11 (*ref*: Duvall et al., 1967). Increased distance from the weld interface causes the maximum temperature to reduce. Hence, a point located 0.010 in. from the weld interface is at 2200°F, and experiences little overall change in ductility because the loss during the heating phase is recovered during cooling. At 0.006 in. the temperature is at 2300°F, ductility reduces to practically nothing at 2250°F and does not recover until the temperature falls off considerably. The length of the zero ductility range (ZDR) is an important indicator from considerations of sensitivity to cracking. Thus, a crack-sensitive alloy has a larger range, while an alloy less prone to cracks has a smaller ZDR. The peak temperature during the tests also affects this value, being longer at higher temperatures.

Changes in phase and precipitation hardening reactions do not occur in solution hardened nickel and cobalt base alloys, making them easier to weld. But cracks and fissures may develop because of the range from liquidus to solidus. Sulphur, phosphorus and other minor elements in appreciable quantity are also known to increase the severity of the problem, mostly because of the increase in ZDR. The effects tend to be additive if more than one of the elements exist. Welding of

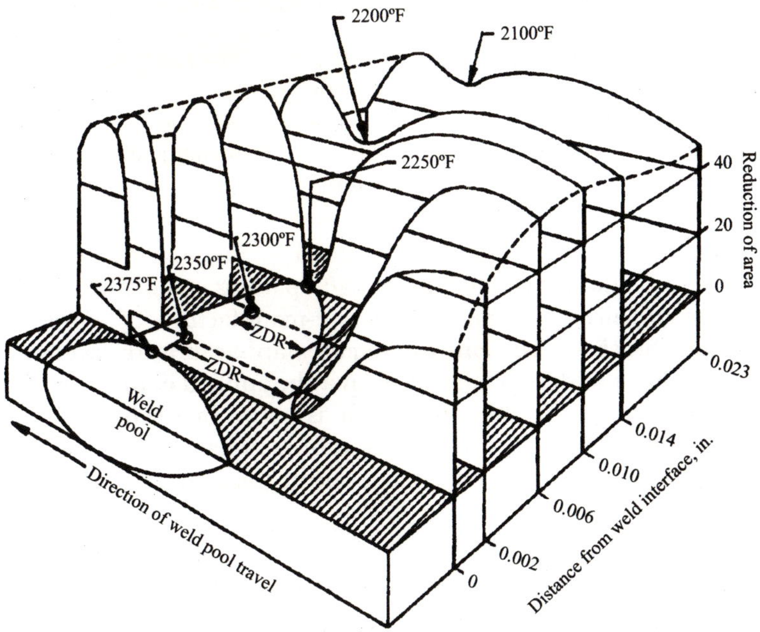

Fig. 18.11 Effects of Peak Weld Temperature on Ductility of Hastelloy X (Duvall, 1967)

nickel based super alloys is good if the titanium and aluminum are present in low levels (*ref*: Prager et al., 1968), as indicated in figure 18.12. Rene 41 and Waspalloy may present difficulties because of cracks during post-weld heat treatment. Casting alloys with high aluminum and titanium (713C and IN-100) are not good candidates for welding because of their low ductility at all temperatures.

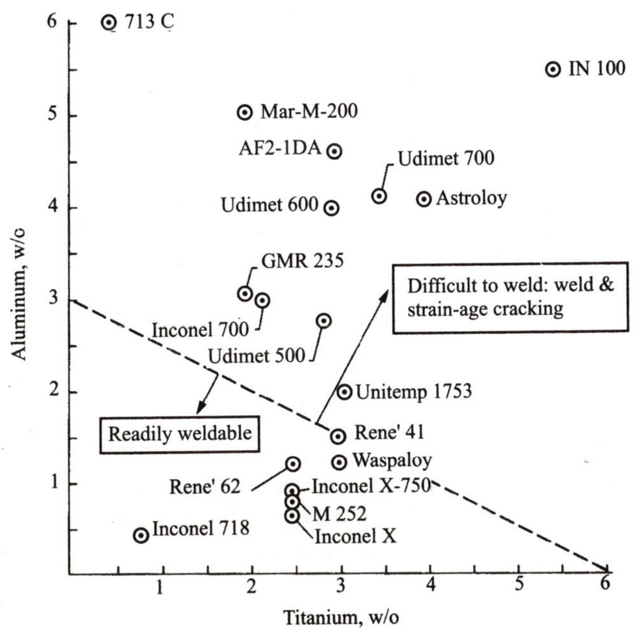

Fig. 18.12 Welding Characteristics of g′ Strengthened Alloys (Prager, 1968)

18.6 BRAZING FOR JOINING NICKEL AND COBALT BASED COMPONENTS

The high-strength characteristics of many super alloys and unacceptable distortion in components with complex geometry are not conducive for welding operations. Many modified wide gap brazing processes have been developed where welding is not practical, such as General Electric's activated diffusion healing method (*ref*: Demo et al., 1992) and Howmet's effective structural repair procedure (*ref*: Wustman et al., 1996). Improvements can be achieved by using the powder metallurgy process for repairing and joining of nickel and cobalt base alloys.

The brazing process by the Liburdi powder metallurgy method has been used for the hot section components of aero and industrial engines (*ref*: Sparling et al., 2002). The procedure has been found useful for repairing casting defects, re-manufacture of shroud segments and reinforcement of mount brackets. The process requires application of layers of high and low melting nickel or cobalt alloy powders. The material may be applied in the form of a tape, paint or paste to the substrate. Addition of components such as boron or silicon aids in depressing the melting point. The mixture is heated to a temperature between the melting points of the two powders. Solid and liquid phase sintering takes place during the heating, accompanied by diffusion bonding and subsequent solidification.

Table 18.2 Composition of Powder Metallurgy Material for Brazing

Element	*Mar M-247 Alloy*	*Mar M 247-7*	*Mar M 247-3*	*Mar M247 Wide gap*	*IN 738 PM*
Ni	59.9	65.26	58.13	65.20	63.32
Cr	8.5	10.11	11.95	11.00	15.40
Co	10.0	7.52	13.00	9.75	8.95
Mo	0.6	0.45	0.42	0.30	1.23
W	10.0	7.52	7.00	5.00	1.82
Ta	3.0	2.26	3.00	1.50	1.98
Hf	1.5	1.13	1.05	0.75	0.00
Al	5.5	4.14	3.85	4.75	3.43
Ti	1.0	0.75	0.70	0.50	2.38
Nb	0.0	0.00	0.00	0.00	0.63
Y	0.0	0.00	0.00	0.00	0.06
B	0.0	0.87	0.90	1.25	0.81

The powder metallurgy process requires a low content of boron for good ductility characteristics, with a wide gap braze having around 1.25 per cent by weight. Since boron has an atomic mass of 11 compared with 59 for nickel, a small change in weight percentage translates into a large percentage by atoms. The composition of some typical powder metallurgy materials are compared with the nominal content of Mar M-247 in table 18.2, together with its wide gap form.

Physical properties of bars subjected to the brazing process are determined by tests on super alloy specimens with slots cut in the middle. Mar M 247-3 and Mar M 247-7 materials are used as filler during the brazing (figure 18.13). The heat-treated test specimens are checked for stress rupture and tension. The ultimate tensile strength of the brazed butt joints is comparable to that of common super alloy repair materials, but the presence of boride particles tends to slightly lower ductility and tensile strength. With Mar M 247-3 the ultimate tensile strength is determined to be

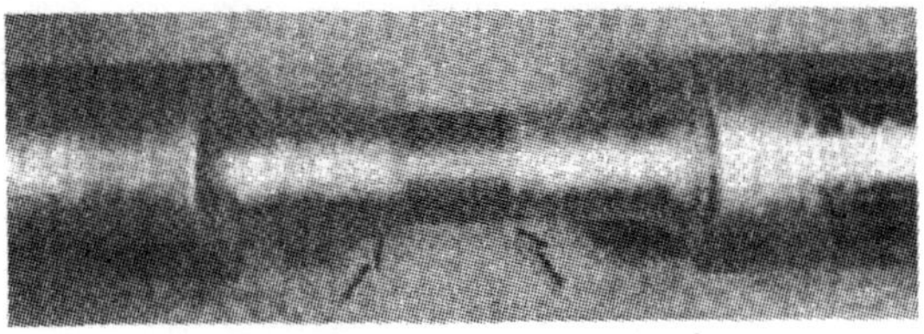

Fig. 18.13 Test Specimen with Butt Joint (Sparling, 2002)

770 MPa at 21°C, reducing to 340 MPa at 927°C. The values with Mar M 247-7 are 640 MPa at 21°C, reducing to 510 MPa at 871°C.

Stress rupture tests are conducted according to ASTM E319 guidelines. Stress, time and temperature are varied to obtain a broad range of Larson-Miller parameters. The results are shown in figure 18.14. The finer equi-axed grain structure of powder metallurgy material tends to lower the creep properties when compared with large grained high strength super alloys. But the materials exhibit superior properties than IN 625, a commonly used filler material for nickel based super alloys.

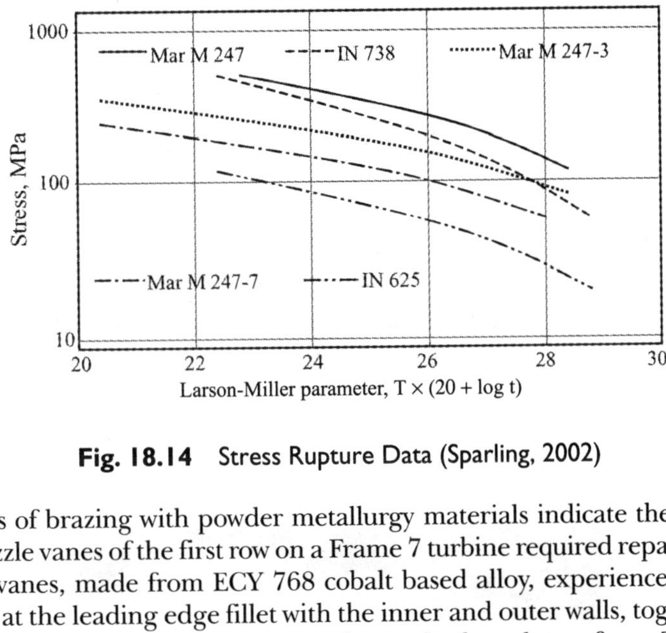

Fig. 18.14 Stress Rupture Data (Sparling, 2002)

Some examples of brazing with powder metallurgy materials indicate the effectiveness of the procedure. The nozzle vanes of the first row on a Frame 7 turbine required repair after 48,000 hours of operation. The vanes, made from ECY 768 cobalt based alloy, experienced extensive thermo-mechanical fatigue at the leading edge fillet with the inner and outer walls, together with oxidation damage of the alloy. Minor impact damage was also noticed on the surfaces. Trailing edges of the vanes suffered thinning during the service, resulting in reduced performance and possible harmonic imbalance due to the increased throat area. After blending and grinding the defect areas in preparation for the repair, Mar M 247-7 powder material is applied in the form of putty. The clay like consistency of the material permits sculpting to fit the repair geometry. The thinned trailing edges and larger areas of oxidation damage are repaired with the tape form of the material, requiring buildup of up to 1.5 mm in thickness. After heat-treatment the nozzles are ground back to nominal contours, and the throat areas are restored to nominal dimensions. Figure 18.15 shows the

Fig. 18.15 Nozzle Vane Repair with Powder Metallurgy Material (Sparling, 2002)

component after the repair procedure, with darker areas representing the area where the powder metallurgy material is applied.

The procedure is also used for closure of the casting holes on the row 1 and 2 blades of Siemens-Westinghouse W501G turbine engine during original manufacture.

Made from Mar M 247 nickel base, the holes are required to maintain position of the casting core. Their larger size precludes closing by welding because of cracking at the directionally solidified grain boundaries. The 6 mm × 25 mm holes are also difficult to close by the conventional wide gap brazing procedure. The powder metallurgy procedure also offers advantages of low porosity and good creep properties. Figure 18.16 shows the tip of the blade before and after the closure of the holes.

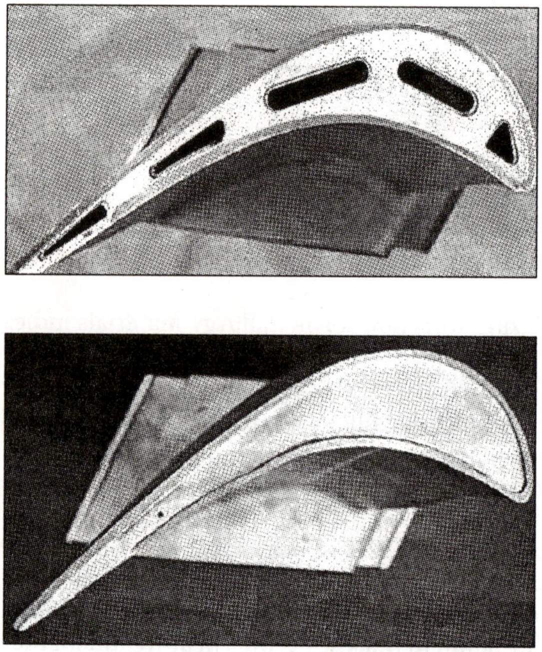

Fig. 18.16 W501G Blade Tip Casting Holes (*upper*); After Closure (*lower*) (Sparling, 2002)

18.7 LASER WELDING TECHNIQUES

Repair of gas turbine blades is mostly limited to the upper portion of the airfoil (tip, angel wings and to a limited extent, down the leading and trailing edges) where operating stresses are generally low. Precipitation strengthening elements that give super alloys their high temperature strength are also fundamentally responsible for the difficulties experienced during weld repair. The limitation is predicated upon the use of low strength welding filler materials and the use of high energy welding processes such as gas tungsten arc welding (GTAW) and plasma transfer arc welding (PTAW). Solid solution strengthened filler materials are employed to assure sufficient ductility is achieved, and cracks and micro fissures from the welding and subsequent heat treatment are minimized. The most commonly employed filler material for repair of IN 738 and GTD 111 blades is IN 625. When the damage is in regions where the stresses are high, the blades are often scrapped or replaced.

A review of various welding technologies indicates that low energy processes such as laser beam welding (LBW) have the potential for repair work in super alloy turbine blades for aircraft engines (*ref*: Gandy et al., 1998). Welding by laser offers the advantage of low heat input, and this promotes shallow penetration, low distortion and minimal residual stress when compared with conventional arc welding processes. Powder filler materials are typically used with the laser process due to the small weld profile (width of bead), but the product is available in forms such as small diameter solid wire. The equipment includes a CO_2 or Nd:YAG laser system, delivery of the beam by hard optics or fibre optic cable, laser welding head to redirect and focus and a filler feed system. The powder is fed with the assistance of a carrier gas through the weld head into the path of the laser beam. The method pre-melts the powder before it is introduced into the molten puddle created on the surface of the substrate by the laser beam. Powder mesh size is selected to assure complete melting and consistent powder flow rate through the feed tubes. The composition and characteristics of the powder must be consistent with the precipitation strengthened substrate alloy. IN 738 and Rene 80 filler materials are selected for the tests.

Development of the procedure calls for evaluation of different welding techniques and parameters to obtain sound weld deposits (*ref*: Frederick et al., 2002). The weld coupons, prepared according to ASTM E139 guidelines, are conditioned by HIP and solution anneal. The shape of the joints (a modified 45° V-groove with overlay buildup) allows for interface between the weld metal and the substrate to be centred. The coupons are sectioned from blades that have logged various service time and operating conditions. A test matrix of travel speed, powder feed rate, shielding and laser power level are used for the evaluation. The weld deposits are screened for lack of fusion, porosity and hot cracking of the weld metal and under-bead cracking by metallographic inspection. The parameters are adjusted with each weld trial to minimize or eliminate the weld defects. An example of porosity and hot cracks in the weld is shown in figure 18.17, and of under-bead cracking in figure 18.18.

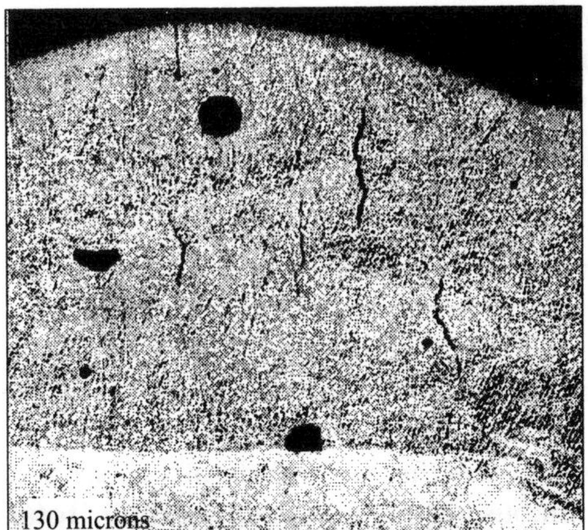

Fig. 18.17 Porosity and Hot Cracks in Weld (Frederick, 2002)

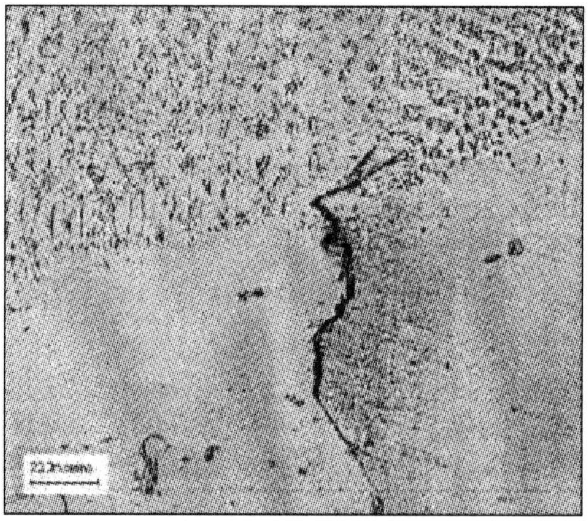

Fig. 18.18 Under-bead Cracking in Weld (Frederick, 2002)

The pattern of the weld and placement of the bead for various joint configurations must be checked during the procedure. The welds comprise of multiple layers of 0.015 – 0.020 in. thickness, and are applied in a controlled manner to obtain an overall level of 1.0 in. for the overlay joint and 0.25 in. for the v-grooved configuration. The coupons are held stationery in the flat position as the laser head is manoeuvred in the horizontal plane to develop the desired weld pattern. After each pass the laser head is retracted and offset by a set distance to avoid stacking of the beads and the potential for inadequate fusion. A final seal weld with a solid solutioning material is intended to cover micro-cracks that may open at the surface. Post-weld heat treatment includes hot isostatic press to optimize the material properties by healing the weld material. The coupons are then sectioned into test specimens for stress rupture and high temperature tensile tests.

Room temperature tensile strength of the GTD111 laser welded coupons ranges between 136.2 ksi and 155.2 ksi. 0.2 per cent offset yield strength is between 127.3 ksi and 130.7 ksi, and elongation ranges between 5.3 and 7.1 per cent. Stress rupture life at 1600°F and 45 ksi for four specimens is between a minimum of 36.5 hours and a maximum of 49.7 hours. Based on these results the selected test conditions may be aggressive. The method, with minor modifications, holds promise for successful repairing of gas turbine blades.

In another example, a shroud ring for a helicopter engine is made of four parts (*ref*: Cantello et al., 1996). The basic ring support made of IN 718 is attached to 0.7 mm thick strips made out of Incoloy 909 and also to a flange made of IN 718. The circular strips are initially welded by the electron beam welding method. The project is intended to evaluate the opportunity for manufacturing cost while fulfilling the quality specifications. A 1 kW solid state Nd:YAG pulsed laser equipped with optical fibre is used for the investigation. The strip at one end of the support ring is welded first, followed by the second ring using the beam incidence angle of 60°, and the flange is welded last. An accurate overlap is required between the strip and the substrate, as also the correct gap between the front ends of the strips. The joining cycle sequence called for spot tacking, continuous tacking and deep welding. Total penetration is 1.3 mm for the strip rings and 3.5 mm for the flange. An advantage of the procedure is that the joining of the two strips to the shroud ring is conducted at the same station, representing a reduction in the cost. The final component assembly of the shroud ring is shown in figure 18.19.

Fig. 18.19 Laser Welded Shroud Ring Assembly (Cantello, 1996)

18.8 GENERATING FIVE-AXIS CUTTER PATH

Production of centrifugal compressor and fan stages with complex and overlapping surfaces requires five-axis numerically controlled machine tools to perform point-milling. The geometry of a typical component is mostly composed of sculptured, ruled and blending surfaces. The desired precision calls for development of correct cutter contact data from the scallop height, over cutting and local gouging of the part surfaces. The possibility of tool interference requires careful checking in order to determine the right cutter location and tool path.

Figure 18.20 schematically illustrates the procedure of five-axis machining. The method is extended to the machining of an impeller to demonstrate its effectiveness (*ref*: Tsay et al., 2001). The cutter is first guided to touch a prescribed contact point $\boldsymbol{r}$ on the surface of the work-piece (figure 18.21). If R is the radius of the cylindrical ball-end cutter and $\boldsymbol{P}$ at its centre then $\boldsymbol{r}$ may be defined by $\boldsymbol{p} = \boldsymbol{r} + R\boldsymbol{n}$, where $\boldsymbol{n}$ is the unit normal vector to $\boldsymbol{r}$. Also, cutter location $\boldsymbol{L} = (\boldsymbol{b}, \boldsymbol{u})$, where $\boldsymbol{b}$ is the position of the cutter tip and $\boldsymbol{u}$ is the unit vector along the cutter axis. Hence $\boldsymbol{b} = \boldsymbol{r} + R(\boldsymbol{n} - \boldsymbol{u})$.

An admissible direction of the cutter axis without encountering interference may be determined as follows. Contact point coordinates on the work-piece are rotated about the z-axis through ϕ, than tilted about the x-axis through θ. The transformation may be expressed in matrix form:

$$[R_{zx}] = [R_z\phi][R_x\theta] \qquad ...(18.1)$$

An identical result is obtained by tilting the work-piece through θ about the x-axis, followed by rotation about the z'-axis through angle ϕ. The transformation for this step is given by:

$$[R_{xz'}] = [R_x\theta][R_{z'}\phi] \qquad ...(18.2)$$

An interference-free tool path can be obtained by placing the cutter's ball end at an offset equal to the radius of the cutter at a specified orientation. The trajectory of the contact points must then

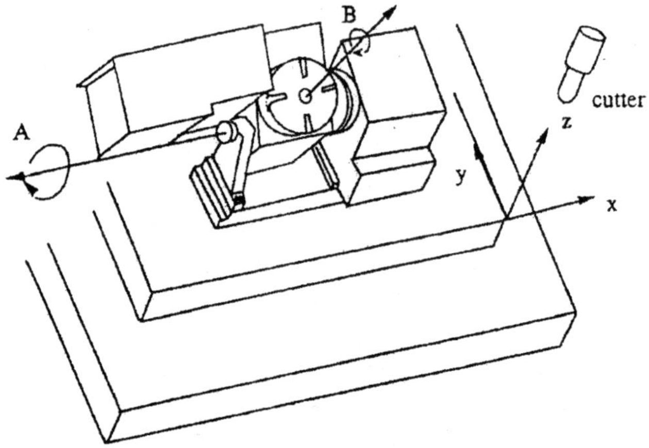

Fig. 18.20 Schematic Illustration of Five-Axis Machining (Tsay, 2001)

lie outside the flank of the cutter. Surface roughness also requires consideration. Two types of roughness are shown in figure 18.22, where the scallop height and chordal deviation must be taken into account. $\boldsymbol{r}(u)$ is the cutter's contact point on an isoparametric tool path in the primary direction $\boldsymbol{u}$ of the machined surface. Chord vector $\boldsymbol{c}$ denotes the bounds between $\boldsymbol{r}(0)$ and $\boldsymbol{r}(1)$. The perpendicular vector $\boldsymbol{d}$ from the chord to the point $\boldsymbol{r}(u)$ of the curve may be written as:

$$\boldsymbol{d} = \boldsymbol{r}(u) - \boldsymbol{r}(0) - \lambda \boldsymbol{c} \quad ...(18.3)$$

where $\lambda = \{[\boldsymbol{r}(u) - \boldsymbol{r}(0)].\boldsymbol{c}\}/\boldsymbol{c}^2$.

The maximum chordal deviation occurs when the tangent vector $d\boldsymbol{r}/dt$ is perpendicular to $\boldsymbol{d}$, or $\boldsymbol{d}.(d\boldsymbol{r}/dt) = 0$, reaching a maximum when $(d^2r/dt^2).\boldsymbol{d} < 0$. Under the maximum tolerant chordal deviation condition, the isoparametric step length u in the primary direction can be determined.

The second component of the surface roughness is along the cross direction $\boldsymbol{v}$ of the machined surface, representing the height of the scallop between two adjacent tool paths of a ball nosed cutter. Both the convex and concave portions are of interest. The diagrams of figure 18.23 provide the geometric relationships of the two contours. For a given ball nose cutter of radius R, the step length s in the convex portion is given by:

$$s = 2\left\{R^2 - \left[\frac{2R\rho - \delta^2 - 2\delta\rho}{2(\delta + \rho)}\right]^2\right\}^{1/2} \quad (18.4)$$

where δ is the scallop height and ρ the radius of curvature of the contour. Similarly, the step in the concave portion is:

$$s = 2\left\{R^2 - \left[\frac{2\delta\rho - \delta^2 - 2R\rho}{2(\delta - \rho)}\right]^2\right\}^{1/2} \quad ...(18.5)$$

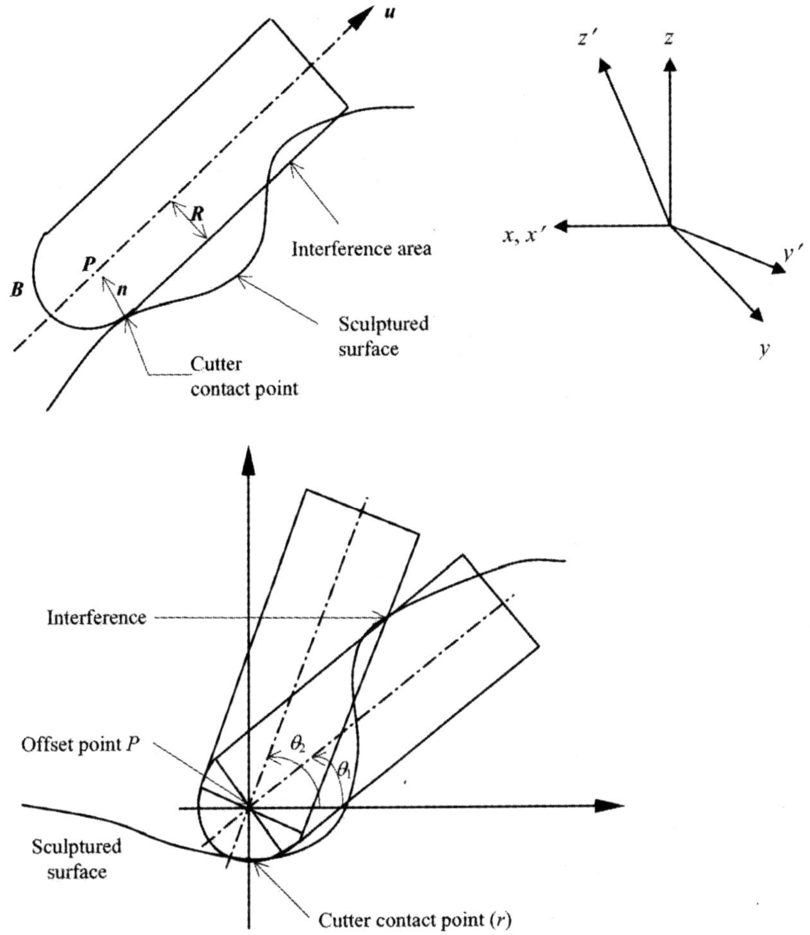

Fig. 18.21 Interference Between Cutter and Work-piece (Tsay, 2001)

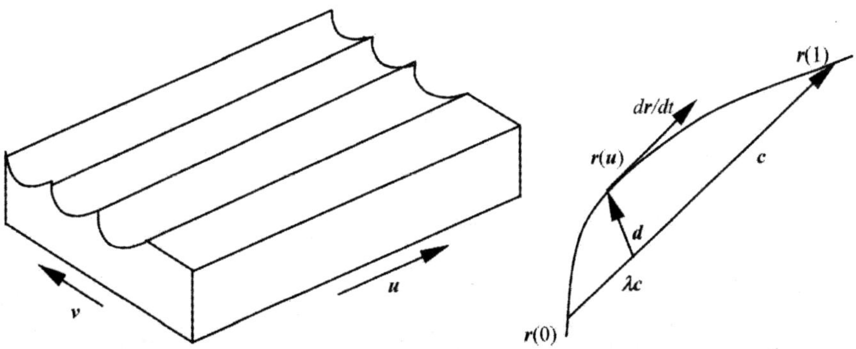

Fig. 18.22 Surface Roughness Composition: Components (*left*); Chordal Deviation (*right*) (Tsay, 2001)

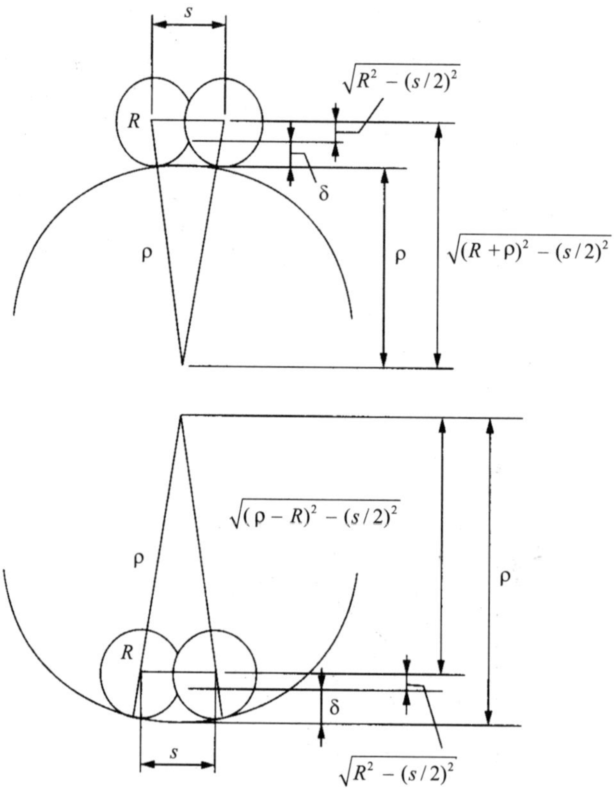

Fig. 18.23 Scallop Height: Convex Surface (*upper*); Concave Surface (*lower*) (Tsay, 2001)

The usefulness and reliability of the approach is demonstrated in the five-axis milling of a centrifugal compressor stage with 15 blades, shown in figure 18.24. Based on the developed algorithm, an interactive computer code in C language is written on a personal computer to implement the procedure. Definition of the geometry is obtained from ruled and camber surfaces. A camber surface is used to form the pressure and suction side surfaces, with offsets applied on both sides. The camber surface is a ruled surface, and is constructed by linearly interpolating between two known boundary curves on the hub and shroud camber-lines. Thus, by extension the pressure and suction surfaces are also ruled surfaces. The hub surface is developed by rotating the hub's (Choi, 1991) is used to construct blends at the root between the blades and the disk on both sides. The blank, made of aluminium alloy 7075, is 118 mm in diameter and 75 mm in thickness. Minimal gap between adjacent blades is 9 mm. The diameter of the selected cylindrical ball ended cutter is 4 mm. For the given conditions the angle ϕ is selected to be 12°. A range of values is used for the tilting angle θ without experiencing interference, although it is limited to a

Fig. 18.24 Machined Impeller (Tsay, 2001)

small zone. Generally, the direction of the cutter's axis should not be collinear with the normal at its contact point, since the tool often leaves marks on the hub surface of the impeller, as shown in figure 18.25 (*upper*). Improvement in the situation is seen in the lower picture by following this guideline.

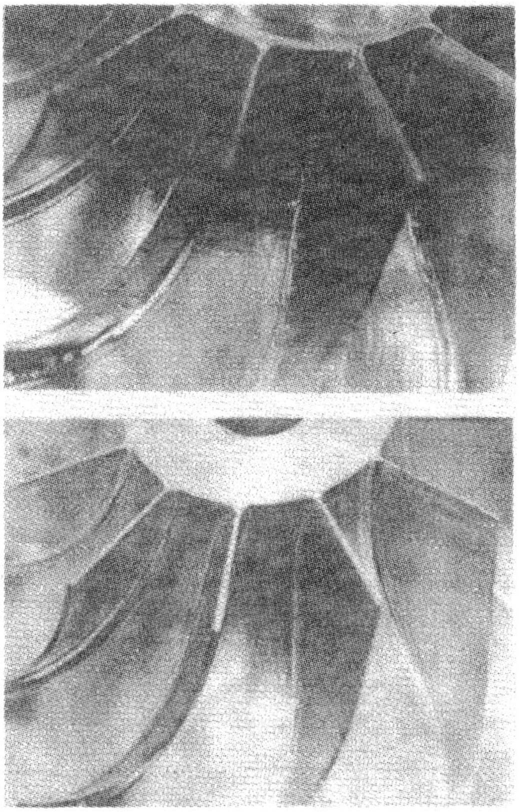

Fig. 18.25 Tool Marks on Hub Surface (*upper*); Improved Surface (*lower*) (Tsay, 2001)

18.9 MACHINING METHODS AND IMPELLER PERFORMANCE

CNC milling of compressor impellers is a mature technology, and is ideal for rapid prototyping and low quantity pre-production, and may sometimes be economical for larger production runs. Casting is cheaper for bigger quantities, but is inferior in accuracy, consistency and structural integrity. Automotive turbo-chargers, for example, have been traditionally cast because of economies of scale. The cost advantage of cast to milled impellers is at least 1:20. For larger high-performance compressors, stress, fatigue and frequency response considerations demand the use of forged billets, and therefore, milling. Milling assures precision and consistency in all directions, irrespective of the size or location.

The manufacturing process results in an inherent surface roughness (figure 18.26). A sand grain form of finish results from the casting process, while machined surfaces display a cusp and a roughness from the cutter's path. Typical blade surface finishes are in the range of 1.6 mm, a de facto industry standard for smoothness, while hub surfaces may vary from 0.8 to 12.5 μm, but may go up to 50 μm. Blades may be flank milled with the side of a cutter or point milled with the tip of a cutter, or milled by a combination of the two. Flank milling takes less time and produces 1.6 μm finish, but is less accurate than point milling. The finish may be improved by light polishing by hand, but adds to the cost. Another parameter of interest is the orientation of the cusps from the cutter path relative to the local flow velocity vector. Figure 18.27 shows examples of cutter path offset from the hub, and with the path oriented relative to an assumed meridional velocity.

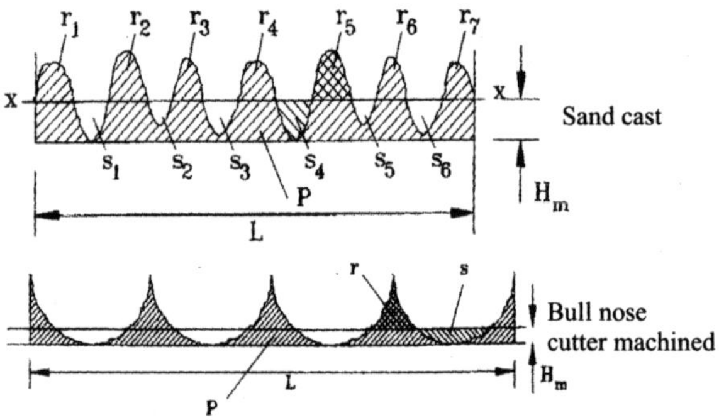

Fig. 18.26 Surface Roughness Examples (Childs, 1997)

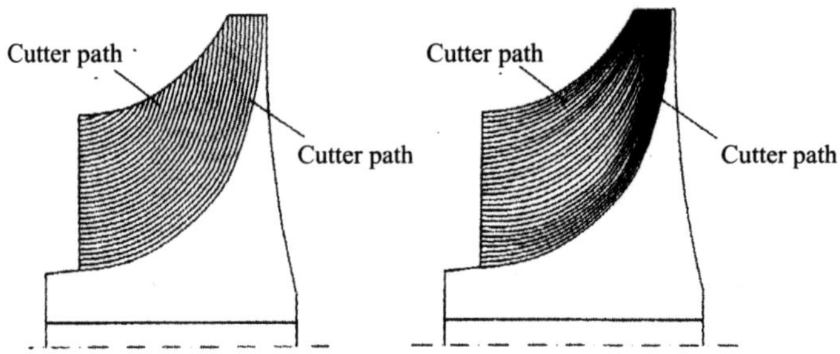

Fig. 18.27 Cutter Path Offset from Hub (*left*), Oriented Relative to Flow Velocity (*right*) (Childs, 1997)

In addition to impeller blade coordinates, a number of critical features come into play. Chief among these are the fillet radius between the blade and the hub, the smallest passage in the impeller and the surface finish. In selecting the blade radius a balance is struck in the design process between aerodynamic performance and stress and vibration resistance. A fixed radius fillet is easier to machine, with the radius matching that ground on the cutter. Variable fillet radii have a limited impact on time for parts in production. Splitter blade length is usually based on flow considerations. The fillet radius between the splitter and the adjacent full blade often defines the limit for the location of the splitter's leading edge. If this fillet radius is held to an appropriate size, it adversely affects production cost, and should be no less than the general fillet radius around the remainder of

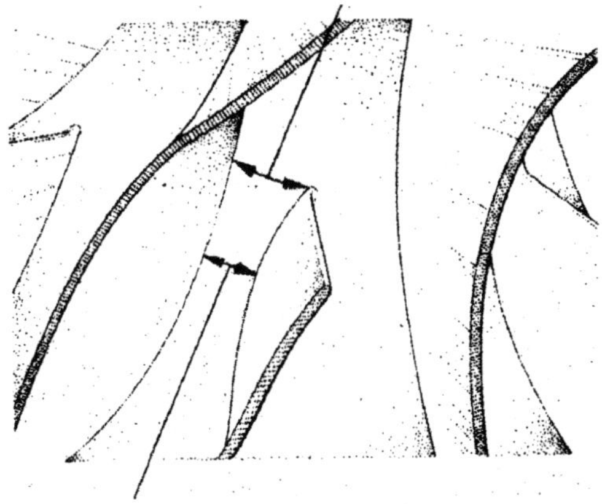

Fig. 18.28 Spacing Between Impeller Main and Splitter Blades (Childs, 1997)

the blade. The smallest distance between consecutive blades is the normal distance from one blade to another (*ref*: Childs et al., 1997), and usually is not in the circumferential direction, as seen in figure 18.28.

Most impellers are designed with straight-line elements to define the hub and the tip, creating ruled surfaces in the process. Flank milling requires the tool to follow the rulings accurately. The dynamics of the machining process introduces chatter on the surface of the blade along the line of contact of the tool with the blade, and may appear in the fillet region. The chatter is sometimes known to leave ridges running from the hub to the blade tip in a direction that may adversely affect the flow. Flank milling also requires accurate tools. Tool inaccuracy, deflection and machine dynamics commonly leave behind striations on blades along curves parallel to the hub line, and may be large enough to impede the flow. Because flank milling employs straight-line elements to machine a curved surface, significant errors are introduced on large impellers and on impellers with considerable twist in the blade.

Point milled blades offer several advantages. Smaller machines of less power may be used, and the cutting force applied is a small fraction of that employed in flank milling. Thus, tool deflection and machining inaccuracy is limited. With such positives going for it, the bad news about higher cutting time may be challenged.

Prediction of compressor performance variation due to surface roughness is based on friction factor for flow in the impeller channel. In one formulation the efficiency is based on Reynolds number (*ref*: Wiesner, 1979):

$$\frac{1-\eta}{1-\eta_t} = a + (1-a)\left(\frac{\text{Re}_t}{\text{Re}}\right)^m \qquad ...(18.6)$$

where η is the efficiency to be determined at Reynolds number Re, η_t is a measured efficiency at a test Reynolds number Re_t, constant a accounts for other losses and m is an empirical constant. Difficulties may be encountered in establishing reliable values for the two constants. Simon et al (1984) suggest an equation based on relative surface roughness:

$$\frac{1-\eta}{1-\eta_t} = \frac{a+(1-a)(f/f_\infty)}{a+(1-a)(f_t/f_\infty)} \quad ...(18.7)$$

where efficiency η is to be determined for friction factor f, η_t is a measured efficiency for friction factor f_t, f_∞ is friction factor for fully turbulent condition and a is an empirical constant. The friction factor is given by:

$$\frac{1}{\sqrt{f}} = 1.74 - 2\log_{10}\left(\frac{2R_a}{b_2} + \frac{18.7}{\mathrm{Re}\sqrt{f}}\right) \quad ...(18.8)$$

where f is the friction factor, R_a is surface roughness, b_2 is impeller exit width and Re is the impeller Reynolds number. The equation requires iteration, or successive substitution, to calculate the friction factor for a given Reynolds number. Casey (1985) proposed another technique:

$$\Delta\eta = -\frac{c}{\mu_0}\Delta f \quad ...(18.9)$$

where $\Delta\eta$ is the change in efficiency resulting from changes in flow or surface conditions, c is semi-empirical constant related to compressor width ratio b_2/D, μ_0 is a work coefficient and Δf is the associated change in friction factor.

Charts for practical use of the procedure are shown in figure 18.29. Baseline efficiency for a compressor of known blade surface roughness must be available. In this figure the curves are plotted for surface roughness of 6.3 μm. Operating condition Reynolds number must be computed. The change in efficiency due to a change in the surface may then be obtained for the Reynolds number.

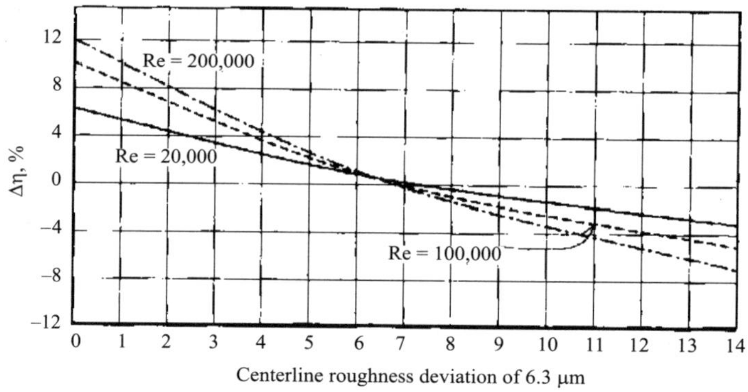

Fig. 18.29 Compressor Efficiency Variation with Blade Surface Roughness (Childs, 1997)

18.10 DIMENSIONAL INSTABILITY IN MACHINING SUPER ALLOYS

The dimensional instability phenomenon is basically a change in dimension with time, unaccompanied by further work on the component. Two probable causes are residual stresses and metallurgical alterations introduced by the machining process. The problem can be controlled by understanding the metallurgy of the material and its plastic deformation characteristics, by controlling the machining parameters and by the effect of process variables on the machined surface and sub-surface.

Highly localized shear stress/strain and temperature in machining of many super alloys has been observed to produce shear instability (*ref*: Komanduri et al., 1986). The general effect of micro-structural changes is to introduce various mechanisms by which the material may undergo dimensional changes as a result of internal changes (*ref*: Marschall et al., 1971).

Plastic deformation characteristics of Inconel 718 and the effects of process variables on a machined surface and dimensional stability in machined jet engine components has been investigated by Subhas et al., 1998. Experimental studies are conducted on ring shaped specimens from compressor disks made of Inconel 718 in a fully heat treated condition. Similar test specimens are also machined out of Ti-6Al-4V alloy and mild steel at identical machining conditions. Dimensional changes are then measured at time intervals up to 220 hours after machining. Titanium alloys are more prone to dimensional instability due to their lower modulus of elasticity.

The ring-shaped specimens have an inner diameter of 55 mm and outer diameter of 76 mm, and are stress-relieved and aged to hardness 44 HRC. The experiments are conducted in two steps. In the first step the machining for the experiment is carried out on a horizontal lathe for studying the effect of machining parameters on plastic deformation characteristics. The test piece is held in a mandrel to avoid clamping pressure. Kennametal brazed carbide tools are used, and the cutting fluid is a soluble oil (1:20). The chips are collected, and optical micrographs of the machined surface and the longitudinal midsection of the chips are obtained. The dimensions are measured on the outer diameter at a predetermined distance on a three-dimensional coordinate measuring machine. Tool flank wear and contact length at the chip-tool interface are measured with a toolmaker's microscope. A portable perthometer measures the surface finish. Cutting forces are determined with the aid of a component tool force dynamometer.

In the second step a hole drilling strain gauge experiment is conducted for determining the residual stresses according to ASTM standard test method E 837. A three-element strain gauge rosette is attached to the machined surface. The gauge grid is connected to a static strain indicator through a switch and balance unit. A precision milling guide is accurately centred over the drilling target on the rosette. After zero-balancing the gauge circuit, a shallow hole is drilled through the centre of the rosette with a carbide cutter powered by a high-speed air turbine unit. Relaxed strains are read for each predetermined incremental depth.

Optical microscopic examination of the longitudinal midsection of the chips produced at speeds ranging between 10 m/min to 38 m/min indicates considerable deformation twinning in the Inconel 718. The effect is observed on the machined surface also, a form that is not generally prevalent in machining other conventional steels and high temperature alloys. Shear instability studies in the machining of the Inconel 718 and Ti-6Al-4V show that the role of limited slip in the latter alloy can be considered to be that of precipitates in the former material. The high resistance to deformation of Inconel 718 probably balances the slightly poor thermal properties of Ti-6Al-4V alloy. The results confirm that the primary deformation mechanism of the two materials is similar, and hence their dimensional stability aspects are also similar.

The results pertaining to the impact of machining parameters on plastic deformation causing residual stresses display some interesting trends. A negative rake angle increases residual stress, while a positive rake angle decreases it. These observations are logically correct, because as the rake angle increases, the shear angle increases, the shear plane length decreases, cutting force decreases and as a consequence the shear stress and strain decrease. Thus, the residual stresses decrease. The chip tool contact length is also of interest. The cutting force decreases as the contact length decreases, and hence reduces the plastic strain and the residual stress. It is also clear that residual stresses in the cutting direction are more tensile than those in the longitudinal direction. At low feeds the surface residual stresses are compressive, but at high feeds they are tensile. The depth of the cut also plays a role, since a larger cut causes a bigger volume of material to be removed. Therefore, the residual stresses at the surface increase with the depth of the cut.

The plastic deformation aspects of Inconel 718 indicate that the machining parameters significantly influence the magnitude of residual stresses, and residual stresses are primarily responsible for the variations in dimensions. Optimization of the machining parameters should be able to control the dimensional stability problem to a manageable level, and hence a procedure is needed to establish the parameter limits. In the design of the statistical experiment the equations for response functions such as residual stress, tool life, surface finish and dimensional stability and their relation with the machining parameters are needed.

Since many of the critical gas turbine components are symmetric about the engine axis in geometry, the turning process is suitable. Here the speed, feed, depth of cut, rake angle, tool nose radius, side cutting edge angle, cutting tool material and cutting fluid have varying degrees of influence on the residual stress. Of these, only the first five play a major role. It may be assumed that the influence of the process variables on the response properties is linear, and hence it suffices to study each variable at two levels only. A half factorial design then leads to 16 experiments.

The effect of the machining parameters on the response functions (residual stress, tool life, dimensional instability, surface finish and material removal rate) is established by multiple regression analysis. The fitted equation is of a first-order, with linear functional relationship. The two levels of independent variables are coded for convenience as – 1 (low) and +1 (high) in the transforming equations. Typical transforming expressions for the variables v are of the form:

Table 18.3 Optimized Machining Parameters

Acceptable Dimensional Instability, Microns	*Independent Parameters*					*Dependent Parameters*				*Composite Desirability*
	v	f	d	α	r	σ_c	σ_l	T	*MRR*	
Group II: 10 to 20	14	0.04	0.4	6	0.4	62.6	8.54	10.2	196	0.296
Group III: 20 to 30	20	0.04	0.5	6	0.4	99.9	10.9	8.4	360	0.312
Group IV: 30 to 50	26	0.04	0.8	6	0.4	152.4	14.8	7.2	780	0.262

$$\frac{2v - (v_{\max} + v_{\min})}{(v_{\max} - v_{\min})} \qquad ...(18.10)$$

$$\frac{2(\ln v - \ln v_{\max})}{(\ln v_{\max} - \ln v_{\min})} + 1 \qquad ...18.11)$$

Expression (18.10) is used for the independent variables when the prediction equations are required in a polynomial form, and (18.11) for the exponential form. The transformation is executed in the computer program to obtain the following equations:

$$\sigma_c = -30.12 + 4.49v + 1477.34f + 76.23d - 10.55\alpha + 2.714r \quad ...(18.12)$$

$$\sigma_l = 2.24 + 0.31v + 1055f + 6.988d - 0.86a + 0.676r \quad ...(18.13)$$

where α = rake angle, degrees, d = depth of cut, mm, f = feed, mm/rev, r = tool nose radius, mm, σ_c = circumferential residual stress, MPa, σ_l = longitudinal residual stress, MPa and v = cutting speed, m/min. Material removal rate is $1000 \times v \times f \times d$. In the evaluation the ratio of lock of fit mean square to pure error mean square at the midpoint of the range is to be determined and compared with the ratio for static conditions.

The method of simultaneous optimization of several variables proposed by Derringer et al (1980) is used to optimize the machining parameters such as dimensional stability, residual stress, surface finish, tool life and material removal rate. The corresponding independent variables v, f, d, α and r may then be used for experimental verification and ready application. In simultaneous optimization the dependent variables are converted into individual desirability functions, which are then combined using the geometric mean to assess the combined desirability of the response.

The tolerance band of modern jet engines is generally classified as shown in table 18.3 for turning operations. Group I (5 – 10) microns is used mostly for jig borings. The tolerance groups decide the maximum and minimum acceptable constraints on the dimensional instability. Similar constraints on the other dependent variables are based on practical experience. The independent variables are increased in discreet steps. Values for the optimized parameters are shown in table 18.3.

Tests are conducted for the tolerance range of 10 to 20 microns, and measured response variables are compared with predicted values. The inspection data indicates that dimensional changes are well within calculated limits. Repeated tests using the same parameters measure a scatter of 5 per cent in the values. Hence, the experiment validates the reliability of the selected independent parameters. Figure 18.30 shows the effects on non-optimized and optimized parameters on dimensional changes. By using the optimized values for the independent parameters, the 400 (+0.02, 0.00) mm locating diameter is machined. Immediately following the machining the dimension measures 400 (+0.010, 0.00) mm. After 360 hours the same dimension is measured 400 (+0.015, 0.00).

18.11 CURVIC COUPLING FOR TURBINE ROTOR

Curvic couplings are used to connect rotating disks and shafts of gas turbine engines for providing precision centering of the components and high load carrying capacity. Considerable disengaging forces develop in the coupling both during steady operations and during start-up and shutdown. Besides accuracy in positioning, the curvic is required to have good indexing capability and high indexing stiffness. Figure 18.31 shows their location on the disk of a turbine rotor assembly and details of the tooth profile on the mating disks (*ref*: Tsai et al., 2002).

The concave and convex tooth profiles are generated by using the outside and inside surfaces of a cylindrical grinding wheel. The machining surface of the grinder is conical, and hence the tooth profiles can be described by parametric equations for the surface of a cone. Since the cone surface is quadric, its generalized form is:

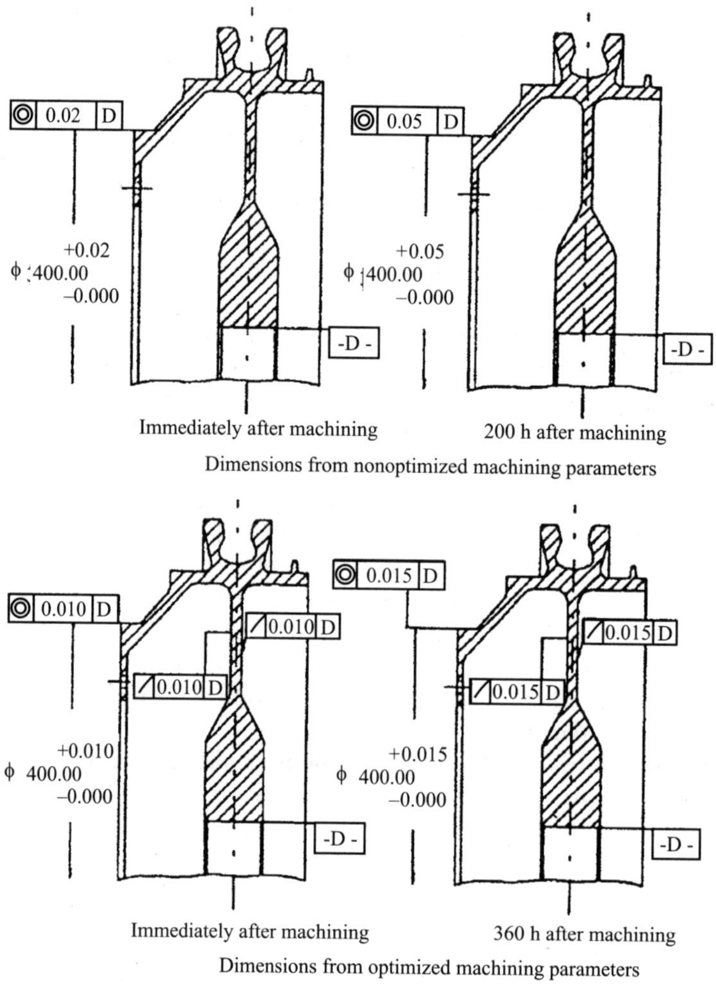

Fig. 18.30 Effects of Machining Parameters on Dimensional Stability (Subhas, 1998)

$$[S]_{gen} = \{x\}[Q]\{x\}^{t} = [x\ y\ z\ 1]\begin{bmatrix} A & D & F & G \\ D & B & E & H \\ F & E & C & J \\ G & H & J & K \end{bmatrix}\begin{bmatrix} x \\ y \\ z \\ 1 \end{bmatrix} \quad ...(18.14)$$

$$= Ax^2 + By^2 + Cz^2 + 2Dxy + 2Eyz + 2Fxz + 2Gx + 2Hy + 2Jz + K$$

$$= 0$$

In the above equation, if $A = B = 1$, $D = E = F = G = H = J = 0$ and $C = -m^2$ ($m = \tan \alpha$, α being the half cone angle) then the equation degenerates to the form:

$$x^2 + y^2 - m^2z^2 = 0 \quad ...(18.15)$$

In the parametric form the cone surface can be described using parameters t and θ by:

Fig. 18.31 Curvic Coupling in Turbine Rotor (*upper*); Tooth Profile (*lower*) (Tsai, 2002)

$$[S(t, \theta)]_{par} = \begin{bmatrix} mt\cos\theta \\ mt\sin\theta \\ t \\ 1 \end{bmatrix} \qquad \text{...(18.16)}$$

where the definition of m remains. Geometrically the above two equations indicate that the vertex of the cone is located on the origin, and the base circle is along the z-axis. But in practice the cone may be anywhere in space, hence a coordinate transformation is required to broadly define the cone's curved surface. The transformation matrix is given by:

$$[T] = \begin{bmatrix} T_{11} & T_{12} & T_{13} & \delta_x \\ T_{21} & T_{22} & T_{23} & \delta_y \\ T_{31} & T_{32} & T_{33} & \delta_z \\ 0 & 0 & 0 & 1 \end{bmatrix} \qquad \text{...(18.17)}$$

Here the T_{11} to T_{33} represent rotations about the fixed axes and δ_x, δ_y and δ_z describe the translations along the fixed axis. Following the coordinate transformation, the generalized and parametric forms of the cone's surface are:

$$[S]'_{gen} = \{x\}(T^t)^{-1}(Q)(T)^{-1}\{x\}^t = 0 \quad ...(18.18)$$

$$[S]'_{par} = (T)\,(S)_{par} \quad ...(18.19)$$

The mathematical model for the machining of the curvic coupling is thus obtained. The manufacturing process is shown in figure 18.32. In the following example the coordinate system origin is located in the centre of the curvic coupling, the z-axis faces up and the x-axis lies in the same direction as the coupling center relative to the center of the grinder. The y-axis is obtained from the right hand rule, and the x-y plane coincides with the pitch plane of the coupling. If the inner and outer diameters (d_i, d_o) of the blank, total number of teeth (N), number of covering teeth (n) and pressure angle (ϕ, generally 30°) are known, the parameters of the grinding wheel can be determined. Some other points deserve attention:

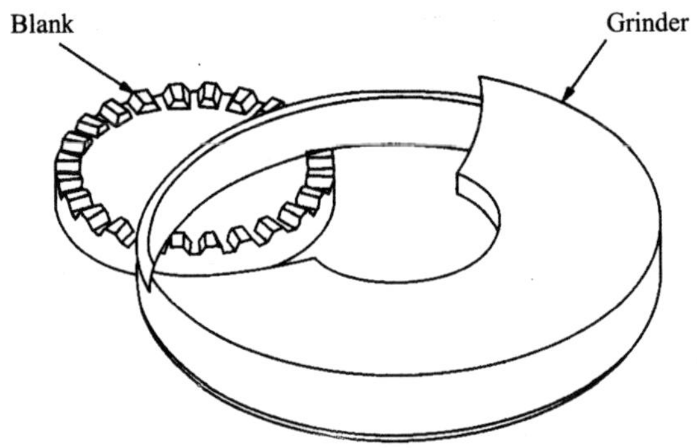

Fig. 18.32 Manufacture of Curvic Coupling (Tsai, 2002)

- On the circle plane formed by the grinder the width of both the convex and concave teeth are the same. On the pitch plane the blank can then be divided into $2N$ teeth.
- The angle covered by the grinder is defined by $\theta = (2n - 1)\pi/N$.
- The circle section of the grinder must be tangential to the section of the tooth profile on the pitch plane. The radius of the grinder is: $R_g = R_{average} \times \tan(\theta_n/2)$, and the centre of the grinder is at: $d = R_{average}/\cos(\theta_n/2)$.

The machining surface of the grinder acts as an important control point. The x, y, z data points on the concave tooth surface can be determined from the generalized and parametric cone surface equations. The contact pattern of the teeth can be predicted by solving for the intersection of the tooth profile.

18.12 VAPOUR DEPOSITION OF THERMAL BARRIER COATING

Thermal barrier coatings (TBC) are applied to separate the hot gases in the turbine from the metal substrate of the components so that the material is subjected to lower temperatures. Air plasma spraying (APS) and electron beam physical vapour deposition (EB-PVD) processes are well established for the deposition of TBC in aircraft engines and industrial gas turbines.

In the EB-PVD method the coating is made from yttria stabilized zirconia (YSZ) that has a columnar structure. This coating material has a long operating life even under severe cycling conditions, when compared with the porous APS coating. Both processes are of the 'line-of-sight' type, meaning that the application of the coating is limited to simple shaped components, unless complex robotic motions are used. Deposition on shadowed regions, for instance on nozzles, poses considerable difficulties. Also, the equipment for EB-PVD is capital-intensive, which has hindered the spread of this technology for industrial gas turbines.

The chemical vapour deposition (CVD) procedure for the deposition of YSZ layers has been developed under a European Commission's Britell/Euram research program for turbine blades and vanes. The lab-scale equipment consists of an evaporator, a deposition reactor and a vacuum system, as shown schematically in figure 18.33. A hot wall reactor with a three-zone resistance furnace is required. The reactor is provided with a stagnation flow arrangement with a quartz tube, with a substrate holder placed in it. Temperature of the deposition is directly measured on the substrate. The system is designed to obtain a homogeneous mass transport to the substrate during the diffusion control portion of the process. The vaporized precursor with the carrier gas is fed through a nozzle into the reactor. Inside the nozzle a thermocouple monitors the gas temperature outside the deposition zone. Preheated oxygen is admitted into the nozzle. The precursors are evaporated separately in a crucible placed in hollow steel blocks with heating cartridges to guarantee quick and uniform heating. The evaporation temperature is recorded. To avoid condensation of the compounds the pipes between the evaporator and the reactor are heated above the evaporation temperature.

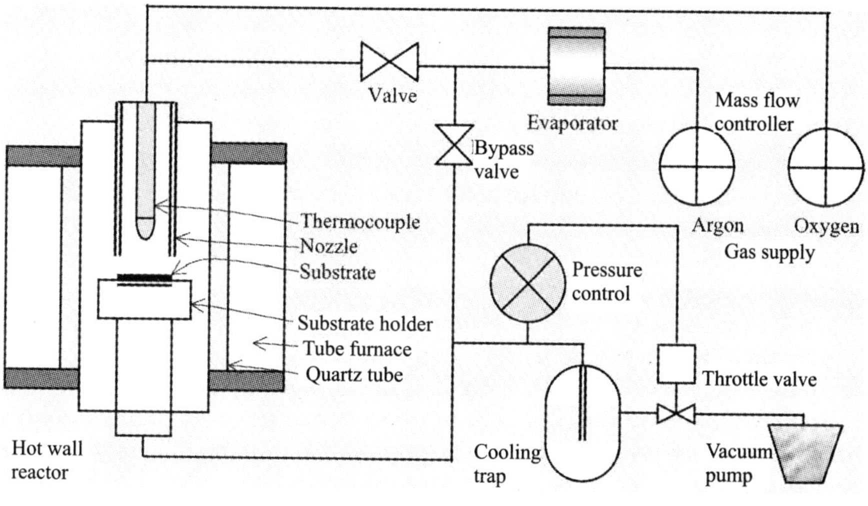

Fig. 18.33 Equipment for Chemical Vapour Deposition (Wahl, 2000)

The β-diketonates Zr and Y are used as source materials for the deposition of YSZ. The two compounds are well suited for the CVD process because of their high vapour pressure at temperatures below 250°C and their stability in the gas phase. Preparation of the source materials calls for conversion of the metal or metal halide into an intermediate alkoxide species, followed by the formation of the desired product. The final compound is isolated by filtration and subsequent trace solvent removal in a vacuum. The purity of the batch is assessed by nuclear magnetic resonance to check for organic components and by inductively coupled plasma for metallic impurities.

Chemical vapour deposition is a process in which the metal-organic precursors are vaporized into a gaseous phase. The gases then react on the surface of the substrate to gradually build up an outer layer. The decomposition of the vaporized compound takes place in the hot zone of the furnace. Deposition temperature is typically in the range of 850 to1200 K. The CVD process occurs under a reduced pressure between 500 and 1000 Pa (*ref*: Hitchman et al., 1993). Gaseous byproducts flow out of the reaction zone through the cooling trap after de-absorption from the surface of the substrate. The overall process may be split into sub-steps:

i. production of vaporized reactants

ii. convective mass transport to the diffusion boundary layer of the substrate

iii. homogeneous reactions in the gas phase to produce gaseous and solid byproducts

iv. de-absorption of the gaseous reaction products

v. diffusion of the gaseous reaction product through the boundary layer

vi. convective discharge of the byproducts from the reactor zone.

The benefits of the CVD process over EB-PVD and plasma spraying arise from its superior shooting power, so it is possible to coat large and complex shaped parts in large batches with good uniformity (*ref*: Stolle, 1997). The evaporation behaviour of the precursors is measured in a micro-balance equipment (*ref*: Pulver et al., 1993). The activation energy of Zr is around 1000 kJ/mol and of Y 117 kJ/mol. The evaporation rate in the stagnation flow equipment is determined by measuring the weight of the precursor before and after the experiment. Concentration level of the gas is important for the structure of the layer and for the rate of deposition. Higher concentration levels increase the deposition rate, reaching up to 100 μm/h. However, the initial columnar grain structure subsequently disappears because of poor adherence of the latter layers to the substrate. At temperatures below 970 K the deposition is controlled by the kinetics of thermal decomposition of the precursors. At higher temperatures the deposition rate is practically independent of the temperature. Hence, diffusion through the gas phase becomes a rate-determined process.

On samples produced by simultaneous deposition of ZrO_2 and Y_2O_3 the main phase detected by x-ray diffraction is cubic. A degree of tetragonal formations are detected. One concern with the CVD process is that at relatively low temperatures the conditions may not be right to form the T′ phase. Nevertheless, given a sufficiently high Y_2O_3 content, formation of the T′ phase should still be possible. The ytttrium composition of the coatings, determined by wavelength dispersion analysis ranged from 2 to 24 per cent by weight. At low temperature, when both Y_2O_3 and ZrO_2 are deposited, the columnar grains are generally shorter and uniform in thickness, have a smooth surface and are pure white in color. With increased deposition temperature the deposition rate increases and also the diameter of the columns increases. But discrete columns extending from the inner to the outer surfaces of the coating are not present. Continuing the increase in deposition temperature results in a growth of discrete columns with a rough outer surface.

18.13 VACUUM PLASMA SPRAYED COATINGS

High velocity oxygen-fuel (HVOF) spraying is capable of providing clean and dense coatings for turbine components. The sprayed MCrAlY (where M represents Fe, Ni and/or Co) coatings may be applied to provide resistance against oxidation and corrosion in the hot components. Diffusion coatings rely on surface enrichment in the form of a layer of oxides of aluminium, chromium and silicon. Overlay coatings, on the other hand, are specifically designed oxidation and corrosion resistant alloys deposited on the substrate. The application of the chemical compositions in many overlay processes requires an inert gas atmosphere in an evacuated chamber. The HVOF procedure, however, is carried out in open air to deliver low oxide content, low porosity and high bonding strength (*ref*: Parker et al., 1994).

Several investigations report achieving near-chamber quality sprayed MCrAlY coatings with HVOF systems in open air (*ref*: Irons et al., 1993). The superior coating performance is primarily a result of the higher particle velocities. Research also indicates that the thicker coatings can be sprayed for improving the residual stress when compared with other thermal spraying systems, although the basic properties such as mechanical properties of HVOF coatings have not been clarified. The properties have been investigated in a research program at Toshiba Corporation (*ref*: Itoh et al., 2000), and compared with the vacuum plasma spray (VPS) method.

Table 18.4 Chemical Composition of Substrate and Spraying Powders

Material	*Chemical Composition (by mass, %)*										
	Cr	*Al*	*Y*	*W*	*Mo*	*Nb*	*Ta*	*Fe*	*Ti*	*Co*	*Ni*
IN738LC	15.8	3.33	-	2.52	1.62	0.99	1.65	0.06	3.32	8.24	Bal.
CoCrAlY	28.9	5.84	0.35	-	-	-	-	-	-	Bal.	0.05
CoNiCrAlY	20.6	7.90	0.40	-	-	-	-	-	-	Bal.	32.5
NiCoCrAlY	17.2	12.6	0.65	-	-	-	-	-	-	21.7	Bal.

A cast nickel based super alloy (IN738LC) is used as a substrate for spraying three commercially available powders: CoCrAlY, CoNiCrAlY and NiCoCrAlY. The chemical composition of the substrate and the powders is shown in table 18.4. The spraying powders are produced by the rapid cooling solidification method using argon as gas atomizer. The particles are mostly spherical, with substantial precipitation of fine metallic compounds.

The procedure does not require preheating. Kerosene is used as the fuel, with combustion taking place at 483 kPa, 703 kPa and 896 kPa. Spraying distance is 400 mm and traverse speed is set at 500 mm/s. The VPS coating requires preheating to 843 K, and the spraying distance is 270 mm in an argon gas atmosphere maintained at 6 kPa. The substrate is roughened by grit blasting after degreasing, and sprayed coating thickness is 2 mm. The bending test specimens are machined out of diffusion heat-treated material (1392 K for 2 hours, 1116 K for 24 hours and argon gas cooled). Mechanical properties such as Young's modulus, Poisson's ratio, bending strength and Vicker's hardness are measured by the four-point bending test using a strain gauge method under a cross head speed of 0.1 mm/min. Figure 18.34 shows the testing apparatus and the test specimen.

One end of the strip shaped substrate is fixed, and deflection at its center on the rear face is measured during the thermal spray with a dial gauge. Changes in temperature at the surface are also measured by a thermocouple during the process. Displacement δ and coating and substrate thickness h_c and h_s permit the calculation of average residual stress by the equation (*ref*: Kuroda et al., 1991):

$$\sigma_R = E'_s \times h_s^2 \times \delta/(3 \times l^2 \times h_c) \qquad ...(18.20)$$

where $E_s' = E_s/(1 - n_s)$, E_s and ν_s are Young's modulus and Poisson's ratio of the substrate and l is the length of the specimen. Following the thermal spray the longitudinal residual stress in the coating is measured with *x*-ray stress measurement equipment. The mechanical properties of the CoCrAlY

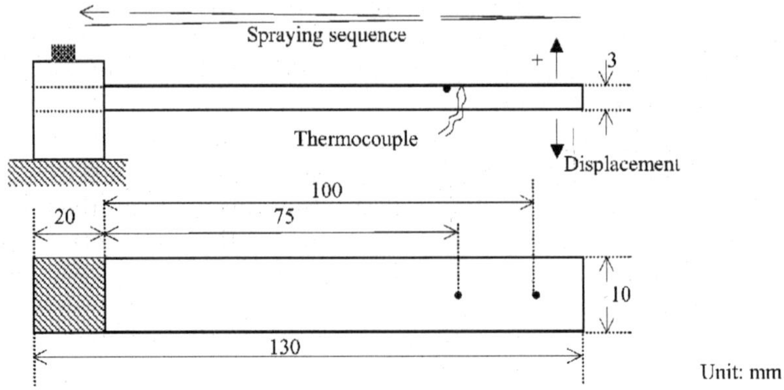

Fig. 18.34 Testing Apparatus (*upper*); Test Specimen (*lower*) (Itoh, 2000)

and NiCoCrAlY coatings using the two methods of application are shown in table 18.5. The bending specimens failed in a brittle mode, with the load versus deflection virtually straight until failure. The failures occur due to the formation of aluminium metallic compounds inside the coating. The strength of the HVOF coatings is improved by the diffusion heat-treatment, mostly because of the melted metal particles filling the crack-like pores. The same trend is noted for the Young's modulus. The Young's modulus of the coating with the HVOF method is inferior by 25 to 30 per cent when compared with the VPS procedure. But the Vickers hardness is substantially higher with the HVOF system. Figure 18.35 (*left*) shows the effect of number of spraying passes and combustion gas pressure for the two spraying methods on the surface temperature, and figure 18.35 (*right*) shows the effect on the displacement of the substrate. The temperatures and displacements increase with increasing gas pressure, which equates to the sprayed particle velocity. Figure 18.36 shows the macrostructure of the HVOF coating for various combustion gas pressures.

Table 18.5 Mechanical Properties of Coatings

Coating	*Method*	*Bending Strength GPa*	*Young's Modulus GPa*	*Poisson's Ratio*	*Vickers Hardness HV300*
CoCrAlY	HVOF	2.01	155	0.30	719
CoCrAlY	VPS	1.85	208	0.26	456
NiCoCrAlY	HVOF	1.63	138	0.33	475
NiCoCrAlY	VPS	1.43	163	0.29	387

Residual stresses computed with the aid of equation (18.20) and the elastic constants shown in table 18.5 are valid for the final exterior coating layer, but the plastic deformation of the substrate and the layer immediately next to it is not taken into account. The residual stress σ_R is represented by the sum of the shrinkage stress σ_S and the thermal stress σ_T induced by the difference of temperature and material constants between the coating and the substrate. The shrinkage stress arises from the solidification of the sprayed particles on the substrate, as shown in figure 18.37. In an HVOF system almost all the sprayed particles dash against the substrate in a poorly melted state. Consequently, compressive stresses $-\sigma_P$ are induced by the peening effect in the substrate and the

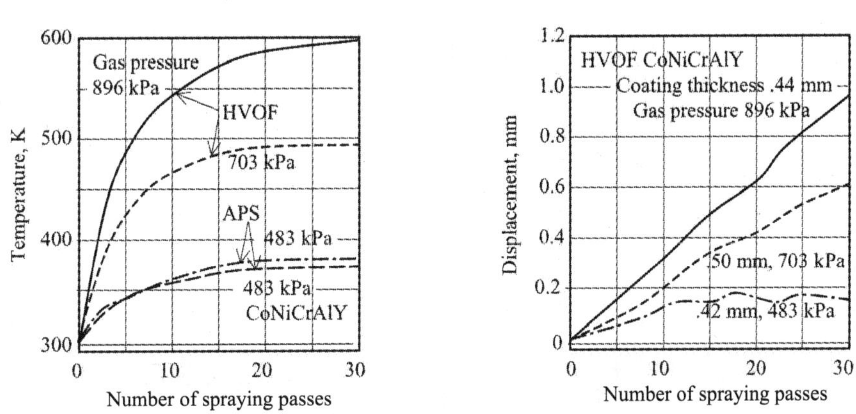

Fig. 18.35 Effect of Spraying Passes on Temperature (*left*), Deflection (*right*) (Itoh, 2000)

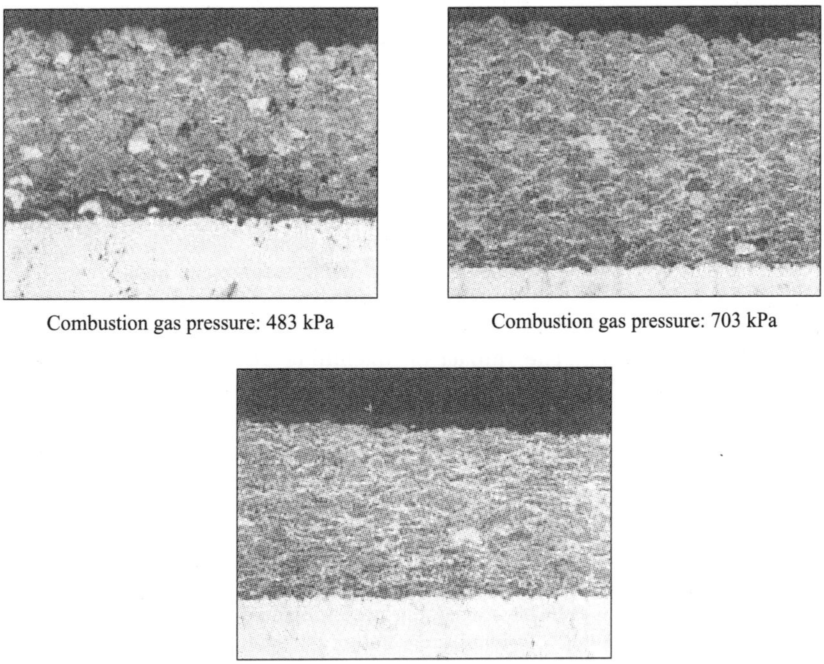

Combustion gas pressure: 483 kPa

Combustion gas pressure: 703 kPa

Combustion gas pressure: 896 kPa

Fig. 18.36 Macrostructure of HVOF Coating with Different Gas Pressures (Itoh, 2000)

layer next to it. The residual stress in the coating is then determined from $\sigma_R = \sigma_S + \sigma_T - \sigma_P$. This effect causes the residual stresses in HVOF coatings to be lower than in the VPS method.

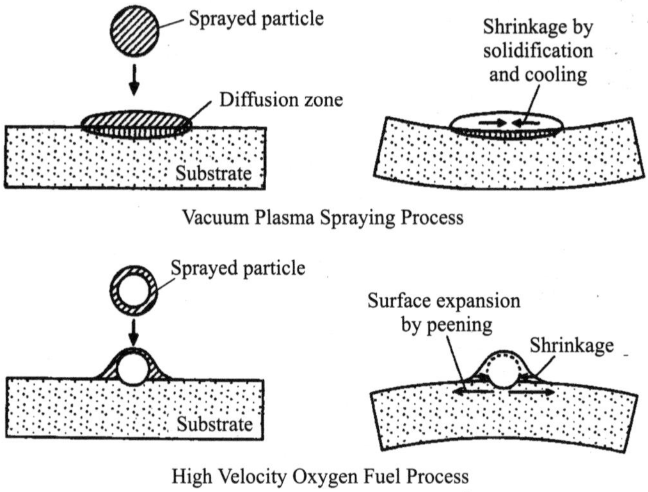

Fig. 18.37 Residual Stress Generating Mechanism During Thermal Spraying (Itoh, 2000)

References & Bibliography

"Annual Book of ASTM Standards", Section 5, Petroleum Products, Lubricants and Fossil Fuels, Volume 05.01, D-56 to D-1660 and Volume 05.02, D-1661 to D-2896", 1983.

Cantello, M., Ricciardi, G. and Gobbi, S.L., "Laser Welding of Super Alloys for the Manufacturing of Aero-engine Components", BRITE/EURAM Programme, Bruxelles, 1996.

Casey, M.V., "The Effects of Reynolds Number on the Efficiency of Centrifugal Compressor Stages", ASME Journal for Engineering Power, volume 107, page 541 – 548, New York, 1985.

Childs, P.R.N. and Noronha, M.B., "The Impact of Machining Techniques on Centrifugal Compressor Impeller Performance", ASME paper # 97-GT-456, New York, 1997.

Choi, B.K., "Surface Modeling for CAD/CAM", Elsevier, Amsterdam, 1991.

Cockell, M.W. and Boyce, K.A.G., "Metallurgy – Powder Review", volume 40 number 3, page 139, 1985.

Demo, W.A. and Ferrigno, S.J., "Brazing Method Helps Repair Aircraft Gas Turbine Nozzles", Advanced Materials and Processes, volume 141, No. 3, page 43 – 45, 1992.

Derringer, G. and Suich, R., Journal of Quality Technology, volume 12, no. 4, page 214 – 219, 1980.

Duvall, D.S. and Owczarski, W.A., Journal of Welding, volume 46, page 423S, 1967.

Ellison, K. A., Lowden, P. and Liburdi, J., "Powder Metallurgy Repair of Turbine Components", ASME paper # 92-GT-312, New York, 1992.

Frederick, G., Gandy, D. and Stover, J. T., "Laser Weld Repair of Service Exposed IN738 and GTD 111 Buckets", ASME paper # GT-2002-30160, New York, 2002.

Gandy, D.W. and Stover, J.T., "Status of Weld Repair Technology for Nickel Base Super Alloy Gas Turbine Blading", EPRI TR-108272, Electric Power Research Institute, Palo Alto, Ca, 1998.

Hitchman, M.L. and Jensen, K.F., "Chemical Vapor Deposition", Academic Press, London, 1993.

Irons, G. and Zanchuk, V., "Comparison of MCrAlY Coatings Sprayed by HVOF and Low Pressure Processes", Proceedings, National Thermal Spray Conference, California, page 191 – 196, 1993.

Itoh, Y., Saitoh, M. and Tamura, M., "Characteristics of MCrAlY Coatings Sprayed by High Velocity Oxygen Fuel Spraying System", Journal of Engineering for Gas Turbines and Power, volume 122, page 43 – 49, New York, January 2000.

Kearns, W.A., "Welding Handbook", volume 5, 7th edition, American Welding Society, 1984.

Komanduri, R. and Schroeder, T. A., ASME Journal of Engineering for Industry, volume 108, no. 2, page 93 – 100, 1986.

Koul, A.K., "Metallurgical Transactions", volume 16A, 17, 1985.

Kuroda, S. and Clyne, T.W., "The Quenching Stress in Thermally Sprayed Coatings", Thin Slid Films, volume 200, page 49 – 66, 1991.

Marschall, C.W. and Maringer R. E., Journal of Institute of Materials, volume 6, no. 2, page 373 – 387, 1971.

Nixon, P. G., "Centrispun High Alloy Steel Castings for Gas Turbine Applications", ASME paper # 87-GT-206, New York, 1987.

Parker, D.W. and Kunter, G.L., "HVOF Moves Into The Industrial Mainstream", Advanced Materials and Processes, volume 146, no. 1, page 31 – 35, 1994.

Prager, M. and Shira, C.S., Welding Research Council Bulletin, volume 128, 1968.

Proceedings of Metal Powder Conference at Movenpick, Zurich, "Metal Powder Report", Shrewsbury, UK, papers 11,16, 17, 1980.

Pulver, M., Wahl, G., Scheytt, H. and Sommer, M., Journal of Physics, volume IV, no. 3, page C3 – 305, 1993.

Reichman, S.H. and Smythe, J.W., "1969 Powder Metallurgy Conference Proceedings", APMI, page 829, New York, 1970.

Savage, W.F., "Welding Design Fabrication", volume 42, page 56, 1969.

Shamblen, C.E., Allen, R.E. and Walker, F.E., "Metallurgical Transactions A", volume 6A, page 2073, 1975.

Simon, H. and Bulskamper, A., "On the Evaluation of Reynolds Number and Relative Surface Roughness Effects on Centrifugal Compressor Performance Based in Systematic Experimental Investigations", ASME Journal for Engineering Power, volume 106, page 489 – 498, New York, 1984.

Sims, C.T., Stoloff, N.S. and Hagel, W.C., "Superalloys II", John Wiley & Sons, New York, 1987.

Sparling, R. and Liburdi, J., "Liburdi Powder Metallurgy New Compositions for High Strength Repair of Turbine Components", ASME paper # GT-2002-30537, New York, 2002.

Standard Test Method for Determining Residual Stresses by Hole-Drilling Strain Gauge Method, Annual Book of ASTM Standards, volume 03-01, E837-85, page 991 – 997, 1987.

Stolle, R., "Kinetische Messung und Simulation der Abscheidung von Bornitrid", CVD/CVI-Prozessen, Braunschweig, 1997.

Subhas, B. K., Bhat, R., Ramachandra, K. and Balakrishna, "Dimensional Instability Studies in Machining of Inconel 718 Nickel Based Super Alloy as Applied to Aero and Gas Turbine Components", ASME paper # 98-GT-469, New York, 1998.

Tien, J.K., Kissinger, R.D. and Nair, S.V., "Super Alloys 1984", TMS-AIME, Warrendale, PA, page 280, 1984.

Tsai, Y.C. and Hsu, W.Y., "A Study on the CAD/CAM of Curvic Couplings", ASME paper # 98-GT-2002-30486, New York, 2002.

Tsay, D.M., Yan, W.F. and Ho, H.C., "Generation of Five-Axis Cutter Paths for Turbo-machinery Components", Journal of Engineering for Gas Turbines and Power, volume 123, page 50 – 56, ASME, New York, 2001.

Union Carbide Corporation, Linde Division, Patented Welding Process.

Vagi, J.J., Meister, R.P. and Randall, M.D., "Weldment Evaluation Methods", DMIC Report # 244, 1968.

Wahl, G., Nemetz, W., Giannozzi, M., Rushworth, S., Baxter, D., Archer, N., Cernuschi, F. and Boyle, N., "Chemical Vapor Deposition of TBC: An Alternative Process for Gas Turbine Components", ASME paper # 00-GT-077, New York, 2000.

19

Economic Considerations

19.1 INTRODUCTION

Engineering projects such as an electric power generating plant require an immense amount of technical effort to bring it into existence. The electric power will be supplied to millions of home and industrial users. The fuel required to power the turbine-generators must be readily available in sufficient quantities and must not produce excessive effluent. Reliability to generate electrical energy continuously is of prime concern. The electricity must be made available to a region encompassing hundreds of square miles and at a reasonable cost.

Analytical techniques are available in engineering economics that permit making decisions concerning capital investment. Because of the immense scope of the project, careful research and analysis is a prime requirement at many different levels prior to, during construction and following start-up of the plant.

Engineering facilities are constructed to meet human or societal needs and priorities. The facility is made up of many diverse components to serve a common purpose. The entire process of planning, design, construction, operation and maintenance of an industrial plant entails many important and often complex decisions. Both technological and economic considerations are essential in undertaking any new project requiring modern technology and involving a heavy commitment of resources.

An engineering system developed to satisfy certain perceived needs must pass tests of technological and economic feasibility under the prevailing institutional and social conditions. It would be a waste of time, effort and resources if the system cannot meet prescribed minimum technical standards. Even if an engineering project is technologically feasible, it may not be worth building if the cost is prohibitively high compared to the benefit derived from the project. Because the advancement of technology is dynamic, the cost of a new system can be greatly reduced through continued technological improvements, and the cost relative to benefit may be rapidly changing. Generally, economic consideration of an engineering project is based on the expected financial return on the investment. However, other institutional and social forces as well as technological developments will tend to influence and modify the economic considerations. So the goal of economic efficiency is often tempered by objectives such as social equity, quality of life or preservation of environment.

In a privately financed enterprise, the failure of a new venture to satisfy perceived needs and desires could result in a loss on investment and its eventual disappearance from the marketplace. In a public sector, however, the failure of a project to meet societal needs is less evident since public welfare cannot be readily measured. Hence, in the development of an engineering project it is important to assess its benefits and costs before making the selection. Although the motives in the initiation of the project are different in private and public sectors, there are enough similarities in the approaches for assessing the merit of the project.

Economic analyses of combined-cycle performance enhancement options normally reveal that heat recovery steam generator duct firing is the best option for the plant. Inlet fogging, evaporative cooling and inlet air chilling are some other available options. However, these analyses were focused on capacity-driven economics resulting from premiums paid for short periods of peak power generation. These economic drivers exist in today's market environment. However, as the market condition changes due to an increase in the installed capacity, escalation of fuel prices and deregulation in the power generation industry, the emphasis will shift to plant efficiency. Thus, plants designed with moderate increase in capacity and high efficiency could provide the highest life cycle profitability.

An economic study using a cost-of-electricity (COE) model is normally done to determine the enhancement option that provides the best overall life cycle benefit for the plant. The model includes both the peak and non-peak performance levels, operating hours, premium paid for power generated during the peak period, cost of fuel, plant capital cost, incremental capital cost of enhancement, and the cost of operating and maintaining the plant. This COE model is then used to determine the best enhancement option for the plant. Most power enhancement alternatives exist in the topping cycle (gas turbine), as opposed to the bottoming cycle (HRSG/steam turbine). The supplementary direct firing within the HRSG is the only power enhancement option for the bottoming cycle. However, most performance enhancement options for the gas turbine result in improvement in the bottoming cycle also due to an associated increase in the exhaust energy from the gas turbine.

19.2 CLASSIFICATION OF PROJECTS

Before a new capital investment is undertaken, an extensive search is required to spot potential investment opportunities. Attractive characteristics of the project must be recognized, particularly their inter-relationships with other potential projects. An investment project is economically independent if it is technically feasible to execute this project by itself and if its expected profit or net benefit will not be affected favourably or adversely by the acceptance or rejection of any other projects; otherwise, the project is economically dependent.

For example, replacing a deteriorating feed water pump with a new one may be regarded as an economically independent project if no other alternative or supporting actions related to replacement are anticipated. In reality, a project is seldom totally economically independent: only if the effects from other possibilities (such as instrumentation) are small, then the project is practically considered independent.

When two projects are dependent on each other, they may either complement or substitute for each other to some degree. In the feed water pump example, a new network of pipelines, valves and accessories to enhance overall improvement may be considered regardless of the acceptance or rejection of the feed water pump proposal, but the replacement pump may be incompatible without the new lines. Then the new network is the prerequisite of the proposed feed water pump. The proposed feed water pump may be driven by a steam turbine or by an electric motor. If the net benefit expected from one form of drive will be completely eliminated by the acceptance of the other, then the two proposals for the drive are mutually exclusive.

In economic evaluation, the potential projects may be arranged in the form of a set of investment proposals so that they are not economically interdependent. Independent proposals are easier to analyse on their own benefit, since the benefit remains unaffected by the acceptance or rejection of other proposals. In mutually exclusive investment proposals, the acceptance of one investment will automatically lead to the rejection of others. The merit of various proposals may then be ranked to select the best among them.

19.3 STRATEGIC ECONOMIC DECISION-MAKING

Project ideas can originate from many different levels in an organization. Procedures may be present or need to be developed to screen new projects and ventures from good ideas. Once identified, the project ideas may be categorized as (*i*) equipment and process selection, (*ii*) equipment replacement, (*iii*) cost reduction, and (*iv*) service or quality improvement.

The classification scheme permits management to address key questions: Can the existing plant, for example, be modified to achieve new production levels? Does the firm have the knowledge base and skill to undertake the new investment? Does the new proposal warrant recruitment of new personnel? The answers to these and related questions help firms screen out proposals that are not feasible, given a company's resources.

Virtually all big businesses sometime face complex engineering situations that require investment decisions from top executives and board of directors. Generally, the larger the investment, the more detailed is the analysis required to support the expenditure. Expenditures required to increase the output of the plant would invariably require a detailed economic justification.

To understand a class of engineering decisions that involve selecting the best course of action when several ways to meet a project's requirements exist, example problem # 19.1 illustrates a process selection situation.

The category of investment decisions considers expenditure necessary to replace worn-out or obsolete equipment. For example, a company may purchase 10 large pumps expecting them to operate for 10 years. After 5 years, however, it finds it necessary to change the process, which requires the pumps to be retired early and purchase compressors. Example 19.2 provides the case of such a situation.

19.4 EQUIPMENT AND PROCESS SELECTION

Investment in capacity expansion alludes to increase in revenue realized by increasing the capacity of the existing plant. The question to be asked is: Should the plant capacity be increased or acquire a new facility? Physical limitations of the present plant's size will play a big role. If space is not a concern, existing facilities for the plant may require limited adjustments to increase the total plant capacity, thus reducing the expenses when compared with establishing a new plant. Additional manpower requirements due to the growth in the plant may also be minimized. On the other hand, if growth in customer base is likely to be in more remote regions, distribution of power and servicing of customers must be adequately addressed.

Investment in a new product leads to increased revenues for the company if a component can be made in a shorter amount of time, thus reducing the downtime for the equipment that uses the component. Example 19.3 illustrates the situation.

A cost reduction project is one that attempts to lower a firm's operating costs. Typically, the question to be answered is should a company buy equipment to perform an operation presently done manually or spend money now in order to save money later. The expected future cash inflows on this investment are savings resulting from lower operating costs.

The examples considered so far relate to the production area, but the decision-making techniques may be extended to improving services to the customer.

An emerging industrial trend is towards the so-called mass customization, permitting the manufacturer to mass-market products one at a time to suit individual needs. For example, a manufacturer providing specially sized mechanic's tools and instruments may take computerized instructions from the customer, requiring him to specify the capacity, size and type of tools. This enables the factory to ship the items directly to the customer for a small shipping fee, but saving the mechanic from making a special trip to purchase the merchandise. It has the potential to eliminate the cost of stocking the items. Eventually, this could lead to reduced manufacturing costs since the company is not mass-producing the item and hoping it sells.

19.5 CAPITAL INVESTMENT CONCEPTS

Investment is defined as a commitment of resources to some economic activity in anticipation of greater returns or benefits in the future. Investment occurs when an entity such as a private corporation or a government agency uses capital from savings out of the capital or financial markets and spends them on acquiring new assets. Savings may flow into investments through the creation of equity or debt. 'Equity' refers to the value of the stock of a corporation, and new assets may be added by either issuing new shares or stocks or by ploughing back corporate earnings. 'Debt' may be incurred either by taking loans from a bank or by issuing bonds to the public which are redeemed at a later date. In the public sector, the government can impose taxes for the purpose of investment.

The difference between an economic evaluation of a project and feasibility study of its financing must be distinguished. The former determines if a capital investment is economically worthwhile, assuming that the resources are available. The latter refers to an investigation to find the means of obtaining necessary funds to acquire a specified capital asset.

19.6 INVESTMENT OBJECTIVES

The primary objective of capital investment is to maximize profit or net benefit within a planning horizon. Given the difficulty and uncertainty involved in estimating benefits and costs, as also the time frame and attractive rate of return, factors other than maximization of profit must be considered in the selection of capital projects. Social, political and environmental factors may be significant in influencing the selection of a project.

Imperfect though it may be, once the concept of maximization of profit is accepted, a decision criterion reflecting the stated objective may be established for the economic evaluation of proposed capital projects. An investment decision criterion consists of two elements: a merit measure and a set of decision rules associated with this measure. Historically, many merit measures and the associated set of rules have been introduced. Two basic types of merit are of special interest in the context of single-period investments: The net future value and the internal rate of return. Real assets are size dependent, hence requiring investments of differing magnitude. For a set minimum attractive rate of return, it is worth investing in a proposed project that will generate a profit greater than this rate of return, regardless of the size of investment. The net future value is a direct measure of the size of profit or benefit at the end of the investment period that the investor would have gained by investing in the proposed project. A comparison between the gross profits resulting from the proposed project and that obtained by the minimum attractive rate of return is already taken into consideration in this measure. The net future value is thus a relative measure of profit, and is dependent on the minimum rate of return.

In the financial world money is considered to be a commodity, just like other goods that are bought and sold. To obtain financing, the cost is established and measured by an 'interest rate', a percentage that is periodically applied and added to an amount (or varying amounts) of money over a specified period of time. When money is borrowed, the interest paid is the charge to the borrower for using the lender's property. The 'time value of money' may appear to be a sophisticated concept, but has to be dealt with regularly. Because money has both purchasing and earning power over time, a dollar received today has a greater value today than a dollar received at a later date.

For convenience in calculations, the elements are represented by the following variables:

A_n = Discrete payment or receipt at the end of an interest period, n

i = interest rate for the period.

N = total number of interest periods.

P = Sum of money at time zero. Also called 'present value' or 'net present worth' (NPW). Expressed mathematically, $P = F/(1 + i)^N = F(P/F, i, N)$.

F = Future sum of money at the end of the period. Also called 'future value' or 'future worth'. $F = P(1 + i)^N = P(F/P, i, N)$

A = An end of period payment or receipt in a uniform series continuing for N periods. This is a special case where $A_1 = A_2 = A_3 = ... = A_n$

V_n = An equivalent sum at the end of the specified period n that considers the time value of money. Thus $V_0 = P$ and $V_n = F$.

19.7 INTEREST CALCULATIONS

Many transactions involve interest – for example, investing money, purchasing money on credit. Interest rate represents the cost or price of money, and is expressed as a percentage of money. Money may be loaned and repaid in many ways. Correspondingly, money can earn interest in similar ways. Mostly the interest is calculated at the end of each interest period on the principal amount according to a specified interest rate.

Two computational schemes for calculating the interest are 'simple interest' and 'compound interest'. Simple interest is earned only on the principal amount during each interest period. For a deposit of P dollars at a simple interest rate of i for N periods, the interest earned is $I = (iP)N$. Then at the end of N periods the future value is given by the expression: $F = P + I = P(1 + iN)$.

In a compound interest scheme the interest earned in each period is determined based on the total amount at the end of previous period. This total amount includes the original amount plus the accumulated interest that is still present in the account. Effectively this increases the deposit amount by the amount of interest, so the new present worth becomes $P + iP = P(1 + i)$ dollars at the end of 1 period. If the principal and interest is reinvested at the same rate i for another period, then at the end of the second period the amount is: $P(1 + i) + i[P(1 + i)] = P(1 + i)(1 + i) = P(1 + i)^2$. Then at the end of 3 periods the balance will be $P(1 + i)^3$, and after N periods the total accumulated value $F = P(1 + i)^N$. Thus, simple interest: $I = iPN$, while compound interest: $I = P[(1 + i)^N - 1]$.

The equivalence of a single amount, P, and its future worth, F, may be expressed by the equations: $F = P(1 + i)^N$. Expressed symbolically, $F = P(F/P, i, N)$. Also, the present worth is determined from the expression: $P = F/(1 + i)^N = F(P/F, i, N)$.

Probably the most familiar category includes transactions arranged as a series of equal cash flows at regular intervals, known as equal payment (or uniform) series. For instance, this method describes cash flow of a common installment loan contract, which arranges repayment of a loan in equal periodic installments. The relation between F, P and A may be expressed as follows:

$F = A[(1 + i)^N - 1]/i$. Expressed symbolically, $F = A(F/A, i, N)$.

$P = A[(1 + i)^N - 1]/[i(1 + i)^N]$. Or, $P = P(A/P, i, N)$

$A = P[i(1 + i)^N]/[(1 + i)^N - 1]$. Or, $A = A(P/A, i, N)$

19.8 PRINCIPLES OF ECONOMIC EVALUATION

Selecting the best investment strategy for a corporation depends on a number of factors. The right investment is a balance of liquidity, safety and return. Liquidity assures money is available to cover financial emergencies. Risk is associated with the safety needs that arise with all financial projects. Safe investments often promise a specified though limited return. Business opportunities having a higher risk associated with them may make (or lose) substantial amounts of money. Both risk and reward are time dependent. The returns associated with higher risk investments could become more stable and predictable over time, thereby reducing the predicted level of risk.

The rate of return that may be realistically expected to earn on an investment is a function of three components: risk-free real return, inflation factor and risk premiums. A risk-free investment (such as US Treasury bills) may yield 2% for a year. If the expected inflation rate is 4% during the time interval and the premium due to a riskier investment is 20%, then the investment should be expected to yield 2 + 4 + 20 = 26%. If a reasonable certainty of having the investment grow at an annual rate of 26% is not present, the investment proposal may not be economically viable.

Risk as it relates to an investment refers to the chance that some unfavourable event will occur. Two forms of risk must be considered: volatility and the effect of changing market conditions, such as a recession, high commodity prices or excessive inflation. Volatility may cause sudden jump up or down of a value. This can cause a big spread in the price paid to purchase and the amount received when sold. Thus, the business must be prepared to see the price drop by the same amount.

Trying to weigh risk and reward can be a challenging task. Investing calls for a trade-off between the opportunity to earn higher returns and the consequences of trying to do so. However, a totally risk-free investment does not exist. The task is to decide what level of risk a corporation is willing to accept and to understand the implications of that choice.

Calculation of the payback period helps in initially identifying projects on the basis of how long it takes for net receipts to equal investment outlays. Projects with a payback period shorter than the maximum acceptable time period permit further consideration. A short payback period assures the corporation of being restored to its initial position within a short span of time. The procedure thus permits focusing information search for the time period in which the firm expects to recover the initial investment. But a number of drawbacks may be present. The principal objection is that it fails to measure profitability, since no profit is assumed during the payback period. The method fails to gauge the earning power of a project, since differences in the timing of cash flows are ignored. By modifying the procedure to consider the time value of money, or interest, one of the shortcomings is addressed, and the modified method is referred to as the 'discounted payback period'.

19.9 ASSET DEPRECIATION

Fixed assets, even when they continue to function and contribute to execution of projects, tend to lose their value. This loss of value due to deterioration and eventual obsolescence is referred to as 'depreciation'. The function of depreciation accounting is to record the cost of fixed assets in a manner that matches their decline in value with the passage of time. The method permits the corporation to stabilize its statement of financial position. On a project level, the practice of depreciating fixed assets influences the investment value of a project. Estimating the allocation of capital costs over the life of a project requires understanding the conventions and practices followed to depreciate assets.

Equipment and real estate are 'fixed assets', and are economic resources acquired to provide future cash flows. The gradual decrease in the utility of fixed assets with use and time may be classified into physical and functional depreciation. The former constitutes reduction in an asset's capability to perform its intended service due to physical impairment. Wear and tear from usage and deterioration from interaction with the environment due to corrosion and chemical changes are examples of physical depreciation. Functional depreciation occurs as a consequence of changes in technology or organization, causing a reduction or elimination of an asset. Inability to meet higher production or quality needs and obsolescence arising from advances in technology can result in a decline of the services performed by an asset.

A brief discussion of the economic principles on which depreciation accounting is based will be useful. Economic depreciation accumulated on an asset may be calculated by subtracting the current market value from the price originally paid for it. The asset need not be sold for this purpose, the market value may be judged without resorting to testing in the market place. The need to have a scheme for recording the continuous decline in the value of an asset is a part of the accounting process.

As in the case of other disbursements, the costs of fixed assets are recorded as an expense on a corporation's balance and income statements. But unlike costs such as material, labour and maintenance, the fixed asset costs are not to be treated in the year they are acquired; rather, the assets are treated as capitalized costs and distributed by subtracting them as expenses from gross income a portion at a time over a number of years. Accounting, or asset depreciation, is thus a systematic allocation of the initial cost of an asset over its depreciable life. The process also provides a basis for determining income taxes associated with the project.

A depreciable property must be used in business or held for production of income, must have a definite service life longer than 1 year and must wear out, become obsolete or lose value from natural causes. The process of depreciating an asset requires information about its cost, its value at the end of useful life, length of its depreciable life and the method of depreciation to be used. Depreciable properties include buildings, machinery and vehicles. Inventories held primarily for sale to customers cannot be depreciated. Land can never be depreciated.

The cost basis of an asset represents the total cost claimed as an expense over an asset's life. This generally includes the actual cost and incidental expenses such as freight and installation.

Mineral properties such as oil, gas, geothermal well or standing timber may be claimed as a deduction as the resource is depleted, because the material is removed and sold. The process of amortizing the investment cost of natural resources in the accounting period is called 'depletion'. The objective of depletion is similar to that of depreciation: to amortize the cost in a systematic way over the asset's life.

'Cost depletion' and 'percentage depletion' methods are commonly used for book and tax purposes to calculate depletion. Cost depletion value depends on the ratio of the adjusted basis of the mineral property and the total number of recoverable units in the deposit. The number of units sold then multiplies this ratio.

$$\text{Thus: Cost depletion} = \frac{\text{(Adjusted basis of mineral property)} \times \text{(number of units sold)}}{\text{(Total number of recoverable units)}}$$

In the percentage depletion method, the depletion allowance is calculated based on a prescribed percentage of gross income from the property during the year. The difference between depreciation and percentage depletion is that in the former the cost is allocated over the useful life, whereas in the latter an annual allowance of a percentage of the gross income from the property is used. Since percentage depletion relies on income rather than the cost of the property, the total depletion may exceed the cost of the property. To avoid this situation, the annual allowance under the percentage method cannot exceed 50% of the taxable income from the property. Allowed percentages for selected mining properties are shown below:

Oil and Gas wells	15%
Sulphur, uranium	22
Gold, silver, copper, iron	15
Coal, lignite, sodium chloride	10
Clay, gravel, sand, stone	5
Most other minerals	14

19.10 DEVELOPING PROJECT CASH FLOWS

A number of variables involved in forecasting cash flows make their projection perhaps the most important and difficult aspect in the analysis of a capital project. The process calls for a number of individuals in the corporation, ranging from engineers to cost accountants and marketing people, to participate in the process. Cash flow elements need to be defined, from which cash flow statements are developed for analysing the economic value of the project.

A full spectrum of activities is initiated by project proposals. Project types differ in many different ways, each having its characteristic set of economic concerns. Possible approaches for relatively straightforward projects may not require much engineering design, but a large project may need thousands of hours of specialized technical effort. Most projects tend to fall between the two extremes, and a combination of methods may be appropriate.

A simple project might need readily available 'off-the-shelf' components that are integrated to form a system. The installed cost is the cost of the items as determined from the vendor's catalogues or quotes and the cost of modifications and utility requirements. A limited amount of design effort is generally required to define the scope of the work for obtaining price quotations from the contractor.

19.11 COST/BENEFIT ESTIMATION

Project benefits are in the form of either new revenue or cost reduction. When the production rate and the selling price are available, the new revenue may be estimated, although these parameters are subject to fluctuations due to supply and demand aspects of the market place. Sophisticated marketing tools are available to gauge the relationship between the price and quantity of an item. The studies precede other activities in order to define the overall size of the capital project. Operation and maintenance costs for the equipment or the facility can be estimated from manufacturer's guidelines and past experience and familiarity with similar processes. A comprehensive estimate may then be drawn from each element of the project.

The estimation of larger projects will follow similar trends, but will be marked by a greater degree of complexity, interaction between the project's components and a tendency towards specialization of processes and equipment. Generally, engineering drawings and specifications are prepared after a commitment has been made to proceed with the project. The design effort may start with conceptualization, from which preliminary and final designs are prepared while keeping in mind the requirements for operating efficiency, environmental impact, safety and minimum cost. The economics of the project must be checked during each phase to confirm the attractiveness and the incentive to continue with the task.

The benefits accruing from the proposal are likely to be well known from the outset. The focus may be either on the price and volume relation or cost reduction arising from process and material efficiency, or a combination of both. Since large and complex projects tend to be designed to operate over several years, forecasts for costs of materials, energy and other resources are subject to high levels of uncertainty.

At the conceptual stage of design, the focus will be on economies of scale, identification and selection from available technology alternatives and their degree of reliability. Preliminary design proceeds to take the most suitable alternative to determine component size, their specifics and overall layout of the equipment and accompanying facilities. Estimation of costs at this stage relies on the number of pieces of equipment and costs associated with similar equipment in use. Actual quotes from vendors are available only when detailed fabrication and construction drawings and specifications are available. Note that as technical requirements and equipment specifications go through greater refinement as the project successively moves from the conceptual to preliminary and final design stages, the accuracy of the cost and benefits estimates is also enhanced. A contingency factor for the cost estimates may be used to account for variations due to uncertainties, varying from between 50 and 100% in the development stage of the project design to 10 and 20% in the detail design stages. Even in the construction and installation stages it may be necessary to have a margin to allow for design modifications as an after-thought.

Investment in a fixed asset such as equipment or a facility is made with expectations of earning a return on the funds in the future. Profitable operation of the asset provides a return at a later date in the form of cash flows. The cash flows occurring as a consequence of the corporation's investment are referred to as differential or incremental cash flows.

The elements of a potential cash flow can take many different forms. Initial investment for purchasing equipment requires a cash outflow. In addition, features such as transportation and installation of the equipment and training of new employees may enhance the initial cash outflow. If the equipment is replacing an older version, proceeds from its sale will reduce the outflow if the transaction results in accounting gain. The gain or loss will depend on the difference between the book value and the salvage value realized.

Extra funds may be needed to support the higher level of production from the new equipment. If the increased investment is in non-depreciable assets, it is often called investment in working capital. A portion of this increase in current assets may come from increased accounts payable, with the remainder coming from permanent capital.

Manufacturing a new product entails expenses attributable to labour, materials and overhead charges. Periodic outlays for repair and upkeep must be taken into consideration when analysing an investment.

Instead of purchasing, a new machinery or facility building may be leased for business use. The expense for leasing is a cash outflow. To avoid rapid technological obsolescence of computers and operating software, many corporations prefer leasing.

Money borrowed to finance a project will require interest expense, as also the return of the principal amount. Cash inflow occurs when obtaining a short-term bank loan or long-term bonds. But repayment of the debt causes cash outflow. Profitable operation of a corporation leads to income taxes, and are treated as cash outflow for budgeting. Depreciating expenses on qualified investments offset part of the taxable income, and are a form of tax shield or saving.

Typical elements of cash inflow are operating revenues, cost reduction and salvage value of replaced assets. When a project approaches termination, inventories are sold off and the items liquidated mostly at cost. Cash flow of working capital is experienced by the corporation at the end of the project, and equals the net investment when the project began.

Cash flow components may also be classified according to the functional characteristics of the company. The elements may be associated with operations, investment activities or project financing. The objective behind this grouping is to glean financial information of the project.

Cash flow from operations includes sales revenues, cost of goods sold, operating expenses and taxes. The effects of cash on transactions entering in the net income calculations are reflected by the operation's cash flows. The interest payment on a loan is a deductible expense, and is a component of the operating activities. Since cash flows based on net income (or with taxes calculated separately) occurring during a year are mostly of interest, they are expressed on a yearly basis. Non-cash expenses (depreciation and amortization) are added to the net income to determine the net cash flow from operation.

Generally, the original investment outflow and working capital occur in year 0 for small projects. If, however, the two do not take place simultaneously but occur gradually over a number of years for larger projects as they get under way, the expenditures may be entered as they occur.

19.12 RISK CONTROL AND MITIGATION

Cash flow from a project has been assumed so far to be known with certainty, with the analysis concerned only with calculating and comparing the economic worth of the projects. Profitability estimates of an investment depend on knowledge of cash flows, which are generally not precise because of a number of uncertainties. Factors such as total market for a product, the market share that a corporation can capture and the selling price limit the accuracy of the revenue estimates for the project. Uncertainties related to the cost of manufacturing the item, labour and overhead expenses also affect the profitability of the investment. Some of the uncertainties may be substantial.

The term 'risk' is used to define an investment project where the cash flows are not known in advance with certainty, but an array of alternative outcomes and their probabilities (or odds) are known. If a range of values for individual cash flow components is possible, then a range of values for the NPW will exist. The analysis requires gauging the odds and reliability of each cash flow component, from which the level of certainty of the overall project's worth may be estimated.

Risk factors may be quantified with numerical probabilities of occurrence in the form of fractions in the range 0.0 to 1.0. A 1.0 probability would mean total certainty of an outcome or an event, while a low value approaching 0 indicates little likelihood of occurrence. The assignment of risk factors to the various outcomes is called 'risk analysis'.

Project risk may be analysed by first determining the uncertainty inherent in the cash flow. The evaluation may be in the form of making informal judgments to calculate complex economic and statistical conditions, and may require sensitivity, breakeven and scenario analyses.

A 'sensitivity analysis' determines the extent of variation in the net present worth in response to a change in a given variable used for estimating after tax cash flow. A convenient way of depicting the results from a sensitivity analysis is by plotting the NPW against the input variable, with the slope of the graphs indicating the extent of sensitivity to changes in the variable.

Decisions related to project management are facilitated by obtaining information relating the extent of reduction in sales below the forecast level before the project begins to lose money. The data is obtained from a 'breakeven analysis'. The technique facilitates studying the effects of variations in output on the project's net worth.

But sensitivity and breakeven analyses have limitations, often due to a lack of precision in the interconnection between a specific variable and other variables and also with the present worth. A 'scenario analysis' considers the sensitivity of NPW to changes in the key variables and also to the range of likely variable values. For instance, the decision may rely on two extreme cases, one assuming low unit sales, low unit price and high manufacturing cost, the other assuming high revenues and low manufacturing expenses. The NPW is calculated for both extreme cases and compared with the expected or base case.

Sensitivity analysis provides answers to questions such as what if incremental sales are only half of that assumed in the base case. The specific variable is then changed above and below the most likely value while holding the other values constant. A new NPW is then determined.

Risk control can be examined within an economic framework by constructing cash flows for available alternatives as investments. A decision situation concerning three alternatives A, B and C can lead to several scenarios. The scenarios for each alternative are identified by the magnitude of their respective NPW's – extremely low, very low, low, good, high, very high and extremely high. Some of the alternatives may have generally smaller returns and smaller spreads than others. Some cases of equal probability distributions, increasing and decreasing likelihood may also be observed. The distributions may be used to introduce various concepts and cases.

19.13 REVIEW PROBLEMS

Example problem # 19.1: A large power generation and transmission company operates a nuclear power plant. The facility was constructed with 44 electrical transformers that use a fluid called polychlorinated biphenyl (PCB) as the liquid cooling component of the transformer insulation system. As illustrated in figure 19.1, the liquid flows around the electrical coils driven by convection currents, the hotter liquid traveling outside the unit to dissipate heat by radiation. The liquid then returns to the coils in the closed loop system. In the event of a fire or explosion in the transformer, PCB's are known to form carcinogenic compounds that are extremely dangerous for humans. Government regulations now ban the production of PCB, so PCB transformers are converted or replaced using alternate coolants. An internal risk assessment study by the company indicated a small probability of an explosion in the existing PCB transformers. The potential cost of cleaning up from a PCB transformer fire would range from $15 to $100 million. The study also identified the transformers generating the greatest risk to individuals in the turbine building because of an open area where contamination would spread in the event of a violent failure.

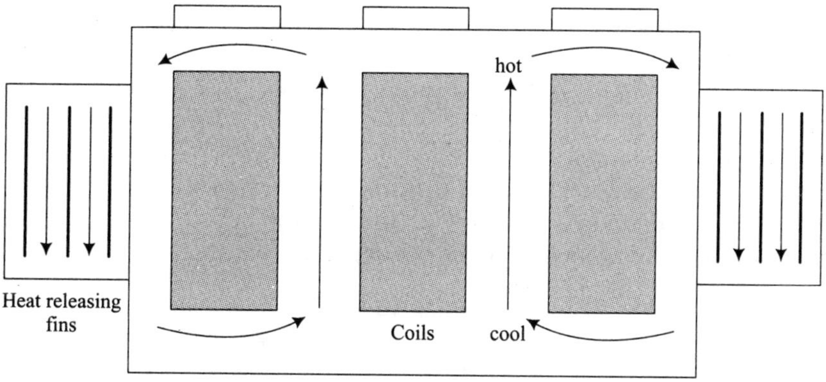

Fig. 19.1 Liquid Cooled Transformer

The company is faced with a decision as to whether it should continue to operate with the well-known PCB related environmental risk, or to reduce the risk by employing one of three options:

- Option # 1. Replace all 44 transformers with non-hazardous liquid filled transformers (LFTs). The high dielectric strength of LFTs provides greater design flexibility in optimizing specific load requirements, and this achieves lower operating costs.
- Option # 2. Replace 24 transformers in the turbine building open area with LFTs and retain the remaining transformers, so only 20 transformers in other areas are exposed to the higher risk.
- Option # 3. Reclassify all the transformers by replacing the PCB fluid with a silicone fluid, which is a non-toxic and is environmentally safe. The challenge associated with this reclassification is the de-rating of the transformers.

Solution: The LFTs cost \$30,000 each and the reclassification requires \$14,000 per unit. Annual maintenance cost including parts and labour is estimated at \$500 per reclassified transformer and \$300 per replaced transformer. A conservative estimate for cleanup in the event of an accident is made at \$80 million. Knowing that an individual transformer failure rate is 0.6% over its lifetime, engineers predict that a strong possibility of one incident would be encountered over a 20-year operating period.

Comments: Expenditure to replace serviceable but potentially hazardous equipment in case of an accident is considered here. Even though the probability is relatively small, the economic consequences of an accident must be included in the analysis. Transformer safety is receiving greater attention. Adverse environmental effects associated with the use, production and disposal of silicon liquid have not been identified. In fact, silicone liquid finds widespread use in domestic products such as skincare products and in food processing industry. As an extra incentive, silicone liquid can be reconditioned and reused.

Example problem # 19.2: Engineers at a power generating station are faced with the problem of cooling 750 gallons of lubrication oil for turbine bearings. The oil returning from the bearings is around 290°F. Circulating air around the oil reservoir tank generally cools the oil to about 125°F during the winter months, which is at the high end of the manufacturer's recommended limit for oil entering the bearings. During the summer the oil from the bearings is close to 340°F, and the temperature in the reservoir is around 210°F. The lubricant's viscosity rapidly deteriorates at the higher temperature, resulting in excessive friction losses in the bearings while also accelerating the rate of wear of the mating surface layer.

Two solutions are available to tackle the problem. One calls for using seawater to cool the bearing oil in a heat exchanger. However, seawater tends to clog the cooling surfaces inside the heat exchanger due to deposits, requiring greater maintenance effort to perform at peak efficiency. The other solution uses desalinated water to cool the oil in the heat exchanger, with the desalinated water cooled in a chiller in a closed loop arrangement. Much of the debate stems from the dramatically different cost structures of the two methods for cooling the lubricant.

Table 19.1

Item	*Seawater Cooled*	*Desalinated Water Cooled*
Heat exchanger cost	\$1.45 million	\$1.30 million
Desalinated water cost	-	\$0.12 million
Chiller cost	-	\$0.59 million
System maintenance cost/year	\$0.22 million	\$0.11 million

The initial cost of the desalinated water system is higher than the alternate seawater cooled system, as shown in table 19.1. However, the recurring cost of maintaining the heat exchanger at scheduled intervals is inherently higher for the seawater-cooled system. Neither system is obviously superior economically. But the salt water-cooled system causes the salty deposits to increase corrosion, thus reducing the useful life of the exchanger relative to the desalinated water system. What then should be the difference in the minimum life expectancy of the systems to make the desalinated water system economically preferable?

Solution: Many factors affect corrosion rate in both systems, which is a prominent factor that will determine the useful life of the equipment. Engineers should also consider cost, efficiency and life expectancy of all major components (chiller, pumps), labour and materials.

Example problem # 19.3: The maintenance department engineers require a rapid prototype system that is capable of designing, duplicating and creating plastic prototypes of complicated parts in just a few hours. A mold is then made from the plastic model for further processing. The procedure allows the engineer to reproduce a complex machine part for replacement of a failed component in less than one week that formerly took 10 to 12 weeks. The complete system costs $385,000, including installation and start-up. Rapid prototyping also saves tens of thousands of dollars per part when compared with purchasing the replacement part from an offshore vendor. In addition, the downtime incurred in procuring the part from the vendor would mean the machinery would be idled for a prolonged period of time, resulting in substantial loss of revenue. In the maintenance department, this technology is a realization of the just-in-time manufacturing concept.

Solution: Here the question is posed: Is it worth spending $400,000 for the machine? How many parts need to be made to recover the investment? The future expected cash flows in this example are the savings experienced from the difference in the cost of the part obtained from the offshore vendor and the manufactured part. Are these savings large enough to warrant the investment in equipment and the cost to make the part?

Example problem # 19.4: An office copying-machine costing $6,000 is expected to be used for 5 years, at the end of which it will have a salvage value of $800. The annual maintenance and operating cost is $1,000, and the benefit generated from the equipment is estimated at $3,000 per year. Assume the receipts and disbursements occur at the end of the year, except the initial cost which is paid at the start of the plan. Describe this proposal in terms of cash flows.

Table 19.2 Cash Flow of Copying Machine Proposal

Year	*Annual Cost*	*Annual Benefit*	*Annual Net Benefit*
0	$6,000	$0	–$6,000
1	1,000	3,000	+2,000
2	1,000	3,000	+2,000
3	1,000	3,000	+2,000
4	1,000	3,000	+2,000
5	1,000	3,800	+2,800

Solution: The series of cash flows may be represented by annual net benefits, i.e., annual benefits in excess of costs, as shown in table 19.2. At the end of year 5 the salvage value is regarded as a benefit and combined with the usage benefit, so total benefit is $3,800.

The cash flows shown above may be conveniently represented in graphic form with a cash flow diagram. Time is represented by a horizontal line, and is marked off with the specified number of time (or interest) periods. Arrows indicate the cash flows at relevant periods. Upward pointing arrows represent positive cash flows (receipts) and downward oriented arrows indicate negative flows (disbursements). The cash flow diagram of example 19.4 is shown in figure 19.2.

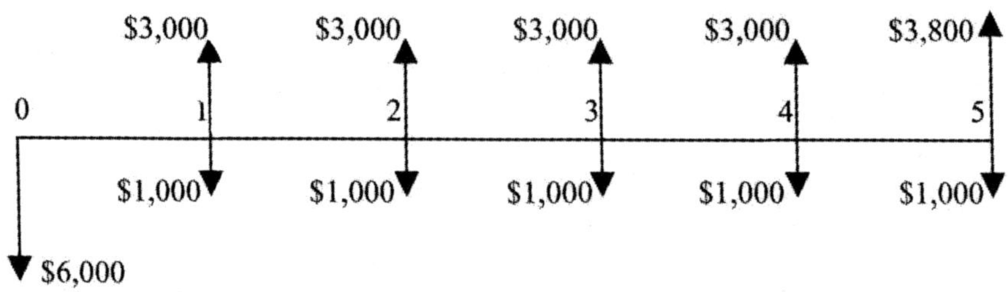

Fig. 19.2 Cash Flow Diagram of Copying Machine Proposal

Example Problem # 19.5: The machine shop of a corporation has just bought a new turning machine at a cost of $105,000 to replace the one that has a salvage value of $20,000. The projected annual savings as a consequence of the improved efficiency is listed in table 19.3, and the payback period is to be determined.

The salvage value of the older machine is of significance in the payback period calculations, since the company has already decided to replace it and thus indicating the new investment cost. Table 19.3 shows the cash flow values and figure 19.3 shows the cash flow and payback period calculation diagrams.

Table 19.3

Period	*Cash Flow*	*Cumulative Cash Flow*
0	–$105,000 + 20,000	–$85,000
1	15,000	–$70,000
2	25,000	–$45,000
3	35,000	–$10,000
4	45,000	$35,000
5	45,000	$80,000
6	35,000	$115,000

Solution: The cash flow is assumed to occur at the end of each year. So the total investment is recovered at the end of fourth year. If cash flow occurs continuously through the year, adjustment is needed. A negative balance exists at the end of third year. If the $45,000 is accrued continuously through the fourth year, then the total investment is recovered in 3.2 years. By including interest amounts, the discounted payback period calculation diagram of this example problem is modified, as seen in figure 19.4.

Payback period evaluation is useful when a company needs to measure the speed of recovering the investment, if the company has a cash flow problem, or when the machine is known to have a short market life. But it is not used exclusively to justify an investment, since it does not indicate profitability. The method is used to supplement other methods of analysis to provide useful information.

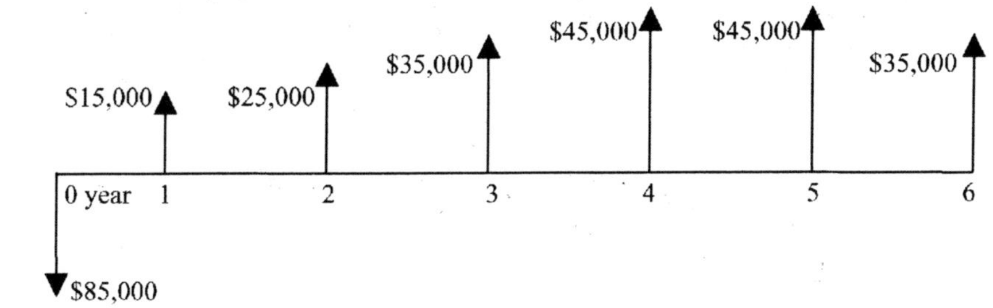

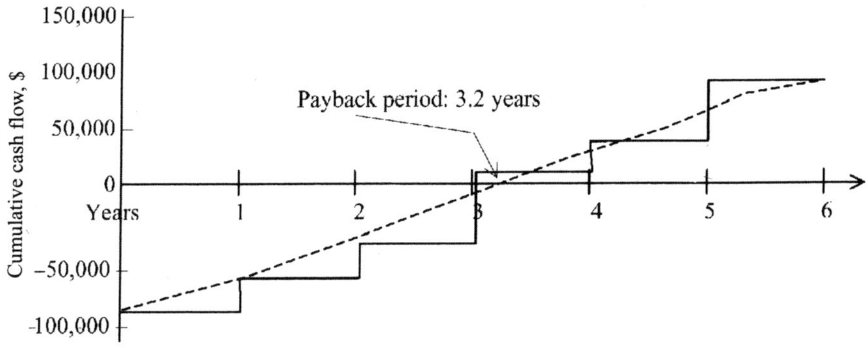

Fig. 19.3 Payback Period Calculation

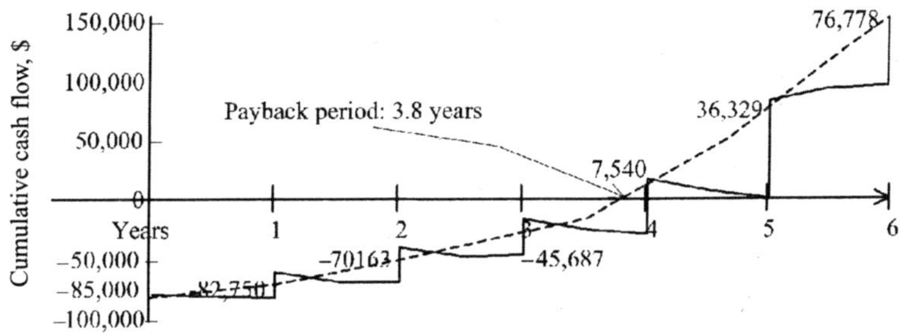

Fig. 19.4 Discounted Payback Period

Example problem # 19.6: A steady electric current of 5,000 amps is to be transmitted a distance of 1,000 ft from a power plant to a sub-station round the clock throughout the year. Copper conductors cost \$8.25/lb. The conductor is expected to have a service life of 25 years and a salvage value \$0.75/lb. Power loss in the conductors is inversely proportional to the cross-section area a. Resistance of copper conductors is 0.8145×10^{-5} ohms per in^2 cross section per linear foot. Energy cost is \$0.05 per kilowatt-hour, interest rate is 9% and density of copper is 555 lbs/ft^3. Calculate the optimum cross section area of the conductor.

Solution: Power loss is a consequence of resistance of the conductor, and follows the rule $R = \rho(L/a)$ where ρ = resistivity, R = resistance, L = length and a = cross-section area of conductor. A consistent set of units needs to be used. Specifically, 1 linear mil = 0.001 in. = 0.0254 mm.

1 circular mil = area of circle 1 linear mil in diameter, and is calculated to be = $(0.5)^2 \times \pi$ mil^2 = 0.7854×10^{-6} in^2. Then 1 in^2 = $1/(0.7854 \times 10^{-6}) = 1.27324 \times 10^6$ cmil. In SI units L is in metres, a in m^2, and ρ in ohm-metres. In SI units copper has $\rho = 1.7241 \times 10^{-8}$ Ω-metre. So: $Ra/L = \rho = 1.7241 \times 10^{-8}$ Ω(1 metre)^2/metre, which equals 1.7241×10^{-8} Ω(39.37 in)^2/3.2808 ft, or = 0.8145421×10^{-5} Ω-in^2/ft. As the current flows through the conductor, power loss due to its resistance is given by:

Power loss = $I^2RT/1000$ kWh, where I = current in amperes, R = resistance in ohms and T = time duration in hours.

$$\text{Energy loss in kWh} = I^2RT/(1000a)$$

$$= (5000^2)(0.008145)(24)(365)/(1000a)$$

$$= 1{,}783{,}755/a \text{ kWh}$$

$$\text{Cost of energy loss} = 1{,}783{,}755 \times \$.05/a = \$89{,}188/a$$

The trade-off is between material cost of the conductor and energy losses. A larger conductor size results in increased copper weight and consequent increased cost. Material weight is calculated to be: $(1000 \times 12 \times 555 \times a)/12^3 = 3{,}854a$. Material cost = $(3{,}854a) \times (\$8.25) = \$31{,}797a$.

Salvage value of the copper conductors is: $(0.75) \times (3{,}854a) = 2{,}890a$. The annual equivalent worth (AE) may be used to measure an investment project by calculating equal payments on a yearly or annual basis. Since a single lump sum cash payment can be readily converted into a series of equal annual payments, the net present worth (NPW) of the original series may be established. This amount may then be multiplied by the capital recovery factor: $AE(i) = PW(i) \times (A/P, i, N)$, where the second expression $(A/P, i, N) = [i(1 + i)^N]/[(1 + i)^N - 1]$

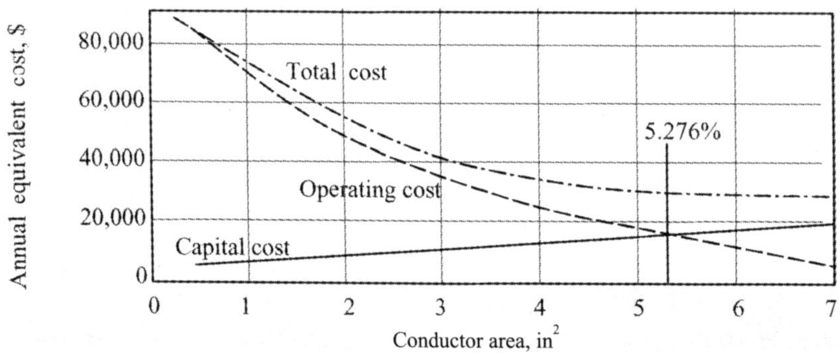

Fig. 19.5 Determine Optimal Cross Section Area of Conductor

Capital recovery cost at 9% is:

$$CR(9\%) = [\$31{,}797a - 2{,}890a](A/P, 9\%, 25) + 2{,}943a(0.09)$$

$$= (2{,}943 + 260)a = 3{,}203a$$

Expressing the total annual equivalent cost of capital and operating costs:

$$AE(9\%) = 3{,}203a + \$89{,}188/a$$

Minimum annual equivalent cost as a function of wire cross section a is:

$$dAE(9\%)/da = 3{,}203 - 89{,}188/a^2 = 0$$

Hence:

$$a^* = \sqrt{(89{,}188/3{,}203)}$$
$$= 5.276 \text{ in}^2$$

Corresponding total minimum annual equivalent cost is determined to be:

$$AE(9\%) = 3{,}203 \times 5.276 + 89{,}188/5.276 = \$33{,}802$$

The relation between costs and conductor size is shown in figure 19.5.

Example problem # 19.7: The energy required to transport a fluid at a given volume flow rate through a pipeline decreases as the pipe size is increased. But the cost of the line increases with the pipe diameter. The most economical pipe size is to be determined for a line running along a length of 600 miles using the data given below. Service life expectancy is 20 years and $i = 10\%$.

$$\text{Pumping power} = 1.33 \times Q \times \Delta P/1{,}980{,}000 \text{ HP}$$

where:

Q = volume flow rate, ft^3/hr,

$\Delta P = 128Q\mu L/g\pi D^4$ = pressure drop, lbs/ft^2

L = pipe length, ft

D = inside pipe diameter, ft

$t = 0.01D$, pipe wall thickness

μ = 8,500 lb/hour-ft, oil viscosity

$g = 32.2 \times 12{,}960{,}000$ ft/hour^2

Power cost = \$0.015 per HP-hour

Oil cost = \$18 per barrel

Pipeline cost = \$1.00 per lb of steel

Pump and motor cost = \$195/HP

Salvage value after 20 years is assumed \$0 due to cost of dismantling.

Solution: The cost of energy to pump the fluid and construction cost through a given pipe size needs to be determined.

$$600 \text{ miles} = 600 \times 5{,}280 = 3{,}168{,}000 \text{ ft.}$$

$$1 \text{ barrel} = 42 \text{ US gallons} = 42 \times 231 \text{ in}^3/\text{gallon} = 42 \times 231/12^3 = 5.6146 \text{ ft}^3$$

$$\text{Density of steel} = 490.75 \text{ lb/ft}^3$$

$$\text{Volume flow rate per hour } Q = 3{,}000{,}000 \text{ barrels/day} \times 5.6146 \text{ ft}^3/\text{barrel}$$
$$= 16{,}843{,}800 \text{ ft}^3/\text{day} = 701{,}825 \text{ ft}^3/\text{hour}$$

$$\text{Pressure drop } \Delta P = 128Q\mu L/g\pi D^4$$
$$= [128 \times 701{,}825 \times 8{,}500 \times 3{,}168{,}000]/[32.2 \times 12{,}960{,}000 \times 3.14159D^4]$$
$$= [1{,}845{,}153{,}595/D^4] \text{ lb/ft}^2$$

$$\text{Pumping power to offset pressure drop} = 1.33 \times Q \times \Delta P/1{,}980{,}000 \text{ HP}$$

$$= \{1.33 \times 701{,}825 \times [1{,}845{,}153{,}595/D^4]\}/1{,}980{,}000$$

$$= [871{,}818{,}975/D^4] \text{ HP}$$

$$\text{Power cost to pump oil} = [871{,}818{,}975/D^4] \text{ HP} \times \$0.015/\text{HP-hour} \times 24 \times 365$$

$$= [\$114{,}557{,}013{,}315/D^4] \text{ per year}$$

$$\text{Pump and motor cost} = [871{,}818{,}975/D^4] \text{ HP} \times \$195/\text{HP}$$

$$= [\$170{,}004{,}700{,}125/D^4]$$

$$\text{Pipe cross sectional area} = 3.14159 \times [(0.051D)^2 - (0.050D)^2]$$

$$= 0.032D^2 \text{ ft}^2$$

$$\text{Total volume of pipe} = 0.032D^2 \text{ ft}^2 \times 3{,}168{,}000 \text{ ft}$$

$$= 101{,}376D^2 \text{ ft}^3$$

$$\text{Total weight of pipe steel} = 101{,}376D^2 \text{ ft}^3 \times 490.75 \text{ lb/ft}^3$$

$$= 49{,}750{,}272D^2 \text{ lbs}$$

$$\text{Pipeline cost} = \$1.00/\text{lb} \times 49{,}750{,}272D^2 \text{ lbs} = \$49{,}750{,}272D^2$$

Annual equivalent cost of material and operating expenses is calculated below, using 20 years service life, no salvage value and i = 10%.

$$\text{Capital cost} = [\$49{,}750{,}272D^2 + \$170{,}004{,}700{,}125/D^4](A/P, 10\%, 20)$$

$$= \$5{,}843{,}648D^2 + \$19{,}968{,}752{,}076/D^4$$

$$\text{Annual power cost} = \$114{,}557{,}013{,}315/D^4$$

Total annual cost:

$$AE(10\%) = \$5{,}843{,}648D^2 + \$19{,}968{,}752{,}076/D^4 + \$114{,}557{,}013{,}315/D^4$$

The best pipe size is obtained by minimizing the annual equivalent cost. Taking the first derivative of $AE(10\%)$ with respect to pipe diameter and equating to zero:

$$dAE(10\%)/dD = 11{,}687{,}297D - 538{,}103{,}061{,}567/D^5 = 0$$

Hence: $$11{,}687{,}297D^6 = 538{,}103{,}061{,}567$$

$$D^6 = 46{,}064.70$$

$$D^* = 5.9868 \text{ ft.}$$

Velocity of the fluid in the pipe should be limited to 10 ft/sec to control friction losses and wear of pipeline and accessories. Checking on flow velocity:

$$Q = \text{Velocity} \times \text{Pipe flow area}$$

$$701{,}825 \text{ ft}^3/\text{hr} \times 1/3600 \text{ hr/sec} = V \times [(3.14159 \times (5.9868)^2]/4$$

$$\text{Hence } V = 6.93 \text{ ft/sec.}$$

which is acceptable. Annual equivalent cost also needs to be checked.

$$\text{Capital cost} = \$5{,}843{,}648(5.9868)^2 + \$19{,}968{,}752{,}076/(5.9868)^4$$

$$= \$224{,}991{,}039$$

Annual power cost = \$114,557,013,315/(5.9868)^4

= \$89,174,911

Total annual cost = \$224,991,039 + \$89,174,911

= \$ 314,165,950

Total annual oil revenue = \$18/barrel × 3,000,000 barrels/day × 365 days/year

= \$19,710,000,000 per year.

Thus adequate revenues are available to offset the capital and operating costs.

Example problem # 19.8: Consider the following financial data for a machine:

Cost basis, I = \$12,000

Useful life, N = 6 years

Estimated salvage value S = \$3,000

Compute the annual depreciation charges and book value using the straight-line depreciation procedure.

Solution: The straight-line depreciation rate = $1/N$ = 1/6. So the annual depreciation charge is:

$$D_n = (\$12,000 - \$3,000)/6 = \$1,500.$$

The depreciation charges and book value during its useful life are shown below:

n	B_{n-1}	D_n	B_n
1	\$12,000	\$1,500	\$10,500
2	10,500	1,500	9,000
3	9,000	1,500	7,500
4	7,500	1,500	6,000
5	6,000	1,500	4,500
6	4,500	1,500	3,000

The values are plotted in figure 19.6.

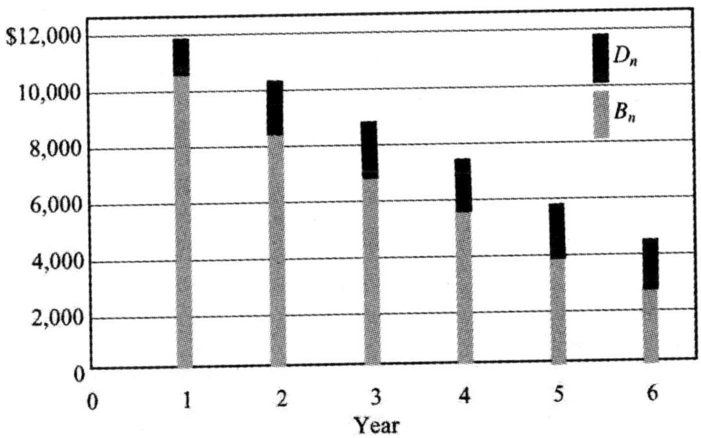

Fig. 19.6 Straight Line Depreciation

Example problem # 19.9: A corporation is thinking of replacing a specialized material handling machine, which is used on two shifts, six days a week. The machine may be either owned or leased. The following data is available. The initial cost is \$35,000, and the machine uses 5 gallons of gasoline per shift (8 hours) costing \$2.00 per gallon. Expected operating life is 5 years (working 2 shifts per day), and engine needs overhauling every 3 years at a cost \$2,750. The machine vendor offers a service agreement with a trained mechanic coming once a month to the site, and costs \$150 per month to keep the machine operating at peak efficiency. The machine is qualified as 5-year property for depreciation purposes, using the sum-of-the-years-digit method, and salvage value of \$7,500 at the end of 5 years. Insurance cost is \$500 per year. The vendor also offers a leasing plan that includes all servicing of the equipment for a minimum of 3 years at a monthly rate of \$1,200. The lease may be cancelled after the minimum period. The company's other similar investments earn 15% after taxes, and their tax rate is 35%. Compare the costs of owning against leasing the equipment.

Solution: Fuel cost for two-shift operation per week is:

(5 gallons/shift)(2 shifts/day) × (\$2 per gallon) × (6 days per week) = \$120 per week, or \$6,240 per year.

Note: However, that the fuel cost will be identical whether the machine is leased or bought, and may thus be ignored for this comparison.

a. The incremental cost of owning over leasing the equipment will be considered.

The maintenance cost is \$150 × 12 months a year = \$1,800 per year, and insurance is \$500 per year, bringing the total overhead and maintenance cost to \$2,300 per year, and its present worth is:

$$P_1 = -\$2{,}300 \times (1 - 0.35)(P/A,\ 15\%,\ 5) = -\$7{,}710$$

Assuming the overhaul cost to be spent all at once, the equivalent present worth of this after tax expense is:

$$P_2 = -\$2{,}750 \times (1 - 0.35)(P/F,\ 15\%,\ 3) = -\$1{,}175$$

Assuming the purchase is financed at 12%, the annual installment for repayment of the debt of \$35,000 is:

$$A = \$35{,}000(A/P,\ 12\%,\ 5) = \$9{,}709$$

The equivalent present worth of this installment is:

$P_3 = -\$9{,}709(P/A,\ 15\%,\ 5) = -\$32{,}547$

Interest payment each year at 12% is as follows:

Year, *n*	Balance	Interest	Payment	Balance
1	\$35,000	\$4,200	\$(9,709)	\$29,491
2	29,491	3,539	(9,709)	23,321
3	23,321	2,799	(9,709)	16,410
4	16,410	1,969	(9,709)	8,671
5	8,671	1,040	(9,709)	0

The combination of interest payments and depreciation will result in tax savings, shown below:

Year, *n*	Depreciation	Interest	Combined Tax Savings
1	\$9,167	\$4,200	\$4,768
2	7,333	3,539	3,805
3	5,500	2,799	2,904
4	3,667	1,969	1,973
5	1,833	1,040	1,006

The equivalent present worth of all the tax credits is:

$$P_4 = \$4{,}678(P/F, 15\%, 1) + \$3{,}805(P/F, 15\%, 2) + \$2{,}904(P/F, 15\%, 3) + \$1{,}973(P/F, 15\%, 4) + \$1{,}006(P/F, 15\%, 5) = \$10{,}483.$$

Net proceeds from the sale of machinery at the end of 5 years is:

Book value = \$35,000 – \$7,500 = \$27,500

Gain (loss) = \$27,500 – \$27,500 = 0

Hence net proceed from sale = \$7,500, and the present value of the amount is:

$$P_5 = \$7{,}500(P/F, 15\%, 5) = \$3{,}729$$

Net incremental cost of owning the machinery through 100% debt financing is thus:

$$PW(15\%)_{buy} = P_1 + P_2 + P_3 + P_4 + P_5 = -\$7{,}710 - \$1{,}175 - \$32{,}547 + \$10{,}483 + \$3{,}729 = -\$27{,}220$$

b. The incremental cost of leasing is calculated next. A tax-deductible expense, the cost per year is:

Annual lease payments = 12 × \$1,200 = \$14,400

Tax = 35% = \$5,040

So annual net cost = \$9,360

Then total net equivalent present incremental cost of leasing is:

$$PW(15\%)_{lease} = -\$9{,}360(P/A, 15\%, 5) = -\$31{,}377$$

Hence, purchase of the machinery through debt financing will save the company \$4,157.

Comments: Lease payments are generally made at the beginning of the period, but here they are assumed to occur at the end of the period. The calculations for the present worth may be modified for the changed timing:

$$PW(15\%)_{lease} = -\$9{,}360 - \$9{,}360(P/A, 15\%, 4) = -\$36{,}083$$

Example problem # 19.10: The maintenance department of a power company is evaluating the purchase of a 5000 lb portable crane. Four alternatives are available for driving the crane – gasoline, diesel, propane gas and rechargeable batteries. Data concerning purchase price, fuel expenses, life expectancy and fixed and variable maintenance costs are shown in the tabulation below. The department operates between 250 to 300 shifts per year of 8 hours each. Income tax is not paid, so depreciation does not enter the comparative analysis. Interest rate of 8% is to be used. Prepare a sensitivity graph to determine how the fuel selection varies as a function of the number of operating shifts per year.

Table 19.4

	Gasoline	Diesel	Propane Gas	Battery
Purchase price	\$75,550	\$79,175	\$77,560	\$92,750
Life expectancy	8 years	8 years	8 years	8 years
Salvage value	\$8,000	\$7,000	\$7,000	\$10,000
Maximum shifts/year	300	300	300	300
Energy consumption/shift	13.2 gallon	9.95 gallon	8.1 gallon	171.1 kWh
Fuel cost	\$1.75/gal	\$1.80/gal	\$1.60/gal	\$0.07/kWh
Fuel cost/shift	\$23.10	\$17.94	\$12.96	\$11.98
Annual maintenance cost:				
Fixed cost	\$2,500	\$3,000	\$3,500	\$2,000
Variable cost/shift	\$5.75	\$4.25	\$7.75	\$4.00

Annual capital cost. Use the capital recovery method.

Gasoline: $CR(8\%) = (\$75{,}550 - \$8{,}000)(A/P, 8\%, 8) + 0.08 \times \$8{,}000 = \$12{,}394$

Diesel: $CR(8\%) = (\$79{,}175 - \$7{,}000)(A/P, 8\%, 8) + 0.08 \times \$7{,}000 = \$13{,}118$

Propane gas: $CR(8\%) = (\$77{,}560 - \$7{,}000)(A/P, 8\%, 8) + 0.08 \times \$7{,}000 = \$12{,}837$

Battery: $CR(8\%) = (\$92{,}750 - \$10{,}000)(A/P, 8\%, 8) + 0.08 \times \$10{,}000 = \$15{,}199$

Annual operating cost. Assume number of shifts = N.

Gasoline: $\$2{,}500 + (\$23.10 + \$5.75)N = \$2{,}500 + \$28.85N$

Diesel: $\$3{,}000 + (\$17.94 + \$4.25)N = \$3{,}000 + \$22.19N$

Propane gas: $\$3{,}500 + (\$12.96 + \$7.75)N = \$3{,}500 + \$20.71N$

Battery: $\$2{,}000 + (\$11.98 + \$4.00)N = \$2{,}000 + \$15.98N$

Total annual cost. Sum of capital and operating costs.

Gasoline: $AE(8\%) = \$14{,}894 + \$28.85N$

Diesel: $AE(8\%) = \$16{,}118 + \$22.19N$

Propane gas: $AE(8\%) = \$16{,}337 + \$20.71N$

Battery: $AE(8\%) = \$17{,}199 + \$15.98N$

Figure 19.7 provides a graph for the total annual equivalent cost as a function of number of shifts N. The battery-operated crane provides the best annual equivalent cost if the number of shifts exceeds 175. Below this value, gasoline powered crane is preferable. Besides sensitivity of variable parameters to the Net Present Worth, project managers are often interested in knowing the minimum level of sales revenue below which the project loses money. The 'breakeven analysis' permits such an evaluation to study the effects of variation in output on a company's NPW or other measure. Computing the PW of cash inflows as a function of an unknown variable x, then:

PW of cash inflows $= f(x)_1$

Also the PW of cash outflows is determined:

PW of cash outflows $= f(x)_2$

NPW is the difference between $f(x)_1$ and $f(x)_2$. For the breakeven value of x, $f(x)_1 = f(x)_2$.

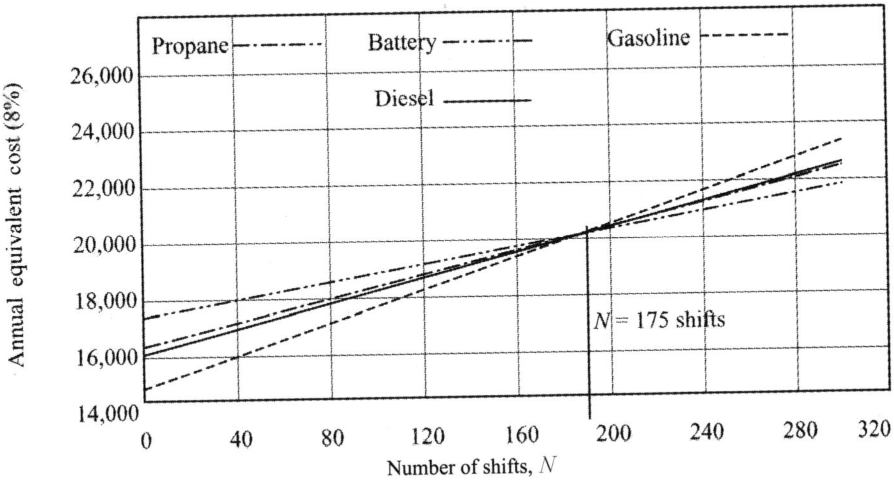

Fig. 19.7 Sensitivity Analysis for Comparison of Crane

19.14 EXERCISES

19.1 A computer system priced at Rs. 750,000 is expected to be used for 6 years, at the end of which it will have a salvage value of Rs. 40,000. The monthly operating cost is expected to be Rs. 3,900, and the benefit generated from the equipment is estimated at Rs. 60,000 per year. If the receipts and disbursements are assumed to occur at the end of the year, except the initial cost, which is paid at the start of the plan, describe this proposal in terms of cash flows.

19.2 A machining centre costing of Rs. 74,50,000 is to replace the one that has a salvage value of Rs. 2,75,000. The projected annual savings in the first 7 years as a consequence of the improved efficiency is projected to be Rs. 12,00,000, Rs. 14,00,000, Rs. 16,00,000, Rs. 18,00,000, Rs. 18,00,000, Rs. 18,00,000 and Rs. 15,00,000. Plot the cash flow values and calculate the payback period.

19.3 Electric current of 1,000 amps is to be transmitted a distance of 2,500 ft from a power plant to a sub-station 12 hours a day. Copper conductors cost Rs. 535/lb. The conductor is expected to have a service life of 20 years and a salvage value Rs. 45/lb. Power loss in the conductors is inversely proportional to the cross-section area a. Resistance of copper conductors is 0.8145×10^{-5} ohms per in^2 cross section per linear foot. Energy cost is Rs. 5.75 per kilowatt-hour, interest rate is 11.5% and density of copper is 555 lbs/ft^3. Calculate the optimum cross section area of the conductor.

19.4 In example problem # 19.7, determine the diameter of the pipe if service life expectancy is 25 years and $i = 9\%$.

19.5 In example problem # 19.7, determine the length of the pipeline if the diameter of the pipe is 54.0 in., service life expectancy is 25 years and $i = 9\%$.

19.6 An investment costing Rs. 6,35,000 has a useful life of 7 years and negligible salvage value. Calculate the average annual depreciation charge using the sum-of-the-years-digit method.

19.7 A forging press machine costing Rs. 13,28,000 is needed for an automobile components manufacturer. The machine may be either owned or leased. The machine uses 150 kWh of electricity per shift of 8 hours, costing Rs. 6.25 per unit. Expected operating life is 8 years (working 2 shifts per day), and needs overhauling every 4 years at a cost Rs. 1,50,000. The machine vendor offers a service agreement with a trained mechanic coming once a month to the site, and costs Rs. 3,500 per month to keep the machine operating at peak efficiency. The machine is qualified as 5-year property for depreciation purposes, using the straight-line method, and salvage value of Rs. 2,50,000 at the end of 5 years. Insurance cost is Rs. 25,000 per year. The leasing plan includes all servicing of the equipment for a minimum of 3 years at a monthly rate of Rs. 42,000. The lease may be cancelled after the minimum period. The company's other similar investments earn 13% after taxes, and their tax rate is 32%. Compare the costs of owning against leasing the equipment.

19.8 Four alternatives are available for an investment proposal. Financial data related to the alternatives is shown in the tabulation below. Depreciation is not a factor to be included in the comparative analysis. Interest rate of 11% is to be used. Prepare a sensitivity graph to determine the most favourable alternative.

	A	B	C	D
Purchase price	Rs. 5,65,000	Rs. 5,82,000	Rs. 5,79,000	Rs. 5,93,000
Life expectancy	8 years	7 years	9 years	8 years
Salvage value	Rs. 27,000	Rs. 24,000	Rs. 25,000	Rs. 28,000
Service expense per year	Rs. 2,29,000	Rs. 2,37,000	Rs. 2,33,000	Rs. 2,42,000

References & Bibliography

Jones, C. and Jacobs, J.I., "Economic and Technical Considerations for Combined Cycle Performance-Enhancement Options," *GE Power Systems,* GER-4200, October 2000.

Wiesner, F. J., "A New Appraisal of Reynolds Number Effect on Centrifugal Compressor Performance", ASME Journal for Engineering Power, volume 101, page 384 – 396, New York, 1979.

Wustman, R.D. and Smith, J.S., "High Strength Diffusion Braze Repairs for Gas Turbine Components", ASME paper # 96-GT-427, New York, 1996.